Acoustical Imaging

Volume 17

Acoustical Imaging

Volume 1	Proceedings of the First International Symposium, December 1967, edited by A. F. Metherell, H. M. A. El-Sum, and Lewis Larmore
Volume 2	Proceedings of the Second International Symposium, March 1969, edited by A. F. Metherell and Lewis Larmore
Volume 3	Proceedings of the Third International Symposium, July 1970, edited by A. F. Metherell
Volume 4	Proceedings of the Fourth International Symposium, April 1972, edited by Glen Wade
Volume 5	Proceedings of the Fifth International Symposium, July 1973, edited by Philip S. Green
Volume 6	Proceedings of the Sixth International Symposium, February 1976, edited by Newell Booth
Volume 7	Proceedings of the Seventh International Symposium, August 1976, edited by Lawrence W. Kessler
Volume 8	Proceedings of the Eighth International Symposium, May 29–June 2, 1978, edited by A.F. Metherell
Volume 9	Proceedings of the Ninth International Symposium, December 3–6, 1979, edited by Keith Y. Wang
Volume 10	Proceedings of the Tenth International Symposium, October 12–16, 1980, edited by Pierre Alais and Alexander F. Metherell
Volume 11	Proceedings of the Eleventh International Symposium, May 4–7, 1981, edited by John P. Powers
Volume 12	Proceedings of the Twelfth International Symposium, July 19–22, 1982, edited by Eric A. Ash and C. R. Hill
Volume 13	Proceedings of the Thirteenth International Symposium, October 26–28, 1983, edited by M. Kaveh, R. K. Mueller, and J. F. Greenleaf
Volume 14	Proceedings of the Fourteenth International Symposium, April 22–25, 1985, edited by A. J. Berkhout, J. Ridder, and L. F. van der Wal
Volume 15	Proceedings of the Fifteenth International Symposium, July 14–16, 1986, edited by Hugh W. Jones
Volume 16	Proceedings of the Sixteenth International Symposium, June 10–12, 1987, edited by Lawrence W. Kessler
Volume 17	Proceedings of the Seventeenth International Symposium, May 31–June 2, 1988, edited by Hiroshi Shimizu, Noriyoshi Chubachi, and Jun-ichi Kushibiki

A Continuation Order Plan is available for this series. A continuation order will bring delivery of each new volume immediately upon publication. Volumes are billed only upon actual shipment. For further information please contact the publisher.

Acoustical Imaging

Volume 17

Edited by

Hiroshi Shimizu,
Noriyoshi Chubachi,
and Jun-ichi Kushibiki

Tohoku University
Sendai, Japan

PLENUM PRESS · NEW YORK AND LONDON

The Library of Congress cataloged the first volume of this series as follows:

International Symposium on Acoustical Holography.

Acoustical holography; proceedings. v. 1–
New York, Plenum Press, 1967–

v. illus. (part col.), ports. 24 cm.

Editors: 1967– . A. F. Metherell and L. Larmore (1967 with H. M. A. el-Sum)
Symposium for 1967– held at the Douglas Advanced Research Laboratories, Huntington Beach, Calif.

1. Acoustic holography—Congresses—Collected works. I. Metherell. Alexander A., ed. II. Larmore, Lewis, ed. III. el-Sum, Hussein Mohammed Amin, ed. IV. Douglas Advanced Research Laboratories, v. Title.
QC244.5.I.5 69-12533

ISBN-13: 978-1-4612-8084-2 e-ISBN-13: 978-1-4613-0791-4
DOI: 10.1007/978-1-4613-0791-4

Proceedings of the 17th International Symposium on Acoustical Imaging
held May 31—June 2, 1988 in Sendai, Japan

Softcover reprint of the hardcover 1st edition 1989
A Division of Plenum Publishing Corporation
233 Spring Street, New York, N.Y. 10013

PREFACE

The 17th International Symposium on Acoustical Imaging was held at Tohoku University, Sendai, Japan, during May 31-June 2, 1988. The symposium was organized by the ultrasonics research group of Tohoku University and the IEEE UFFC Society, Tokyo Chapter. Of the 128 papers submitted, 88 were presented during the symposium, which comprised 144 researchers from 13 countries.

This volume contains 81 papers as the record of the symposium and is classified into the following sections: (1) Acoustic Microscopy and its Applications; (2) Non-Destructive Evaluation; (3) Signal Processing of Images; (4) Acoustic Measurements and Physical Acoustics; (5) Medical Ultrasonic Diagnostics; (6) Acoustic Sensors; (7) Acoustic Holography and Tomography; (8) Seismic Exploration; and (9) Imaging Instrumentation and Other Techniques.

A number of the papers submitted were associated with medical ultrasonic diagnostics and acoustic microscopy, reflecting a major activity in acoustical imaging at Tohoku University. Accordingly, two invited talks were focused on this area: acoustic microscopy by Dr. G. A. D. Briggs of the University of Oxford and medical ultrasonics by Prof. M. Tanaka of Tohoku University. In light of the history of research in this field at our university, we are delighted to have had the opportunity to host the 17th symposium.

We would like to express our gratitude to many individuals and to several organizations. First of all we would like to thank the authors for their significant contributions. We appreciate the cooperation of the IEEE UFFC Society, the Acoustical Society of Japan, the Japan Society of Ultrasonics in Medicine, and the Institute of Electronics, Information and Communication Engineers. Sendai City, Sendai Convention Bureau, and many companies also supported the symposium in a variety of ways, including financial contributions. Special thanks are extended particularly to K. Yamada, T. Sannomiya, Y. Sugawara, and A. Hashimoto for their important assistance in the arranging and functioning of the symposium. Furthermore, we are grateful to E. Kimura, S. Inoue, and J. Yamaguchi for editing this proceedings.

The 18th International Symposium on Acoustical Imaging will be held during September 18-20, 1989 at the Red Lion Resort, Santa Barbara, California, U.S.A., under the Chairmanship of Profs. Glen Wade and Hua Lee.

Hiroshi Shimizu
Noriyoshi Chubachi
Jun-ichi Kushibiki

CONTENTS

ACOUSTIC MICROSCOPY AND ITS APPLICATIONS

NON-DESTRUCTIVE EVALUATION

SIGNAL PROCESSING OF IMAGES

ACOUSTIC MEASUREMENTS AND PHYSICAL ACOUSTICS

MEDICAL ULTRASONIC DIAGNOSTICS

ACOUSTIC SENSORS

ACOUSTIC HOLOGRAPHY AND TOMOGRAPHY

SEISMIC EXPLORATION

IMAGING INSTRUMENTATION AND OTHER TECHNIQUES

ACOUSTIC MICROSCOPY OF OLD AND NEW MATERIALS

G.A.D. Briggs, C.M.W. Daft[1], A.F. Fagan, T.A. Field, C.W. Lawrence, M. Montoto[2], S.D. Peck, A. Rodriguez[2], and C.B. Scruby

Department of Metallurgy and Science of Materials
University of Oxford, Parks Road, Oxford OX1 3PH, England

[1]Department of Electrical and Computer Engineering
University of Illinois, 1406 W. Green Street, Urbana
IL 61801, USA

[2]Departamento de Geologia, Universidad de Oviedo
33005 Oviedo, Spain

INTRODUCTION

The large number of papers on acoustic microscopy in this symposium testifies both to the range of techniques available to the user, and to the range of applications that are being found. Imaging is now routinely possible at frequencies ranging from close to those used in conventional nondestructive testing and medical ultrasound to 2GHz[1,2], with contrast arising from scattering by internal defects or factors affecting surface wave propagation. Quantitative measurements can be made by time resolved impulse measurements[3] or by the V(z) technique with larger pulses. For very accurate measurements the line focus beam technique has become well established[4], offering unprecedented capabilities especially for surface wave measurements on anisotropic materials. In a growing number of disciplines it is being discovered that the acoustic microscope can provide new information, specifically about the elastic properties of specimens and how acoustic waves interact with them. This paper offers a small selection of applications, ranging from natural materials in the form of rocks and teeth, to new man-made materials in the form of ceramics and hard-metals.

ROCKS

The oldest materials that have been studied by acoustic microscopy are rocks[5,6]. Figure 1 shows a geological specimen of granite imaged by two optical microscopy techniques. Figure 1a is a polarized light micrograph, which is suitable for studying the textural characteristics; it shows two quartz grains, with a number of microcracks in the right hand grain. The most sensitive optical technique for the detection of cracks like these uses a fluorescent dye penetrant; such an image is shown in figure 1b. In the fluorescence image a crack roughly parallel to the

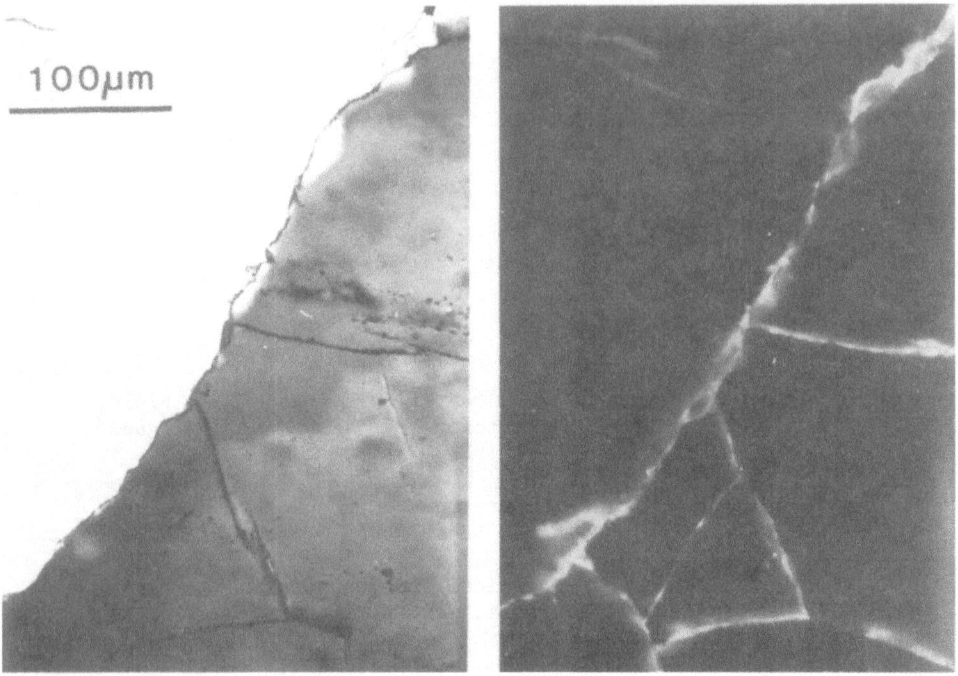

Fig.1. A polished thin section of granite observed by light microscopy: [a] polarized light; [b] fluorescent dye penetrant. The images should be rotated about 30° anticlockwise for comparison with the following images of the same specimen.

grain boundary appears in the quadrilateral near the bottom; this crack is more difficult to see in the polarized light micrograph. Both the images in figure 1 were originally in colour; they are reproduced here in black and white. They should be rotated by about 30° anticlockwise to relate them to the following images of the same specimens. The fluorescent dye penetrant technique can of course be used in this form only in transparent materials.

When the same specimen was examined in a scanning electron microscope at progressively higher magnification the images of figure 2 were obtained. These show, at successively higher magnifications, the structure of microcracks. The crack in the quadrilateral is visible along much of its length, but rather higher magnification is necessary to find some of the very fine cracks such as the one seen diagonally in figure 2c.

The acoustic microscope is able to reveal both the grain structure and the microcracks, including the very finest ones. In [a] and [b] the frequency was 1·6GHz; the two grains are distinguished by the reversal of the contrast at different defocus[7]. The crack in the quadrilateral is seen quite well, especially in [b]. The crack seen at high magnification in the s.e.m. is visible in both these images, and also at much lower frequency in [c] where the defocus has been increased to optimise the Yamanaka fringes due to Rayleigh wave scattering by the cracks[8]. Comparison with figure b gives some indication of just how fine a crack can be and yet still give plenty of contrast acoustically.

The series of acoustic micrographs also shows some new boundaries in

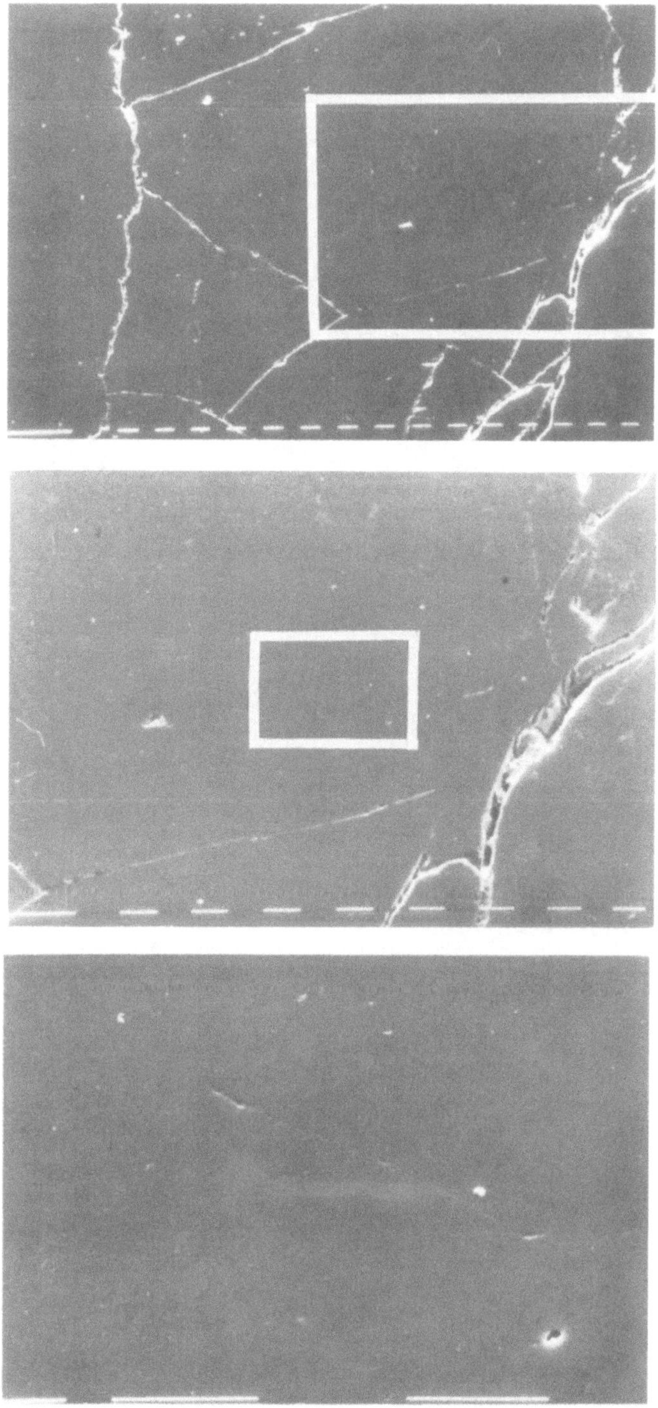

(a)

(b)

(c)

Fig.2. The same granite specimen as in figure 1, viewed in a scanning electron microscope at progressively higher magnifications. The bars represent 10μm in each case. The rectangles in [a] and [b] indicate the areas.

(a)

(b)

(c)

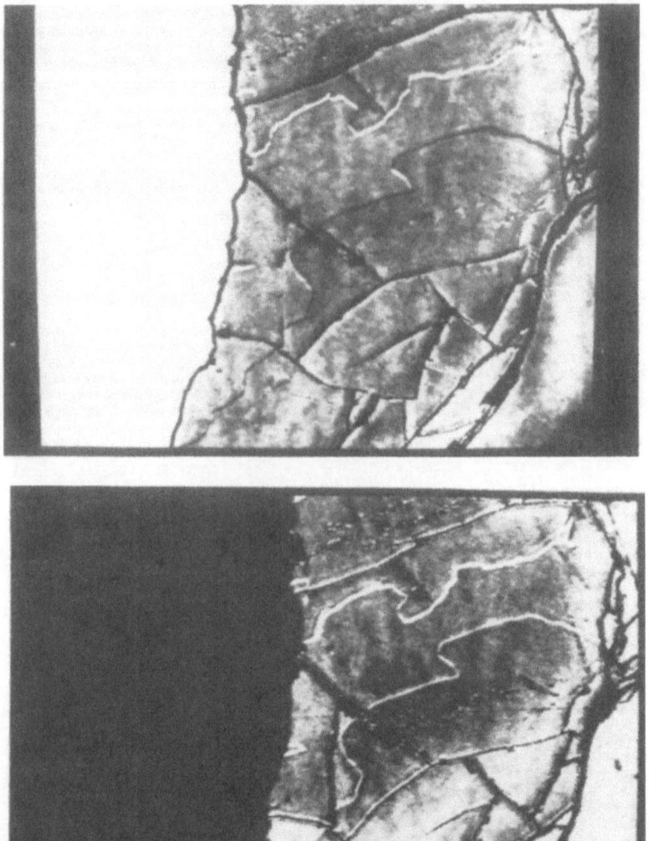

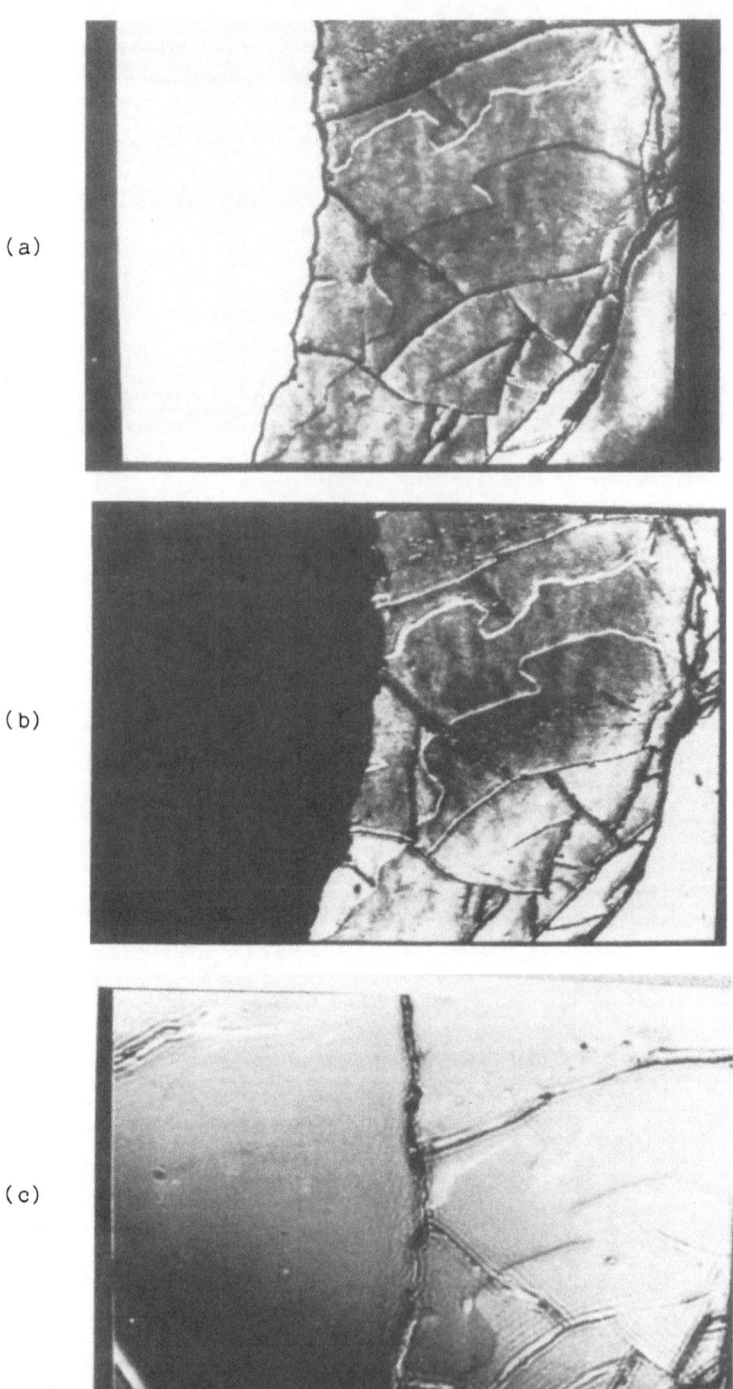

Fig.3. Acoustic micrographs of the granite specimen, under various imaging conditions: [a] f = 1·6GHz, z = -3·1μm; [b] f = 1·6GHz, z = -5·8μm; [c] f = 370MHz, z = -10μm.

the right hand quartz grain. These are subgrain boundaries, associated with plastic deformation of the rock early in its history (before the cracks formed). Simple V(z) measurements at the points indicated in figure 4 enable an assessment of the mean surface wave velocities to be made. These are $3174 ms^{-1}$, $3471 ms^{-1}$, $3019 ms^{-1}$ for an average of four measurements at each of the three points. This might suggest that the parent grain has the same orientation above and below the middle subgrain, but that the middle subgrain has a different orientation. More accurate measurements could be made using an angle resolved microspectroscopy system[9].

TEETH

Incipient caries lesions in teeth appear as white spots, because they scatter light more than healthy enamel does[10]. Sections through caries lesions can be studied in transmission by polarized light microscopy and by microradiography, and in reflection by acoustic microscopy. Once again, acoustic microscopy shows morphology that is recognisably the same as that shown by longer established techniques, but with additional information[11]. Dental enamel is almost the only biological material with high enough velocity of sound to give Rayleigh wave contrast in the acoustic microscope.

The variation of contrast along a given line in an acoustic image can be displayed as V(x,z), i.e. when the horizontal axis represents position along a line parallel to the specimen surface and the vertical axis represents displacement of the lens perpendicular to the specimen surface[12]. The two scales need not be the same; indeed the range of z displacement will usually be much smaller. The z displacement should, as usual, be thought of not as imaging below the surface (as would be displayed in a medical scan), but rather as altering the interference conditions for the contrast from surface waves.

An acoustic micrograph of a section through a caries lesion is shown

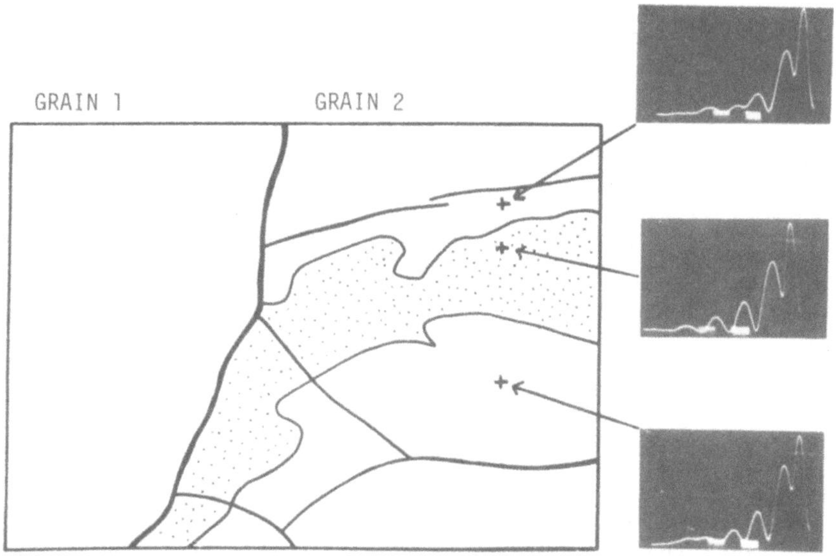

Fig.4. V(z) curves at 370MHz recorded at three of the points in figure 3 [c].

(a)

(b)

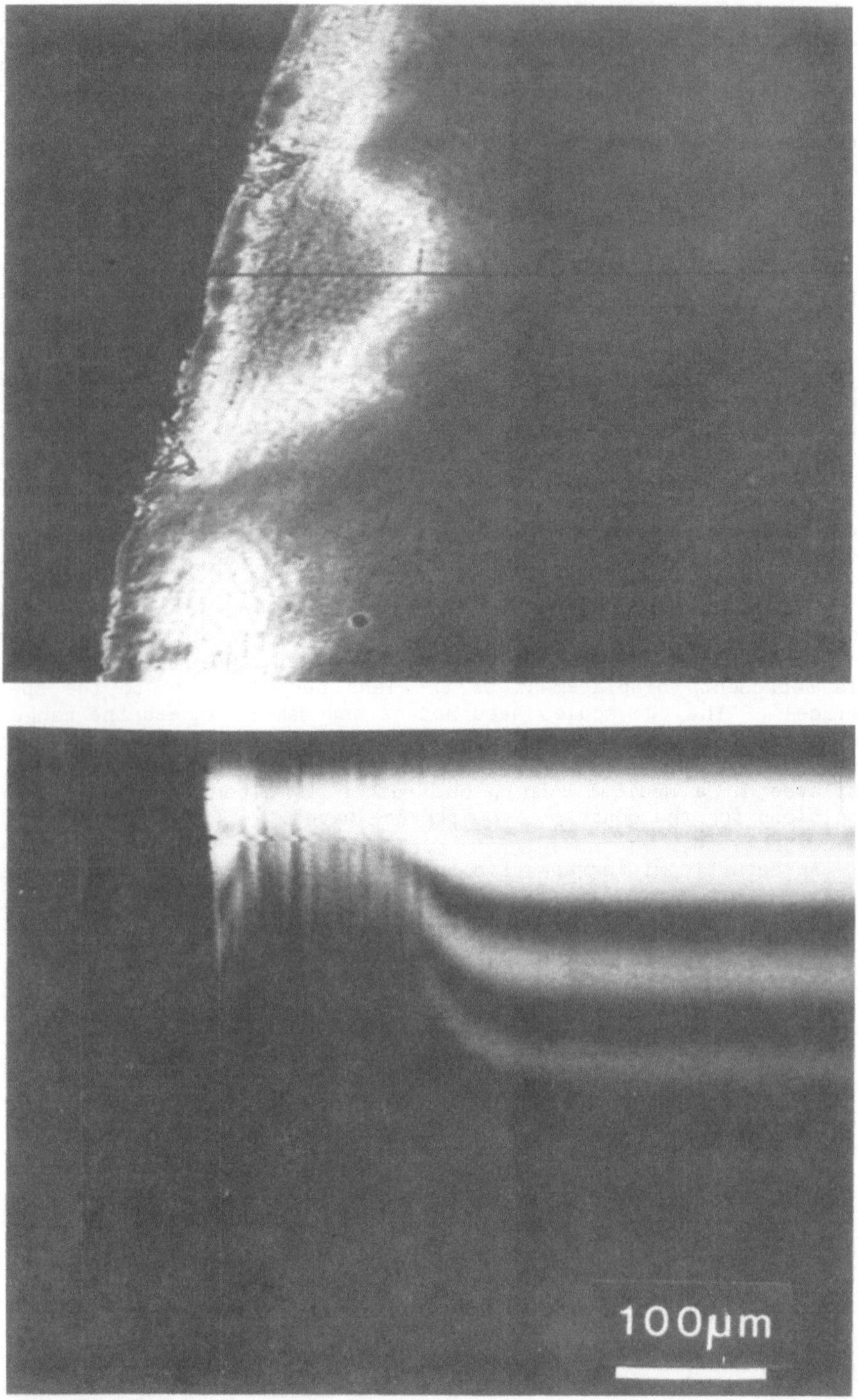

Fig.5. An acoustic micrograph of a section through human dental enamel containing a caries lesion (f = 370MHz, z = -4µm), together with V(x,z) recorded along the line indicated at 250MHz (the total range of defocus was 172µm).

in figure 5a. Along the indicated line V(x,z) was recorded, and is shown in figure 5b. Each vertical line in the V(x,z) can be thought of as a V(z) curve displayed in a brightness modulated form, and can therefore be interpreted in just the same way as any other V(z) curve[13 14]. The fringes correspond to the oscillations in V(z) due to Rayleigh wave interference. In the lesion they are closer spaced and they decay faster, indicating reduced Rayleigh velocity and increased attenuation[15].

Another image of caries lesion is shown in figure 6[16]. Here discrete V(z) curves have been measured at 15μm intervals along the line indicated in [a]. The V(z) curves were analysed by Fourier techniques to yield the Rayleigh wave velocity and attenuation, as plotted in [b]. In a caries lesion the density of the enamel may be 25% or so less than that of sound enamel, and by itself this would lead to an increase in acoustic wave velocity. Therefore, since in fact a decrease in velocity is observed there must be a reduction in elastic stiffness considerably greater than the reduction in intensity.

An alternative way to observe Rayleigh wave phenomena in the acoustic microscope is to use very short broadband pulses[3 17]. As the microscope is defocussed, so the normal reflection and the Rayleigh reflection separate in time, the normal reflection arriving first. By measuring the delay and strength of the Rayleigh reflection, the Rayleigh wave velocity and attenuation can be deduced. Such a measurement of a cross-section of a tooth is presented, together with the parameters deduced from it, in figure 7[18]. The line scan is presented in figure 7a as a V(y,t). Here the vertical axis represents position of the lens along a line parallel to the specimen surface. The horizontal axis represents time so that each horizontal line across the picture is like an oscilloscope trace, except that the signal is represented by brightness so that a positive value appears white and a negative value appears black. In some regions, such as the core of the lesion, there are strong fluctuations in the Rayleigh pulse (as indeed there are in figure 5b), and so measurements there should be treated with caution. Nevertheless, it is interesting to compare the Rayleigh wave parameters in the enamel in the vicinity of the lesion in figure 7b with the results by conventional V(z) analysis in figure 6b.

TISSUE SECTIONS

The time resolved system used for figure 5 was originally developed for the measurement of thin sections of biological tissue, in order to overcome some of the ambiguity found in interference measurements (using conventional gated c.w. excitation) of tissue sections[19]. A study of a transverse section of mouse muscle is presented in figure 8[20]. Conventional optical and acoustic micrographs are shown in [a] and [b]; these images show the soleus and gastrocnemius muscles, and the margin between them. V(y,t) is shown in unprocessed form in [c], and after Wiener filtering to sharpen up the pulses in [d][21]. The method of recording and displaying V(y,t) is the same here as with the tooth enamel, but the significance is quite different. There is no Rayleigh signal, and the two reflected pulses correspond to the top and bottom of the tissue section, which in this case had been cut in a freeze-microtome to a nominal thickness of 14μm. The time of the first pulse gives the height of the top of the section, and since it is on a flat glass slide this enables the thickness to be deduced (stability of the temperature of the water is crucial for this, since a small error in the velocity in the water leads to a large error in the calculated section thickness). The amplitude of the first echo enables the acoustic impedance to be calculated. The time interval between the first and second echoes (since the thickness is already known) gives the velocity, and hence also the

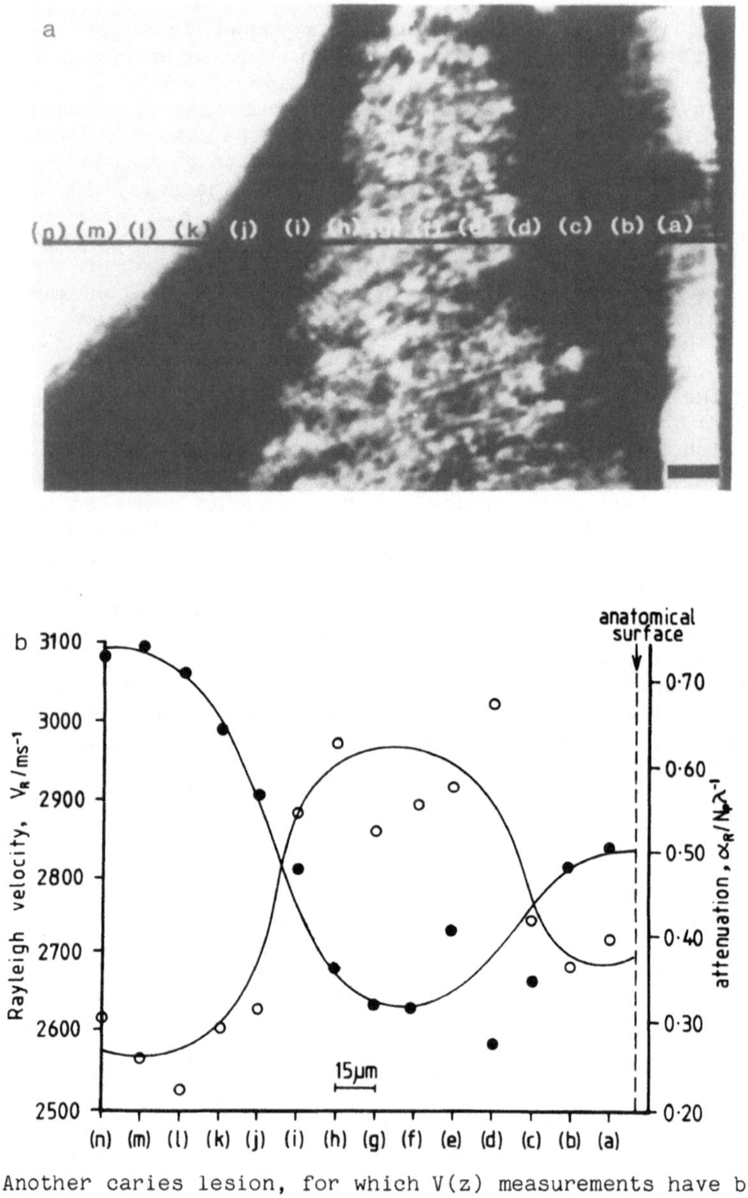

Fig.6. Another caries lesion, for which V(z) measurements have been made at points (a)-(n). The Rayleigh wave velocity and attenuation at each point were deduced from Fourier analysis of each V(z) curve.

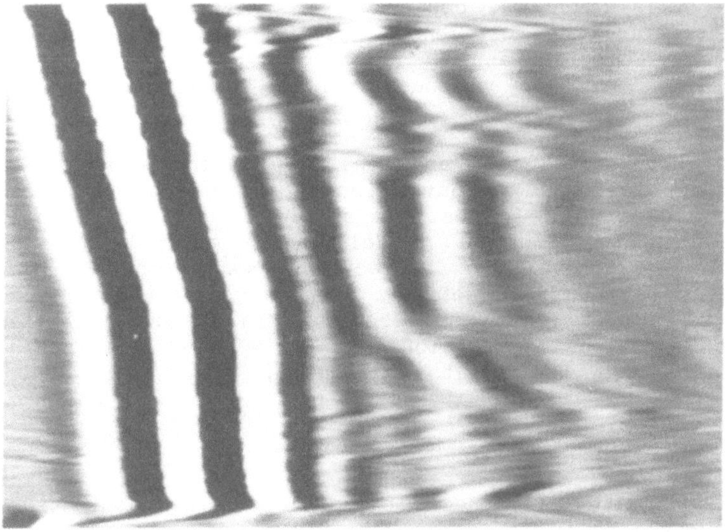

(a) B-scan of tooth - image width is 38 ns; vertical scan is 870 um.

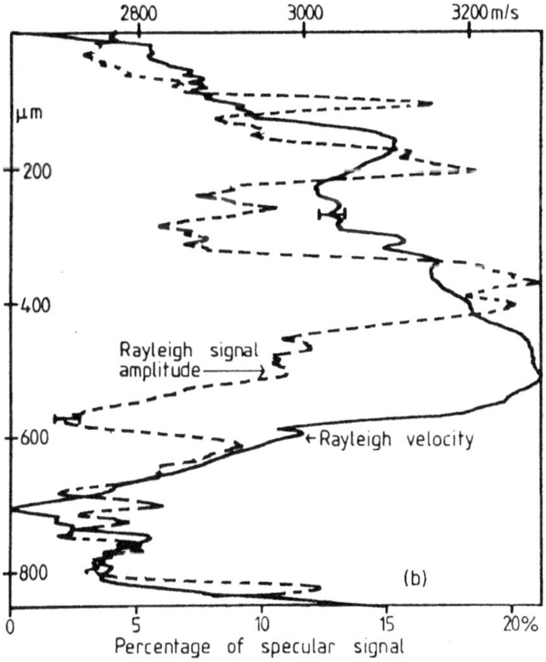

(b) Elastic properties of tooth

Fig.7. [a] V(y,t) for a caries lesion. The total horizontal scan corresponds to 38ns, and the vertical scan is 870μm. [b] Elastic properties of the tooth deduced from V(y,t).

density. Finally, the amplitude of the second echo gives the attenuation, and, using spectral analysis, its frequency dependence. These quantities are shown in figure 9, deduced from analysis of V(y,t) in figure 8. In particular, the collagenous material between the muscle fibres has a higher velocity and attenuation than is typical for tissue. At this resolution, the frequency dependence of the attenuation appears to behave anomalously[20].

Time-resolved measurements such as these help interpret images of tissue sections obtained in a conventional gated c.w. acoustic microscope, as in figure 9a. In most images of tissue, the acoustic attentuation dominates the contrast. Collagenous tissue, however, produces a far larger tissue/water reflection because of its high impedance. This can lead to novel effects such as contrast reversal[22] when the lens is focused at top of the section. In the present time-resolved experiment, the lens is positioned at $z = +7\mu m$ to minimize the effects of such focussing on the results.

CERAMICS AND HARD METALS

Turning to modern materials, one of the most powerful applications of acoustic microscopy lies in its great sensitivity to surface cracks. Indeed, it is proving invaluable in simple measurements of the length of radial cracks from hardness indentations for measurement of fracture toughness[23]. The Rayleigh wave fringes from such cracks tend to stop at a well defined length, thus removing some of the ambiguity in finding the tip of the crack by light or electron microscopy.

The ability to detect microcracks is being exploited in research into the micromechanisms of creep and fracture in low ductility materials[24]. Acoustic micrographs of the vicinity of crack tips in two hard-metal specimens (WC-6%Co) are shown in figure 10. Both specimens were precracked and then subjected to stress at 850°C (about 50°C above the transition temperature at which the behaviour ceases to be linear elastic). In [a] the specimen was loaded at temperature for a very short time (less than one minute). The crack is straight with a well defined tip, and little creep has occurred. The situation in [b] is quite different. The specimen was held under load at 850°C for 12 minutes. The crack has grown by void nucleation and coalescence, and secondary microcracking associated with creep damage is apparent. Considerable ingenuity has been expended in combining s.e.m. and s.a.m. observations using image processing and analysis techniques in order to make such observations more quantitative, and to enable the sizes of the relevent zones to be measured[24].

In components made of low ductility materials such as engineering ceramics, the critical defect size that can be tolerated is usually much smaller than in conventional alloys, because the limited capacity for plastic deformation leads to low fracture toughness. Therefore if such components are to be inspected nondestructively for cracks and other defects, a much shorter wavelength is necessary than is common in conventional ultrasonic n.d.t. Figure 11a is an acoustic micrograph of a Hertzian crack in the surface of a specimen of silicon nitride; the crack was formed by dropping a load on the surface. An actual component, a silicon nitride ball-bearing, is shown in figure 11b. This ball bearing had never been used in service, but it was one of a type that often exhibited premature failure. The acoustic image (which shows circular artifacts due to the curvature of the specimen surface) reveals serious cracks in the ball bearing, which could well grow catastrophically when it was subjected to load.

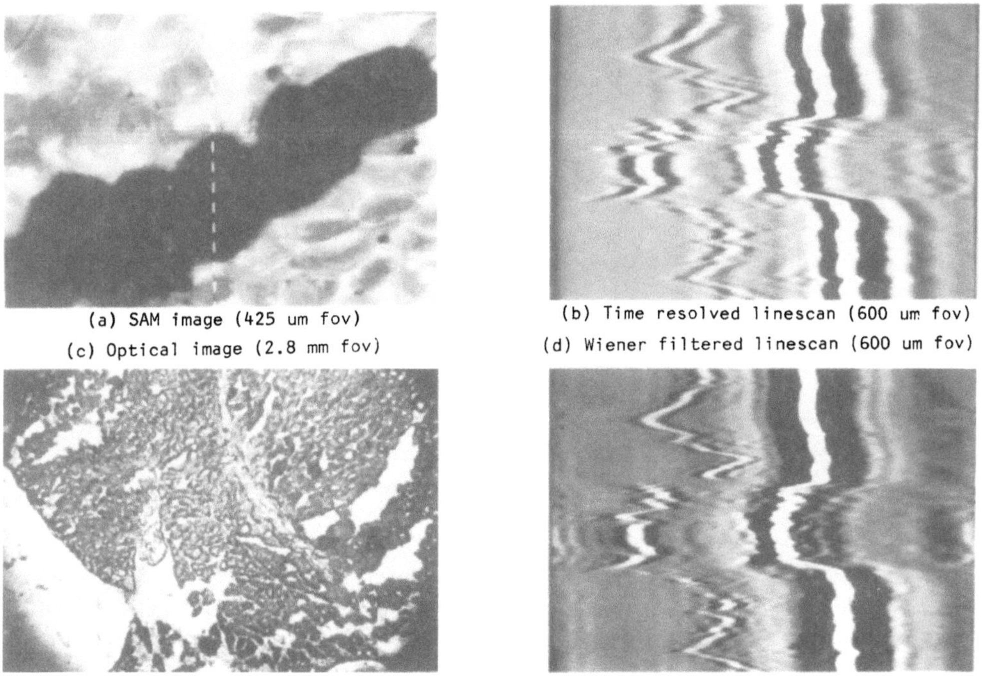

(a) SAM image (425 um fov)
(b) Time resolved linescan (600 um fov)
(c) Optical image (2.8 mm fov)
(d) Wiener filtered linescan (600 um fov)

Fig.8. Images of a transverse section of mouse muscle.

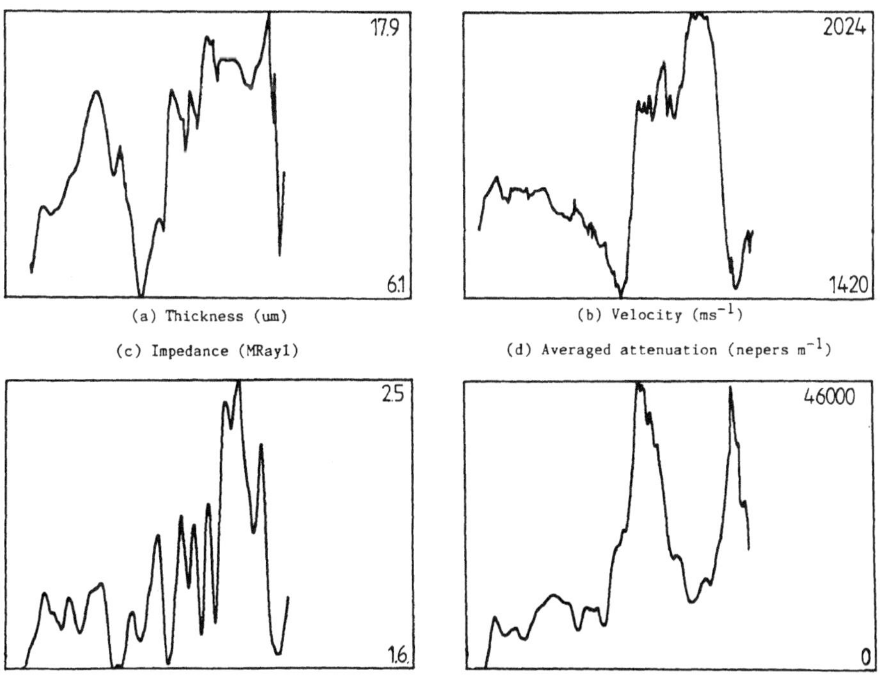

Fig.9. Elastic properties of the transverse section of mouse muscle.

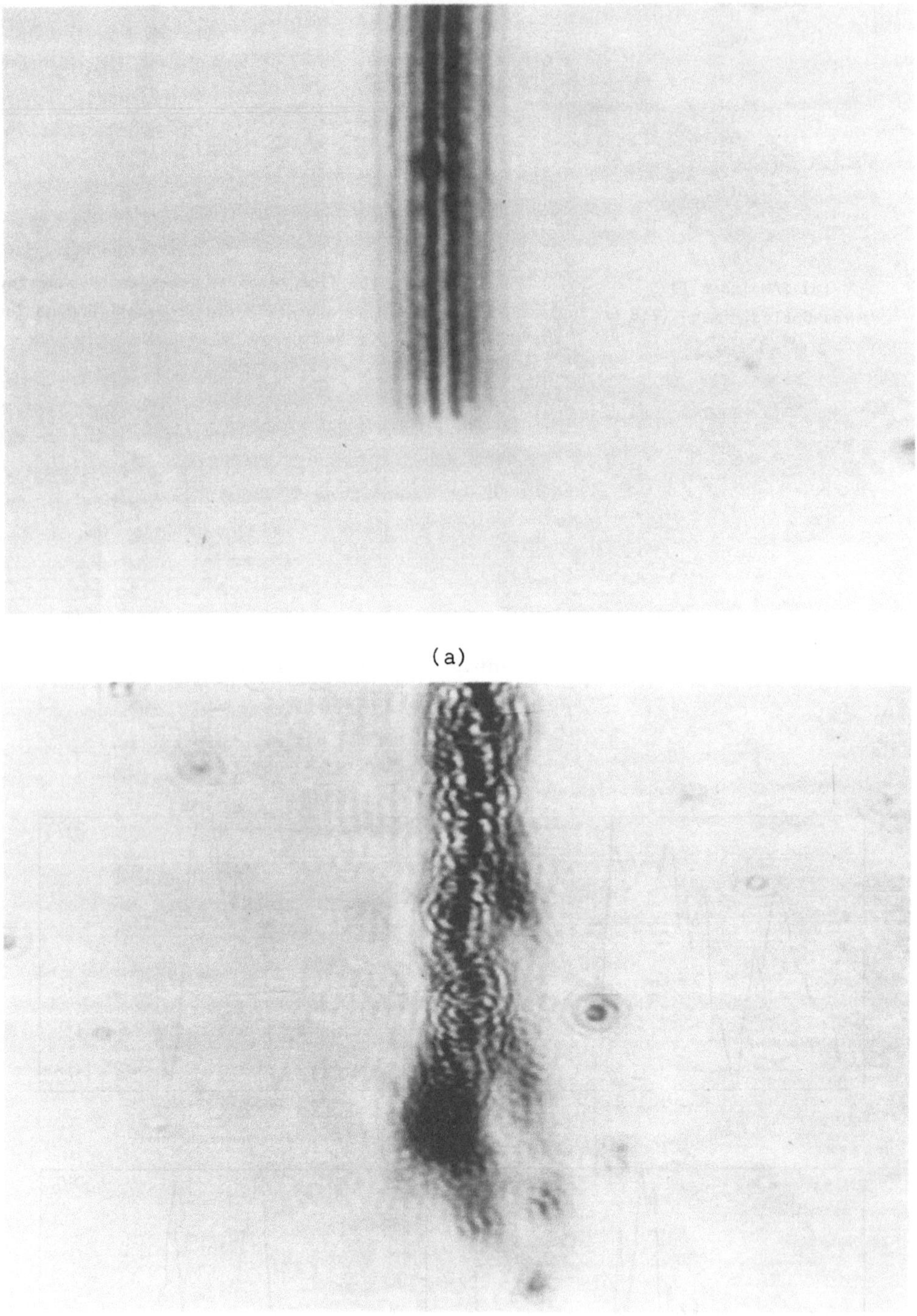

(a)

(b)

100μm

Fig.10. Specimens of WC-6%Co containing precracks subjected to stress at 850°C for: [a] less than 1 min (f = 370MHz, z = -25μm); [b] 12 min (f = 250MHz, z = -39μm).

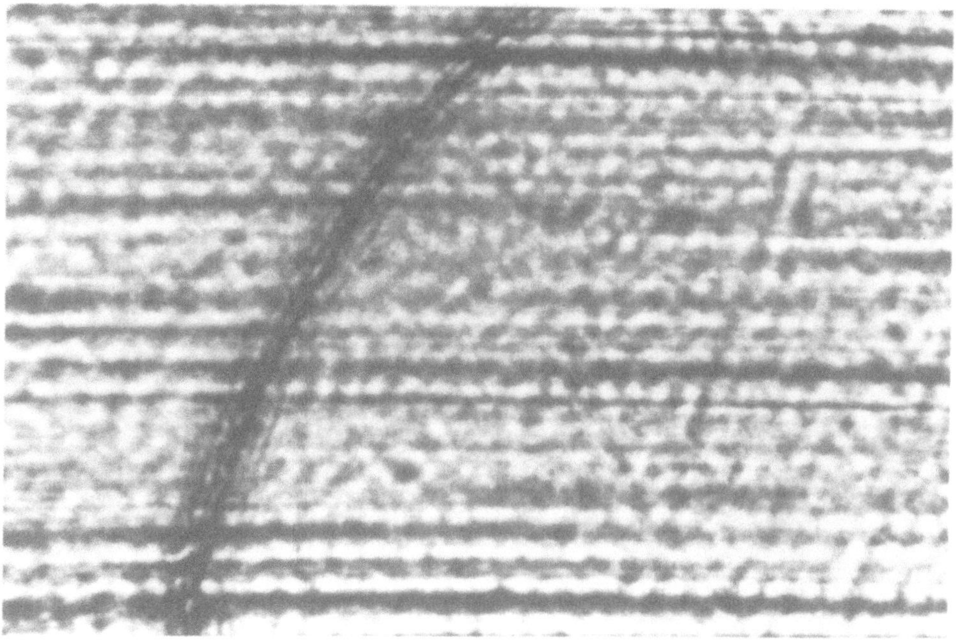

(a)

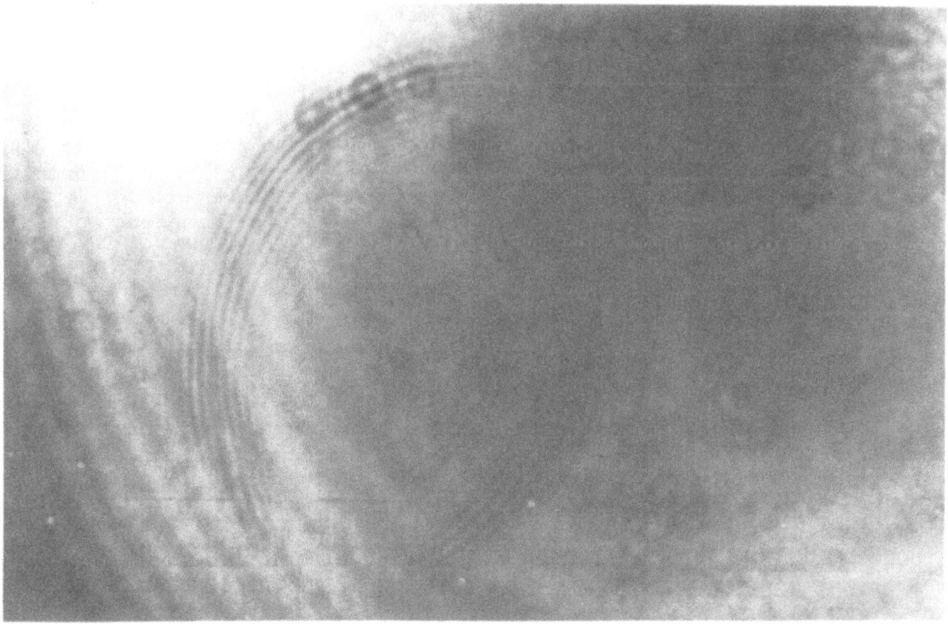

(b)

200µm

Fig.11. [a] Acoustic micrograph of a crack in a block of Si_3N_4 (f = 250MHz, z = -144µm). [b] Cracks in a Si_3N_4 ball (f = 250MHz, z = -180µm (variable, because the surface is curved).

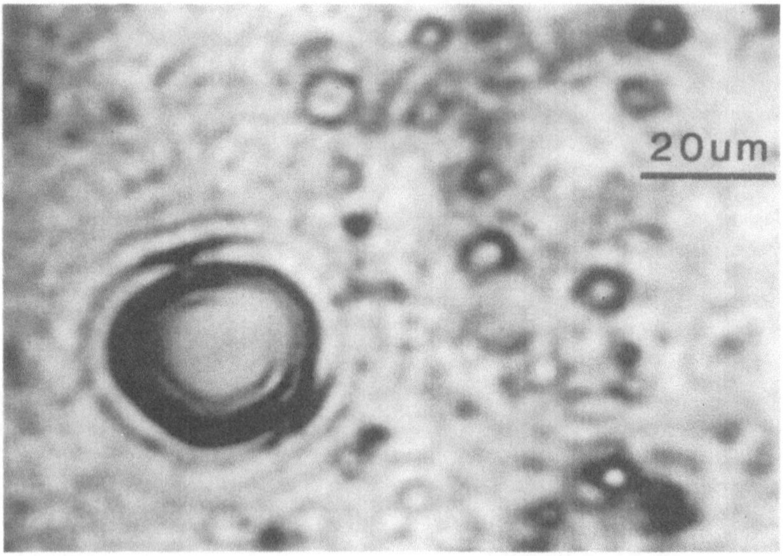

Fig.12. A section through a single fibre specimen of Pysic (f = 370MHz, z = -15μm).

As well as examining simple cracks, the acoustic microscope can give contrast from interfaces in composite. An example is shown in figure 12[25], which is a micrograph of a section through a specially prepared single fibre specimen of Pysic, a silicon carbide fibre-reinforced composite with a pyrex glass matrix. The contrast varies around the interface, indicating that the interface is not uniform. Curiously enough, in the series of experiments from which this example was taken the greatest contrast at the interface was found with specimens prepared so as to give the best adhesion. But the acoustic microscope does not image adhesion as such directly, in the sense of the strength that a bond will show under load. Rather, the acoustic microscope images elastic properties, such as for example the extent of true contact, and interpretation of such contrast requires detailed knowledge of how this might relate to fracture behaviour for the particular system of materials being examined.

WHAT ELSE?

The range of applications of acoustic microscopy is growing so fast that the kind of results illustrated here are scarcely even representative. There are, for example, well established applications in the fields of cell biology on the one hand[26], and in semiconductor manufacturing, at all stages from the raw wafer through the etching process to examination of lines on finished circuits, on the other. A common theme in many of these applications is the need for quantitative analysis.

For imaging purposes, commercial instruments are now available that offer the highest resolution that is likely to be available in the foreseeable future with convenient coupling fluids, and sophisticated image processing and analysis packages can be connected to these. For accurate measurement of surface acoustic wave parameters on isotropic and anisotropic surfaces the line-focus-beam technique has been well demonstrated for some time. Remaining needs include better established methods for wave parameter measurement with a point focus lens as robust as those that exist for the line-focus-lens[27], and also techniques for measurement of dimensions, such as the width of lines below the resolution limit, or the depth of cracks.

REFERENCES

1. R. S. Gilmore, K. C. Tam, J. D. Young and D. R. Howard, Acoustic microscopy from 10 to 100MHz for industrial applications, Phil. Trans. R. Soc. Lond. A320, 215-235 (1986).
2. A. Atalar and M. Hoppe, A high performance acoustic microscope, Rev. Sci. Inst. 57, 2568-76 (1987).
3. K. Yamanaka, Surface acoustic wave measurement using an impulsive converging beam, J. Appl. Phys. 54, 4323-29 (1983).
4. J. Kushibiki and N. Chubachi, Material characterization by line-focus-beam acoustic microscope, IEEE Trans. SU-32, 189-212 (1985).
5. M. Hoppe and J. Bereiter-Hahn, Applications of scanning acoustic microscopy, IEEE Trans. SU-32, 289-301 (1985).
6. C. Ilett, M. G. Somekh and G. A. D. Briggs, Acoustic microscopy of elastic discontinuities, Proc. R. Soc. Lond. A393, 171-83 (1984).
7. M. G. Somekh, G. A. D. Briggs and C. Ilett, The effect of anisotropy on contrast in the scanning acoustic microscope, Phil. Mag. 49, 179-204 (1984).
8. K. Yamanaka, Surface acoustic wave measurements using an impulsive converging beam, J. Appl. Phys. 54, 4323-4329 (1983).
9. N. Chubachi, Ultrasonic microspectroscopy via Rayleigh waves, in: Rayleigh-wave theory and applications, E. A. Ash, E. G. S. Paige, eds., Springer-Verlag, Berlin, pp291-297 (1985).
10. B. Augmar-Mansson and J. J. ten Bosch, Optical methods for the detection and quantification of caries, Adv. Dent. Res. 1, 14-20 (1987).
11. S. D. Peck and G. A. D. Briggs, A scanning acoustic microscope study of the small caries lesion in human enamel, Caries Res. 20, 356-360 (1986).
12. V. B. Jipson, Acoustic microscopy at optical wavelengths, Ph D Thesis, Stanford University (1979).
13. C. F. Quate, A. Atalar and H. K. Wickramasinghe, Acoustic microscopy with mechanical scanning - a review, Proc. IEEE 67, 1092-1114 (1979).
14. J. Kushibiki, A. Ohkubo and N. Chubachi, Effect of leaky SAW parameters on V(z) curves obtained by acoustic microscopy, Electron Lett. 18, 668-70 (1982).
15. S. D. Peck and G. A. D. Briggs, The caries lesion under the scanning acoustic microscope, Adv. Dent. Res. 1, 50-63 (1987).
16. S. D. Peck, J. M. Rowe and G. A. D. Briggs, Quantitative acoustic microscope studies on sound and carious enamel, (submitted to J. Dental Res.) (1988).
17. K. Liang, S. D. Bennett, B. T. Khuri-Yakub and G. S. Kino, Precise measurement of Rayleigh wave velocity perturbation, Appl. Phys. Lett. 41, 1124-6 (1982).
18. J. M. R. Weaver, C. M. W. Daft, S. D. Peck and G. A. D. Briggs, Applications of broadband scanning acoustic microscopy, (submitted to IEEE Trans. UFFC) (1988).
19. C. M. W. Daft, J. M. R. Weaver and G. A. D. Briggs, Phase contrast imaging of tissue in the scanning acoustic microscope, J. Microsc. 139, RP3-4 (1985).
20. C. M. W. Daft, G. A. D. Briggs and W. D. O'Brien, Frequency dependence of tissue attenuation measured by acoustic microscopy, (submitted to J. Acoust. Soc. Am.) (1988).

21. Y. Murakami, B. T. Khuri-Yakub, G. S. Kino, J. M. Richardson and A. G. Evans, An application of Wiener filtering to nondestructive evaluation, Appl. Phys. Lett. 33, 685-7 (1978).
22. C. M. W. Daft and G. A. D. Briggs, Wideband acoustic microscopy of tissue, (IEEE Trans. UFFC, in the press) (1988).
23. K. Yamanaka, J. Kushibiki and N. Chubachi, Anisotropy detection in hot-pressed silicon nitride by acoustic microscopy using the line-focus-beam, Electron Lett. 21, 165-7 (1984).
24. H. G. Schmid, The mechanisms of fracture of WC-11wt%Co between 20°C and 1000°C, Materials Forum 10, 184-197 (1987).
25. D. G. P. Fatkin, C. B. Scruby and G. A. D. Briggs, Acoustic microscopy of low ductility materials, (submitted to J. Mater. Sci.) (1988).
26. J. Bereiter-Hahn, Scanning acoustic microscopy visualizes cytomechanical responses to cytochalasin D, J. Microsc. 146, 29-39 (1987).
27. J. M. Rowe, Quantitative acoustic microscopy of surfaces, D Phil Thesis, Oxford University (1988).

DUAL-BEAM DIFFERENTIAL AMPLITUDE CONTRAST SCANNING ACOUSTIC MICROSCOPY

M. Nikoonahad and E. A. Sivers

Bio-Imaging Research, Inc.
425 Barclay Blvd., Lincolnshire, IL 60069, USA

1. INTRODUCTION

In the reflection acoustic microscope {1}, the large impedance mismatch at the coupler/object interface can lead to a large, but relatively constant, signal. This is because, despite the large reflectivity at this interface, in some specimens, the *variations* in object impedance and, therefore, in the reflectivity may be quite small. For example, it has been shown that the grain structure in certain solids results in only a few percent change in reflectivity {2}. In such situations, the large background signal can overshadow the signal which arises from the reflectivity variations, leading to a poor image contrast. These issues become more serious in gas medium {3} and cryogenic {4} systems to the extent that, in these forms of microscopy, a major part of the image contrast is purely due to topography. It is clear that, to attain the largest image contrast, only the changes in the signal have to be displayed. Also, it is clear that, ideally, one requires a system which yields no signal for a perfect reflector. In such a system – a differential amplitude microscope – the difference in the amplitudes from adjacent points of the sample constitute the imaging signal.

We previously reported a differential phase contrast acoustic microscope based on a dual-beam lens {5,6}. Also, recently we described a differential amplitude microscope {7} and a high resolution flow measuring system {8} based on the same lens concept. In the past, the design methodology for the dual-beam lens has been geometrical acoustics. Although this approach provides a starting point, it cannot adequately describe the behavior of the dual-beam lens. The aims of this paper are to: (a) introduce a differential amplitude microscope; (b) present a diffraction formulation for the dual-beam lens; and (c) demonstrate experimentally a differential amplitude microscope at 10 MHz.

The basic elements of the dual-beam, differential amplitude microscope are shown in figure 1. The dual-beam lens produces two adjacent foci on the surface of the specimen. The signals received from these two foci are first amplitude detected and then brought to baseband by means of two sample-and-hold modules, figure 1. The electronic subtraction of the two signals leads to the differential amplitude imaging signal.

In section 2, a diffraction analysis for the dual-beam lens is presented. Section 3 deals with the experimental results and the images obtained at 10 MHz. The conclusions are presented in section 4.

2. DIFFRACTION ANALYSIS FOR THE DUAL–BEAM LENS

In this section, we present a two–dimensional diffraction analysis for the dual–beam lens. This analysis is based on a scalar field theory and is accompanied by a number of simulations for a 10 MHz microscope.

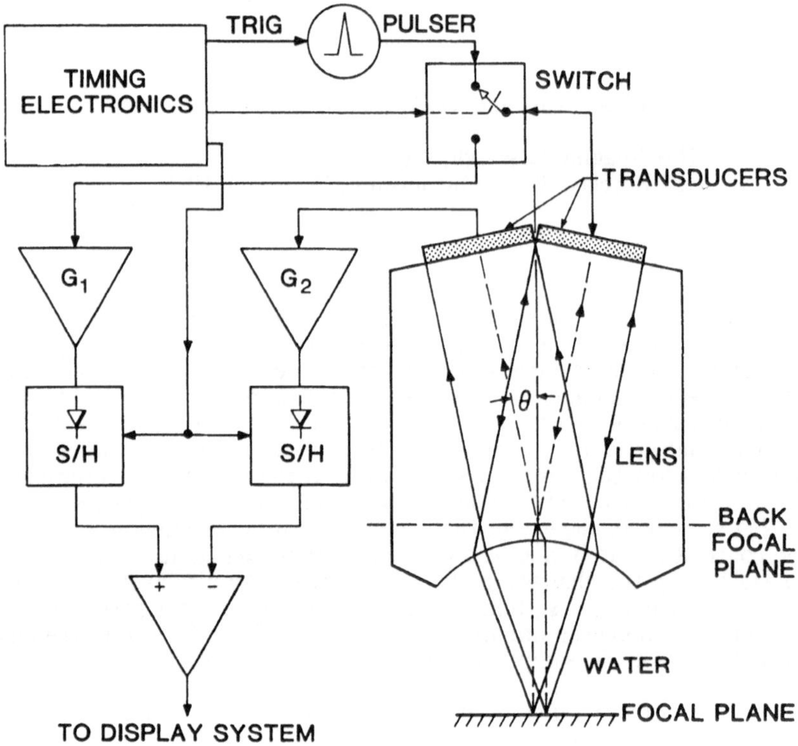

Figure 1. Dual–beam differential amplitude microscope.

2.1 Theory

Based on geometrical acoustics, the tilt angle and the dimensions of the lens are designed so that the central ray from each transducer passes through the intersection point of the back focal plane and the lens axis, figure 2. Once the two central rays traverse the lens, they travel parallel to the lens axis and, hence, we have two off–axis foci separated by $|\Delta x|$ where:

(1) $\Delta x \simeq 2\frac{f}{n}\tan\theta$

$f \equiv nR/(n-1)$ is the paraxial focal length for a velocity ratio of n across the lens and θ is shown in figure 2. It is clear that the maximum width of each transducer w is dictated by:

$$(2) \quad w = 2\left(\ell - \frac{f}{n}\right)|\sin\theta|$$

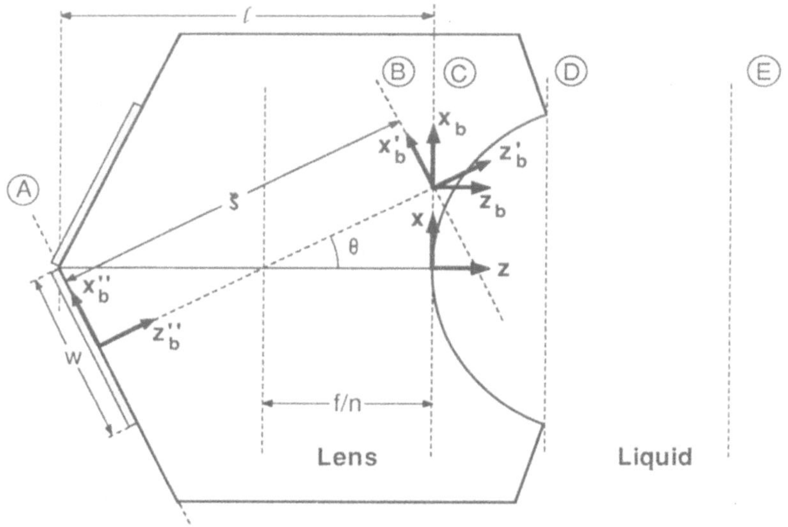

Figure 2. Various parameters and the coordinate systems used in the formulation.

We assume that the ultrasonic disturbance can be described by a monochromatic scalar field U(x,z), and we ignore the effects of reflections and mode conversion within the lens. The derivation presented here is for the bottom transducer only, figure 2. The corresponding relationships for the top transducer are obtained by replacing θ by $-\theta$ in the equations. There are four coordinate systems, figure 2, used in our formulation; (x''_b, z''_b), (x'_b, z'_b), (x_b, z_b) and (x,z). The transformations between these systems are given by:

(3–a) $x''_b = x'_b$ (3–b) $z''_b = z'_b + \zeta$

(4–a) $x'_b = x_b \cos\theta - z_b \sin\theta$ (4–b) $z'_b = z_b \cos\theta + x_b \sin\theta$

(5–a) $x_b = x - \Delta x/2$ (5–b) $z_b = z$

Given the transducer field $U^A(x''_b,0) = \mathrm{rect}(x''_b/w)$, the field at plane B can be readily calculated by either Fourier optics or by an appropriate diffraction analysis. It turns out that, in the majority of cases, the lens is at the far field of the transducer aperture. Using the Frounhofer approximation, the field at B can be written as {9}:

$$(6)\quad U^B(x''_b,z''_b) = \frac{e^{[j k_s z''_b]}}{2\pi j z''_b}\, e^{\left[\frac{j k_s x''^2_b}{2 z''_b}\right]}\, k_s w \,\mathrm{sinc}\left[\frac{k_s w x''_b}{2\pi z''_b}\right]$$

where k_s is the wave number in the solid lens. Given equation (6), the field distribution can be found at plane C in the (x,z) system by taking the following steps: (i) find the field in the (x'_b,z'_b) system using equation (3); (ii) transform the field to the (x_b,z_b) system using equation (4); and (iii) express the field in plane C in the (x,z) system by using equation (5).

$$(7)\quad U^C(x,z) = \frac{e^{[j k_s (z\cos\theta + (x - \Delta x/2)\sin\theta + \zeta)]}}{2\pi j (z\cos\theta + (x - \Delta x/2)\sin\theta + \zeta)}$$

$$\cdot e^{\left[\frac{j k_s ((x - \Delta x/2)\cos\theta - z\sin\theta)^2}{2(z\cos\theta + (x - \Delta x/2)\sin\theta + \zeta)}\right]}$$

$$\cdot k_s w\,\mathrm{sinc}\left[\frac{k_s w((x - \Delta x/2)\cos\theta - z\sin\theta)}{2\pi(z\cos\theta + (x - \Delta x/2)\sin\theta + \zeta)}\right]$$

We assume that the lens imposes a spherical (circular in 2-D) phase curvature on the incident field. Furthermore, we ignore any diffraction within the lens. In this respect, the lens is assumed to act as a "thin lens". We have:

$$(8)\quad U^D(x,0) = \mathrm{rect}(x/A)\; e^{j(k_l - k_s)\sqrt{R^2 - x^2}}\; U^C(x,0)$$

where A and k_l are the lens aperture and the wave number in the liquid respectively. Given the field just after the lens, we can now use standard Fourier optics to compute the field at any arbitrary plane E in front of the lens:

$$(9)\quad U^E(x,z) = \int_{-\infty}^{\infty} A^D(f_x)\, e^{jz\sqrt{k_1^2 - (2\pi f_x)^2}}\, e^{2\pi j f_x x}\, df_x$$

where, for a spatial frequency f_x, the angular spectrum $A_D(f_x)$ is given by:

$$(10)\quad A^D(f_x) = \int_{-\infty}^{\infty} U^D(x,0)\, e^{-2\pi j f_x x}\, dx$$

Alternatively, one could use equations (1) through (7) to find the field at the back focal plane and then use the Fourier transform properties of a thin lens. We have deliberately used an angular spectrum approach in order to study the formation, separation and overlapping of the two beams at various planes in the vicinity of the focus.

2.2 Simulations

The lens dimensions used in the simulations are : $\ell = 32$ mm, w = 3.7 mm, R = 5 mm, A = 9 mm and $\theta = 3.5^o$. The lens is assumed to be aluminum (velocity = 6.35 mm/μsec) focusing in water (velocity = 1.5 mm/μsec). Using these parameters, n = 4.2, f = 6.55 mm and Δx = 190 μm. The simulations have been carried out for a 10 MHz system. With these dimensions, the lens is located at 5.7 times the Fresnel distance from the transducers and the far field approximation is valid. Figure 3 shows the field distribution just before the lens, computed from equation (7). We see that the lens aperture of 9 mm extends over most of the mainlobe of the *sinc* function.

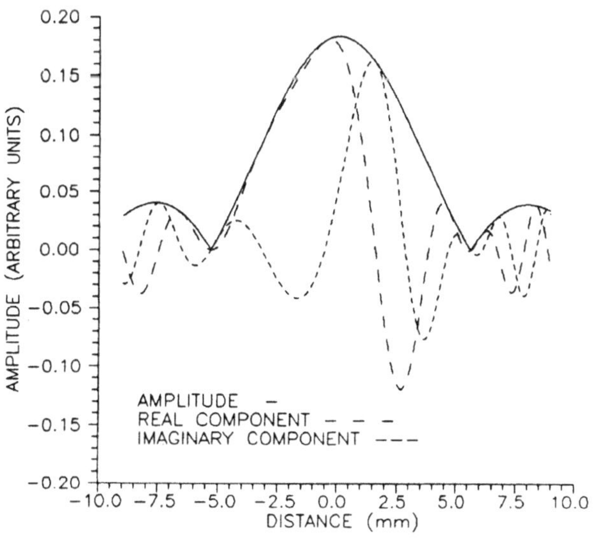

Figure 3. The field distribution just before the lens in the (x,z) coordinate system (bottom transducer).

Such an apodized lens aperture, of course, leads to a reduced sidelobe level at the focal plane. Also, the asymmetry of this field, as illustrated by the real and imaginary components, is the result of the tilt angle and coordinate transformations. It is this asymmetry which consequently leads to the formation of an off-axis focus. Figures (4-a) through (4-i) show the field distributions in front of the dual-beam lens at nine different values of z, 5.0 mm < z < 8.0 mm, including the paraxial focal plane at z = f = 6.55 mm. We see that when z < f, there is a considerable amount of interference in each beam. As z approaches f, the two beams start separating from each other. The separation between the two beams at the paraxial focal length is 184 μm, which is in excellent agreement with the design based on geometrical acoustics. Also, since this is a wide angle lens, we see that the field amplitudes are higher, the mainlobes are narrower and the sidelobes are higher when z is shorter than the paraxial focal length. These observations are in good agreement with what has been reported previously for single beam lenses {10}.

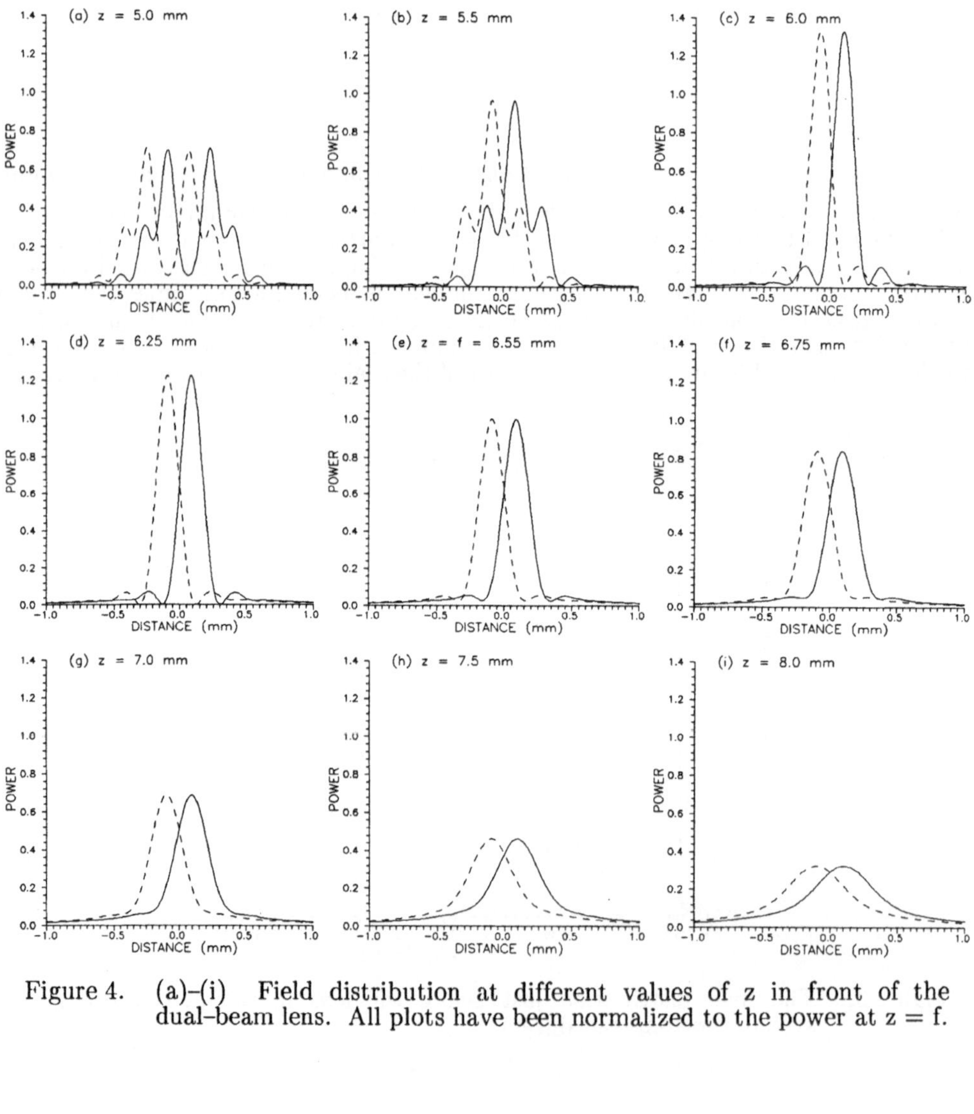

Figure 4. (a)-(i) Field distribution at different values of z in front of the dual-beam lens. All plots have been normalized to the power at z = f.

3. EXPERIMENTS

Using the lens dimensions given in the last section, a lens was fabricated and two 10 MHz transducers were bonded to the sloped surfaces. The point sensitivity function of the dual-beam lens was measured by scanning the lens over a wire stretched between two points in water. We have experimentally measured {5,6} a Δx of 195 μm, which is in excellent agreement with the simulations. We previously reported that, by using the signal emitted from one transducer and collected by the other, only one driving pulser need be used {6,7}, figure 1. To perform differential amplitude imaging, the gains of the two pre-amplifiers, G_1 and G_2, are adjusted so that, for a perfect reflector, the output of the differential amplifier is at the noise level. This ensures a maximum imaging sensitivity. Figure 5 shows a set of line scans obtained from a 10 μm ($\lambda/15$) step in stainless steel. The differential nature of the microscope is demonstrated by this result. As far as topography is concerned, we assume that, for a height change of Δz from the focus, the transducer outputs vary as $\exp(-(\Delta z/z_0)^2)$ (with z_0 being a measure of the depth of focus). It can easily be shown that, for a small Δz, the differential amplitude signal is proportional to $(\Delta z)^2$. The measured SNR for this step was 40 dB in a 100 Hz bandwidth. It is clear that, if topography is the only source of contrast, the microscope, with its present electronics, is sensitive to height changes down to 0.1 μm. The electronics can be improved if higher sensitivities are required. As far as height imaging is concerned, physical intuition suggests that phase contrast imaging {6,7,11,12} may offer a higher sensitivity. A one-to-one comparison requires a complete system analysis and depends on the total input powers. However, one must point out that, regardless of any topography, the differential amplitude signal can arise only from impedance differentials encountered by the two beams.

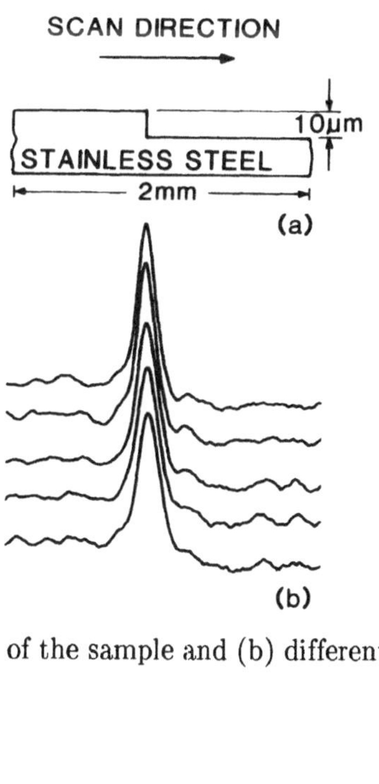

Figure 5. (a) Schematic of the sample and (b) differential amplitude line scans.

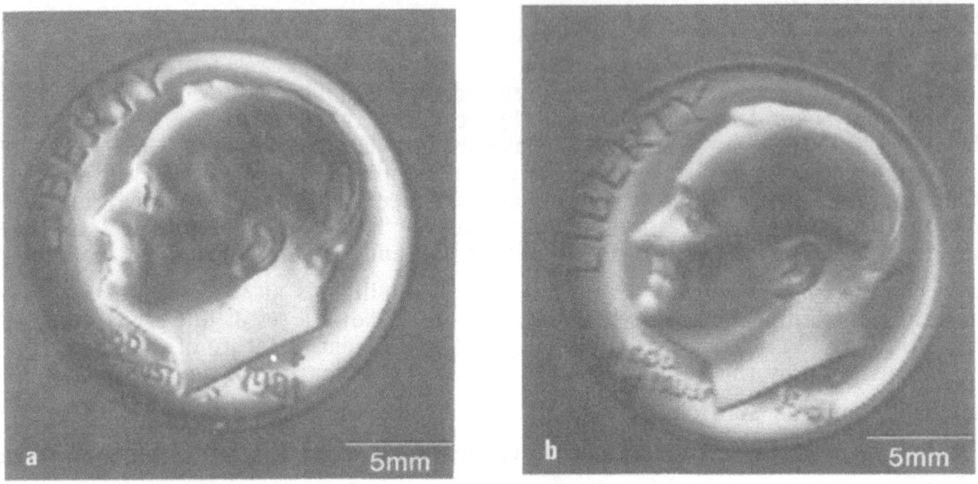

Figure 6. Differential amplitude images obtained from a coin. (a) Two foci scanned along horizontal direction and (b) two foci scanned along vertical direction.

Figure 6(a) shows an image obtained from a coin. In this image, the two beams have been scanned along the horizontal direction and it is evident that the image is mainly differential along this direction. Figure 6(b) illustrates an image taken with the lens rotated by 90°. We see that the differentiation is more pronounced along the vertical direction. Figure 7 shows an image obtained from a slice of zinc–selenide crystal. The interest in this crystal is its use in solid state visible lasers. The face of the crystal was polished and etched, which has created a variation in its surface topography. A mechanical profiler showed that the peak–to–peak topography in this sample was 0.25 μm. The contrast in the image is mainly due to this variation.

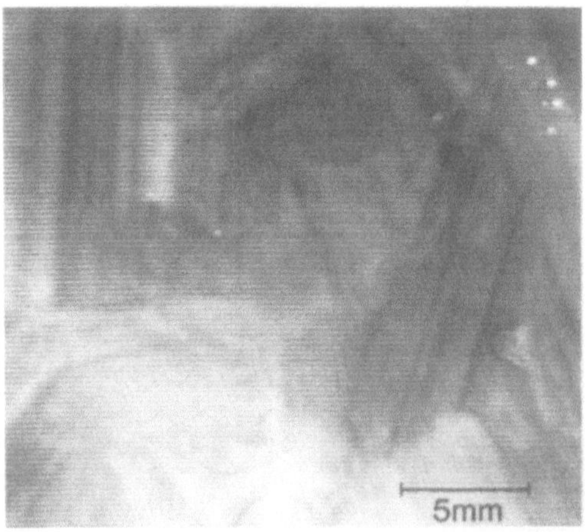

Figure 7. Differential amplitude image obtained from grains in a ZnSe crystal.

4 . CONCLUSIONS

A dual-beam differential amplitude contrast acoustic microscope was described. We presented a diffraction formulation to analyze a dual-beam lens and demonstrated a differential amplitude microscope by a range of images obtained at 10 MHz using such a lens. We have observed an excellent agreement between the simulations and the experimentally measured performance of the dual-beam lens. It is believed that a differential amplitude microscope provides a means to increase the sensitivity in situations where the reflectivity changes in the specimen are small.

REFERENCES

{1} C. F. Quate, A. Atlar and H. K. Wickramasinghe (1979), "Acoustic Microscopy with Mechanical Scanning - A Review", *Proc. of IEEE,* 67(8), pp 1092–1113.

{2} Q. R. Yin, D. Ilett and G. A. D. Briggs (1982), "Acoustic Microscopy of Ferroelectric Ceramics", *Journal of Materials Science,* 17, pp 2449–2452.

{3} B. Hadimioglu and J. S. Foster (1984), "Recent Developments in Superfluid Helium Acoustic Microscopy", *Proc. of IEEE Ultrasonics Symposium,* 84CH2112-1, vol 2, pp 593–597.

{4} F. Faridian (1985), "Gas Medium Acoustic Microscopy at 160 MHz", *Proc. of IEEE Ultrasonics Symposium,* 85CH2209-5, vol 2, pp 759–762.

{5} M. Nikoonahad (1987), "Differential Phase Contrast Acoustic Microscopy Using Tilted Transducers", *Elect. Lett.,* 23, pp 489–490.

{6} M. Nikoonahad (1987), "New Techniques in Differential Phase Contrast Scanning Acoustic Microscopy", *Acoustical Imaging,* vol 16, Plenum, in press.

{7} M. Nikoonahad (1987), "Differential Amplitude Contrast in Acoustic Microscopy", *Appl. Phys. Lett.*, 51(21), pp 1687–1689.

{8} M. Nikoonahad and F. Li (1988), "High Resolution Ultrasound Traverse Flow Measurement", *Elect. Lett.,* 24(4), pp 205–207.

{9} See, for example, J. W. Goodman, (1968), Introduction to Fourier Optics, McGraw–Hill, San Francisco.

{10} M. Nikoonahad and E. A. Ash, (1982), "Ultrasonic Focussing in Absorptive Fluids", *Acoustical Imaging,* vol 12, pp 47–60.

{11} I. R. Smith and H. K. Wickramasinghe (1982), "Dichromatic Differential Phase Contrast Microscopy", *IEEE Trans. Sonics and Ultrasonics,* SU 29(6), pp 321–326.

{12} K. K. Liang, S. D. Bennett, B. T. Khuri–Yakub and G. S. Kino (1985), "Precise Phase Measurements with the Acoustic Microscope", *IEEE Trans. Sonics and Ultrasonics,* SU 32(2), pp 266–273.

WHAT CAN SCANNING ACOUSTIC MICROSCOPY TELL ABOUT ANIMAL CELLS AND TISSUES ?

J. Bereiter-Hahn, J. Litniewski, K. Hillmann, A. Krapohl, and L. Zylberberg[1]
Cinematic Cell Research Group, Johann Wolfgang Goethe-University, D 6000 Frankfurt/M, Fed. Rep. Germany
[1]Université Paris VII, Laboratoire d'Anatomie Comparéé. Equipe de recherche "Formations squelettiques" (CNRS UA 041137), Paris, France

INTRODUCTION

Scanning acoustic microscopy (SAM) is now becoming a widely used method for the investigation of solid materials while its application in the fields of biology and medicine where soft materials are prevalent is still in its infancy. The reasons for this situation are twofold
- the physiological significance of mechanical properties of cells is just in an early stage of being appreciated by the scientific community
- our knowledge on the interaction of cellular components with ultrasound in the GHz range is very limited.

The future of SAM in cell biology will depend on the importance of cellular mechanics for our understanding of cell physiology. A first summary of this topic has been compiled by one of the present authors (Bereiter-Hahn et al. 1987). Some observations point to an outstanding role of cell mechanics in the control of macromolecular synthesis among those are the close relationship between adhesion to a substrate and proliferation (Folkman and Moscona 1978), between tension in a cell layer and thymidine incorporation (Iwig et al. 1981) or mechanical stress and differentiation of keratinocytes (Görmar et al. 1988) and the intimate relation of intermediary filaments to nuclear pore complexes (Penman et al. 1983). In addition cytoskeletal elements seem to be involved in microcompartmentation of the cytoplasm (Bereiter-Hahn 1988, Clarke et al. 1983, Clegg 1984, Masters and Wilson 1981, Masters 1984, Tillmann and Bereiter-Hahn 1986). It is the cytoskeleton which can be supposed to determine the mechanical properties of cells which can be revealed by SAM.

Cytoskeletal structures are filamentous, so they are primarily tension resisting, bending resistivity (stiffness) is reached by lateral connections between the filamentous elements. Anchorage of the fibrils to cellular membranes is the structural basis for the generation of tension and cell shape. Parts of the cytoskeleton (actin fibrils and microtubules) are highly dynamic structures undergoing rapid assembly and disassembly processes which cannot be followed in living cells by conventional light microcope methods. The resolving power of SAM operated in the range of 1 to 2 GHz is sufficient to demonstrate some thick stress fibres although it does not allow to resolve single microtubules in living cells. However, direct visualization may not be required to follow the dynamics of supramolecular interaction of the type described above. SAM can be regarded as a non invasive method for the investigation of living cells (Hildebrand and Rugar 1984, Hoppe and Bereiter-Hahn 1985) therefore it seems to be suited for studying living cells in particular and thus revealing properties which by means of any other technique become apparent.

In the present contribution we first demonstrate some capabilities of SAM for studies of biological hard tissues. Most of the presentation is dedicated to the problems arising by quantification of acoustic properties of living cells and the physiological correspondence of acoustic impedance and attenuation.

HARD TISSUES

Teeth

SAM allows the investigation of polished surfaces of grounded teeth without demineralization. The various regions, enamel, dentin and pulpa can easiliy be distinguished (Fig. 1). In the dentin tubules are obvious (Fig. 1 b,c,d). They appear to be surrounded by a zone of higher reflectivity (Fig.1d). Whether this zone reveals peritubular dentin of high mineral and a low organic matrix content cannot be decided on the basis of the present matrial. The dark structures inside the tubules are assumed to be remnants of odontoblast processes (Tomes fibres). These structural details of totally mineralized teeth cannot be obtained by light microscopy.

Fish scales. The skeletal core of elasmoid scales of fish is composed of three layers the basal plate, the external layer and the outer limiting layer. These layers differ in the orientation of collagen fibres representing an important part of the organic matrix where the mineral is deposited. The basal layer remains unmineralized (for details see Zylberberg et al. 1988). The internal organization can be used to distinguish the different layers by polarization microscopy, in SAM these layers are discerned primarily by their different extent of mineralization (Fig. 2). The clearest image is obtained with frozen sections placed on plastic surfaces with the focal plane of the acoustic lens coincident with the upper surface of the section (ca. 10µm thick). In this case the reflections from the plastic surface are very small compared with those from the section surface, thus primarily variations in acoustic impedance of the tissue are shown. Treatment with the calcium chelating agent EDTA reveal the significance of mineral deposition for the reflection properties in the different layers. The basal layer contains densly packed and highly ordered collagen, while the organic matrix in the external layer is composed of a few unoriented collagen fibres and a loose isotropic proteoglycan matrix. Therefore demineralization results in a considerable loss of reflectivity in the outermost

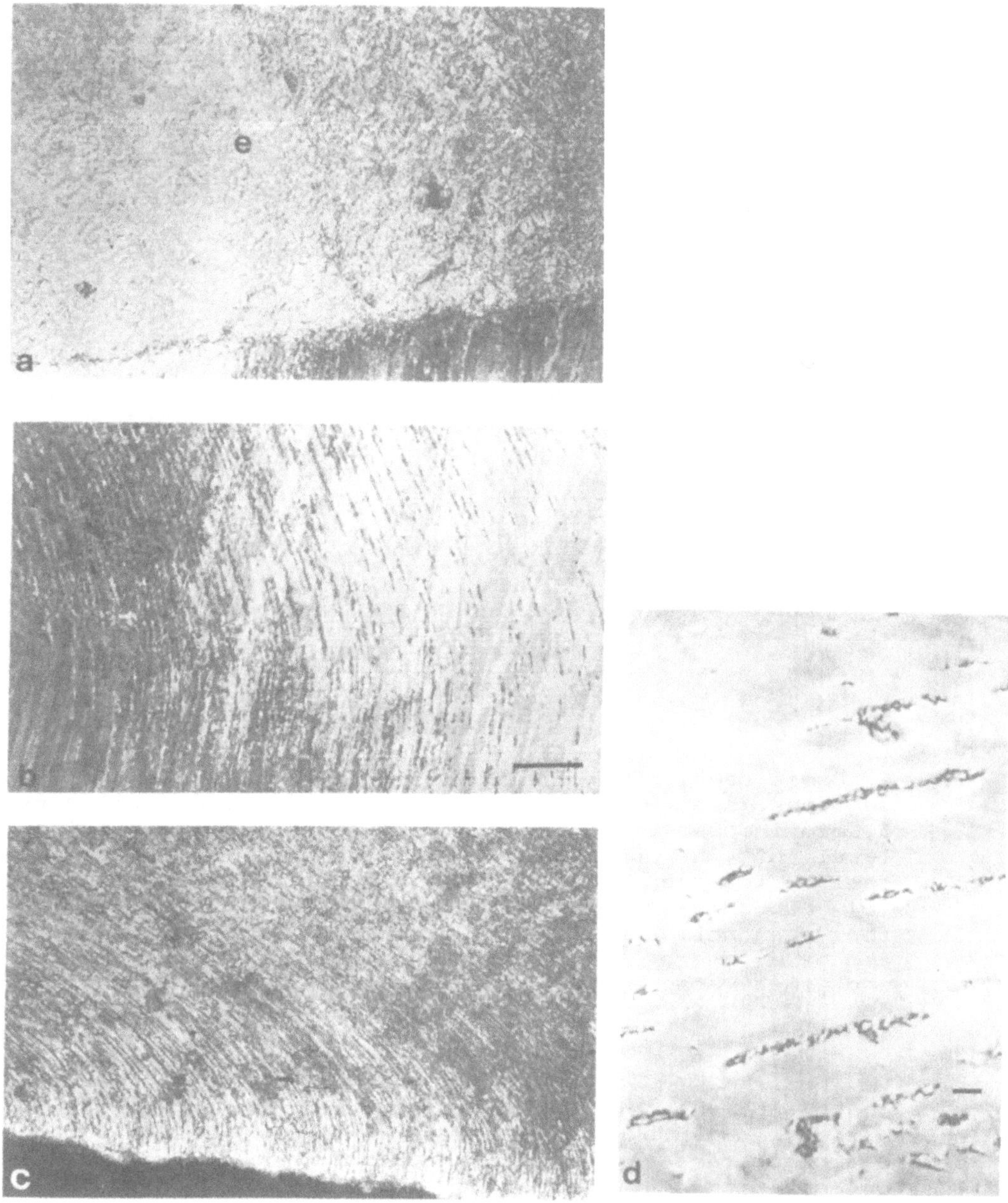

Fig. 1. SAM images of the surface of a grounded and polished human premolar taken at 1GHz. (a) Enamel (**e**) / dentin boundary. The enamel is characterized by an irregular structure, no prismatic organization is visible. (b) Dentin with Tomes fibres (dark). The larger overlying structures are scratches from grinding. (c) Dentin at the margin of the pulpa (black area). Note the brighter reflection at the pulpa boundary and the bending of dentin tubules. (d) Tomes fibres from the central dentin at higher magnification. Bar a-c: 100 µm, d: 10 µm.

layer of the scale while the basal layer remains nearly unaffected. The external layer behaves in an intermediate fashion. A more detailed description and interpretation will be given in a forthcoming publication.

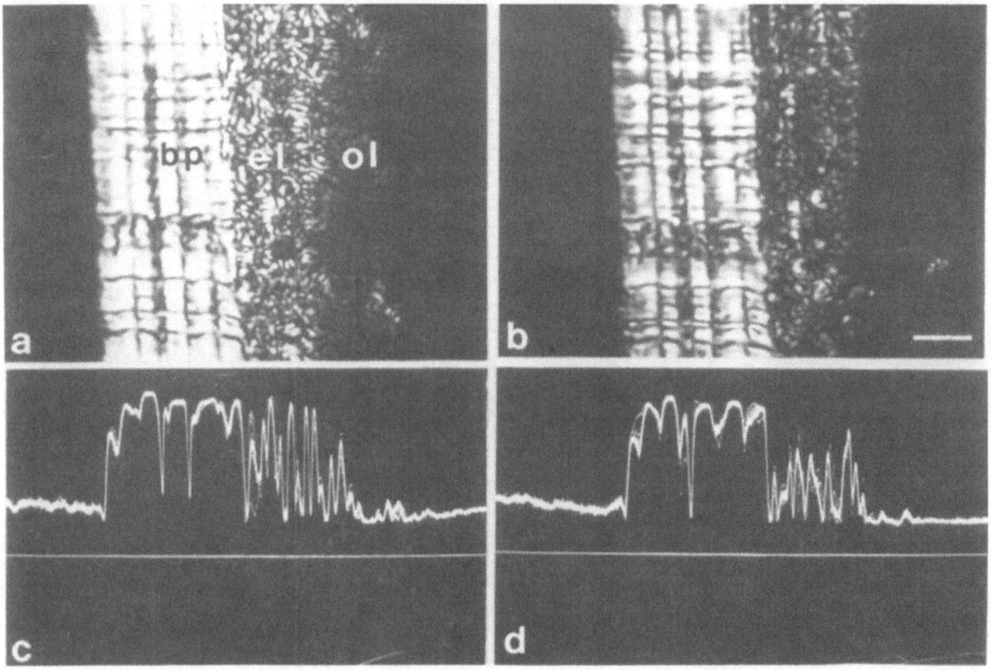

Fig. 2. Cross section through Carassius auratus scales, frozen sections, seen with SAM at 1.0 GHz. (a) before, (b) after 30 min incubation in 0.1M EDTA to remove calcium. Focus level of the acoustic lens is on upper surface of the scale. (c) and (d) line scans corresponding to (a) and (b) respectively. Reflectivity in the bright basal plate (**bp**) does not change, while removal of mineral is indicated by a loss of reflection in the external layer (**el**) and the very faintly visible outer limiting layer (**ol**). Bar: 10 µm

ANIMAL CELLS

In the optimal case SAM reveals acoustic impedance and attenuation and thickness of a cell region. The significance of these parameters for an interpretation of the supramolecular organization of cytoplasm of living or fixed cells can be deduced only by model experiments with purified cytoskeletal components and on the basis of our knowledge (hypothesis) of the factors determining the mechanical properties of cytoplasm and controlling cell shape and locomotion.

Destruction of F-actin and microtubules

Microtubules and actin fibrils are those components of the cytoskeleton which exhibit the most pronounced assembly disassembly cycles. Actin fibrils are intimately associated with the plasmamembrane forming a cortical meshwork which is supposed to be responsible for the maintainance and generation of a pressure difference between the cytoplasm and the extracellular space (Strohmeier and Bereiter-Hahn 1987). Microtubules form an intracellular scaffold interacting with other filamentous components, the endoplasmic reticulum membranes and vesicles which are moved along them. Both these components can be assumed to add considerably to the viscoelasticity of cytoplasm (Bereiter-Hahn et al. 1987), and both are sensitive to an increase of cytosolic calcium concentration which causes disassembly (Stolz and Bereiter-Hahn 1988a,b).

Experimentally, the increase of cytoplasmic calcium is mediated by the addition of a calcium specific ionophore (10µM A23187) in the presence of normal culture medium (calcium concentration: 1.36 mM). Fig. 3 gives an impression of the changes in cell shape and its interaction with ultrasound evoked by this treatment. Most apparent is the fact that the cells become more "translucent" on the destruction of actin fibres and microtubules, thus acoustic attenuation can be supposed to be decreased considerably. This can be seen most clearly in the peripheral regions (compare Figs. 3a and 3c, Figs. 3b and d). The second parameter which obviously changed is the cell volume as can be followed from the number and course of interference lines seen in Figs. 3b and d. The swelling is restricted to the "cell body", it terminates at the proximal margin of the lamellar regions which are used to be limited by strong actin fibres arranged in parallel to each other delineating the boundary with the lamella. Due to the swelling process this boundary is now more prominent (Fig. 3c, arrows).

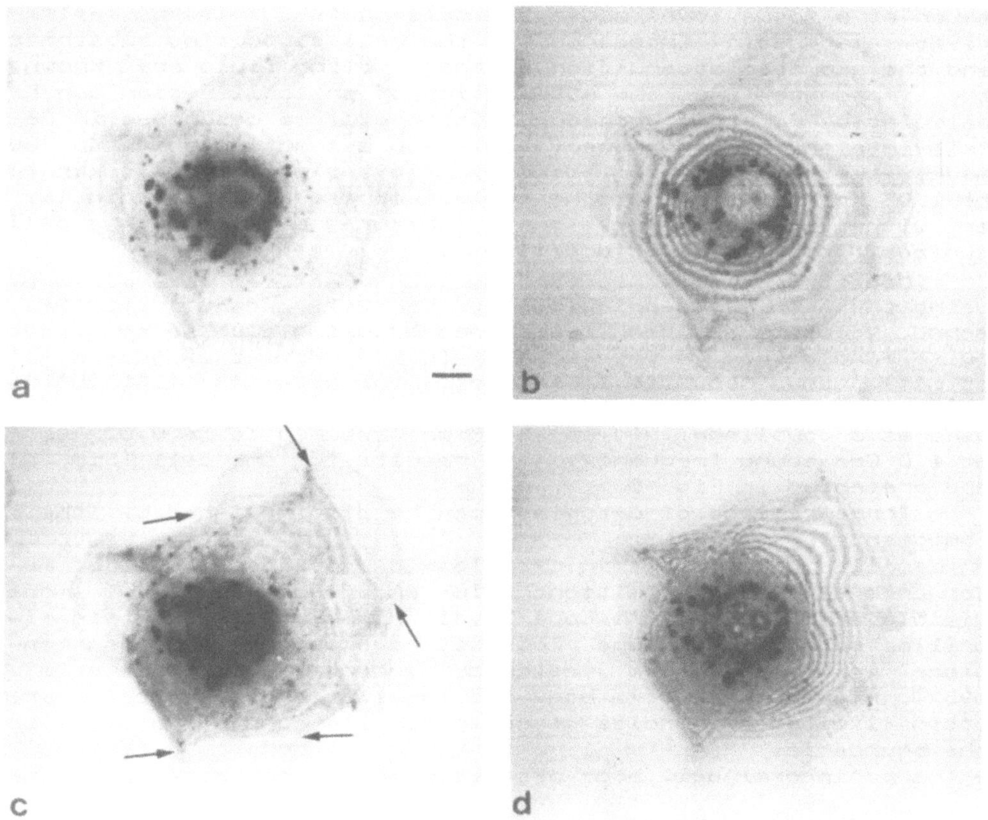

Fig. 3. SAM images of an endothelial cell in culture taken at 1 GHz, focus level is either on substrate (a) and (c) or 3µm above (b) and (d). The culture was treated for 10 min (continued)

These observations support the hypothesis that cell shape is controlled by tension forces modelling the cytoplasm against a swelling force resulting from an osmotic imbalance between the extra- and the intracellular spaces (Strohmeier and Bereiter-Hahn 1987). In addition the apparent decrease in acoustic attenuation demonstrates the contribution of the polymerized cytoskeletal elements to this acoustic parameter.

Acoustic impedance and attenuation of cultured cells, alive and after fixation

A qualitative description of changes of acoustic properties of cells and their relation to cytoplasmic activities can be based just on the observation of SAM images. A deeper understanding of the factors controlling image generation by SAM and of the significance of mechanical properties of cells for their physiological activities requires quantification of the parameters. In general, SAM images contain information of thickness, acoustic impedance and of attenuation of a cell or a cellular region. Complications arise from the fact that the relative influence of acoustic attenuation and impedance on the brightness at a given point are difficult to assess. Hildebrand and Rugar (1984) tackled this problem by using model calculations and by taking two pictures of the same cell at different focus levels. We have tried to solve the problem by evaluating intensities of interference lines in a single image taken at a focus level above the cell surface. If the reflectivity (acoustic impedance) of the cell supporting substrate and the acoustic attenuation of the coupling fluid are known, the impedance and the attenuation of any cell region can be calculated. For this procedure, which will be described in detail in a forthcoming paper, only two assumptions have to be made, that of a constant density of cytoplasm; it is taken to be 1.03 kg m^{-3}, and that the cells comprise a single thin layer with two boundaries only (culture medium/upper cell surface, lower cell surface/glass).

Density of the supporting glass has been determined by using the Archimedes principle to be 2472 kg m^{-3} (±0.2%), sound velocity has been determined by using a phase sensitive pulse/echo method in the range of 30-100 MHz. It is 5869 m s^{-1} (±1.4%), thus acoustic impedance of the glass is 14.51 10^6 kg m^{-2} s^{-1} (±1.4%). The attenuation coefficient of Hanks saline used as a coupling fluid at 28 °C was assumed to be 0.02 μm^{-1} at 1.0 GHz sound frequency. The results for one cell (Fig. 4) are presented in Fig. 5.

Three regions of cytoplasm can be distinguished by their components:

- Lamellae which are thin cytoplasmic processes devoid of major organelles as are mitochondria and vesicular inclusions visible on a light microscope level. The lamellae contain fibrillar material and some ribosomes enveloped by a plasmamembrane which may form vesicular indentations (Small et al. 1982). Towards the cell body the lamellae of single cells are often limited by bundles of actin fibrils running parallel to the boundaries.
- The peripheral cell body cytoplasm which contains all the

→ with 10 µM A23187 (c,d), an ionophore which makes cellmembranes permeable to calcium. Extracellular calcium concentration: 1.36 mM. The arrow indicates the cell body lamella boundary. Bar: 10µm.

well known organelles and surrounds the perinuclear region.
- The nuclear and perinuclear region of the cell body which contain the nucleus, all the other membranous organelles and in many cases a large amount of vacuoles and granular cytosomes in the immediate vicinity of the nucleus.

Due to the granular inclusions the basic assumption of a single thin layer is not valid for the nuclear and the perinuclear regions. Therefore these areas have not been considered for calculations of their acoustic properties (open fields in Figs. 5 a-c). Fig. 5a shows the thickness profile along the scan line of the cell seen in Fig. 4, the lamella regions are more or less distinct from the rest of the cell (see arrows). On fixation of the cells with 2.5% glutardialdehyde in phosphate buffered saline (pH 7.2) the profile remains unaltered in the limits of resolution. Acoustic impedance in living cells is slightly higher in the broader lamella than in other parts of cytoplasm (Fig. 5b), irregular but major changes are observed in the cell body/lamella transition region. These variations in acoustic impedance can be ascribed to changes in the velocity of the longitudinal waves because no variations in density are expected throughout the cytoplasm regarding a protein content between 10 and 20% and a specific density of protein close to that of water. Acoustic attenuation (Fig. 5c) of living cells ranges from 1.5 to 2.3 times that of water and exhibits a clear minimum at the lamella/ cell body transition region, rising as well towards the perinuclear region as towards the lamella periphery. In fixed cells the general distribution of these parameters remains, however, their variation between different cell areas increases considerably (broken lines in Figs. 5 b and c). After fixation an increase in sound velocity in the lamella is most obvious indicating an increase stiffness induced by cross linking the densely packed fibrillar material of the lamella.

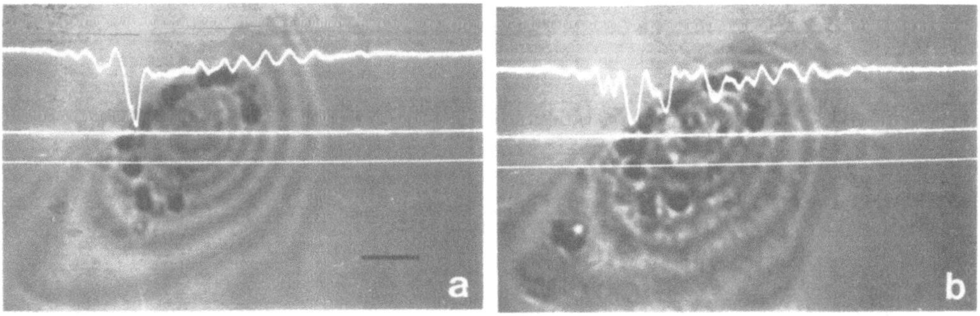

Fig. 4. SAM image of a frog endothelial cell in culture taken at 1 GHz, focus 7 μm above glass surface. (a) living cell, (b) the same cell after fixation in 3% glutardialdehyde in 0.1M phosphate buffer, pH 7.2. Small changes in surface topography are obvious, however the general outline remains unaltered. The line scans provide the data needed for the calculation of acoustic impedance and attenuation shown in Fig. 5. Bar: 10 μm.

The minimum of attenuation at the lamella/cell body transition region is not understood at the moment. The relative high attenuation in the lamella can be due to a high degree of polymerized macromolecules, that in the cell body could result from scattering at the various boundaries provided by the membrane bound organelles. An approach to the influence of supramolecular organization on acoustic impedance is presented in the following chapter.

Relation of sound attenuation to the amount of protein traversed

For the cell biologist an answer to the question whether acoustic attenuation is proportional to the amount of cellular

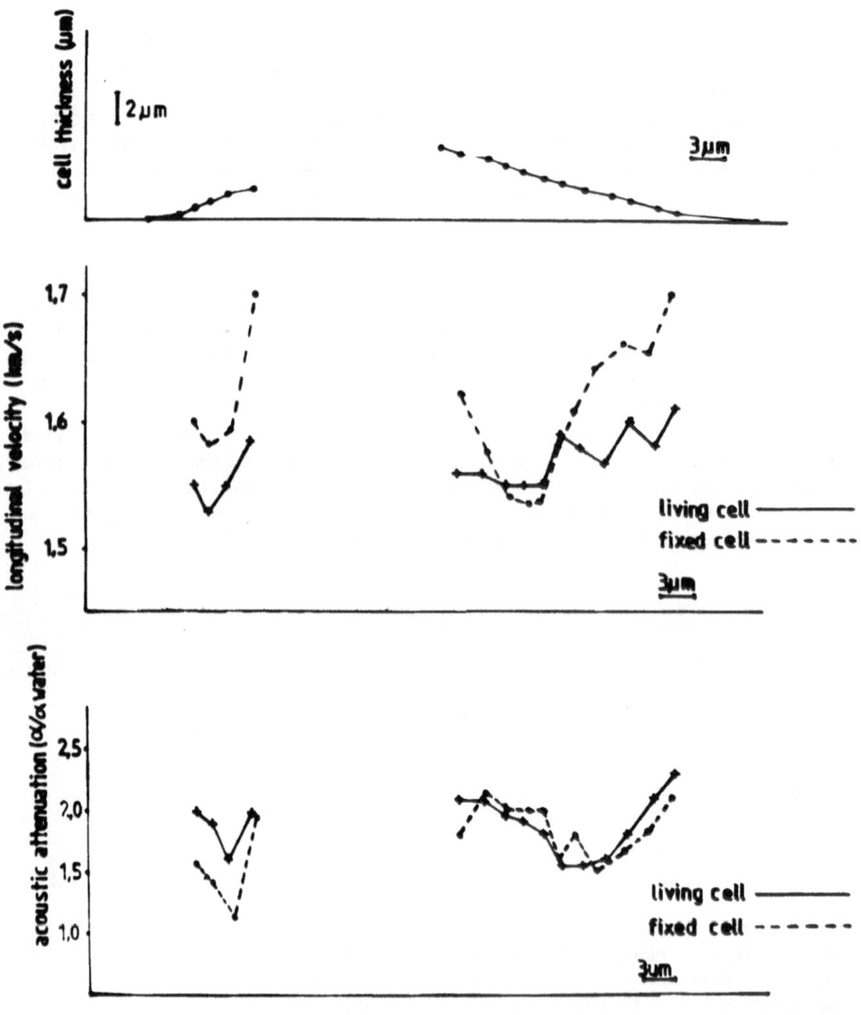

Fig. 5. Evaluation of acoustical properties of the cell shown in Fig. 4. (a) surface profile along the scan line; magnification is higher in y-direction than in x-direction to enhance comprehensiveness. (b) Velocity of longitudinal sound waves: position of data points coincides with position on cross section shown in (a). (c) Attenuation coefficients of cytoplasm at the various sites (corresponding to (a)) related to that of water. Full lines connect values taken as long as the cell was alive, broken lines after fixation.

dry mass traversed or whether it provides information of the supramolecular structure of cytoplasm is of crucial importance. Acoustic attenuation was determined by comparison of the sound intensity reflected from a glass surface without penetrating a cell and that reflected from such a surface onto which a cell is adhering. In these cases the lens was focussed on the glass surface and the influence of interference fringes was eliminated by taking the mean between adjacent minima and maxima. The percentage of signal decrease was then related to the optical path difference at the same site of the same cell after fixation. The optical path difference between the retrieved cell and its environment was determined by means of an interference microscope of the Mach-Zehnder type (E. Leitz, Wetzlar, F.R.G.) operated at 578nm. Microinterferometry allows calculation of the dry mass (DM) in any given area of a cell from the optical path difference (Γ). Dry mass and opical path difference are related to each other by the specific refraction increment α (a constant giving the change in refraction number on changing solute concentration by 1%):

$$\Gamma = d \times \delta n \qquad (1)$$

and

$$DM = \Gamma \times A/100\alpha + \delta n \times V/100\alpha \qquad (2)$$

where DM is the dry mass, A is the area of the specimen with the thickness d and the volume V (A x d), Γ is the optical path difference between specimen and surroundings, α is the specific refraction increment, δn is the difference of the refractive indices of culture medium and cytoplasm. For small differences in refractive indices and for thin objects the second term becomes negligibly small, thus Γ is directly proportional to the amount of matter present in a certain area of the microscopic field which can be related to the acoustic attenuation in this area. As shown in Fig. 6 in the cell periphery (lamella region) attenuation increases conside-

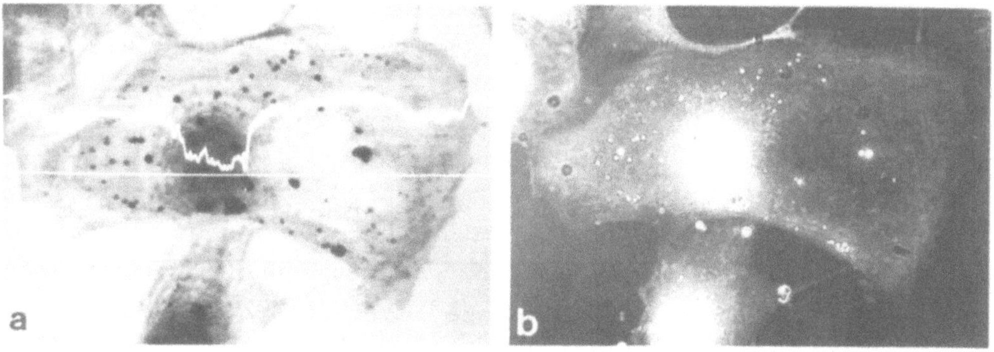

Fig. 6. Comparison of SAM image (a) with light microinterferometric picture of the same cell (b). SAM image was focused on glass surface and taken at 1.3GHz. (a) Living endothelial cell, (b) Interference contrast image: Distance between the interference lines is adjusted to infinity, thus grey levels are a direct marker of optical path differences between the cytoplasm and a reference specimen. (for more detailed description of the procedure see [Bereiter-Hahn 1985]).Bar: 10 µm.

rably more with increasing optical path difference (amount of protein traversed) than in the more central parts of the cytoplasm. A more detailed description has been given in previous reports (Bereiter-Hahn 1987, Bereiter-Hahn and Buhles1987).

CONCLUSIONS

All these observations point to the significance of the organization of cytoskeletal elements for their acoustic properties. The most impressive variations in soft cells are found for attenuation which seems to be related to cytoplasmic viscosity and to scattering by cellular organelles. Further studies of cells under different metabolic conditions like those presented in this contribution are promising to deepen our understanding of mechanical parameters involved in generation of cell shape and in the control of cell metabolism which by means of no other method can be revealed with comparative ease. However, we are at the very beginning of the development of methods for the evaluation of SAM images.

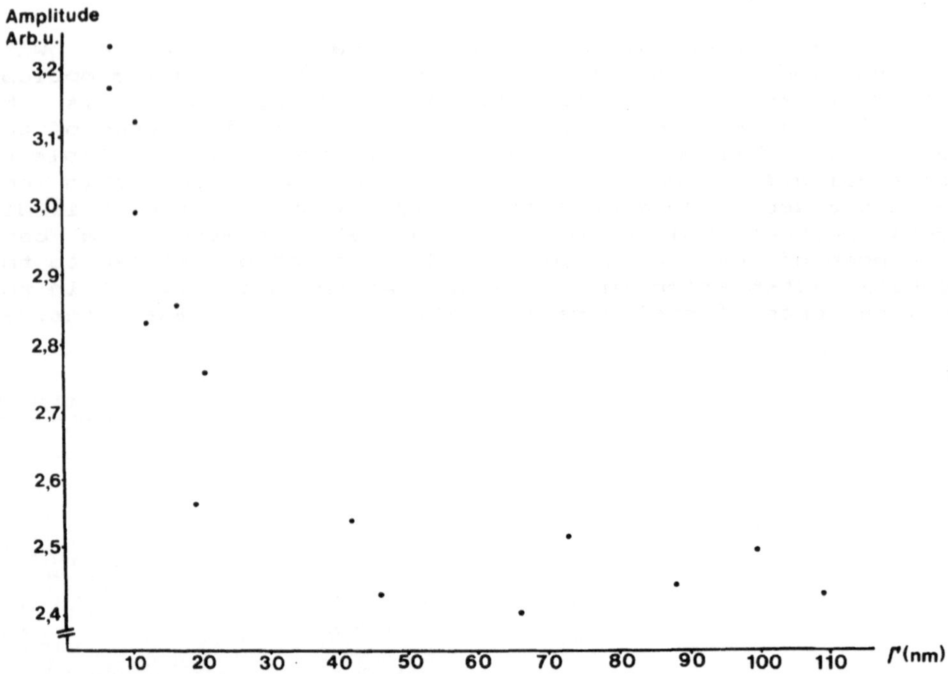

Fig. 7. Relation of optical phase shift (Γ) to acoustic attenuation (ordinate) obtained from a living endothelial cell. The small Γ values correspond to very thin cytoplasmic layers, mostly to lamella regions, while the larger values are found in the central parts of the cells. Attenuation is given relative to a constant signal.

For studies of hard tissues SAM offers a unique method which easily reveals areas of different degree of mineralization.

ACKNOWLEDGEMENTS

The tooth shown in Fig. 1 was kindly provided by Dr. Löst, Universitätszahnklinik, Mainz. We thank Prof. Dr. Grill (Frankfurt/M) for the opportunity to perform the determinations of acoustic impedance of glass and plastic materials in his laboratory with the help of Mr. T.J. Kim which is also gratefully acknowledged.

The study has been supported by a grant from the Deutsche Forschungsgemeinschaft (Be 423/13) and by the CNRS/DFG exchange convention program provided to L.Z..

REFERENCES

Bereiter-Hahn J., 1987, Comparison of the appearance of cultured cells observed using scanning acoustomicroscopy with that obtained by interference and fluorescence microscopy., Scanning Img. Techn.SPIE, 809: 162-165

Bereiter-Hahn J., Anderson O.R., Reif W.-E., eds., 1987, "Cytomechanics. The mechanical basis of cell form and structure." Springer Vlg., Berlin, Heidelberg, New. York

Bereiter-Hahn J., 1988, Involvement of microcompartmentation in the regulation of cell proliferation, in: "Microcompartmentation," D. Jones, ed., CRC Press Inc. (in press)

Bereiter-Hahn J., Buhles N., 1987, Basic principles in interpretation of scanning acoustic images obtained from cell cultures and histological sections, in:"Imaging and visual documentation in medicine", 537-543, Wamsteker K., ed., Elsevier Publ., Amsterdam.

Clarke F., Stephan P., Morton D., Weidemann J., 1983, The role of actin and associated proteins in the organisation of glycolytic enzymes, in:"Actin", 249-258, D. Remedios, J.A. Barden, ed., Acad. Press Sydney, N.Y..

Clegg J.S., 1984, Intracellular water and the cytomatrix, J. Cell Biol. 99: 167s-171s

Folkman J., Moscona A., 1978, Role of cell shape in growth control., Nature, 273: 345-349

Görmar F., Bernd A., Holzmann H., 1988, Differentiation of human keratinocytes in cell culture induced by cyclic mechanical, Europ. J. Cell Biol., 46 Suppl.22: 22

Hildebrand J.A., Rugar D., 1984, Measurement of cellular elastic properties by acoustic microscopy, J. Micr., 134: 245-260

Hoppe M., Bereiter-Hahn J., 1985, Applications of scanning acoustic microscopy - survey and new aspects, IEEE Transactions on sonics ultrasonics, SU-32: 289-301

Iwig M., Glaesser D., Bethge M., 1981, Cell shape-mediated growth control of lens epithelial cells grown in culture. Exp. Cell Res. 131:47-55

Masters C., 1984, Interactions between glycolytic enzymes and components of the cytomatrix, J. Cell Biol., 99: 222s-225s

Masters C.J., Wilson J.E., 1981, Interaction between soluble enzymes and subcellular structure, CRC Critical Rev. Biochemistry II., 2: 105-141

Penman S., Capco D.G., Fey E.G., Chatterjee P., Reiter T., Ermish S., Wan K., 1983, The three three-dimensional structural networks of cytoplasm and nucleus: Function in cells and tissue. Modern Cell Biol. 2: 385-415

Small J.V., Rinnertaler G., Hinssen, H., 1982, Organization of actin meshworks in cultured cells: The leading edge, Cold Spring Harbor Symp. Quant. Biol. 56: 599-611

Stolz B., Bereiter-Hahn J., 1988a, Calcium sensitivity of microtubules changes during the cell cycle of Xenopus laevis tadpole endothelial cells, Cell Biol. Intern. Reports, 12: 313-320

Stolz B., Bereiter-Hahn J., 1988b, Increase of cytosolic calcium results in formation of F-actin aggregates in endothelial cells, Cell Biol. Intern. Reports, 12: 321-329

Strohmeier R., Bereiter-Hahn J., 1987, Hydrostatic pressure in epidermal cells is dependent on Ca-mediated contractions, J. Cell Sci., 88: 631-640

Tillmann U., Bereiter-Hahn J., 1986, Relation of actin fibrils to energy metabolism of endothelial cells, Cell Tissue Res., 243: 579-585

Zylberberg L., Bereiter-Hahn J., Sire J.-Y., 1988, Cytoskeletal organization and collagen orientation in the fish scales, Cell Tissue Res. 245 (in press)

ADVANCED FEATURES FOR THE QUALIFICATION OF COATINGS BY MEANS OF REFLECTIVE ACOUSTIC MICROSCOPY

H. Vetters, E. Matthaei, Th. Lübben, A. Schulz, and P. Mayr
Institut für Werkstofftechnik (Institute for Materials Science and Engineering)
Bremen, FRG

ABSTRACT

The main object of the investigation described in this paper is the usefulness of the V(z) response for the quantitative examination of the interfacial microstructures in coated parts. By means of the numerically calculated functions the experimental conditions for obtaining contrasted indepth patterns can be predicted. For this purpose the difference-characteristics of the calculated V(z) functions have been derivated as a quantitative tool for the estimation of the materials contrast conditions.

The adaptation of the theoretical functions with the experimentally measured V(z) curves is proofed by variation methods. The dependence of the changes in the response for different classes of material parameters with defocusing adjustment is demonstrated.

INTRODUCTION

A number of methods have been derived to characterize material structures such as subsurface defects like pores, glass fibres and adhesion failures in material bondings with scanning acoustic microscopy (SAM) [1,2]. At least in biological materialography the acoustic microscope features the examination of organic coatings such as the enamel of teeth [3]. Coatings are increasingly being used to protect the surfaces of engineering components. Although a wide variety of coatings are employed, including metals, ceramics and polymers.

The reflection of sound waves depends on specific features such as sound velocities, attenuation, density in layer and substrate. Therefore the V(z) functions, which depict all these material specific characteristics, are used in acoustic imaging to describe the contrast behaviour. Namely, the varieties in the layer composition, such as partial inhomogeneities due to delaminations, adhesion failures or the change in thickness, can influence the amplitude response and the periodicity of the V(z)-characteristics [4].

The V(z) curves which exhibit the specific imaging behaviour of layers and substrate can be calculated numerically by means of a theoretical model [5] connected with the reflection function of bulk or layered materials [6], specially developed in a computer program. For the case of a circularly shaped acoustic lens it depends only on the polar angle. Based on this method the acoustic parameters of bulk materials can be determined by adaptation with measured values. The developed program can handle the prediction of the contrast and focusing condition for given material combinations.

EXPERIMENTAL

Instrumentation

The scanning acoustic microscope has been described in detail elsewhere [1,7]. Some technical data are listed in table 1. The instrument is equipped with several additional features. The temperature of the stage can be varied from room temperature up to 80° C, to reduce the attenuation of the coupling liquid. Furthermore a reflected light optical microscope incorporated into the instrument facilitates the positioning of the specimen to be examined by SAM. A variable time gate features the discrimination of the superposed responses of the lens.

Calibration of the V(z) data

The demodulated output signal V_{out} is responsible for the determination of the acoustic materials signature V. This signal is defined as following [8] :

$$V\ [dB] = 20\cdot\log\ (|V_{out}|/|V_{in}|) + c_v \qquad (1)$$

with V_{out}: demodulated integral output signal, V_{in}: amplitude of the input signal, c_v: instrumental parameter.

The contrast conditions to be obtained are based on this signal. For the experimental determination of the V(z) characteristics and the correlation with theoretically calculated curves an adaptation of these values is necessary. In practice the measured V(z) values depend on some instrumental parameters. By means of glass platelets, without any materials structures, these functional relations can be found.

On the other hand, theoretical V(z) curves can only be calculated with exception of one constant factor. In the case of spherical lenses with an aperture angle of Θ_A, the output can be expressed in a simplified integral [5,8] described as

$$V(z) = 20\cdot\log\ \left|\Pi \int_0^{\Theta_A} U^2(\Theta)\cdot R(\Theta)\cdot \sin(2\Theta)\cdot \exp(i2k_L^{(1)}\cos\Theta\cdot z)\ d\Theta\right| \qquad (2)$$

with $U(\Theta)$: incident field, $R(\Theta)$: reflection function, $k_L^{(1)}$: wave number in coupling fluid.

Table 1. Technical data of the acoustic microscope (type UH2, OLYMPUS Optical Co., Japan)

Acoustic lenses					
frequency (MHz)	aperture angle (deg.)	working distance (mm)	resolution lateral (μm)	resolution indepth (μm)	penetration depth (max.) (μm)
100	60	1.25	12	20-50	0-2000
200	120	0.31	6	10-40	0- 350
200	60	0.50	6		
200	60	1.25	6		
400	120	0.31	3	4-10	0- 50
400	60	0.50	3		
Scanning areas:			0.25 x 0.19 mm 0.50 x 0.38 mm 1.00 x 0.76 mm 2.00 x 1.53 mm		

Taking all these experimental and theoretical conditions into consideration, the relation between calculated {V(z)} and measured {$V_E(z)$} signals is represented by the following equation

$$V(z) = a \cdot V_E(z_M + z') + k \quad (3)$$

This function describes a linear relation between both values with a slope "a", which includes mainly electronic parameters of the microscope. The value "k" is formed by further parameters, for example by the amplitude of the incident sound field and the focal length of the lens. Additionally it depends on the instrumental set-up. Therefore, the set-up conditions during the measuring should remain constant.

Furthermore, in eq. (3) the transformation of the z-axis, defined in theory, is included with $z = z_M + z'$. z_M represents the material-dependent shift of the signal maximum relative to the theoretical maximum.

For a general correlation of V(z) and $V_E(z)$ the parameters a and k have been determined for constant experimental conditions. For this, some bulk materials, like epoxy resin, glass, fine-grained nickel, were investigated and their $V_E(z)$ curves were registered. Assuming the general validity of the theoretical V(z)-model and the smallness of the errors of measurements a good agreement of V(z) and $V_E(z)$ should be expected, if the theoretical curve is calculated with the exact material parameters, describing the test material. But there exist always uncertainties in determing these data, coming from literature or from measuring. This can be taken into account by the procedure of the variation method.

For that, in a computer program the measured $V_E(z)$ values were directly compared with the corresponding theoretical V(z) values. The degree of similarity between theory and experiment can be expressed with the correlation coefficient r due to linear regression (least-square-method) with eq. (3). Therefore, under variating the material parameters around a given mean value, this coefficient has to reach a maximum. The principle scheme for variation of the parameters to get maximum correspondence is [8] :

1) variation of sound velocitites within defined regions
2) variation of attenuation
3) variation of density
until in each case the maximum of r is reached, then repeating of No. 1) to 3) with smaller variation steps.

The systematic investigations of several bulk materials, mentioned above, connected with optimizing the correlation, yield the mean parameters $\bar{a}$ and $\bar{k}$ with $\bar{a} = 1.21 \pm 0.06$ and $\bar{k} = (-34.89 \pm 0.78)$ dB.

The comparison of experimentally determined $V_E(z_M+z')$ data of nickel versus the calculated V(z) values (fig. 1) yield a correlation coefficient of r = 0.989. The results from linear regression with V and V_E (broken line) and the plot which is determined with the mean values a and k (straight line), mentioned above, show good correspondance. Table 2 shows the parameters for nickel resulting of this variation analysis.

In the case of an isotropic bulk medium, considered in the reflection function, the variation of 5 parameters is possible, without taking into consideration the data of the coupling fluid. To reduce the time-consuming procedure of calculation it is useful to know the relevance of each parameter influencing the V(z) behaviour due to amplitude and periodicity [9]. Taking into account the aperture angle of the acoustic lens, 3 classes of materials can be distinguished corresponding to their acoustic parameters in relation to the criticial angles of total reflection and the coupling medium (table 3).

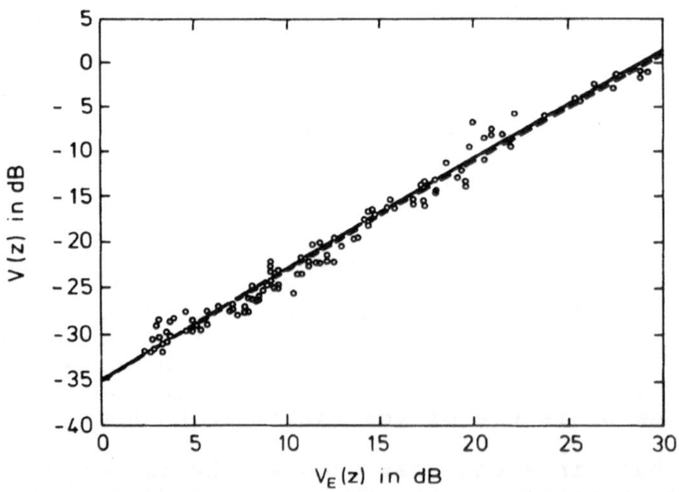

Fig. 1 Comparison of experimental data $V_E(z)$ with calculated V(z) values for nickel (200 MHz)
Solid line: calibration line ($V = \bar{a} \cdot V_E + \bar{k}$)
Broken line: regression line of the measured data

Table 2. Parameters resulting of the variation analysis with V(z) and $V_E(z)$ for nickel (200 MHz)

parameter	unit	value
density	(g/cm^3)	8.78
sound velocity long.	(km/s)	5.63
attenuation long.	(1/mm)	1
sound velocity transv.	(km/s)	2.94
attenuation transv.	(1/mm)	5.4
shift z_M	(μm)	4.5
corr.coeff. r		0.989
a		1.201
k	(dB)	-35.13

Table 3. Classification of materials due to acoustic parameters and lens geometry.

classes	condition
class 1:	no critical angle within the aperture angle $v_L^{(2)} < v_L^{(1)}/\sin\Theta_A$
class 2:	one critical angle within the aperture angle $v_L^{(2)} > v_L^{(1)}/\sin\Theta_A$ and $v_S^{(2)} < v_L^{(1)}/\sin\Theta_A$
class 3:	Rayleigh-angle within the aperture angle $v_R^{(2)} > v_L^{(1)}/\sin\Theta_A$ with $v_S^{(2)} > v_L^{(1)}/\sin\Theta_A$
with	Θ_A: aperture angle of the lens v_L: sound velocity of longitudinal waves v_S: sound velocity of shear waves v_R: sound velocity of Rayleigh waves (1): in coupling fluid (2): in solid

The periodicity Δz of the V(z) curve can be expressed with the equation [5]

$$\Delta z = \lambda_L^{(1)}/(2\cdot(1-\cos\Theta_C)) \tag{4}$$

with $\lambda_L^{(1)}$: wave length in coupling fluid, Θ_C : the relevant critical angle (with C = A, L, or R, see above [8]).

According to the materials class the relevance of each of the parameters for changing the V(z) behaviour is different. So, for example, for materials of class 2 a variation of the velocity of shear waves shows only small effects in changing the periodicity. Here the main contribution comes from the longitudinal waves. The variation of their velocity alters the periodicity.

Contrast characteristics due to layers

For the materials contrast recorded under equivalent conditions the difference-characteristic D(z) between the two response signals $V_1(z)$ and $V_2(z)$ can be achieved as

$$D(z) = V_1(z) - V_2(z) \tag{5}$$

With this model material contrast can be described between two regions with lateral different components. To determine the lateral distribution of the image contrast $D(x,y;z)$ a specific reflection function $R(x,y)$ (compare function $R(\Theta)$ in eq. (2)) characterizing the local distribution of the materials parameters must be evaluated. The contrast in the vicinity of surface cracks can be described theoretically as a function of defocusing [10]. A general analytical solution to describe objects including different materials, phase boundaries and discontinuities is due to its complexity not yet to realize.

For the localization of subsurface material structures and material failures the prediction of material contrast formation is modeled without the influence of discontinuities, because artificial structures around such discontinuities, such as fringes, can easily be distinguished in the acoustic image while defocusing [11]. But furthermore, it is necessary to estimate the valid focusing parameter z in which determinable subsurface structures can appear sufficiently contrasted.

Under the assumption that the material parameters of the object show no lateral variation within sufficient large volume elements, a model can be designed (fig. 2). By means of this the composites to be examined by SAM can be described as an array of several "columns" which characterize the local distribution of the layered components. Then, the reflection function $R(\Theta)$ can be calculated for a halfspace without lateral boundaries.

To examine the penetration depth and the $V_E(z)$ behaviour a special test sample has been developed. A mesh grid mounted as a sublayer indicates the lateral resolution and the contrast conditions while changing the focusing adjustement (fig. 3). The sample is embedded in epoxy resin.

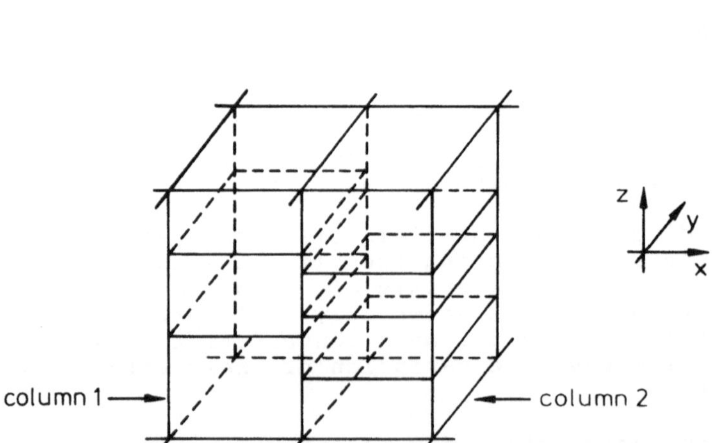

Fig. 2 Model to characterize layered objects with different "columns"

Thus, the resulting model-system can be characterized by two "columns":

aluminium - nickel and
aluminium - epoxy resin.

The recorded $V_E(z)$ curves of the two components (fig. 4) show nearly equivalent characteristics down to a value of z = -80 µm due to the dominating response of the aluminium layer (30 µm thick). Exceeding z to lower values the increasing influence of the different subsurface structure materials can be figured out by the distinctive characteristics. Then the difference plot of the V(z) curves (fig. 5) indicates the contrast conditions. The positive values are to be interpreted as bright subsurface structures, and negative values as dark subsurface structures, respectively. The contrast inversion, resulting from the first change in periodicity, is distinguishable at a focusing of z = -80 µm. This has been proofed by sequential examination of the micrographs obtained at the indicated z-values.

In theory, delaminations can be simulated with a system "coating superposed to vacuum (air)" to characterize the enhancement of sound reflection on a solid/gas interface [12]. The calculated values for a 7 µm thick aluminium foil above air compared with the experimental data (fig. 6) show the same characteristic sequence in maxima and minima, but a slight shift within z-position. This can be caused either by the uncertainties of the experimental calibration, or of incorrect material parameters, such as the sound velocities.

The loose in adhesion of a molybdenum boride layer (thickness: 10.8 µm) can be analysed by means of this technique (fig. 7). The calculated plots of V(z) for the adherent layer on molybdenum in comparison with the delaminated zones on the system molybdenum-boride-layer above vacuum let expect an enhancement in amplitude and periodicity for the defective areas. The difference curve D(z) (fig. 8) shows also a periodic form, and the inversion points and the extremals become obvious.

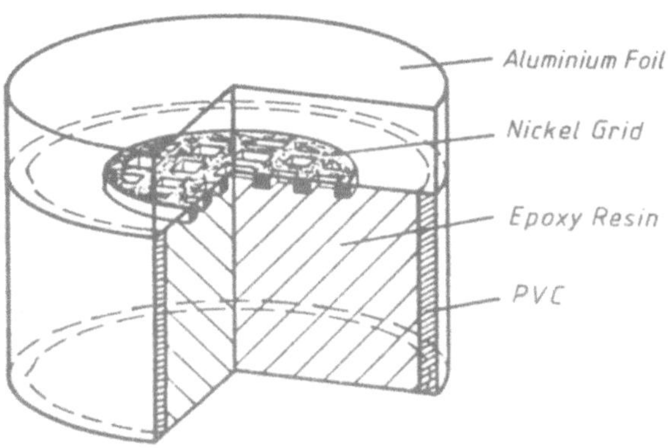

Fig. 3 Test sample for subsurface imaging

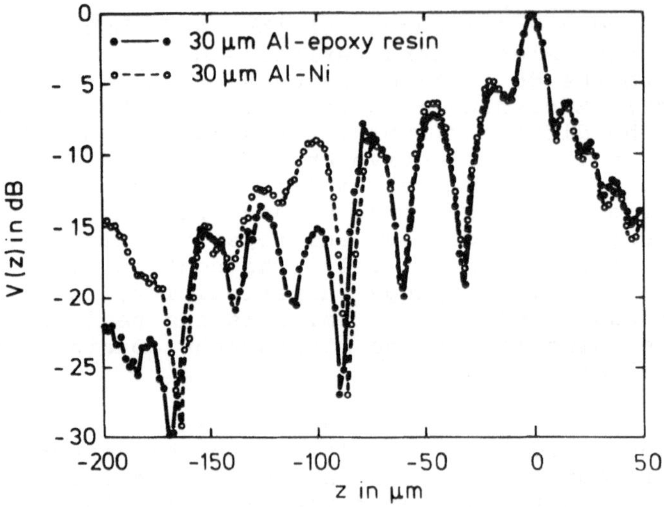

Fig. 4 Measured $V_E(z')$ curves on different regions on the test sample with a 30 µm aluminium layer (200 MHz)
Solid line: column aluminium - epoxy resin
Broken line: column aluminium - nickel

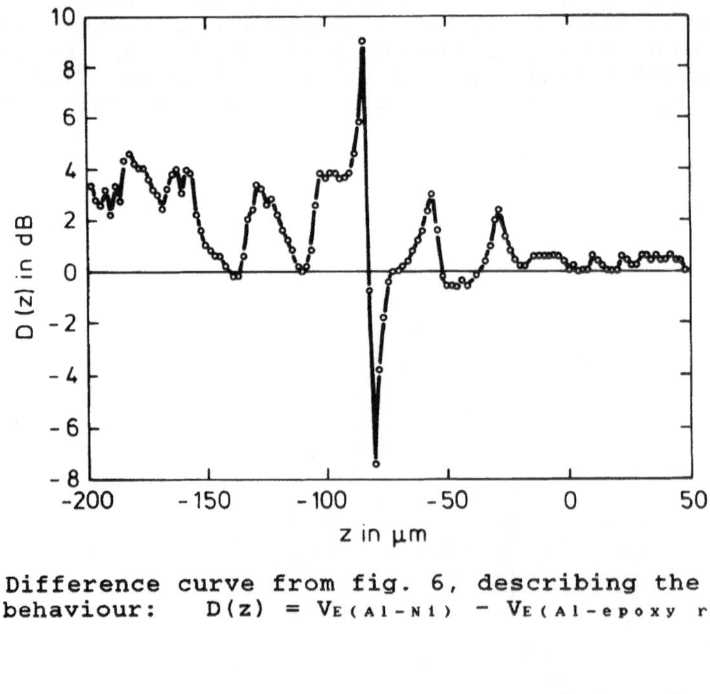

Fig. 5 Difference curve from fig. 6, describing the contrast behaviour: $D(z) = V_{E(Al-Ni)} - V_{E(Al-epoxy\ resin)}$

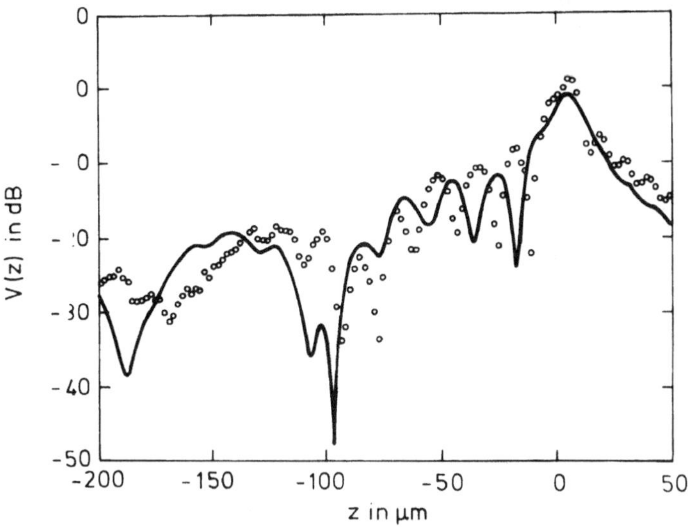

Fig. 6 Comparison of calculated and measured V(z) values for a 7 µm aluminium foil above air (200 MHz)
Solid line: calculated values
Dots: measured data

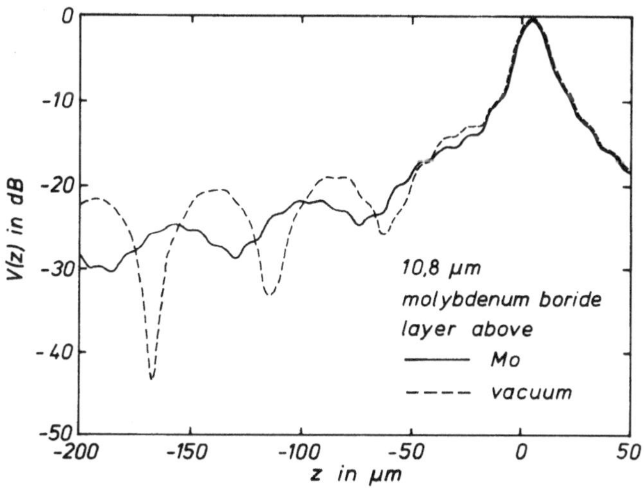

Fig. 7 Calculated V(z) curves for two material columns molybdenum boride above molybdenum (solid line) and molybdenum boride above vacuum (broken line) to simulate local delaminations (200 MHz)

In practice, the acoustic micrograph to be obtained at the specific z-position of -150 µm (fig. 9) shows the delaminated areas with bright contrast. While comparing theory and experiment the shift z_M, defined above, should always be taken into consideration. Surface cracks are seamed with the typical fringes [11]. These fringes are raised by interactions with surface acoustic waves. Therefore the surface cracks can be distinguished from subsurface cracks.

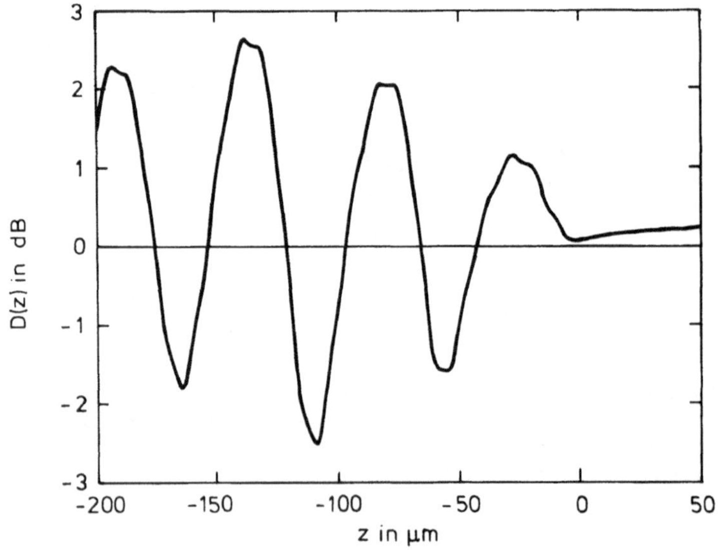

Fig. 8 Difference curve D(z) for the two columns of fig. 8

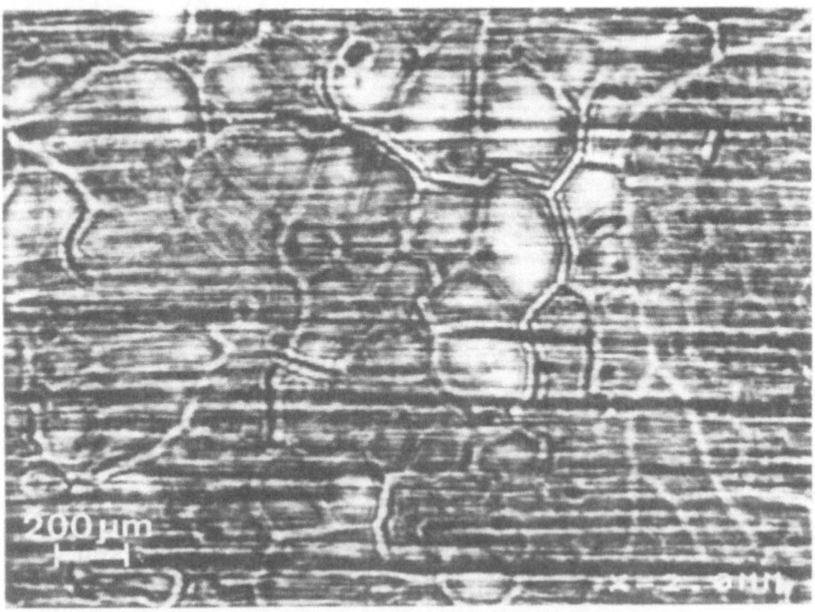

Fig. 9 Acoustic micrograph of a defective layer of molybdenum boride (10.8 µm) above molybdenum (200 MHz, z' = -150 µm)

When measuring $V_E(z)$ curves the distances between the ultrasonic probe and discontinuities in the sample should be sufficiently extended to avoid artificial patterns caused by the fringe-effect and additional scattering processes. For the experimental conditions to be used the ultrasonic field on the object surface spreads to a radius of about 310 µm at the lowest working distance.

CONCLUSIONS

The SAM-detection of subsurface failures under quantitative aspects needs calibration procedures for the correlation of experimental and theoretical V(z) values. As shown by practical experiments, the constants of materials can be determined by linear regression and variation analysis.

For this, the knowledge of relevance of each acoustic parameter on influencing the V(z) formation is useful to reduce the time-consuming variation calculations. Therefore, the materials to be examined by SAM have been classified in three types: 1) where no critical angle, 2) where one critical angle, 3) where Rayleigh angle are within the aperture. By variation of individual parameters, the relevant data which characterize each class can be estimated. Regarding equivalent experimental conditions, the characteristics of different materials can now be compared under quantitative aspects.

For predicting the material contrast for a given material combination the application of the difference function $D(z) = V_1(z)-V_2(z)$ is useful. By means of this tool, preferable z-values, where the different materials in layered media appear contrasted, can be selected. By implementing a special model, the compositions of various layered media can be calculated by an array of columns which represent the variety of the sublevels of the composite. The contrast due to variations in thickness or composition, the focusing conditions to detect subsurface flaws can then be predicted by means of the numerically calculated difference curve. For an automatical preselection-device the signal output data from the microscope are fed into a computer coupled with the instrument.

REFERENCES

1. Quate C.F., The acoustic microsope: a concept for microscopy using waves of sound, Nav. Res. Rev.(USA), 33(1) (1980-81), pp. 24-32
2. Yamanaka K., Enomoto Y., Characterization of PVD films with scanning acoustic microscope, Proc. 7. Int. Conf. on Vacuum Metallurgy (ICVM), 26.-30.11.1982, Tokyo, Japan pp. 157-163
3. Peck S.D., Briggs G.A.D., A scanning acoustic microscope study of the small caries lesion in human enamel, Caries Res., 20 (1986), pp. 356-360
4. Weglein R.D., SAW dispersion and film thickness measurements by acoustic microscopy, Appl. Phys. Lett., Vol. 35, No. 2 (1979), pp. 215-217
5. Atalar A., A physical model for acoustic signatures, J. Appl. Phys. Vo. 50, No. 12 (1979), pp. 8237-39

6. Brekhovskik L.M., "Waves in layered media", 2nd Ed., Academic Press, NY (1980)
7. Scanning Acoustic Microscope (Report from Olympus Corp. of America, NY), Solid State Techn., Oct. 1982, pp. 119-120
8. Lübben Th., V(z)-curves in acoustic microscopy, diploma thesis, university of Bremen, 1986, in german
9. Kushibiki J., Ohkubo A., Chubachi N., Effect of leaky SAW parameters on V(z)-curves obtained by acoustic microscopy, Electron. Lett., Vol. 18, No. 15 (1982), pp. 668-670
10. Briggs G.A.D., Somekh M.G., Acoustic microscopy of surface cracks: theory and practice, Solid Mechan. Res. for Quant. Nondestr. Eval., Proc. Conf., Evanston Illinois, USA, 18.-20. Sept. 1985, Martinus Nijhoff Publishers, Netherlands (1987), pp. 155-169
11. Yamanaka K., Enomoto Y., Observation of surface cracks with scanning acoustic microscopy, J. Appl. Phys., Vol. 53, No. 2, Febr. 1982, pp. 846-850
12. Bray R.C., Quate C.F., Calhoun J., Koch R., Film adhesion studies with the acoustic microscope, Thin Solid Films, 74 (1980), pp. 295-302

ACOUSTIC MICROSCOPE STUDY OF A TITANIUM DIFFUSION BOND

R. D. Weglein

6317 Drexel Avenue, Los Angeles, CA 90048, USA

INTRODUCTION

The need for more reliable and cost-effective nondestructive inspection methods is never ending. In particular, the ability to locate and possibly to quantify the microscopic strength-determining imperfections in diffusion bonds suggests that ultra-high frequency (UHF) ultrasonic techniques might render an appropriate tool. However, the investigation of diagnostic techniques of diffusion bonds via the acoustic microscope with micron dimensional resolution capability has been surprizingly sparse[1]. In addition, the metrology mode[2] of this versatile inspection tool has to date not been invoked in this context to ascertain the elastic properties in the immediate vicinity of the joined surfaces of a diffusion bond.

The metrology mode employs Rayleigh waves that are generated in the wide-angle acoustic lenses, exclusively used in acoustic microscopy, to produce a periodic acoustic material signature (AMS) from which quantitative Rayleigh- and, on occasion, other surface-propagating mode velocity information is retrieved. Because Rayleigh waves may be excited from VHF through the microwave frequency range with this technique, the consequent pixel resolution for imaging and elastic inspection of micron-sized areas, quickly and nondestructively, would seem nearly ideal for the quantitative characterization of diffusion bonds.

In this paper, the results of a preliminary investigation into the nature of the elastic material characterization of a titanium (Ti) diffusion bond is described. Acoustic images and material signatures at 200 MHz are presented with an attendant surface area and depth resolution of less than 15 μm. Although the new generation of acoustic microscopy instrumentation features some form of metrology-mode capability, the present study was conducted with an instrument of earlier vintage[3] devoid of this option. Therefore, the necessary data for the AMS generation was obtained on a point-by-point basis as will be described below.

PRIOR WORK

A cursary review of the published literature on the bulk elastic parameters of titanium metal and titanium alloys reveals some variations. No data on elastic surface properties, such as Rayleigh waves appears to have been of interest to date.

Considerable effort has been devoted to diagnostic studies of acoustic bulk wave propagation in titanium alloys commonly considered in joining parts by the diffusion-bonded process[4,5,6,]. In these efforts, the effect of the gas contamination, inherently involved in the joining process, on the longitudinal velocity and other measurable propagation properties of the titanium alloys was considered. The measured results from these investigations, conducted at 5 MHz, are not totally in agreement with each other. In one instance[4], a fractional increase of 3% was noted in the longitudinal velocity with contamination of 0.9 wt% of oxygen. Contradicting evidence[5,6] is also reported, where a longitudinal velocity decrease of 0.7% was determined with 0.3 wt% of oxygen increase in Ti-6211 titanium alloy.

In addition, the effect of oxygen contamination up to 30 wt % on lattice spacing along the "a" and "c" directions has been carried out in fundamental studies that revealed the formation of an intermediary δ-phase, containing higher-order oxides of titanium[7].

Table I - Elastic properties of bulk titanium[a].

Density	4.50 Kg/m^3
Longitudinal velocity	$6.07 < v_l < 6.13$ mm/µsec
Shear velocity	$3.12 < v_{sh} < 3.18$ mm/µsec
Poisson's ratio[b]	$0.325 < \sigma < 0.311$
Rayleigh velocity[b]	$2.91 < v_r < 2.96$ mm/µsec
Acoustic attenuation [c] (α)	6.33 dB/µsec
Longitudinal velocity[c]	6.32 mm/µsec

Notes:
a. polycrystalline unless noted;
b. derived from bulk velocities.;
c. Single crystal, along "a" at 1 GHz with $f^{1.4}$ frequency dependence from 0.5-2.0 GHz [8].

Finally, the results of a brief literature search that revealed some variations in the elastic properties, particularly bulk velocities of the polycrystalline titanium metal is summarized in Table 1. The acoustic velocity appears to be subject to an uncertainty of 1% and a consequent 2 % variation in the value of the corresponding modulus. The additional information for crystalline titanium with natural hexagonal anisotropy (hcp) is supplied to suggest that one possible explanation for the variation in published material properties lies the large anisotropy factor of titanium[9].

From the above, it would seem warranted to add some recent measurements on the surface-oriented properties of this important metal in the context of the determination of the elastic properties in the vicinity of the diffusion bond specimen joining two titanium parts. In addition, the published literature does not offer diagnostic studies of Ti diffusion bonds employing Rayleigh waves.

THE DIFFUSION-BONDED SPECIMEN

The specimen represents a diffusion-bonded joint of Ti-alloy containing 6% Al and 4% V that has been oxygen-contaminated in the diffusion process near the bond plane. The diffusion-bonded specimen has been cut normal to the bond plane and polished for metallurgical diagnostic studies. Indentation marks were provided to bracket the diffusion bond plane as is shown in the sketch of Figure 1. Optical views of this region are shown in the images of Figure 2. In Figure 2b the sample has been metallurgically etched to reveal the oxygen-enriched bond region that appears brighter than the host material with no more than 0.2 wt% oxygen. Figure 2a is devoid of contrast but does exhibit phase boundaries that suggest the presence of α- (hcp) and β- (bcc) phases, known to exists in Ti-O systems[7].

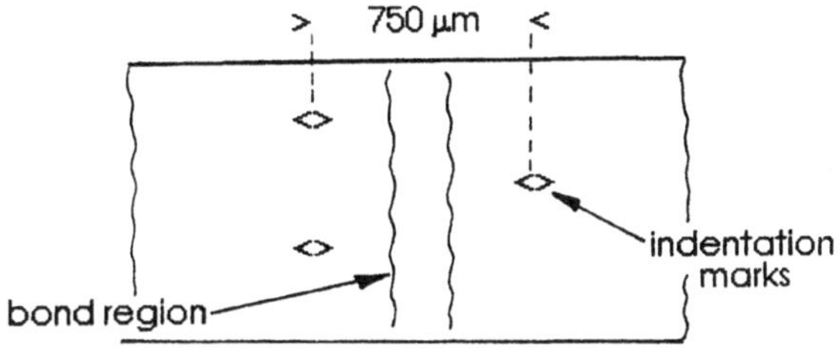

Figure 1. The diffusion-bonded region of the specimen; the bond plane extends normal to the paper.

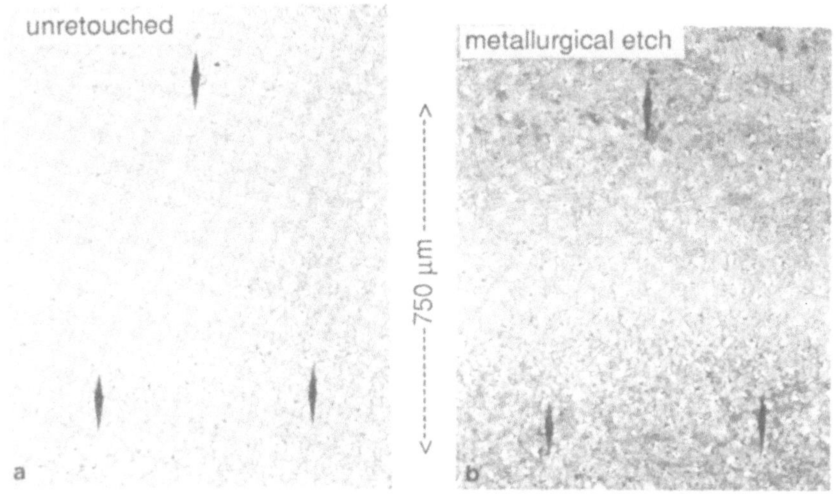

Figure 2. Optical micrographs of the bond region; a) unretouched surface; b) metallurgically-etched surface that shows brighter image contrast in the oxygen-rich bond plane.

ACOUSTIC IMAGING

The presence of different metallurgical phase distributions between the bond- and host-material regions and the attendant variation in their respective elastic properties suggest that the bond region might be more clearly identified for further study with the aid of acoustic microscopy. The large image contrast that results from spatial acoustic impedance variations is well documented in the published literature[10]. A comparison of representative acoustic and optical images of the bond region, shown in Figure 3, clearly illustrates that this is indeed the case.

The acoustic diagnostic investigation was carried using a prototype acoustic microscope at 200 MHz[3]. At this frequency, the wavelength of Rayleigh waves in titanium is 14.5 μm. One should keep in mind that this is also the approximate depth of penetration of Rayleigh waves in the specimen and, therefore, depth variations within this region will not be evident if the frequency is held constant. Such depth variations are not likely to exist in this specimen.

The elongated dark area in Figure 3, essentially bisecting the area between indentation marks, is evidence of signifiant changes in the elastic properties (acoustic impedance changes) brought about by the oxygen enrichment. The semi-circular artifact in the lower left region of the acoustic image is to be disregarded, as it is the result of an unfortunate accident during the inspection, in which the acoustic lens was inadvertently driven into the specimen, leaving its imprint.

Large contrast variations of this type are the result of a two dimensional acoustic material signature (AMS) that constitutes the metrology mode. Provided that the gray scale of the acoustic microscope is calibrated, these contrast variations may furnish a quantitative spatial elastic map of the diffusion bond area, as has been suggested earlier[11]. The calibration procedure did not take place in this preliminary study and therefore, the images in this report are of a qualitative nature. However, quantitative information in the form of AMS-derived Rayleigh velocities is still retrieved at specified locations, as will be described in what follows.

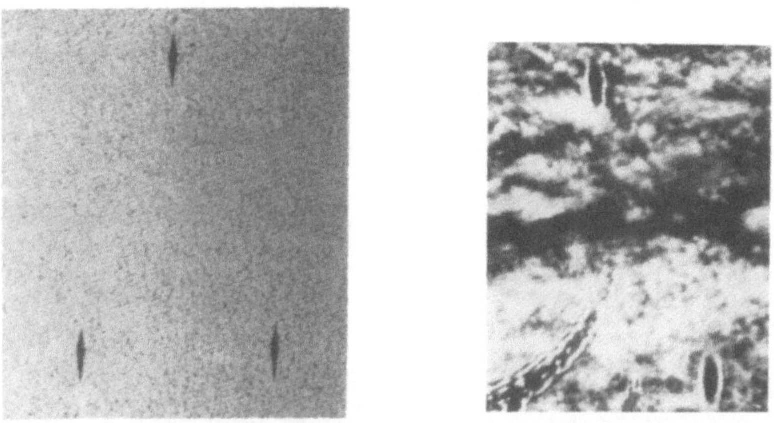

Figure 3. Optical (100x) and acoustic (80x) micro-images of Ti diffusion bond; the black region bisecting the bracketed region is evidence of large acoustic impedance change in the bond region.

ACOUSTIC MATERIAL SIGNATURE ACQUISITION

The prototype acoustic microscope used in this study did not feature an automatic AMS acquisition mode from which elastic material data via the Rayleigh velocity is obtained. However, this information was still obtainable in the following manner. In its simplest form, the AMS records the variation of the detected transducer output magnitude as the specimen is translated from the focal plane toward the acoustic lens. The focusing control of this instrument is a micrometer in which the smallest repeatable increment is one micrometer (μm). The predicted AMS curve for titanium is of the form shown in Figure 4, that corresponds to a frequency of 200 MHz. The Rayleigh velocity v_R is then calculated from the period Δz_N at which the minima (dark image) or maxima (bright image) repeat, as has been described previously[2]. Figure 4 has been drawn in this manner by using the Rayleigh velocity value obtained on the host material.

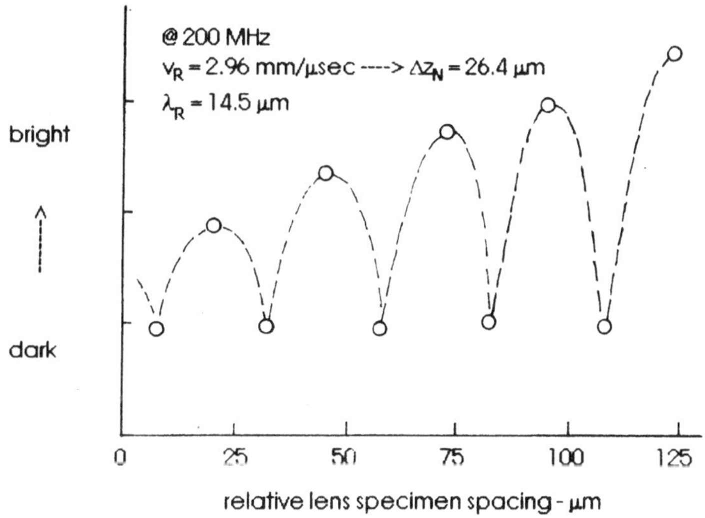

Figure 4. Predicted acoustic material signature (AMS) for titanium.

Following this procedure, two sets of acoustic images were taken by recording the micrometer readings that corresponded to the minima (dark) in both the diffusion bond area (Figure 5) and also in the host material region (Figure 6). In each of the respective images the micrometer reading is shown that corresponds to the minimum brightness for that particular set. In Figure 7, a partial AMS has been reconstructed from the data of the minima micrometer readings. It appears that the period Δz_N that corresponds to the diffusion bond area is reduced as compared to that of the host material region. Finally, much of the obtained data is summarized in Table II. The average value of the AMS period Δz_N in the bond and bulk regions of the specimen is indicated. From this data the corresponding Rayleigh velocity, that is a function of both frequency and Δz_N [12], is also listed in the table.

Table II - AMS data and derived results

Position (μm)	Bulk (μm)	Bond (μm)
1st min	38	23
2nd min	64	43
3rd min	90	64
4th min	≈114	85
Δz_N (μm)	26	21
v_R (mm/μsec)	2.91	2.60

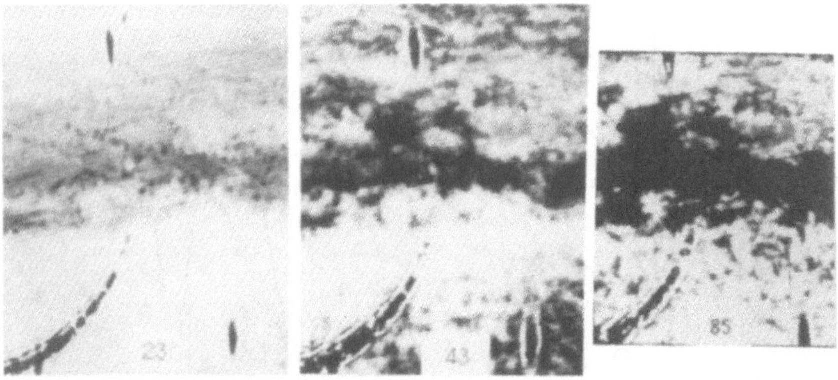

Figure 5. Acoustic images of Ti-diffusion bond;
(taken at minimum brightness in area of bond plane).

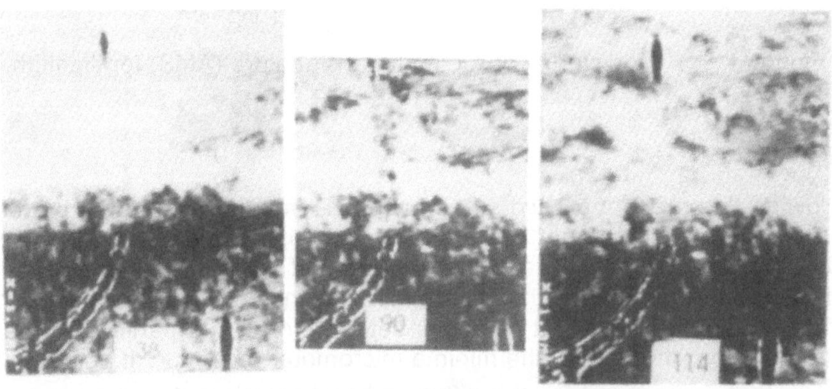

Figure 6. Acoustic images of Ti-diffusion bond;
(taken at minimum brightness in host material area).

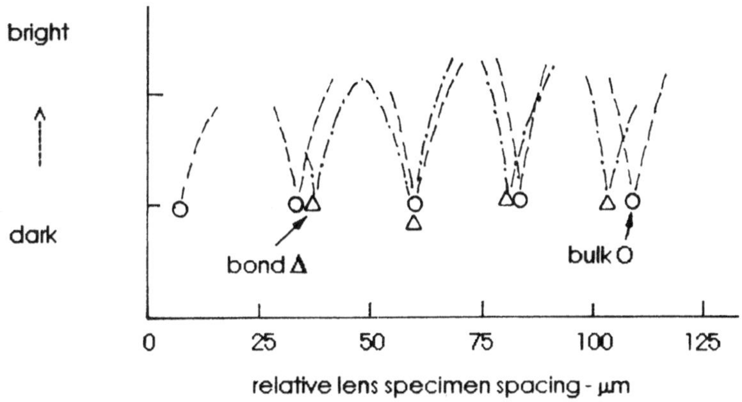

Figure 7. Reconstructed, partial AMS on Ti-diffusion bond specimen.

DISCUSSION

The substantial image contrast variations in the diffusion-bond area is undoubtedly the result of changes in the spatially-varying elastic properties. Quantitative estimates of these spatial variations are not easily determined unless an absolute gray-scale calibration of the contrast range has been established prior to image acquisition. In this preliminary study, this somewhat time-consuming procedure was not implemented. However, some quantitative determination of the gross properties of the specimen was effected.

The accuracy of the AMS aquisition procedure in this case rests on the assumed procedure relying essentially on the mechanical stability of the mechanical linkage between the acoustic lens and the measuring micrometer. There was no evidence of backlash in the mechanical movement and the least significant step for recording the axial specimen movement of one micrometer assures an maximum inaccuracy of 4 % in the measured period Δz_N between brightness minima. Seasoned users of metrological optical microscopes will agree that the actual accuracy in such a case is probably much better than that.

Measurements in the diffusion-bond region and elsewhere in the bulk region of the specimen indicate that a significant reduction in Rayleigh velocity v_R of about 11 % has occured in the former. It is possible to obtain the desired estimate of the probable changes that have taken place in the bulk modulus or bulk strength in the bond region. It was found, in earlier studies[13] that neither the density ρ nor the Poissons ratio σ of the oxygen-contaminated regions were affected in a significant way. Following these assumptions and also that the specimen is largely isotropic, the effective Ti bulk velocities vary then directly as v_R and the probable reduction in bulk modulus of the two regions are then related solely to σ and v_R, as follows:

$$v_{shear} = [(1 + \sigma)/(0.87 + 1.12\,\sigma)]\, v_R,$$

and

$$v_{shear}/v_{long.} = f(\sigma) = [(1-2\sigma)/2(1-\sigma)]^{1/2}.$$

Substitution from Table II yields a bulk modulus reduction factor ΔB in the bond as compared to the bulk region of

$$\Delta B = 1-(v_R'/v_R)^2 \approx 21\%.$$

Based on this result, the effect of the oxygen contamination is to significantly compromise the integrity of the specimen.

CONCLUSIONS

The acoustic microscope study has shown that diffusion-bonded regions in titanium may be identified and imaged with large contrast and good gray scale. The region of the diffusion-bond stands out significantly from the surrounding bulk, adjacent to the bond plane. Quantitative information of the elastic properties in and away from the bond plane were derived from the acoustic material signature (AMS) that was obtained manually at axial positions of minima in the acoustic images. The resultant Rayleigh velocity data shows a substantial reduction along the bond plane as compared to the remainder of the Ti specimen. The lower Rayleigh velocity suggests correspondingly reduced bulk moduli and consequently reduced strength in the bond region.

ACKNOWLEDGEMENT

The author is grateful to Dr. R. C. Addison of Rockwell International Science Center who suggested the topic for this study.

REFERENCES

1) B. Derby, G.A.D. Briggs and E. R. Wallach, "Non-destructive testing and acoustic microscopy of diffusion bonds", J. Materials Science, vol. 18, I983.

2) R. D. Weglein, "Acoustic Micro-Metrology", IEEE Trans. on Sonics and Ultrasonics, vol. SU-32, No. 2, pp. 225-234, March, I985.

3) kindly made available through the courtesy of the Olympus Corporation of America.

4) N. Hsu and H. Conrad," Ultrasonic Wave Velocity Measurements on Titanium- Oxygen alloys", Scripta Metallurgica, vol. 5, pp. 905-908, Pergamon Press, I971.

5) See, for example, O.P. Arora, et al, "Proceedings of the 1st workshop on nondestructive evaluation (NDE) of titanium alloys", Report No. DTNSRDC-SME-CR-14-82, 12 contributions, Naval Research Laboratory, Wash., DC 20375, June I982.

6) S. R. Buxbaum and R. E. Green, Jr., "Ultrasonic Characterization of Oxygen Contaminated Titanium 6211 Plate", Proc. Ultrasonics International '83, pp.91-96, I983.

7) E. S. Bumps, H. D. Kessler, and M. Hansen, Trans. ASM, vol. 45, pp. 1008-1028, l953.

8) M. T. Wauk II, "Attenuation in microwave acoustic transducers and resonators", PhD Thesis, Stanford University, July l969.

9) B. A. Auld, ACOUSTIC FIELDS AND WAVES IN SOLIDS, vol. 1, App. 2, John Wiley & Sons, N.Y. 1973.

10) Special Issue on Acoustic Microscopy - IEEE Trans. Sonics & Ultrasonics, vol. SU-32, No. 2, March l985.

11) R. D. Weglein and A. K. Mal, "Toward quantitative imaging with the acoustic microscope", ACOUSTICAL IMAGING, eds. A. J. Berkout, J. Ridder and L. F. van der Wal, pp. 387-396, vol. 14, l985, Plenum Press, N. Y.

12) W. Parmon and H. L. Bertoni, "Ray interpretation of the material signature in the acoustic microscope", Elect. Lett., vol. 15, No. 21, pp. 684-686, Oct. l979.

13) S. R. Buxbaum and R. E. Green, Jr., "Ti-Alloy Acoustic Data", pp.15-36, in reference 5.

APPLICATION OF SCANNING ACOUSTIC MICROSCOPE TO FERROELECTRICS

Seiji Kojima

Institute of Applied Physics, University of Tsukuba
Tsukuba, Ibaraki 305, Japan

INTRODUCTION

For the characterization of inhomogeneity in ferroelectric compounds, various types of instruments have been used up to the present. An optical polarizing microscope has been mostly applied to observe the elastic inhomogeneity indirectly through the piezo-optic effect, whereas scanning acoustic microscope (SAM) is useful to observe the elastic inhomogeneity directly and it is also useful to determine the elastic properties quantitatively.[1]

The most important feature of an acoustic focusing lens is the excitation of leaky Rayleigh waves (LRW) which propagate along the interface between a specimen and liquid coupler. The interference effect due to the propagation characteristics of LRW can afford the precise information about elastic properties. One phenomenon resulting from this effect is the periodic variation of V(z) curves in the region $z<0$, where V denotes the output voltage of a lens and z is measured from the focal point and negative values of z correspond to lens-sample spacing shorter than the focal length. It has been confirmed that a phase velocity v_R of LRW was determined by the periodicity Δz. Further the development of a cylindrical lens enables the determination of the directional dependence of v_R in an anisotropic material.[2] Another phenomenon caused by the LRW effect is the interference fringe patterns around a crack. Such a fringe is formed owing to the reflection of LRW and the spacing $\Delta \ell$ is equal to a half of a wavelength λ_R of LRW.[3]

The present work has extended the applications of SAM to ferroelectric and the novel phenomena have been investigated to develop a new method of obtaining various information in a very small region.

EXPERIMENTAL

All measurements were performed with scanning acoustic microscopes manufactured by Olympus Optical Co., Ltd., at the reflection mode. The frequencies of the soundwave bursts are around 0.2 GHz and 0.4 GHz. All acoustic lenses used were made of sapphire, on which ZnO piezoelectric films were sputtered. For the measurements of V(z) curves, the line-focus lens with a cylindrical plane was also used. The surface of the specimen to be objected was lapped with #4000 green carborundum and was finally polished with 0.05 μm colloidal alumina (Linde B).

STUDY OF FERROELECTRIC CRYSTALS

Ferroelectric materials have been widely used in the field of ultrasonics, electronics and opto-electronics. Ferroelectricity was observed in a large number of compounds. The various textures of ferroelectrics are classified into several categories from crystallographic symmetry. In the present paper the two typical cases, namely, uniaxial and biaxial ferroelectrics are investigated.

Barium Titanate. Barium Titanate $BaTiO_3$ (BT) is one crystal of most technological importance. At room temperature, a BT crystal is uniaxial with a tetragonal system and a spontaneous polarization exists along the c-axis. Tetragonal ferroelectrics have two types of domain, namely, a 90° domain and a 180° domain, where the angles denote the change of the polarization direction on both sides of a domain wall. Generally in a uniaxial ferroelectrics the propagation characteristics of LRW on both sides of a 180° domain wall are equivalent and the contrast among different domains cannot be expected in acoustic images. It is very important to observe 180° domains on an unetched surface. However, in fact, such a observation has not yet succeeded. Therefore in the present work, 90° domains and 90° domain walls were investigated. As to the mechanism of the contrasts among domains and domain walls was already, analyzed elsewhere[4] and the two phenomena were studied as follows.

Identification of domain configuration. The geometrical patterns of domains and domain walls have been observed by an optical polarizing microscope (OPM) through the change of optical birefringence. However, in fact, the determination of crystallographic orientations in each domain is impossible. Fig. 1 shows the inclined boundary in an unpoled BT crystal with the size $7 \times 7 \times 10$ mm^3. The shape was observed by OPM as shown in Fig. 1. For such a boundary inclined to the surface, four possible cases are derived from the crystallographic consideration as shown in Fig. 2. Firstly, the acoustic images were investigated in the neighborhood of the inclined boundary as shown in Fig. 3. When the surface of the sample was located in the immediate vicinity of a focal plane, the acoustic micrograph (Fig. 3(a)) shows only a bright line along the positions where the surface and the boundary cross. With decreasing z, the interference fringe appears around the boundary. However the contrast between both sides of the boundary did not appears even in a defocused region. Consequently it is concluded that the crystallographic orientations of both sides are equivalent and the possibility of the cases (a) and (b) in Fig. 2 is denied.

Secondly, to distinguish the two cases (c) and (d) the V(z) curves were measured accurately by utilizing a usual spherical lens. To obtain the standard V(z) curves of each orientation, the another poled crystal was used. Fig. 4 shows the V(z) curves obtained on both a (100) plane and a (001) plane: the averaged velocities $\bar{v}_R$ of LRW on a (100) plane and on a (001) plane were 3.76 km/s and 3.37 km/s respectively. Next, in order to identify the orientations of each side of the inclined boundary V(z) curves were measured on both sides. The line shapes of the curves were essentially equivalent to that of a (100) plane. The obtained values of $\bar{v}_R$ are 3.75 km/s on the right hand side and 3.76 km/s on the left hand side. The values of both sides are equal to that of a (100) plane within experimental uncertainty. Therefore the domain configuration of the present sample was identified uniquely to the case (d) of Fig. 2.

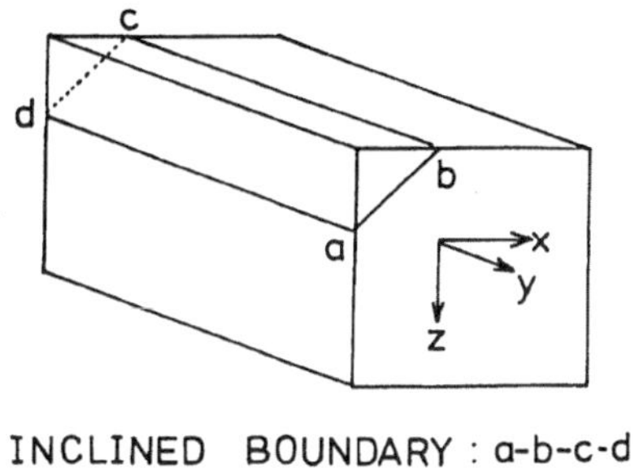

Fig. 1. The schematic illustration of the inclined boundary in a barium titanate crystal.

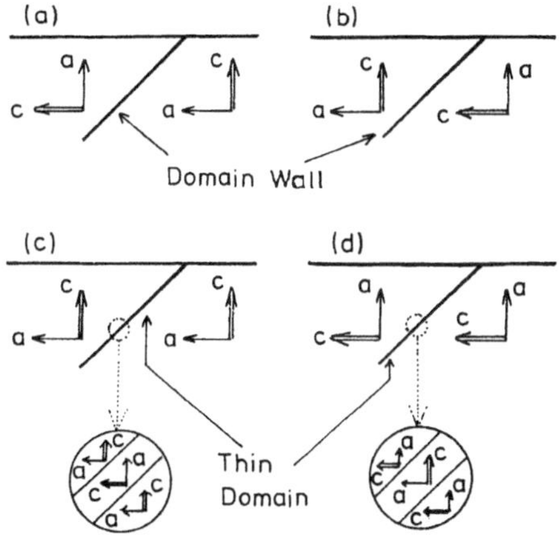

Fig. 2. The schematic illustration of four possible cases in domain configurations.

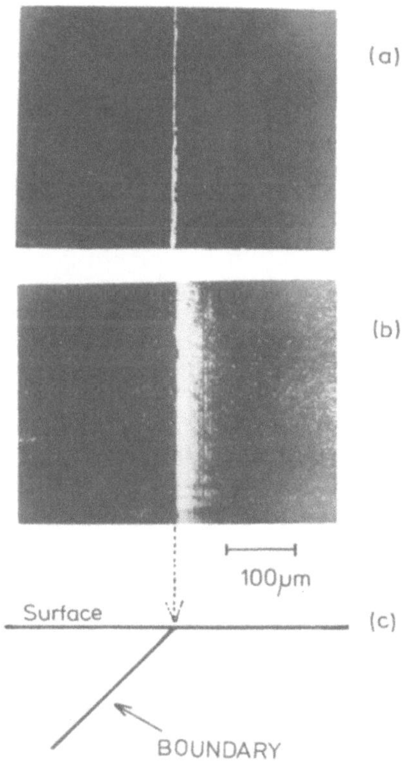

Fig. 3. Acoustic micrographs for two different focusings ((a)z= -0.1μm, (b)z=-100μm) and (c) the schematic cross-sectional drawing of a barium titanate crystal.

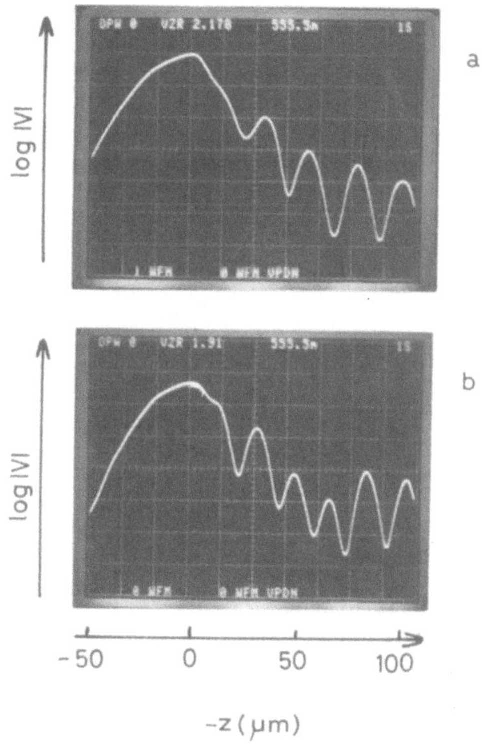

Fig. 4. V(z) curves obtained by a spherical lens in a poled barium titanate crystal. The operating frequency is 0.42GHz. (a) The curve on a (100) plane. (b) The curve on a (001) plane.

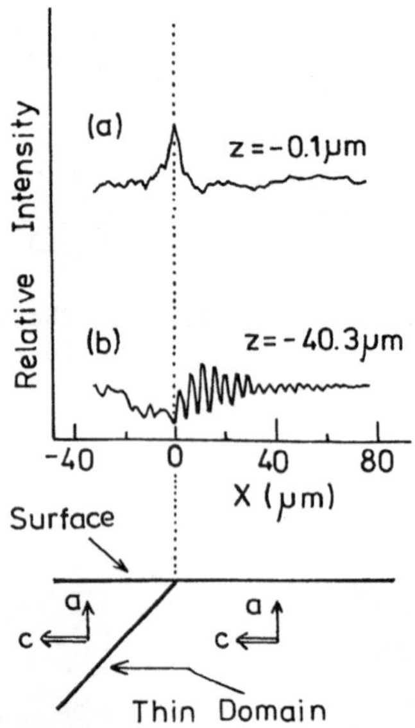

Fig. 5. Line scan profiles of acoustic signals of the asymmetric fringe patterns in a barium titanate crystal. Operating frequency is 0.4GHz. (a) z=-0.1μm, (b) z=-40.3μm.

Table 1. Values of $\Delta\ell$, $\lambda_R/2$ determined from the interference fringes and the V(z) curve of a cylindrical lens. The operating frequency is around 0.42GHz.

Method	$\Delta\ell$(μm)	$\lambda_R/2$(μm)	v_R(km/s)
Interference fringe	4.1 ± 0.1	—	3.44 ± 0.08
V(z) curve	—	4.15 ± 0.02	3.49 ± 0.02

As described above, the measurements of V(z) curve are very important to determine the crystallographic orientations in a very small area on a multidomain crystal or a polycrystal.

The Asymmetric fringe around the inclined boundary. As to the crack situated perpendicular to a surface, interference parallel fringes appear around the crack. For such a fringe, the mechanism of the appearance has been considered to be the interference due to the equivalent reflections of excited LRW on the both sides of the crack. By the consideration of the phase difference, it is concluded that the fringe spacing $\Delta \ell$ is equal to $\lambda_R/2$ as reported previously.[3,5]

In contrast, the interference fringe around the inclined boundary, the intensity of the fringes is not symmetric on both sides of the boundary as shown in Fig. 3(b). In order to determine the spacing accurately, the line scan profiles were measured by the wave analyzer with a 1024 times averaging as shown in Fig. 5. The obtained value of spacing $\Delta \ell$ was 4.1 μm. The measurement of V(z) curve by utilizing a cylindrical lens was done to determine the wavelength λ_R of LRW, of which wave plane is parallel to the line where the surface and the boundary cross. Fig. 6 shows the obtained curves. The obtained values from these experiments are listed in Table 1. The value $\Delta \ell$ is in good agreement with $\lambda_R/2$ within experimental uncertainty. So it is concluded that the mechanism of the fringe is due to the interference of LRW.

Next, a simple model is proposed for the interpretation of the asymmetry. In this model only three types of rays are considered as shown in Fig. 7. One is the geometrically reflected beam with the normal incidence (beam, a, d). The two others with the incident angle θ_R (the Rayleigh cristal angle) are converted to LRW; they are then reflected at the boundary and reradiate the waves to the fluid with incident angle θ_R and finally reach the transducer (beam, b, c, e, f). The difference of the output intensity of these beams is mainly dependent on both the propagation loss of LRW and the reflectivities at the boundary. Due to propagation loss of LRW the intensity of the beam b is negligible in comparison with that of the beam c. In the same way, the intensity of the beam f is negligible in camparison with that of the beam e. Therefore, the asymmetry can be caused by the difference between the interference of two beams, a, c, and d, e.

As to the reflectivity of LRW or B.G. waves at the coners of various angle was already studied in detail in ref. 7, 8. These results showed that the intensity of the beam e is fairly larger than that of c and the interference of the beam d, e to be strong in comparison with that of the beam a, c. Therefore it is concluded that the interference fringes appear strongly on the right side of the boundary.

Another experiment was also made by the author on a inclined twin boundary of a calcite crystal. A similar asymmetric fringe was observed and such asymmetry strongly supports the above discussion. Asymmetric fringes are thus important sources to obtain the knowledge about the subsurface inclination of any kind of the boundary.

Gadolinium Molybdate

Gadolinium molybdate β-$Gd_2(MoO_4)_3$ (GMO) is one of the famous ferroelectrics. At room temperature, a GMO crystal is biaxial with a orthorhombic system and a spontaneous polarization exists along the c-axis. The special feature of GMO is that the phase velocity of shear waves in lower than that of usual dielectric crystals. So the condition $\theta_R > \theta_{lens}$ holds, where θ_{lens} means the half opening angle of an acoustic lens. In this case LRW cannot be excited and the periodicity of V(z) curves is determined by

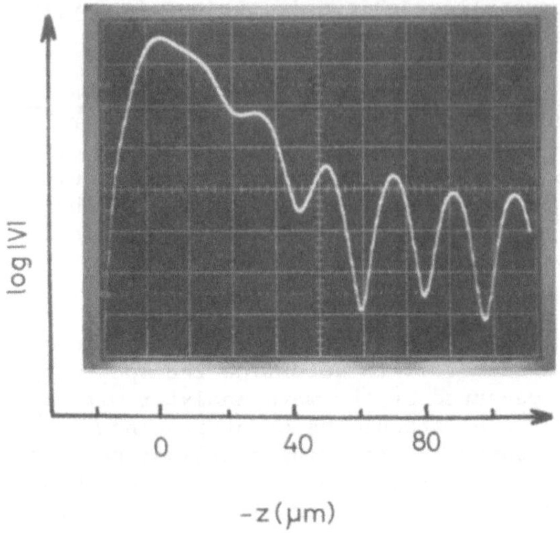

Fig. 6. The V(z) curve measured by utilizing a cylindrical lens. The operating frequency is about 0.42GHz. The periodicity gives the value v_R of LRW, of which wave plane is parallel to a (001) plane of a barium titanate crystal.

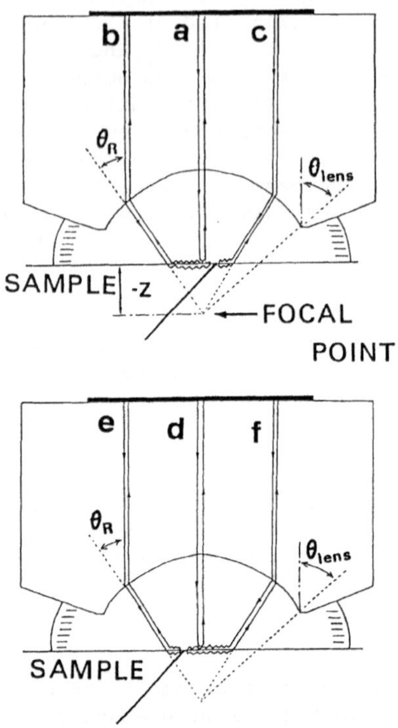

Fig. 7. The schematic illustration of the reflection of LRW at the very thin domain which is inclined to the surface.

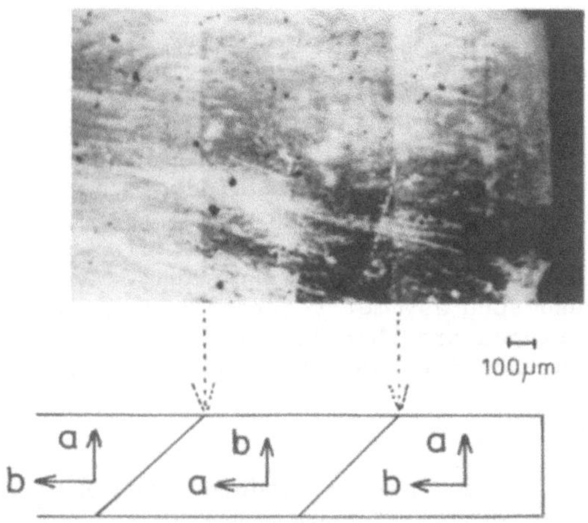

Fig. 8. The acoustic micrograph of three domains of an unpoled GMO crystal. The operating frequency is 0.2GHz.

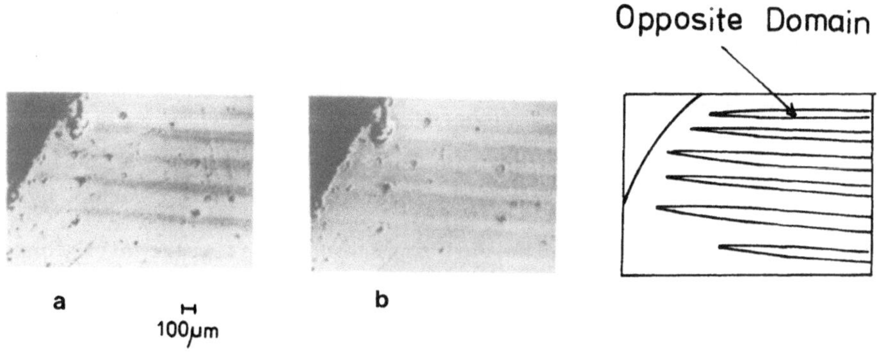

Fig. 9. Acoustic micrographs of a GMO crystal on a (001) plane. The operating frequency is 0.2GHz. The sharp lines correspond to 180° domain walls. (a) z=-60µm, (b) z=-80µm.

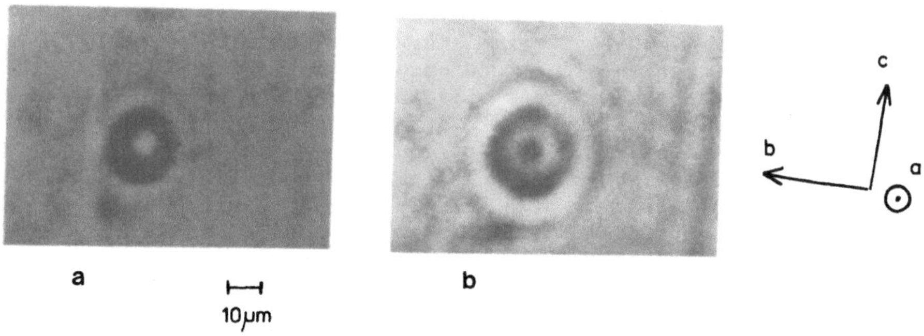

Fig. 10. Elongated interference rings around a very small holl on the (100) plane of a GMO crystal. The operating frequency is 0.2GHz. (a) z=-20µm, (b) z=-40µm.

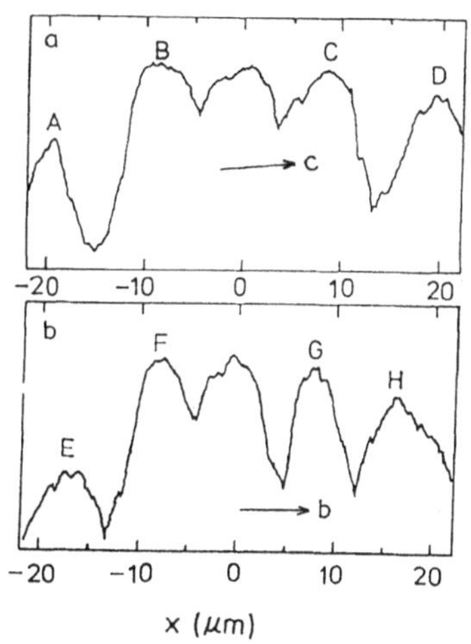

Fig. 11. Line scan profiles of the ouyput intensity in the vicinity of the elongated interference rings,(a) along the c-axis and (b) along the b-axis. The zero point of abscissa is adjusted to the center of the interference rings. A-H denote peaks of the interference rings.

the phase velocity of surface skimming compressional waves (SSCW).[8] It is therefore predicted that the output signals of a lens strongly reflect the propagation characteristics of SSCW.

The contrast due to SSCW. In acoustic images of GMO, the appearance of the contrast due to SSCW is expected in the defocused region. The observation was made on a (100) plate of an unpoled crystal, which contains one opposite domain. When the surface of the plate was adjusted at a focal plane, domains were scarcely distinguished. On the other hand, the contrast among domains appeared clearly with decreasing the lens-object spacing as shown in Fig. 8. The dark area located at the center corresponds to an opposite domain as shown in the bottom of Fig. 8.

The contrast between a domain and a domain wall is also predicted due to the reflection of SSCW at a wall. The observation was made on a (001) plate of an unpoled crystal. The contrast was clearly appeared as shown in Fig. 9.

These results shows that even if $\theta_R > \theta_{lens}$ holds the contrast due to SSCW plays a similar role to that due to LRW.

Elongated Interference Rings. As to a straight line reflector, the interference fringe appears as parallel lines with the periodicity $\lambda_R/2$ as described previously. Whereas a small crylindrical or hemispherical hole reflects acoustic waves in all directions on a surface. It is then expected that the interference rings will appear around such a small hole and that their spacing will vary with directions owing to elastic anisotropy. A high reflectivity is also required for the sharp contrast of such a ring. For a hemispherical hole the maximum reflectivity occurs in the condition that the radius of a hole is nearly equal to the wavelength. The fabrication of a precise small hemispherical hole is difficult and the hole to be studied was selected from holes peeled off in the process of polishing a surface. Fig. 10 shows acoustic micrographs obtained in defocused regions around a hole on a (100) plane. The elongation of the rings along the c-axis was clearly observed. For the exact determination of a ring spacing, line scan profiles were carefully measured along both the c and b-axis as shown in Fig. 11. On the assumption that the ring spacing is equal to a half wavelength, the phase velocities v_b and v_c along the b and c-axis, respectively, were determined.

The value of v_b and v_c were also determined from the V(z) curves of a cylindrical lens. These values obtained by the two methods are presented in Table 2. By the comparison among the values, it is concluded within experimental accuracy that the phase velocities determined from the interference rings are in good agreement with those from the V(z) curves. Therefore the directional dependence of ring spacing is considered to be strongly connected to the propagation characteristics around a hole. Such a property of elongated interference rings can be of great importance in obtaining the knowledge about local elastic anisotropy.

CONCLUSION

Scanning acoustic microscopy has been applied to the two typical ferroelectric crystals. In a study of barium titanate crystals. It is shown that the V(z) curve is useful to determine the crystallographic orientations in a very small area and that the asymmetry of the intensity of the fringes is caused by the subsurface inclination of a boundary. In a study of gadolinium molybdate crystals, it is shown that the contrast due to SSCW is very important when the equation $\theta_R > \theta_{lens}$ holds and the elongated interference fringes are strongly connected to the elastic anisotropy.

Table 2. Values of v_b, v_c determined from the interference ring and the V(z) curves of a cylindrical lens, where the operating frequency of V(z) is 0.42GHz.

Method	v_b (km/s)	v_c (km/s)
Interference ring	3.7 ± 0.2	4.7 ± 0.2
V(z) curve	3.82 ± 0.02	4.45 ± 0.02

ACKNOWLEDGEMENT

This work was partially supported by both the University of Tsukuba Project Research and Grant-in-Aid of Special Project Research on Ultrasonic Spectroscopy and its Application to Material Science from the Ministry of Education, Science and Culture of Japan. The author is grateful to Prof. T. Suzuki for his valuable discussion about SAM and Prof. T. Nakamura for supplying with the specimen. The author is thankful to Mr.F.Uchino and Dr.H.Tateoka, Olympus Optical Co., Ltd., for their technical supports of SAM.

REFERENCES

1. R.A. Lemon and C.F. Quate., Appl. Phys. Lett., 24: 163 (1974).
2. J. Kushibiki, A. Ohkubo and N. Chubachi., Electron. Lett., 17: 534 (1981).
3. K. Yamanaka and Y. Enomoto., Electron. Lett., 17:638 (1981).
4. S. Kojima., Proc. 6th Int. Meet. Ferroelectricity, Kobe, 1985, Jpn. J. Appl. Phys. 24-2: 553 (1985).
5. S. Kojima., Proc. 4th Symp. Ultrasonic Electronics, Tokyo, 1983, Jpn. J. Appl. Phys. 23-1: 203 (1984).
6. I. A. Victrov., 1967, "Rayleigh and Lamb Waves," Plenum Press, New York.
7. P. Kielczynski and W. Pajewski., J. Appl. Phys. 60: 78 (1986).
8. R.D. Weglein., Appl. Phys. Lett., 34: 179 (1979).

CONTINUOUS WAVE REFLECTION SCANNING ACOUSTIC MICROSCOPE (SAMCRUW)

A. Kulik, G. Gremaud, and S. Sathish

EPFL - Institut de Génie Atomique
CH-1015 - Lausanne
Switzerland

INTRODUCTION

Most of the reflection acoustic microscopes use pulse echo technique for imaging and for V(**z**) measurements. Amplitude of the surface echo of the sample is recorded as a function of the **xy** position (imaging) or as a function of the **z** (distance between the lens and the sample). Such a technique has two inherent weak points - usually phase information is lost and the signal to noise ratio is poor due to the wide bandwidth of the receiver electronics.

For the surface acoustic wave (SAW) measurements in dispersive media (for ex. layered structures) short (wide bandwidth) pulses are smeared out, which makes the measurements difficult. Working frequency is not well determined.

We propose a reflection continuous wave setup, which gives us vector measurements at a single frequency, with the usual drawback of coherent radiation. Most of the electronics used in our system are commercially available.

REALISATION OF THE SAMCRUW

Construction of our microscope uses mostly commercially available elements (fig. 1).

Ultrasonic Part of the System

As transmitter, receiver and measuring system, we use a Network Analyzer (HP 8753A), together with the S-parameter Test Set (HP 85046A). On the input of the system, acoustic lens is directly connected. The Analyzer measures S_{11} parameter, defined as the ratio of the reflected to incident wave, as a function of the frequency. The Analyzer is equipped with the time domain option (opt 010). Such a Fourier Transform processor allows us to calculate equivalent response in the time domain (pulse response) of the system, with excellent time resolution (fig. 2).

We use different types of lenses (200 MHz and 1 GHz) from OLYMPUS and LEITZ, which have a wide bandwidth and are well electrically matched.

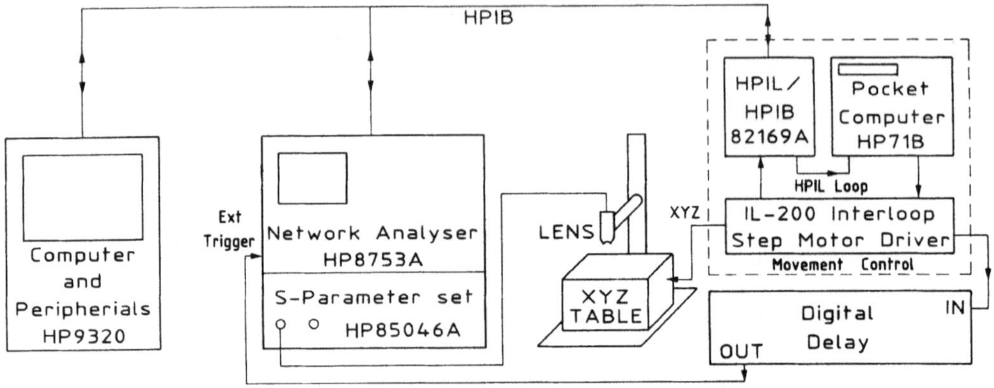

Fig. 1. Block diagram of the continuous wave scanning acoustic microscope SAMCRUW

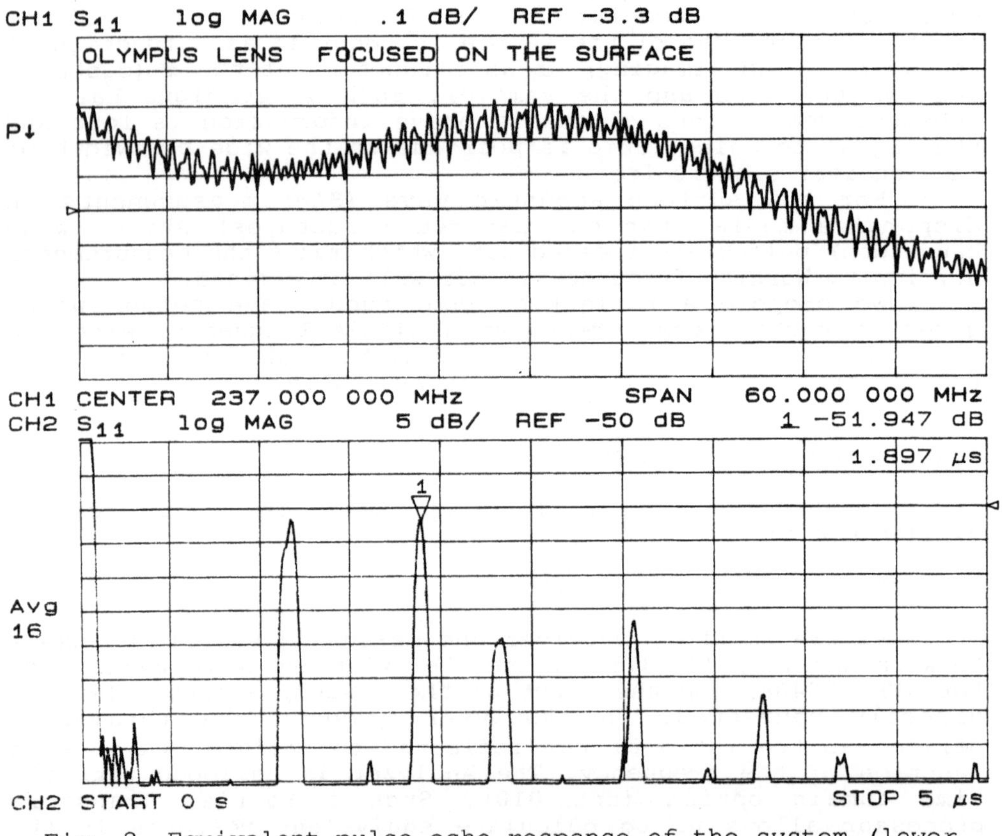

Fig. 2. Equivalent pulse echo response of the system (lower trace) calculated from frequency response (upper trace). Sample : Tungsten carbide.

The Mechanical Scanning System

Movement in the **xyz** directions is performed by a standard optical measuring microscope table from Microcontrole, which uses step motors for all three axis. The resolution chosen was 1 μm/step for the **x** and **y** direction and 0.1 μm/step for **z** axis.

Acoustic microscope for research needs maximum flexibility. Therefore a two computer system was chosen: one for the movement and scanning control, and the other for data processing and imaging.

Movement is controlled by an HP 71B pocket computer which in turn, controls an interloop # 200 Step Motor Driver, using an HPIL interface. For data exchange with the main computer, an HPIL/IB interface is provided. The keyboard of the pocket computer is redefined, allowing easy dialogue with the user. A program written in HP-BASIC offers a full choice of the absolute movements, scanning **xy** for imaging, scanning **z** for V(**z**) measurements. Synchronizing signals for the ultrasonic part can be delayed, as needed, in order to avoid measurements during the acceleration phase of the movements.

Data Processing and Imaging

For the data processing and imaging, we use an HP-9320 workstation running HP-UX (UNIX) system. The computer has 6 MB of RAM memory and 571 MB of the system disc. Another 55 MB disc with integrated cartridge backup allows us to archive the images off line. Screen has 256 gray levels and a 768 x 1024 resolution.

Computer is connected with our microscope, using a dedicated HPIB interface. Data for imaging in the form of consecutive lines is stored on working files for both imaginary and real parts. The resolution and number of lines can be chosen from 201 up to 1601, as determined by the setup of the Network Analyzer. Therefore, data is scaled and stored in the appropriate form for the NewView Image Processing Library[1]. NewView is a particularly convenient tool for research with the following facilities :

- accepts the images of any reasonable format, aspect and size
- can simultaneously display a lot of images on the screen as overlapping windows
- offers all the darkroom operations facilities
- has most of the usual imaging operations such as: storage, pixel manipulation, filtering, histogram, quantization, statistics, panning, zoom, rotation and dyadic operations too
- includes facilities for hard copy preprocessing
- offers a learning (script) facility

A coherent user interface using the mouse and pop-up menus for control of the microscope, is under development and will be incorporated into NewView.

THE WORKING MODES OF THE SAMCRUW

The Continuous Wave Acoustic Microscope can be used in three different modes.

Continuous Wave V(**z**) Measurements (CW V(**z**))

The working frequency can be freely chosen inside the useful bandwidth of the lens transducer. Therefore, the Network Analyzer is set into CW-TIME mode, which measures the real and imaginary part of the s_{11} parameter as a function of time at constant frequency.

The lens is defocused completely - sample is almost touching the lens. Then the **z** movement is activated and at the same time CW-TIME measurement is done. On the Analyzer's screen V(**z**) curve is directly displayed (fig. 3). The Fourier transform of V(**z**) is calculated in the Analyzer ($\mathbf{z}^{-1}$ domain) which allows us to find the principal periodicities of V(**z**).

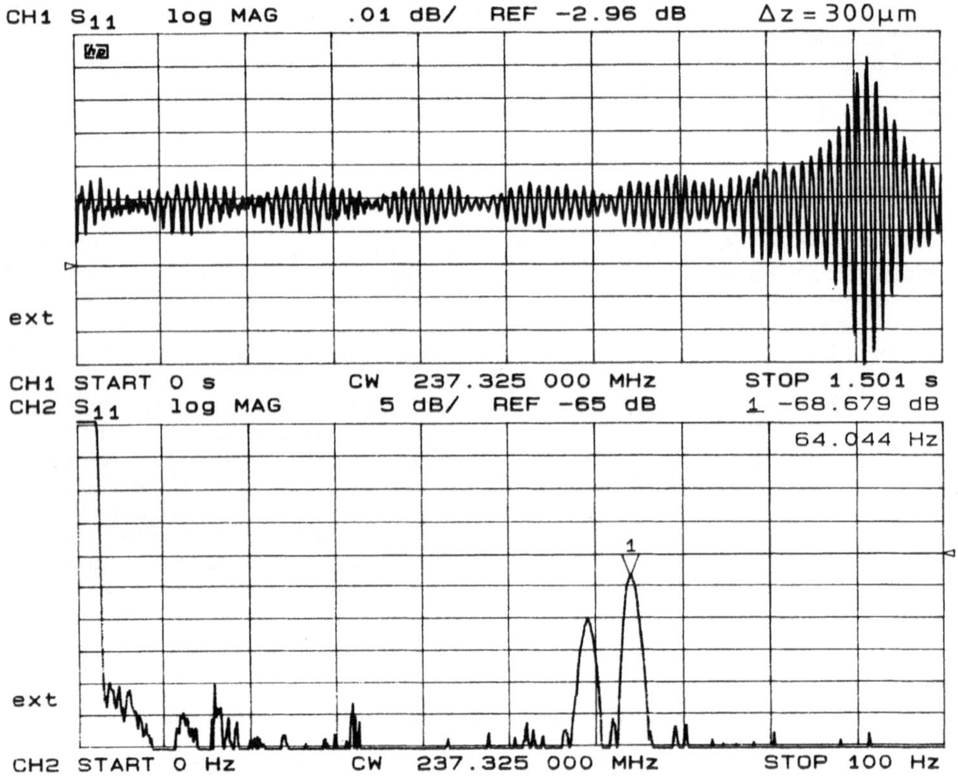

Fig. 3. Continuous wave V(**z**) curve (upper trace) and its Fourier transform (lower trace). Sample : Tungsten carbide.

Continuous Wave Imaging (CW Imaging)

CW imaging is similar to CW V(**z**) measurements. The lens is focused on the surface using quick V(**z**) measurement. The Analyzer measures real and imaginary parts of the s_{11} parameter as a function of the position at each line of the scan. Consecutive lines data are transferred to the main computer, scaled and stored as a standard file which can be read using the NewView[1] Image Processing Library. A typical result is presented in fig. 4. Real and imaginary parts images can clearly be distinguished - (see corresponding histograms underneath).

Equivalent Pulse Echo V(**z**) Measurements or Imaging

We calculate equivalent pulse echo response of the lens-water-sample system using the time domain option of the Analyzer (Fig. 2). With cursor readout we measure the amplitude of the sample's echo at each distance **z** between the sample and the lens. Note almost perfect similarity between the Equivalent Pulse Echo V(**z**) curve (Fig. 5) and the envelope of the CW V(**z**) (Fig. 3). Moreover, the measured Equivalent Pulse Echo V(**z**) matches well with pulse echo V(**z**) measurements obtained with a classical microscope, using the same type of lens and samples.

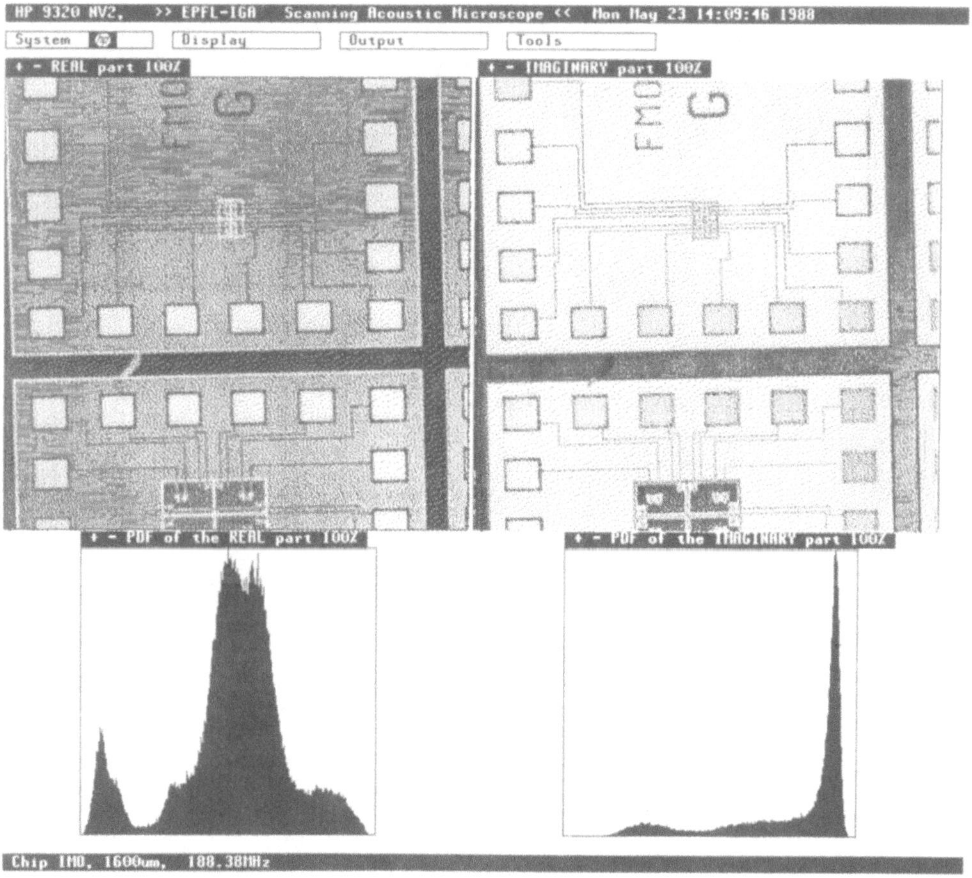

Fig. 4. Typical Continuous Wave images of the real and imaginary part of the s_{11} (Floyd-Steinberg diffused for hard-copy). Underneath corresponding PDF histograms.

This means that imaging possible in this mode will correspond to the classical pulse echo microscope, giving in addition phase information. The actual drawback of the imaging is the time necessary to calculate a Fourier transform : ~1 sec. per point.

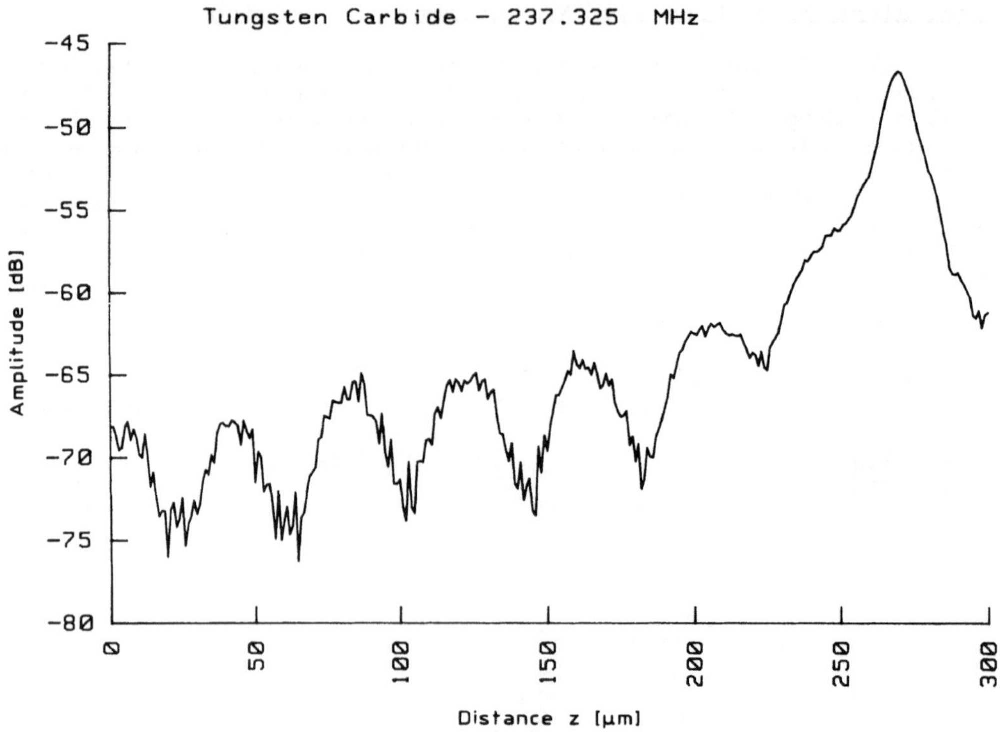

Fig. 5. Equivalent pulse echo V(**z**) response measured in the same setup as presented in fig. 3.

DISCUSSION OF THE RESULTS

Continuous Wave V(z)

CW V(**z**) curves can be understood as a series of standing waves in water (as in CW ultrasonic interferometer for liquids) modulated in amplitude by the classical V(**z**) curve (Fig. 3). By the Fourier transform we obtain main peak, allowing us to determine sound velocity in the coupling fluid and one or two sidelobes related to the V(**z**) amplitude modulation. From the distance between the main peak and sidelobe, we can calculate the mean value of the V(**z**) oscillation period and in turn Rayleigh wave velocity in the sample.

Such measurements need high sensitivity equippment since the full amplitude of the CW V(**z**) is usually smaller than 0.1 dB. Their advantage over the classical measurements lies in the fact that the working frequency is well determined and can be easily chosen anywhere inside the lens transducer bandwidth. Such a method seems to be well suited to the study of the dispersion relations in the layered media[2].

Continuous Wave Imaging

Despite of the high sensitivity of the method for the sample's topology variations, due to the oscillatory behaviour of the CW V(**z**) response of the system, lens-water-sample at one frequency can be compared with the CW ultrasonic measurements[3]. In principle, knowledge of the real and imaginary parts of the s_{11} parameter should allow the calculation of the complex reflectance of the sample. But topology variations give interference fringes, similar as these described in [4] with an equipment using mixed pulse echo and CW method. This means that a more detailed analysis should be performed in order to correctly interpret the contrast.

Comparable vector measurements were already done using a pulse echo apparatus[5,6], but they are technically more difficult to realize and are limited to a rather low frequency range.

Equivalent Pulse Echo V(**z**)

The calculated pulse response from the frequency response for each distance **z**, usually gives better time resolution than the classical pulse echo method. Unfortunately, we lose the advantage of working at constant frequency, so Equivalent Pulse Echo V(**z**) must be understood as measured on the mean frequency of the frequency sweep.

CONCLUSIONS AND PLANNED DEVELOPMENTS

Continuous Wave Scanning Acoustic Microscope has been constructed. Its main advantages are in a well defined working frequency, good Signal to Noise Ratio and phase sensitive measurements. The most important practical application can be seen in the local elastic measurements, using CW V(**z**) curve.

In the future, our project will consist of developing measurement procedures and algorithms which will automatically produce images with physical content, i.e. the topology, the Rayleigh wave velocity, attenuation, etc..

ACKNOWLEDGEMENTS

The authors are grateful to HP Laboratories and HP Geneva for the software and hardware support, and to LEITZ and OLYMPUS for making the lenses available.

This work was partially supported by the Swiss National Science Foundation.

REFERENCES

1. Courtesy HP Laboratories. Actually available from : Northwest Digital Research, Discovery Park, 3700 Gilmore Way, Vancouver B.C., Canada V5G4M1.

2. J. Kushibiki, T. Ishikawa, N. Chubachi, Precise measurement of film thickness by line-focus-beam acoustic microscope, in: Ultrasonic International 87, Butterworth, Guildford (1987).

3. D.I. Boleff, J.G. Miller, High frequency Continuous Wave Ultrasonics, in: Physical Acoustics VIII, W.P. Mason, R.N. Thurston, ed., Academic Press (1971).

4. M. Poirier, M. Castonguay, C. Neron, J.D.N. Cheeke, Nonplanar surface characterization by acoustic microscopy, J. Appl. Phys., 55: 89 (1984).

5. J.A. Hildebrand, K. Liang, S.D. Bennett, Fourier transform approach to materials characterization with the acoustic microscope, J. Appl. Phys. 54: 7016 (1983).

6. K. Liang, S.D. Bennet, B.T. Khuri-Yakub, G.S. Kino, Precise Phase Measurements with the Acoustic Microscope, IEEE Trans. SU, 32: 266 (1985).

VISUALIZATION OF ULTRASONIC WAVES PROPAGATING IN SCANNING ACOUSTIC MICROSCOPE

Hyo Ung Li and Katsuo Negishi

Institute of Industrial Science
University of Tokyo
Minato-ku, Tokyo 106, Japan

Introduction

In the study of scanning acoustic microscope(SAM), it is very important to know wave phenomena induced by focusing the ultrasonic beam at an interface between coupling water and solid sample. In addition to the direct reflection of focused ultrasonic beam, it is known that leaky surface acoustic waves(LSAW) excited by the incident beam play an important role in a visual contrast of acoustic image and also in material evaluation through V(z) curves.[1] Nevertheless, the wave phenomena occurring at a liquid-solid interface in SAM have so far been imagined only theoretically.

Recently, we succeeded in simultaneous visualization of ultrasonic pulses in both liquid and solid using the technique of stroboscopic photoelasticity together with the schlieren method. In this paper are presented some new photographs which reveal various modes of waves induced at the liquid-solid interface by focused ultrasonic waves in a large-scale model of SAM.

Experimental

Since the acoustic lens of SAM is too small to visualize ultrasonic waves in it, a large cylindrical lens made of optical glass (BK-7) is used as a model of microscopic lens. As shown in Fig.1, the lens is 80 mm wide, 60 mm height and 15 mm thick. The radius of curvature is 20 mm and its focal length is 27 mm in water. A piezoceramic transducer of 40 mm in width is attached to the top surface of the lens. Fundamental frequency of this transducer is 5.5 MHz. The lens is immersed in water to about 3/4 of its height and a large glass block is positioned near its focus as a specimen.

The block diagram of the experiment is shown in Fig.2. A stroboscopic light source, which delivers light flashes with the duration of 180 ns, makes it possible to visualize ultrasonic waves in still or slow motion by controlling a delay time between the transmitting pulse and the stroboscopic light. In order to visualize various modes of ultrasonic waves in the glass, a circularly polarized light is used in the photoelastic visualization. Because the polarizer-analyzer system is not perfect, there exists a weak leakage light which can be used for the schlieren visualization of ultrasonic waves in water. The schlieren method is realized by setting a slit just in front of the camera lens through which only the zero-th order diffracted light is passed. Then, the ultrasonic waves in glass are visualized as bright images and those in water as dark images in a halftone background.

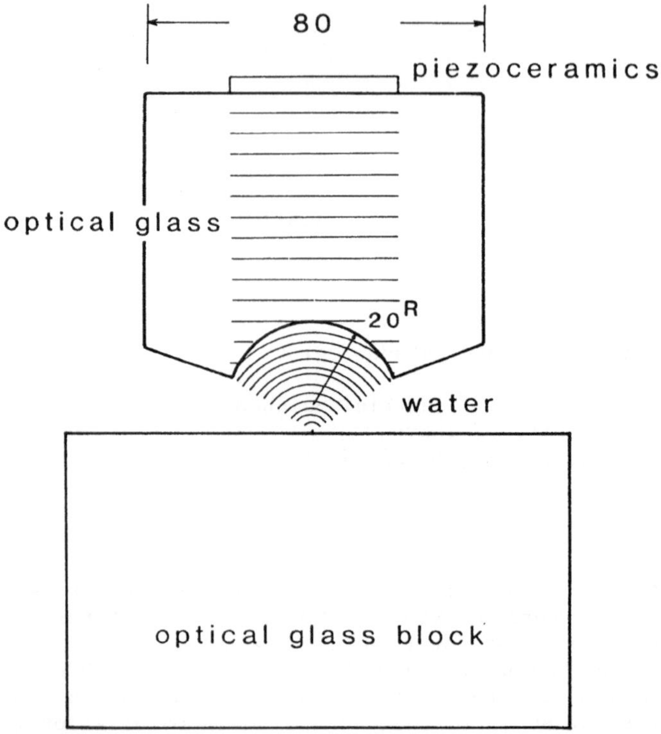

Fig.1. Geometry of cylindrical lens and solid sample.

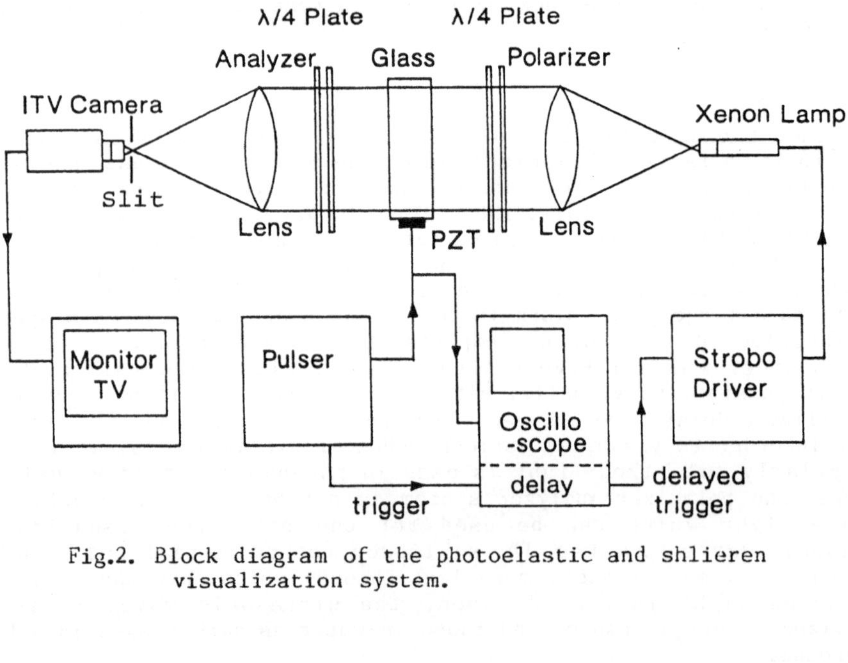

Fig.2. Block diagram of the photoelastic and shlieren visualization system.

Results and Discussion

A time sequence of ultrasonic pulse of 1.5 MHz propagating in the cylindrical lens is shown in Fig.3. In Fig.3 (b), the leading wavefront 1 is the longitudinal waves. The shear waves 2 also are excited from both edges of the transducer. In Fig.3 (c), the longitudinal wave 1 is converted into shear waves 3 by reflection at the side boundary of the glass. The faint longitudinal wave 4 is produced by double mode conversion between the front and rear surfaces, being longitudinal-shear-longitudinal conversion.[2] When the longitudinal waves 1 reflect from the cylindrical surface, in addition to longitudinal waves, shear waves 5 are excited by the mode conversion. In Fig.3 (e), longitudinal waves 1 arrive near the top surface after one round trip.

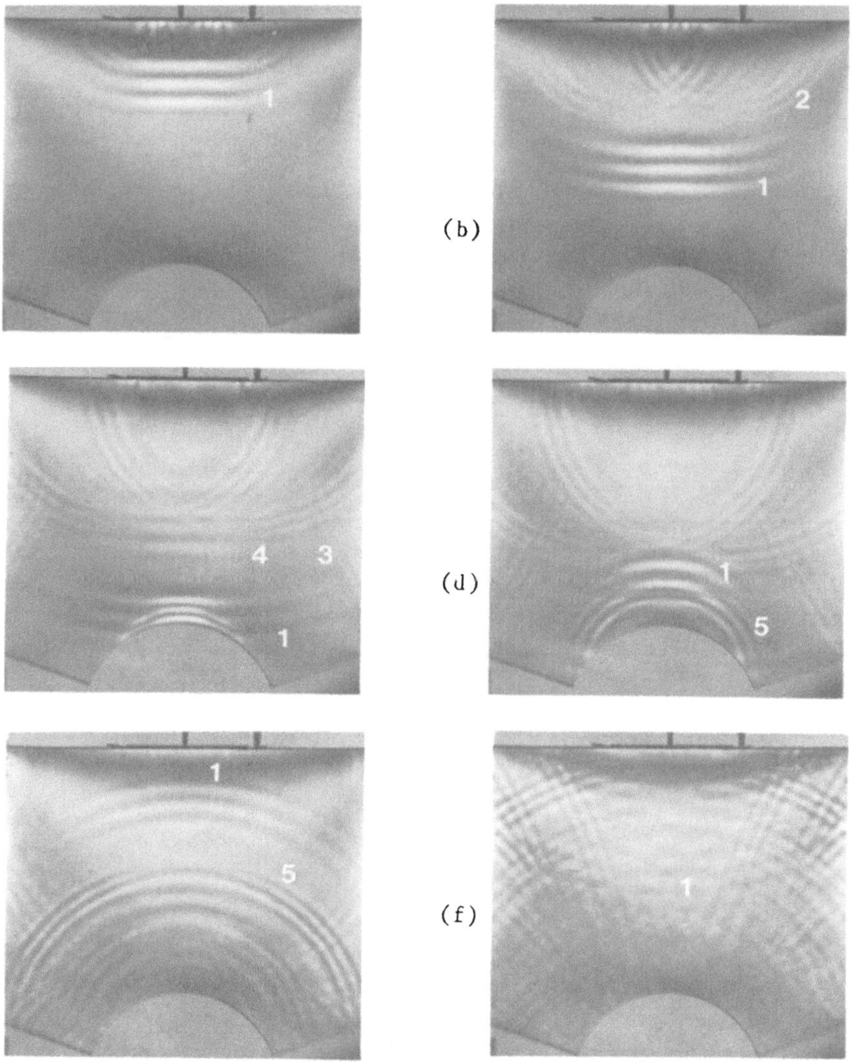

(a) (b) (c) (d) (e) (f)

Fig.3. Propagation of ultrasonic pulse in the cylindrical lens. The pulse is excited by sinusoidal waves of 1.5 MHz with the duration of 3 cycles.

Sharp focusing of a 5.5 MHz ultrasonic beam upon a glass block is shown in Fig.4 (a). There appears a diffraction field radiated from the transducer in the lens. A little later, reflection and transmission occur from the water-glass interface as shown in (b). Broadening of the beam waist and striations on both sides of the reflected beam are evident. Two bright beams appearing in the glass block are interpreted as transmitted shear waves. Transmitted longitudinal waves are too weak to be seen.

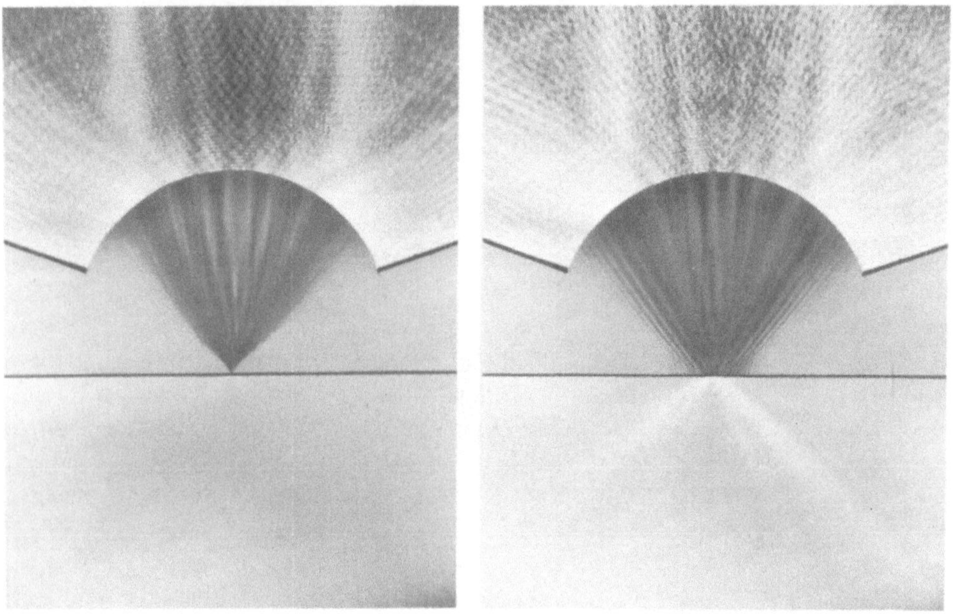

(a) (b)

Fig.4. Visualization of focusing (a), and reflection and transmission (b) of 5.5 MHz ultrasonic waves.

Figure 5 shows a time sequence of ultrasonic pulse focusing and reflecting from the glass surface. The carrier frequency of the pulse is 5.5 MHz and the width 5 μs. Here, the striations become clearer than those in Fig.4 (b). The reflection pattern shown in Fig.5 (e) remains unchanged when the lens is moved vertically.

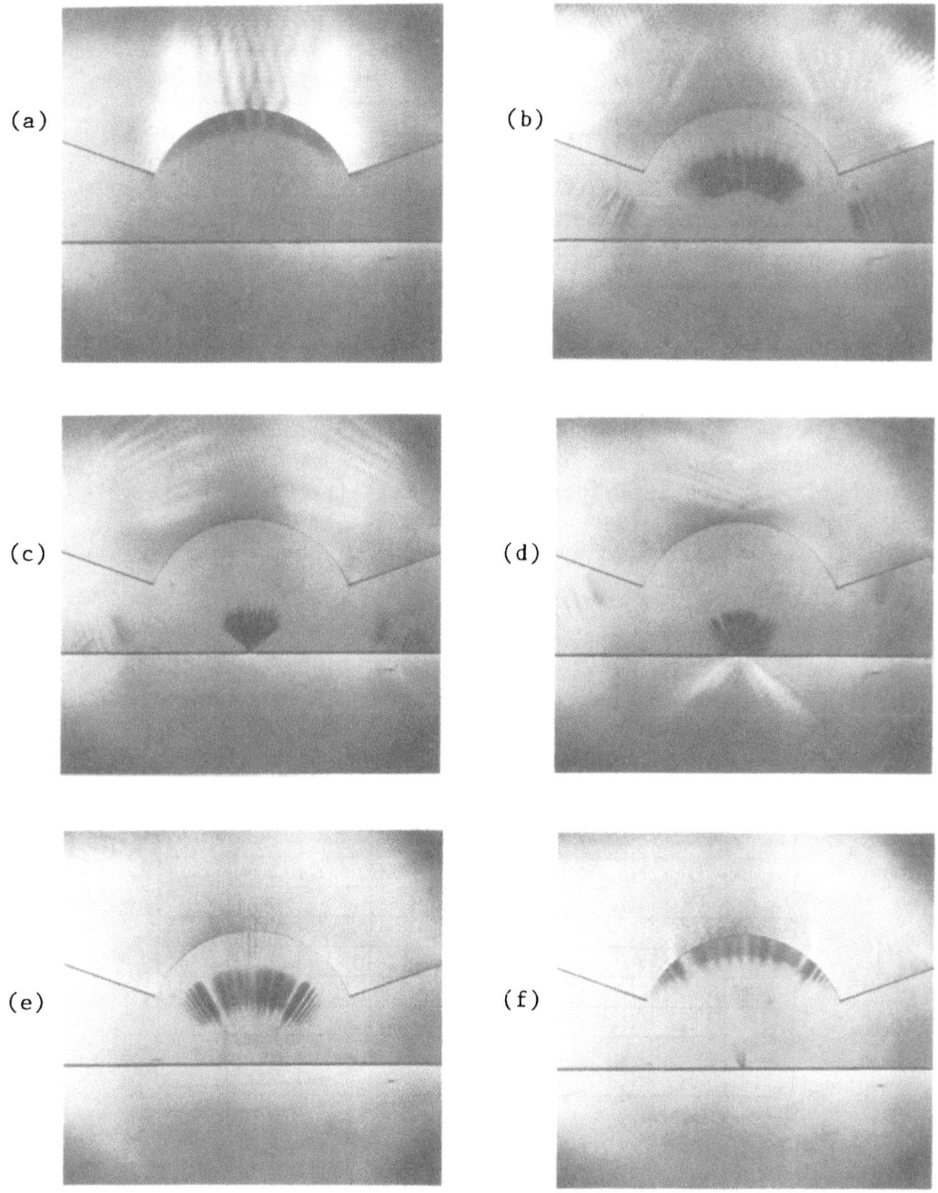

Fig.5. A time sequence of focusing and reflection of 5.5 MHz ultrasonic pulse.

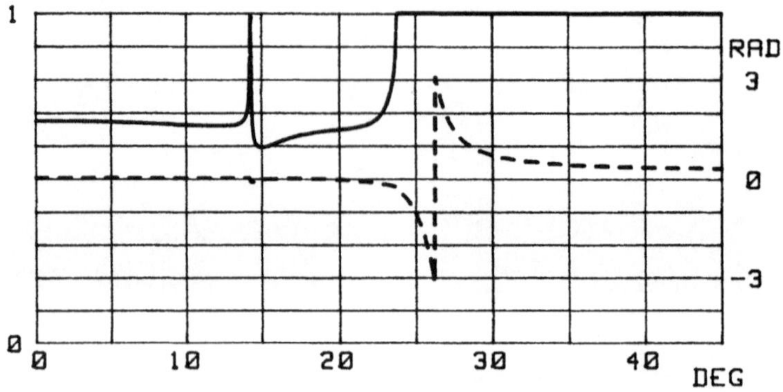

Fig.6. Theoretical intensity reflectivity (solid line) and its phase (broken line) at water-glass interface as a function of incidence angle.

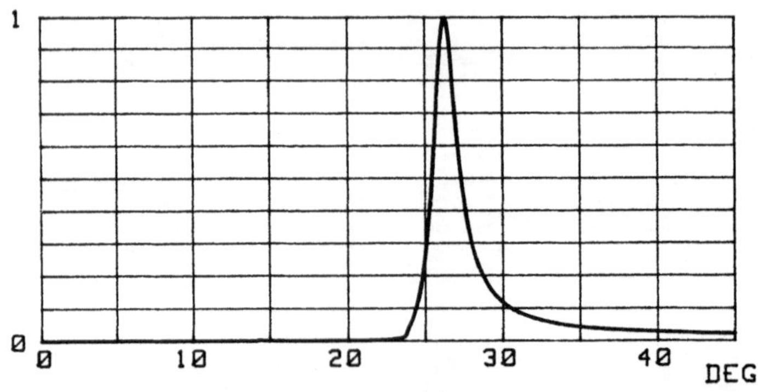

Fig.7. Theoretical reradiation intensity from water-glass interface as a function of incidence angle.

These striations are not due to the angular dependence of reflectivity from the glass surface, because total reflection occurs above the shear critical angle of 23.8° as shown in Fig.6. There exists, however, a large phase variation near Rayleigh critical angle of 26.3°. This causes LSAW to propagate along the glass surface and simultaneous reradiation of plane waves in water as shown in Fig.7. Then, broadening of the focal waist and the appearance of striations can be explained by LSAW.

Coexistence of reradiated plane waves and specularly reflected waves diverging cylindrically results in interference fringes as shown in Fig.8. Striations appearing in Figs.4 and 5 are nothing but interference fringes of those two components of waves. In other words, the existence of interference fringes is a direct evidence of LSAW generation.

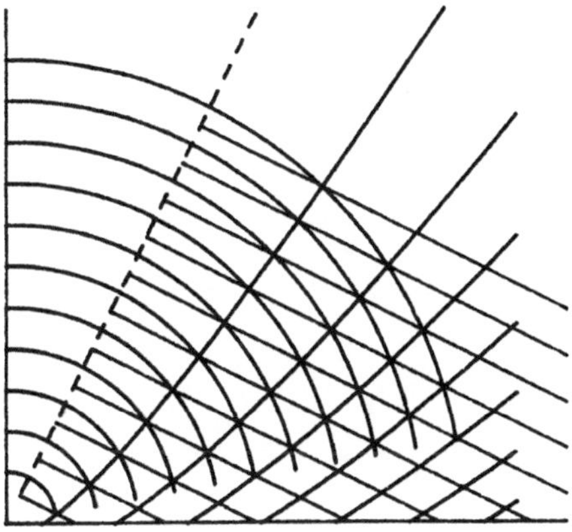

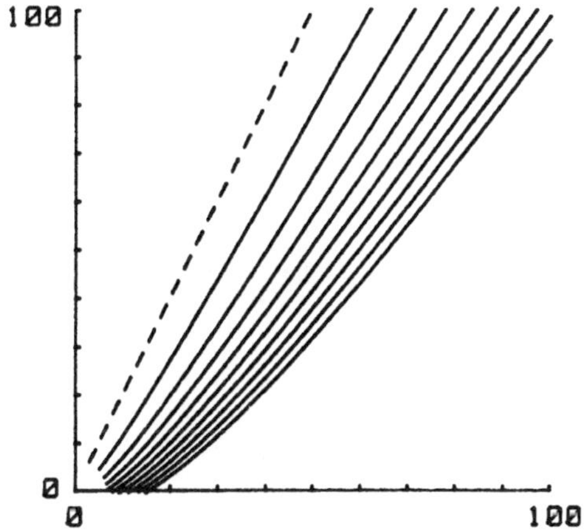

(a) (b)

Fig.8. (a) Interference of circular reflection with reradiated plane wave. (b) Resulting interference fringes over the range of 100 wavelength. Broken lines show Rayleigh criltical angle.

To confirm this interference, a lower ultrasonic frequency is needed for the visualization of wavelength in water. In Fig.9, the ultrasonic frequency is lowered to 1.2 MHz. By adjusting the slit width, ultrasonic waves in water are visualized as bright images. Cylindrically reflected and transmitted waves can be seen together with Rayleigh waves just below the glass surface and reradiated plane waves in water which interfere with the cylindrically reflected waves. In this case, theoretical decay length of LSAW is 4.85 wavelength.

In Fig.4 (b) and Fig.5 (d), faint interference fringes similar to those appearing in water can be found in the glass block. This proves the existence of shear waves reradiated from leaky surface-skimming compressional waves(LSSCW) interfering with cylindrically transmitted shear waves.

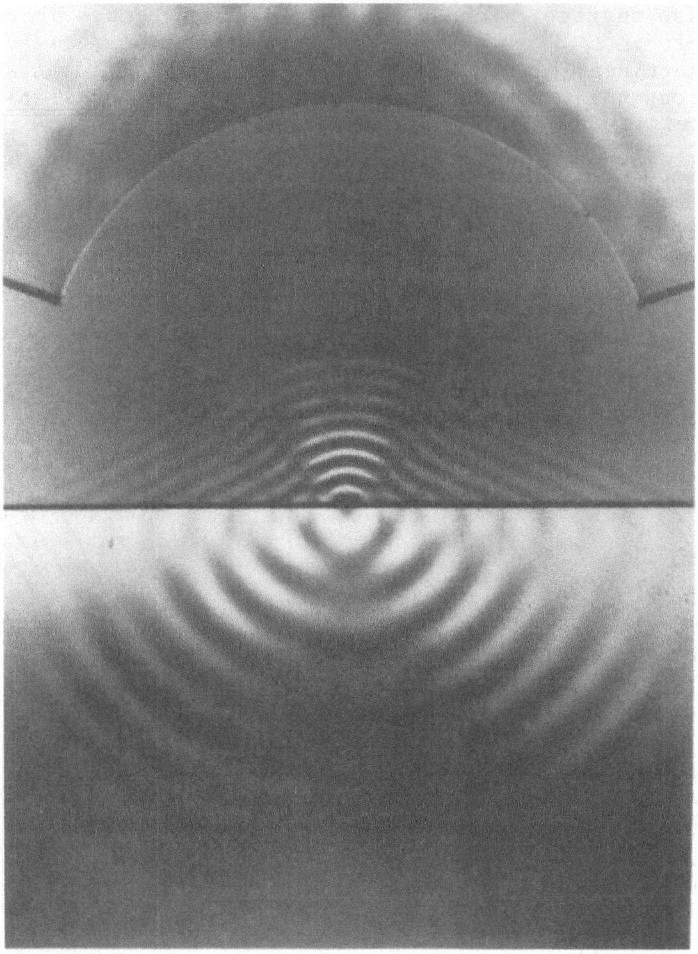

Fig.9. Reflection and transmission of 1.2 MHz ultrasonic pulse focused on the glass surface.

In conclusion, the visualized images of reflected, transmitted and reradiated ultrasonic waves in water, in glass and their interface have verified the theoretical assumption of V(z) curve analysis. It has also been shown that simultaneous visualization of ultrasonic pulses in liquid and solid is not only a tutorial means but also a powerful tool to clarify complicated wave phenomena caused by interaction between various wave modes in liquid and solid and at their interface.

Reference

1. For example, J. Kushibiki and N. Chubachi, " Material Characterization by Line-Focus-Beam Acoustic Microscope(invited) ", IEEE Trans. SU-32:189 (1985)

2. K. Negishi and Y. Tsuboi, " Visualization of Ultrasonic Pulse in Glass by the Technique of Photoelasticity ", Jpn. J. Appl. Phys. Suppl.22-3:19 (1983)

TIME-RESOLVED ACOUSTIC MICROSCOPY OF POLYMER COATINGS

A.M. Sinton, G.A.D. Briggs, and Y. Tsukahara*

Department of Metallurgy and Science of Materials
Oxford University, Oxford OX1 3PH, England

* Technical Research Institute, Toppan Printing Co. Ltd.
Sugito-machi, Kitakatsushika-gun, Saitama 345, Japan

1. INTRODUCTION

As an alternative to the standard gated CW (toneburst) mode for the scanning acoustic microscope, the lens may be excited by short impulses instead of gated bursts of CW [e.g. Yamanaka 1983]. The short temporal extent of the impulse excitation allows separate echo pulses from the top and bottom of a thin specimen on a slide, or from separate interfaces in a layered specimen, to be identified in the received signal (see Fig. 1). The scanning acoustic microscope at Oxford University (here called the OXSAM) has been used in this time-resolving mode to measure the elastic properties of soft and hard biological tissues on a microscopic scale [Daft and Briggs 1988a, 1988b, Daft et al. 1988] and to characterise surface acoustic waves and surface-breaking cracks in other hard materials [Weaver et al. 1988].

This paper describes the OXSAM's time-resolving capability to the characterisation of layered specimens consisting of a thin polymer coating on stone-finish steel sheet. The thickness of the coating and the presence or absence of bonding between the coating and the steel have been measured, for polypropylene (PP) and polyethylene teraphthalate (PET) coatings of 15um to 40 um thickness.

2. THE MICROSCOPE

The components of the OXSAM in its time-resolving configuration are shown in Fig. 2. The lens is made of sapphire, with a spherical cavity of 500um radius, 60 degree half-angle, and a matching layer of chalcogenite glass. The lens may be fast-scanned in the x-direction at 50Hz, and slow-scanned relative to the specimen in y and z.

The lens is excited by short electrical impulses generated from a step-recovery diode, at a repetition rate of 12.5kHz. The electrical impulses are bandpass-filtered by a shorted micro-strip to match their spectral properties to the frequency response of the lens. The impulses are then amplified and applied to the lens via a SPDT switch. The echo pulses received from the specimen via the lens, SPDT switch, and receive amplifiers, are sampled by a Tektronix sampling oscilloscope, averaged to reduce background noise and switching transients, and digitised for processing by a minicomputer [Daft 1987 chapt. 4].

The time resolution available from the system, and hence the resolution of vertical structure in specimens, is determined by the temporal width of individual received pulses. Figure 3a shows the received pulse from a glass microscope slide at focus. The pulse has significant value over an approximately 15ns interval. The spectral

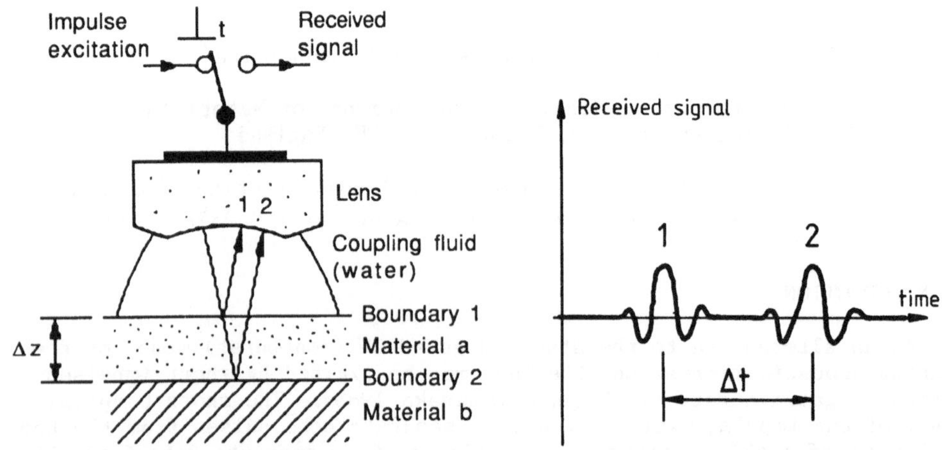

(a) Geometry (b) The received signal

Figure 1. Time resolution of vertical structure in an acoustic microscope specimen using impulse excitation. Pulse 1 returns from boundary 1 and pulse 2 from boundary 2.

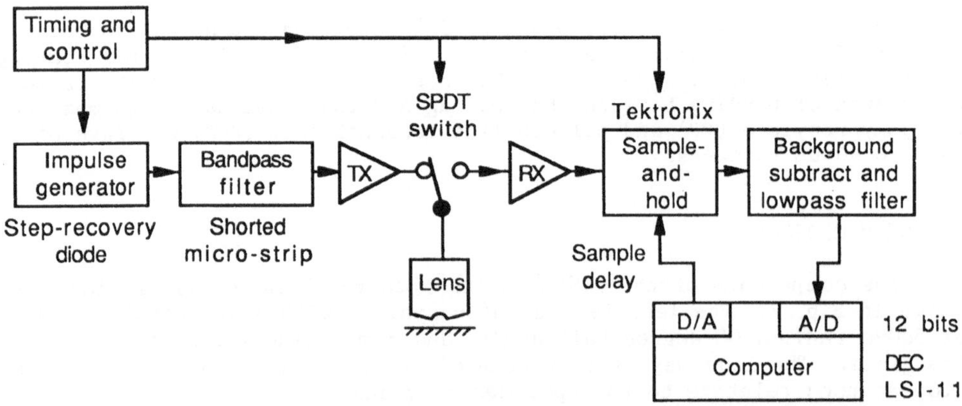

Figure 2. Block diagram of the OXSAM in its time-resolving configuration. Impulses are generated at a repetition rate of 12.5kHz, sampled, averaged to reduce noise, and digitised for computer processing. The sample-and-hold block is a Tektronix sampling oscilloscope comprising a 7613 mainframe, 7T11 timebase unit, 7S11 sampling sweep unit, and S-2 sampling head.

magnitude of the pulse is shown in Fig. 3b. The centre frequency is 230 MHz and the half-power bandwidth is 110MHz. The shape of this spectrum is governed almost entirely by the frequency response of the acoustic components of the microscope (transducer, lens, matching layer, and coupling fluid), since the electronic components of the microscope have broad bandwidths compared with Fig. 3b. The reciprocal of the bandwidth provides a measure of the effective width of the received pulse, which from Fig. 3b is approximately 10ns, corresponding to a vertical resolution in polypropylene (2530 m/s) of 12um. The spectral magnitude of the received pulses from polymer-coated specimens at positive defocus, as discussed in the following sections, is similar to that shown in Fig. 3b. (Note that the useful frequency range of the lens with CW excitation extends considerably beyond the half-power bandwidth, to nearly two octaves.)

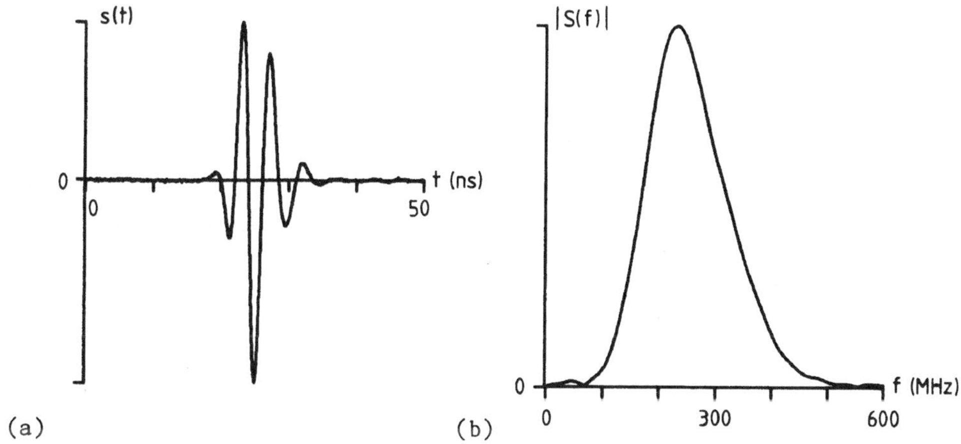

Figure 3. Received pulse from a glass microscope slide at focus, shown in the time domain (a) and in the frequency domain (b). These figures illustrate the time resolution and frequency response of the OXSAM.

3. MEASUREMENTS OF LAYER THICKNESS

The received signal from the coated specimens contains two pulses, a first pulse from the water/polymer boundary and a second pulse from the polymer/steel boundary. These are well resolved from each other for each of the coating types shown in Fig. 4. We have found the shapes of the two pulses to be relatively insensitive to the position of the specimen in z when the topmost part of the specimen (i.e. the water/polymer boundary, which generates the first received pulse) is positioned with positive defocus of z >= 30um. All of the polymer coating measurements reported in this paper were made with +40um defocus.

To measure the coating thickness, the temporal separation Δt between the two pulses is determined and, with an assumed sound propagation velocity v in the coating, the thickness Δz is computed from the equation

$$\Delta z = \frac{\Delta t.v}{2}$$

(cf. Fig. 1).

Since the pulses are of known shape (observe that in Fig. 4, and also in Figs. 8 and 9c, the pulses are all of similar shape), a simple pattern-matching algorithm may be used to locate the two pulses in the received signal. The pulse-finding algorithm implemented on the OXSAM for the work reported here is described in Fig. 5. It is applied separately to left-hand and right-hand regions of the received signal to locate the two received pulses. It returns for each pulse the pulse's temporal position and peak amplitude.

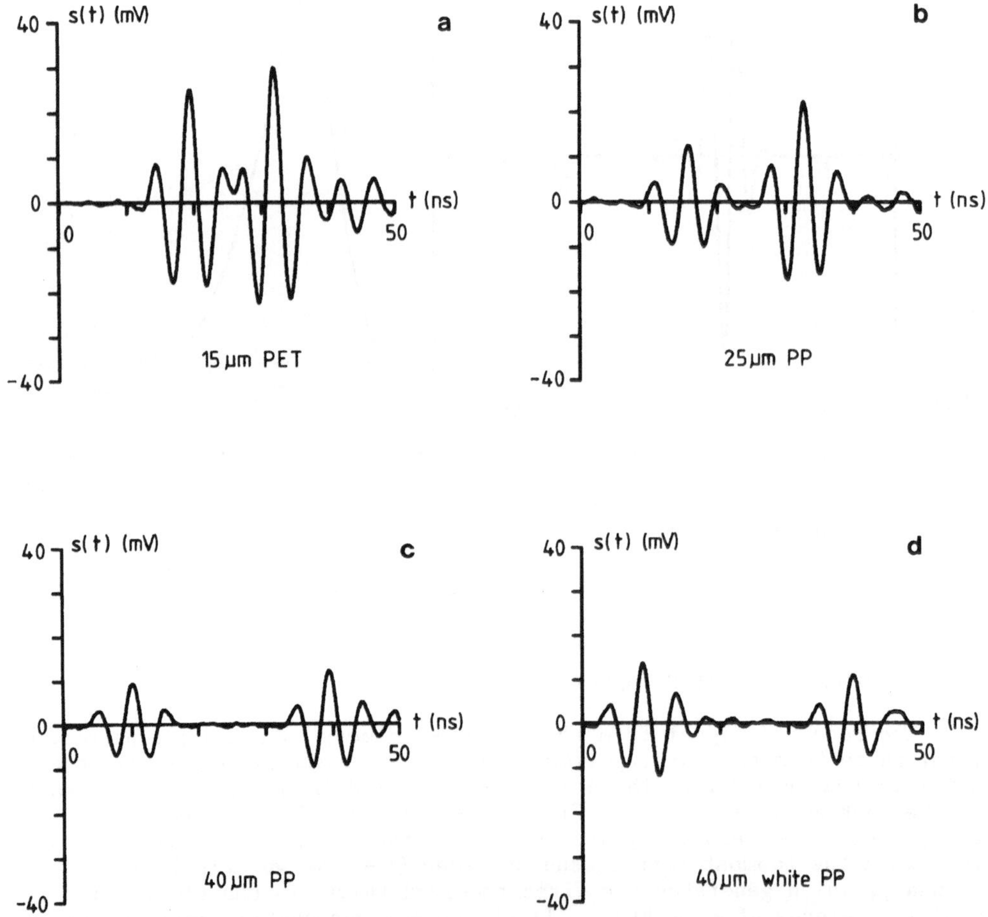

Figure 4. Received signals from impulse excitation of polymer-coated sheet steel specimens. The coatings are 15um thick PET (a), 25um PP (b), 40um PP (c), and 40um PP containing 0.3um TiO_2 particles (d). The top of the polymer coating was positioned at z = +40um defocus in each case.

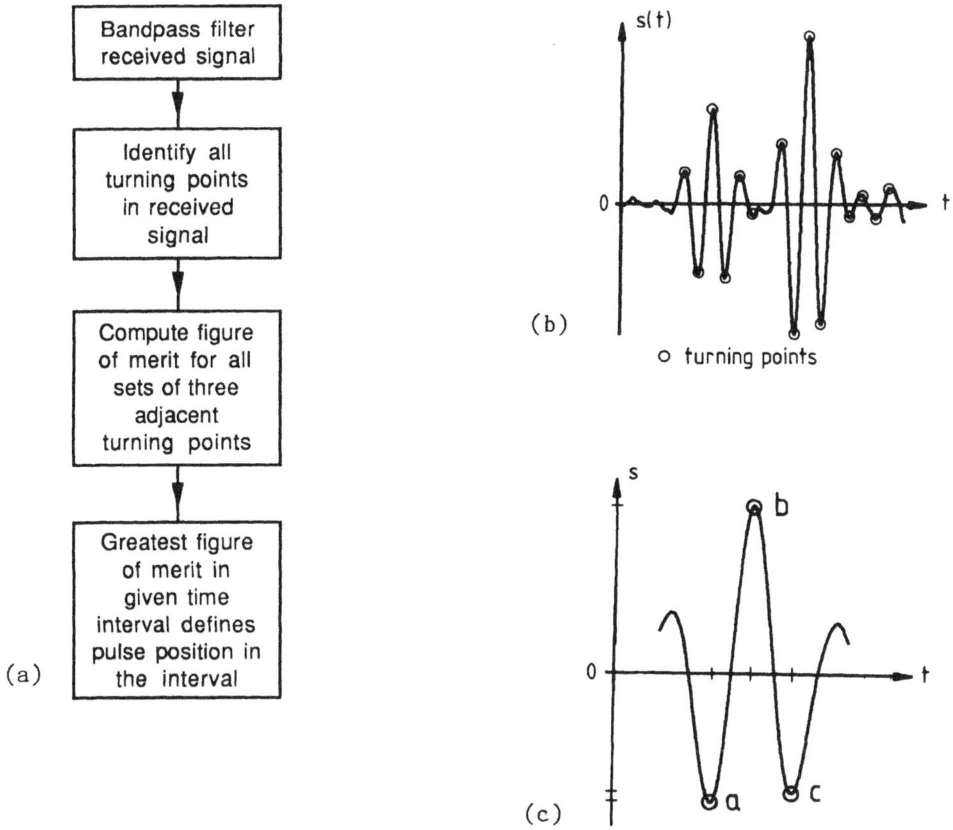

(d)

Pulse position = t_b

Pulse size = s_b

Time asymmetry $A_t = \dfrac{(t_c - t_b) - (t_b - t_a)}{t_c - t_a}$

Size asymmetry $As = \dfrac{s_c - s_a}{s_b}$

Figure of merit $F = |s_b| \cdot (1 - \alpha|A_t| - \beta|As|)$, with $\alpha = 1$, $\beta = 2$

Figure 5. Computer algorithm for detecting individual pulses in the received signal. (a) Flow diagram of the algorithm. (b) Turning points in the received signal, marked by circles. Adjacent turning points are required to exceed a minimum magnitude, to be separated by a minimum time, and to have second differences of opposite sign. (c) A pulse defined by the three turning points a, b, c. (d) Definition of the figure of merit for the pulse in (c).

Table 1. Statistics of measured coating thicknesses for specimens with coating types and thicknesses as in Fig. 7, measured by (a) the OXSAM with impulse excitation and (b) the leaky-pseudo-Sezawa-wave method.

Material	Nominal thickness	(a) OXSAM N	(a) OXSAM Mean (um)	(a) OXSAM Std dev. (um)	(b) Sezawa N	(b) Sezawa Mean (um)	(b) Sezawa Std dev. (um)
PET	15um	15	14.3	0.8	30	14.9	0.8
PP	20um	9	18.8	1.2	18	20.0	1.3
PP	25um	6	24.5	1.7	10	24.9	1.0
PP	40um	5	40.2	1.5	10	39.7	1.3
White PP	40um	5	36.8	1.2	10	40.5	2.2

The algorithm of Fig. 5 differs from the pulse-finding algorithms for biological tissue developed by Daft [1987 sect. 3.4.3, Daft and Briggs 1988b sect. 4] in being tailored specifically to the empirically observed symmetry of shape of the received pulses from the polymer-coated specimens and to the need to identify inversions of polarity of the received pulses (Sect. 4 below). Daft's algorithms, which can recognise pulses of dissimilar shapes, are more general but computationally more expensive than the algorithm described here.

Figure 6 shows plots of received pulse positions (scaled from time t to vertical position z using v = 2530 m/s) and coating thickness (the difference of the positions) along a 2mm scan in y across a typical specimen's surface. The thickness plot agrees well with the nominal coating thickness of 25um specified by the manufacturers of the laminate

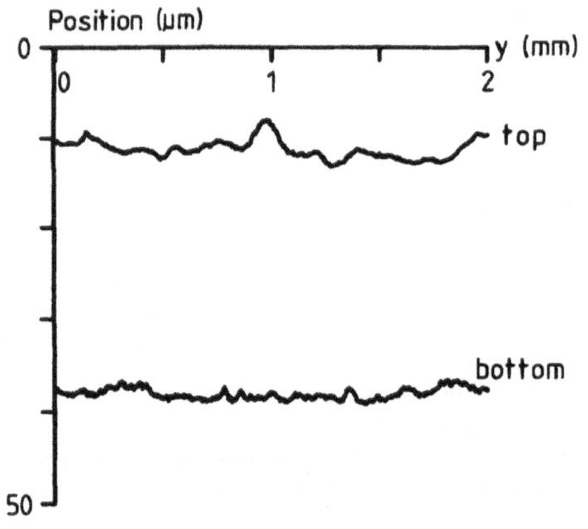

(a) Coating boundary positions

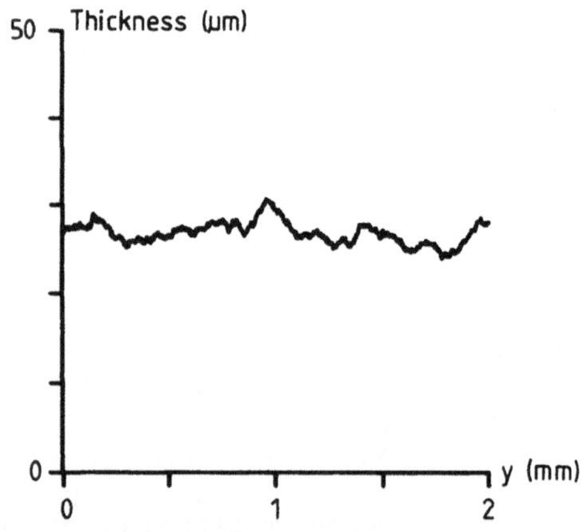

(b) Coating thickness

Figure 6. Polymer coating top position, bottom position, and thickness in z along a 2mm y scan of a specimen with 25um PP coating. The position plots are displayed according to the usual convention of larger z being further from the lens (the zero is arbitrary).

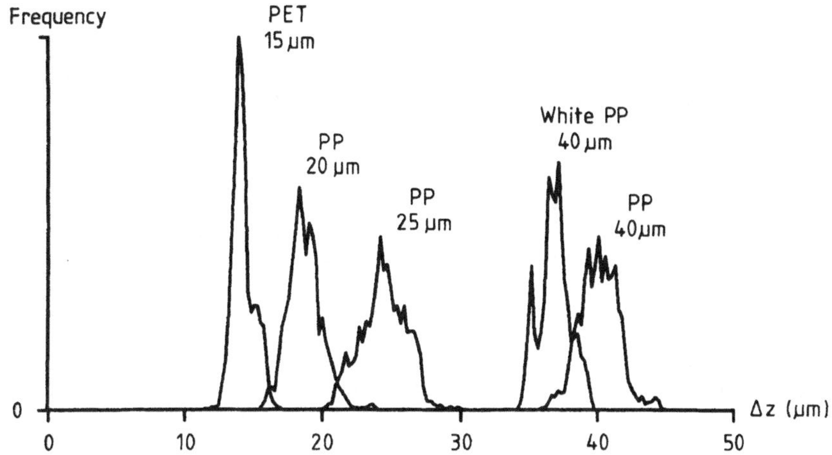

Figure 7. Frequency histograms, normalised to unit area, of polymer coating thickness for specimens with coatings of various nominal thicknesses. Measured acoustic velocities of 2309 m/s (PET), 2530 m/s (PP), and 2560 m/s (white PP) were used in computing Δz from Δt.

from which the specimen was taken. From data such as in Figure 6b, statistical descriptions of the coating thickness may be compiled. Figure 7 shows frequency histograms of coating thickness for specimens with a variety of coating types. Measurements of coating thickness of the same specimens by the leaky-pseudo-Sezawa-wave method [Tsukahara et al. 1988] yield results in good agreement with the OXSAM results (see Table 1).

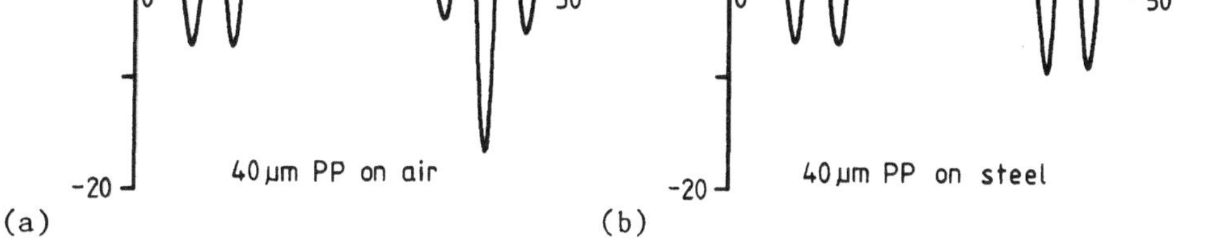

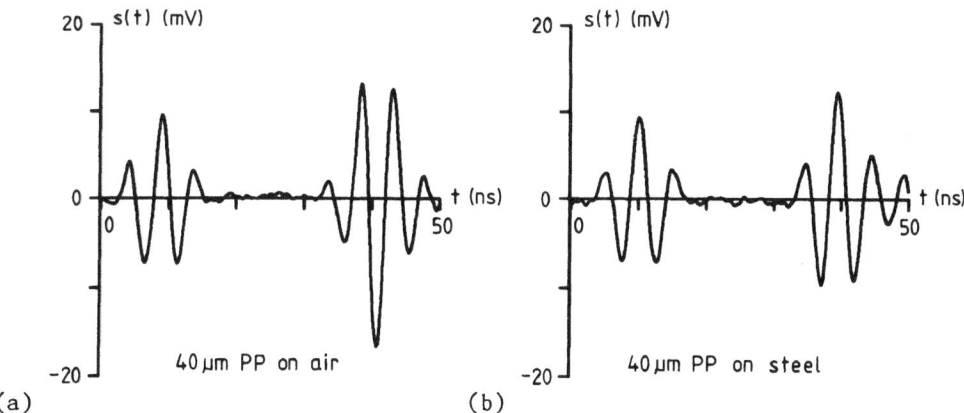

Figure 8. Received signal from a 40um thick layer of polypropylene with water (the coupling fluid) above and air below (a). The material is similar to the polymer coating of Fig. 4c, for which the received signal is shown again in (b) for comparison. The polarity of the second pulse is reversed in (a).

4. MEASUREMENTS OF BOND QUALITY

The acoustic impedance of the polymer in the coatings considered in this paper is intermediate between that of water and steel [cf. Kaye and Laby 1966, Hung and Goldstein 1983]. Hence, with the top and bottom boundaries of the coating both at positive defocus, the first and second received pulses both have the same polarity (as in Fig. 4). If the polymer coating is not bonded to the steel, so that the bottom boundary is between the polymer and a material of lower acoustic impedance than the polymer (e.g. water or air), the polarity of the second received pulse is reversed (see Fig. 8). The polarity reversal allows regions of bond failure between the polymer coating and the underlying sheet to be detected acoustically. Peak voltages of the first and second received pulses along a 2mm y-scan of a specimen showing small regions of disbonding between the coating and the underlying steel are shown in Figure 9. The first pulse size is uniform across the scan, indicating that the acoustic properties of the polymer are uniform. The second pulse size shows the polarity inversions characteristic of disbonding.

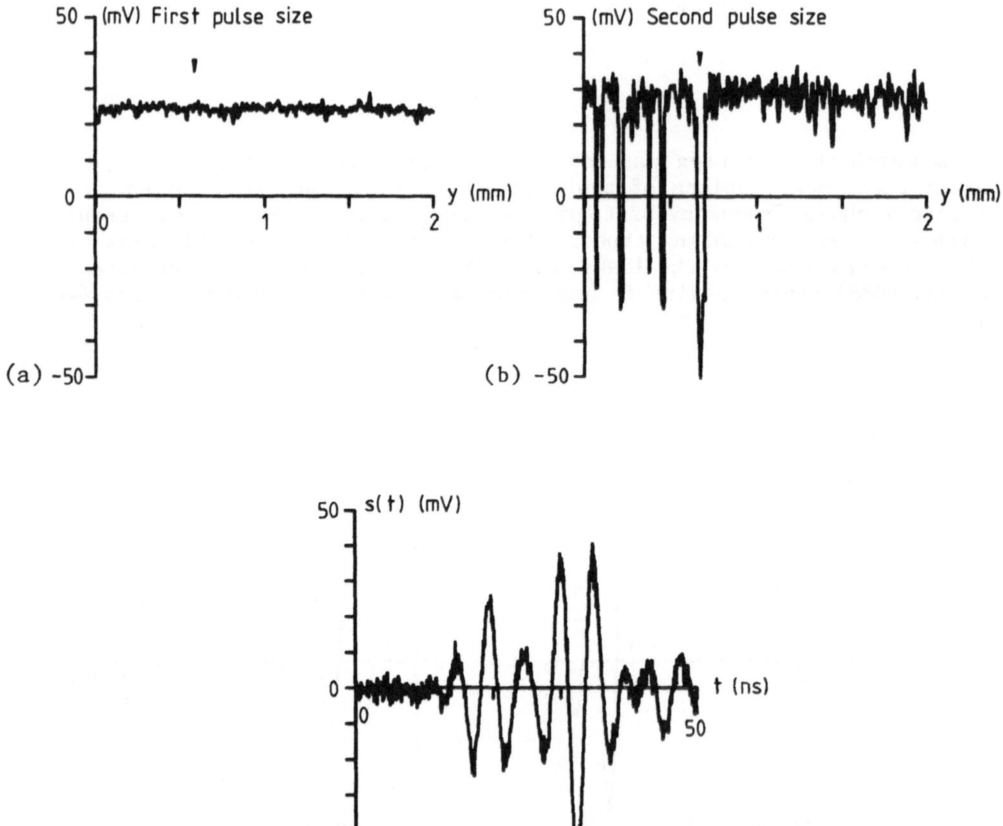

Figure 9. First (a) and second (b) received pulse sizes along a 2mm y-scan of a specimen with 15um PET coating on steel, showing disbonded regions (polarity reversals in the second pulse size). The received signal for the position marked by arrowheads is shown in (c) (cf. Fig. 8a).

5. CONCLUSIONS

The scanning acoustic microscope, operated with short impulse excitation, can measure the thickness and the bonding integrity of polymer coated sheet steel on a microscopic scale. In principle any similar layered structure may also be measured. The OXSAM easily resolved polymer layers of 15 to 40 um thickness. The minimum resolvable layer thickness is determined by the bandwidth of the acoustic microscope. The maximum resolvable layer thickness is limited by signal-to-noise ratio in the received signal, which worsens as the thickness is increased due to increasing attenuation and geometrical spreading of the acoustic energy reflected from the bottom of the layer. The operating frequency and lens geometry of the microscope may be chosen to suit a given range of layer thicknesses.

ACKNOWLEDGEMENTS

We gratefully acknowledge the assistance of Professor C.M.W. Daft and Dr. J.M.R. Weaver, who designed and constructed the time-resolving extension to the OXSAM, and of Dr. J. Kushibiki of Tohoku University, who supplied the lens used in this work.

REFERENCES

Daft, C.M.W., 1987, "Acoustic Microscopy of Biological Tissues", D. Phil. thesis, Oxford University, Oxford.

Daft, C.M.W. and Briggs, G.A.D., 1988a, The elastic microstructure of various tissues, to be submitted to J. Acoust. Soc. Am.

Daft, C.M.W. and Briggs, G.A.D., 1988b, Wideband acoustic microscopy of tissue, to be submitted to IEEE Trans. Ultrasonics, Ferroelectrics, and Frequency Control.

Daft, C.M.W., Briggs, G.A.D., and O'Brien Jr., W.D., 1988, Frequency dependence of tissue attenuation measured by acoustic microscopy, to be submitted to J. Acoust. Soc. Am.

Hung, B. and Goldstein, A., 1983, Acoustic parameters of commercial plastics, IEEE Trans. Sonics and Ultrasonics, SU-30:249.

Kaye, G.W. and Laby, T.H., 1966, "Tables of Physical and Chemical Constants," 13th edition, Longmans, London.

Tsukahara, Y., Ohira, K., Saito, M., Briggs, G.A.D., 1988, Evaluation of polymer coatings by ultrasonic spectroscopy, in Acoustical Imaging Vol. 17, Plenum, New York.

Weaver, J.M.R., Daft, C.M.W., Peck, S.D., and Briggs, G.A.D., 1988, Applications of broadband scanning acoustic microscopy, to be submitted to IEEE Trans. Ultrasonics, Ferroelectrics, and Frequency Control.

Yamanaka, K., 1983, Surface acoustic wave measurements using an impulsive converging beam, J. Appl. Phys., 54:4323.

BROADBAND ACOUSTIC MICROSCOPY: SCANNED IMAGES WITH AMPLITUDE AND VELOCITY INFORMATION

R.S. Gilmore, R.A. Hewes, L.J. Thomas III, and J.D. Young

GE Research and Development Center
P.O. Box 43, Schenectady, NY, 12301, USA

INTRODUCTION

Since the introduction of the mechanically scanned acoustic microscope by Lemons and Quate,[1] a world-wide effort has taken place to make this device [2] one of the most widely used tools for materials characterization and development. With the exception of work by Tsai,[2,3] and the pulse compression acoustic microscopy by Nikoonahad, Yue, and Ash [2] most studies have utilized acoustic pulses containing many wavelengths, resulting in narrow bandwidth systems. Calculations for material properties therefore require amplitude and phase measurements at different heights of the acoustic transducer above the sample. Algorithims for the use of these $V(z)$ data have become highly sophisticated as reported in the work of Weglein, Kushibiki, Chubachi, Bertoni, Kino, Laing, Kuri-Yakub, Ash, Wickramasinghe, and others.[2] This work continues to develop methods for time and frequency domain observations of broadband acoustic pulses. These observations permit material properties to be determined with the acoustic transducer at a constant height above the sample. The necessary data can be acquired during uninterrupted mechanical scanning by digitizing the reflected waveforms from the sample. One of the advantages of this approach is that scanned broadband systems are, and historically have been, widely used for industrial quality control.

Ultrasonic images for industrial inspection were first suggested by Sokolov and Pohlmann, as described by Berger.[4] Early images were attempted using Pohlmann cells and Sokolov tubes until C-Scanning was introduced. The first industrial C-Scan images by Buchanan and Hastings [4] were made in 1955 and recorded amplitude data, usually flaw echoes, that exceeded an amplitude threshold. These early hard copy images were binary, almost always at 1X magnification, and they consisted of black areas representing flaws and white areas without flaws. The threshold amplitude was usually set to some fraction of the amplitude from a reference reflector. The reference reflectors, still the most commonly used sensitivity references for industrial NDE, were typically machined slots, flat bottomed holes or side drilled holes. They were machined into the same test material to assure that the reference echo occurred in a coherent acoustic noise field from an equivalent microstructure. The noise, however, was excluded by the threshold setting.

C-Scan imaging was a significant advancement in materials evaluation. It provided a hard copy record; it simplified detection by plotting black patches representing flaws on the white field of the recording paper and it provided spatial correlation of the multiple signals usually produced when scanning a single small flaw with overlapping beams. The spatial correlation provided by a C-Scan, in turn gave a higher probability of detection than A-Scan observations, the most used method of signal observation.

Mechanically scanned acoustic microscopes (SAMs) were introduced by Lemons and Quate in 1973. These early systems operating at GHz frequencies produced images with optical microscope resolution and initiated a world wide research effort in acoustic microscopy.[2] However, with the exception of the previously mentioned pulse compression work, all of the acoustic imaging work at GHz frequencies was being done with tone burst systems with only a few percent bandwidth. In industrial ultrasonic testing, the emphasis, historically, has been toward broader and broader bandwidths. Because of the short time duration of the pulses, the entry surface is obscured for only one to two round-trip wave periods. Little surface wave scanning was done because of the sensitivity of sharply inclined beams to finish roughness, finish direction, and to flaw orientation. Until acoustic microscopy, ultrasonic surface wave testing has been considered one of the least dependable methods of interrogating surfaces. Acoustic microscopy is not affected by the directionality of either flaws or surface finish. Its contribution to surface inspection cannot be overemphasized. In addition, the sharply focused beams used to produce $V(z)$ images can also produce time-resolved surface waves (Fig. 1) using broadband industrial ultrasonic systems.

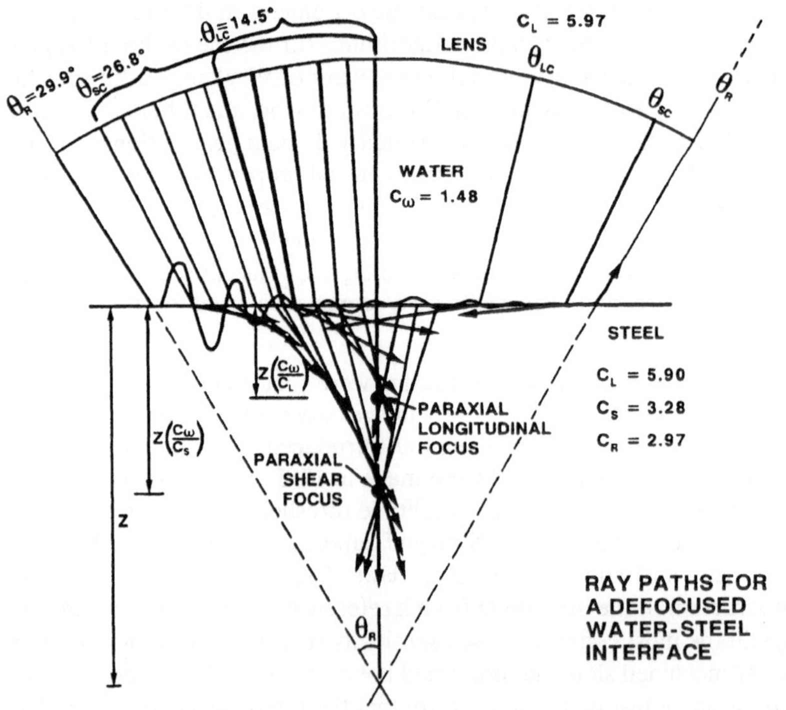

Fig. 1. The angular spectrum for a water-steel interface showing the critical angles for longitudinal and shear waves, and the Rayleigh critical angle.

Acoustic images also contain substantial information for materials characterization. Grain size can be measured with standard observational techniques applied to the surface wave gray scale images. Liang et al. [5] have published powerful analytic methods by inverting $V(z)$. Similar time and frequency domain calculations can be developed for broadband systems where the bandwidth of the beam is used to provide information provided by vertical transducer displacement in the $V(z)$ method. The purpose of this discussion is to report on some of these developments and to indicate future work.

AMPLITUDE IMAGING OF SURFACES AND VOLUMES

This discussion reviews expressions for the lateral resolution of time-resolved surface wave images and for beams focused in a beam-forming medium such as water and their focus in a higher velocity substrate. When the cone of focus (Fig. 1) is great enough to include the Rayleigh critical angle, usually subsurface information from both the longitudinal and shear components is included in the image. Often the relative amplitude of the pixels displaying a subsurface feature will indicate the nature of the energy forming the image. Lower amplitudes often indicate surface waves because of the pulse transmission nature of this imaging mechanism. Higher amplitudes often indicate longitudinal or shear wave echoes from features below the reach of the surface waves. Features with low amplitude edges around a high amplitude center can indicate a near surface feature interacting with the entry circle of a surface wave but reflecting both longitudinal and shear energy as those much smaller focal zones begin to interact with the feature.

Lateral Resolution for Time-Resolved Surface Waves

The entry circle schematic in Fig. 2 and Fig. 4 shows that any discontinuities on the surface that interrupt, or change direction of, any of the converging bundle of surface wave rays, changes the amplitude received by the piezoelectric element. The mechanism by which this occurs can be clarified by considering the two opposing 60-degree segments of that entry circle, and again to consider only those rays propagating from left to right. The 60-degree segment depicts a converging ray bundle, point-focused at the center of the entry circle, and then diverging to the opposite side of the circle perimeter. What is in fact displayed is a surface wave transmission acoustic microscope that is focused by the curvature of the entry surface circle. Entry circles of three to four surface wavelengths are required to time-resolve the leaky-Rayleigh wave from the direct reflection. However, unlike subsurface scanning, this does not obscure any portion of the entry surface material. In isotropic materials, the surface wave originates at an entry circle, and provides a 360-degree cylindrically convergent, pulse transmission, surface wave inspection method that can detect surface features with equal probability regardless of the direction of their surface strike. Utilizing broadband pulses with center frequencies at 1.0 MHz extends the depth of interrogation of the broadband systems to three surface wavelengths or 9.0 mm in steels and equivalent velocity media. This surprising depth is due to the low-frequency components of the broadband pulse.

Two parameters describe the resolution of a time-resolved surface wave imaging system. These, summarized in Fig. 2, are the diameter of the entry circle (D_R) and the spot size of the surface wave focal zone or crossing zone at the center of the circle. The spot diameter (ε_R) can be calculated from the amplitude point spread function of the double cylindrical lens system placed on the entry surface by the entry circle. The point spread function produced by a cylindrical lens has been shown by Born and

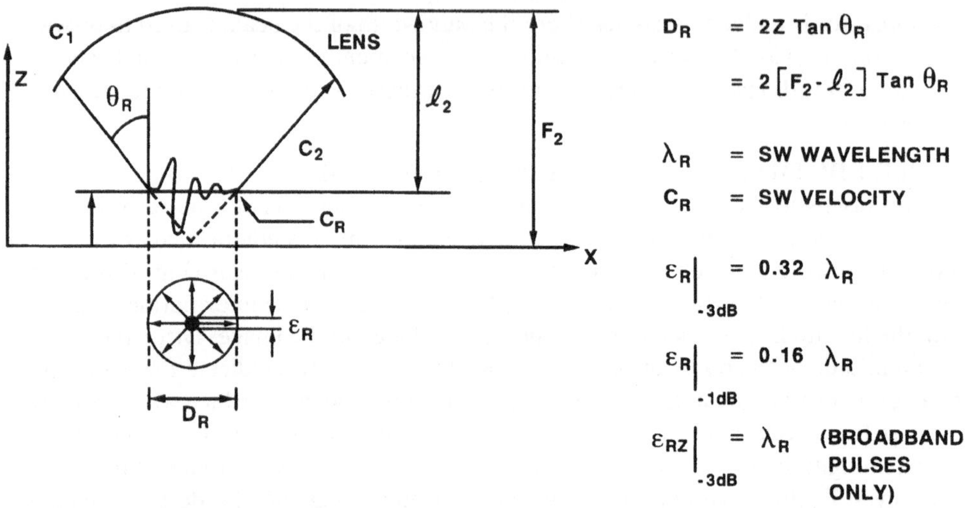

Fig. 2. Summary of the equations and parameters for the lateral resolution of a time-resolved surface wave image.

Wolf,[6] and Kino, [7] to be of the form (sinX/X)**2 for the double lens system. Gilmore, et al. [8] used these results to calculate the -20 dB, -3 dB, and -1 dB diameters for the point spread function for entry circle focused surface waves. The transmitter/receiver 180 degree segments are, in fact, F/0.5 lenses, which produce very sharply focused spot sizes . Discontinuities of one-third to a full surface wavelength in diameter can cause -3 dB to -20 dB changes in received amplitude. The surprisingly small -3 dB diameter (0.32 λ_R) of the center focus explains the consistently surprising resolution of all mode-converted surface wave imaging techniques that involve the formation of an entry circle and that develop the cylindrically convergent surface wave.

The general rules for broadband surface wave imaging are summarized in Fig. 2. The entry surface circle must typically be three wavelengths in diameter in order to time-resolve the surface wave arrival from the entry surface reflection. Therefore, D_R is typically nine times the -3 dB diameter of the focus at its center and three times the -20 dB size. Cracks, seams, and other linear discontinuities having little width, but with lengths that exceed D_R, are imaged as having a width that approaches D_R. Features much smaller than D_R, however, will produce detectable changes in amplitude only when they interact with the zone of focus at the center. They will, however, produce some decrease in amplitude as soon as they become included in the entry circle.

In specifying a surface wave scan, the -1 dB spot diameter, 0.16 λ_R, should be used for the line-to-line spacing and the pulse-to-pulse spacing along the line. This will limit the amplitude ripple in the scanned acoustic field to 1 dB or less. For 50-Mhz surface wave images in most materials, this spacing is 0.01 mm or less. Clearly, an image

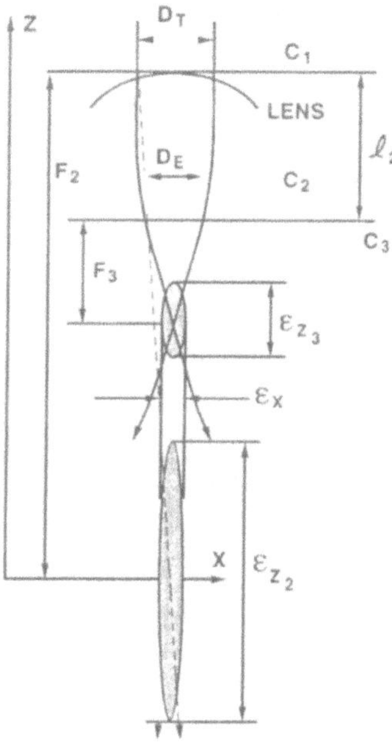

SPHERICAL FOCUS $F/D \geq \frac{C_3}{C_2}$

$$\varepsilon_X \Big|_{-3\text{dB}} = 1.03\,\lambda_2 \frac{F_2}{D_T} = 1.03\,\lambda_3 \frac{F_3}{D_E}$$

$$\varepsilon_X \Big|_{-1\text{dB}} = 0.4\,\lambda_2 \frac{F_2}{D_T}$$

$$\varepsilon_{Z_2} \Big|_{-3\text{dB}} = 4\,\lambda_2 \left[\frac{F_2}{D_T}\right]^2$$

$$\varepsilon_{Z_3} \Big|_{-3\text{dB}} = 4\,\lambda_2 \left[\frac{F_2}{D_T}\right]^2 \frac{C_2}{C_3} = 4\,\lambda_3 \left[\frac{F_3}{D_E}\right]^2$$

$$F_3 = [F_2 - \ell_2]\,\frac{C_2}{C_3} \qquad \lambda_3 = \frac{C_3}{C_2}\,\lambda_2$$

$$D_E = \frac{F_2 - \ell_2}{F_2}\,D_T$$

$$\frac{F_3}{D_E} = \frac{F_2}{D_T}\,\frac{C_2}{C_3}$$

Fig. 3. Summary of the equations and parameters for the lateral resolution of a focused beam in water or a high-velocity substrate.

scanned at -1 dB requires an index between lines and pulses of half of the -3 dB beam, but this assures that the Nyquist spacing of the pixels is compatible with the production of the -3 dB beam resolution in the image.

Lateral Resolution for Beams Focused in Water or High-Velocity Substrates

Volume inspection also requires that an acoustic field be scanned to a uniform amplitude throughout the length, width, and depth of the inspected material. In order for this to be accomplished in an economical number of scans, the interrogating beam must have a depth of focus (Fig. 3) that is, if not equal to, at least an appreciable fraction of, the material depth. For many industrial parts this could be 1 to 10 cm or considerably greater. If the material is to be scanned to a uniformity of 1 dB, peak to peak, then the 1 dB diameter and the 1 dB depth of focus must be again established for the interrogating acoustic beam. For very small flaws there may be no economical solution. A beam diameter small enough to detect the flaws may require excessive time to complete a scan of the material volume.

The gain required to detect the critical flaw size is also a key design parameter for the nondestructive evaluation of a volume of material. This may be calculated from a ratio of the critical flaw area to the area required to block the beam, Krautkramer.[9] Using work in Krautkramer, Gilmore, et al. [8] has shown (Fig. 3) that a circular reflecting flaw or disk, perpendicular to and centered on the axis of symmetry of an acoustic beam focused at a distance F in a beam-forming medium with longitudinal velocity C_2, by a lens of diameter D_L, will totally block the beam if the back reflected beam angle from the disk is equal to the angle of convergence of the focused beam. If C_2 and λ_2 are the velocity and wavelength of the beam forming medium, and the convergent angle of the inspecting beam is well within the first critical angle $\sin\theta_L = C_2/C_3$ of the longitudinal velocity (C_3) in the material, then

$$Df = 2.4\,\lambda_2 \frac{F}{D_L} \tag{1}$$

The insonification, by the focused beam across the face of disk *Df*, does not have the uniformity assumed in the derivation, but Krautkramer shows that measurements of the reflected amplitude from flat bottomed holes, with *Df* diameters, usually fall very close to the amplitude from a flat backwall reflection. Therefore, Equation (1) does give a good blocking diameter.

With an expression for blocking diameter, Fig. 3 gives expressions for the lateral resolution and depth of focus. A scanning plan can now be established for the inspection of volumes of material. The calculation for the -3 dB depth of focus in Fig. 3 is not as accurate as that published by Laing, et al.,[5] but it gives values about 10% above Liangs for focal length to diameter ratios of 1.0 or higher and is more convenient.

MEASUREMENT OF VELOCITY

The time domain photographs in Fig. 4 show surface wave pulses at two different path lengths in a highly isotropic material, tungsten carbide. In forming the surface wave arrival, each ray proceeds from the piezoelectric element (1) to the lens (2). Following specifically those rays that are refracted at the Rayleigh critical angle (θ_R), they proceed in a converging truncated cone of rays to intersect the specimen surface (3) where they intersect it to form the entry circle. The energy then mode converts to surface waves and propagates across on a circle diameter to the far side (4) where it reconverts and proceeds back to the lens (5) and the piezoelectric element (6). For a buffer-rod of length *L*, a lens of radius *K*, focusing at an axial distance *F*, where it is defocused to an axial distance Z_A, the time (t_1) required for the directly reflected pulse following the ray path on the lens axis to travel from the piezoelectric element and return is

$$t_1 = \frac{2L}{C_1} + \frac{2(F - Z_A)}{C_2} \tag{2}$$

where C_1 and C_2 are the velocity in the lens and coupling liquid, respectively. The round-trip travel time for the pulse traveling along the ray incident at the Rayleigh critical angle is

$$t_2 = \frac{2L}{C_1} + \frac{2(1 - \cos\theta_L)K}{C_1} + \frac{2(F - Z_A) - 2K(1 - \cos\theta_L)}{C_2 \cos\theta_R} + \frac{2\,Z_A \tan\theta_R}{C_R} \tag{3}$$

where θ_L, θ_R, and C_R are the lens angle, the Rayleigh critical angle, and the Rayleigh velocity. From Fig. 1 and Snell's law, the diameter (D_R) of the entry circle, where the incident pulse in the liquid is mode converted into a leaky-Rayleigh wave and (θ_R) are

$$D_R = 2\,Z_A \tan\theta_R \quad \text{and} \quad \theta_R = \sin^{-1}\frac{C_2}{C_R} \tag{4}$$

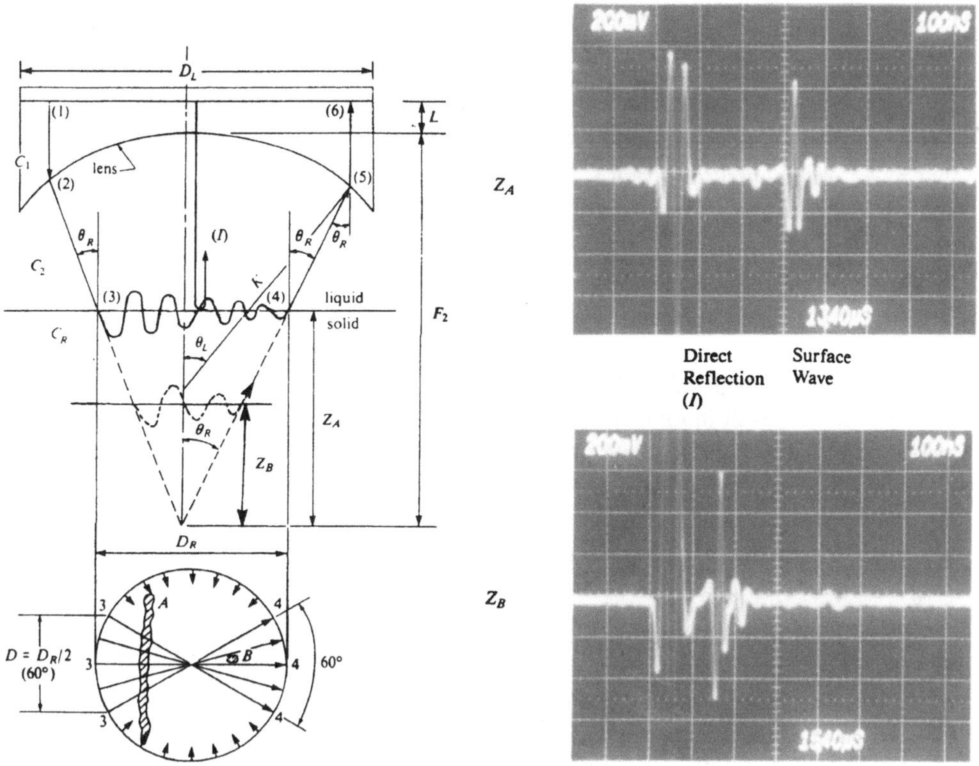

Fig. 4. Schematic showing surface wave arrivals (the 50-MHz signals are from tungsten carbide) at water path distances Z and Z, 60 degrees of the entry circle ray geometry producing these signals is shown with two different sized flaws, "A" and "B."

Writing the differences in round-trip travel time for the direct reflection (Δt_1) and the surface wave (Δt_2) at the two defocus distances Z_A and Z_B and combining Equations (2), (3), and (4) and solving for C_R gives

$$C_R = \sqrt{\frac{(C_2)^2}{1 - \left(\frac{\Delta t_2}{\Delta t_1}\right)^2}} \tag{5}$$

Equation 5 permits the Rayleigh group velocity to be determined from time measurements at two water path settings where both the velocity in the water and the material are determined. However, for temperature-stabilized water baths, C_2 is known and invariant. Therefore, for a constant water velocity, a flat sample, and an invariant waterpath during scanning the time delay of the surface wave arrival behind the direct reflection $t_2 - t_1$ can give the surface wave velocity directly.

Contrast to the Surface Wave Path Length

The difference in amplitude of the surface wave signals produced by the two water path distances $(F - Z_A)$ and $(F - Z_B)$ in Fig. 4 is inversely proportional to the distance each pulse travels across the surface of the tungsten carbide. Some loss is due to material attenuation, but most of the drop on amplitude is due to the continuous radiation of elastic energy, characteristic of leaky-Rayleigh waves, back into the water during propagation. The received amplitude decreases as the length of the entry-exit path increases. Therefore, when the path length D_R is specified, this also specifies the energy lost back into the coupling fluid for a given frequency or range of frequencies that make up the ultrasonic pulse. Rewritten in terms of the surface wave velocity D_R becomes

$$D_R = 2Z \frac{C_2}{\sqrt{(C_R)^2 - (C_2)^2}} \qquad (6)$$

At constant Z, high values for C_R result in small values for D_R; low velocities, on the other hand, result in long entry-exit paths. Now consider two adjacent regions with velocities C_{R_1} and C_{R_2}. For many engineering materials C_R is in the vicinity of 2.7 to 3.9 (mm/microsecond) or at least twice the velocity in water (1.48 mm/microsec). Equation (6) evaluated for this range in velocity shows that for the fixed lens-surface water paths used in scanned surface wave imaging, the distance D is inversely proportional to the the surface wave velocity.

Higher frequencies, however, lose proportionately more amplitude than low frequencies for a given surface wave path length. The loss appears to vary per wavelength of propagation path for a given liquid-solid interface. Consider a situation in which the surface wave changes from region 1 to region 2 such as in grain size images at constant density. If the amplitude is proportional to

$$A \approx \frac{\lambda_R}{D_R} \quad \text{then} \quad \frac{A_1}{A_2} = \frac{C_{R_1}}{C_{R_2}} \cdot \frac{D_{R_2}}{D_{R_1}} \qquad (7)$$

Fig. 5 shows D_R and amplitude as functions of C_R for constant transducer height and density. An increase in C_R of 3% produces an amplitude increase of 5%. Fig. 6 shows surface wave arrivals wave in fused quartz and tungsten carbide. Note that quartz shows the greatest amplitude loss because of its much better acoustic match to water. Equation 7 therefore only works for small changes in velocity rather than for the range used in Fig. 5.

ANISOTROPIC REGIONS WHERE THE VELOCITY CHANGES WITH THE DIRECTION OF THE SURFACE STRIKE

When the surface wave velocity is directionally variable in the plane of the surface, such as for images of grain size where the grains are anisotropic, then the travel time along each entry-exit path is different, and the received amplitude is dependent on both length of each path and the summation of all of the arrivals back at the transducer with respect to both their amplitude and phase. In the case of an image of a polycrystalline material, if the maximum, minimum, and average velocities for each of the grains on the surface can be uniquely related to the orientation of the surface with respect to the crystal axes of the grain, then the received amplitude (and therefore the gray scale at which each grain is displayed) can be related to the crystallographic orientation. The interaction between anisotropic grains and single frequency signals has also been considered by Somekh. [10] In the broadband case consider a stress pulse

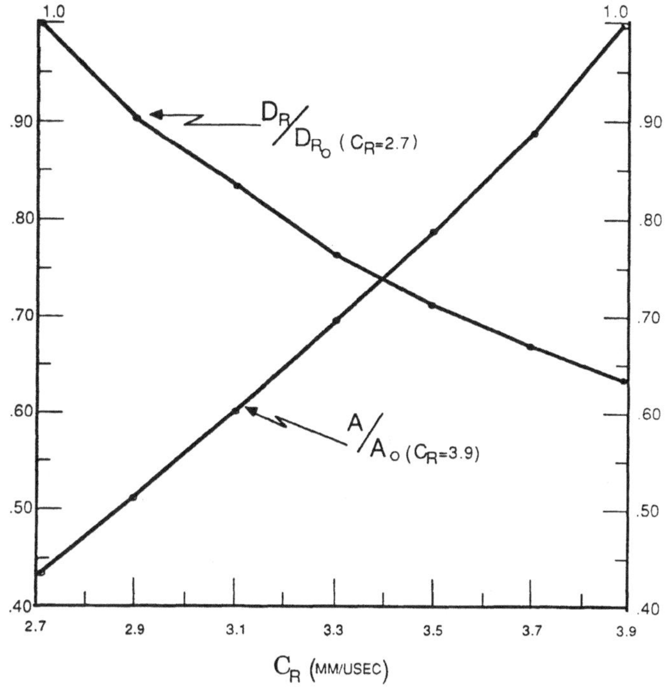

Fig. 5. Amplitude and entry circle diameter as a function of surface wave velocity for constant water path (Z) and density such as in a single phase grain size image.

of carrier frequency F_0. If the pulse is N wavelengths long, it can be represented in time as

$$P_t = \begin{cases} P_0 \cos 2\pi f_0 t & t \leq \dfrac{N}{2 f_0} \\ 0 & t > \dfrac{N}{2 f_0} \end{cases} \tag{8}$$

Various authors including Fitting and Adler [11] have shown that the Fourier transform is written in frequency as

$$F_P(f) = \frac{\sin N\pi(1 - f/f_0)}{N\pi(1 - f/f_0)} \tag{9}$$

Consider now an azimuthal velocity variation on the surface such that the maximum propagation delay is t_2 and the minimum delay is t_1 and the difference between the two is $t_2 - t_1$. Such a variation could be produced by an anisotropic grain at the surface (Fig. 7). The surface wave velocity on such a grain has been estimated for titanium as a function of orientation (Table 1), by numerically evaluating Kristoffels equations for that crystal structure. The Fourier transform for the time difference produced by such a variation can be written [11,12] as

$$F_{t_2 - t_1}(f) = \frac{\sin \pi f(t_2 - t_1)}{\pi f(t_2 - t_1)} \tag{10}$$

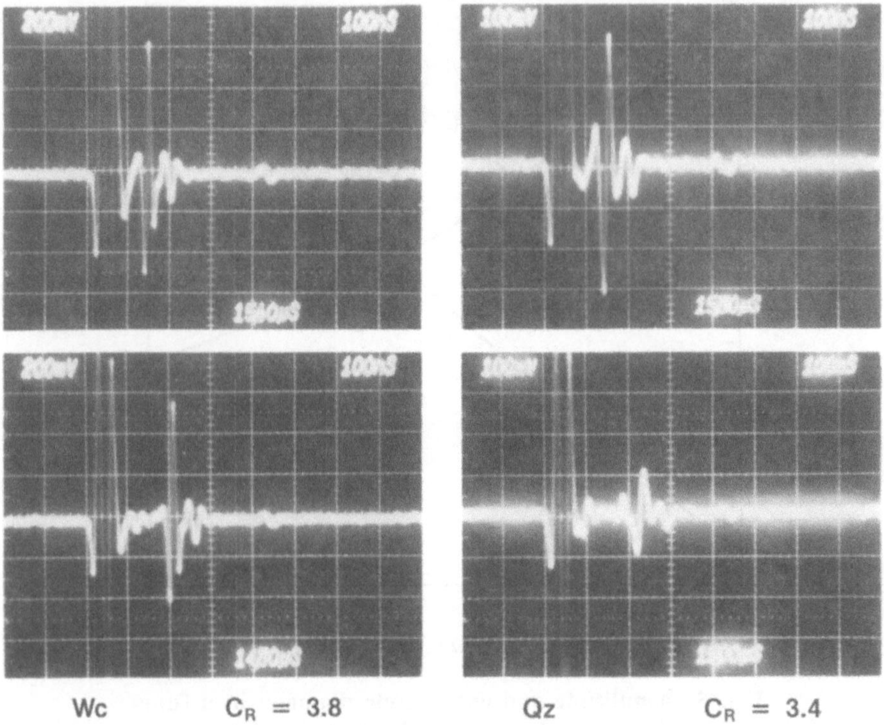

Fig. 6. Surface wave arrivals in fused quartz and tungsten carbide. Note that the pulse suffers little degradation in the carbide because of its very high surface wave acoustic impedance; the loss in quartz is much greater.

The effect of the velocity variation on the transmitted amplitude can now be investigated by convolving these two spectra to give

$$F_T(f) = F_P(f) \cdot F_{(t_2 - t_1)}(f) \tag{11}$$

A schematic representation of such a convolution is shown in Fig. 8. The effect of such a process on the amplitude and information content of the time domain signal is shown for two grain orientations in the titanium 6-4 alloy in Fig. 9. The greater the difference in arrival time produced by the surface, the lower the amplitude. Therefore for a consistent waterpath the image contrast will be determined by the range of the differences in time for each zone encountered by the surface wave.

Fig. 10 shows images of fully annealed titanium 6-4 sample from which all data were taken. Note that each titanium grain shows a consistent shade of gray and therefore a consistent amplitude (wave form). From grain to grain, however the Ti signals vary in amplitude over a factor of 0-156 (7+ bits).

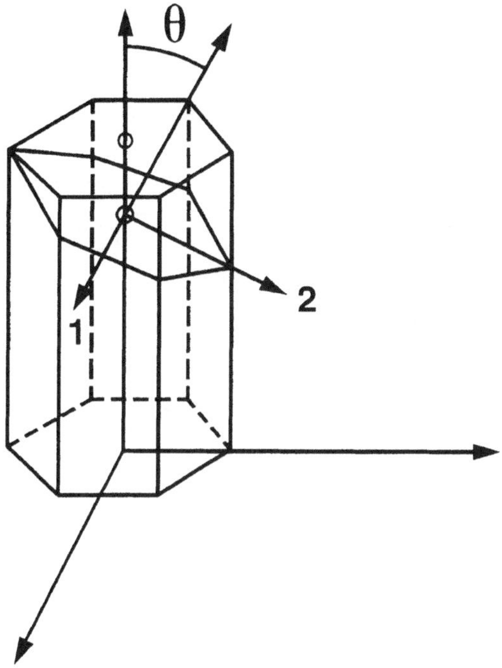

Fig. 7. Schematic showing a grain surface of arbitrary orientation in a hexagonal structure.

Table 1. Titanium at room temperature.

degrees	mm/μsec	mm/μsec	mm/μsec	
θ	$C_R\|_1$	$C_R\|_2$	ΔC_R	$\frac{\Delta C_R}{\bar{C}_R}$
0.0	2.998	2.998	0.000	0.000
5.0	2.996	3.002	0.006	0.002
10.0	2.987	3.012	0.024	0.008
15.0	2.974	3.027	0.053	0.018
20.0	2.956	3.045	0.090	0.030
25.0	2.933	3.065	0.132	0.044
30.0	2.907	3.084	0.177	0.059
35.0	2.877	3.100	0.222	0.074
40.0	2.846	3.110	0.264	0.089
45.0	2.813	3.114	0.301	0.101
50.0	2.780	3.110	0.330	0.112
55.0	2.747	3.099	0.352	0.120
60.0	2.717	3.083	0.366	0.126
65.0	2.689	3.062	0.374	0.130
70.0	2.664	3.040	0.376	0.132
75.0	2.644	3.019	0.374	0.132
80.0	2.629	3.001	0.372	0.132
85.0	2.620	2.990	0.369	0.132
90.0	2.617	2.986	0.368	0.132

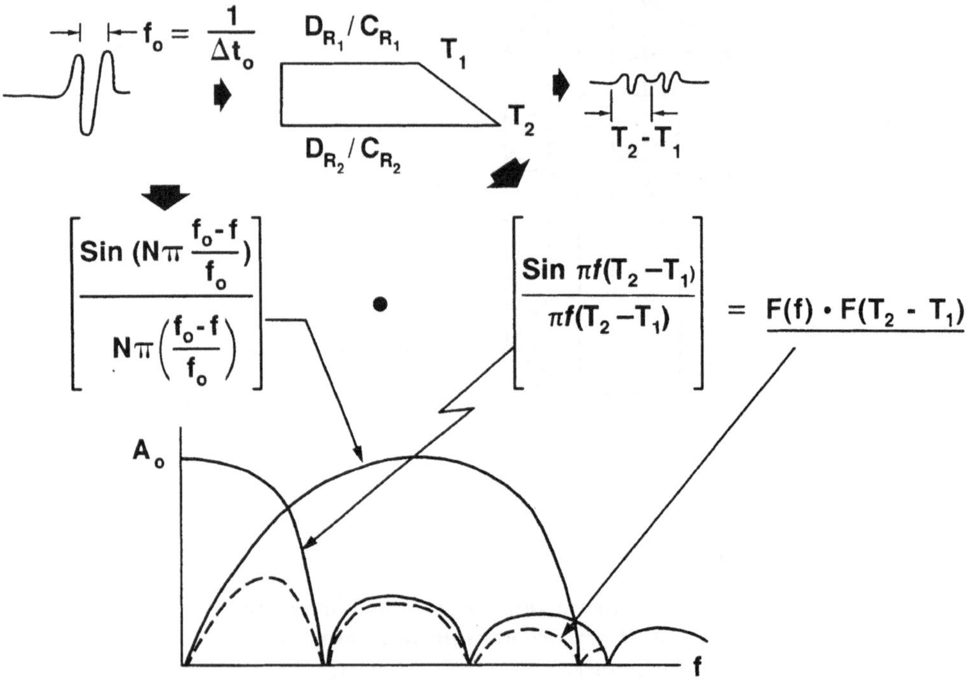

Fig. 8. A schematic showing the modulation of the amplitude spectrum of a broadband pulse by a velocity variation.

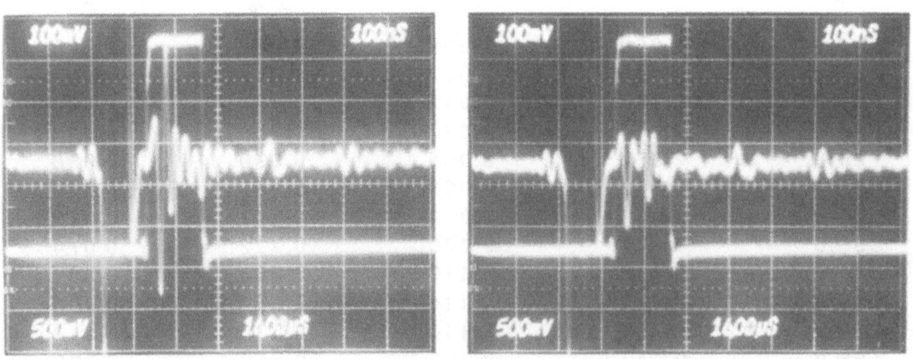

Fig. 9. Time domain signals from the fully annealed titanium grain structure image shown in figure 10. Gate shows extent of waveform used to image surface. The left waveform is from a relatively isotropic region; the right region is so anisotropic that the two arrivals are time-resolved.

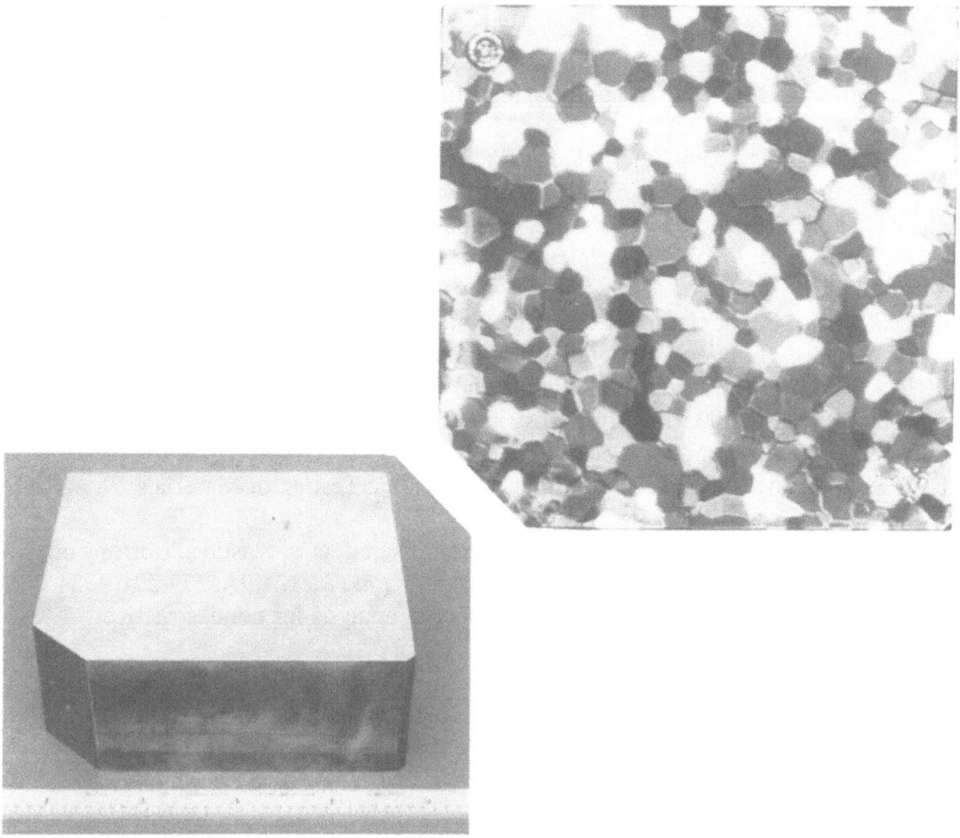

Fig. 10. 20-MHz surface wave image of a fully annealed grain structure; the titanium 6-4 sample is shown.

CONCLUSION

Future progress will almost certainly be driven by the dramatic growth in availability and power of computers for image processing and image analysis. Image data based on waveform capture is memory-intensive, but permits the operators developed for ultrasonic spectroscopy to be used in analysis of the data as well as the image processing operators such as Wiener filters and fast Fourier transforms. It seems reasonable to suggest that all of the analytic procedures published by Weglein,[13] Kushibiki and Chubachi [14] and Liang, Kino, and Khuri-Yakub [5] have corresponding operators which should be developed for amplitude-frequency analysis of broadband ultrasonic images.

REFERENCES

[1] R.A. Lemons and C.F. Quate, "Acoustic microscopy," Physical Acoustics, Ed. W.P. Mason and R.N. Thurston, Academic Press, NY, 1979.

[2] *IEEE Transactions on Sonics and Ultrasonics*, Special Issue on Acoustic Microscopy, SU-32, No.2, pp. 130-375, March 1985;

[3] C.S. Tsai, S.K. Wang, and C.C. Lee, "Visualization of solid Material Joints using a Transmission Acoustic Microscope," *Appl. Phys. Lett*; 31: 791-793, December 1977.
[4] H. Berger, "Ultrasonic Imaging Systems for Nondestructive Testing," *JASA*, 45: 859-867, April 1969.
[5] K.K. Liang, G.S. Kino, and B.T. Kuri-Yakub, "Material Characterization by the Inversion of V(z)," *IEEE Trans. on Sonics and Ultrasonics*, Special Issue on Acoustic Microscopy, SU-32, No.2, pp. 130-375, March 1985.
[6] M. Born and E. Wolf, "Principles of Optics," 6th Ed. London: Pergamon, 1980.
[7] G.S. Kino, "Fundamental of Scanning Systems," in Scanned Image Microscopy, ed E.A. Ash, London, Academic Press, 1980.
[8] R.S. Gilmore, K.C. Tam, J.D. Young, and D.R. Howard, "Acoustic Microscopy from 10 to 100 MHz for Industrial Applications," *Phil Trans. R. Society,* A320: 215-235, London, 1986.
[9] J. Krautkramer and H. Krautkramer, "Ultrasonic Testing of Materials," Springer-Verlag, 1977, pp. 75-79.
[10] M.G. Somekh, G.A.D. Briggs, and C. Ilett, "The effect of elastic anisotropy on contrast in the scanning acoustic microscope," *Phil. Mag.*, 49A: 179-204, 1984.
[11] D.W. Fitting and L. Adler, Ultrasonic spectral analysis for nondestructive evaluation, Plenum Press, NY, 1981.
[12] R.S. Gilmore, G.J. Czerw, and L.B. Burnet, "The Use of Radiation Field Theory to Determine the Size and Shape of Unknown Reflectors by Ultrasonic Spectroscopy," *Materials Evaluation,* 1, January 1977.
[13] R.D. Weglein, "Acoustic Micro-Metrology," *IEEE Trans. Sonics and Ultrasonics,* SU-32, No. 2, pp. 225-235.
[14] J. Kushibiki, and N. Chubachi, "Material Characterization by the Line-Focus-Beam Acoustic Microscope," *IEEE Trans. on Sonics and Ultrasonics*, Special Issue on Acoustic Microscopy, SU-32, No.2, pp. 130-375, March 1985.

NON DESTRUCTIVE EVALUATION OF COMPOSITE MATERIALS USING ACOUSTIC MICROSCOPY*

B. Nongaillard, J.M. Rouvaen and N.E. Imouloudene

Laboratoire d'Opto-Acousto-Electronique (U.A.832 CNRS)
Université de Valenciennes - Le Mont Houy
59326 Valenciennes Cédex - France

INTRODUCTION

The effectiveness of the technique of scanning acoustic microscopy has been clearly demonstrated since some time, in various application domains like biology, non destructive evaluation of materials, bounds and microstructure analysis.

This technique has only been, in its initial stage, devoted to a qualitative analysis, but, since a few years, an evolution towards quantitative evaluation methods is performed, due to numerous contributions to the theoretical foundations of acoustic microscopy and to the understanding of the significant factors for the image contrast. Particularly, in the majority of cases, the generation of interface waves at the boundary between the material and the coupling medium (1,2,3) consitutes an important source of contrast.

For inhomogeneous or stratified materials, the interference between the different waves generated on all boundaries constitutes a major source of contrast. This phenomenon is particularly important if the ultrasonic pulse duration is sufficiently higher than the time delay between the different structures in the material and if the angular spectrum of the incident ultrasonic field is sufficiently large. Both conditions are generally satisfied for the non destructive evaluation of composite materials using acoustic microscopy.

The aim of the present work is to try to explain the mechanism responsible for the contrast in the images of composite samples obtained using acoustic microscopy.

Problems encountered in the testing of composite materials

The problems of non destructive evaluation of composite materials is very specific. Opposite to classical materials, which consist in a single solid phase, the composite materials are heterogeneous and aniso-

* This work is supported by the Direction des Recherches, Etudes et Techniques FRANCE

tropic, and comprise several solid phases. Their elaboration processes involve moreover numerous stages and is easily meant that variations of the texture, polymerization or fiber orientations may induce very different defects.

The minimum requirement is to detect these defects, but the actual trend, for companies involved in composite material elaboration, is to be able to identify them also : this implies the obtention of their sizes, depths, forms and natures. The typical dimensions of the defects searched in composite materials is around 0.1 mm, so that, owing to the mean ultrasonic velocity in these materials of, say, 2500 m/s, leads to an operating frequency near 25 MHz.

Like in the majority of ultrasonic non destructive applications, a tradeoff must be realized between the desired spatial resolution and the exploration depth inside the material. The heterogeneity of composite materials induces parasitic phenomena during ultrasonic wave propagation, like multiple scattering, responsible for an increased attenuation and a significant wavefront distortion. This restricts the available exploration depth to 2 or 3 mm for operating frequencies between 10 and 25 MHz, according to the composition of the composite materials under test.

Our main interest is in the non destructive evaluation of carbon epoxy composite plates. This work is an extension of the studies performed by B.F. Khuri Yakub (4) and the Laboratory of the "Aerospatiale" (5) in Suresnes, France, using low frequency acoustic microscopy inside composite materials. These experiments, mainly performed at a 10 MHz operating frequency, using spherical piezoelectric transducing cups with a wide angular aperture (numerical aperture near unity), demonstrated the very high sensitivity of this technique, which allows the observation of the defects in the composite materials and also their texture.

Convenient geometrical parameters for an ultrasonic probe structure with a delay line are first deduced in the following study. The limitations given by the parasitic echoes propagating in the delay line over the receiving dynamic range are also presented. Finally, modelizations of the focalization process in the anisotropic multilayers composite materials and of the ultrasonic propagation are proposed in order to interpret the sources of the contrast in the acoustic images.

Definition of the probe geometry

The geometry of the ultrasonic probe is shown in Fig. 1 below, where T is the plane piezoelectric transducer, L is the delay line length, e is the thickness of the water coupling medium, and d is that of the composite sample.

The first conditions to be verified correspond to a round trip transit time in water sensibly less than in the composite, and that in the delay line sensibly higher than the sum of the two previous ones. If v_w, v_d and v_c stand for the ultrasonic wave velocities in, respectively, water, delay-line medium, and composite medium, these conditions may be written as :

$$e > \frac{d}{n_1} \qquad L > (e + \frac{d}{n_1}).n_2 \qquad (1)$$

Next, the radius of curvature R of the spherical acoustic lens must be chosen lower than a limit value in order to focalize inside the composite sample. Using the paraxial model, and neglecting any sample inhomogeneity and anisotropy, the following relations apply :

$$f = e + n_1 . d \qquad R = \frac{(n_2 - 1)}{n_2} . (e + n_1 . d) \qquad (2)$$

from which a typical maximum value for R appears to be nearly 6 mm for a 3 mm thick sample.

The angular aperture of the acoustic lens must be sufficiently large in order to optimize the spatial resolution. The practical limit is attained when the incident angular spectrum on the composite sample extends up to the total reflection angle for longitudinal waves, which is nearly equal to 37 degrees for a typical v_c value of 2500 m/s.

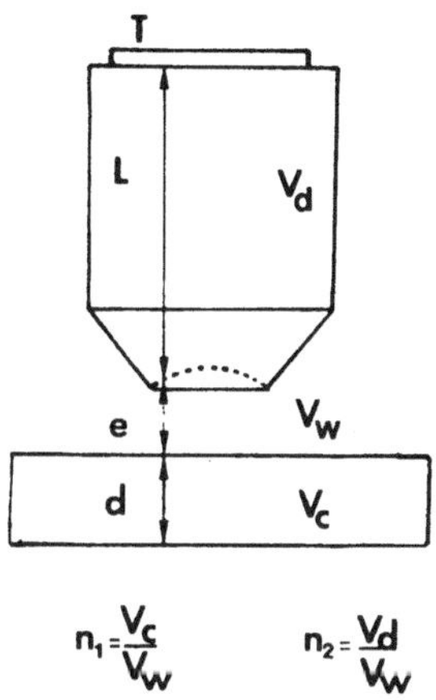

Fig. 1 Geometry of the ultrasonic probe.

Another problem arises from the echoes propagating in the delay line medium, bound to the mode conversion during the successive reflections on the lens and to the multimode (longitudinal and transverse) generation by the transducer. The later is unavoidable, due to the high longitudinal to transverse acoustic power isolation needed in the experiment (around 60 dB). However, at the used frequencies, a significant thickness of water may be used without an excessive ultrasonic attenuation, enabling an easy discrimination of the useful echoes from the parasitic ones.

The photographs in Fig. 2 below illustrate the interest of phase contrast imaging. The operating frequency is 100 MHz. The images represent a 5 μm width line engraved in a fused silica slab, a dimension well below the acoustic wavelength of 60 μm. For the first photograph, amplitude only imaging is used and the line is poorly detected in a very noisy background. Phase contrast between the echo reflected on the line and a fixed parasitic echo is used in the second photograph, where the line is neatly resolved.

For composite samples, the contrast and detection limit are also very good, despite of their heterogeneous and anisotropic nature. In what follows, we will try to explain this feature using a modelization of the ultrasonic behaviour of the composite.

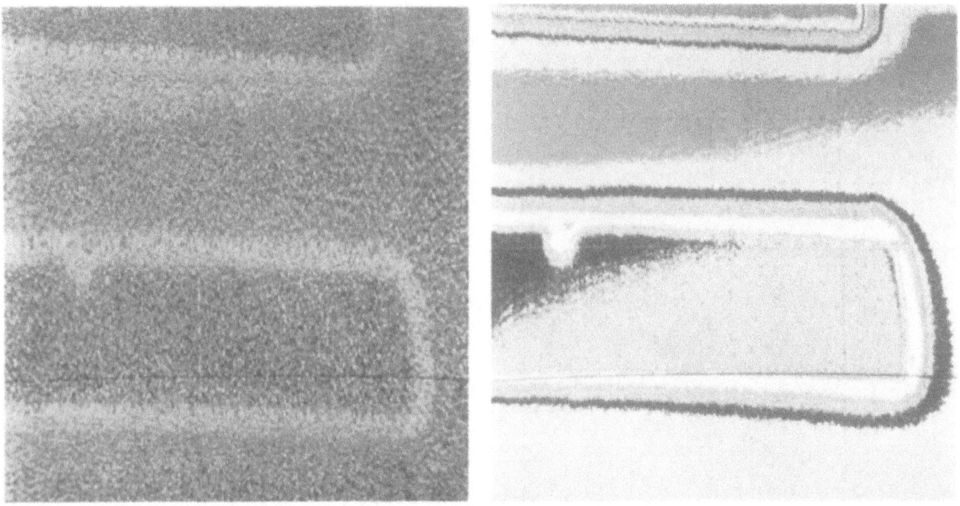

a b

Fig. 2 Image of a test line in a fused silica sample.
a) amplitude contrast only.
b) phase contrast image.

Theoretical problems

- Spatial resolution inside a composite material.

The focalization of an acoustic beam inside a composite material has been modelized in order to evaluate the performance of an acoustic microscope used for non destructive evaluation purposes. Attention has been given to carbon-epoxy composites, whose mechanical properties are very anisotropic.

The acoustic field inside the composite sample has been computed assuming water as a coupling medium, for acoustic lenses with various angular apertures and focal lengths inside water, by several approaches like evaluating the anisotropic diffraction integral (6,7), using the angular spectrum diffraction technique (8) or the geometrical theory of diffraction (9). The results derived from all these methods agree significantly well. The geometry of the anisotropic diffraction problem is shown in Fig. 3 below. The radiating surface σ is assumed plane here. The acoustic ray (vector r) PM from the source P to the observation point M is along the energy flux direction, in general non collinear to the wavevector (k).

The acoustic field potential distribution inside the sample may be written as :

$$A(M) = C.\int_{\sigma} a(P).\cos\beta.\frac{e^{j(\omega t - kr)}}{r}.d\sigma \qquad (3)$$

where C is some normalization constant.

For these computations, mechanical isotropy has been assumed in a plane transverse to the stratification direction (normal to the plies). The matrix of the effective elastic constants is then identical to that pertaining to the hexagonal cristallographic classes. In the following calculations, the following approximate relations for the wave velocity $v(\psi)$ and energy flux deviation angle ϕ are used :

$$v^2(\psi) = v_1^2.(\cos^2\psi + p.\sin^2\psi) \qquad p = v_2^2/v_1^2$$

$$\Theta = \psi + \phi \qquad \psi = tg^{-1}\left(\frac{tg\,\theta}{p}\right) \qquad (4)$$

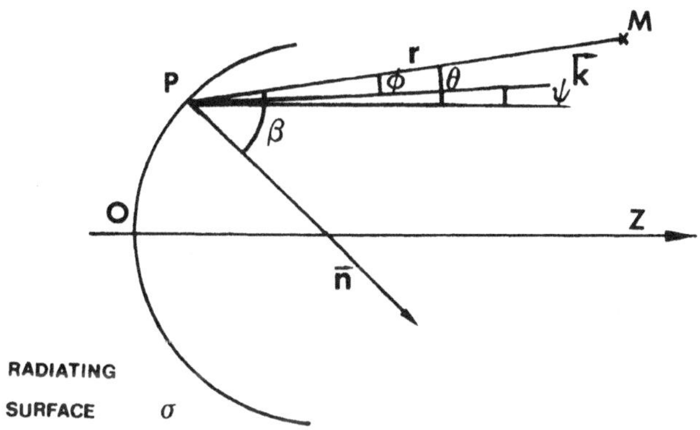

Fig. 3a Geometry of the anisotropic diffraction problem

where v_1 stands for the longitudinal wave velocity along the stratification direction and v_2 for that in the composite isotropy plane (tissue), and p is an anisotropy parameter. For carbon fiber composites typical values are v_1 = 2600 m/s and v_2 = 6000 m/s. The focal length of the acoustic lens is assumed to insure a free focalization in water at a distance 3 mm away behind the position where the sample surface will be localized. If the sample was isotropic, with a velocity v_1, the focalization depth inside the sample would be 1.7 mm. In Fig. 3, the computed focalization depth (normalized to that, F_0 = 1.7 mm, for an isotropic sample with velocity v_1) is shown versus the value of the anisotropy parameter p. With the value expected in our case for p (around 4), it is seen that the focalization depth is severely reduced with respect to that for an isotropic sample, and this feature must be taken into account by accordingly increasing the focal length of the acoustic lens.

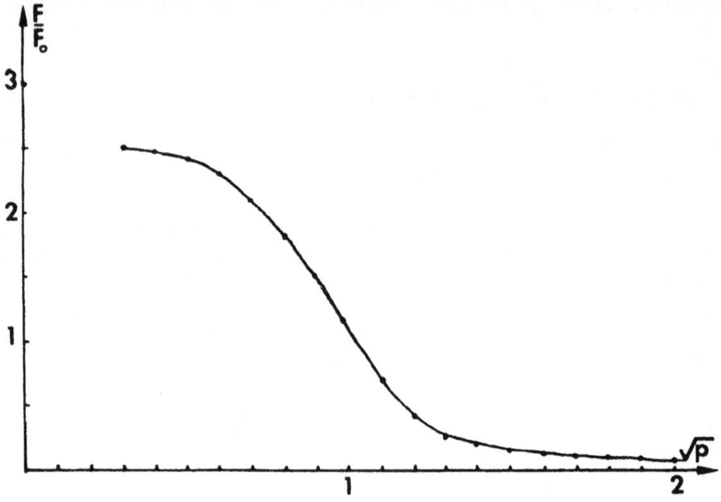

Fig. 3b Variations of the focalization dept versus anisotropy.

The half-width w (expressed in wavelengths for the acoustic velocity v_1) is also given versus p in Fig. 4. The degradation of the spatial resolution due to the anisotropy is clearly evidenced here for values of p greater than 1.5. This feature is responsible for a lowering of the contrast, due to the noise increase produced by the diffusion of a larger acoustic beam on the heterogeneities.

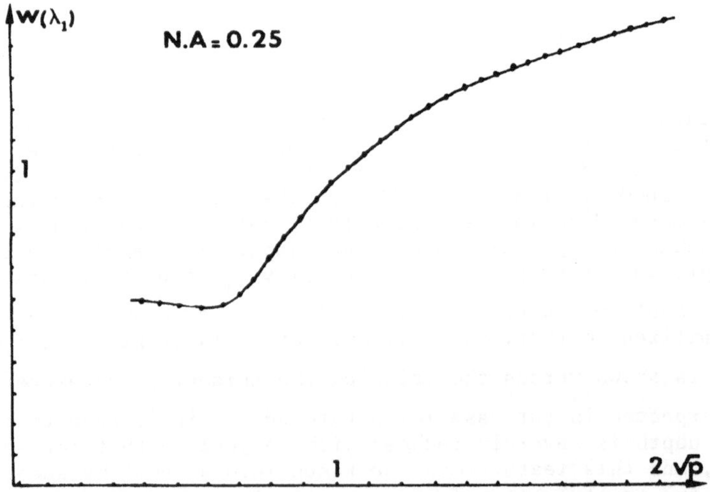

Fig. 4 Variations of the focal spot width versus anisotropy.

Interpretation for the contrast in the interference mode

The contrast observed in the images of composite samples may arise from the interferences between the part of the wave impinging under nearly normal incidence upon a fiber of a given ply and those other parts reaching the transducer after being converted to other (longitudinal, transverse, surface, cylindrical) propagation modes on a fiber of the same or another ply. This may be illustrated by considering the simple model of Fig. 5. An equiphase spherical surface is considered here and the interference occurs between rays A B C D E F (suffering from a mode conversion on fiber p) and S M P M S (direct reflection on fiber n).

It is easily shown that the time delay for the following rays may be written as :

$$t(ABCDEF) = \frac{2R}{v_0} - \frac{2Z\cos\Theta_0}{v_0} + \frac{2d_p\cos\Theta_1}{v_1} \tag{5}$$

$$t(SMPMS) = \frac{2(R - z)}{v_0} + \frac{2d_n}{v_1} \tag{6}$$

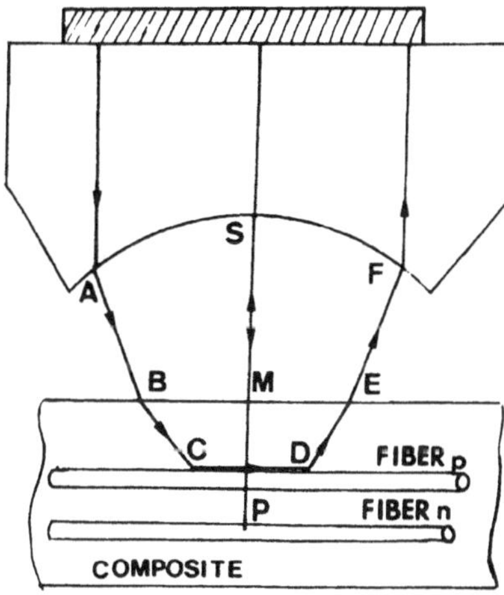

Fig. 5 Geometry for the interference calculation.

where v_0 and v_1 are the wave velocities inside, respectively, the coupling medium and the normal to the plies of the composite, d_n and d_p are the depths of the considered fibers from the sample surface, Θ_0 and Θ_1 are the incidence and refraction angles (whose dependance is given by SNELL'S law) at the sample surface for ray ABCDEF, R is the radius of curvature of the equiphase surface, and z is the offset of the sample surface from the focal plane position in the coupling medium. For the values v_0 = 1500 m/s, v_1 = 2600 m/s, Θ_0 = 5°, Θ_1 = 8°, and a sample thickness of 2 mm, one sees that the delay between all these rays is always lower than 1.6 μs, so that in the one to ten MHz operating frequency range, interference may occur between a number of directly reflected or mode converted rays. A very good phase contrast may result in the imaging mode when the acoustic probe is scanned over the sample and some of the previous rays alternatively exist or disappear according to the material microstructure. The observed response is then analogous to the classical V(z) curves, and may be computed from a relation like (10) :

$$V(Z) = \frac{\int P^2(\frac{r}{f_0}).\ R(\frac{r}{f_0}).\ \exp\left[-2jkz\sigma(r)\right]\frac{rdr}{\sigma(r)}}{\int P^2(\frac{r}{f_0}).\ \frac{r\,dr}{\sigma(r)}} \qquad (7)$$

where $\sigma(r) = \sqrt{1 - (r/f_0)^2}$

and where K is a normalization constant, P is a pupil function (which is generally assumed to be near unity), R is a reflectance function at the sample surface, and z is the offset between the sample surface and the free focal plane of the acoustic lens (inside water). In our case, the reflectance R is to be replaced by the sum of two or more interfering echoes. By recording the variations of the output voltage when scanning over the composite sample surface, informations about the internal structure (mean ply thickness, ply orientation using a cylindrical lens, the carbon tissue mesh period ...) may be retrieved from the V(z)-like curves.

Conclusion

Some problems encountered in the non destructive evaluation of composite materials have been studied. The geometrical design of the acoustic lens has been particularized to this kind of sample, but without taking anisotropy effects into account (this would lead to intricate aspherical acoustic lens shapes). The influence of the anisotropy on the focalization and the working in phase contrast mode have been investigated using simple modelizations of the mechanical behaviour of the composite.

These models explain the high resolution and contrast observed by us in the images of carbon-epoxy composite samples. By refining the models, the present work will also be developped towards the characterization of dimensional and physical properties of the tissue.

Acknowlegments

We greatly acknowledge the "Laboratoires de l'Aérospatiale" in Suresnes, France, for their contribution in the elaboration of the carbon-epoxy composite samples and fruitful discussions about these materials.

References

(1) J.I. KUSHIBIKI, and N. CHUBACHI, "Material Characterization by Line-Focus-Beam Acoustic Microscope", I.E.E.E. Trans. Sonics Ultrasonics, Vol. SU-32, p. 189, 1985.
(2) J.M.R. WEAVER, M.G. SOMEKH, A.D. BRIGGS, S.D. PECK, and C. ILETT, "Applications of the scanning Reflection Acoustic Microscope to the Study of Materials Science", I.E.E.E. Trans. Sonics Ultrasonics, Vol. SU-32, p. 302, 1985.
(3) K. LIANG, B.T. KHURI-YAKUB, S.D. BENNETT, and G.S. KINO, "Phase Measurements in Acoustic Microscopy", I.E.E.E. Ultrasonics Symp. Proc. (n° 0090-5607/83), p. 599, 1983.
(4) P. REINHOLDTSEN, W.W. HIPKISS, and B.T. KHURI-YAKUB, "Low-Frequency Acoustic Microscopy", Review of Progress in Quantitative N.D.E., San Diego, July 1984.

(5) B.T. KHURI-YAKUB, P. REINHOLDTSEN, "Acoustic Imaging of Subsurface Defects in Composites and Samples with Rough Surfaces", I.E.E.E. Ultrasonics Symp. Proc. (n° 0090-5607/85), p. 746, 1985.
(6) M.G. COHEN, "Optical Study of Ultrasonic Diffraction and Focusing", J. Appl. Phys. Vol. 38, p. 3821, 1967.
(7) N.R. OGG, "A Huyghen's Principle for Anisotropic Media", J. Phys. A (Gen. Phys), Vol. 4, p. 382, 1971.
(8) M.S. KARUSHI, and G.W. FARNELL, "Diffraction and Beam Steering for Surface Waves Comb Structure on Anisotropic Crystals", I.E.E.E. Trans. Sonics Ultrasonics, Vol. SU-18, p. 35, 1971.
(9) W.A. RADASKY, and G.L. MATTHEI, "Fast Computation of Diffraction in General Anisotropic Media by Use of the Geometrical theory of Diffraction" I.E.E.E. Trans. Sonics Ultrasonics, Vol. SU-30, p. 78, 1983.
(10) K.K. LIANG, G.S. KINO, and B.T. KHURI-YAKUB, "Material Characterization by the inversion of V(z)", I.E.E.E. Trans. Sonics ULtrasonics, Vol. SU-32, p. 213, 1985

ACOUSTIC MICROSCOPY: DEEP FOCUSING INSIDE MATERIALS AT GIGAHERTZ FREQUENCIES

J. Attal, A. Saied, J.M. Saurel, and C.C. Ly

LAMM, USTL, Place Eugène Bataillon
34060 Montpellier Cedex, France

ABSTRACT

Another way of imaging inside materials using transverse modes generated at the liquid-sample interface is shown. These modes can be highly focused for a given depth by means of aspherical lenses which apodize the efficiency conversion. Calculations of the profile of these lenses are reported and experiments are performed at 500 MHz through standard silicon wafers. This technique is extended to other types of material and structures such as metal-metal and ceramic-metal bondings.

INTRODUCTION

One of the most significant features of acoustic microscopy is focusing inside thick samples with spatial resolution as high as possible. However, serious difficulties arise when a convergent beam strikes a plane object : high reflectivity due to acoustic impedance mismatch, scattering of the beam into transverse, longitudinal and surface modes, aberrations induced by the plane liquid-object interface, diffraction and diffusion by the surface and the volume of the object itself.

The problem has been studied by a number of workers and different approaches have been undertaken (1-8). Reducing the surface artefacts is the first goal, which can be achieved by acoustic impedance matching the liquid to the object (1). The easiest way to do it is to use heavy liquids such as mercury or gallium. However, these cause wetting problems and they attack most metals. Applying a plastic antireflection coating of appropriate thickness is also a technique for reducing the surface echo. If the matching layer could be part of the instrument itself, it could be considered as a NDE technique. Some results have been carried out with organic thin film membranes at 20 MHz (2).

For deep penetration, with very short pulse and time gating detection we can hope to eliminate the surface echo. However, due to the low level of the tracking edge, the signal will be partially masked by the surface echo.

The second problem is that solids have a substantially higher velocity than most liquids and severe aberrations occur with a plane solid-liquid interface. The acoustic energy is spread on the axis of the lens from the paraxial focus to the surface of the object. The outer rays from the lens are totally reflected which lowers the efficiency of the wave transmission and implies a substantial degradation of the imaging performance. Some work has been devoted to design specific lenses to improve the stigmatism inside the object. We know that longitudinal and shear waves generated at the solid-liquid interface can both be focused inside a solid. Imaging with longitudinal waves is easier to do since it corresponds to paraxial focusing (3).

However, there are some advantages to use shear waves because of the reduction, by a factor of about two, in the velocity of waves in solids with a corresponding increase of the resolution and a possibility of imaging deeper for the same lens aperture.

Combining all these improvements by using liquid metals and aspherical lenses adapted to transverse waves, we show that it is possible to focus deeply in solids with a frequency practically limited by the absorption in the sample. Frequencies as high as 1.5 GHz can be achieved for a penetration depth of 0.4 mm in silicon for example. The corresponding resolution should be less than 10 microns.

COMPUTATION OF THE LENS PROFILE

Geometrical ray tracing provides a good insight into various situations using Snell's law. The field distribution is obtained by summing the plane waves contributions. The lens profile can be determined using Fermat's principle which equalizes the transit time of any rays coming from the ideal focus inside the sample back to the incident planar wave front located at the back face of the lens (fig.1).

The parametric equations defining the lens profile is given by:

$$x(r) = N \frac{(D + nz - nz/\cos r)\cos(\operatorname{asin}(n \sin r)) - D}{\cos(\operatorname{asin}(n \sin r) - N}$$

$$y(r) = z \operatorname{tg} r + (D - x(r))\tan(\operatorname{asin}(n \sin r))$$

where $N = V1/V2$ and $n = V2/V3$ are respectively the relative index of the lens-liquid and liquid-object (4). The other parameters are shown in figure 1.

If we fix the penetration z to be attained, the distance D between the surface of the object and the top of the lens, N and n, we can sketch the profile of the lens.

VIZUALISATION INSIDE SILICON USING MERCURY AS COUPLING FLUID

Though gallium seems to be the best candidate for inside examination with its very low absorption and high acoustic velocity, this liquid wets the surface of samples too much and makes cleaning difficult Mercury exhibits high surface tension which allows acoustic transmission without wetting except for some metals. In that case a thin protective layer of plastic, for instance, is necessary for examination.

Figure 2 shows the partition of the acoustic energy inside silicon versus the incidence angle when using mercury. We can notice that almost 100% is

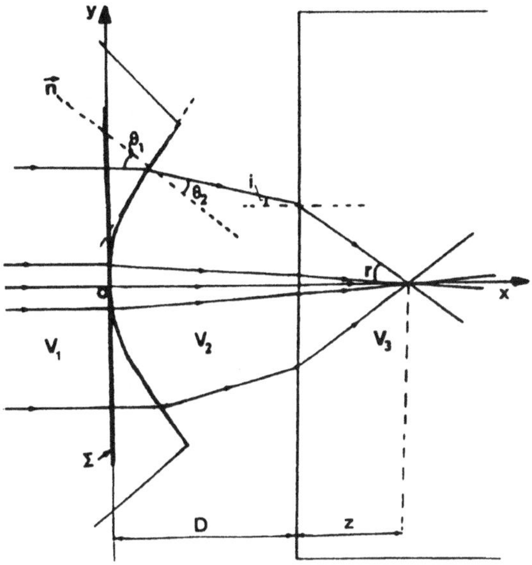

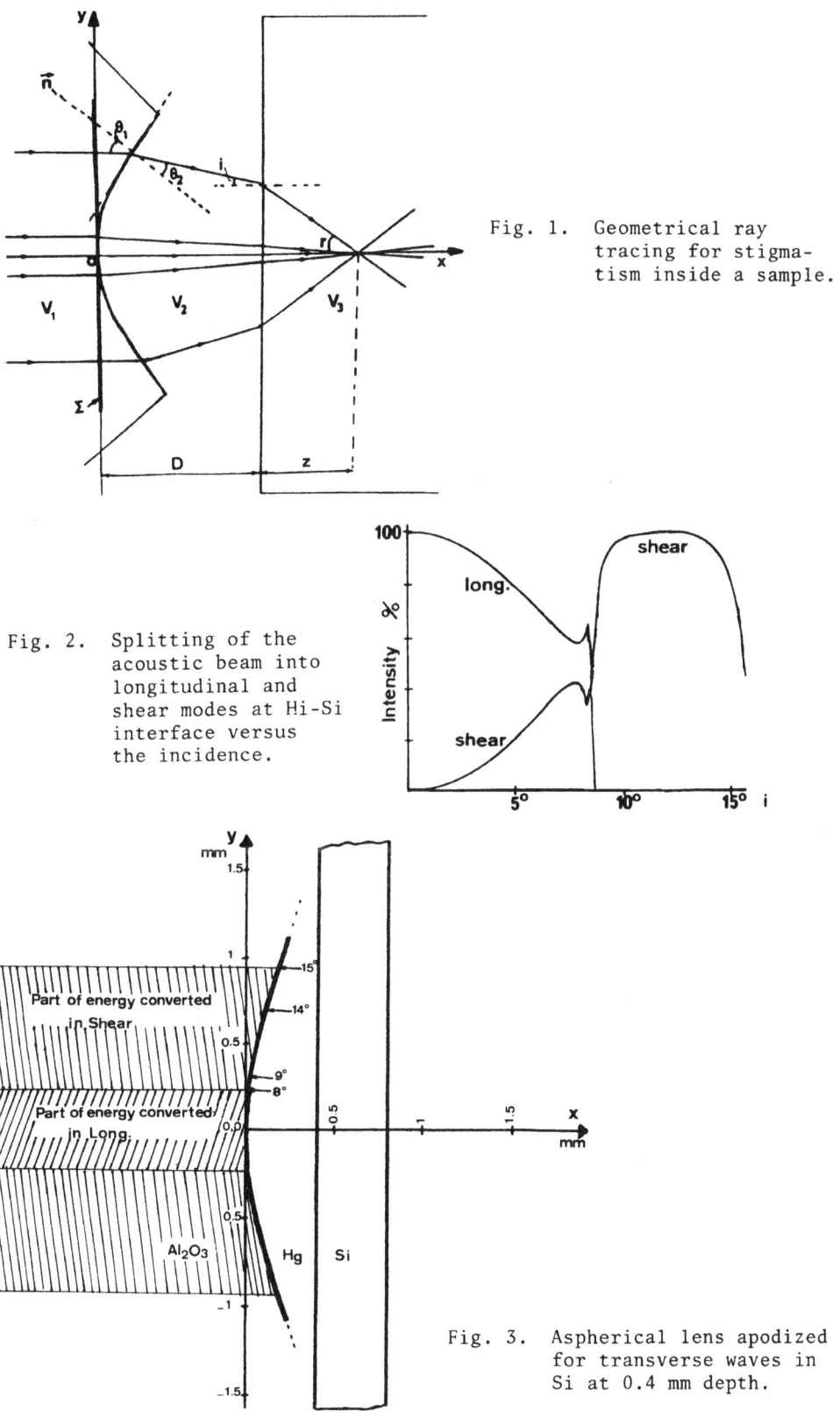

Fig. 1. Geometrical ray tracing for stigmatism inside a sample.

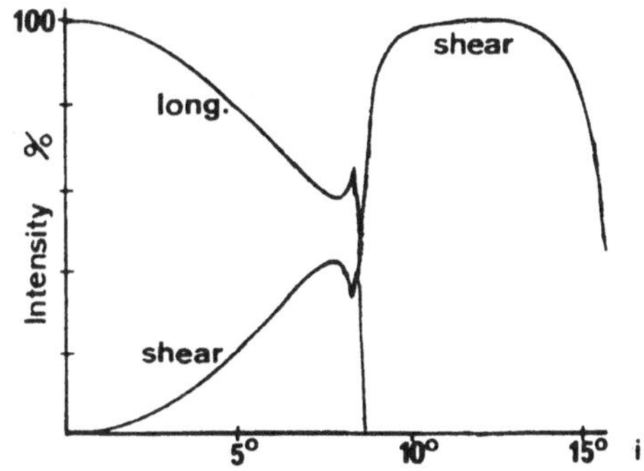

Fig. 2. Splitting of the acoustic beam into longitudinal and shear modes at Hi-Si interface versus the incidence.

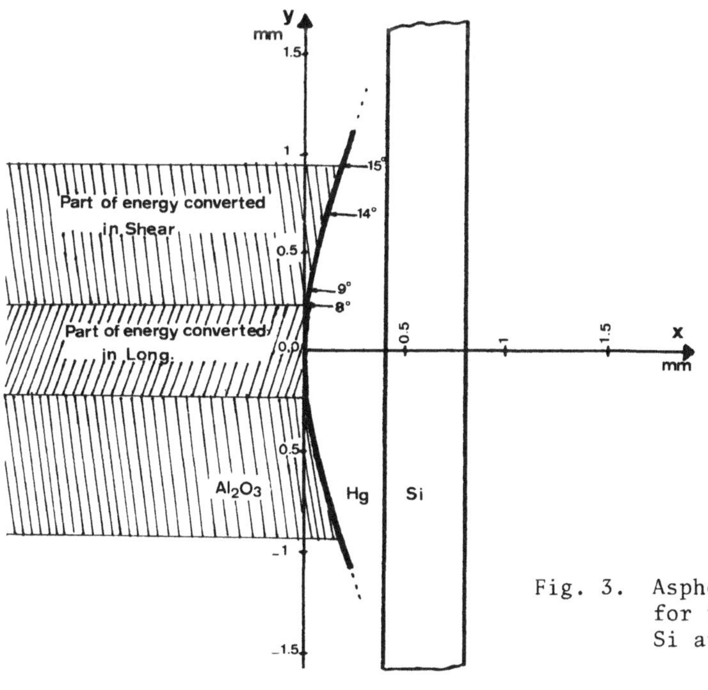

Fig. 3. Aspherical lens apodized for transverse waves in Si at 0.4 mm depth.

transmitted into longitudinal modes in normal incidence. Within the critical angles (which are 8° for the longitudinal modes and 15° for transverse) almost 100% of the energy is converted into transverse. Beyond 15°, all the energy is reflected back if we neglect the acoustic surface generation. In order to minimize the waste of acoustic energy, the incident angles on Si must not exceed 15°, and we can see in fig. 3 the part of the incoming energy which will be transformed into longitudinal and transverse modes. If we assume a uniform illumination of the lens, the ratio transverse/longitudinal is of the order of 12 dB. This value can be increased when taking into account the absorption due to different paths in the liquid since the absorption is more significant as the frequency is increased. If we also assume that the only cause of limitation in increasing frequency is absorption in mercury which is around 50 dB/mm at 1 GHz, we can evaluate the maximum frequency attainable with this type of lens, according to the following formula :

$$F_{max} = \left(\frac{A}{\alpha} \frac{1}{\langle d \rangle}\right)^{\frac{1}{2}} \quad \text{in GHz}$$

where A is the one way loss acceptable in the liquid

α is the absorption per unit length

$\langle d \rangle$ is the average distance between lens and object for the path of the transverse waves .

If we fix 15 db for A, 0.2 mm for lens-object distance which gives $\langle d \rangle$ around 0.13 mm, we find Fmax = 1.5 GHz.

GEOMETRICAL ABERRATIONS AS A FUNCTION OF DEPTH

In order to evaluate in what range the aberrations in silicon vary when we move the object along the axis direction, we have plotted these

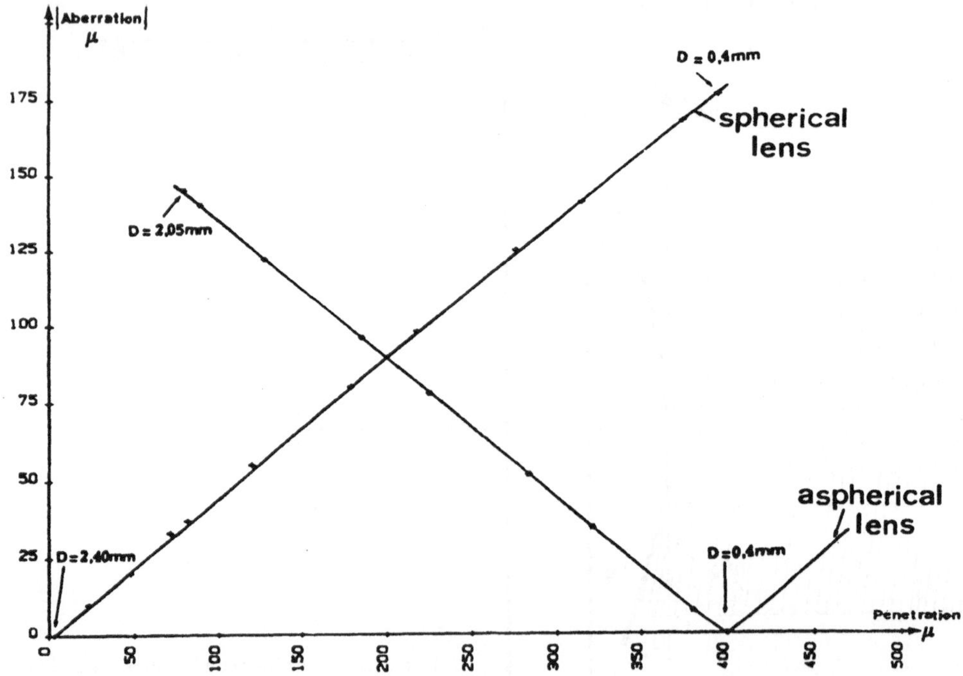

Fig. 4. Comparison between the aberrations of spherical and aspherical lenses versus the penetration.

variations (fig. 4) with spherical lens and aspherical lens apodized for a depth of 400 microns. The amplitude of the aberrations is limited by the contour of the efficient part of the beam for transverse waves. This is bounded between 9° and 14° with respect to the normal of the object. Obviously we have a sharp focusing at the surface for a spherical lens but the aberrations reach 175 microns at 400 microns of depth. The reverse situation occurs with an aspherical lens which gives a good rejection of sharp structures located near the surface. We can also notice that we get rid of the V(z) signature by increasing the transit of the lossy surface waves, especially at such high frequencies.

ASPHERICAL LENSES FOR EXAMINATION INSIDE METALS WITH TRANSVERSE WAVES

If we collect data on transverse waves velocities in most metals, we can notice that all the values are at ±10% close to three specific values, so we can split them in three groups (Table I). The first group includes Al, Fe, Ni, W with a mean velocity of 3000 m/s, the second group is Cu, Zn and alloys with 2300 m/s. The third one is soft metal such as Au, Sn, Ag with 1450 m/s. The last one does not of course require aspherical lenses for sharp focusing since there is matching in terms of velocity. For the other two classes, we can define a specific profile of aspherical lenses which can minimize the aberrations for a given deepness.

TABLE 1

Metal	Density g/cm^3	Long. velocity m/s	Trans. Velocity m/s	Average trans. Velocity m/s
Al	2.70	6420	3040	
Fe	7.90	5950	3240	
Ni	8.90	6040	3000	3000
Tungsten	19.30	5220	2890	
Dural.175	2.90	6320	3130	
Cu	8.93	4760	2325	
Zn	7.10	4210	2440	2300
Brass	8.60	4700	2110	
Au	19.70	3240	1200	
Ag	10.40	3650	1610	1450
Sn	7.30	3320	1670	

EXPERIMENTAL RESULTS

Different aspherical lens profiles have been realized for focusing in silicon and metal at different depths, and frequencies lying between 100 and 500 MHz. The large aperture of the lens makes difficult to correctly tune the transducer at frequencies above, due to the high value of the capacitance and rather small radiation resistance.

First, we have checked at 500 MHz the resolution and penetration through a wafer of Si focusing on the back side where a film of gold was deposited through an electron microscopy grid. For easy measurements, two thicknesses of film have been selected (100 Å and 1000 Å) and placed edge to edge. In figure 5, we see very clearly the patterns of the 1000 Å thickness grid and we are able to distinguish the 100 Å patterns which

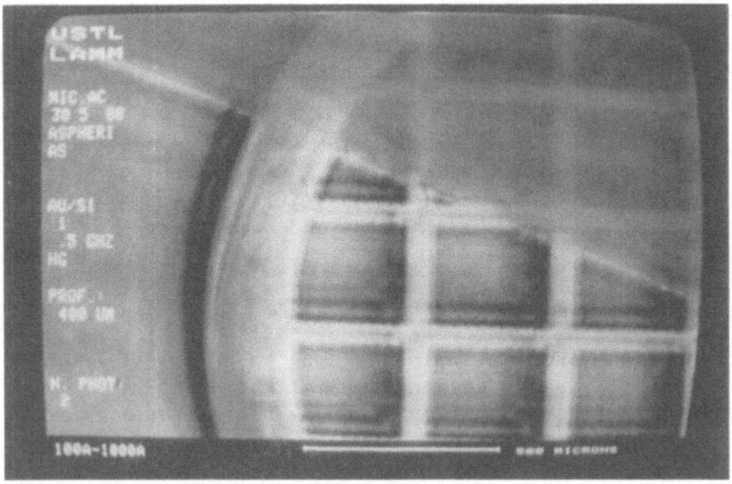

Fig. 5. Resolution in depth at 500 MHz inside 0.4 mm of Si with gold films deposited on the back side.

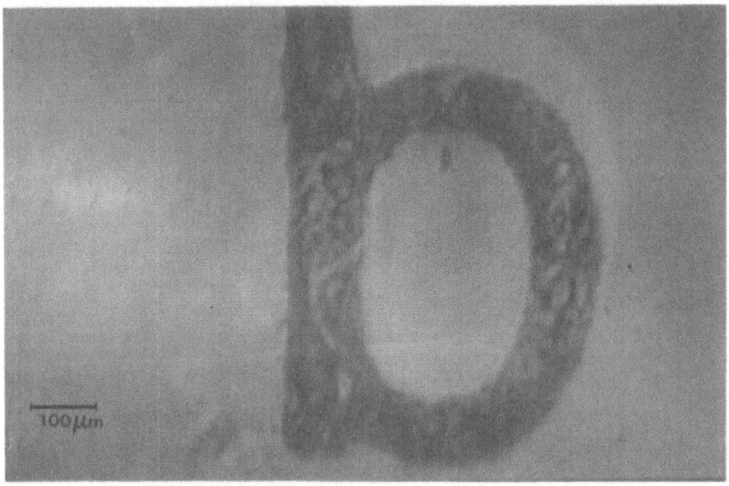

Fig. 6. Lateral resolution check at 1 GHz through 0.4 mm of Si.

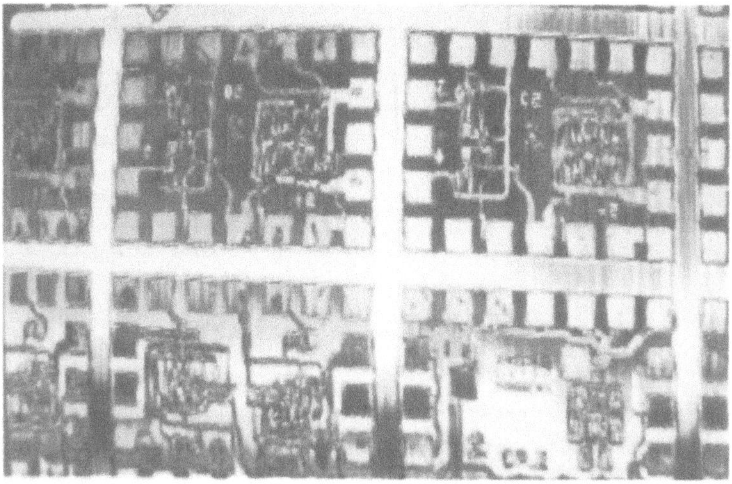

Fig. 7. Visualization of an integrated circuit on GaAs through its substrate at 500 MHz.

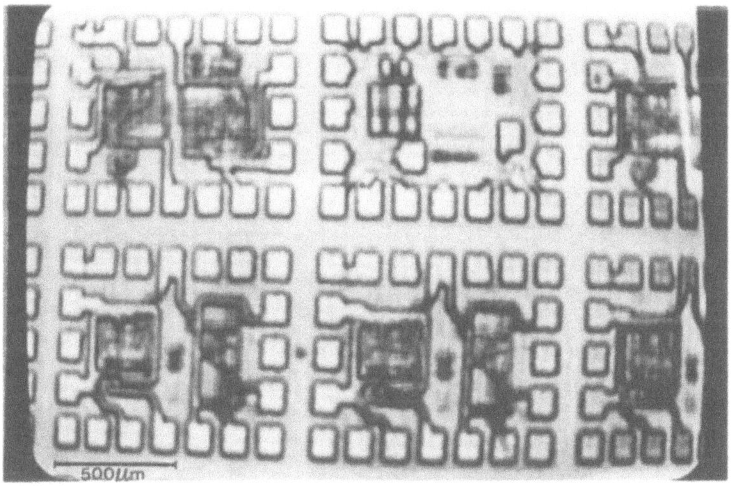

Fig. 8. Same field as fig. 7 but direct observation of the circuit with water at 600 MHz.

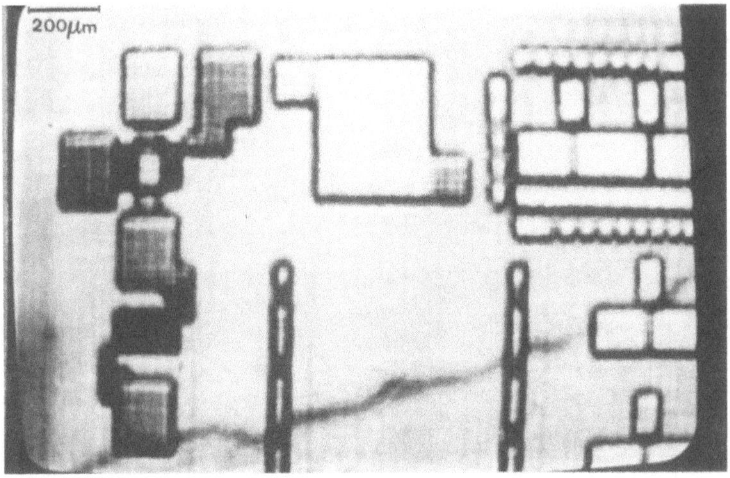

Fig. 9. Zooming of a part of fig. 7.

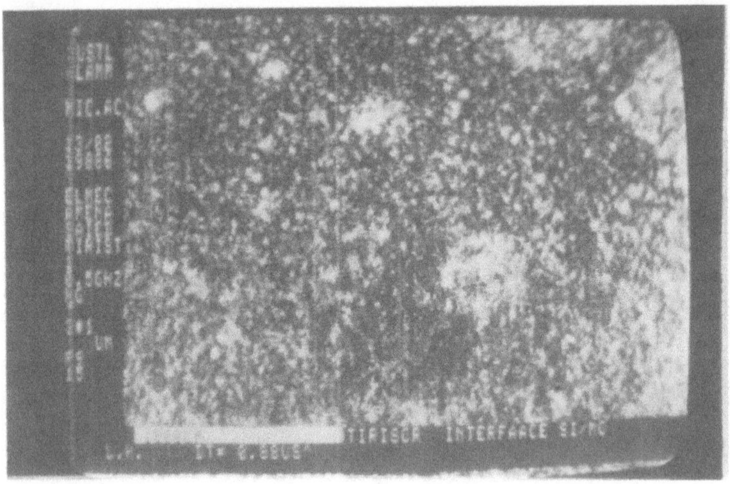

Fig. 10. Visualization of a Silicon-Molybdenum interface bonding.

fixes the resolution in thickness at something slightly less. The best lateral resolution is shown in figure 6 where the operating frequency is 1 GHz. Here, the pattern is deposited on the back side of a standard Si wafer. The resolution should be better than 10 microns.

The next two figures are the visualization of integrated circuits through their GaAs substrate with adequate aspherical lenses operating at 500 MHz (fig. 7). Comparison with the image of the front surface using spherical lens at 600 MHz and water has been made (fig. 8). The corresponding wavelengths are 6 microns for transverse waves at 500 MHz in GaAs and 2.5 microns at 600 MHz. It seems that the resolution inside the crystal is equivalent to that obtained at the surface at 100 MHz which is five times worse.

We have also investigated many devices with these aspherical lenses in order to check the quality of the bonding between materials. As an example we show in figure 10 a Si-Mo interface used for power device application which reveals large areas where bonding can be suspected. This may cause failure or reduction of lifetime of the device due to a lack of efficient heatsink.

EFFECT OF ANISOTROPY IN SAMPLES

The strong anisotropy in many crystals affects the quality of the stigmatism with transverse waves. For instance depending on the orientation, we note a change in velocity from 2480 to 3350 m/s for GaAs and from 4680 to 5850 m/s for Si. We clearly see this effect in figure 5 which produces a superimposition of images of the same pattern shifted in the plane.

CONCLUSION

Combining the use of liquid metals and aspherical lenses, we can optimize the focusing depth and the resolution of the acoustic microscope which can be extended to the gigahertz range. These performances tend to be limited only by the intrinsic absorption in the samples which is much more satisfactory. Even if mercury still has the problem of harmfulness, it is beginning to be used in industrial environments for specific applications where no other choice is available.

REFERENCES

1 J. Attal, G. Cambon and J.M. Saurel, Imaging interior planes by acoustic microscopy, Acoustic Imaging, vol. 10, p. 803-815, 1982.
2 A. Umeda, H. Nikoonahad and E. Ash, New sub-surface imaging technique using scanning acoustic microscopy. Ultrasonics symposium, p. 750-754, 1985.
3 F. Pino, D.A. Sinclair and E. Ash, Scanning acoustic microscopy of solid objects using aspheric lenses, Acoustical Imaging Vol. 11, p. 1-19, 1983.
4 A.SAIED Conception and realization of a new type of lens to improve the depth resolution of an acoustic microscope. (Thesis). May 1985. Université des Sciences et Techniques du Languedoc Montpellier France.
5 Atalar, Penetration depth of the scanning acoustic microscope, IEEE Trans. Son. Ultr., Su-32 p. 164-167, 1985;
6 B.T. Khuri-Yakub, P. Reinholdtsen and C.H. Chou, Acoustic imaging of subsurface defects in composites and sample with rough surface, Ultrasonics symposium, p. 746-749, 1985.

7 B. Nongaillard, J. Grosmaire and J.M. Rouvaen, Zooming lens for acoustic microscopy and analysis of non destructive testing images, Ultrasonics symposium, p. 772-776, 1985.
8 Ian, R. Smith, R.A. Harvey and D.J. Fathers, An acoustic microscope for industrial applications, IEEE Trans. Son. Ultr. SU-32 p. 274-288, 1985.

DATA ACQUISITION PHASE ERROR ESTIMATION AND CORRECTION IN TOMOGRAPHIC ACOUSTIC MICROSCOPY

Hua Lee and Richard Y. Chiao

Department of Electrical and Computer Engineering
University of Illinois at Urbana-Champaign
Urbana, Illinois 61601, USA

Abstract - This paper considers the problem of phase estimation and correction in tomographic acoustic microscopy. Recently, a quadrature receiver has been implemented for the Scanning Laser Acoustic Microscope (SLAM) for the purpose of holographic and tomographic imaging. Because the real and imaginary components are detected sequentially, a relative phase error is introduced. In a recent paper, Lin and Wade suggested that the relative phase error is constant throughout the image, and they proposed a single-sideband approach for the removal of the phase error. In this paper, we show that the phase error is a spatial variation. In addition, we point out that the single-side band technique removes only the effect of the phase error in the image intensity profile, and the phase error remains in the wavefield data samples. Here we present a new correction technique for the complete removal of the phase error. This technique has been applied to experimental data from the SLAM, and the results are included in this paper.

Introduction

In recent years, considerable effort has been devoted to the conversion of conventional acoustic microscopy into holographic and tomographic operating modes [1-3]. With the addition of quadrature detectors, acoustic microscopy is currently capable of high-resolution subsurface holographic imaging. During the data acquisition process, the real and imaginary parts of the complex wavefield are detected separately [4]. As a result, a relative phase error is introduced in the data samples which contributes to the degradation of the reconstructions in the form of a fixed-frequency background fluctuation. The phase error becomes an even more dominant degradation factor in tomographic superposition. Therefore, accurate estimation and correction of the phase error are important subjects in acoustic tomographic microscopy. A recent paper by Lin and Wade proposed to remove the degradation by eliminating one of the spectral side-bands [5].

In this paper, we point out that the single side-band technique removes only the effect of the degradation. The phase error still exists in the remaining side-band, which makes this technique infeasible for tomographic reconstruction. Here we present an alternative approach to the estimation and correction of the phase error. In addition, the experimental results also indicate that the phase error is a spatial variation instead of a constant as it was assumed in the recent publication by Lin and Wade. This technique has been applied to experimental data from the SLAM and the results are presented.

Phase Error and Degradation Effects

Figure (1) illustrates the structure of the SLAM data acquisition system. The unknown object is insonified by plane waves, and the scattered wavefield is detected by a raster-scanned laser. A quadrature receiver with an electronically-generated coherent reference is used to demodulate the resultant signal to obtain the complex wavefield distribution [4]. We denote the two-dimensional unknown object distribution as

$$a(x,y) = |a(x,y)|\, e^{j\phi(x,y)}. \tag{1-a}$$

For simplicity, we use the one-dimensional format for the analysis

$$a(x) = |a(x)|\, e^{j\phi(x)}, \tag{1-b}$$

where x is the scanning direction of the laser beam. Then, the planar object is modulated by the illumination wavefield $u(x)$ and can be written as

$$p(x) = a(x)u(x). \tag{2}$$

Specifically, when the illumination wavefield is a unit-amplitude plane wave with spatial frequency f_s, the scattered wavefield is given by

$$p(x) = a(x)e^{j2\pi f_s x}. \tag{3}$$

Ideally, the quadrature receiver detects the real and imaginary parts of the scattered field

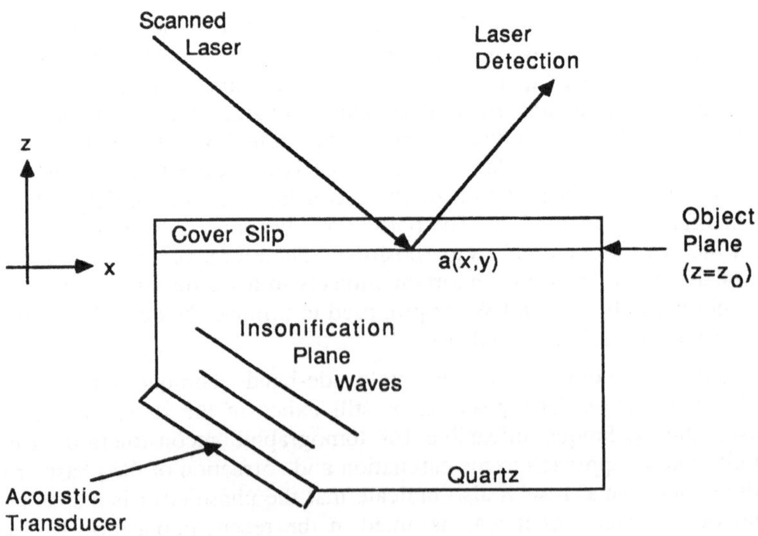

Figure 1. Basic setup of the SLAM.

$$p_r(x) = Real\ \{a(x)\ exp\,(j\,2\pi f_s x)\} \tag{4-a}$$

$$= |a(x)|\ cos\,(2\pi f_s x + \phi(x))$$

and

$$p_i(x) = \text{Im}\ \{a(x)\ exp\,(j\,2\pi f_s x)\} \tag{4-b}$$

$$= |a(x)|\ sin\,(2\pi f_s x + \phi(x)).$$

In practice, due to a relative phase error $\Delta\theta(x)$, the two output components of the quadrature receiver become

$$p'_r(x) = |a(x)|\ cos\,(2\pi f_s x + \phi(x) - \Delta\theta(x)/2) \tag{5-a}$$

and

$$p'_i(x) = |a(x)|\ sin\,(2\pi f_s x + \phi(x) + \Delta\theta(x)/2). \tag{5-b}$$

As a result, the wavefield actually recorded is

$$p'(x) = p'_r(x) + jp'_i(x) \tag{6}$$

$$= cos\,(\Delta\theta(x)/2)\ a(x)\ exp\,(j\,2\pi f_s x) + sin\,(\Delta\theta(x)/2)\ a^*(x)\ exp\,(-j\,2\pi f_s x).$$

The phase error results in an additional spectral concentration centered at $-f_s$.

The degradation effect of the phase error can be observed by examining the intensity of the recorded wavefield

$$|p'(x)|^2 = |a(x)|^2 + |a(x)|^2\ sin\,(\Delta\theta(x))\ sin\,(4\pi f_s x + 2\phi(x)). \tag{7}$$

The degradation due to the phase error is a background fringe pattern with spatial frequency $2f_s$. In tomographic reconstruction, several projections are combined during the image formation process [2]

$$\hat{p}(x) = \frac{\sum_{k=1}^{M} u^*_k(x) p_k(x)}{\sum_{k=1}^{M} |u_k(x)|^2}, \tag{8}$$

where the subscript k denotes different projection angles, and $u(x)$ and $p(x)$ are the insonifying and scattered wavefields, respectively. The effect of the phase error is more detrimental in this case due to the complex superposition process. Correct phase relationship between the terms in the summation is essential for a quality reconstruction.

The single-sideband approach has been proposed for the removal of the phase error [5]. However, this technique removes only the *degradation effect* of the phase error in terms of background fringe patterns. The phase error is still embedded in the complex image and consequently remains a problem for the tomographic case. In the next section, we will review the single-sideband correction method for comparison purposes.

Single-Sideband Approach

The phase correction method proposed by Lin and Wade is to remove one of the sidebands of the detected signal. Taking the upper sideband of the signal given in Eq. (6), we have

$$p'_+(x) = cos\,(\Delta\theta(x)/2)a\,(x)exp\,(j\,2\pi f_s x). \tag{9}$$

We assume $\Delta\theta(x)$ is a narrowband variation. Equivalently, we can assume that f_s is sufficiently large so that the two sidebands do not overlap. The intensity of the single-sideband image is

$$|p'_+(x)|^2 = \cos^2(\Delta\theta(x)/2)\;|a\,(x)|^2. \tag{10}$$

We note that the fringe patterns with spatial frequency $2f_s$ have now been removed. However, the intensity is still modulated by an undesirable $\cos^2(\Delta\theta(x)/2)$ factor which is less noticeable than the sinusoidal component. This method is useful for holographic imaging when the phase error is a constant, and is not feasible for tomographic reconstruction because the phase error is still embedded in the complex signal. In the next section, we present an algorithm for the complete removal of the phase error.

Phase Error Estimation and Correction

In this section, we present an alternative approach to the phase error problem. We first estimate the phase error at each spatial position and then make the correction accordingly. Because a relative phase difference cannot be estimated from two real numbers, we cannot use the outputs of the quadrature receiver directly to estimate the phase error. Instead, we take the single-sideband of the quadrature receiver outputs. Then Eq. (5) becomes

$$p'_r(x) = \frac{1}{2}e^{-j\Delta\theta(x)/2}\;a\,(x)\;exp\,(j\,2\pi f_s x) + \frac{1}{2}e^{j\Delta\theta(x)/2}\;a^*(x)\;exp\,(-j\,2\pi f_s x) \tag{11-a}$$

and

$$p'_i(x) = \frac{1}{2}e^{j\Delta\theta(x)/2}\;a\,(x)\;exp\,(j\,2\pi f_s x) - \frac{1}{2}e^{-j\Delta\theta(x)/2}\;a^*(x)\;exp\,(-j\,2\pi f_s x). \tag{11-b}$$

The upper-sideband signals are

$$p'_{r+}(x) = \frac{1}{2}e^{-j\Delta\theta(x)/2}\;a\,(x)\;exp\,(j\,2\pi f_s x) \tag{12-a}$$

and

$$p'_{i+}(x) = \frac{1}{2}e^{j\Delta\theta(x)/2}\;a\,(x)\;exp\,(j\,2\pi f_s x). \tag{12-b}$$

Subsequently, the phase error can be estimated

$$e^{j\Delta\theta(x)} = \frac{p'_{i+}(x)}{p'_{r+}(x)}. \tag{13}$$

The lower sidebands can also be used to give a phase error estimate.

To derive the correction operation, consider the data

$$p'_r = A\ cos\,(\alpha-\beta) \tag{14}$$

$$p'_i = A\ sin\,(\alpha+\beta),$$

where α and β are constants. We wish to obtain the corrected data in the form of

$$\hat{p}_r = A\ cos\,(\alpha) \tag{15}$$

$$\hat{p}_i = A\ sin\,(\alpha).$$

Rearranging Eqs. (14) and (15), we obtain the following matrix relationship

$$\begin{bmatrix} p'_r \\ p'_i \end{bmatrix} = \begin{bmatrix} cos\,(\beta) & sin\,(\beta) \\ sin\,(\beta) & cos\,(\beta) \end{bmatrix} \begin{bmatrix} \hat{p}_r \\ \hat{p}_i \end{bmatrix}. \tag{16}$$

The correction matrix operation is in the form of

$$\begin{bmatrix} \hat{p}_r \\ \hat{p}_i \end{bmatrix} = \begin{bmatrix} cos\,(\beta) & sin\,(\beta) \\ sin\,(\beta) & cos\,(\beta) \end{bmatrix}^{-1} \begin{bmatrix} p'_r \\ p'_i \end{bmatrix} \tag{17}$$

$$= \frac{1}{cos\,(2\beta)} \begin{bmatrix} cos\,(\beta) & -sin\,(\beta) \\ -sin\,(\beta) & cos\,(\beta) \end{bmatrix} \begin{bmatrix} p'_r \\ p'_i \end{bmatrix}.$$

Letting $\beta=\Delta\theta(x)/2$, we can obtain the corrected signal for each position x

$$\begin{bmatrix} \hat{p}_r(x) \\ \hat{p}_i(x) \end{bmatrix} = \frac{1}{cos\,(\Delta\theta(x))} \begin{bmatrix} cos\,(\Delta\theta(x)/2) & -sin\,(\Delta\theta(x)/2) \\ -sin\,(\Delta\theta(x)/2) & cos\,(\Delta\theta(x)/2) \end{bmatrix} \begin{bmatrix} p'_r(x) \\ p'_i(x) \end{bmatrix}. \tag{18}$$

Here we also utilize the narrowband assumption for the phase error distribution for the formulation of the algorithm. We wish to point out that the accuracy of the error estimation and correction depends largely on the variation of the phase error. For phase error variation with wider bandwidth, the error may not be completely removed in a single step. Therefore, we modify this approach into an iterative process, whereby the estimation and correction process is applied to the data samples recursively. We perform error reduction at each iteration, and we terminate the algorithm when the error reaches an acceptable level. Fig.(2) shows the block diagram of the recursive correction algorithm.

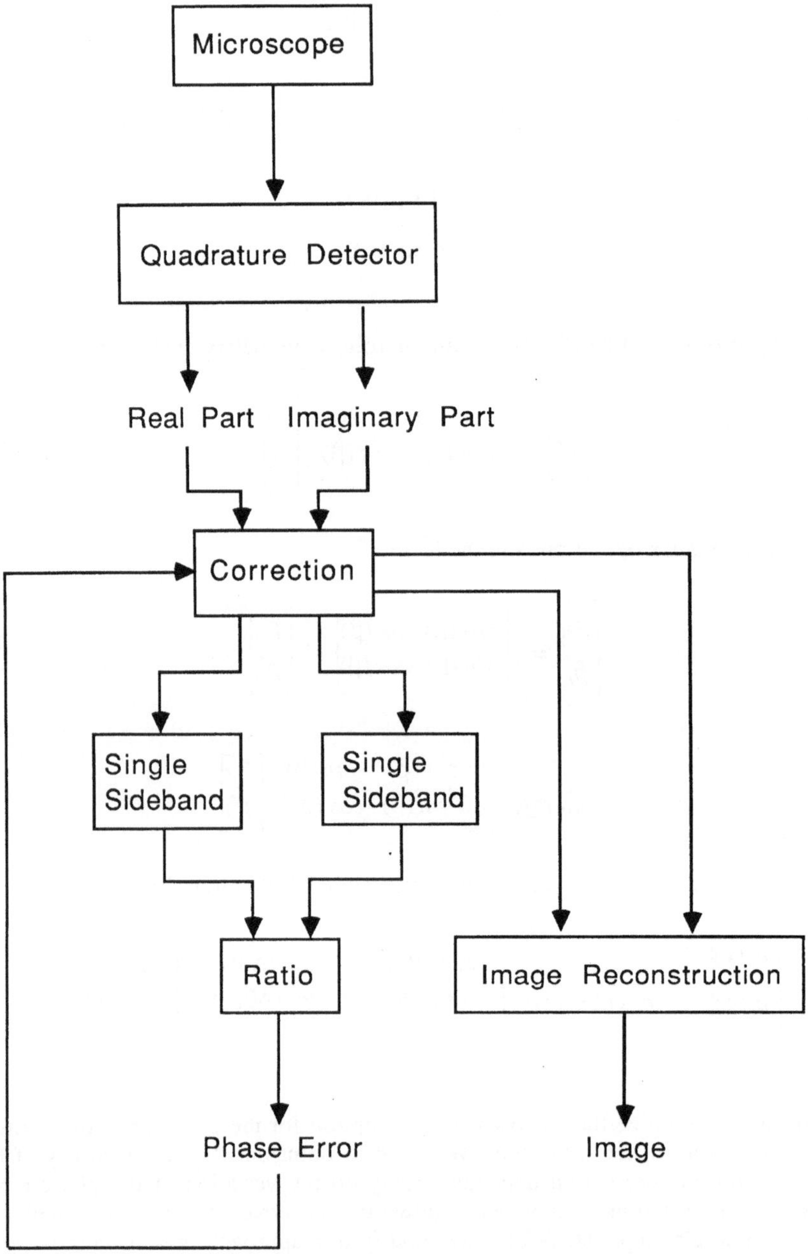

Figure 2. Block diagram of the algorithm.

Experimental Results

Fig.(3) shows the magnitude of a complex data sequence from the SLAM, and Fig.(4) shows the magnitude of the spectrum of the complex sequence. The small peak in the lower-sideband is caused by the phase error. Fig.(5) shows the estimate of the phase error distribution of the wavefield data sample sequence. We wish to point out that the spatial phase error distribution is not a constant as was assumed by Lin and Wade. Figs.(6-a), (6-b), and (6-c) are the phase error distributions after 1, 3, and 5 iterations, respectively. Fig.(7) illustrates the gradual reduction of the phase error after recursive corrections. The error does not converge to zero largely due to the presence of noise in the data samples.

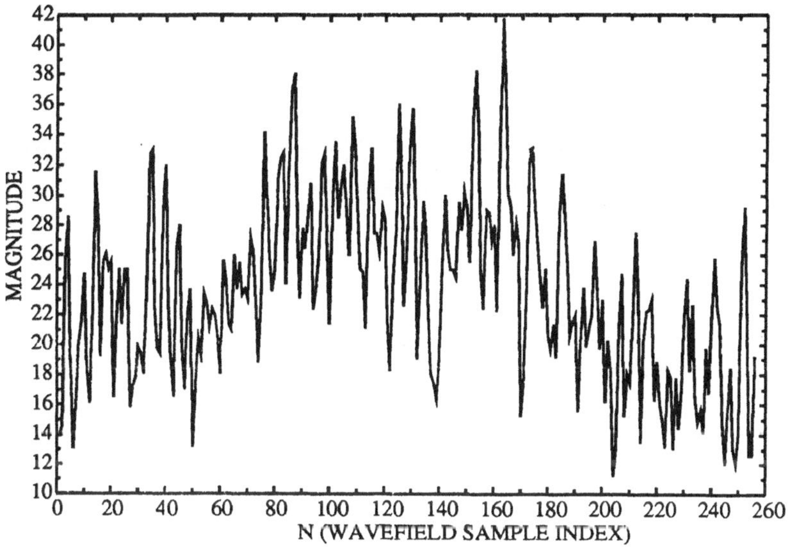

Figure 3. Magnitude of complex data sequence.

Conclusion

In tomographic acoustic microscopy, the image quality is very sensitive to the relative phase error in the projections. In this paper, we propose a technique for phase error estimation and correction which is effective and computationally efficient. This approach is superior to the method previously proposed by Lin and Wade because (1) the phase error is generalized to a spatial variation instead of a constant, and (2) it removes the phase error instead of removing only the degradation effects. Experimental data samples are used as examples to demonstrate the effectiveness of this approach. We have demonstrated phase error estimation and correction, but it is also possible to include amplitude error estimation and correction in this algorithm with minor modifications.

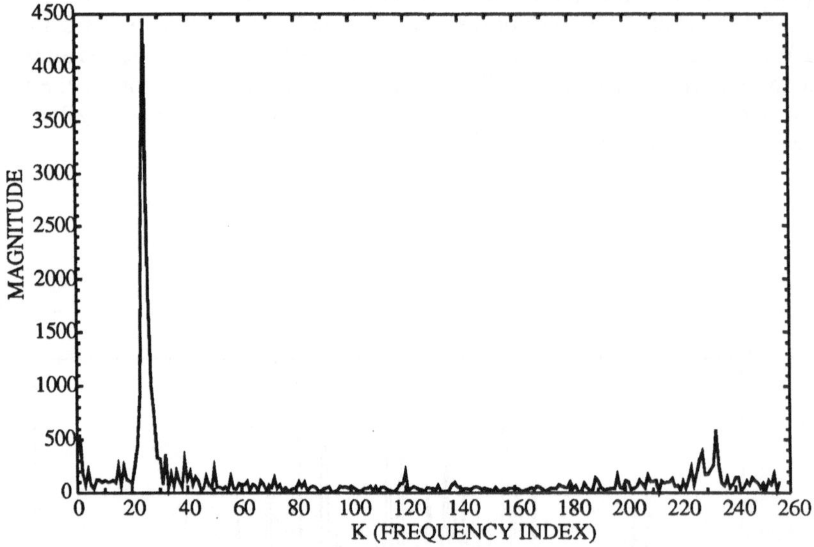

Figure 4. Magnitude spectrum of data.

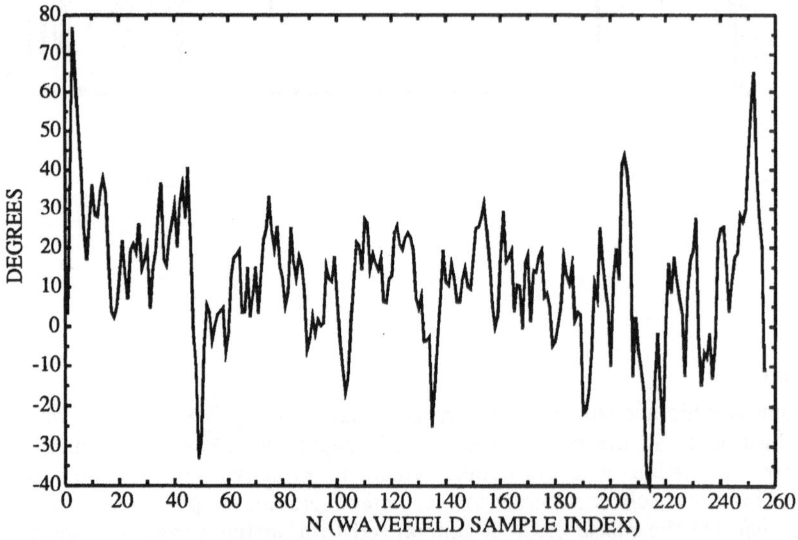

Figure 5. Phase error distribution.

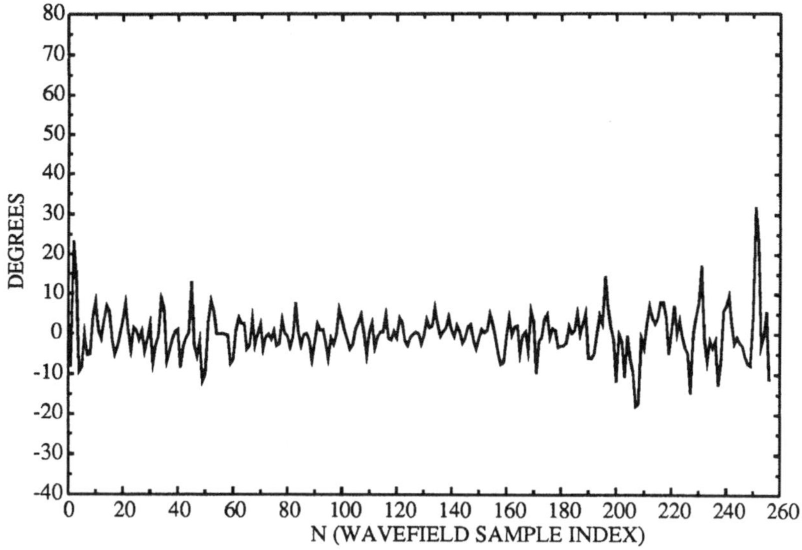

Figure 6-a. Phase error distribution after one iteration.

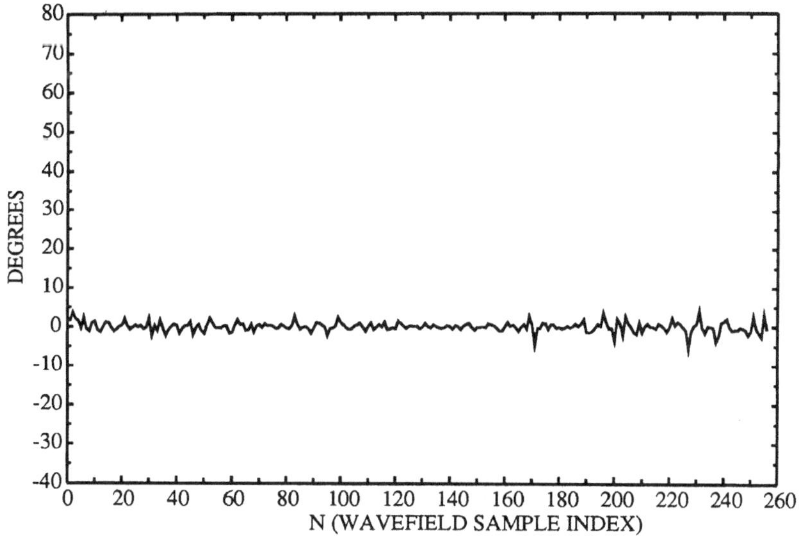

Figure 6-b. Phase error distribution after three iteration.

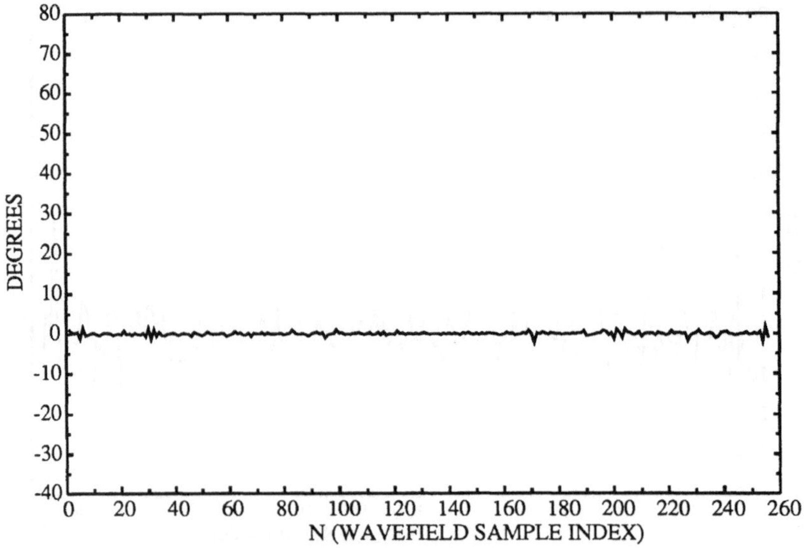

Figure 6-c. Phase error distribution after five iterations.

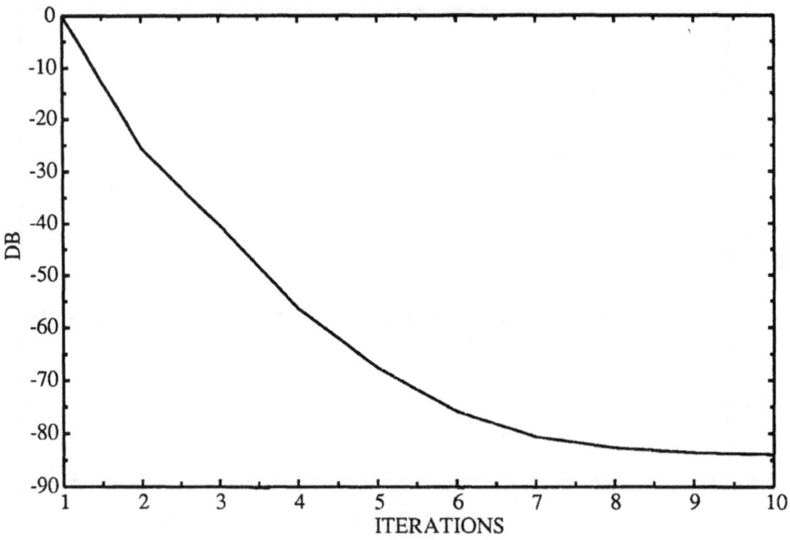

Figure 7. Error power as a function of iterations.

Acknowledgment

This research is supported by the National Science Foundation under Grants ENG-8451484 and ECS-8406511.

References

[1] Z. C. Lin, H. Lee, and G. Wade, "Back-and-Forth Propagation for Diffraction Tomography," *IEEE Trans. Sonics Ultrason.*, vol. SU-31, pp. 626-634, 1984.

[2] Z. C. Lin, H. Lee, and G. Wade, "Scanning Tomographic Acoustic Microscopy: A Review," *IEEE Trans. Sonics Ultrason.*, vol. SU-32, pp. 168-180, Mar. 1985.

[3] Z. C. Lin, H. Lee, G. Wade, M. G. Oravecz, and L. W. Kessler, "Holographic Image Reconstruction in Scanning Laser Acoustic Microscopy," *IEEE Trans. Ultrason. Ferroelec. Freq. Contr.*, vol. UFFC-34, pp. 293-300, May 1987.

[4] Hua Lee and Carlos Ricci, "Modification of the Scanning Laser Acoustic Microscope for Holographic and Tomographic Imaging," *Applied Physics Letters*, 49(20), pp. 1336-1338, November 1986.

[5] Zse Cherng Lin and Glen Wade, "Signal Processing and Image Reconstruction in Microscopic Holography," *IEEE Transactions on Acoustics, Speech, and Signal Processing*, 35(7), July 1987, pp. 1037-1045.

IN-SITU MEASUREMENT OF CATHODIC DISBONDING OF POLYBUTADIENE COATING ON STEEL

R.C. Addison, Jr., M.W. Kendig, and S.J. Jeanjaquet

Rockwell International Science Center
1049 Camino Dos Rios
Thousand Oaks, CA 91360

INTRODUCTION

Corrosion of organic polymer-coated steel involves the formation of a corrosion cell consisting of the metal substrate, water, oxygen, and a layer of conducting electrolyte in contact with both reaction sites.[1,2] Initially the anodic and cathodic reactions occur at proximate sites (Fig. 1A). The principal reactions are the anodic oxidation of the iron and the cathodic reduction of oxygen. As the corrosion product forms and oxygen and electrolyte concentration gradients are established, the reaction sites separate and localize (Fig. 1B). Finally, as the volume of corrosion product increases, the site of the cathodic reaction shifts to the periphery, leading to an increase in the local pH, and through several possible mechanisms, to a loss of adhesion of the polymer coating, thereby propagating coating failure (Fig. 1C). The proposed mechanisms for this cathodic disbonding invoke chemical attack of the coating[3] or metal oxide,[4] attack of a metal/polymer interphase,[5] or alteration of the surface energy of the metal.[6] Two extensive reviews of the mechanism are available.[7,8]

A standard test[10] is frequently used to evaluate the sensitivity of a coating/substrate system to cathodic disbonding. The coating is scratched through to the substrate and the panel is then exposed in a salt spray cabinet. We are using a modification of this procedure that moves the anodic reaction away from the disbonding site, allows precise control over the potential between the electrolyte and the substrate, and allows nondestructive monitoring of the propagation of the disbond. The scribed substrate is placed in a 0.5 M NaCl solution containing a platinum counterelectrode; the potential between a Ag/AgCl reference electrode and the substrate is maintained at a constant value using a potentiostat. A small region near one of the scribe marks is then monitored using the scanning acoustic microscope. We have found that once the potential is applied, the disbond front is marked by a series of microblisters that appear with high contrast in the acoustic micrographs as the disbond grows. In addition to being nondestructive, this method of monitoring the growth of the disbond permits a time and spatial resolution for an _in-situ_ examination that has not been possible using previous techniques. Although optical microscopy might be considered for monitoring the cathodic disbonding of polymer coatings, we believe that the sensitivity of acoustic microscopy to changes in the polymer and the polymer/substrate interface makes it superior for detection of features that mark the disbond front such as zones of polymer swelling, weak fluid boundaries and microblistering. Further, contrast changes in an optical micrograph depend on changes in the index of refraction of the polymer that are generally small in the visible portion of the spectrum. On the other hand, contrast changes in an acoustic micrograph depend on changes in the elastic properties of the polymer and

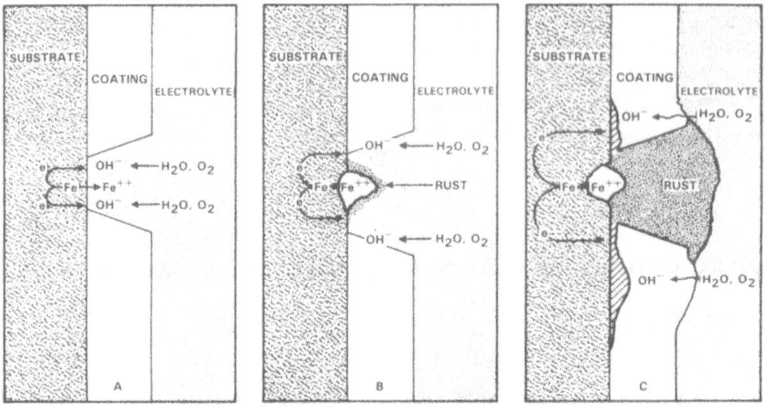

Fig. 1. The corrosion at a pinhole in an organic polymer coating is depicted for three phases of its development. In the initial stage (A), the iron is oxidized and the oxygen is reduced at proximate sites. As corrosion proceeds (B), a small rust deposit forms, oxygen and electrolyte gradients are formed, and the reaction sites separate. Finally the cathodic reaction is moved to the periphery (C) where there is an increase in the local pH. The mechanisms producing the subsequent disbonding of the polymer coating are the subject of active research. (After Dickie and Smith[9]).

interface. These are readily detectable since various types of surface acoustic waves, whose energy is concentrated in the polymer film and the interfacial region, are generated by the scanning acoustic microscope (SAM). Thus small changes in the elastic properties can produce large changes in the image contrast. An example of the difference seen in optical and acoustic micrographs is shown in Fig. 2. This paper describes the experimental technique and reports the preliminary results of the use of scanning acoustic microscopy to make virtually instantaneous in-situ measurements of the rate and morphological details of cathodic disbonding of a hydroxy-terminated polybutadiene film from a carbon steel substrate.

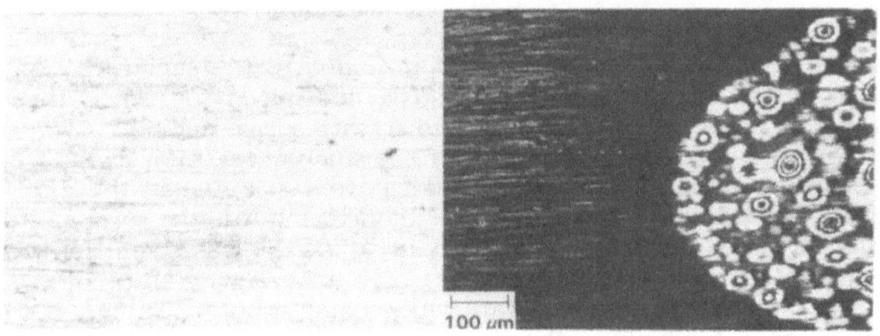

Fig. 2. Illustration of difference in observed contrast of a cathodically disbonded region as viewed in the optical brightfield micrograph on the left and the 1.0 GHz acoustic micrograph on the right.

EXPERIMENTAL

Apparatus

The SAM used for these measurements was the commercially available ELSAM made by E. Leitz Inc. This unit can obtain either an acoustic or optical micrograph of the same field of view. Most of the acoustic micrographs were obtained at

1.0 GHz, providing a resolution of about 1 μm. The magnification was typically the lowest possible (field-of-view of about 0.8 mm × 1.0 mm). During the experiments, acoustic micrographs were recorded at intervals of 2 to 15 mins which was compatible with the 35 sec interval required to record a micrograph.

An in-situ cell which contained the electrolyte and maintained a constant potential was designed to fit on the stage of the SAM. Figure 3 shows a schematic of this cell and the acoustic lens. A Pt ring counterelectrode, an Ag/AgCl reference electrode with a salt bridge of PVC tubing containing a ceramic frit, the coated specimen as the working electrode, and an insulating backplate comprised the in-situ electrochemical cell. The electrolyte used for the tests was 0.5 M NaCl equilibrated with the laboratory environment. The potential between the specimen and the reference electrode was maintained constant with a potentiostat. Care had to be taken to maintain the proper electrolyte level (0.3 cm) in the cell so as not to contact the anodized aluminum of the acoustic lens heater assembly. As an additional precaution, the potentiostat and all cell electrodes were electrically isolated from the acoustic microscope.

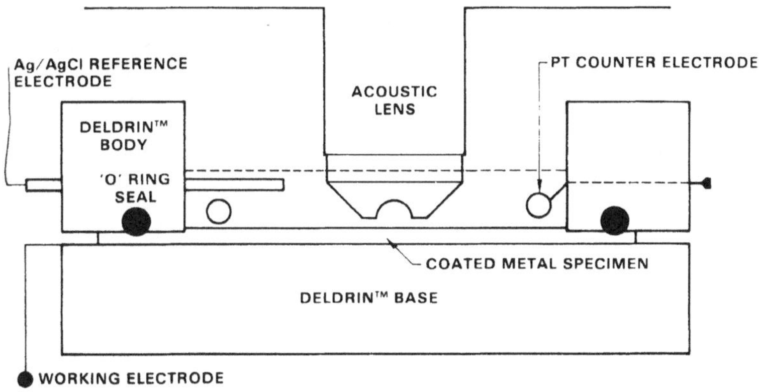

Fig. 3. Schematic of the in situ cell used to monitor the cathodic disbonding process.

Specimen Preparation

A solution of four parts of hydroxy-terminated polybutadiene (PB-OH) in three parts mineral spirits containing 0.1% by weight vs solids of the active ingredient Modiflow™ (Monsanto Co.) for leveling was spin coated onto polished and degreased 7.6 cm × 7.6 cm and 0.055 cm thick steel Q-panels™ (The Q-Panel Co.). The coating solution was passed through a 0.2 μm membrane filter before coating to remove any dust or gel particles. The steel specimens were polished dry with 600 grit paper, wiped with a towel moistened with acetone and degreased at room temperature in hexane before applying the coating. Wet-coated samples were air-dried for 12 h and oxidatively crosslinked in an oven reaching 205°C for 9 min. The resulting coatings had a nominal 4 μm thickness. Individual specimens were prepared for polarization in the in-situ cell, each at a given cathodic potential. A few specimens were given a finer polish to eliminate the background clutter in the acoustic micrographs. These specimens were polished with 1200 grit SiC followed by polishing in a kerosene medium using 15, 3 and 1 μm diamond paste successively. The resulting sample was then degreased and coated as the other, more roughly polished specimens.

Procedure for Data Acquisition

Using a SiC tool, two 2.5 cm long scribes intersecting at right angles were placed in the coating just before each test. For the 600 grit polished substrates, one line of the scribe was placed parallel and the other perpendicular to the 600 grit polishing marks. The specimen was then mounted in the in-situ cell and oriented so that the scribe marks were parallel to the scan directions of the SAM. The acoustic micrographs were typically taken adjacent to the scribe mark that was oriented parallel to the polishing marks of the 600 grit polished specimens. This meant that the disbond front being monitored had to propagate normal to the polishing marks. The speed of propagation of the disbond front was noticeably slower if it was moving normal to the polishing marks. This confirms the observation made by Leidheiser[7] that the tortuousity of the path for lateral diffusion between the substrate and the coating slows the rate of disbonding.

Shortly after the cell was filled with electrolyte, the focus of the acoustic lens was adjusted to provide the best contrast for the features in the disbonded region; the potential was applied, and an acoustic micrograph was recorded. This field of view was monitored continuously and acoustic micrographs were recorded as necessary to provide a complete record of the rate of disbonding. Depending on the specific experimental variables - potential, surface treatment of substrate, film thickness - the disbonding current and an acoustic micrograph were recorded at intervals ranging from 2 to 15 mins. The focal position of the lens was stable over the several hour period necessary to complete an experiment. Once the initial focus was determined, it was only necessary to change it to accommodate changes in the image caused by variations in the polymer properties. The propagation of the disbond was monitored until it had traveled a distance of about 0.5 mm from the scribe mark. The potential was then turned off and the monitoring was continued until the disbond stopped propagating or for about 1 hr, whichever occurred first. The electrolyte was removed from the cell, the surface of the polymer was dried, and Scotch tape was applied to the scribe marks to remove the disbonded film. An optical micrograph was then made of the same field of view that had been monitored acoustically showing the region where the film had been removed by the Scotch tape.

EXPERIMENTAL OBSERVATIONS

As an example of the data acquired with the SAM, consider the results obtained from a PB-OH coated steel specimen polarized at -850 mV vs Ag/AgCl. Three acoustic micrographs of the region being monitored are shown in Fig. 4 as they appeared at 1.5 min, 35.3 min, and 100.3 min after the potential was applied. The scribe mark appears at the bottom of these micrographs. The disbond front is marked by the advancing field of microblisters that appear in the micrographs. In addition there are several other features that can be seen in the acoustic micrographs including the polishing marks, the grain boundaries of the steel substrate (seen in the left micrograph), and various heterogeneities in the polymer film. Here the dark regions seen at the top of the left and middle micrograph are attributed to heterogeneities in the polymer film. These heterogeneities, which were observed in nearly all of the specimens, seem to have a subtle effect on the size and distribution of the microblisters that form in these regions as seen in the right micrograph. Note also that the background region surrounding the microblisters in the right micrograph has darkened indicating a general change in the polymer in that region, possibly due to swelling. The advance of the disbond front is measured by drawing a "best fit" straight line through the forward edges of the most advanced microblisters and measuring its average distance from the scribe mark. The fit of the line representing the disbond front is currently done by eye. This data is then plotted vs $\sqrt{t}$ as shown in Fig. 5. We have found that the data can be represented by an equation of the form:

$$d = 0 \qquad t <= t_0 \qquad (1A)$$

$$d = K_d\sqrt{t - t_0} \qquad t > t_0 \qquad (1B)$$

This behavior is consistent with the model proposed by Thornton[11] and the observations of Leidheiser and Wang.[7] Leidheiser and Wang show a linear dependence of the area of the circularly symmetric region with time (representing a $\sqrt{t}$ dependence of the radius). The experimentally determined dependence of the rate constant, K_d,

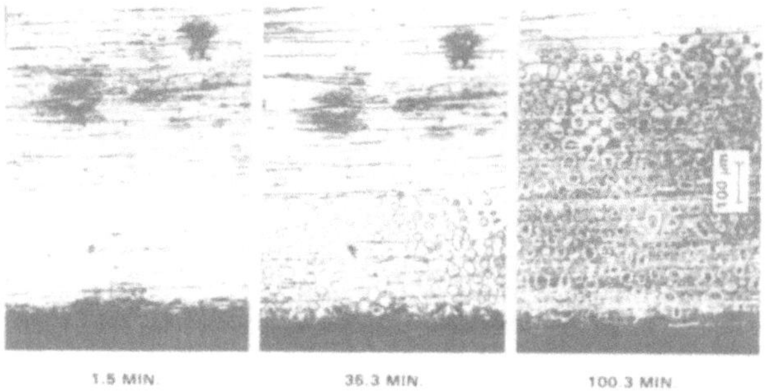

Fig. 4. Acoustic micrographs of a hydroxy-terminated polybutadiene coated steel specimen in 0.5 M NaCl polarized at -850 mV vs Ag/AgCl. The micrographs were made at the indicated times after the potential was applied.

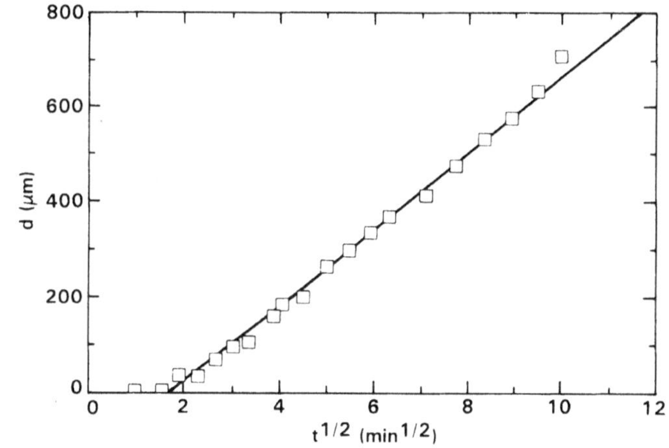

Fig. 5. The distance of the disbond, d, from the scribe mark as a function of the square root of time measured in (minutes)$^{1/2}$ for the specimen shown in Fig. 4.

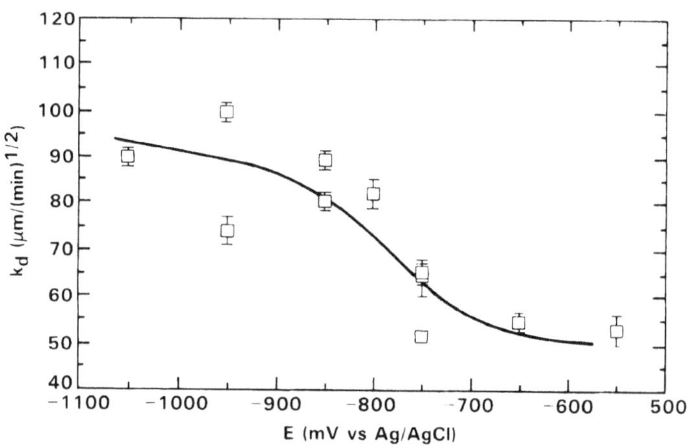

Fig. 6. Disbond rate constant, K_d, as a function of potential in mV vs Ag/AgCl.

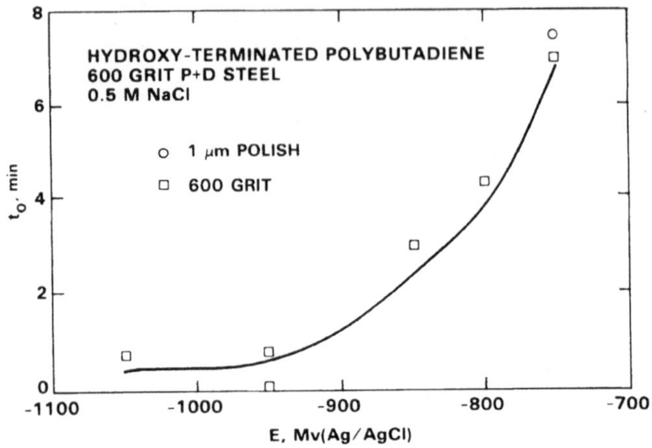

Fig. 7. Initiation time, t_0, as a function of potential in mV vs Ag/AgCl.

on the applied potential is shown in Fig. 6. It increases with decreasing (more cathodic) potential, displaying a rapid increase between -700 mV and -950 mV. The initiation time, t_0, decreases with decreasing potential, becoming constant for potentials more negative than -950 mV. (Fig. 7).

The precision with which the disbonding rate constant, K_d, can be measured using this technique is far greater than that available from other testing techniques. We believe that the scatter in the data points (Figs. 6 & 7) is caused by variations in the samples rather than in the technique for measuring the disbond rate. The acoustic micrographs of the polymer films frequently display many heterogeneities and small variations in thickness. The SAM is particularly sensitive to changes in the elastic properties of a layered system because of its ability to detect small changes in the velocity of surface waves that propagate in these systems. Kushibiki[12] has been able to measure surface wave velocities at lower frequencies with an accuracy of better than 0.5% using a derivative of the SAM known as a line focus microscope. In the micrographs made at 1.0 GHz, the minimum detectable change in the surface wave velocity is estimated to be about 10%.

The precision of the in-situ SAM measurements allows a sensitive comparison of the rates of cathodic disbonding between various specimen and surface treatments. The measured disbonding rates and initiation times for six different treatments are summarized in Table 1. The treatment of the surface with the organotitanate slows the disbonding rate by a factor of two compared to the polished and degreased surface.

Table 1. Effects of Different Sample and Surface Treatments on Cathodic Disbonding

Specimen ID	Specimen/Surface Treatment*	K_d ($\mu m/\sqrt{min}$)	t_o (min)
65	P+D	82±3	4
96	Polished < 1 μm	108±4	25
53	P+D/HNO_3	68±2	71
41	P+D/inh. HCl	52±2	26
45	P+D/LICA in polymer	52±1	2
46	P+D/LICA on surface	41±2	27

*Glossary of Terms

P+D	polished and degreased
HNO_3	steel oxidized with HNO_3
inh. HCl	steel cleaned with HCl containing inhibitor
LICA	Kenrich Organotitanate Coupling Agent #38

DISCUSSION

Microblisters

The appearance of the microblisters in the disbonded region was unexpected. Their presence has not been reported in the literature to our knowledge. It was our initial belief that they were the result of local swelling of the polymer and electro-osmotic influx of electrolyte that caused the surface of the polymer to be raised. For this explanation to be correct, microblisters would be optically detectable using a microscope with Nomarsky contrast or by examining the polymer surface with an interferometric microscope. Both of these techniques have been tried, but neither of them was able to detect any surface distortion of the polymer that could have been associated with the microblisters. In addition, the microblisters remain after dry-out and partial readhesion of the coating and after storage for 24 hours under vacuum.

Our current hypothesis is that an irreversible localized change in the elastic properties of the polymer occurs. Possibly the polymer is being hydrolyzed by the high pH solution that is present in the boundary layer of the disbonded region.[11] To evaluate this possibility and to ascertain the effect of concentrated sodium hydroxide on the polymer, we placed a drop of 1 M NaOH on the polymer surface for 1 min. The region below the drop turned a straw color. The region was wiped with a dampened Q-tip and the straw colored layer was removed leaving no optical evidence that any change in the polymer had occurred. However, an examination with the SAM clearly showed a contrast change in the region that had been exposed to the NaOH (Fig. 8). This simple experiment demonstrates that a high pH solution is capable of causing irreversible elastic changes in the polymer. Two additional observations made during the experiments support the alkaline hydrolysis hypothesis. The fringes seen in the microblisters mark changes in the acoustic path-length. For a given thickness of polymer, there is always the same maximum number of fringes for the larger microblisters. Thus once the hydrolysis has occurred over the full polymer thickness, there is no further vertical change in the microblister although lateral spreading and coalescence may continue. Further evidence is provided by Fig. 9 which shows a specimen that, by chance, contained a microblister-like feature in its as-fabricated condition. Presumably this feature was created during the polymer curing process. There was no growth of this microblister until the disbond front reached it. After the disbond front passed, this microblister grew at the same rate as the others. These three observations support the hypothesis that the high pH solution in the boundary layer of the disbonded region hydrolyzes the polymer. We do

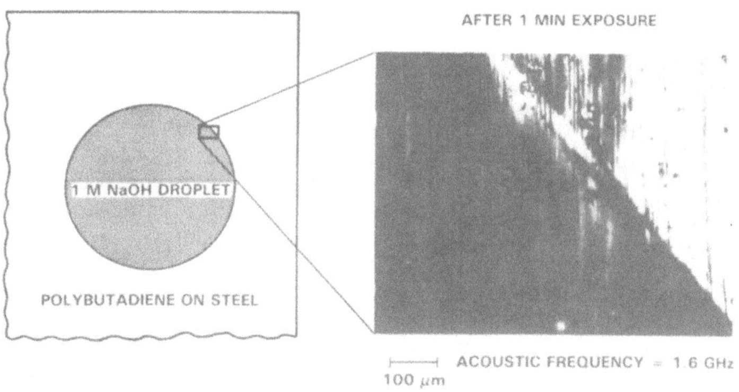

Fig. 8. Acoustic micrograph of specimen after exposure to 1 M NaOH for 1 min. The dark region on the left has been exposed. The light region on the right is unexposed.

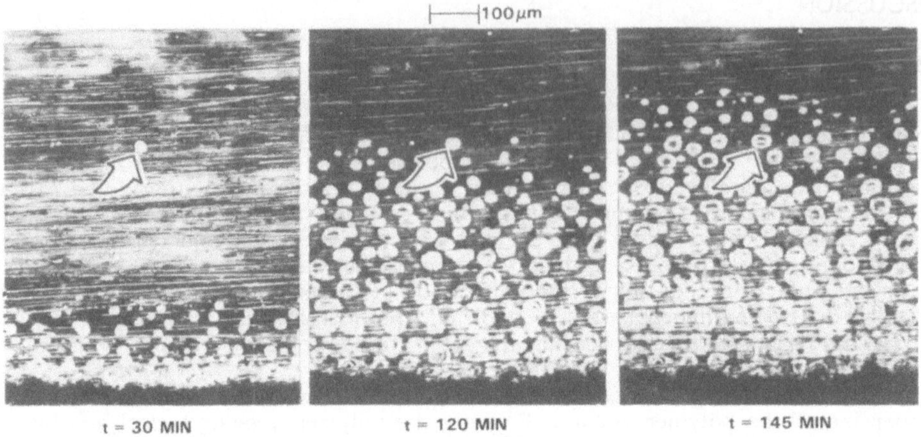

Fig. 9. Series of acoustic micrographs showing that a microblister (marked by arrow) does not grow until the disbond front reaches it. The scribe mark is at the bottom in these micrographs.

not yet have a quantitative explanation for the presence of microblisters rather than a spatially uniform hydrolysis of the polymer. Others have previously considered the heterogeneity of crosslinking in organic coatings to cause a variation in the modulus, swelling, and electrical properties.[13 14] Most likely a similar explanation can be made for heterogeneous rates of hydrolysis. Qualitatively it appears that there are local regions of the polymer that are more susceptible to hydrolysis than others. Once the hydrolysis has started, the process is accelerated because the diffusion of the reactants through the polymer film is greater at that point, particularly for these thin 4-μm films.

Acoustic Response

We have only approximate values for the elastic parameters of the polybutadiene. However, when these values are used to calculate the response of the microscope to the layered structure as a function of its spacing from the specimen (V(z) curve), we obtain a curve with a maximum response that occurs at a z spacing other than the focal length of the lens (Fig. 10). This strongly suggests energy trapping within the polymer such as that obtained with a Sezawa wave.[15] The sensitivity of the characteristics of this wave to small changes in the ratio of the shear wave velocity in the polymer to that in the substrate probably explains the strong contrast changes in the acoustic micrographs that arise from the microblisters. It may also explain the significant variations in the response that are observed among nominally identical specimens.

Cathodic Disbonding

Figure 11 summarizes our current hypothesis concerning how the disbonding process occurs. The cathodic reaction is driven by the potential between the electrolyte and the steel substrate. Castle and Watts[5] have previously shown that an iron oxide/polymer interphase exists between the bulk polymer and the metal oxide substrate for this particular coating system. The diffusion of the reactants into the interphase region at the leading edge of the disbond is consistent with the observation of a $\sqrt{t}$ time dependence. These reactants H_2O, O_2, and Na^+ diffuse in from the scribe mark and also through the polymer. A set of rate equations for these reactants has been formulated by Thornton et al.[11] They solved the equations under a simplifying set of assumptions showing that the rate of advance of the disbond was proportional to $\sqrt{t}$. The electrochemical reactions reduce oxygen to form OH^- whose charge is balanced by the influx of hydrated Na^+. Charge transport by diffusion of the hydrated cation to the interphase due to osmosis provides the mechanical pressure to rupture the interfacial bonds and drive the disbond along the interface.

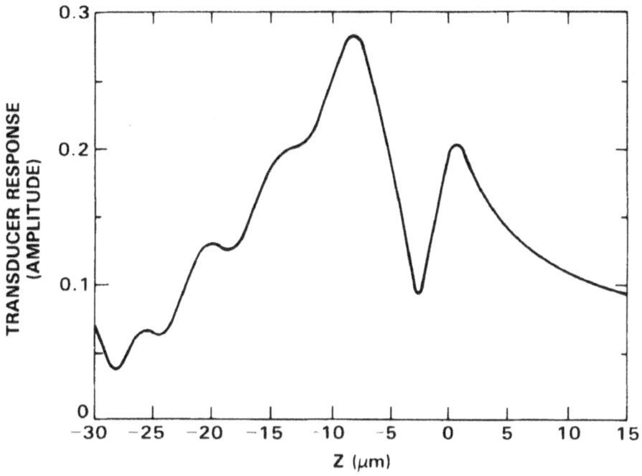

Fig. 10. Response of SAM to polybutadiene/ steel model as a function of the z spacing of the lens (V(z) curve). The focal point of the lens corresponds to Z=0.

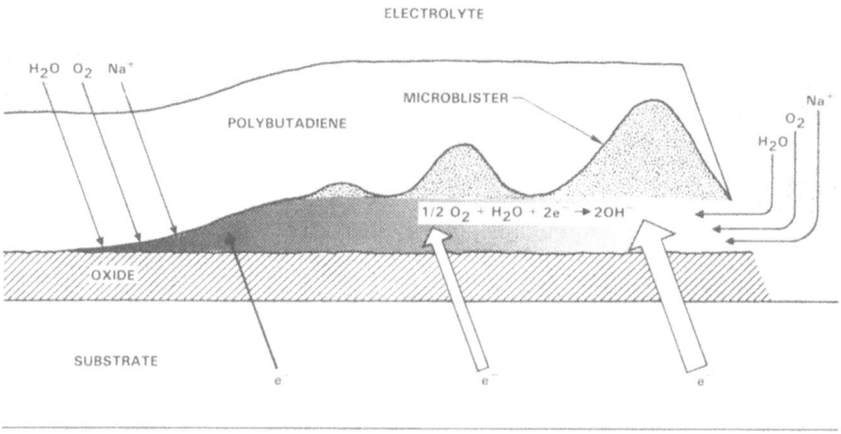

Fig. 11. Model for cathodic disbonding process based on experiments with SAM.

The presence of high pH regions at the coating/metal interface may ultimately lead to the dissolution of the metal oxide, particularly under reducing conditions as shown by Ritter,[4] and may produce alkaline attack on the polar groups of the polymer as observed here. However, it is unclear whether these are necessary conditions for cathodic disbonding. Rather it is suspected that they result from attack of the freshly exposed interfaces produced after rupture of the interfacial bonds by the mechanical pressure induced by the electroosmotic swelling of the coating at the interface.

SUMMARY

We have described a new nondestructive, precise, and rapid method for the in-situ monitoring of cathodic disbonding. It provides an excellent technique for comparing the rate of disbonding of organic coating/substrate systems that have been subjected to a variety of treatments to retard disbonding. The acoustic micrographs clearly show differences in the polymer for nominally identical coatings. Future work will be directed toward the study of cathodic disbonding kinetics for more anodic potentials as well as for different combinations of polymers and surface treatments.

ACKNOWLEDGEMENT

This work was supported in part by the Office of Naval Research under Contract N 00014-87-C-0075.

REFERENCES

1. M.W. Kendig and H. Leidheiser, Jr., "The Electrical Properties Of Protective Polymer Coatings As Related To Corrosion Of The Substrate," J. Electrochem. Soc., 123:982-989 (1976).
2. F. Mansfeld and M.W. Kendig, "Electrochemical Tests For Protective Coatings," ASTM Special Testing Publication, No. 866, pp. 122-142, American Society For Testing And Materials, Philadelphia, (1985).
3. J.S. Hammond, J.W. Holubka, J.G. DeVries, and R.A. Dickie, "The Application Of X-Ray Photo-Electron Spectroscopy To A Study Of Interfacial Composition In Corrosion-Induced Paint De-Adhesion," Corr. Sci., 21:239-253 (1981).
4. J.J. Ritter, "Ellipsometric Studies On The Cathodic Delamination Of Organic Coatings On Iron And Steel," J. Coating Technology,. 54:51-57 (1982).
5. J.E. Castle and J.F. Watts, "Interface Chemistry Of Stoved Organic Coatings," I&EC Product Res. & Dev., 24:361-369 (1985).
6. E.L. Koehler, "The Mechanism Of Cathodic Disbondment Of Protective Organic Coatings - Aqueous Displacement At Elevated pH," Corrosion, 40:5-8 (1984).
7. H. Leidheiser, Jr., W. Wang, and L. Igetoft, "The Mechanism For The Cathodic Delamination Of Organic Coatings From A Metal Surface," Prog. in Org. Coatings 11:19-40 (1983).
8. H. Leidheiser, Jr., "Cathodic Delamination Of Polybutadiene From Steel - A Review," J. Adhesion Sci. Tech., 1:79-98 (1987).
9. R.A. Dickie and A.G. Smith, "How Paint Arrests Rust,"Chemtech, pp. 31-35, (1980).
10. "ASTM Standard G8-72,"Annual Book of ASTM Standards, American Society For Testing And Materials, Philadelphia, 27:869 (1979).
11. J.S. Thornton, R.E. Montgomery, and J.F. Cartier, "Failure Rate Model For Cathodic Delamination Of Protective Coatings," NRL Memorandum Report 5584 (1985). Available from DTIC as report AD-A157 483.
12. J. Kushibiki and N. Chubachi, "Material Characterization by Line-Focus-Beam Acoustic Microscope," IEEE Trans. Sonics Ultrason., SU-32:189-212 (1985).
13. E.M. Kinsella and J.E.O. Mayne, "Ionic Conduction in Polymer Films," Br. Polym. J. 1:173 (1969).
14. J.E.O. Mayne and J.D. Scantlebury, "Ionic Conduction in Polymer Films - II. Inhomogeneous Structure of Varnish Films," Br. Polym. J., 2:240-243 (1970).
15. B.A. Auld, Acoustic Fields and Waves in Solids, John Wiley & Sons, N.Y., 1973.

APPLICATION OF ACOUSTIC MICROSCOPY IN DENTAL RESEARCH

S. Kasahara, K. Yoshida, J. Kushibiki*, and N. Chubachi*

First Department of Prosthetic Dentistry, School of Dentistry and *Department of Electrical Engineering, Faculty of Engineering, Tohoku University, Sendai 980, Japan

INTRODUCTION

In the dental research fields, there are many fundamental problems to be investigated on the elasticity. For example, studies on the elastic properties of dental materials for dental treatments and further the changes of elastic properties of dental hard tissues associated with the progress of caries should be most interesting subjects in clinical dentistry. Recently, the technology of acoustic microscopy has been demonstrated in these fields [1,2]. This system has been expected to be a new and useful research tool which can measure directly the elastic properties of dental materials.

In this paper, primary caries lesion in human enamel and dentin are investigated with a reflection-type point-focus-beam acoustic microscope at a frequency of 400MHz. The usefulness of this system in dental research fields are discussed comparing with images obtained with transmission type optical microscopy as well as microradiography.

MATERIALS AND METHOD

An extracted human incisor with a primary caries in enamel was taken for a typical carious sample for this study. The sample configuration is shown in Fig. 1. The procedure of sample preparation is as follows: The surface of tooth to be observed is supported with a dental composite resin

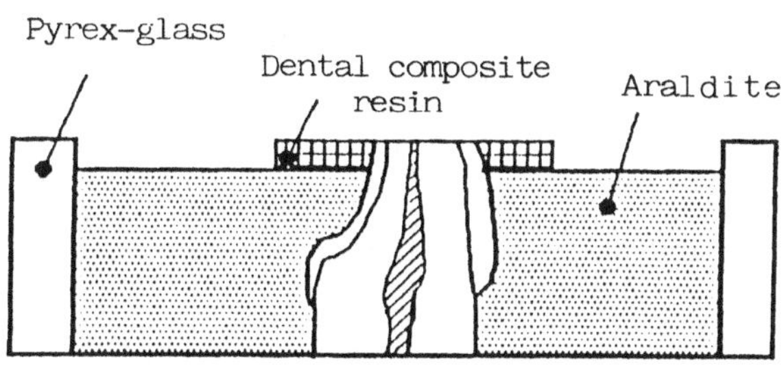

Fig. 1. Sample configuration of human incisor.

(Photo Clearfil, KURARAY, Japan) which is actually used to repair human decayed teeth. This material is very useful for obtaining a smooth surface with sharp edges, because it has almost the same hardness as dental hard tissues. The sample with the composite resin is embedded in a glass case with epoxy resin (Araldite: CY230 resin with HY951 hardener). The surface of specimen to be imaged is polished with emery papers and lapping films. A point-focus-beam acoustic microscope operating in a reflection type with a spatial resolution of 3μm at 400MHz was employed to image. After the investigation of the acoustic microscope, the sample was sectioned about 200μm thick and parallel for imaging with a transmission type optical microscope and microradiograph.

RESULTS

Primary Caries in Enamel

A schematic representation of primary caries in enamel is shown in Fig. 2(a). Caries in enamel originate from the subsurface of enamel and usually spread along the enamel rods and the striae of Retzius. An optical micrograph of primary caries in enamel is shown in Fig. 2(b). In this figure, carious enamel region is observed as a darker portion than sound enamel region. This is because the optical light is scattered in carious region due to changes of histological structure by demineralization and destruction in carious region. A microradiograph for the same area is shown in Fig. 2(c). In this figure, carious enamel region is observed as a bright portion in which mineral contents are lessen by caries to increase the radiolucency of tooth. An acoustic image of the same area is shown in Fig. 2(d). Comparing these three images, more detailed structures are observed in the acoustic image, especially in the carious region, than in the images of optical micrography and microradiography. The conventional images obtained with transmission types include whole informations of the structures consisting throughout the thickness of the section, so that it is difficult to image in detail within a shallow layer in subsurface of specimen.

Dentin Adjacent to Pulp Cavity

A schematic representation of dentin adjacent to the dental pulp cavity is shown Fig. 3(a). In this region, dentin has a complicated histological structure. An optical micrograph of dentin is shown in Fig. 3(b). The bright portion near the pulp cavity is called "transparent dentin", in which the optical translucency of dentin increased because of calcification in dentinal tubules. A microradiograph of the same area is shown in Fig. 3(c). The radiolucency is not uniform in this area because the degree of mineralization are uneven in dentin. In images obtained with both optical microscopy and microradiography, the concentric patterns due to the periodical changes of optical translucency and radiolucency in dentin are observed. It is considered that these periodical changes are related to the changes introduced in mineralization and in structures during the growth of dentin. However, as optical and microradiographic images taken in the transmission type are also affected by the thickness of specimen, it is difficult to explain the relationship between the changes in the fine structures or in the degree of mineralization and the concentric patterns in these images. An acoustic micrograph of the same area is shown in Fig. 3(d). This is the first image of dentin which was observed with an acoustic microscope. It is a very interesting image that indicates the changes of elastic properties in structure under the subsurface of dentin, which can not be imaged in the above optical and microradiographic images. To understand this acoustic image, further investigations with acoustic microscopy should be carried out comparing with the investigations on fine

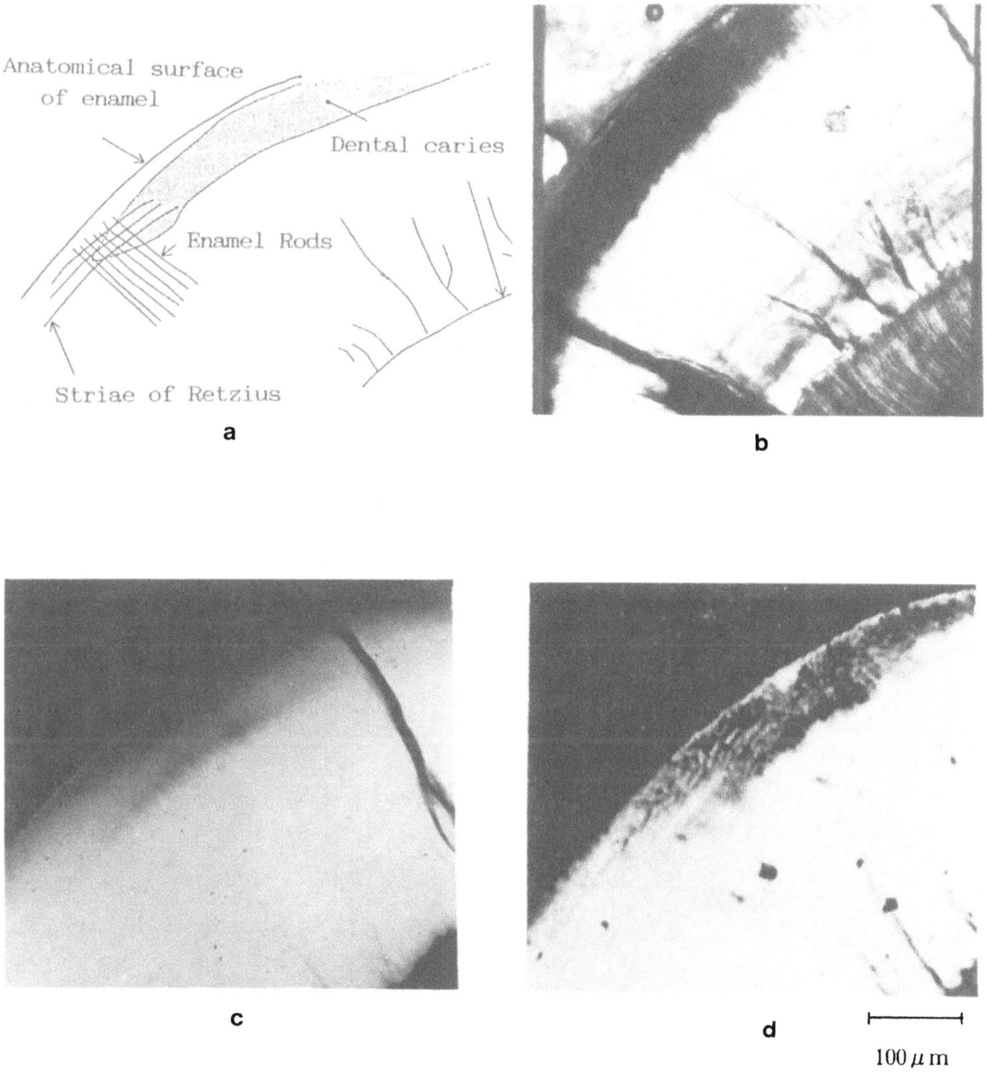

Fig. 2. Images of human enamel.
(a) diagrammatic presentation. (b) optical micrography.
(c) microradiography. (d) acoustic image.

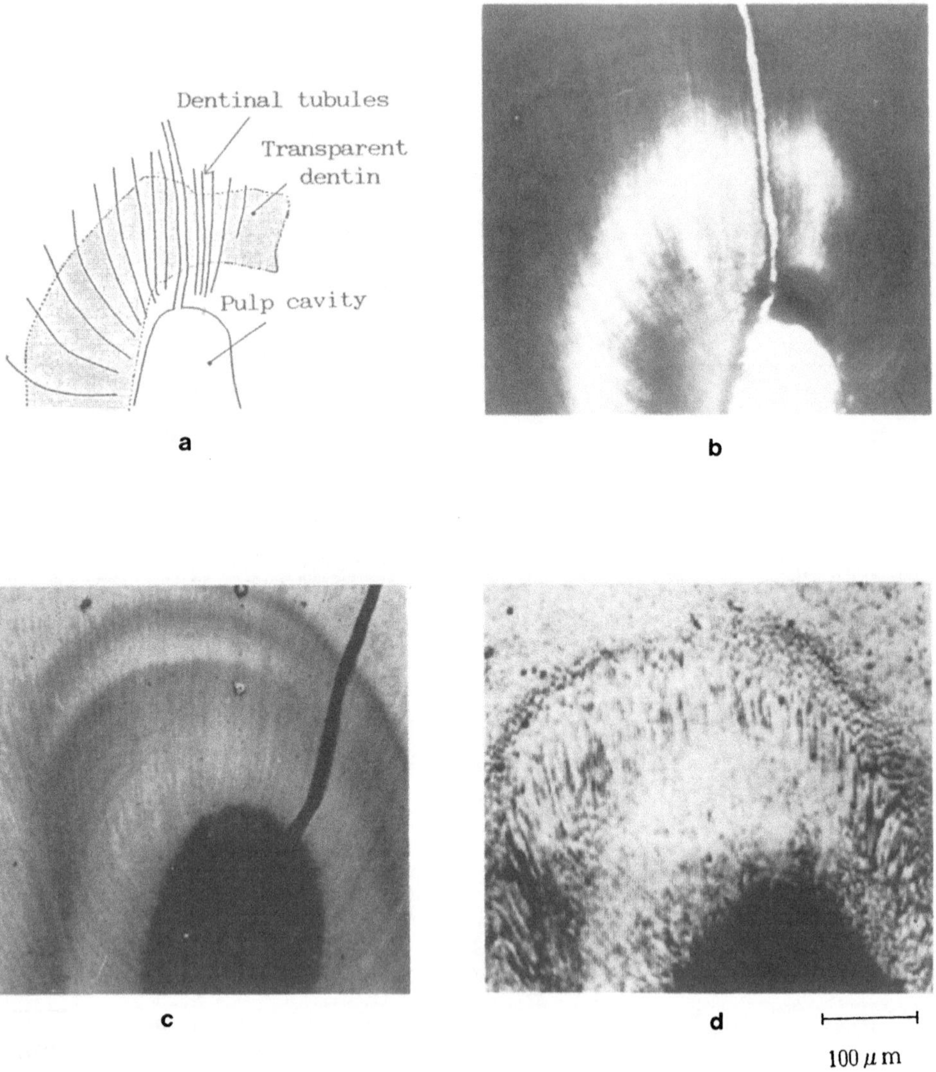

Fig. 3. Images of dentin adjacent to pulp cavity.
(a) diagrammatic presentation. (b) optical micrography.
(c) microradiography. (d) acoustic image.

structures with electron microscopy or on mineral contents with x-ray micro-analyzer.

CONCLUDING REMARKS

In this paper, acoustic images of an extracted human incisor are observed with a point-focus-beam acoustic microscope at 400MHz. These images on enamel and dentin were compared with conventional images obtained with a transmission type optical microscope and microradiograph. In imaging with an acoustic microscope, detailed structures in enamel and drastic changes in elastic properties in dentin are observed. Especially, a acoustic image associated with dentin is the first image obtained with acoustic microscope. New informations on elastic properties in teeth are expected by analyzing this image, because the images like this had not been provided with conventional method, namely optical microscopy and microradiography. The studies on elastic properties of tooth are important and interesting subjects in clinical dental research fields. There were some researches on the changes in hardness of tooth using microhardness testing [3,4,5]. The hardness of tooth is varied in carious region because of demineralization and destruction of tooth substances. However, as the hardness values with this measurement method are dependent upon both the structures of materials and measurement conditions [6], the relationship between the degree of caries progression and the changes in hardness is remained unsolved. The hardness values are related to elastic properties closely, so that the direct measurements of elastic properties with acoustic microscopy is expected to resolve these clinical problems.

In basic dentistry, namely histology and biochemistry, a study on mineral content is an important subject, because the elastic properties are associated with mineral contents of teeth [7]. Acoustic microscopy can measure the elastic properties directly, so that it is expected to be a useful tool not only in basic but also in clinical dentistry.

ACKNOWLEDGMENTS

The authors are very grateful to T. Sannomiya, K. L. Ha and H. Kato, Department of Electrical Engineering, Tohoku University, for their helpful discussions and technical assistance on experiments, to Prof. M. Kagayama and Prof. K. Kindaichi for their useful discussions on histology and M. Eguchi for his cooperation in measurements with microscope and microradiograph, School of Dentistry, Tohoku University, and to M. Ishibashi, Dental Hospital, Tohoku University, for his cooperation in preparation of specimen. This work was supported in part by Research Grant-in-Aids from Japan Ministry of Education, Science & Culture, and by Takeda Science & Technology Grants.

REFERENCES

1. J. Kushibiki, K. L. Ha, H. Kato, N. Chubachi and F. Dunn, Application of Acoustic Microscopy to Dental Materials Characterization, Proc. IEEE Ultrason. Symp., 837 (1987).
2. S. D. Peck and G. A. D. Briggs, The caries lesion under the scanning acoustic microscope, Adv Dent Res 1:50 (1987).
3. K. Okuse, Relationship between hardness, discoloration and organismal invasion in carious dentin, The Journal of the Japan Stomatological Society 31:187 (1964) (in Japanese).
4. R. G. Craig and F. A. Peyton, The microhardness of enamel and dentin, J Dent Res 37:661 (1958).

5. R. G. Craig and P. E. Gehring and F. A. Peyton, Relation of structure to the microhardness of human dentin, J Dent Res 38:624 (1959).
6. F. Nishimura, K. Okazaki, Y. Kono, K. Komori and S. Nomoto, Compressive behavior and micro Vickers hardness of human enamel and dentin, J J Dent Mat 5:449 (1986) (in Japanese).
7. C. L. Davidson, J. Arends and I. Hoekstra, Density changes in enamel after decalcification, J. Biomech 9:81 (1976).

LINE FOCUS ACOUSTIC LENS WITH WEDGE TRANSDUCER SOURCE

J. David N. Cheeke, J.-O. Fossum*, and N. Rene

Department of Physics, University of Sherbrooke
Sherbrooke, Quebec, Canada J1K 2R1

ABSTRACT

Results are reported for a new concept in line focus acoustic microscopy. We use a wedge shaped lithium niobate transducer as the source for a standard cylindrical lens. The resonant frequency varies with position down the axis so that the frequency can be used to spatially vary the focal point position along the focal line.

Several experimental results are given confirming the performance of this new lens, particularly regarding the spacial resolution and the process of selective Rayleigh wave generation in the specimen.

INTRODUCTION

Scanning acoustic microscopy (SAM) as developed by CF Quate and co-workers (1) relies on the use of mechanical scanning for the obtaining of an acoustic image. The ultrasonic beam is focussed to a diffraction limited point on the specimen surface which is then raster scanned over the field of view. This system has several advantages including simplicity and the retention of diffraction limited resolution, which would be difficult to achieve in more complex systems. However it is inherently slow, with the exception of possible developements in very high speed mechanical scanning (2). In most systems a typical imaging time is of the order of a minute. For many applications, particularly on line non destructive evaluation,this is too long and for such work it would be highly desirable to have a real time scanning system, even at the expense of loss fo some of spatial resolution. The present paper describes recent developements in our laboratory in this direction.

* Centre for Materials Science MIT Cambridge Mass. USA.

The present work is based on the use of a line focus beam, a technique which is used extensively in ultrasonic C scan imagery and in acoustic microscopy (3). A Quate type lens is used but the cavity is now a cylinder, typically a few mm long. Although one cannot produce an acoustic image in such a system it is possible to obtain quantitative information on the Rayleigh wave velocity and attenuation by the so-called V(z) phenomenon (3) and this as a function of direction by a rotation of the acoustic lens. This system is useful for characterising homogeneous materials and defects with directional properties but the loss of spatial resolution is a serious limitation.

In the wedge transducer assembly we replace the constant thickness ultrasonic transducer by a wedge transducer with constant transversal thickness and the long axis parallel to the cylindrical lens axis. In typical experiments so far we have operated with a 20 to 30% bandwidth. If we conceptually consider a local element of the transducer as an effective resonator with resonant frequency inversely proportional to the thickness then there is a continuous variation of the resonant frequency along the length of the transducer. We can thus vary the position of the focal spot along the cylindrical lens axis by a simple variation of the frequency over the transducer bandwidth. Evidently this is a highly oversimplified description. In practice such effects as the consequences of finite wedge angle, diffraction and substrate broadening of the intrinsic transducer resonance must play a very important role. In fact the aim of the preliminary work shown here is to determine the real device potential of this lens assembly.

EXPERIMENTAL RESULTS; SPATIAL RESOLUTION

The wedges were 2 x 6 mm^2 plates of 36° Y cut $LiN6O_3$ supplied by Valpey Fisher corporation. For the first set of experiments to be described (figs 1-3) the fundamental bandwidth was approximately 20-30 MHZ and the third harmonic 65-85 MHZ as seen in fig. 1. The general transducer response was considerably better for the third harmonic so this was used for the results given here. The fact that the fifth harmonic was also clearly identified indicates that a good compromise between resonant frequency and bandwidth was obtained far this transducer. The latter was bonded with vacuum grease to an aluminium block which had a 1 mm diameter cylindrical lens ground and polished in the opposing face. The lens was about 4 mm long. Water was the coupling medium and the experiments were done with a Matec 6000 pulse echo system.

The basic experiment is essentially a resolution test of the lens on the edge of a glass slide as shown in fig. (2). We monitor the height of the reflected echo as a

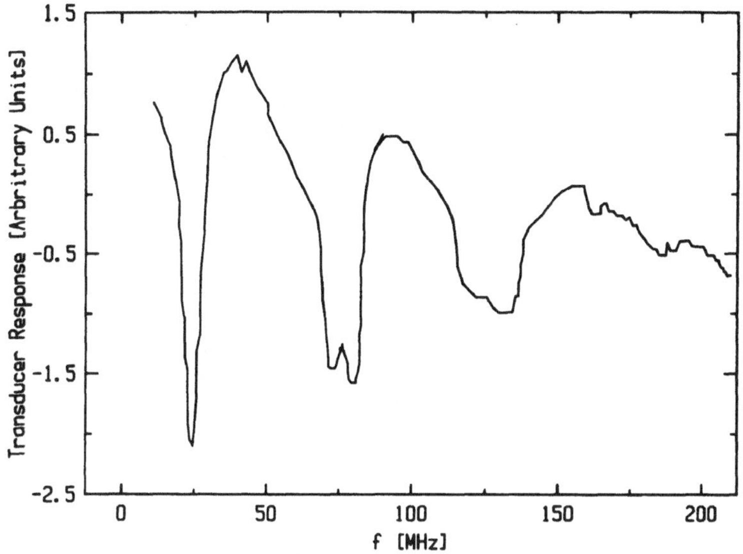

Fig. 1. Frequency response of wedge transducer with fundamental bandwidth approximately 20-30 MHZ.

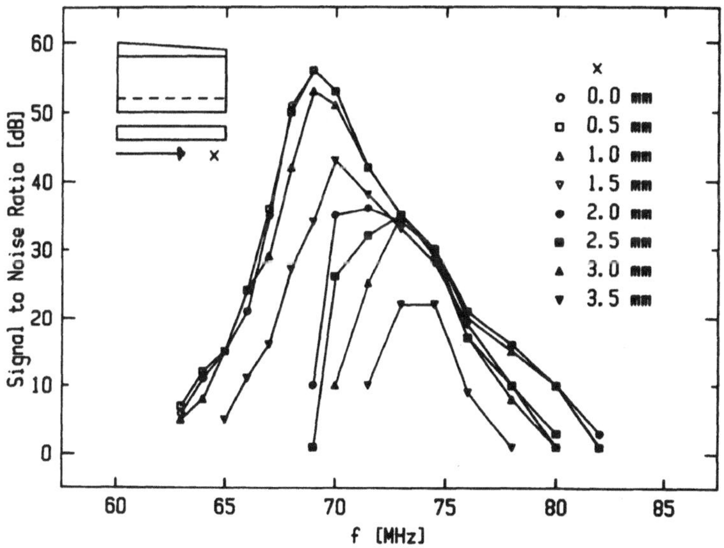

Fig. 2. Reflection signal for wedge transducer lens assembly from a glass plate as a function of plate displacement. The geometrical configuration is shown in the inset.

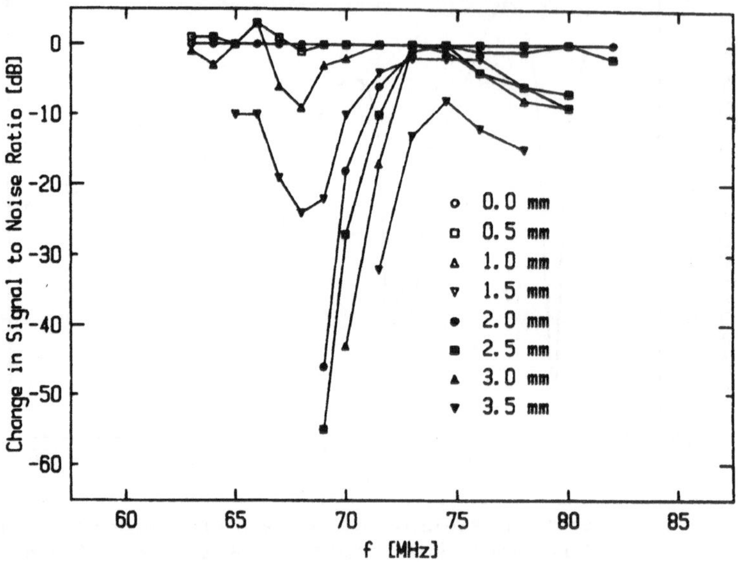

Fig. 3. The data of fig. 2 normalised to the results for x = 0.

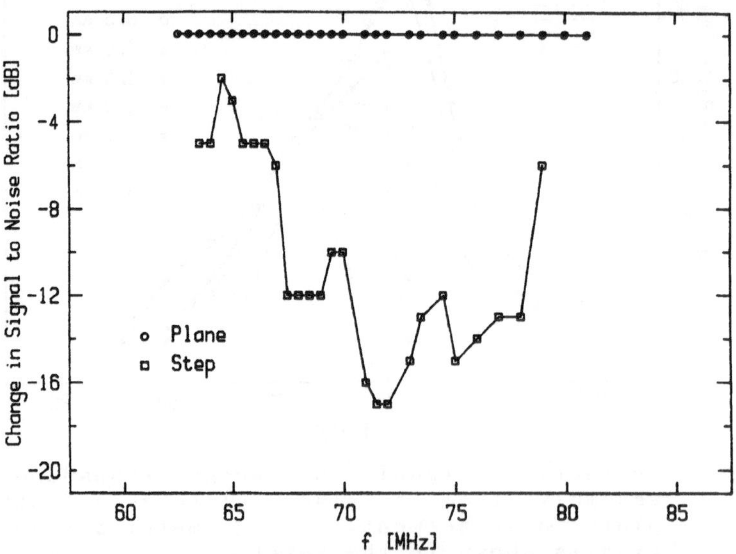

Fig. 4. Normalised data for reflection from two superposed 0.25 mm glass slides forming a step.

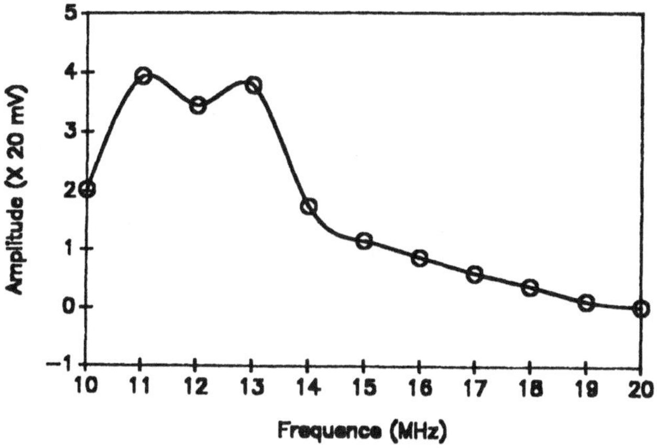

Fig. 5. Reflection signal for wedge lens assembly as a function of frequency for aluminium reflector at the focal plane.

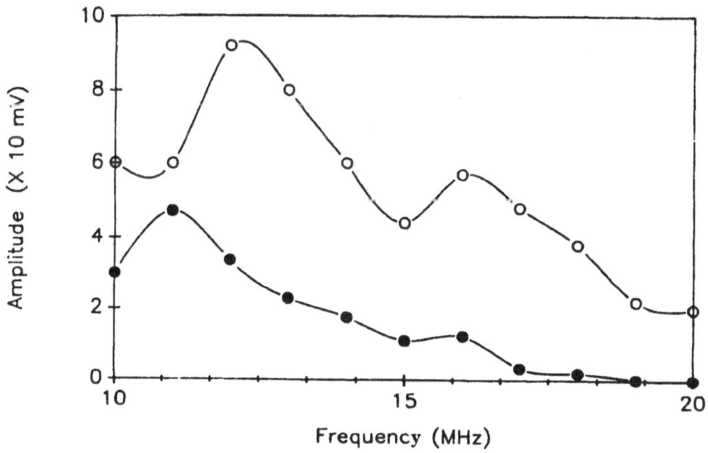

Fig. 6. Results for the assembly of fig. 5 for the deflector in two unfocussed positions: o 0.5 mm • 1 mm.

function of frequency for a fixed position of the glass slide. The latter is then displaced along the lens axis by a small amount and the experiment repeated. For the set up shown in fig. 2 we expect to progressively lose the low frequency content of the signal as the plate is displaced to the right and this is definitely seen to be the case for $x > 1$ mm. The result can be seen more clearly by normalising the data to that at $x = 0$; the latter corresponds to the overall transducer-lens response if we neglect edge effects. The result is shown in fig. 3 and it is seen that the data suggests for this lens a usable field of view of about 2 mm and a resolution in the range 100-500 microns. Similar behavior was seen for displacement of the plate in the opposite direction.

An extension of this experiment was made by repeating it for the case of two glass slides superposed to form a step, with the lens and target in a fined position and varying the frequency. The normalised data is shown in fig. (4) where the minimum in the curve corresponds to the known position of the step. The structure at $\pm$ 5 MHZ of the step is thought to be due to beam spreading.

EXPERIMENTAL RESULTS: RAYLEIGH WAVE GENERATION

A second experiment concerns selective Rayleigh wave generation. Two identical wedge transducers are arranged symmetrically as sources for a cylindrical lens with their long axes perpendicular to the cylinder axis. We excite one transducer with an RF tone burst and receive with the other. By varying the frequency we excite different sections over the length of the transducer which should lead to a selective variation of the incidence angle from zero out to the maximum angle possible for the lens, here chosen to be considerably greater than the Rayleigh angle for the reflector water combination.

Figure 5 shows the reflected signal as a function of frequency for an aluminum reflector at the focal position. This curve basically gives a measure of the product of the lens angular response and the transducer frequency response. The signal is largest far roughly normal incidence as expected and then falls off monotonically. The interesting result is shown in fig. 6 for the reflector in two defocussed positions. We clearly see an extra peak near 16 MHZ, which is shown in greater detail in figure 7. For this particular transducer and the aluminium reflector in water this is exactly the frequency expected for Rayleigh wave generation. While further experiments are needed this first result is very encouraging.

POTENTIAL APPLICATIONS

These preliminary results are presently being continued with the immediate objective of obtaining an acoustic image with electronic scanning in one direction. For many applications this is quite sufficient e.g. an assembly line where movement in one direction is already pre-

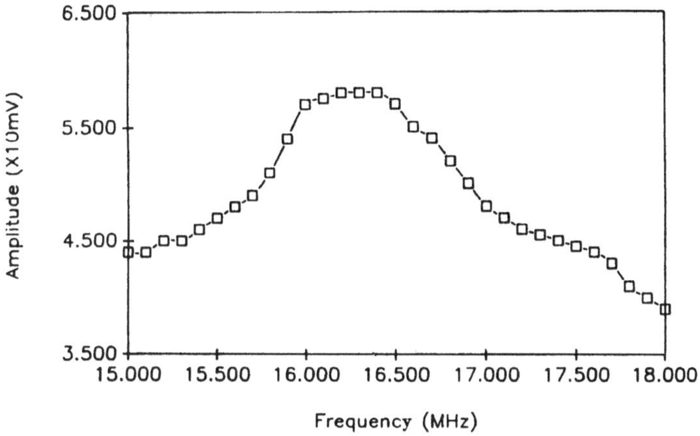

Fig. 7 Detailed results for the peak of the 0.5 mm defocus data.

sent for the processing of the product. When scanning in only one direction is required essentially real time imaging should be easily accessible. Further work is also underway to see if it is possible to obtain directional variations of Rayleigh wave velocity without mechanical scanning.

At this point it is useful to make contact with other work on wedge transducers. There has been a certain interest in the subject from the point of view of obtaining very wide band operation (and hence good time domain response) without the axe of lossy backing layers. Alphonse (4) has approached the problem using a wedge shoped face plate in front of a conventional PZT transducer. He obtained a 71% bandwidth with much lower insertion loss than is typically obtained with backed transducers. An interesting study has been carried out by Tomikawa et Al (5) on tapered piezoelectric ceramic transducers for both longitudinal and shear wave generation. Particular attention was paid to measuring the radiation pattern, which, as expected was displaced from the normal to the front face. Using a double wedge configuration it was found that the direction of radiation could be controlled electronically by driving either one or the other of the tapered sections. Finally a theoretical treatment of wedge transducers using a staircase model has recently been presented (6). Complete convergence of the modelled results was obtained for N 500. This approach is identical to that adopted in our qualitative interpretation given in the present work and so it would appear to be suitable for qualitative modelling of the wedge line focus lens assembly at a later date.

CONCLUSION

In conclusion the present work is compatible with previous studies of wedge transducers as regards frequency and time domaine response. In addition these preliminary results suggest interesting perspectives for real time imaging and direct determination of Rayleigh wave velosity variations.

ACKNOWLEDGEMENTS

One of the authors (JDNC) wishes to thank M. Poirier and B. Chick for stimulating discussions.

This work was supported by the GRSD (Groupe de Recherche sur les Semiconducteurs et les Dielectriques), the CRPS (Centre de Recherche en Physique du Solide) and the National Sciences and Engineering Research Council of Canada.

REFERENCES

1. R. A. Lemens and C.F. Quate "Acoustic microscopy" in Physical Acoustics (ed. W.P. Mason and R.N. Thurston) 14, 1-92, Academic Press, London (1979).

2. C.F. Quate "Acoustic Microscopy: Recollections", IEEE Trans. SU-32, 132-135 (1985).

3. J. Kushibiki and N. Chubachi "Material characterisation by line focus beam acoustic microscope" IEEE Trans. SU-32, 189-212 (1985).

4. G.A. Alphonse "The wedged transducer - a transducer designed for broad band characteristics" ultrasonic imaging, 1, 76-88, (1979).

5. Y. Tomikawa, H. Yamada and M. Onoe "Wide band ultrasonic transducer using tapered piezoelectric ceramics for non-destructive inspection", Jpn Jour App. Phys. 23,supp. 23-1, 113-115 (1984).

6. P.G. Barthé and P.J. Benkeser "A staircase model of tapered piezoelectric transducers" IEEE Ultrasonics Symposium, Denver, 697-700 (1987).

IMAGE OF ION BEAM EXCITED ACOUSTIC MICROSCOPE OF THE TEETH

Hiroto Tateno, Youichirou Iwashita*, Kazunori Kawano*
and Takenori Noikura*

Department of Physics, Faculty of Science
Kagoshima University
Kagoshima 890, Japan
*Department of Dental Radiology
Kagoshima University Dental School
Kagoshima 890, Japan

ABSTRACT

The ion beam excited acoustic microscope (IAM) is a new microscope which provides visual representations of the elastic waves caused by localized chopping ion energy. Topographs and images of ultrasonic attenuation and local elastic variation were obtained by measuring frequency and amplitude at a composite resonant point between transducer and sample. Acoustic images of the teeth were measured by IAM. The elasticity of the hydroxyapatite crystal of the teeth was examined.

INTRODUCTION

Photoacoustic (PA) signals have been measured by various exciting sources such as laser or electron beams. However, ultrasonic signals by ion excitation have not yet been measured. This paper describes the system of ion beam excited acoustic microscope (IAM) and IAM images of human teeth.

IAM SYSTEM

Figure 1 shows a block diagram of the IAM system developed in the authors' laboratory[1]. The instrument was constructed by adding a beam blanking apparatus, a sample holder containing a PZT resonator and an extractor electrode to an ion microanalyzer (IMA-2AZ) produced by Hitachi Corp. To measure large samples like teeth, the sample holder was able to move on X and Y axes by pulsemotors with an accuracy of 0.416μm/pulse, and the ion beam was excited at 25% duty by applying a 600V pulse to a deflector electrode. All systems were controlled by microcomputer system. The ultrasonic waves in the sample were detected by the PZT resonator, and the phase difference ϕ between acoustic amplitude Va and beam blanking pulse was obtained by a lock-in amplifier (NF LI-575). A special resonance point tracking circuit was developed to maintain the phase angle constantly, because resonator sensitivity was higher at the resonance point due to pumping.

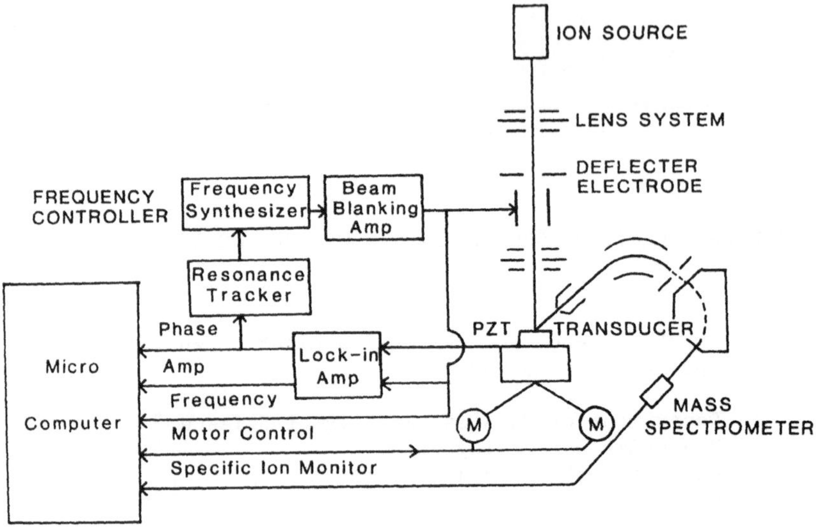

Fig. 1. Block diagram of IAM.

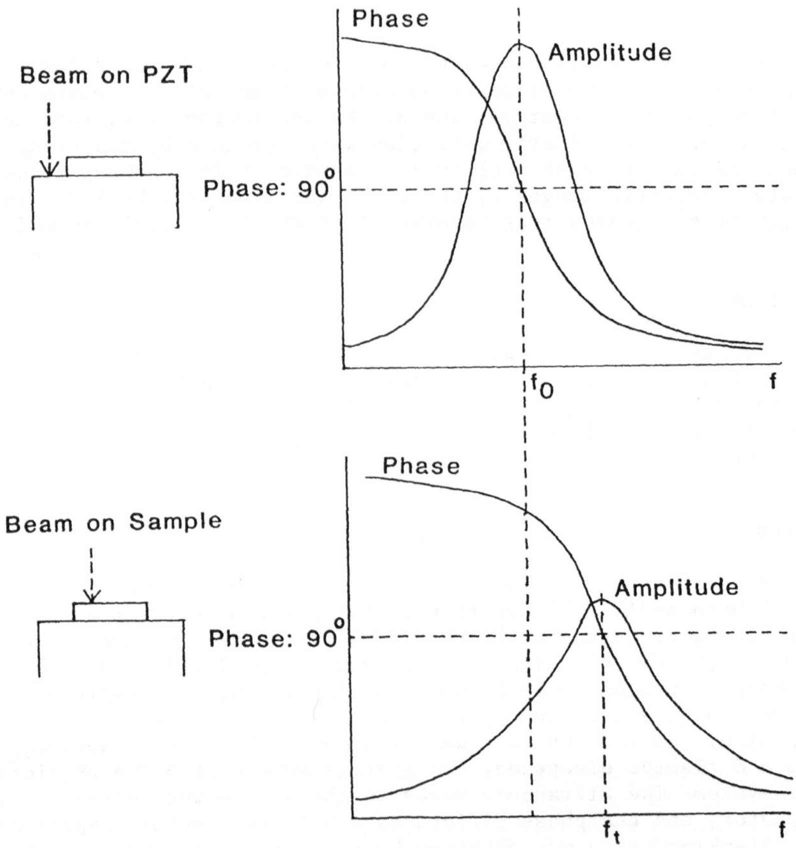

Fig. 2. Frequency dependence of IAS signal.

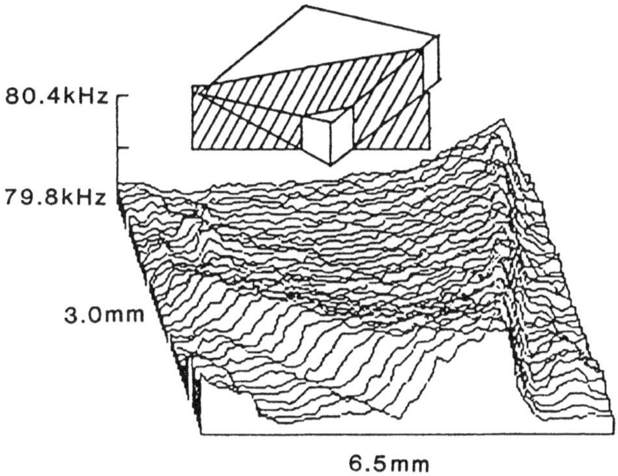

Fig. 3. Frequency topograph in tapered aluminum by IAM. Scanning area is shaped.

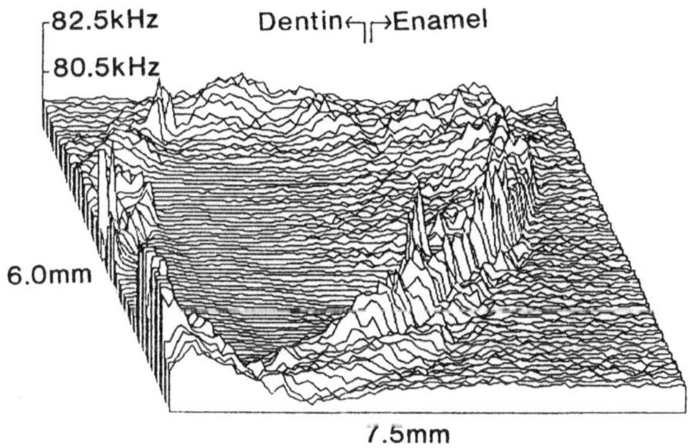

Fig. 4. Frequency topograph of human tooth.

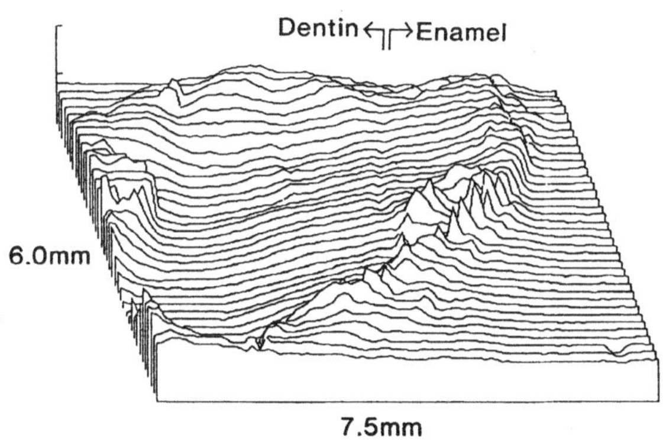

Fig. 5. Attenuation topograph of human tooth.

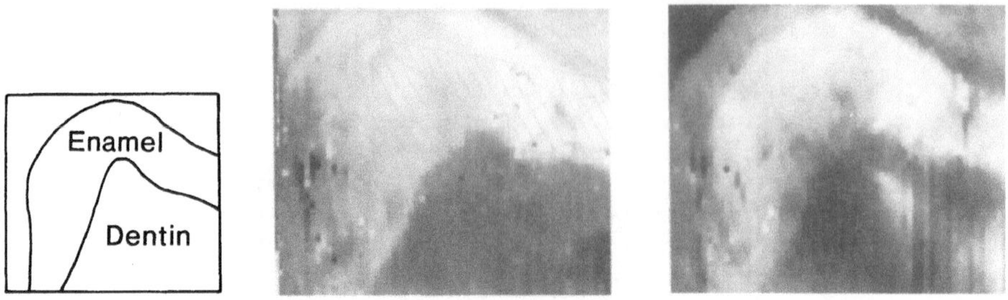

Attenuation image Frequency image

A. Matured third molar tooth. Attenuation of enamel is higher than that of dentin.

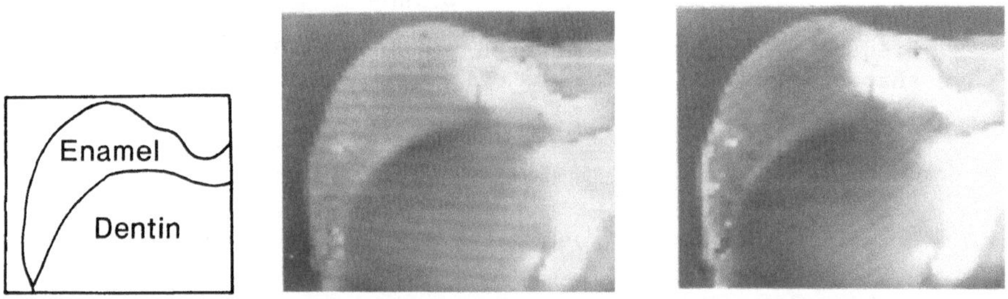

Attenuation image Frequency image

B. Unerupted third molar.

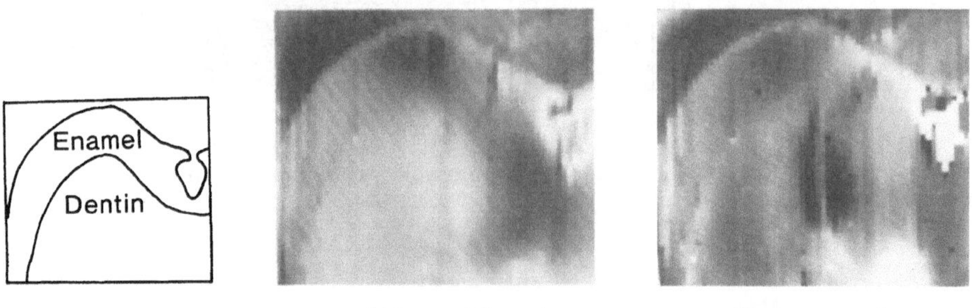

Attenuation image Frequency image

C. Unerupted third molar. Attenuation of enamel is slightly lower than that of dentin.

Fig. 6. Acoustic images of third molar teeth.

PRINCIPLE OF IAM

The acoustic waves in the sample, caused by ion beam excited at 25% duty, were detected by the PZT resonator, and the acoustic amplitude Va and the phase difference o between acoustic wave and ion beam pulse was obtained by lock-in amplifier. (Fig. 1)

When the beam was injected on a sample, resonance frequency (ft) shifts as shown in Fig. 2. From tapered aluminum data, we obtained empirical formula,

$$ft = (L/v)\ fo^2 + fo$$

where the thickness of sample is L, acoustic velocity is v, proper frequency of the PZT is fo. When the thickness of sample L and density ρ are constant, the Young's modulus image E(x,y) is given by

$$E(x,y) = \rho\ fo^4\ L^3 \ /\ (\ ft - fo\ )^2.$$

Figure 3 shows a frequency topograph of the IAS signal observed with an aluminum wedge of thickness varying from 50 µm to 600 µm. The sample was glued to PZT by Araldite and measured by a chopping ion beam under the condition of chopping frequency of 80kHz.

IAM IMAGE OF THE TEETH

The teeth are composed of hydroxyapatite $Ca_{10}(PO_4)_6(OH)_2$ which resembles a crystal of $M^{2+}{}_{10}(R^{5+}O_4)_6X^-{}_2$. The developmental mechanism of calcification with different crystals of hard enamel and elastic dentin has not yet been clarified.

Figure 4 shows frequency topograph of a human tooth. Higher frequency in enamel than in dentin is seen. The density of enamel was higher than that of dentin, but difference of the elastic constant seemed to be small. Figure 5 shows attenuation topograph of a human tooth. Higher in enamel than that in dentin is seen.

Acoustic images of human matured molar teeth in Fig. 6 showed that the attenuation of enamel is higher than that of dentin.(Fig. 6. A, B) However, some of the unerupted third molars showed different images, in which the attenuation of enamel is slightly lower than that of dentin. (Fig. 6. C) These results suggest the possibility of structural change and/or characteristic change of crystal during amelogenesis.

Reference

1) H. Tateno et al: Proc. 6th Symp. on Ultrasonic Electronics, Tokyo 1985, Jpn. J. Appl. Phys. 25(1981) Suppl. 25-1, 188.

AMPLITUDE AND PHASE ACOUSTIC MICROSCOPY

AND ITS APPLICATION TO QNDE

B. T. Khuri-Yakub, P. Reinholdtsen, C-H Chou,
P. Parent, and C. Cinbis

Edward L. Ginzton Laboratory
Stanford University
Stanford, CA 94305-1502

INTRODUCTION

We have two amplitude and phase measuring acoustic microscopes, one at low frequency (3-10 MHz) which is used for measurements in metals and composites, and the other operating at frequencies of up to 200 MHz which is used for higher resolution measurements. The added dimension of having phase information allows us to use image processing for a variety of applications. We have demonstrated the following applications with these two microscopes: V(z) inversion for reflectance function calculations, depth determination for delaminations in composite materials, slowness curve measurements in anisotropic materials, image enhancement of subsurface defects, measurements of depth of trenches in aluminum samples (scaled problem of silicon trenches), and measurement of visco-elastic properties of thin surfactant films on a water surface.

- - - - - -

Acoustic microscopy has become a major nondestructive evaluation tool for imaging, characterizing, and detecting defects in structural materials. The literature and this conference contain ample evidence of this fact. Except for our work at Stanford University, amplitude-only acoustic microscopes are used. We believe that the added image of phase variation across a sample gives a major advantage in image interpretation, signal processing, and material characterization. Thus, we have developed two acoustic microscopes that measure amplitude and phase, and that operate in the frequency range of 1-200 MHz .

The following is a list of problems that we have, over the last few years, solved by using amplitude and phase acoustic microscopy: (1) V(z) inversion for reflectance function evaluation [1,2]; (2) thin film thickness measurements with a resolution corresponding to 0.1 degree phase detection sensitivity; (3) surface wave velocity variations due to anisotropy, surface residual stress, and thin film loading of a half space; (4) thickness determination of thick films such as delamination in composite materials [3]; (5) contrast enhancement by spatial filtering which is superior to the same process using amplitude-only information [4]; (6) surface roughness removal to image subsurface defects, such as imaging the back side of a coin through the front surface roughness at 10 MHz for a U.S. quarter and without time gating the back surface echo; (7) depth resolution enhancement by two-dimensional spatial deconvolution; and (8) measurement of the surface viscosity of thin films on the surface of the ocean [5].

Several parts of the above-mentioned work have been published; we will concentrate in this paper on the last two examples mentioned above.

A schematic diagram of the electronic set-up for amplitude and phase measurement with the low-frequency acoustic microscope is shown in Fig. 1. The major feature in this measurement scheme is the use of the divide by n and n + 1 chip, which allows us to step the phase of the signal by 360/n degrees at a time with respect to a fixed-phase reference. Thus, we use n mixings to determine the amplitude and phase of the signal at a point in the scanning field. This type of measurement enhances the accuracy and the sensitivity of the measurement by removing electronic offsets in the various components of the system. The high-frequency measurement, shown in Fig. 2, is essentially the same, except that due to the unavailability of very high-frequency divide-by chips, we mix the phase-shifted signal with a local oscillator to arrive at the frequency of operation. The mechanical scanning, display, and computer control of the microscope is conventional and merits no further mention in this paper.

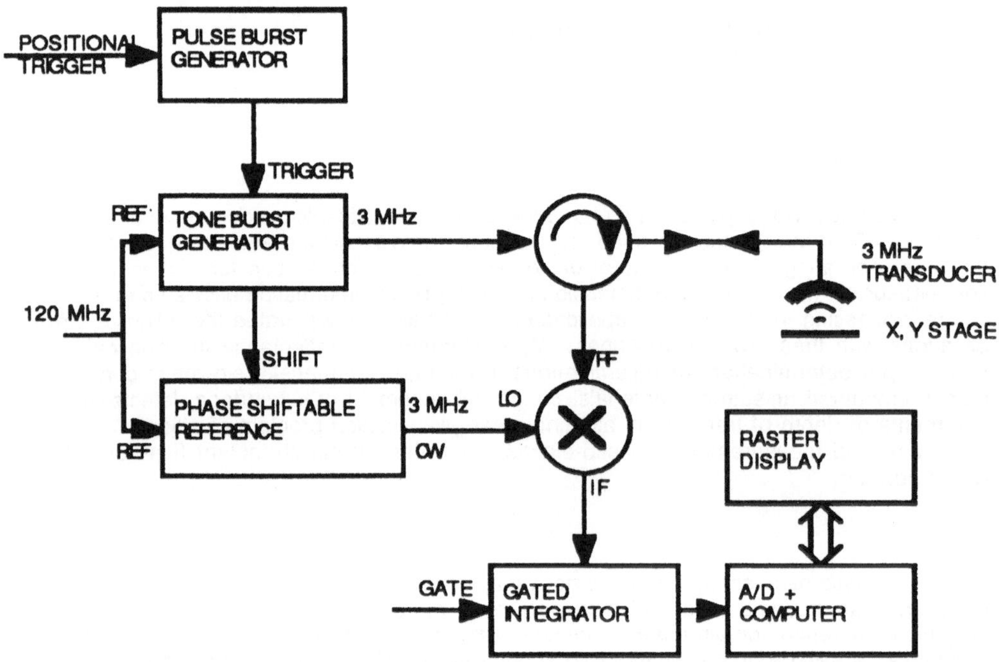

Fig. 1. Schematic diagram of amplitude and phase low-frequency acoustic microscope.

THREE-DIMENSIONAL RESOLUTION ENHANCEMENT

For objects embedded in a uniform medium, the image formed as the microscope scans in all three dimensions is a convolution of the object with the microscope's three-dimensional impulse response.

$$g(\vec{x}) = h(\vec{x}) *** f(\vec{x}) \tag{1}$$

where g is the image, h is the microscope's impulse response, and f is the object. In general the transducer's impulse response is less than ideal; it may have a broad central lobe or relatively large sidelobes. Figure 3a shows the line spread function of an F2 transducer operating at 3 MHz . This was experimentally determined by scanning a

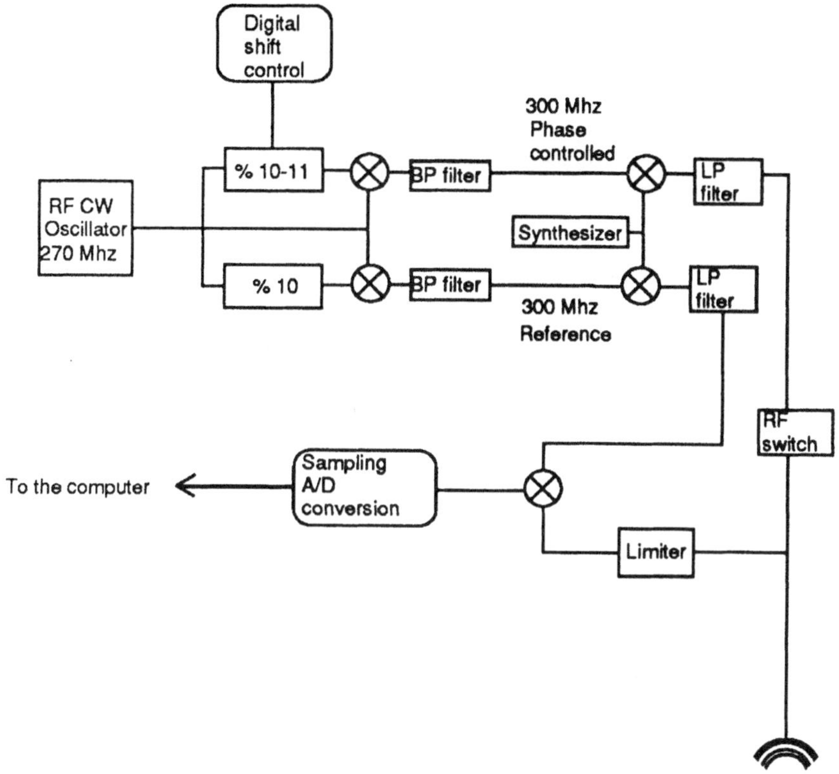

Fig. 2. Schematic diagram of amplitude and phase high-frequency acoustic microscope.

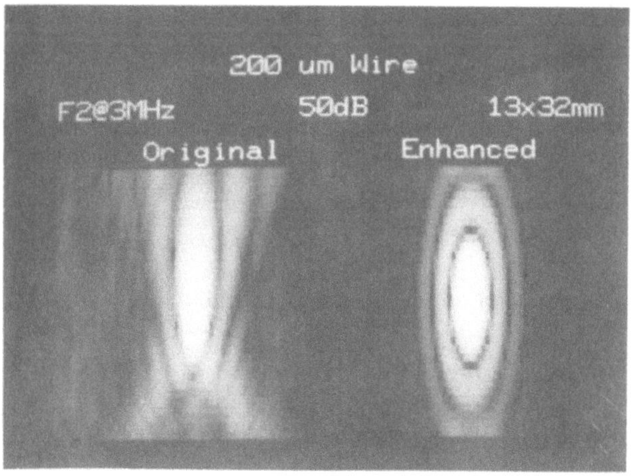

(a) (b)

Fig. 3. (a) Two-dimensional line spread function of F2 transducer. (b) Enhanced line spread function of same transducer.

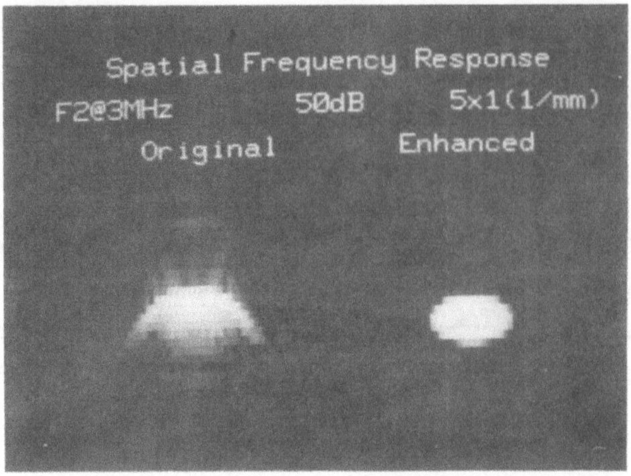

(a) (b)

Fig. 4. (a) Spatial frequency response of F2 transducer. (b) Enhanced spatial frequency response of same transducer.

200 μm wire. Notice the large diagonal sidelobes and that the width of the main lobe in the z direction is greater than 27 mm . The non-ideal characteristics of the microscope's impulse response degrade the quality of the acoustic images.

If a transducer can be characterized, there is a simple technique for numerically enhancing the effective transducer response of an acoustic microscope. It is possible to numerically tailor the transducer's effective impulse response to have a narrower central lobe, giving better resolution, or giving it smaller sidelobes.

First, we Fourier transform the images into the spatial frequency domain to get the image spectrum:

$$G(\vec{k}) = H(\vec{k})F(\vec{k}) \tag{2}$$

where G is the Fourier transform of g , H is the transform of h , and F is the transform of f . H is called the spatial frequency response (SFR) of the microscope.

$$G(\vec{k}) = \Im\{g(\vec{x})\} \tag{3}$$

This relationship can be inverted via an inverse Fourier transform.

$$g(\vec{k}) = \Im^{-1}\{G(\vec{x})\} \tag{4}$$

An image spectrum corresponding to a desired SFR is given by:

$$G_{Opt}(\vec{k}) = H_{Opt}(\vec{k})F(\vec{k}) \tag{5}$$

The desired image spectrum is obtained from the measured image by dividing Eq. (5) by Eq. (2) and cancelling terms:

$$G_{Opt}(\vec{k}) = \frac{H_{Opt}(\vec{k})}{H(\vec{k})} G(\vec{k}) \qquad (6)$$

The enhanced image is obtained by inverse Fourier transforming G_{Opt} into the space domain. This amounts to convolving the experimental image with an "enhancing" filter whose spectrum is H_{Opt}/H .

Figure 4a shows the SFR of our transducer. Notice that it is non-zero only for a finite region in the spatial frequency domain. Therefore, Eq (6) is only valid in that range. This range is called the base of support for the SFR. In order for Eq. (6) to be valid, the base of support for H_{Opt} must be a subset of the base of support for H .

Figure 4b shows the enhanced SFR that was chosen. The base of support is an ellipse where the enhanced SFR is constant in amplitude and phase except the edge has a cosine squared taper.

Figure 3b shows the enhanced line spread function. Notice that the diagonal sidelobes have been eliminated and the width of the main lobe in the z-direction has been greatly decreased. The transverse resolution has been increased noticeably, but the depth resolution has been doubled, as shown in Fig. 5.

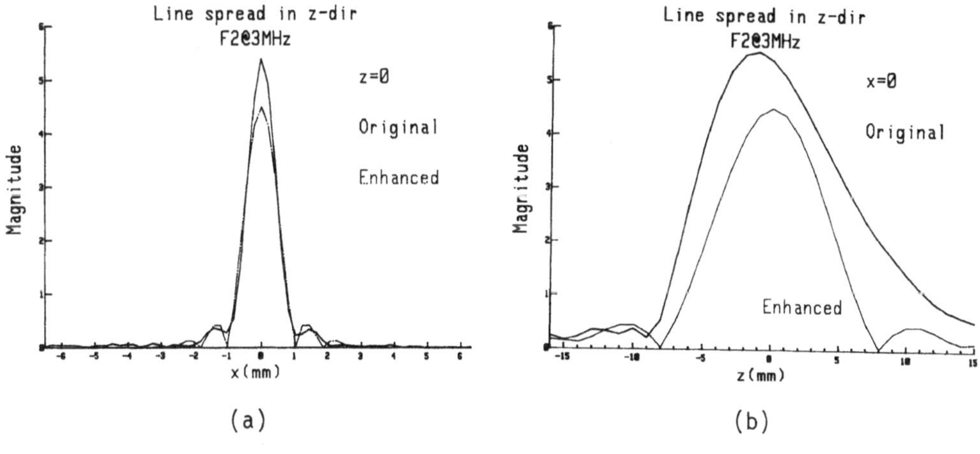

Fig. 5. (a) Line spread function of F2 transducer in x-direction. (b) Line spread function of F2 transducer in z-direction.

This dramatic increase in the z-direction can be better understood by looking in the spatial frequency domain. Figure 6 shows that the spectrum falls off rapidly as kz differs from k , but that the spectrum is still well above the noise floor. Thus, the effective width of the useful spectrum can be numerically doubled.

EXPERIMENT: DEEP TRENCHES IN ALUMINUM

We applied this technique to measuring the profile of a deep trench (2 mm wide by 5 mm deep) using the F2 transducer. The sample is show schematically in Fig. 7.

The trench was designed to simulate a vertical capacitor in an integrated circuit (2.5 μm wide by 6 μm deep). This technique could be scaled up in frequency to characterize integrated circuits with an optical microscope that measures amplitude and phase, or with an acoustic microscope operating at 2.5 GHz .

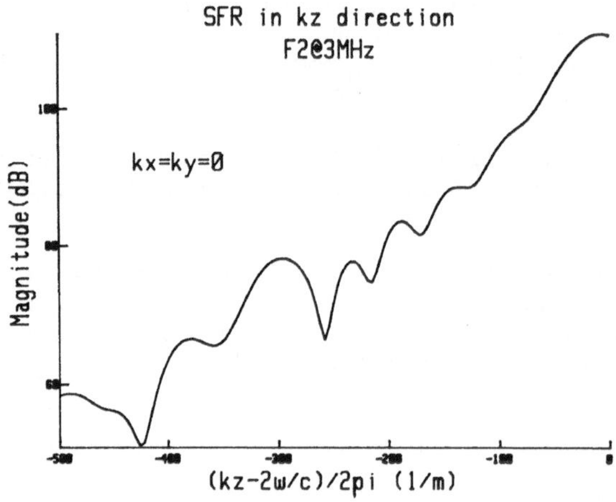

Fig. 6. SFR in kz direction of F2 transducer.

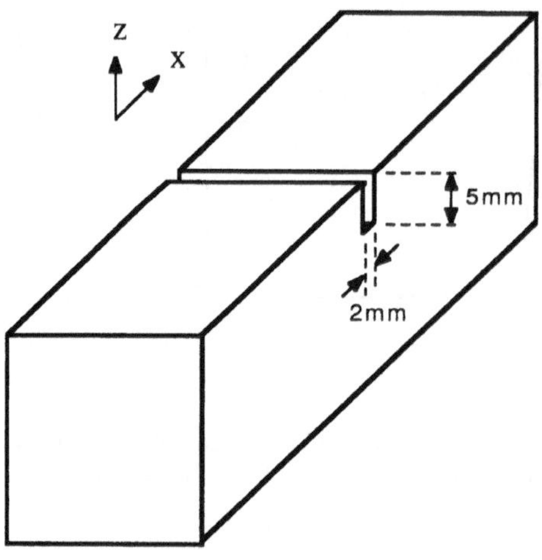

Fig. 7. Schematic of trench in Al.

Figure 8 shows the original acoustic image compared to the enhanced image. The enhanced image qualitatively looks much sharper, showing the location of the bottom of the trench much more clearly. Comparing line scans in Figure 9 shows that the quantitative improvement has been dramatic. The transverse profile of the trench, when focused on the bottom, has been improved noticeably. The most dramatic improvement is

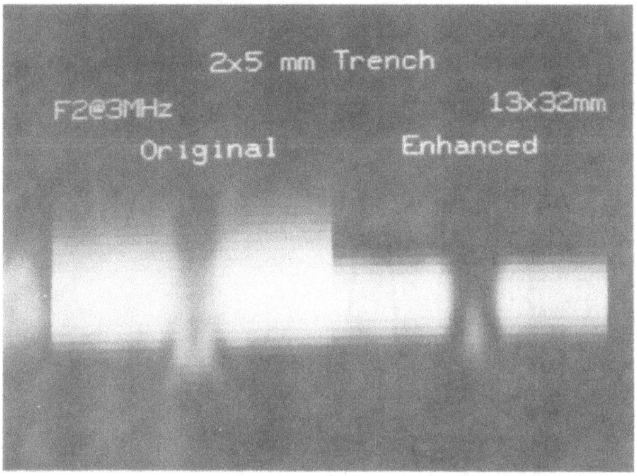

Fig. 8. X-Z plane image of trench in Al.

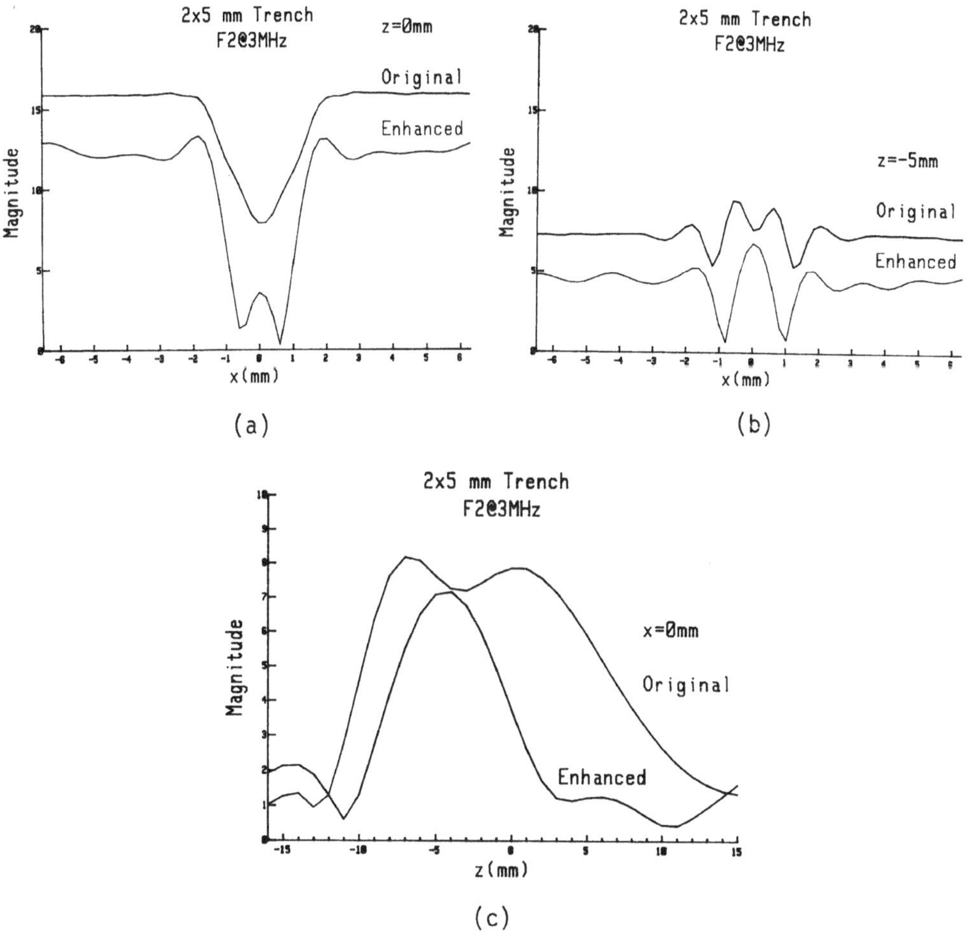

Fig. 9. (a) Original and enhanced line scans of trench in Al at Z = 0.
(b) Original and enhanced line scans of trench in Al at Z = -5 mm
(c) Line scan through Al trench at x = 0.

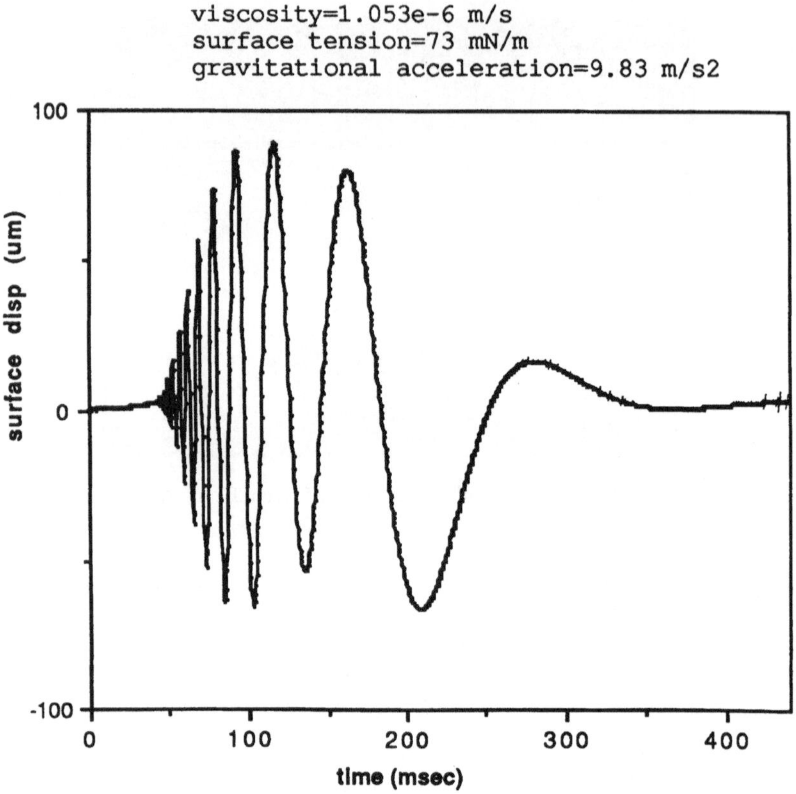

Fig. 10. Transient capillary wave evolution of a Gaussian dimple.

shown in the depth scan of the bottom of the trench. In the original image, the peak corresponding to the bottom occurs at $z = -7$ mm and there is another peak corresponding to the top surface. The enhanced image locates the bottom closer to the correct location at $z = -5$ mm with little interference from the top surface.

MEASUREMENT OF SURFACE TENSION

The ability of the acoustic microscope to measure phase allows us a novel application for acoustic microscopes, that is the measurement of ripples or capillary waves on the surface of water. Capillary waves are short-wavelength ripples on the surface of water which occur due to wind interaction with the water. Capillary waves are controlled by surface tension which acts to restore the disturbed surface of the water to its original position. We use the acoustic microscope to measure a controlled capillary wave packet that we excite with an acoustic transducer.

The schematic of the set-up is shown in Ref. 17. The first transducer excites a high-amplitude ultrasonic wave whose radiation pressure lifts the surface of the water and evolves as a transient capillary wave with a broad frequency spectrum. The second ultrasonic transducer which makes part of the phase measuring acoustic microscope is used to measure the amplitude of the capillary wave. The measured waveform is then analyzed to yield the surface tension of the water. This method allows us to measure the effect of surfactants on wind-wave interactions.

Figure 10 shows the result of a theoretical calculation of the capillary wave packet due to single Gaussian excitation 500 microns wide. The packet is measured at a distance of 4 cms from the source, and the surface tension of the water is taken to be

73/m ; surface viscosity is taken to be 1.053 μm/s . The results of measurements on sea water samples agree very well with the theory of Fig. 10 and demonstrate the variation of the surface tension as a function of time, which is due to the rise of surfactants to the surface. The results indicate that we are indeed capable of measuring the properties of surface films on water.

CONCLUSIONS

Amplitude and phase measuring acoustic microscopes offer a major advantage in applying the microscope to material characterization problems. We demonstrated the improvement in most areas of application of acoustic microscopes with the added advantage of phase measurement. We also show a technique for making the phase measurement that is fairly simple, reliable, and which does not reduce the speed of operation of the instrument.

ACKNOWLEDGMENT

This work was supported by the Department of Energy under Contract No. DE-FGO3-84ER45157 and by the Office of Naval Research under Contract No. N00014-86K-0433.

REFERENCES

1. K. K. Liang, G. S. Kino, and B. T. Khuri-Yakub, IEEE Sonics and Ultrasonics SU-32 (2), 213 (March 1985).
2. P. Reinholdtsen and B. T. Khuri-Yakub, "Acoustic Microscopy Using Amplitude and Phase Measurements, Review of Progress in Quantitative Nondestructive Evaluation, Vol. 6A, Eds: D. O. Thompson and D. E. Chimenti, 543-551 (Plenum Press, New York, 1987).
3. P. A. Reinholdtsen, C-H. Chou, and B. T. Khuri-Yakub, "Quantitative Acoustic Microscopy Using Amplitude and Phase Imaging," Proc. IEEE Ultrasonics Symp., October 14-16, 1987, Denver, Colorado (1987).
4. P. A. Reinholdtsen and B. T. Khuri-Yakub, "Removing the Effects of Surface Roughness from Acoustic Images," Review of Progress in Quantitative NDE, Eds: D. O. Thompson and D. E. Chimenti (Plenum Press, New York, 1988).
5. G. Ewing, "Slicks, Surface Films, and Internal Waves," J. Marine Res. 9, 161-187 (1950).
6. W. D. Garrett and J. D. Boltman, "Capillary-Wave Damping by Insoluble Organic Monolayers," J. Colloid Sci. 18, 798-801 (1963).
7. R. S. Hansen and J. A. Mann, "Propagation Characteristics of Capillary Ripples: I. The Theory of Velocity Dispersion and Amplitude Attenuation of Plane Capillary Waves on Visco-Elastic Films" J. Appl. Phys. 35, 152-158 (1964).
8. G. D. Crapper, Introduction to Water Waves, Ellis Harwood, Chichester (1984).
9. J. W. Goodman, Introduction to Fourier Optics, McGraw-Hill (1968).
10. P. C. D. Hobbs, R. L. Jungerman, and G. S. Kino, "Proc. of SPIE, Vol. 565, Micron and Submicron Integrated Circuit Metrology (1985).
11. F. J. Harris, Proc. IEEE 66 (1), 51 (January 1978).
12. R. A. Lemons and C. F. Quate, Appl. Phys. Lett. 24, 163 (1974).
13. C. F. Quate, A. Atalar, and H. K. Wickramasinghe, Proc. IEEE 67, 1052 (1979).
14. A. Atalar, J. Appl. Phys. 45, 5130-5139 (October 1979).
15. See IEEE Trans. on Sonics and Ultrasonics, Special Issue on Acoustic Microscopy SU-32 (2) (March 1985).
15. K. Liang, S. D. Bennett, B. T. Khuri-Yakub, and G. S. Kino, "Precision Phase Measurements with the Acoustic Microscope," IEEE Trans. on Sonics & Ultrasonics SU-32 (2), 266-273 (March 1985).

16. B. A. Auld, Acoustic Waves and Fields in Solids, Wiley-Interscience (New York, 1973).
17. B. T. Khuri-Yakub, P. A. Reinholdtsen, C-H. Chou, J. F. Vesecky, and C. C. Teague, Appl. Phys. Lett. 52 (19), 1571 (9 May 1988).

CONSIDERATIONS FOR QUANTITATIVE CHARACTERIZATION OF BIOLOGICAL TISSUES BY SCANNING ACOUSTIC MICROSCOPY

N. Akashi, J. Kushibiki, N. Chubachi, and F. Dunn*

Department of Electrical Engineering, Tohoku University
Sendai 980, Japan
*Bioacoustics Research Laboratory, University of Illinois
1406 W. Green St., Urbana, IL 61801, USA

INTRODUCTION

The acoustic microscope has received much attention as a new research tool to measure acoustic properties of biological tissues on a microscopic scale. A reflection-type acoustic microscope system for the use was reported at the previous symposium [1]. In that system, acoustic velocity and attenuation were determined approximately from measured phase and amplitude, using the simplest model analysis for which acoustic multiple reflections and viscosity in a tissue section, layered on a substrate, were neglected.

In order to improve the accuracy of measurements, the processing and analytical procedures have been investigated, and are discussed in this paper. A revised model, taking effects of the acoustic multiple reflections and viscosity into consideration, is proposed, and a fitting method is developed to determine the velocity, attenuation, density, and thickness of tissue specimens. Numerical calculations are performed to clarify the differences between acoustic properties obtained with the revised model and those obtained with the simplest model.

REVISED MODEL

Sample configuration for the ultrasonic characterization is shown in Fig. 1, in which a tissue section is mounted on a substrate. In order to determine the acoustic properties of tissues, the amplitude and phase of reflected waves from a sample, in reference to those from the substrate alone, are measured. For theoretical consideration, a plane wave model is used considering that the focused acoustic waves near the focal plane can be approximated as plane waves.

The reflection coefficients, R_S for the sample and R_W for the exposed substrate alone, are defined as the ratios of the reflective waves to the incident waves at the boundary a-a'. Acoustic multiple reflection occurs in a sample because the thickness of a sample is finite. From transmission line theory, the reflection coefficints, R_S and R_W, are, respectively, given as follows:

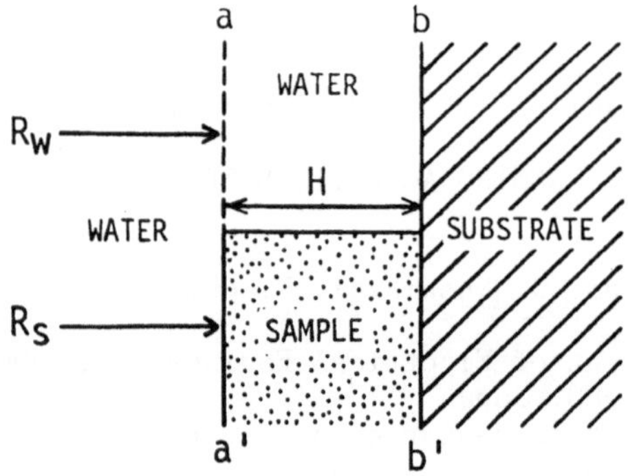

Fig. 1. Sample configuration and definition of reflection coefficients R_s and R_w.

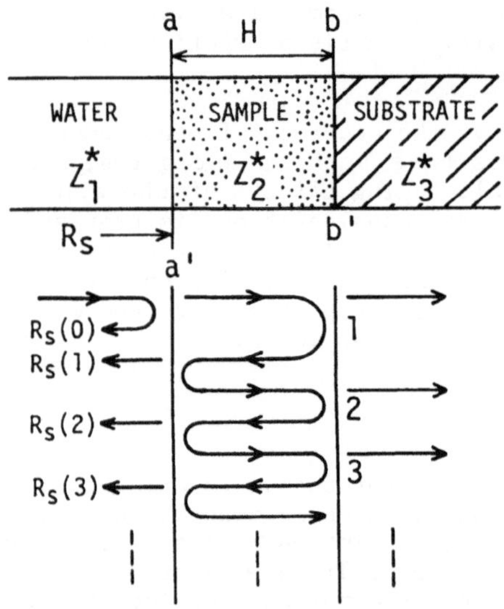

Fig. 2. Acoustic transmission line model and definition of components $R_s(n)$ in R_s.

$$R_s=R_s(0)+\sum_{n=1}^{\infty}R_s(n)$$

$$=R_{12}+\sum_{n=1}^{\infty}(1-R_{12}^2)(-R_{12})^{n-1}R_{23}^n\exp(-2n\gamma_2 H) \quad (1)$$

$$R_w=R_{13}\exp(-2\gamma_1 H) \quad (2)$$

$$R_{ij}=\frac{Z_j^*-Z_i^*}{Z_j^*+Z_i^*}, \qquad \gamma_i=\alpha_i+jk_i, \qquad k_i=\frac{\omega}{V_i}$$

where Z_1^*, Z_2^*, Z_3^* are the complex acoustic impedances of water, sample, and substrate, respectively. H is the thickness of the sample, ω is the angular frequency, and α_i, k_i, and V_i are the attenuation coefficient, wave number, and velocity of medium i, respectively. α_i and V_i are dependent on the acoustic frequency because of the viscosity of tissues [2]. The complex acoustic impedance Z_i^*, according to the Voigt model, is given by

$$Z_i^*=\sqrt{\rho_i(c_i+j\omega\eta_i)} \quad (3)$$

where ρ_i, c_i, and η_i, are the density, stiffness constant, and viscosity constant of medium i, respectively.

The acoustic properties of a sample are obtained by measuring the frequency characteristics of both the phase $\phi=\arg(R_s/R_w)$ and the amplitude $|R_s/R_w|$ of the normalized reflection coefficient R_s/R_w. The velocity, attenuation, density, and thickness of a sample are then calculated by computer fitting.

Figure 3 shows the calculated results of the normalized phase $\phi/(fH)$ as a function of acoustic frequency for the four velocities of 1450, 1550, 1650, and 1750 m/s, with the other acoustic parameters as listed in Table 1. According to the literature [3,4], the velocities of biological soft tissues have a range of 1450 to 1750 m/s where the lowest velocity corresponds to that of fat and the highest velocity corresponds to that of tendon. ζ_i is defined as $\zeta_i=\eta_i/c_i$ [2]. The published data [3,4], and our measured data, show that the attenuation coefficients of fresh tissues, around 100 MHz, are several to ten times greater than that of water, while those of fixed tissues are a few tens times greater than that of water. In the numerical calculations, $\zeta_2/\zeta_1=10$ is taken for fresh tissues. The

Table 1. Acoustic parameters used for numerical calculations. The biological tissue is considered as liver.

Medium 1 (water)	Medium 2 (tissue)	Medium 3 (fused quartz)
$\rho_1=9.982\times10^2$ kg/m^3	$\rho_2=1.05\times10^3$ kg/m^3	$\rho_3=2.2\times10^3$ kg/m^3
$V_1=1483$ m/s	$V_2=1600$ m/s	$V_3=5973$ m/s
$\alpha_1/f^2=2.53\times10^{-14}$ s^2/m	$\zeta_2/\zeta_1=10$	$\alpha_3/f^2=1.3\times10^{-16}$ s^2/m
	H=10 µm	

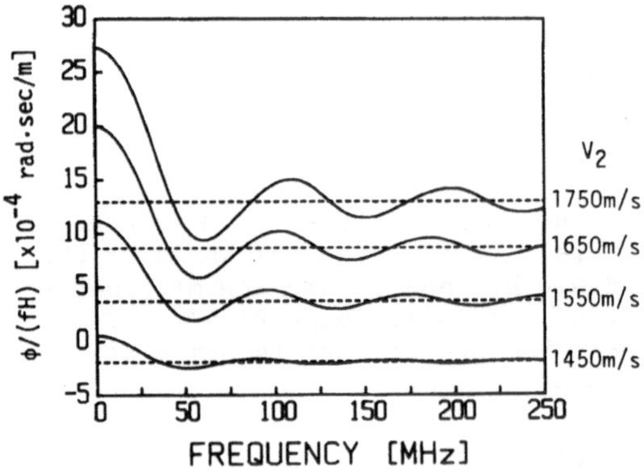

Fig. 3. Calculated results of phase $\phi/(fH)$ of normalized reflection coefficient R_s/R_w for V_2=1450, 1550, 1650 and 1750 m/s.

Curves are for $R_s=\sum_{n=0}^{\infty} R_s(n)$, and dashed lines for $R_s=R_s(1)$.

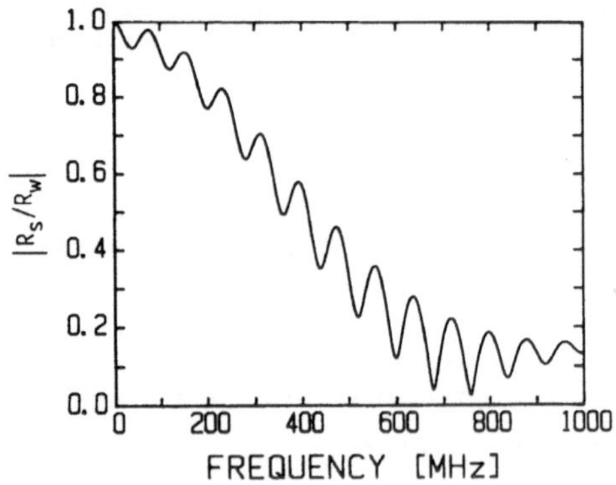

Fig. 4. Calculated ratios of amplitude $|R_s/R_w|$ of normalized reflection coefficient R_s/R_w.

dashed lines, being parallel to the horizontal axis in Fig. 3, correspond to the phase values obtained for $R_s=R_s(1)$. Intervals among the "cross" frequencies, at which the phase curves intersect the lines, approximates $V_2/4H$. If media are assumed to be acoustically lossless, the intervals are exactly equal to $V_2/4H$. The density of a sample influences the interference amplitude of $\phi/(fH)$.

The specimen parameters V_2, ρ_2, and H can be obtained by the fitting procedure using the phase versus frequency characteristics.

On the other hand, the attenuation is determined from measurements of amplitude $|R_s/R_w|$. Figure 4 shows the calculated curve of the frequency characteristics of the amplitude, determined using the acoustic parameters listed in Table 1. The attenuation α_2 is obtained by calculating the frequency characteristics best-fitted for a measured curve with the velocity, density, and thickness obtained through the phase measurements.

DISCUSSION

Numerical calculations are carried out to show how multiple reflection and viscosity of tissues affect the measurements, as regards the revised model proposed here. The differences between the revised and simplest models are discussed in relation to the accuracy of measurements.

Multiple Reflection

The amplitude of each component $R_s(n)$ of the multiple reflection components in Eq.(1) is first examined, considering acoustic impedances of a variety of biological soft tissues. Figure 5 shows the ratios of relative amplitudes of the components $R_s(n)$ to that of $R_s(1)$ as a function of the acoustic impedance ratio, Z_2/Z_1, where Z_i (i=1, 2, 3) is the acoustic impedance of each medium given by assuming $\eta_i=0$ in Eq.(3). In the figure, all of the relative amplitudes become smaller as Z_2/Z_1 approaches unity or as the number n increases. The values of Z_2/Z_1 for biological soft tissues have approximately a range of 0.9 to 1.3 according to the literature [3,4]. The values of 0.9 and 1.3 typically correspond to the impedance ratios for fat and tendon, respectively.

When fused quartz with an acoustic impedance ratio Z_3/Z_1 of 8.9 is used as the substrate in the sample configuration shown in Fig. 1, the transmission coefficient T_{12} at the water/sample boundary is very large, while the transmission coefficient T_{23} at the sample/substrate boundary is very small. Thus, $R_s(1)$ is the largest $R_s(n)$, and $R_s(0)$ is the second largest of the components. The relative amplitudes of $R_s(2)$, $R_s(3)$, and $R_s(4)$ are -20dB, -41dB, and -61dB, respectively, even at $Z_2/Z_1=1.3$.

To examine the multiple reflection effect, the cross frequencies defined previously were calculated for the cases in which all or some of the components were considered, using the acoustic parameters listed in Table 1. A number of cross frequencies are seen in Fig. 3. The cross frequencies around 120 MHz are investigated further, defining the frequency deviation df as df=100x(f'-f)/f (%), where f' is the cross frequency obtained for the limited components, and f is that for all of the components. The deviations were obtained as -0.4%, -0.02%, and -0.002% when the first one, two, and three components of the multiple reflection components were taken into account, respectively.

Thus, it is understood that only a few components of the multiple reflection components contribute significantly to the determination of the acoustic properties.

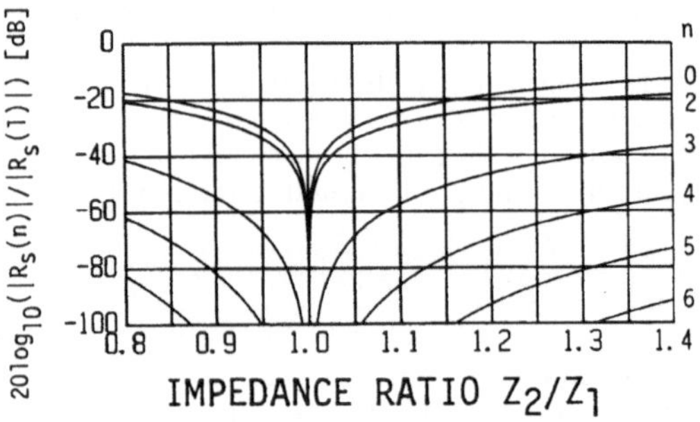

Fig. 5. Calculated results of relative amplitudes of components $R_s(n)$ to that of $R_s(1)$ versus acoustic impedance ratio Z_2/Z_1.

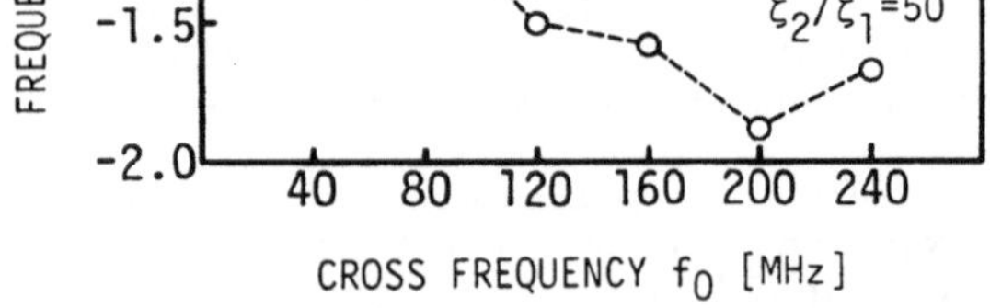

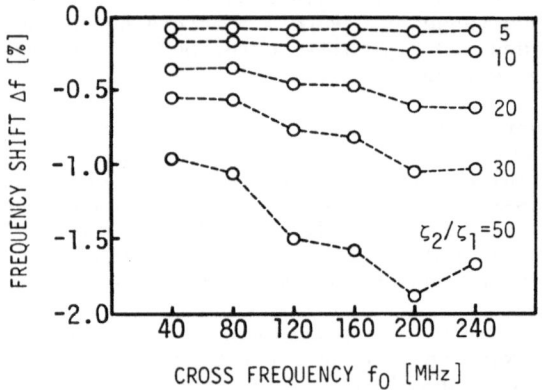

Fig. 6. Calculated results of effect of viscosity on cross frequency f_0.

Viscosity

Biological tissues are viscoelastic media. Viscoelastic properties in media result in velocity dispersion and attenuation in wave propagation, and phase shifts occur in reflection and transmission at the boundary between different media.

In order to discuss the effect of viscosity on the measurements, the cross frequencies were calculated with the acoustic parameters in Table 1 as a function of ζ_2/ζ_1, using the revised model. For this purpose, the frequency shift Δf is defined as $\Delta f=100x(f-f_0)/f_0$ (%), where f_0 is the cross frequency calculated for lossless media and f is that for lossy media. Figure 6 shows the calculated results of the frequency shifts Δf versus f_0. In the calculations, ζ_2/ζ_1 was taken as 5, 10, 20, 30, and 50. In the figure, the values of Δf are negative, resulting mainly from the phase shifts in reflections and transmissions at boundaries, rather than from velocity dispersion [2]. Thus, it is seen that the effect of viscosity becomes significant for values of ζ_2/ζ_1 greater than ten and at the higher operating frequencies.

Comparison between Revised and Simplest Models

In order to compare the acoustic properties determined by the revised model with those by the simplest model, numerical calculations were made, using the acoustic parameters in Table 1, in the frequency range from 100 to 200 MHz, which corresponds to the range employed in the previous study [1]. The calculated results of the phase $\phi/(fH)$ and the amplitude $|R_s/R_w|$ of the normalized reflection coefficients for both models are shown in Figs. 7 and 8, respectively. The percent differences of $\phi/(fH)$ and $|R_s/R_w|$ for the simplest model, referring to their frequency characteristics for the revised model, are also shown. The differences are the uncertainty in determining the acoustic properties using the simplest model. The maximum difference in $\phi/(fH)$ is about 18%, while the maximum difference in $|R_s/R_w|$ is about 5%. The difference in $\phi/(fH)$ of 18% corresponds to the velocity uncertainty of 2%, according to the following equation [1] for the determination of the velocity in the simplest model,

$$V_2= 1/(1/V_1-\phi/4\pi fH) \ . \tag{4}$$

All of the multiple reflection components are taken into account in the revised model, while only the $R_s(0)$ and $R_s(1)$ components are included in the simplest model for determining the attenuation. The interference amplitudes for the frequency characteristics of $|R_s/R_w|$ depend on the number of multiple reflection components included in the models. However, the difference in $|R_s/R_w|$ of 5% is not directly related to uncertainty of the attenuation coefficient α_2 to be measured. Since α_2 is determined with the predetermined values of V_2, ρ_2, and H for a tissue sample, it could be estimated that uncertainty of α_2 is the same order of a few percent as that of the other acoustic parameters.

CONCLUDING REMARKS

A revised model for the quantitative characterization of biological tissues by scanning acoustic microscopy has been proposed. Herein, the effects of acoustic multiple reflection and viscosity of the sample have been taken into account, and a fitting method has been developed to determine the velocity, attenuation, density, and thickness of tissues with reasonable accuracy. It has been shown from numerical calculations that the first two or three components of multiple reflection contribute most significantly to the determination of acoustic properties, and that the

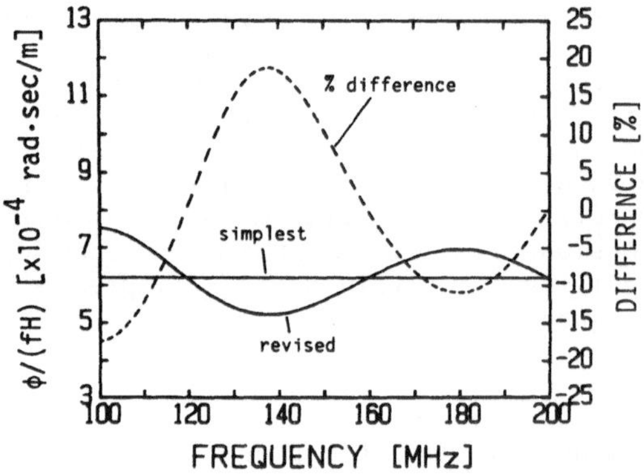

Fig. 7. Calculated frequency characteristics of phases of normalized reflection coefficient R_S/R_W for revised and simplest models, and their difference(dashed curves).

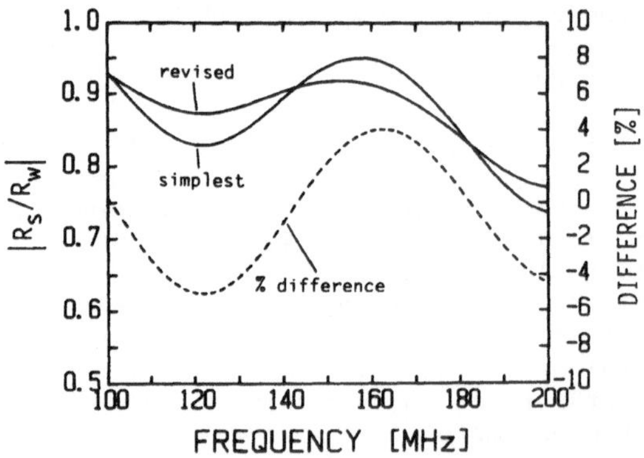

Fig. 8. Calculated frequency characteristics of amplitudes of normalized reflection coefficient R_S/R_W for revised and simplest models, and their difference(dashed curves).

effect of viscosity of biological soft tissues cannot be neglected at the high frequencies at which acoustic microscopes operate.

ACKNOWLEDGMENTS

The authors are very grateful to T. Sannomiya for his helpful discussions. This work was supported in part by the Research Grant-in-Aids from Japan Ministry of Education, Science & Culture.

REFERENCES

1. N. Chubachi, J. Kushibiki, T. Sannomiya, N. Akashi, M. Tanaka, H.Okawai, and F. Dunn, Scanning acoustic microscope for quantitative characterization of biological tissues, in: "Acoustical Imaging Vol.16," Plenum, New York (in press).
2. N. Akashi, J. Kushibiki, and N. Chubachi, Fundamental study on quantitative measurements of acoustic properties of biological tissues in acoustic microscopy — Theoretical and experimental consideration using plane wave model —, Technical Reports of IECE, US86-33:1(1986) (in Japanese).
3. S. A. Goss, R. L. Johnston, and F. Dunn, Comprehensive comparison of empirical ultrasonic properties of mammalian tissues, J. Acoust. Soc. Am., 64:423 (1978).
4. S. A. Goss, R. L. Johnston, and F. Dunn, Compilation of empirical ultrasonic properties of mammalian tissues II, J. Acoust. Soc. Am., 68:93 (1980).

QUANTITATIVE DISPLAY OF ACOUSTIC PROPERTIES OF THE BIOLOGICAL TISSUE ELEMENTS

Hiroaki Okawai[1)], Motonao Tanaka[1)], Floyd Dunn[2)], Noriyoshi Chubachi[3)], and Keisuke Honda[4)]

1) Research Institute for Chest Diseases and Cancer Tohoku University, Sendai 980, Japan, 2) Bioacoustics Research Laboratory, University of Illinois, Illinois 61801, USA, 3) Faculty of Engineering, Tohoku University 4) Honda Electric Company, Toyohashi 441-31, Japan

Introduction

Though the acoustic properties in the frequency range 1 to 10 MHz, have been considered by many investigators for the purpose of tissue characterization (TC) [1,2,3], TC at the microscopic level has not yet been studied in detail. For TC at the microscopic level, it is necessary to investigate the acoustic properties and the structural features of tissues.

In this paper, a measurement method is described which is capable of yielding such data, together with a discussion of the significance of a measurement system and a method for quantitative two-dimensional display of the distribution of attenuation constant and sound speed at the microscopic level.

METHOD

(a) Instrumentation

In order to measure the acoustic properties of tissue elements at the microscopic level, a measurement system shown in Fig.1 has been developed. The main function at the instrument quantitative display of acoustic properties of tissue elements.

The ultrasonic frequency is variable over the range 100 to 200 MHz of spectroscopic measurement. The transducer is equiped with a lens for which the dimensions of diameter, radius of curvature and aperture angle are 1.25 mm, 1.25 mm and 60 degrees, respectively, and having a -3 dB beam width in water (20°C) between 5 μm (at 200 MHz) and 10 μm (at 100 MHz). Such beam is scanned over a 2 mm width with the mechanical scanner and the ultrasound is transmitted for every 4 um interval. The number of sampling points are 480 in one scanning line and 480x480 points makes one frame. In the analogue signal processor, the detector of both amplitude and phase is fixed, and an A/D converter (6 bits) is employed, in which the error of quantization is approximately 2 m/s for a 10 μm thick tissue at the ultrasound frequency of 130 MHz. The original image produced in the analogue signal processor can be displayed in the display unit directly. However, the original image data, by amplitude and by phase detection, do not have real values of attenuation and speed, respectively, so that the image processor composed of the memory for 16 frames, a computer, and a display controller was introduced to obtain real data. A display controller provides the unique image in color scales quantitatively.

(b) Investigation of components of reflected wave

The specimens were tissues, formalin fixed and paraffin embedded, cut in a sections approximately 10 μm in thickness and mounted on glass slides in order to measure the acoustic properties with precision. The paraffin was removed just before the experiment.

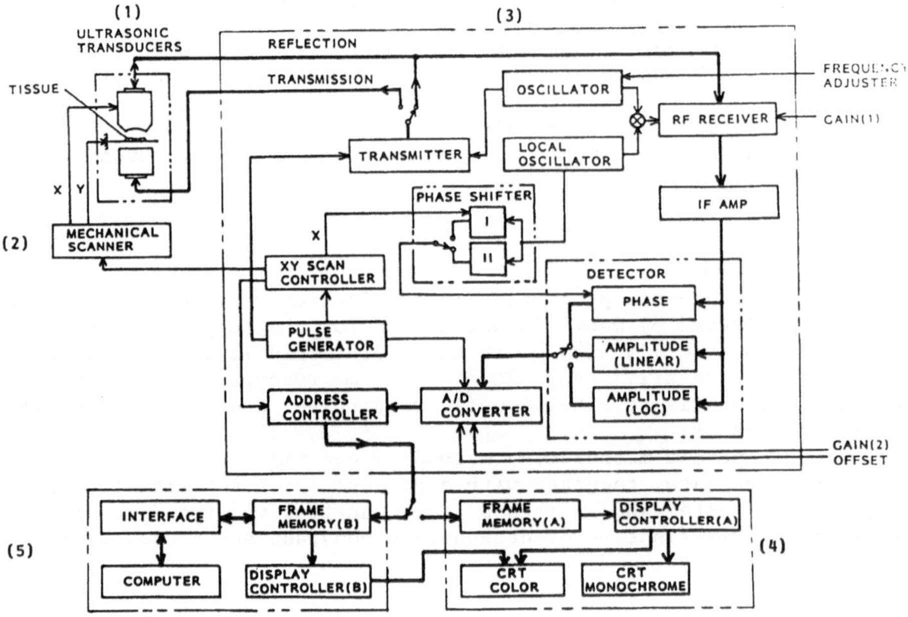

Fig.1 Block diagram for the measurement system.
(1) ultrasonic transducers, (2) mechanical scanner, (3) analogue signal processor, (4) display unit, and (5) image processor.

The reflection method using the focused wave was employed. For the purposes of imaging and quantitative measurement, the components of the reflected wave are as shown in Fig.2. The oblique incident wave, the waves reflected from the front surface, y_{2a}, y_{2b} and the round trip waves, y_{1a}, y_{1b} are received. However, the more than two times round trip waves, y_{1a}, y_{1b}, are outside the region where the wave is received. Consequently, the round trip wave and the wave reflected from the front surface become major components as illustrated in Fig. 2 by y_1, y_2 [4,5].

The following equations describe detected amplitude and phase:

$$L=-(10 \log y^2 - 10 \log y_3^2)$$

$$\phi=\arg(y/y_3)$$

where y is y_1+y_2.

(c) Determination of specimen thickness

When the attenuation constant and speed of sound are determined by transmitted ultrasound, it is essential to have knowledge of the specimen thickness. The problem is that the specimen thickness cannot be measured by a contact method such as with a micrometer. This problem has been resolved by developement of a non-contact method [4,5], described as follows.

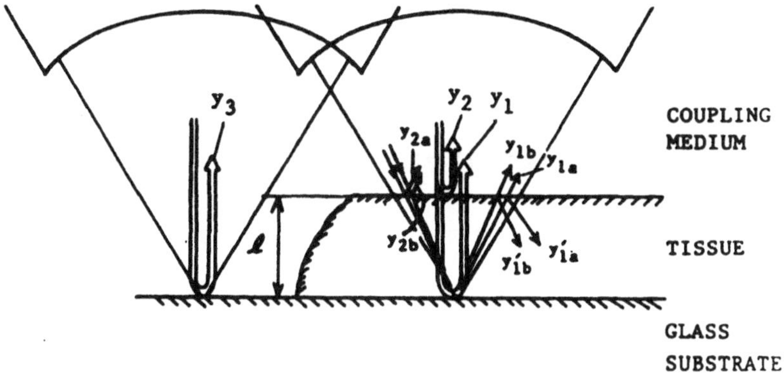

Fig.2 Acoustic model to discuss acoustic components of reflected wave and the relationship between acoustic beam and a tissue.

The amount of decrease in amplitude and of shift in phase of the wave reflected from a thin tissue mounted on a glass slide exhibits undulating curves in the frequency characteristics, as shown in Figs.3 (described as attenuation) and 4, respectively. Though L and Ø do not express real attenuation or sound speed of a tissue, respectively, the traces of these quantities have similar characteristic patterns as the levels of attenuation and sound speed, the same positions of maxima and minima, and the attenuation slopes in the calculation and in the measurement. Therefore, specimen thickness can be determined by comparing these characteristic patterns. Once specimen thickness is determined, the attenuation constant and the sound speed in a tissue can be determined from the straight lines in the figures. The values on the straight lines correspond to those of the round trip wave y_1. The attenuation constant is determined by normalizing the values on the straight line in the attenuation graph by twice the thickness value. Sound speed is calculated from the values on the straight line in the phase graph and from the thickness.

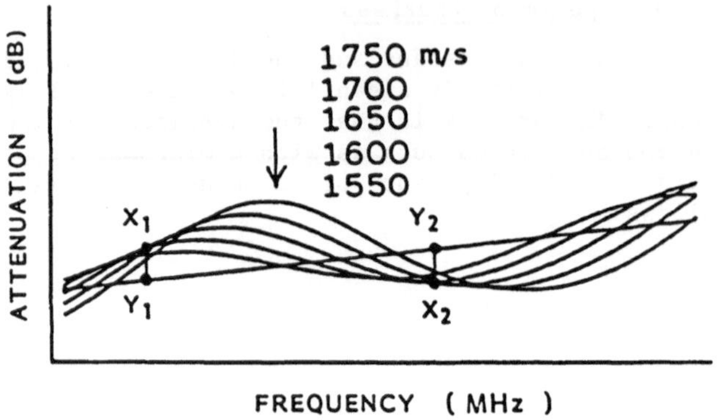

Fig.3 Attenuation vs. frequency.
The undulating curves are calculation of y_1+y_2, while a straight line is calculation of y_1.

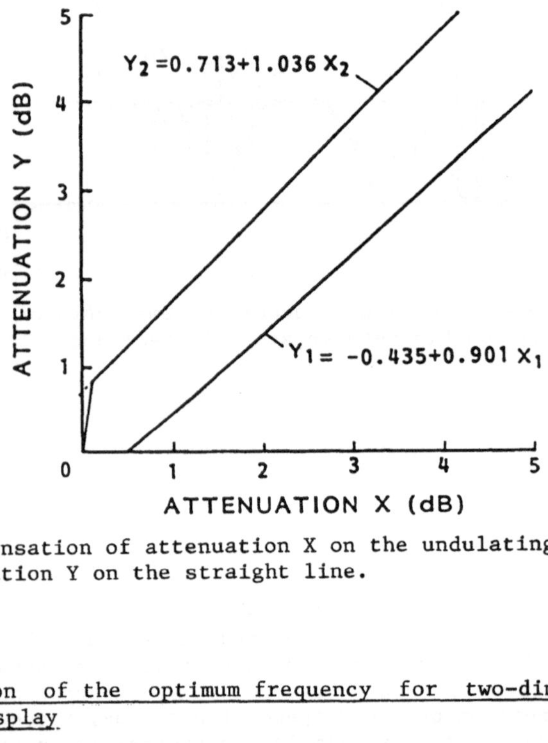

Fig.4 Compensation of attenuation X on the undulating curves to the attenuation Y on the straight line.

(d) Determination of the optimum frequency for two-dimensional quantitative display

In order to display the attenuation constant and sound speed in two-dimensions, it is necessary to choose an ultrasonic frequency and to calculate these two quantities for each pixel using a computer. The method for choosing the frequency for displaying the attenuation constant image is shown investigated in Fig.3. The curves depending upon sound speed intersect at X_1 and X_2. The values at these points are insensitive to sound speed so that the frequencies at these points are optimum for expressing the attenuation constant. In Fig.3 the values on the straight line are the amount of reduction of amplitude in the round trip wave, so that it is appropriate to use the values Y_1 or Y_2 by compensating the values X_1 or X_2, respectively. This compensation is carried out by using the relation in Fig.4, which was calculated by the

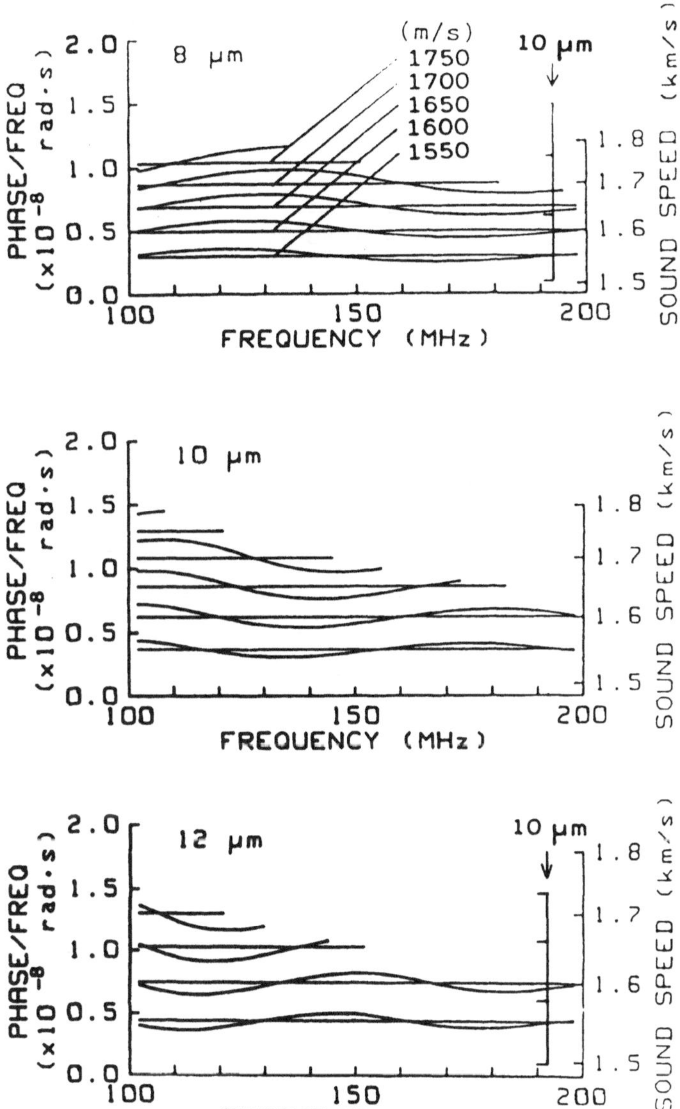

Fig.5 Relationship between phase shift and sound speed depending upon the specimen thickness. The speed scale on the right side is for each thickness, while the speed scale labeled "10μm" is determined under the consideration that the thickness is 10 μm.

regression for five relations between X and Y. In this way actual attenuation in the tissue is obtained, and attenuation constant obtained by normalizing by twice the thickness, with an uncertainty of 10%.

The choice of frequency for determinig sound speed is shown in Fig.5. Note the points in which the curved lines, obtained experimentally, intersect the straight lines. The values at these intersections correspond to the phase shift of the round trip wave. The intersections exist in the frequency range 120 to 130 MHz for 10 μm in thickness of the tissue; the optimum frequency in this range.

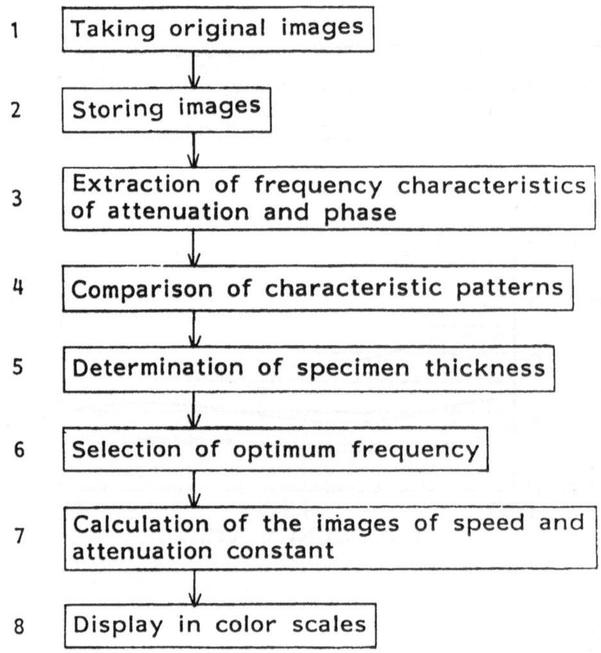

Fig.6 Proceeding flow of processing of the image of attenuation constant and sound speed.

For the case that specimen thickness is less than 10 μm, the values on the curved lines are larger than those on the straight lines in the frequency range 120 to 130 MHz. The scale labeled "10 μm" is speed scale calculated by substituting 10 μm, although the actual thickness is 8 μm. It is noted that, in the range 120 to 130 MHz, the values on the curved lines fit well the scale labeled "10 μm". For the case that the thickness is larger than 10 μm, the values on the curved line are less than those of the straight line in the range 120 to 130 MHz, so that the values on the curved line fit well to the speed scale labeled "10 μm".

Thus, in determining the speed of sound, the phase values in the range 120 to 130 MHz was converted to speed values by the scale labeled "10 μm", within the uncertainty of 10 m/s.

Table 1 Values of sound speed and attenuation constant.

(a) sound speed and attenuation constant are divided in 30 m/s and 0.25 dB/mm/MHz, respectively,

(b) sound speed and attenuation constant are divided in 20 m/s and 0.2 dB/mm/MHz, respectively.

a

SPEED	units:m/s	ATTENUATION	units:dB/mm/MHz
red	(≧1765)	red	(≧ 2.375)
magenta	(1750 ± 15)	magenta	(2.25 ±0.125)
orange	(1720 ± 15)	orange	(2.0 ±0.125)
brown	(1690 ± 15)	brown	(1.75 ±0.125)
yellow	(1660 ± 15)	yellow	(1.5 ±0.125)
green	(1630 ± 15)	green	(1.25 ±0.125)
olive green	(1600 ± 15)	olive green	(1.0 ±0.125)
cyan	(1570 ± 15)	cyan	(0.75 ±0.125)
royal blue	(1540 ± 15)	royal blue	(0.5 ±0.125)
blue	(1510 ± 15)	blue	(0.25 ±0.125)
black	(<1495)	black	(< 0.125)

b

red	(≧1690)	red	(≧1.9)
magenta	(1680 ± 10)	magenta	(1.8 ±0.1)
orange	(1660 ± 10)	orange	(1.6 ±0.1)
brown	(1640 ± 10)	brown	(1.4 ±0.1)
yellow	(1620 ± 10)	yellow	(1.2 ±0.1)
green	(1600 ± 10)	green	(1.0 ±0.1)
olive green	(1580 ± 10)	olive green	(0.8 ±0.1)
cyan	(1560 ± 10)	cyan	(0.6 ±0.1)
royal blue	(1540 ± 10)	royal blue	(0.4 ±0.1)
blue	(1520 ± 10)	blue	(0.2 $^{+0.1}_{-0}$)
black	(<1510)	black	(<0.2)

The attenuation constant and sound speed can be determined by relatively simple processes, and, so that these values for all pixels in a frame can be calculated with a computer. Consequently, two-dimensional computer processed images can be obtained by the procedure shown in Fig.6.

(e) Proceeding flow for quantitative display

Figure 6 shows an image processing method to obtain the images of actual attenuation constant and sound speed. The original images of amplitude and phase within 16 frames are stored. The frequency characteristics of amplitude and phase for some small portions on the original image were obtained. By comparing these characteristics with the model pattern of the frequency characteristics obtained by the theoretical calculation, the specimen thickness, attenuation constant and sound speed are determined as shown in section of method (c). The next procedure is choosing the ultrasonic frequency in order to obtain the images of actual attenuation and speed as shown in section of method (d). Then the actual attenuation and sound speed for every pixel is calculated and the processed image is displayed in color scales.

Table 1 shows the relationship between the color and the value range of attenuation constant and sound speed. In Table 1(a), sound speed and attenuation constant are divided into 30 m/s and 0.25 dB/mm/MHz intervals, respectively. In Table 4(b) these values are divided into 20 m/s and 0.2 dB/mm/MHz intervals, respectively.

RESULT

Figure 7 is an example of human liver cirrhosis. In the acoustic images, the left figure is a speed image and the right figure is an attenuation constant image. The upper two images are shown in the scale of Table 1(a) and the lower two are shown in the Table 1(b). The optical image is an optical micrograph of a stained specimen from a section next to the one for the acoustic measurement. The portion P, pseudolobules, has sound speed in the range blue to cyan (1510 to 1570 m/s) and attenuation constant in the range cyan to yellow (0.75 to 1.75 dB/mm/MHz), in the scales shown in Table 1(a). The portion Q, fibrosis, has sound speed in the range cyan to red (1570 to 1780 m/s) and attenuation constant in the range cyan to red (0.75 to 2.5 dB/mm/MHz). The fibrosis portion shows greater values than the pseudolobule portion by 130 m/s and 0.35 dB/mm/MHz.

DISCUSSION

The measurement by the image processing shown in Fig. 6 could contain an uncertainty of approximately 10% in attenuation constant and 10m/s in sound speed. However, it is acceptable because it enables to measure acoustic properties with structural features.

With respect to the acoustic properties, the portions of both P and Q exibit variation in the values of attenuation constant and sound speed, particularly fibrosis portion exibits wide variation. Judging from comparison between the acoustic image and the optical image, it is suggested that the variation in acoustic properties is due to the variation in distribution of tissue elements: the portion in which the tissue elements are densely distributed has greater attenuation and speed. This suggests the consistency with the idea that, at the macroscopic level, the attenuation constant and sound speed increases with the protein content [2].

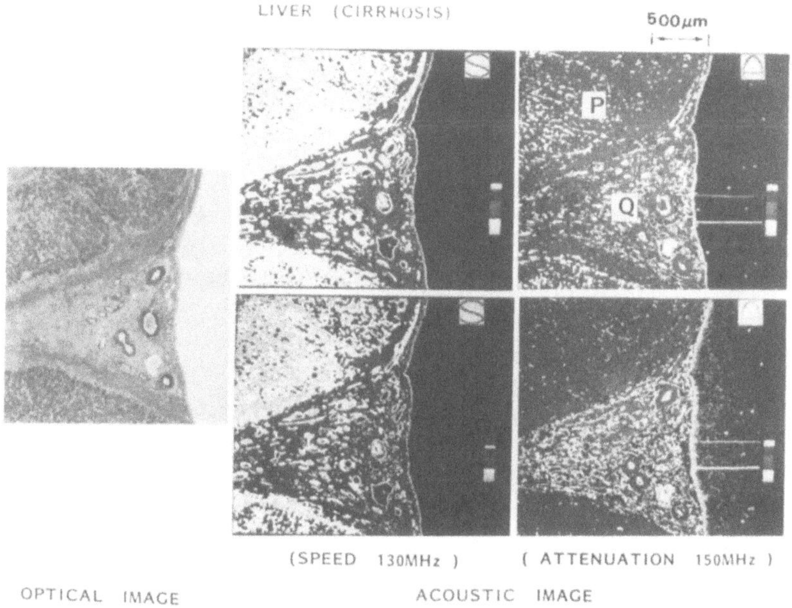

Fig.7 Example of the quantitative display of the processed images of sound speed and attenuation constant.

REFERENCES

1) J.G.Miller, J.E.Parez, and B.E.Sobel;"Ultrasonic characterization of myocardium." Progress in cardiovascular Diseases, Vol.28, No.2, pp.85-110,1985
2) R.L.Johnston, S.A.Goss, V.Maynard, J.K.Brady, L.A. Frizell, W.D. O'Brien Jr., and F.Dunn;"Elements of tissue characterization. Part I", Ultrasonic tissue characterization, M.Linzer, ed., National Breau Standards, Spec Publ. 525. pp.19-27, 1971.
3) S.A.Goss, R.L.Johnston, V.Maynard, L.Nieder, L.A.Frizell, W.D. O'Brien Jr., and F.Dunn;"Elements of tissue characterization. Part II." Ultrasonic tissue characterization, M.Linzer, ed., National Breau Standards, Spec Publ. 525. pp.43-51,1971.
4) H.Okawai, M.Tanaka, N.Chubachi and J.Kushibiki;"Non-contact simultaneous measurement of thickness and acoustic properties of biological tissue using focused wave in a scanning acoustic microscope."JJAP Suple 26-1, pp. 52-54, 1986.
5) H.Okawai, M.Tanaka and F.Dunn," A non-contact acoustic method for the simultaneous measurement of thickness and acoustic properties of biological tissues." IEEE Trans. UFFC (submitting).

CONTINUOUS WAVE OPERATION OF ACOUSTIC MICROSCOPE USING DOPPLER DOPPLER MODULATION

M.G. Somekh, D. Zhang, and R.K. Appel

Dept. of Electronic Engineering, University of London
Torrington Place, London WC1E 7JE, UK

1. INTRODUCTION

The applications of scanning acoustic microscope technology may be divided in two categories, subsurface bulk wave and near surface Rayleigh wave. Subsurface imaging of optically opaque materials is the most well-established area of application. Surface and near-surface imaging above 1GHz using excitation and detection of surface waves (V(z) mode) is still handicapped by the ability to obtain accurate quantitative measurements. On the other hand, there has been considerable success in using this contrast mechanism in non-imaging mode at lower frequencies as a tool for local acoustic measurements[2]. The V(z) effect can be used to determine the surface wave velocity of materials with considerable accuracy. This has several applications such as determination of crystallographic orientation, film thickness[3] thin film mechanical properties[4] and residual stress[5]. As instrumentation improves there will undoubtedly be many more application areas for both cylindrical and spherical lenses.

There are several difficulties with the conventional pulse echo implementation of the V(z) method that limit its accuracy and the upper frequency over which it may be accurately used. (i) It is difficult to obtain phase information, (ii) the signal is not truly monochromatic as the pulse used typically has a fractional bandwidth of one part in 80, (iii) the pulses must be well separated so the method has not been accurately operated above a few hundred MHz, and (iv) the pupil function of any one lens is fixed. The difficulties outlined in (i) and (ii) can be overcome by extracting a single Fourier component of the received signal.[6 7]

We described a simple method which we believe overcomes many of the problems associated with the pulsed operation of V(z). In particular we will argue that the method is particularly suited for operation at high acoustic frequencies where other methods tend to break down. In this paper we present preliminary results using a line focus lens operating close to 200MHz. This lens was chosen for the preliminary tests both because it allowed ready comparison with pulse echo measurements and because in some ways it provides a more severe test of the method than operation in the GHz region.

2. IMPLEMENTATION

2.1. Theoretical Aspects

It is necessary to separate the signal reflected from the sample surface from the very much larger spurious signals arising from electrical breakthrough and reflections from the lens water interface. In conventional pulse echo methods this separation is achieved by time gating. In this work we separate the sample reflection by frequency shifting. The sample is periodically vibrated through a small amplitude at frequency f_m so that the reflected incident acoustic energy (frequency f) contains frequency components at f, $f+f_m$ and $f-f_m$. By extracting the information from one of these sidebands the amplitude and phase of the reflected signal may be extracted.

Before discussing the practical implementation of our system it is perhaps necessary to summarise the important aspects of the pulse echo based V(z) method. It is well known that the utility of the V(z) method depends on interference between a normally incident ray and one suffering mode conversion to a surface wave on the material.[8] In this mode of operation the microscope may be thought of as an acoustic interferometer. In the pulse echo mode of operation, care has to be taken to ensure that the output one detects really is a true interference between the Rayleigh and the specularly reflected ray. For instance, the different times of flight means that the two returning pulses do not perfectly overlap, or to put it another way, the finite bandwidth of the pulses means that caution has to be exercised when one specifies the centre frequency at which the interference takes place. This issue has been explicitly addressed by Liang et al[16] and Weaver[7] where they used heterodyning methods to downconvert the RF signals thus enabling the use of narrowband filters to eliminate all but the desired centre frequency. These systems are, to the authors' knowledge, the only ones giving true single frequency interference. This does not imply that curves obtained by other authors are erroneous merely that without precisely defining the electrical detection circuitry the exact operation frequency cannot be certain. All theoretical formulations used for interpretation of curves rely on a single frequency analysis. The method we present here is another single frequency technique since the only carrier of information is the doppler shifted reflection from the sample surface.

We will now show how the V(z) formulation given by Sheppard and Wilson[9] is modified in the presence of an oscillating sample. Their analysis implicitly assumes a true single CW frequency source which is of course valid for our system. The analysis shown below applies to a cylindrical lens but may be easily modified for a spherical lens[10]

V(z) for conventional system:

$$V(z) = \int_0^\infty P(\theta)\, R(\theta)\, \cos(\theta)\, \exp\{j2kz\cos(\theta)\}\, d\theta \qquad (1)$$

where θ is the angle of incidence, $P(\theta)$ is the lens pupil function taking into account both transmission and reception. $R(\theta)$ is the reflection coefficient of the sample, and k is the wave number in the coupling fluid. Note that the limits of integration are allowed to vary for 0 to infinity because the pupil function $P(\theta)$ truncates the integral. The time dependence $\exp j2\pi ft$ is assumed throughout, so that the V(z) is an RF signal with angular frequency $2\pi f$.

If the sample oscillates sinusoidally around the average defocus position, z, with amplitude δz at angular frequency $2\pi f_m$ the expression for V(z) is modified to:

$$V(z) = \int_0^\infty P(\theta)R(\theta)\cos(\theta)\exp(j2kz\cos(\theta))\exp(j2k\delta z\cos(\theta)\cos2\pi f_m t)d\theta \tag{2}$$

The Fourier series representation of the second exponential means that the signal returning to the transducer contains a series of sidebands separated by the modulation frequency. It can be seen that only signals incident on the sample will be frequency shifted so that any signal other than the carrier can be used to acquire a V(z) curve from the sample surface. If we now represent the component returning at frequency $f+nf_m$ as $V_n(z)$ it can be readily seen that equation 2 is equivalent to:

$$V_n(z)=\int_0^\infty P(\theta)J_n(2k\delta z\cos(\theta))R(\theta)\cos(\theta)\exp(j2kz\cos(\theta))\, d\theta \tag{3}$$

where J_n is the nth order Bessel function of argument $2k\delta z\cos(\theta)$. We note therefore that the effect of modulating the sample is effectively to modify the pupil function by the appropriate Bessel function. This is not simply a multiplication factor on the pupil function but actually modifies its angular variation. Large index modulation will enable synthesis of a wide range of pupil functions with either maxima or minima at the Rayleigh wave critical angle thus providing a novel means of enhancement[11] or suppression[12] of Rayleigh wave contrast. Whether it will prove possible to produce such high modulation indices without unduly disturbing the couplant is the subject of present investigation.

It can be readily shown that if the vibration amplitude is small compared to the wavelength (ie, $k\delta z \ll 1$) that only the centre frequency and the two first order sidebands are significant. The expression for the terms at $f\pm f_m$, $V_{\pm1}(z)$, then reduces to:

$$V_{\pm1}(z)=k\delta z\int_0^\infty P(\theta)R(\theta)\cos^2(\theta)\exp(j2kz\cos(\theta))d\theta \tag{4}$$

2.2. Mechanical and Electronic Realisation

A simplified schematic diagram of the system is shown in figure 1. A piezoelectric pusher was mounted under the specimen stage, which was driven by a sinusoidal drive signal whose nominal frequency was 600Hz. The amplitude of vibration was measured with an optical interferometer and was found to vary linearly with drive amplitude. The peak to peak drive amplitude chosen throughout these experiments was between 100 and 300 Angstrom, which was sufficiently small not to disturb the coupling fluid. This was confirmed by the linear response of the system as a function of drive voltage, if any turbulence were introduced into the water one would expect the response to be erratic. Furthermore fairly accurate values of the sample displacement could also be found by comparing the ratio of the first and second acoustic sidebands. Figure 2 compares the results obtained for the vibration displacement against voltage using both the ratio of the acoustic sidebands and the more accurate optical interferometry.

The output signal from the system contains the carrier and the doppler shifted sidebands. The output from the circulator was passed into mixer A whose local oscillator was driven by the signal from the source. The output from the mixer at frequency f_m, $O_A(z)$ can be expressed as:

$$O_A(z)=-\cos(\phi-\beta_A)\ k\delta z\cos(2\pi f_m t)\ V_1(z) \tag{5}$$

where the $\exp j2\pi ft$ dependence is no longer assumed and ϕ is the phase of the V(z) signal and β_A is the phase of the drive to the local oscillator.

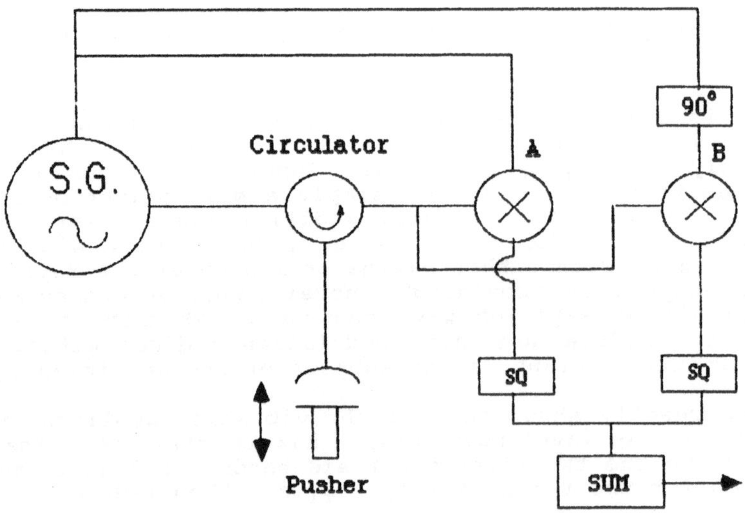

Figure 1. Schematic Diagram of simple system used for extraction of CW V(z) curve.

The premultiplying cosine term arises because both the upper and lower sidebands are downconverted to the same frequency. The amplitude of this output therefore oscillates with a periodicity in z of a quarter the water wavelength. There are several ways of overcoming this problem such as using heterodyne methods as described in the discussion. Here, however, we simply used another mixer B whose local oscillator was driven by a signal 90 degrees out of phase with the drive to mixer A (ie, $\beta_A = \beta_B \pm \pi/2$). Simple algebra can then be used to show that the premultiplication factor is then $\sin(\phi-\beta_A)$. The outputs from each mixer were rectified (in our case with a lock-in amplifier) and sent to the computer via two A/D channels. These two outputs were then squared and added to yield the true V(z) curve. Slight imbalance in the two channels leads to a small amplitude modulation whose periodicity is a quarter the wavelength in water.

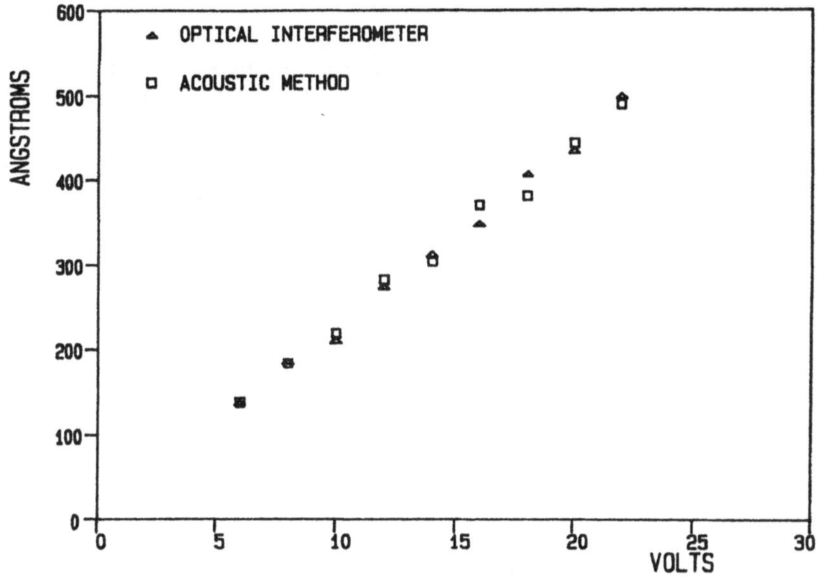

Figure 2. Input voltage to piezo pusher against peak to peak vibration amplitude measured acoustically and with an optical interferometer.

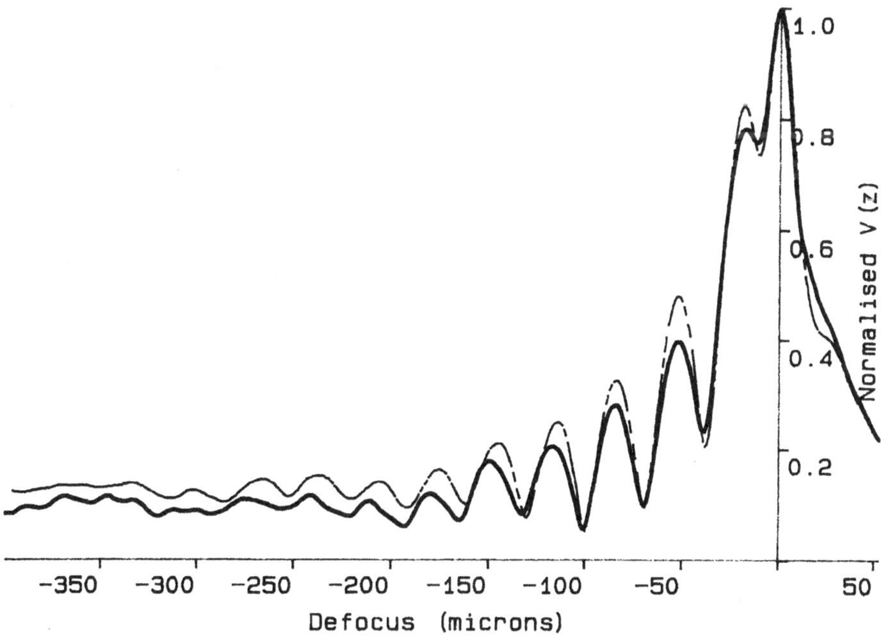

Figure 3. Experimental V(z) curves on glass surface, continuous curve was obtained with doppler modulation system and dashed curve was taken using pulse echo electronics. Frequency 202.46MHz.

2.3 Results

Figure 3 shows two V(z) curves taken on the same glass sample using a line focus lens operating at 202MHz. The continuous curve was taken with the modulation technique and the dashed curve was obtained using conventional pulse echo electronics. It can be seen that the two curves are very similar exhibiting the same periodicity. There are, however, detailed differences in the shape of the curve. It is worth pointing out that the slight difference in the amplitude of the Rayleigh ripple is due to the variation in modulation index with angle, that is the extra $\cos\theta$ term in equation 4.

3. DISCUSSION

The system we have described is fairly simple to implement and can be used to reproduce good V(z) curves. We would not claim that the method is ideal for the frequency range we have presented here and much of this discussion will be devoted to a detailed consideration of the areas where we believe this method will be particularly useful.

3.1. Some problems with the method

The method relies for its successful operation on the ability to separate the desired reflection from the spurii. There are,

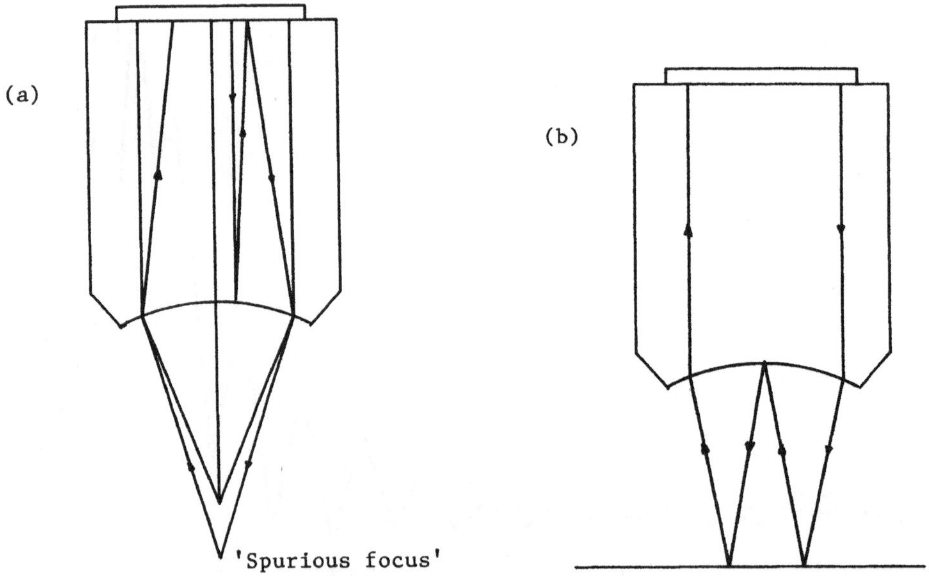

Figure 4a. Ray diagram showing multiple reflections in the buffer rod leading to false focus.

Figure 4b. Ray path showing multiple reflections in the coupling fluid giving rise to high frequency ripple.

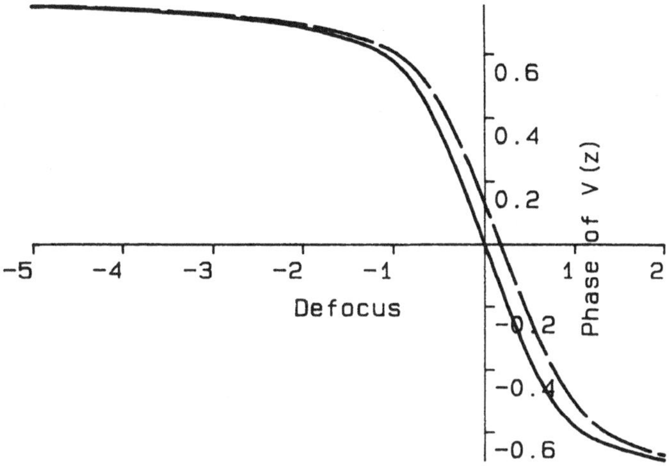

Figure 5. Diagram showing phase of V(z) for a perfect reflector subtracted from 2kz for a cylindrical lens. Solid curve is with no spurious reflections and dashed curve arises when the multiple reflection shown in figure 4a has equal amplitude to the "direct" path.

however, two main reflections that the system does not eliminate whose effect needs to be evaluated both at the frequency of operation used here and at higher frequencies. One is a signal whose path is shown in figure 4a. This ray is initially reflected from the lens/water interface then reflected from the top face of the buffer rod prior to interacting with the sample. These rays will suffer the same frequency shift as the rays which pass through the buffer rod and thus cannot be distinguished by their differing frequency. These rays effectively form an imperfect focus a small distance below the true focus, so that the pupil function of the transmitted rays is modified somewhat. We have modelled the effect of this extra ray passage by adding two V(z) responses. It has been shown [13, 14] that an accurate V(z) will only be obtained in regions where the rate of phase change of the V(z) from a perfect reflector is exactly 4π per wavelength defocus. Figure 5 shows the deviation of the V(z) from this value both for an ideal situation with no spurious reflection present and for the situation where a spurious reflection whose amplitude is equal to the wanted reflection which is far worse than the real situation. It can be seen that this modifies the response in regions close to the focus where the V(z) is in any case not accurate, but a modest defocuses there is no observable error in the phase V(z). Similar arguments can be applied for the phases of the Rayleigh contribution. We can thus conclude that provided the unwanted reflections are no greater than this value that our method will yield accurate values of surface wave velocity. By operating the lens in pulsed mode we observed that the reflection following this ray path was in fact 20dB down so that we need not worry about this effect. It is obvious that a good matching layer is helpful in minimising this effect. Furthermore the relative size of the signal should be smaller with a spherical lens.

A second ray path that can lead to spurious reflections is shown in figure 4b. This results from multiple reflections in the coupling fluid. Since the modulation index is small most of the reflected signal is not frequency shifted so that reflection off the lens surface onto the sample can lead to a signal returning the transducer at the same frequency as the wanted signal. The rate of phase change of this signal with defocus will be 8π per wavelength so that when it beats with the single sample reflection to give a ripple with periodicity of half the water wavelength. This may be readily filtered out in the same way as the corresponding signal in a conventional V(z). The relative size of this signal is reduced at high frequencies due to increased water attenuation.

The two paragraphs above allow us to conclude that accuracy of the V(z) curve for determination of surface wave velocity is not impaired by operating in the manner described in this paper.

There is another difficulty that continuous wave operation introduces and this we believe is crucial at low frequencies but becomes less so as the frequency is increased. We mentioned earlier that there are several signals present at the output port of the lens. The presence of spurious signals reduces the amount of preamplification that can be applied to the receiver signal prior to down conversion in the mixer. This can be the most serious limitation on the signal to noise ratio achievable with this method at low frequencies. The performance then depends critically on the effectiveness of the circulator, in the experiments performed at 200MHz the circulator eliminated approximately 6dB of the breakthrough signal. At frequencies above 500MHz non-reciprocal ferrite devices improve considerably allowing close to 30dB elimination of electrical breakthrough thus allowing sufficient preamplification to optimise the signal to noise ratio.

3.2. The effect of increasing the frequency

The main theme of this paper emphasises the advantages of this method at high acoustic frequencies. In the previous subsection we argued that more effective electronics could be incorporated above 500MHz. We will now consider the differences in acoustic performance. One limitation is the amount of frequency shifting we can safely introduce before turbulence in the fluid becomes a problem. The relative size of the sidebands is given by $J_n(2k\delta z)$ which for small index modulation is $(k\delta z)^n/n!$. It follows that for a given maximum "safe" value of δz that the energy in the frequency shifted signal returning to the transducer will increase with frequency.

A fundamental limitation with pulse echo methods is the need to maintain adequate separation of pulses. As the frequency increases the water attenuation increases as f^2 so that the relative bandwidth of the transducer must increase with frequency. This places very severe constraints on the high frequency acoustic lens, particularly if accurate quantitative measurements are required. Firstly, the efficiency of the transducers will deteriorate, so reducing the signal to noise ratio and secondly the water path must be large enough to separate the pulses and prevent multipath reflections in the buffer rod reaching the transducer at the same time as the return signal. Our Doppler method, on the other hand, allows a more relaxed approach to the design of high frequency lenses. Temporal separation is no longer necessary, so high efficiency, high Q transducers may be used. The path length in water may be determined solely be considerations of adequate focussing and acoustic loss rather than pulse separation.

3.3. Improved Electronic system and phase acquisition

The electronic system shown in figure 1 to extract the V(z) curve is not ideal requiring accurate 90 degree phase shift between the drives to the local oscillators. A far more satisfactory scheme is to extract information from one doppler shifted sideband only, by mixing the output from the transducer with a phase locked reference signal. A system has recently been constructed in which the drive to the piezo is derived from the difference frequency between the two sources. This system is presently being evaluated for the extraction of both amplitude and phase information.

The ability of the doppler method to extract phase information is particularly useful. The advantages of extracting phase in acoustic microscopy have been reported elsewhere and our method provides a particularly simple way of acquiring it. There is, however, another very important advantage in using phase information in amplitude only V(z) curves. In order to interpret these curves the water temperatures must be accurately monitored since it effects the velocity of sound. Even with very small thermocouples one cannot be certain that the temperature of the region examined is identical to that in the acoustic path. If the phase of the return signal is measured simultaneously with the amplitude of the V(z) the wave velocity in the fluid through which the sound is passing may be continuously monitored.

4. CONCLUSION

This paper describes preliminary results obtained with a new CW technique. The merits and problems of the method have been examined in some detail. In the immediate future the method will be extended to higher frequencies where the technique should be of greatest use.

ACKNOWLEDGEMENTS

This work was funded by the Science and Engineering Research Council. We would also like to acknowledge the British Council and the Chinese National Education Committee for funding DZ. We particularly wish to thank J Kushibiki of Sendai, Japan for provision of the line focus lens used in the experiments. We also wish to thank Drs John Rowe and John Weaver for significant and stimulating discussions.

REFERENCES

[1] I R Smith, R A Harvey and D J Fathers, IEE Trans on Sonics and Ultrasonics, Vol SU-32, p 274, 1985.

[2] J Kushibiki and N Chubachi, IEEE Trans. on Sonics and Ultrasonics, Vol SU-32, p 189, 1985.

[3] G M Crean, A Golanski, J C Oberlin, Appl. Phys. Lett., Vol 50, p 74, 1987.

[4] G M Crean, M G Somekh, A Golanski and J C Oberlin, Proc. IEEE Ultrasonics Symposium, Denver, Co., p 843, 1987.

[5] "Stress measurement by a line-focus-beam acoustic microscope", by H Shimada, p 50-57 in "Ultrasonic Spectroscopy", Editor-in-chief Y Wada, Project report for work supported by Grant in Aid for Scientific Research from the Japanese Ministry of Education, Science and Culture, 1984-86.

[6] K Liang, G S Kino and B T Khuri-Yakub, IEEE Trans. on Sonics and Ultrasonics, Vol. SU-32, p 213, 1985.
[7] J M R Weaver, submitted to Rev Sci Instr.
[8] W Parmon, H L Bertoni, Electron. Lett., Vol. 34 (3), p 179, 1979.
[9] C J R Sheppard and T Wilson, Appl. Phys. Lett., Vol. 31 (12), p 791, 1977.
[10] M G Somekh, G A D Briggs and C Ilett, Phil. Mag., Vol. 49A, p 179, 1984.
[11] A Atalar, Proc. IEEE Ultrasonics Symposium, Denver, Co, p 791, 1987.
[12] P Sivaprakaspilla, M Nikoonahad and E A Ash, Proc. IEE Ultrasonics Symposium, Atlanta, Ga., p 632, 1983.
[13] H L Bertoni, IEEE Trans on Sonics and Ultrasonics, Vol. SU-31, p 104, 1984.
[14] H L Bertoni and M G Somekh, Proc. IEEE Ultrasonics Symposium, San Francisco, CA, p 715, 1985.

A POSSIBILITY FOR ULTRASOUND FLOW MEASUREMENT WITH MICROSCOPIC RESOLUTION

M. Nikoonahad

Bio-Imaging Research, Inc., 425 Barclay Boulevard
Lincolnshire, Illinois 60069, USA

1. INTRODUCTION

In diagnostic ultrasound, Doppler {1} and the newly emerging time domain correlation techniques {2,3} provide powerful means for flow characterization. In Doppler, the motion of a volume of scatterers along the ultrasound beam causes a shift in the frequency of the return signal. In time domain correlation, one essentially seeks to find the time taken for a volume of scatterers to travel a short distance along the beam. It is clear that both Doppler and time domain correlation are sensitive primarily to axial (parallel to beam) components of the flow. These techniques are relatively insensitive to transverse (parallel to transducer) flow components. As far as the complete characterization of the flow is concerned, the transverse components can potentially also play a significant role.

Recently in a letter we reported a novel technique for measuring the *transverse* flow components with high lateral resolution {4}. The principal operation of this method resides primarily on a dual-beam acoustic lens, which was previously reported for differential acoustic microscopy {5,6}. The two foci generated by this lens are positioned in the region where the flow is to be measured, Figure 1. In the presence of flow, the scattered ultrasound from these two foci leads to two quasi-identical baseband signals shifted in time. The time shift between these two signals is measured by cross correlation, resulting in the flow velocity.

This technique can potentially offer a lateral measurement resolution which, for a given wavelength and aperture, is comparable to that attainable with scanning acoustic microscopy {7,8}. It is, therefore, conceivable that this method could pave the way to very high resolution flow imaging applications. The lateral resolution is, to first order, determined by the separation between the two focal spots and by the beamwidths.

The aim of this paper is to demonstrate this technique by means of results obtained at 4.5 MHz using a laboratory flow setup. Description of the system and experimental results are presented in Section 2. Section 3 deals with conclusions.

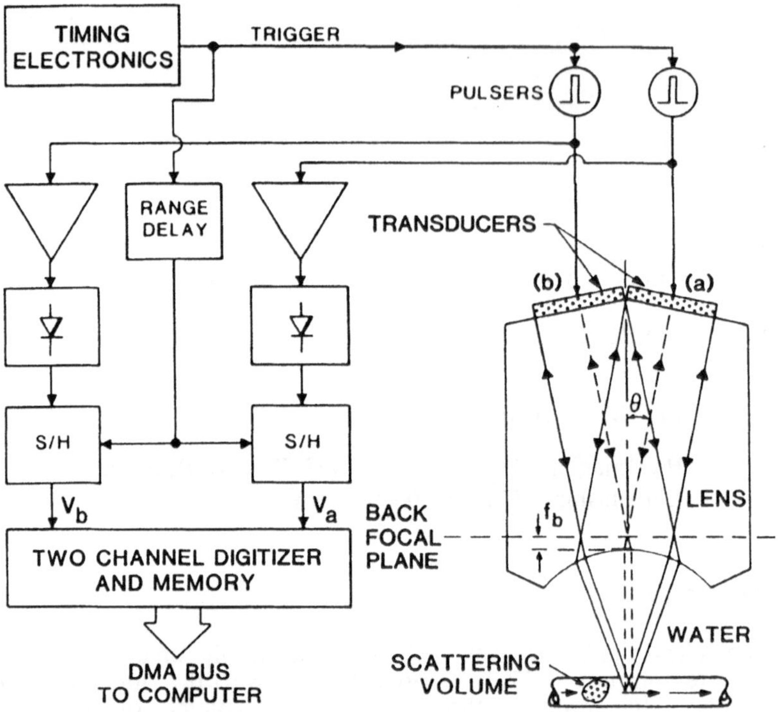

Figure 1. Basic elements of the flow measurement system.

2. EXPERIMENTS

For a given speed-of-sound ratio, n, across the lens, the separation between the two focal spots is given by $\Delta x=2f_b\tan\theta$, Figure 1. For a lens of radius R, the back focal length is given by $f_b=R/(n-1)$. To calculate the flow velocity, we first measure the time taken for a volume of scatterers to travel the distance, Δx, prescribed by the lens.

Aiming for Δx=2 mm, a 13 mm radius lens was fabricated with tilt angles of 14°. The lens material was aluminum (n=4.2) leading to front and back focal lengths of 17 mm and 4 mm respectively. The point sensitivity function and the separation between the spots for this lens were measured by scanning the lens over a 0.1 mm diameter steel wire stretched between two fixed points in water. The scattered signal amplitudes were simultaneously recorded, Figure 2. We observe an excellent agreement between the designed and experimentally measured values for Δx.

To measure flow, the two transducers are excited by two separate pulsers with a PRF of 10 kHz. After suitable amplification the two received signals are amplitude detected and fed into two sample-and-hold (S/H) modules. As the scatterers traverse the foci, two baseband signals, v_a and v_b, are formed at the output of the S/Hs. In the ideal case in which over Δx the scattering characteristics are unchanged, v_a and v_b are identical but shifted by Δx/v. In practice, v_a and v_b are not quite identical and we determine the Δx/v by seeking the peak location of the cross correlation function of v_a and v_b. The two signals v_a and v_b must, therefore, be recorded for a reasonable duration of time. This time duration depends on the dominant frequency in v_a and v_b, which in turn is determined by the size of each focal spot and the average size of the scattering agent. The output integration time for each S/H is 1 ms, which with $\Delta x=2$ mm sets a maximum velocity of 200 cm/s as being measurable with the present electronics. For higher velocities the output integration time can be reduced.

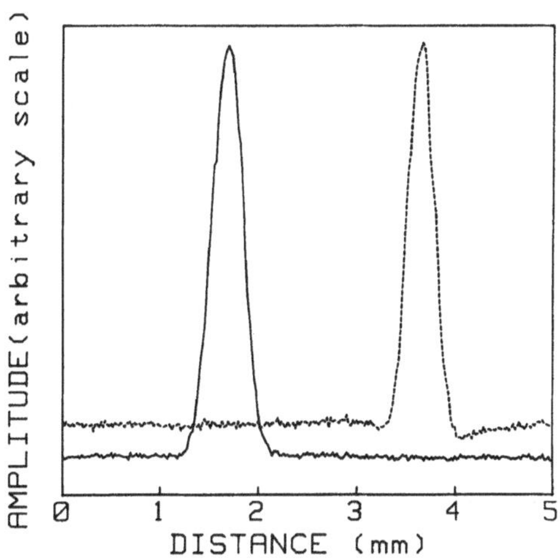

Figure 2. The point sensitivity function of the dual-beam lens measured by scanning the lens over a 0.1 mm diameter wire.

In our experiment, alumina particles (up to 0.6 mm diameter) suspended in water provided the means for scattering. The water reservoir was raised above the measurement table so that under gravity a steady flow could be set up in a 6.4 $\pm$ 0.2 mm nylon tube. We independently measured a volume flow rate of 29 cc/s. No safeguards were used to achieve a particular flow profile. Assuming a uniform flow profile, and based on the volume flow measurements, the flow velocity was estimated to be in the 85 to 96 cm/s range, depending on the exact diameter of the tube. The two voltages v_a and v_b were sampled at 10 kHz and recorded for 100 ms. Figure 3(a) shows the v_a and v_b recorded from the center of the tube. Identifiable features have been marked with identical letters on both traces. The cross-correlation function of these two sequences gives a peak at 2.2 ms, Figure 3(b), which with $\Delta x=2$ mm results in a flow velocity of 91 cm/s. This is within the estimated range of velocities based on independent volume flow measurements.

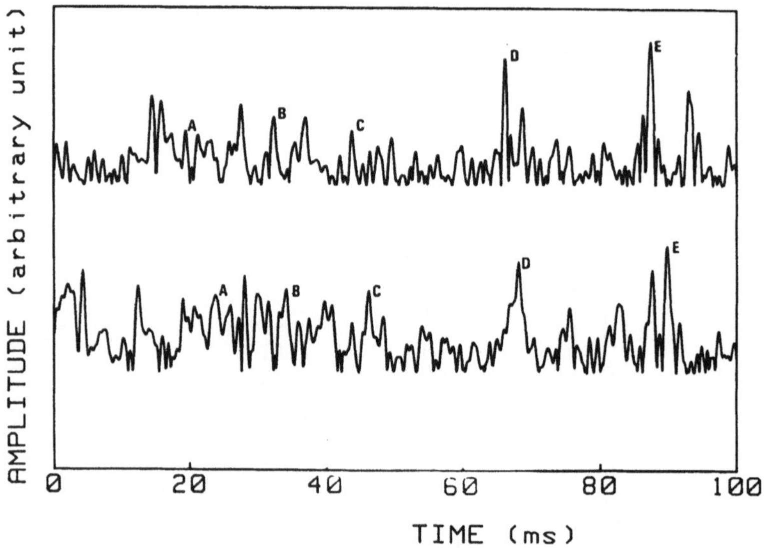

(a)

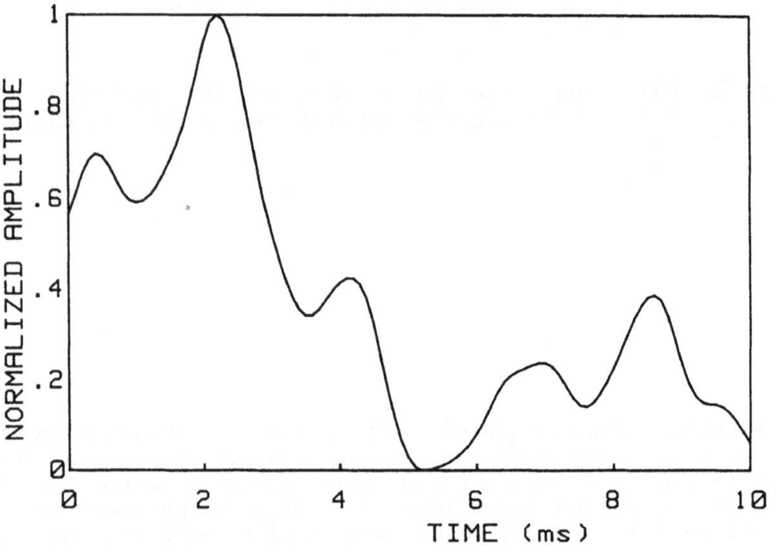

(b)

Figure 3. (a) Two baseband signals collected from the middle of the tube: upper trace (upstream) and lower trace (downstream). (b) Cross-correlation of the two waveforms in (a).

The range delay was then positioned at time locations corresponding to several points across the diameter of the tube. For each point, v_a and v_b were recorded and a cross correlation was performed resulting in the velocity profile across the tube, Figure 4. It is seen that the flow has a maximum at the center of the tube and falls to lower values at the sides.

All the velocity values reported here were calculated with a single measurement of, and cross correlation between, v_a and v_b. For higher accuracies, an average delay can be obtained from a series of sequences. Other factors such as beamwidths, depth of focus, particle size, and spot separation also have a direct effect on the measurement accuracy. Such issues are currently under investigation.

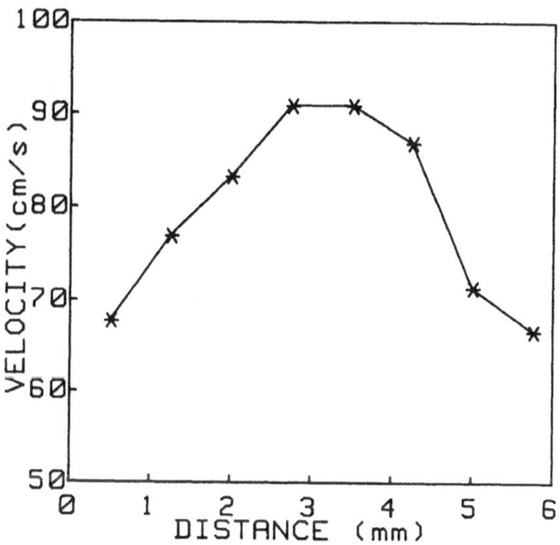

Figure 4. The velocity profile measured across the diameter of the tube.

3. CONCLUSIONS

We have described a high resolution ultrasound transverse flow measurement system; the lateral resolution at present is 2 mm. The principal operations of this system were demonstrated by results obtained from flow in the 50-100 cm/s range. The lateral resolution can be further improved by increasing the frequency and reducing the separation between the two focal spots. From Figure 2 it is clear that Δx can be reduced to 1 mm before scattering cross-talk between the two beams could pose a limitation on flow measurements. Beyond this, to first order, we can scale down Δx with increasing frequency. It is conceivable that at 45 MHz a lateral resolution of 0.1 mm could be achieved. Such a system could, for example, provide a potential capability for characterizing flow in small blood vessels in planes near the skin surface. It should also be possible to aim for mildly focused dual-beam probes at an intermediate frequency (about 15 MHz) for macroscopic transverse flow characterization.

REFERENCES

{1} See for example P. Atkinson and J. P. Woodcock, "Doppler ultrasound and its use in clinical measurements", Academic Press, London, 1982.

{2} P. M. Embree and W. D. O'Brien Jr., "The accurate ultrasonic measurement of the volume of blood by time domain correlation", Proc. IEEE Ultrasonics Symposium, pp 963-966,1985

{3} O. Bonnefous, P. Pesque and X. Bernard, "A new velocity Estimator for color flow mapping", Proc. IEEE Ultrasonics Symposium, pp 855-860, 1986.

{4} M. Nikoonahad and F. Li, "High resolution ultrasound transverse flow measurement", Electron. Lett., 24(4), pp 205-207, 1988.

{5} M. Nikoonahad, "Differential phase contrast acoustic microscopy using tilted transducers", Electron. Lett., 23(10), pp 489-490, 1987.

{6} M. Nikoonahad, "Differential amplitude contrast in acoustic microscopy", Appl. Phys. Lett., 51(21), pp 1687-1689, 1987.

{7} C. F. Quate, A. Atalar and H. K. Wickramasinghe, "Acoustic microscopy with mechanical scanning - a review", Proc. of IEEE, 67(8), pp 1092-1113, 1979.

{8} M. Nikoonahad, "Recent advances in high resolution scanning acoustic microscopy", Contemp. Phys., 25(2), pp 129 - 158, 1984.

NON-CONVENTIONAL PIN SCANNING ULTRASONIC MICROSCOPY

Jerzy K. Zieniuk and Antoni Latuszek*

Institute of Fundamental Technological Research
Polish Academy of Sciences, 00-49 Warszawa
*Institute of Physics, Technical University
02-542 Warszawa; Poland

INTRODUCTION

In a classical scanning ultrasonic microscope using ultrasonic lenses, lateral resolution is defined by wavelength and F# of the lens. Development of tunnelling microscopy[1] gave us the idea of exploring the possibility of building an ultrasonic microscope with a pin instead of a lens to scan the object[2,3]. It is interesting to note that a similar - to some extent - approach has been adopted to develop near-field optical scanning microscope[4]. In this case an aperture (an opening in a opaque layer) of dimensions smaller than optical wavelength is scanned over a sample forming the image. In our case using a pin probe of physical dimensions of the tip much smaller than Λ we can obtain resolution better than with diffraction limited lenses. Unfortunately, because of the very small aperture dimensions of the tip, some problems arise connected with the low level of transmitted energy.

Our basic idea was to build a probe of a conical shape with a small angle of opening and with an ultrasonic transducer attached to the broader and of the pin. The probe can then be scanned over a carefully leveled sample. In such a case the lateral resolution is determined by the aperture of the tip of the probe. The opening angle of the probe is an importent factor. Dürr, Sinclair and Ash[5] found that in the case of their probe having the opening angle of around 120° and being irradiated by bulk wave, in the probe not only bulk waves but also surface waves propagate. Because of that, a single impulse of the bulk wave produces, at the receiving transducer, a number of impulses coming with different time of delay. It makes the interpretation of the impulse shape difficult. In case of opening angle not greater than 20° in a probe, a clean signal of bulk waves propagate 20 dB higher than any other signal.

EXPERIMENTAL SETUP

We used our self-built ultrasonic scanning microscope which is equipped with interchangable lenses (see Figures 1 &

2). Figure 1 shows the mechanical part of the microscope already in position. A special adaptor was used to place a center of the tip of the probe on the axes of the lens.

The electronics of our microscope give a maximum amplification of 80 dB on receiver side, with the maximum level of the output of the generator, 15V. It turns out, that because of the very small surface of the tip, the received signal was too low to be used with the computer to produce the image. For our present study we have used a defectoscope with which we were able to check the possibility of assessing lateral resolution as well as checking the topography of the surface.

As we wished to scan the object with the tip located as close to the sample surface as possible, the crucial point was very exact leveling. As can be seen in the photo (Figure 1) our microscope is equipped with a rather typical, precisely built stage, giving the possibility of leveling along the X and Y axes. To perform leveling of the stage itself or with metal sample already in position, we developed an electric system with which an electrically isolated sharp tip closes a circuit when touching the stage or sample. A simple algorithm makes it possible to asses needed correction. After two or three succeeding repetitions of the procedure it is possible to have an error no bigger than 1 μm. The same is done in the case of the second axis. After that, the auxiliary tip is removed and the pin probe put into position. We can move our lens or pin probe in the Z direction controlling its location with exactnees to 0.5 μm.

Figure 1. Ultrasonic scanning microscope with pin probe

As has been mentioned above, our pin probe consists basically of conical needle made of quartz, metal or other suitable material. Its opening angle (see Figure 3) is 20°. Both ends are flat and parallel. At the broader end of the pin a piezoelectric transducer is located. We use mostly $LiNbO_3$ 36°Y cut crystals polished and gold plated. Its resonance is at 35 MHz.

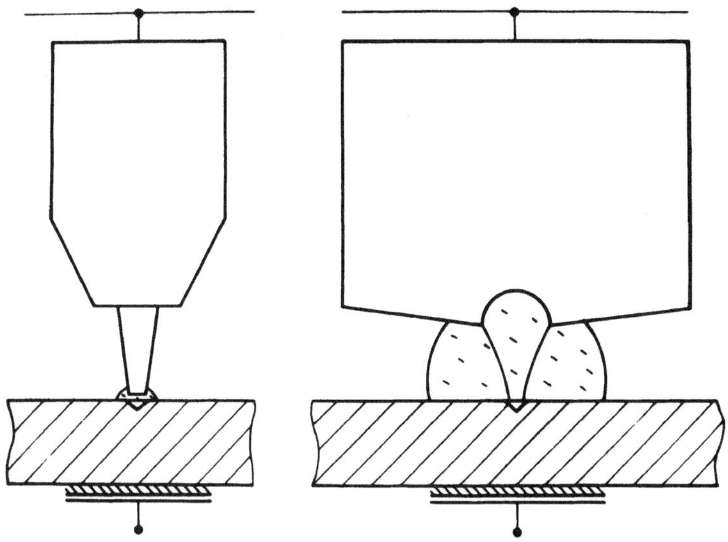

Figure 2. Ultrasonic microscope working in transmission mode. Left: arrangement with pin probe, right: ultrasonic lens.

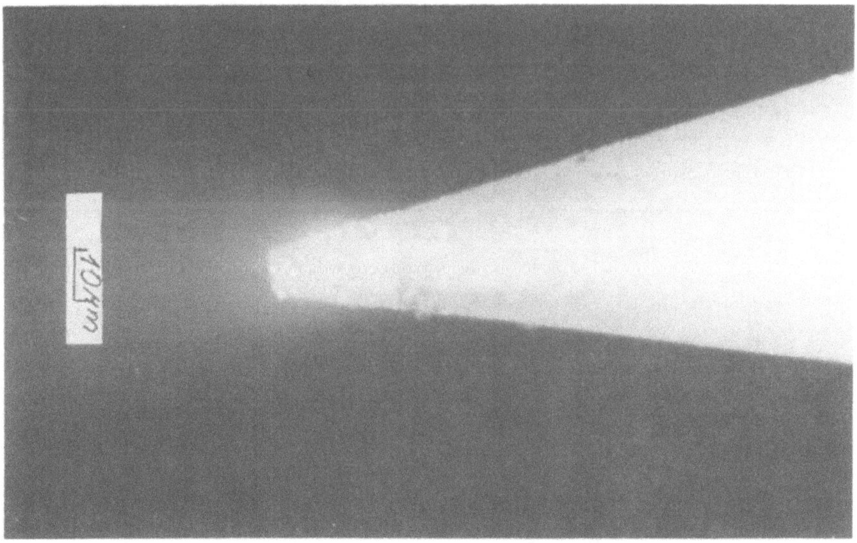

Figure 3. Micrograph of the pin. Diameter of the flat tip is equal to 10 µm.

As the pin probe itself is small and difficult to handle, the transducer is sandwiched between the needle and supporting rod which serves as a backing medium for transducer as well. There has been developed also a system helping to avoid accidentally crushing the needle against sample or stage. As both ends of the conical needle should have flat and parallel surfaces, a special procedure for preparing and polishing needles has been developed.

The diameter of the needle currently in use is equal to 10 μm. It means, that the surface of the tip has only $7.8.10^{-5}$ square milimeter. It is the reason for the need of special electronics.

EXPERIMENTS

To explore the possibility of pin scanning microscopy a series of experiments has been performed. We concentrated on the transmission mode pin probe being the receiver. Two main goals were: i) to see the possibility of fine topographic investigations of the sample surface, and ii) to check the resolution which can be obtained with the system.

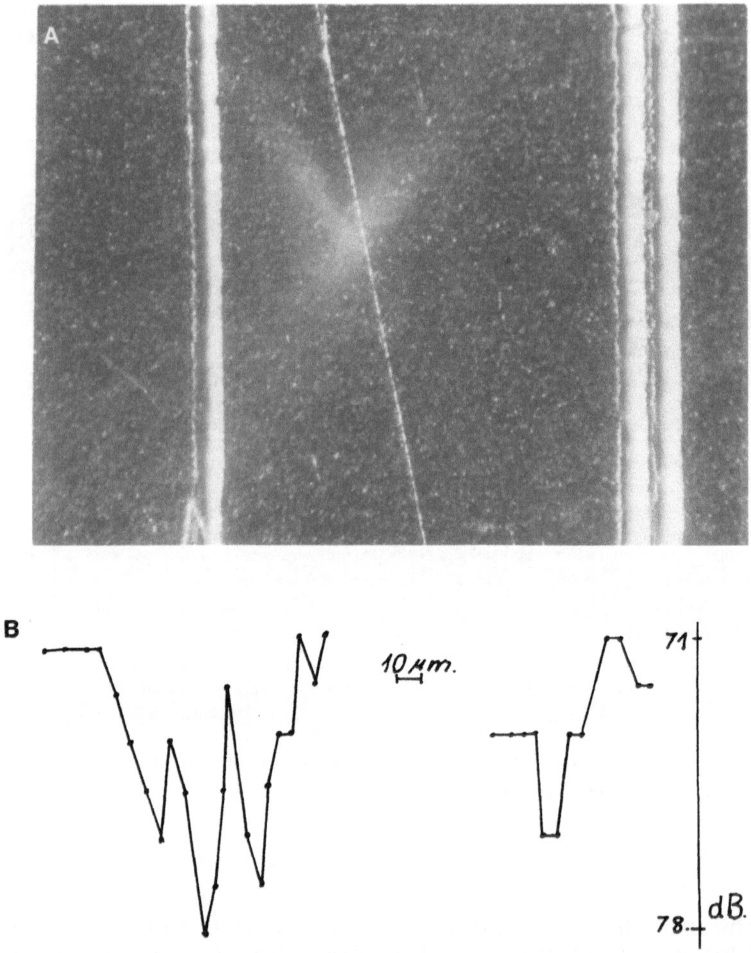

Figure 4A and B. A: micrograph of the surface of the test block with scratches. B: fluctuations of the signal caused by scratches.

Checking the topography. In the preliminary experiments we used a drop of oil as a coupling medium. It was smeared over the surface to form a layer as thin as possible. Lowering the probe, we obtained the first signal when the needle touched the surface of the oil. Further lowering caused some fluctuations in signal level until the needle eventually touched the surface of the sample. We assessed the thickness of such "thin" layer as of the order of 100 μm. It was not possible to get any significantly thinner layer this way.

As we wanted to scan the sample very close to the surface, a method of producing a thin film of the coupling medium was developed. A solution of 10 drops of ethylene glycol in 0.9 mililiter of acetone was used to wet the carefully cleaned surface of the sample. Acetone evaporates quickly, leaving thin film of glycol. Ethylene glycol has good physical and acoustical properties. Its surface tension is much lower than water (44 instead 72 N/m), its boiling temperature is twice that water so it evaporates much slower. Also its acoustical impedance is two times higher than that of water.

After carefully leveling the sample surface and covering it with liquid film, we found experimentally that we are able to detect very fine differences in level. Moving the tip of the probe only 1 μm in the Z direction from the surface causes a 10 dB drop in the signal.

Checking Resolution. A rectangular bar of aluminium with very finely polished surface and with transducer attached to the opposite parallel surface was used as a test block. The thickness of this bar was chosen so that the upper surface was in the far field of the transducer. A diamond tip was used to make a series of parallel scratches. Figure 4A shows a micrograph showing the scratches. Figure 4B shows a fluctuations in the received signal level. It is interesting to compare those two Figures. One can see the correlation between line distribution and significant signal drop. The pin probe with the tip located very close to the upper surface of the test bar was scanned along the line normal to the scrathes' axis. Every 10 μm the level of the signal was measured. Far from the scratches signal level fluctuated with changes no bigger than 1.5 dB. When the probe approached scratches some bigger fluctuations of the signal were observed. It is perhaps a result of superposition of acoustical surface waves generated at the edge of the cratch with bulk wave. It is interesting to note that in case of twin scratches, biggest drop of signal can be observed between scratches.

Comparision of micrograph and a diagram shows that the lateral resolution is around 10 μm, just the same as crossection of the tip. The lateral resolution which would be achieved with the lens would be in this case 50 μm.

CONCLUSIONS

It has been experimentally prooved that the ultrasonic microscope equipped with pin probe can be used to investigate fine details of the sample surface and its topography with the resolutions exceeding resolution of the ultrasonic lens.

ACKNOWLEDGMENTS

JKZ wishes to acknowledge financial support for this project by the Institute of Physic of the Academy of Sciences (Project CPBP 01.04). We greatly appreciate professors L. Filipczyński and J. Auleytner encouragement and support. Many thanks to D. Wytrykowski for his help and I. Głębocka for suggesting ethylene glycol.

REFERENCES

1. C. F. Quate,Vacuum tunelling: A new technique for microscopy. Physics Today, 1, 8 (1986)
2. J. K. Zieniuk, A. Latuszek, A conception of an acoustic pin scanning microscope. Royal Microscopical Society Proccedings. 21: S13 (1986)
3. J. K. Zieniuk, A. Latuszek, Ultrasonic pin scanning microscope a new approach to ultrasonic microscopy, in: "1986 Ultrasonic Symposium Proceedings" IEEE (1986)
4. U. Ch. Fisher, U. T. Dürig, D. W. Pohl, Near-field optical microscopy in reflection. Appl. Phys. Lett. 52: 249 (1988)
5. W. Dürr, D. A. Sinclair, E. A. Ash, High resolution acoustic probe, Electronics Letters. 16: 805 (1980)

ACCURATE RECONSTRUCTION OF FLAWS USING A NEW IMAGING TECHNIQUE

D.K. Mak

Metals Technology Laboratories
Canada Centre for Mineral and Energy Technology
Ottawa, Canada

ABSTRACT

A method which requires two transducers to collect data for ultrasonic imaging is described. Each transducer is used as a combination transmitter/receiver. Then one transducer is used as a transmitter and the other as a receiver. The computer software searches for the echoes which arise from the same defect. Experiments have been performed with both linear and circular apertures. Noise was pre-filtered from the signals to eliminate most of the artefacts. The accuracy obtained in imaging the defects was better than one wavelength in both the longitudinal and lateral direction. This method has also been extended to three-dimensional space where three transducers were used. Some data have been collected.

INTRODUCTION

Various techniques have been used for ultrasonic imaging. Usually, the rf signal is recorded, the data processed and the image displayed. Other methods[1-3] record the peaks of the pulses and use them as indicators of the pulse arrival time; this simplifies the process of data collection and image processing. In all these methods, a single set of data was collected for each transducer scan position.

A different method is described here which uses two transducers, and three sets of data are collected for each transducer-pair scan position. The time of the first arrival of each echo is registered and used in image processing.

Method

This method uses two transducers for acoustical imaging. It is assumed that the speed of sound is constant in the material inspected. When a single transducer is used as a transmitter/receiver, a point defect lying within its beam width will scatter acoustic waves back to the transducer. The point defect lies on a circular arc, centred at the transducer position, with radius equal to the travel distance between the defect and the transducer.

If the point defect lies within the beam width of another transducer acting as a transmitter/receiver, the point defect can be located at the intersection point of the two circles within the material. If there is more than one point defect within the region of interest (Fig. 1), each

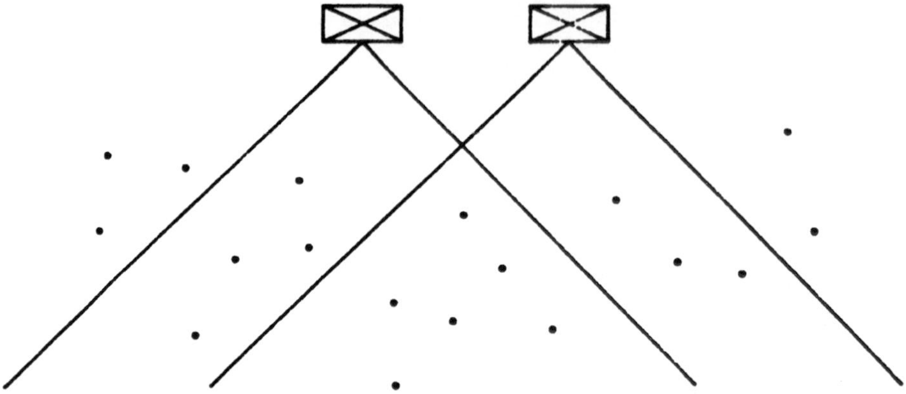

Fig. 1. Defect points in the beam insonification regions of the transducers.

transducer will detect a number of rf backscattered echoes, some of which arise from point defects that lie within the common acoustical zone of both transducers. If echoes from the same point defect can be identified in the two A-scans, the defect can be located accurately at the intersection point of two circles. When two transducers are used as a transmitter-receiver system, a point defect that lies within their common beam width will lie on the arc of an ellipse with foci at the two transducer positions.

Let t_j', t_k'' be the travel time of the echoes detected by the two transmitter/receivers. Let t_i be the travel time of the echoes detected by the same two transducers used as a transmitter-receiver system.

$$t_1', t_2', --, t_j', -- t_m'$$

$$t_1'', t_2'', -- t_k'', -- t_n''$$

$$t_1, t_2, -- t_i, -- t_p$$

A travel time that satisfies the following equation can be used to calculate the locations of the point defects.

$$t_i \pm \delta t = 1/2\ (t_j' + t_k'') \qquad \begin{array}{l} i = 1 ---- p \\ j = 1 ---- m \\ k = 1 ---- n \end{array} \qquad (1)$$

where δt is the experimental error in travel time. In our experiment, δt was chosen to be the travel time corresponding to approximately half a wavelength.

It has been found experimentally that the travel time from the centre of the transducer crystal to a point defect corresponds to the first arrival time of the echo and not to the peak of the echo. The rf waveform in these experiments was sampled above the Nyquist rate and then rectified by appropriate software. The travel time of the peak was measured and then used in Eq 1. The time of first arrival of each echo was measured also; actually it was the time of the zero crossing of an echo below a threshold set to be slightly above noise level (Fig. 2). This was used to locate the defect. When the signal-to-noise ratio was low for a certain echo, the zero crossing below the preset threshold did not correspond to the first arrival time, and an error would show up during the location of the defect. Thus, the peak time is used in Eq 1 because it is not affected by the threshold level setting.

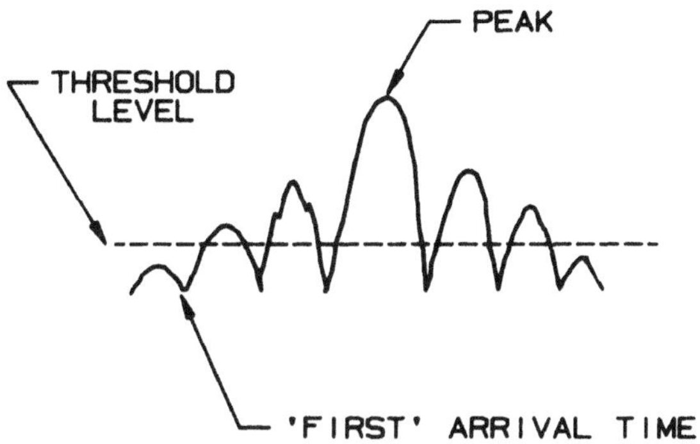

Fig. 2. Locating the peak and 'first' arrival time of a rectified signal.

In ultrasonic testing, a number of detected signals are not caused by sound waves scattered directly from the defects but are caused by geometric reflections, mode-converted signals, multiple scattering between defects or grain noise[3-6]. The most common method of eliminating some of these signals is to choose an arbitrary threshold level and display image points which correspond to peak amplitudes above the set threshold. Another method is to plot peak transit time versus scan position and extract the flaw signal by certain interactive programs. A flaw signal will trace out a hyperbolic curve[4] for a linear aperture and a cosine-like curve for a circular aperture[3]. Different flaw curve extraction programs have been written for linear aperture[4,5,7]. The interactive method that has been used in our work is the direct approach of choosing the transit-time point which is considered to be a flaw signal. This prefiltering can eliminate almost all the artefacts in the final image. However, some artefacts can still be generated by our imaging method (Fig. 3).

Experiment and Results

An experiment was performed on an aluminum block 3.8 cm wide, 5.0 cm high and 2.5 cm thick with a side-drilled hole 1.5 mm in diameter. Two

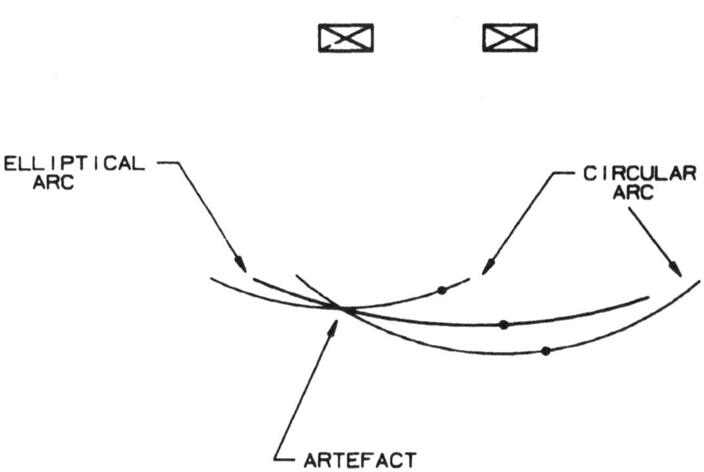

Fig. 3. Appearance of an artefact.

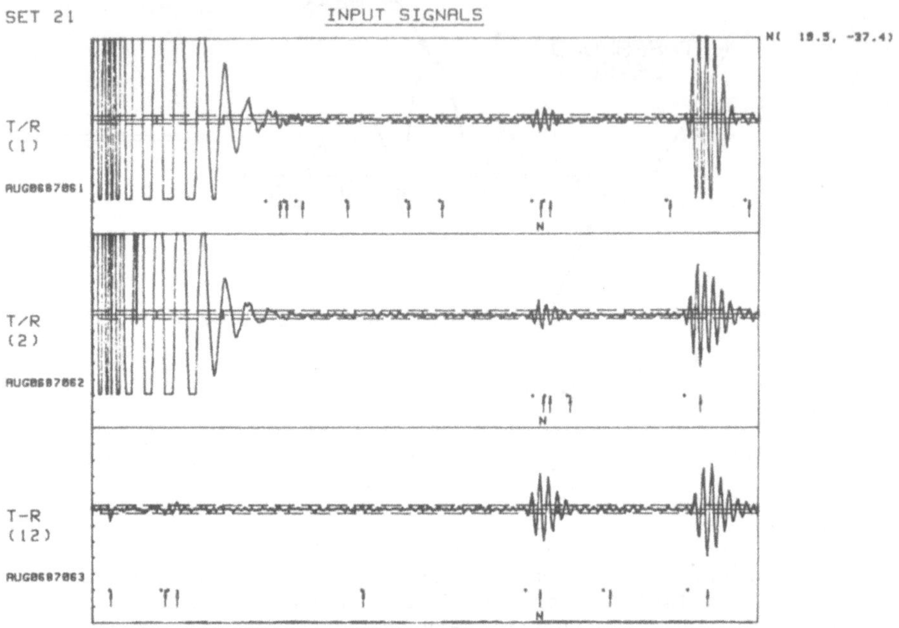

Fig. 4. The three rf waveforms recorded at one of the scan position pairs in linear scanning.

5 MHz 6.35 mm diameter transducers were spaced 10.2 mm apart centre to centre. They were scanned automatically along the surface of the block in scan steps of 0.5 mm. At each of the forty-one scan position pairs, data were recorded. The rf waveforms from one of the scan position pairs are illustrated in Fig. 4. The peak travel times versus scan position for transducers 1-2 are plotted in Fig. 5. The threshold was set slightly above the noise level. It can be seen from the plot that quite a number of other signals apart from ones from the hole were detected. Using these raw data for transducers 1, 2 and 1-2, and applying the method described above, the image field was displayed in Fig. 6. A number of points clustered at the top of the hole. However, a number of spurious points were also displayed.

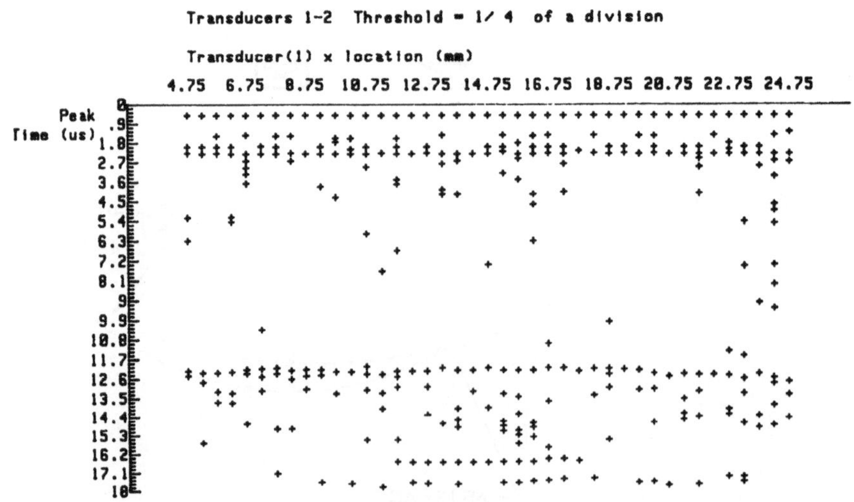

Fig. 5. Peak transit time versus scan position for transducers 1-2.

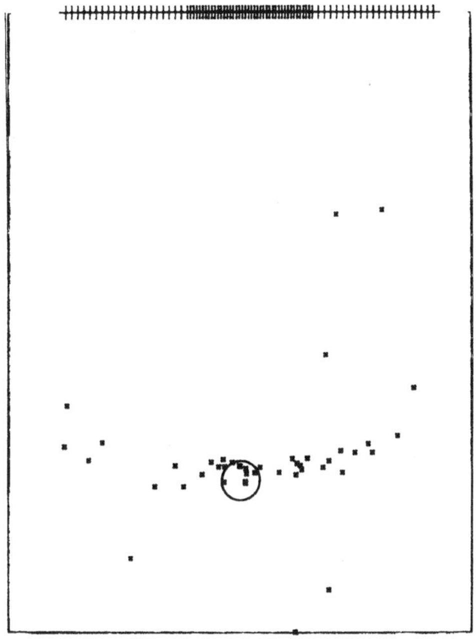

Fig. 6. Image points using the raw data. +'s denote transducers positions. The boundaries of the block and the holes are drawn in for comparison.

Prefiltering was then performed on the transit time plots. Points on the hyperbolic curves were sorted out. The filtered transit time plot for transducers 1-2 was displayed in Fig. 7. Using these filtered data, the image field was replotted and shown in Fig. 8. Image points corresponding to the top of the hole could be observed.

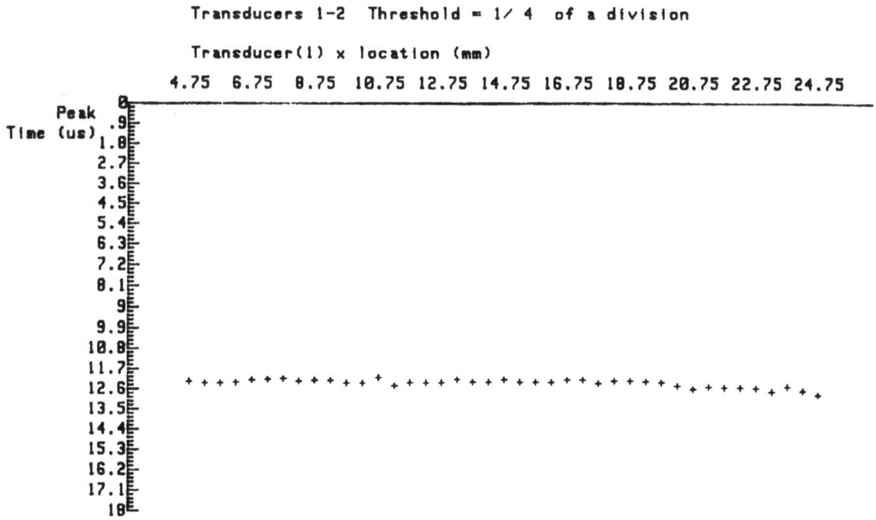

Fig. 7. Filtered data of Fig. 5.

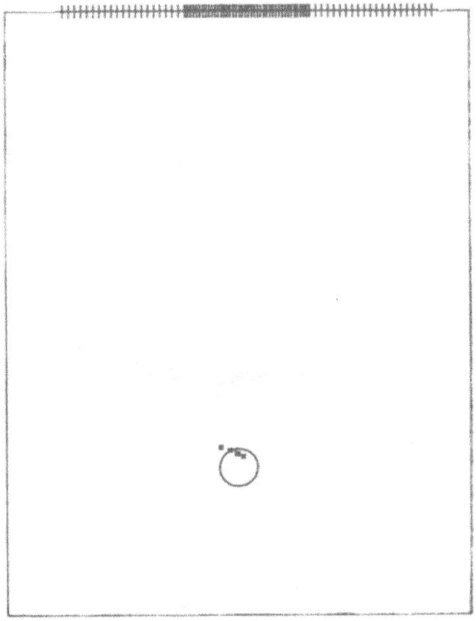

Fig. 8. Image points corresponding to the top of the hole using filtered data.

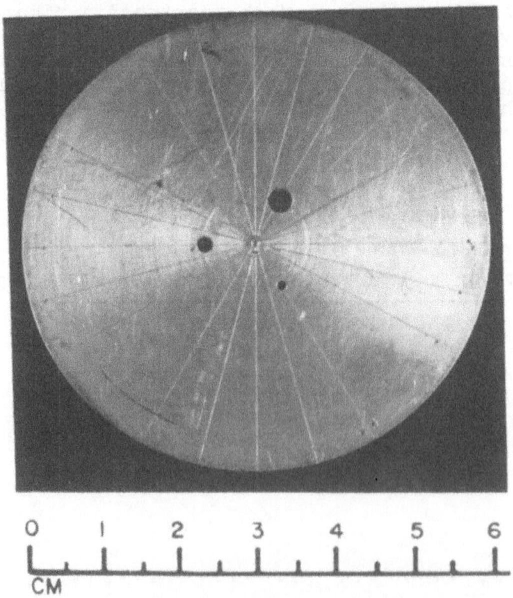

Fig. 9. An aluminum test object 6 cm in diam with 1 mm, 2 mm and 3 mm holes.

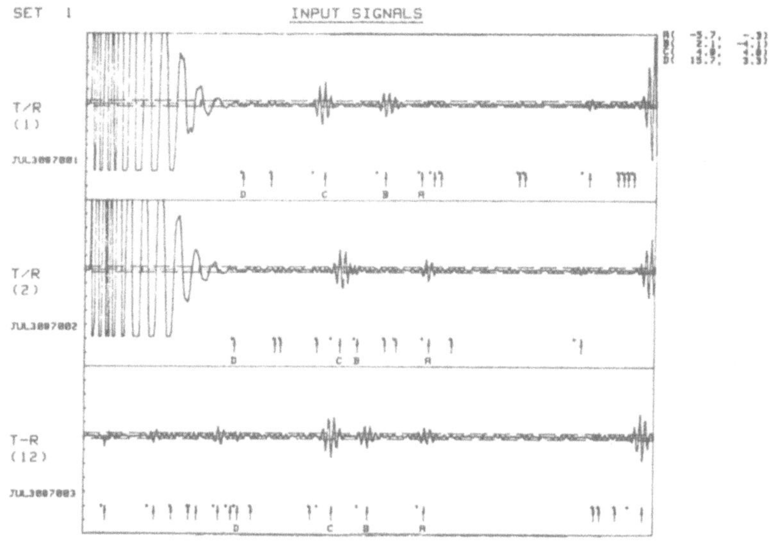

Fig. 10. The three rf waveforms recorded at one of the scan position pairs in circular scanning.

A linear scan along the top surface of the block only allows the imaging of the top part of the hole. For holes in a cylinder, the boundaries of the holes can be imaged by using circular apertures, i.e., scanning the transducers along the cylindrical surface. Three holes 3 mm, 2 mm and 1 mm in diameter were drilled in an aluminum block of radius 3 cm. The holes were centred at a radius of 6.5 mm and spaced at equal angles. A picture of the cylinder is shown in Fig. 9. The cylinder was put on a computer-controlled turntable, with its centre lining up with the centre of the turntable. Two 5 MHz 12.7 mm diam transducers were held on the same horizontal plane against the surface of the cylinder, subtending an angle of 30° at its centre. The turntable was rotated through one turn, in steps of 6°, i.e. through 60 positions. Three rf signals recorded at one position are displayed in Fig. 10. Transducer one was located at a polar angle = 30° and transducer two at a polar angle = 0°. "A" denoted signals from the 2-mm hole, "B" denoted signals from the 1 mm

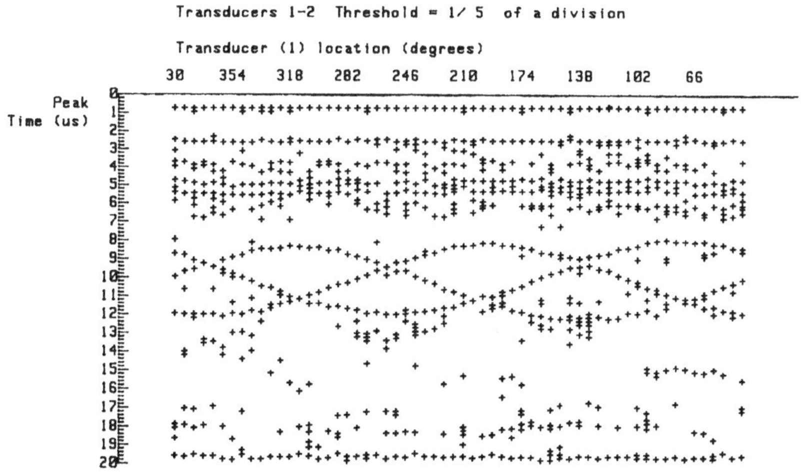

Fig. 11. Peak transit time versus degrees for transducers 1-2.

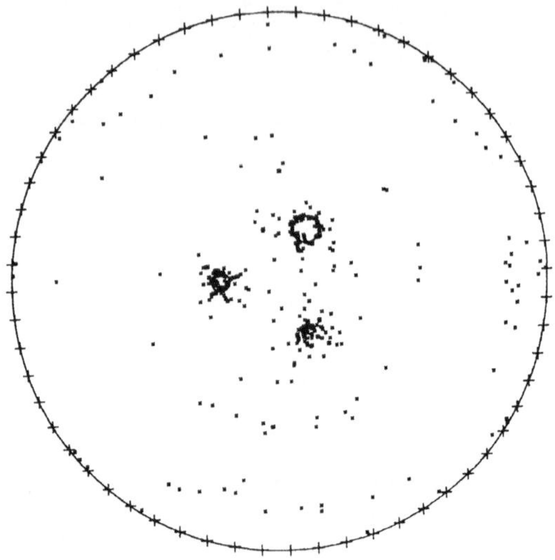

Fig. 12. Image points using the raw data. +'s denote transducers positions. The boundaries of the cylinder and the holes are drawn in for comparison. Beam width is ±19° for both transducers.

hole and "C" denoted signals from the 3-mm hole. Other signals were also detected. "D" will form an artefact in the image. Peak travel time versus scan position for transducer 1-2 were plotted in Fig. 11. The image field shown in Fig. 12 was constructed using raw data from transducers 1,2 and 1-2 and applying the method described above. A condition was imposed to constrain the points to lie within the beam width angles of ±19° for both transducers. This angle was calculated assuming that the transducer was generating a parallel beam. The curved surface of the cylinder acted as a diverging lens and the parallel beam diverged inside the specimen[2,8,9]. The diverging wavefront was approximated by circles at the

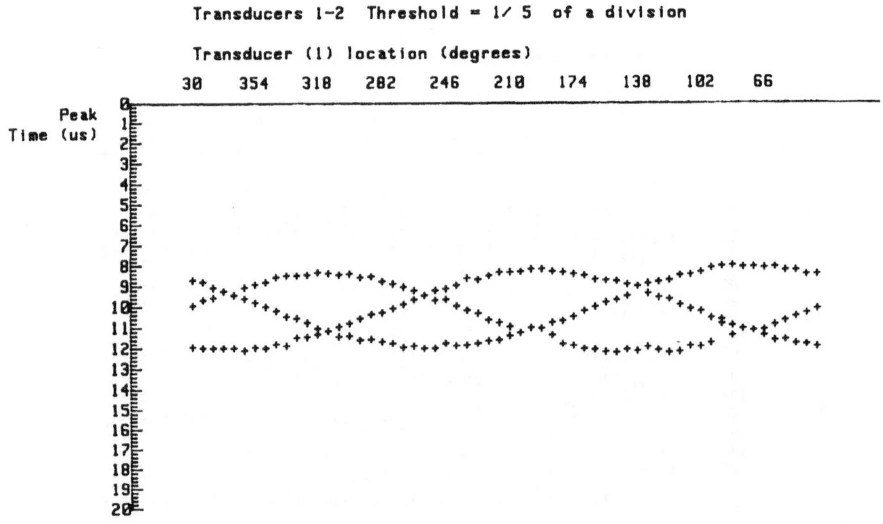

Fig. 13. Filtered data of Fig. 11.

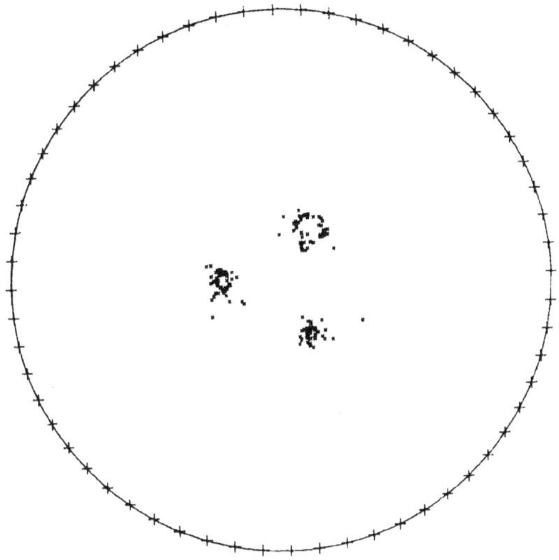

Fig. 14. Image points using the filtered data. Beam width is ±19° for both transducers.

centres of the transducer crystals. The image points described the boundaries of the holes quite well. However, spurious points also appeared in the image. Points on the cosine-like curves which correspond to the holes were then sorted out from the transit time plots. An example was displayed in Fig. 13 where filtered data for transducers 1-2 were plotted. Using these data, the image field was replotted in Fig. 14. A sharper image of the hole could be seen. Even with the constraints of the beam width removed, only a few more spurious points appeared in the image (Fig. 15).

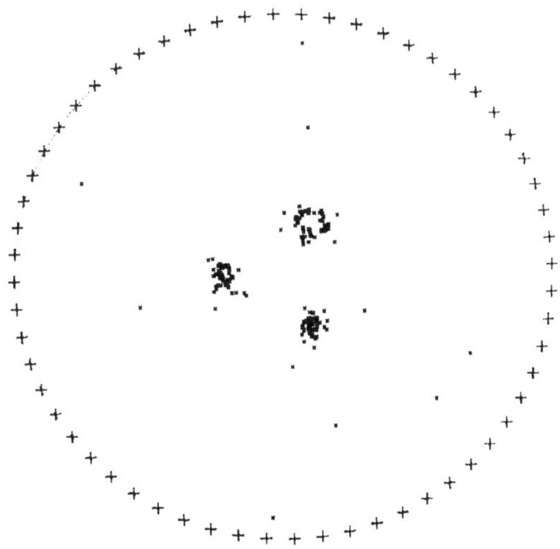

Fig. 15. Image points using the filtered data. No beam width constraint was applied.

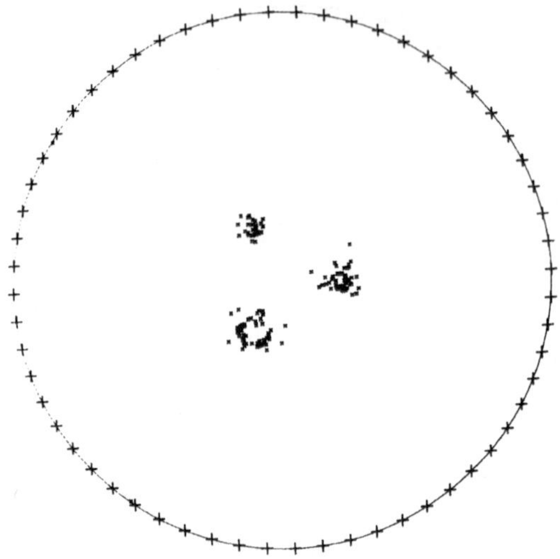

Fig. 16. Image points using the raw data. Threshold level was raised to twice that of the noise level. Beam width is ±19° for both transducers.

A threshold operation on unfiltered data was also attempted. The threshold level was set to be twice that of the noise level, and only pulses with peak amplitude above this level were used for imaging. Beam width constraint was imposed. The image was displayed in Fig. 16. Only a few spurious points were present.

The method was extended to a three-dimensional case. An Al block, 80 mm x 50 mm x 25 mm with a 2 mm diam vertical hole 10 mm long was used as a test block. Three 5 MHz, 6.35 mm diam transducers mounted on 45° (in Al) wedges were put on top of the block. They radiated shear waves toward the tip of the hole. Each transducer was used as transmitter/receiver. Then each two transducers were used as a transmitter-receivers. The travel time corresponding to the tip of the hole was sorted out for each transducer. The location of the tip of the hole was calculated as the intersection point of three spheres inside the block. The centre of each sphere corresponded to the location of each transducer and its radius corresponded to its travel distance to the tip of the hole. The calculation

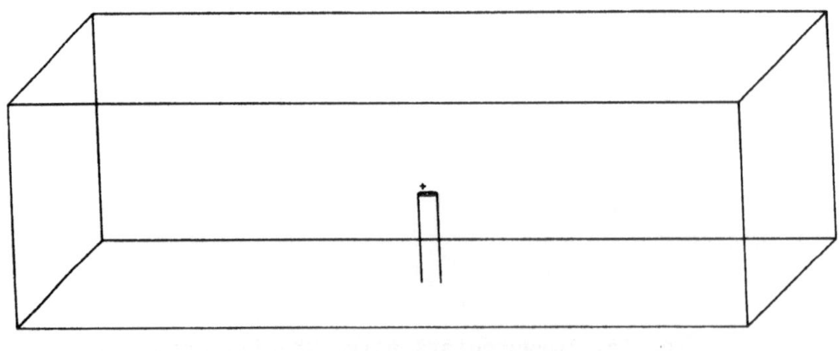

Fig. 17. + denotes the calculated intersection point corresponding to the top of the hole in the block.

yields a result of (40.38, 25.03, -14.24) in millimetres. This can be compared with the measurement of a vernier caliper which yields (40.10, 25.04, -15.06) in millimetres. The calculated point and the hole was pictured in Fig. 17.

DISCUSSION AND CONCLUSION

An acoustical imaging method which searches for echoes scattered from the same point defect from different transducer positions has been described. Accuracy in locating the defect point was better than a wavelength. A circular aperture provides a useful scanning method for producing images as the defect is viewed from different angles. Defect points cluster around the boundary of the defect and can easily be recognized even in the presence of other spurious points. Prefiltering or threshold operation can enhance the image by eliminating most of the artefacts. Linear scanning provides a limited aperture where the defect is viewed. Prefiltering of the signals is necessary to eliminate spurious points in the image. The method can be applied to imaging in three-dimensional space. Three transducers are required

ACKNOWLEDGEMENT

The author would like to thank Mr. I.R. Somerville and Mr. W. Kung for their assistance in collecting the data.

REFERENCES

1. P.D. Hanstead, Fast digital ultrasonic imaging, in: "Conf Proc of Ultrasonics International 81", pp 62-66 (1981).
2. M. Moshfeghi and P.D. Hanstead, Ultrasonic reflection tomography of cylindrical rods, Ultrasonics 23:206-214 (1985).
3. M. Moshfeghi, Ultrasonic reflection - mode tomography using fan-shaped-beam insonification, IEEE Trans on Ultrasonics, Ferroelectrics and Frequency Control, UFFC-33:299-314 (1986).
4. B. Grohs, O.A. Barbian, W. Kappes, H. Paul, R. Licht and F.W. Hoh, Characterization of flaw location, shape and dimensions with the ALOK system, Materials Evaluation, 40:84-89 (1982).
5. T.A. Slesenger and G.B. Hesketh, Ultrasonic defect positioning by non-linear least squares curve fitting, Acoustical Imaging, 14:547-558 (1985).
6. L.N.J. Poulter, Signal processing methods applied to PWR inlet nozzles, NDT International, 19:141-144 (1986).
7. O.A. Barbian, B. Grohs, W. Kappes and C. Hullin, Inspection of thick-walled components by ultrasonics and evaluation of the data by the ALOK-technique, in: "New Procedures in Nondestructive Testing," pp 133-150, Springer-Verlag (1983)
8. J. Krautkramer and H. Krautkramer, "Ultrasonic Testing of Materials," third revised edition, p 331, Springer-Verlag (1983).
9. D.K. Mak, Acoustical imaging with minimized scanning, Conference Proceedings of Ultrasonic International 87, pp 26-31 (1987).

ACOUSTICAL IMAGING USING A DAIS TECHNIQUE

FOR NON-DESTRUCTIVE TESTING

J.Y. Guigné, P.G. Williams, and D.K. Mak*

C-CORE, Centre for Cold Ocean Resources Engineering
Memorial University of Newfoundland, St. John's, Canada
*Physical Metallurgy Research Laboratories, Centre for
Mineral and Energy Technology, Ottawa, Canada

INTRODUCTION

The extraction of oil from offshore sites off the East Coast of Canada will probably be achieved using steel and concrete platforms comparable to those used in the North Sea. The welded steel nodes of these structures are susceptible to fatigue damage which occurs because of dynamic loading under the wave action and stress concentrations at the weld[1]. This fatigue damage leads to surface breaking cracks which propagate through the chord wall and cause a rapid decrease in stiffness. It is therefore important that these surface cracks be detected through inspection at the earliest opportunity after crack initiation. Fracture mechanics based models can then be used to assess the fatigue crack growth rate and the remaining life of the joint. Repairs can thus be effectively conducted as necessary. Laboratory studies[2,3] of such nodes have shown that the time period between initiation and breakthrough of such a crack is of the order of one to six months and this is extended for the in-service situation.

A technique known as Dynamic Acoustic Intensity Scanning (DAIS) has been developed which makes use of the energy flow in a structure to provide an early indication of fatigue cracking[4,5]. Most current non-destructive testing (NDT) techniques such as ultrasound or magnetic particle inspection, rely on the defect to reflect an introduced pulse or field within finite spatial and temporal resolution limits. Acoustic Emission is also limited as an NDT technique as it relies on the continuous monitoring of the transmission emanating from the defect and on the sensitivity in tuning the receiving instrumentation to the transmitted frequency.

In contrast DAIS captures the flow of acoustic energy in the near field of a steel structure, mapping sources and sinks which relate directly to the work taking place. Intensity vectors are measured using phase matched microphones, and the resulting energy flow patterns are dynamically displayed using a combination of vector plots and contour plots. From these, a unique Residual Factor (R_f) is calculated which expresses the perturbation in energy flow as a response to the changes which occur in the structure as a function of time. It is this change in energy flow, as quantified by the acoustic images, which identifies clearly the behaviour and location of the defect.

A significant aspect of this passive approach is that the receivers can be microphones, hydrophones or accelerometers. Since measurements can be made without surface contact and in most fluid media, the method is not affected by surface finish or environment, and yet it is highly sensitive[4]. This technique has been verified on large scale welded tubular T-joint and T-plate specimens[4]. This paper presents the results of experiments using the DAIS technique on flat steel plates with weld beads and machined notches.

INTENSITY AND DAIS

Intensity

The classic definition for sound intensity is a vector quantity which describes the net amount, and direction of flow, of acoustic power at a given point in space[4], and has units W/m^2. In a medium without bulk movement or flow, the intensity vector can be written simply as the time-averaged product of the instantaneous pressure and the corresponding particle velocity:

$$\vec{I} = \lim_{T\to\infty} \int_{-T/2}^{+T/2} p(t).\vec{V}(t)\, dt$$

and hence the intensity vector component in a given direction is

$$I_r = \lim_{T\to\infty} \int_{-T/2}^{+T/2} p(t).V_r(t)dt$$

In practice, intensity can be measured using two closely spaced phase matched microphones by applying Eulers equation and a finite difference approximation to obtain the particle velocity[3].

$$I = \overline{\frac{P_a + P_b}{2\rho\ \Delta r} \int (P_b - P_a)dt}$$

where ρ is the density of the propagation medium, Δr is the microphone spacing, and P_a and P_b are the pressures measured at the two transducers. The bar denotes time averaging. This technique allows for the measurement of the projection of the intensity vector along the line of the receivers, which provides sufficient information for sound power and source location calculations.

Dynamic Acoustic Intensity Scanning

It is necessary to measure the intensity vector by simultaneously acquiring its three spatial components if a complex sound field is to be defined. For stationary or periodic fields such as those seen over vibrating structures this instantaneous time restriction is relaxed. The DAIS logic[4,5] is designed to collect and analyse sound fields by measuring the three spatial components of the intensity vector. The use of real time digital filtering along with phase matched microphone pairs and expert software allows a dynamic interrogation of the energy flow to be performed.

A combination of a vector map for the z-plane component, and a contour map for the z-direction (Fig. 3(a)) illustrates the data in the three spatial dimensions. Alternatively, the three directional components making up the energy flow can be displayed separately as contour maps (Fig. 4(a)), where the dashed lines represent 'negative' flow and the solid lines 'positive' flow relative to the probe orientation.

The Residual Factor (R_f) is the relative difference between the magnitude of the intensity levels (in dB) measured over the same area. This 'energy difference' is then displayed as a contour map for each component, as shown by Figure 6. The R_f processing and data display are carried out through software.

LABORATORY INVESTIGATION

Objective

A laboratory investigation has been undertaken to validate the use of the DAIS Residual Factor (R_f) processing. This has been performed on flat steel specimens with machined notches and weld beads introduced into the specimens and referenced against the no-defect situation.

Test Specimens

The specimens were cut from 19mm thickness 1024 steel plate, and had dimensions of 100mm x 600mm (Fig. 1). A triangular extension was left at one end to provide a location for an accelerometer. Three specimens were fabricated with notches or weld beads as follows:

- Plate 1: No defect
- Plate 2: 10mm wide weld bead run across width of plate, 25mm from centre line (see Fig. 1)

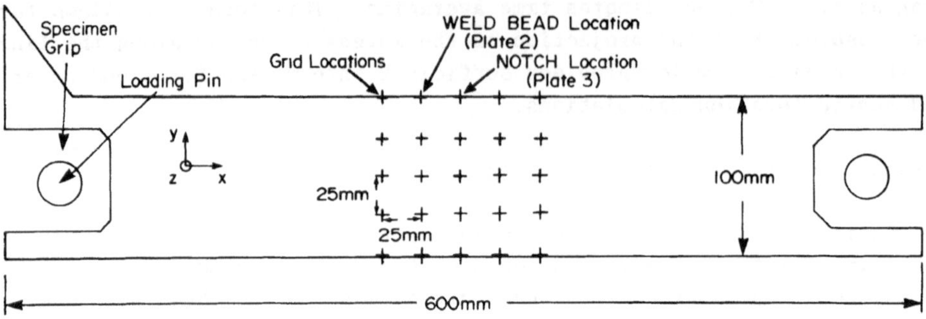

Fig. 1 Steel Test Specimen Dimensions.

- Plate 3: 1.6mm wide X 6.25mm deep notch machined across width of plate at centre line (see Fig. 1)

The specimens were placed in a servo-hydraulic test frame (see photograph 1) and loaded in tension to a static load of 75 KN. On top of this static load, a dynamic loading with a maximum amplitude of 0.25 KN was applied using a pseudo-random noise source. The pseudo-random noise had a bandwidth of 10 kHz and a sequence length of 32767 events. This provided a stationary acoustic energy source for the plate.

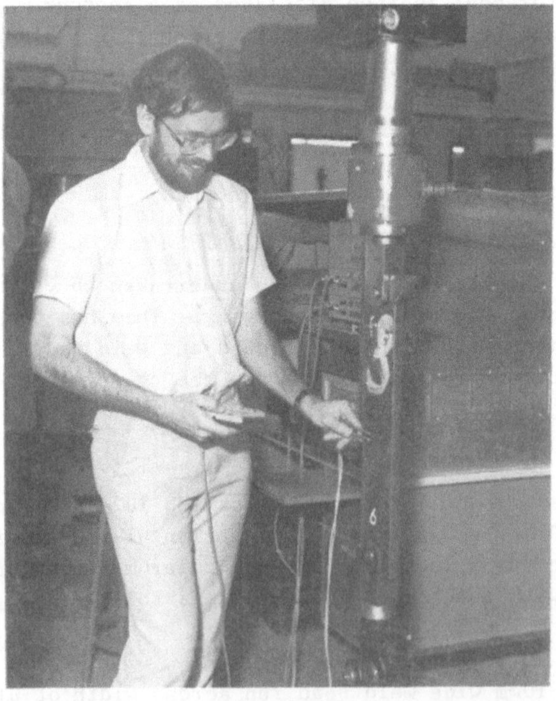

Photograph 1 Specimen Mounted in Test-Frame.

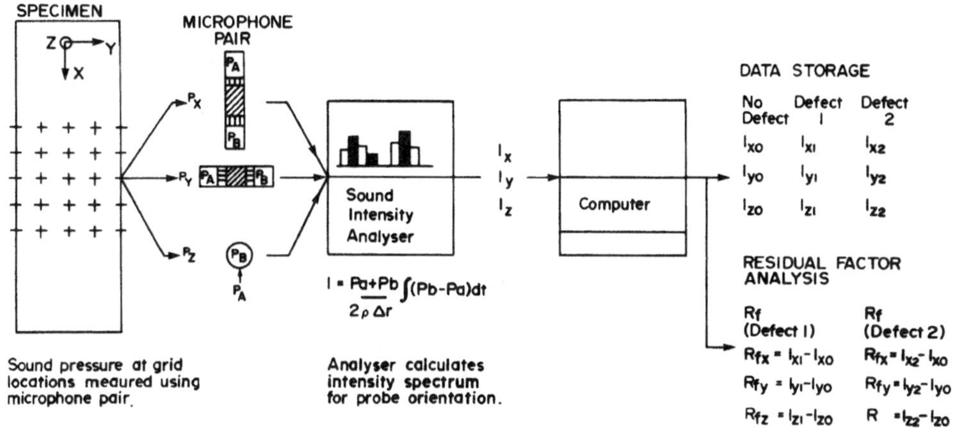

Fig. 2 Flow Logic of DAIS.

Measurement Technique

The acoustic intensity measurements were made using a 1/3 octave sound intensity analyzer. A bank of digital filters measures the intensity spectrum in real time and all measurements were made using linear averaging over a period of 1 second. A single pair of phase-matched microphones, held manually and rotated to the three required orientations was effectively employed since the stationary nature of the vibrations allowed the finite time between measurements to be ignored during reconstruction of the data. Figure 2 illustrates the flow of the logic used in the experiment.

Photograph 2 Intensity Spectrum Measurement (x-component).

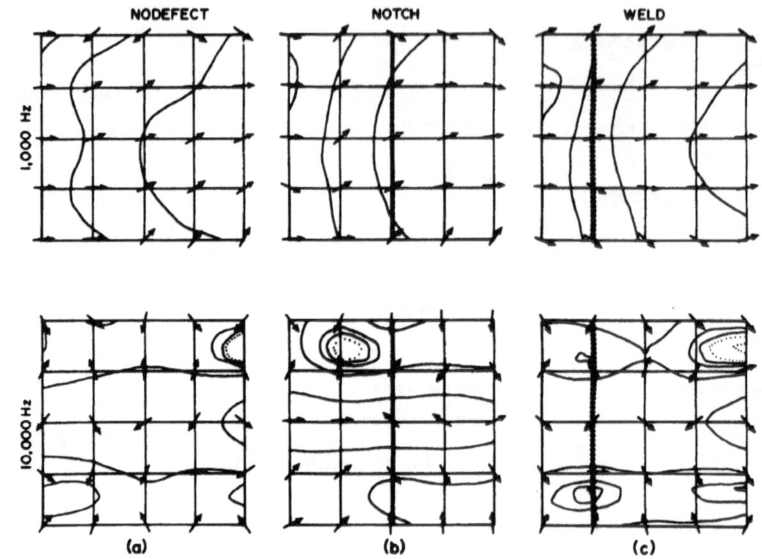

Fig. 3 Combined Vector and Contour Plots 1000 Hz and 10000 Hz Intensity Measurements.

An imaginary grid was established on each specimen and the intensity spectrum for the three directions was measured at each grid intersection (Photograph 2). The spectra were transferred to a desk top computer for processing. Two sets of data were taken for each specimen configuration in order to provide a control for the validation of the data.

RESULTS

Intensity Measurements

The results of the experimentation on the three control plates are synthesized in Figures 3 to 5. The pertinent frequencies were detected by the DAIS as being 1000 Hz and 10000 Hz and these are presented here. Figures 3(a), (b) and (c) show the combined vector and contour map plots for specimen configurations 1, 2 and 3 respectively (see section "Test Specimens" for defect dimensions). Figures 4 and 5 represent the same data displayed for 1000 Hz and 10000 Hz respectively using separate contour maps for the x, y and z directions.

Residual Factor Analysis

The Residual Factor (R_f) was calculated at 1000 Hz and at 10000 Hz for specimens 2 and 3 with respect to the no defect condition represented by specimen 1. Figures 6(a) and 6(c) show the R_f between specimens 1 (no defect) and 2 (notch) at 1000 Hz and at 10000 Hz respectively. The data are exhibited by contour maps for the x, y and z directions. Figures 6(b) and 6(d) are of the R_f between specimens 1 (no defect) and 3 (weld), again at 1000 Hz and 10000 Hz respectively.

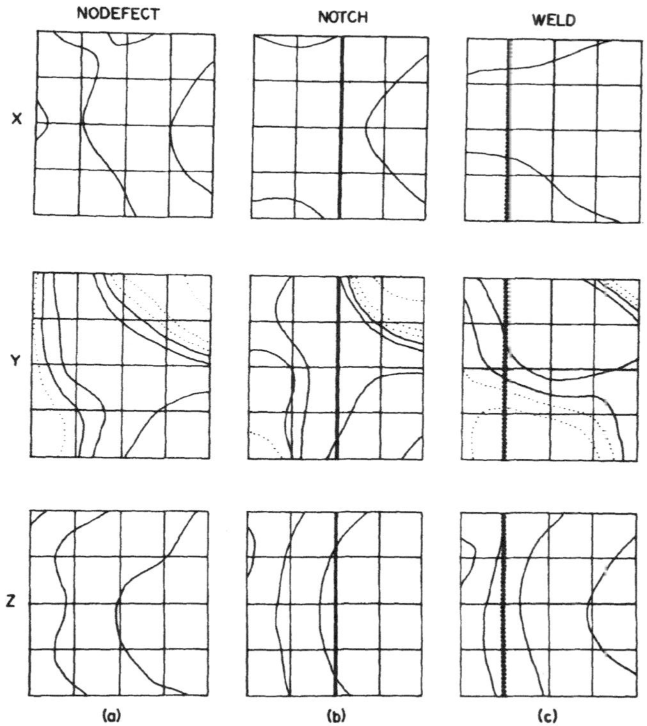

Fig. 4 Contour Plots of x,y and z Components of Intensity Measurements at 1000 Hz. 6 dB Between Contours

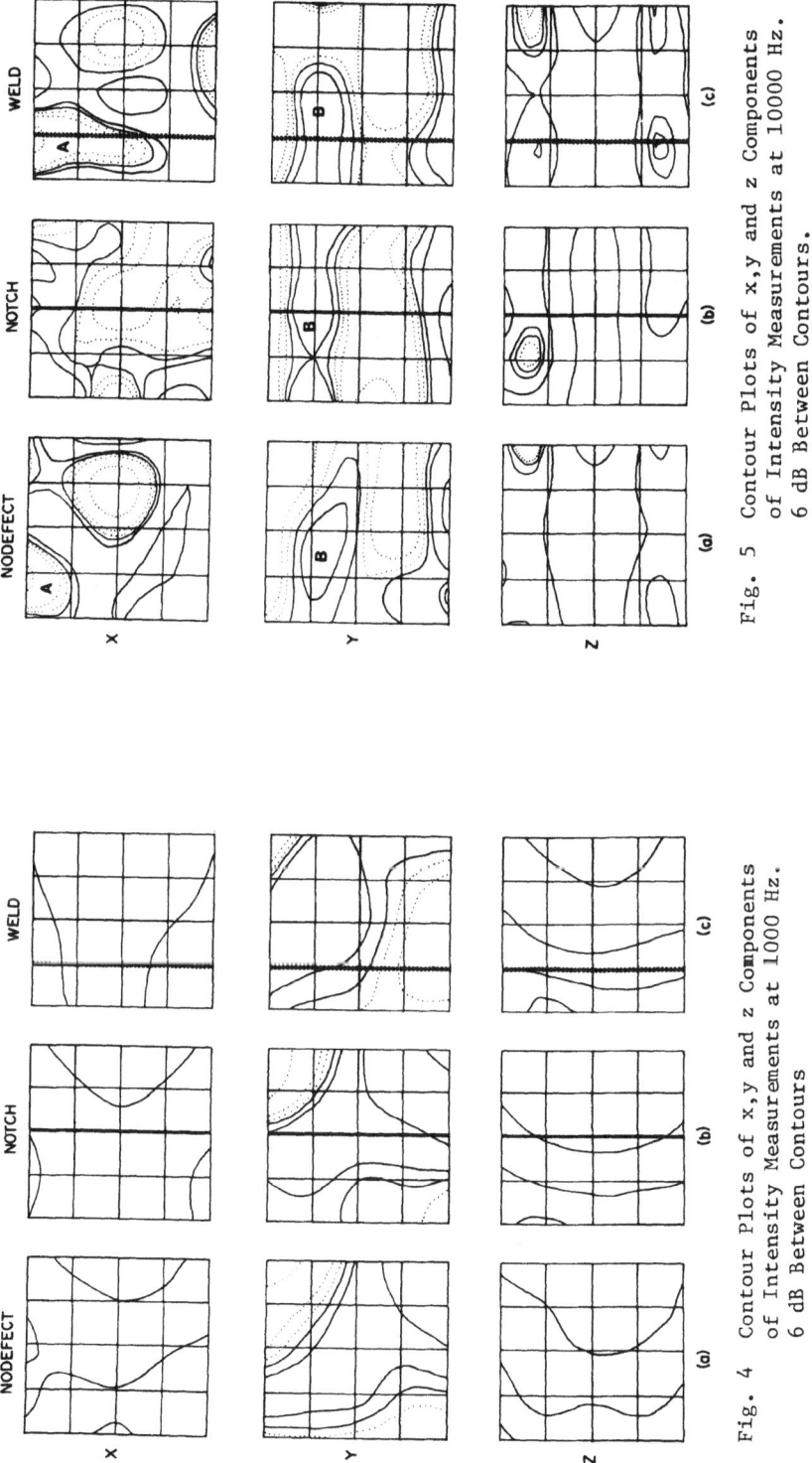

Fig. 5 Contour Plots of x,y and z Components of Intensity Measurements at 10000 Hz. 6 dB Between Contours.

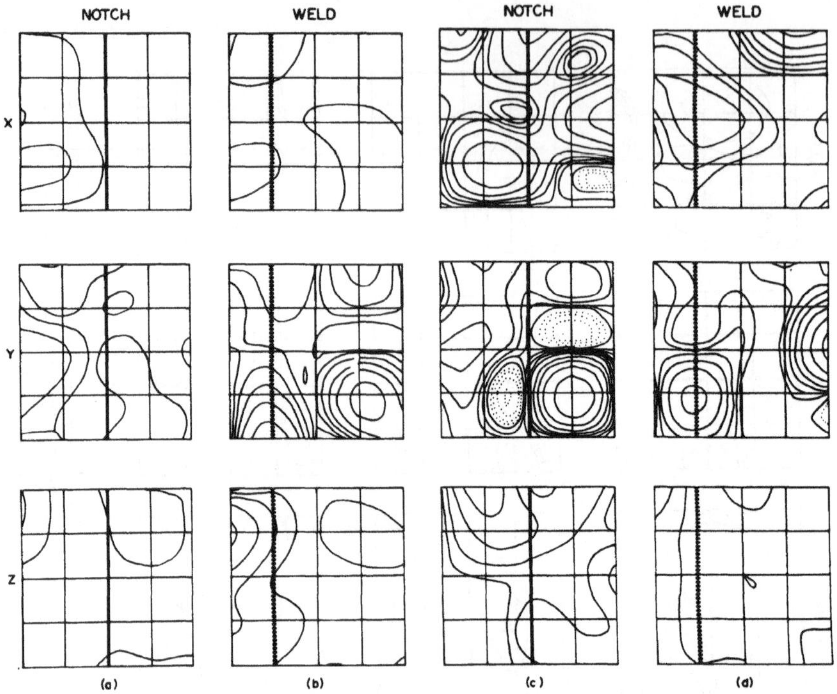

Fig. 6 Contour Plots of x,y and z Components of Residual Factor (a), (b) at 1000 Hz & (c), (d) at 10000 Hz. 2dB Between Contours.

INTERPRETATION

From inspection of the data presented in Figure 3, where the z-plane component of the vector is plotted over the z-direction contours, we can see that the energy flow patterns for the no defect, notch and weld specimens at the 1000 Hz frequency are similar and that information pertaining to the defect is not revealed explicitly. This holds true for the 10000 Hz frequency data base although anomalies are noted. The reason for the subdued exhibition of differences in the plot is that the vector descriptors mask the information held in the x and y spatial components.

By separating the vectors into isoline representations of their three spatial components (Figures 4 and 5), it becomes clearer that they demonstrate distinct flows of energy, associated with the unique defects in each of the specimens. In particular the 10000 Hz frequency reveals dominant trends of change.

The A zone in the x-isoline plot for the no defect specimen (Fig. 5(a)) is transformed into an elongated feature along the weld bead (Fig. 5(c)). The damping effect that the weld has on the plate is thus highlighted. In contrast, the notched specimen (Fig. 5(b)) redistributes the energy flow in the x-direction in a more symmetrical pattern along both sides of the notch. This reflects the behaviour of minor standing waves in the plate as caused by the notch.

In the y-isoline plots the B zone in all three specimens is indicative of the physics attributed to the plates. For example, where the weld dampens the plate the B zone is foreshortened towards the weld (Fig. 5(c)) compared to the no defect pattern (Fig. 5(a)). In comparison, the notch (Fig. 5(b)) extends the zone B to again give a symmetrical pattern on either side of the notch.

The z-direction represents the energy component radiating normal to the plate and this is less affected by the weld bead damping or by the reverberation at the notch. Minor pattern changes are, however, still to be observed, and these are indicative of the sensitivity of the technique.

The residual factors are noted in Figure 6. The isolines represent the relative difference between the magnitude of the intensity levels. The energy flow differences for the notch and for the weld are contrasted by their x and y component isoline representations, especially at 10000 Hz (Fig. 6(b) & (d)). At 1000 Hz, the distinctions between the two specimens are found mainly in the y-isoline representation of the R_f. The Residual Factor approach therefore directly illustrates the effect of defects, related to damping (Fig. 6(a) & (b)) and reverberation (Fig. 6(c) & (d)), on the flow of energy. These patterns could thus be used for image processing in non-destructive testing applications.

CONCLUSIONS

The flow of acoustic energy in the near field of three steel plates, was captured using the DAIS technique. Two of the three specimens contained calibrated defects. The resultant intensity vectors were differentiated into their orthogonal components. This approach characterized the perturbation of the energy flow by the defects in a definitive manner which cannot be obtained directly from the vector images.

The Residual Factor (R_f) process further removed ambiguities associated with the defect free plate. The final R_f images exhibited damping and reverberation flow patterns in concordance with the weld and notch defect.

These R_f patterns are ideally suited to robotic pattern matching techniques. An attempt is now in place to store and retrieve the R_f information using a neural network architecture. This should provide the DAIS with an ability to handle problematic patterns, similar to the type of analysis executed in a biological brain. Matching complex and "blurry" images would be feasible using the Neural Network's adaptive rules, complementing and extending the DAIS' already broad dynamic range for non-destructive testing as seen in this paper.

ACKNOWLEDGEMENTS

Thanks are due to C-CORE for supporting this research, to NORDCO Ltd. and Dr. A.S.J. Swamidas (Memorial University of Newfoundland) for their interest, and to Dr. Robert Thompson and CANMET for encouragement and financial backing of the program. Thanks are also due to Desiree King for the diagrams and Norma Matthews for typing the paper.

REFERENCES

1. W.D. Dover and L.J. Bond. Weld Crack Characterization on Offshore Structures Using AC Potential Difference and Ultrasonics. NDT International Vol. 19, No. 4, (1986).

2. A.A. Laurenssen and O.O. Dijkstra. Fatigue Tests on Large Post-Weld Heat Treated and as Welded Tubular T-Joints. 14th Annual Offshore Technology Conference, Houston, (1982).

3. M. Hara, Y. Kawai, A. Narumoto and S. Matsumoto. Corrosion Fatigue Strength of 490 MPa Class High Strength Steels Produced by the Thermo-Mechanical Control Process. 18th Annual Offshore Technology Conference, Houston (1986).

4. J.Y. Guigné, P.G. Williams and V. Chin. A Concept for the Detection of Fatigue Cracks in Welded Steel Nodes, Marine Technology, Vol. 18, No. 4, (1987).

5. J.Y. Guigné, P.G. Williams and V. Chin. Analysis of the Deformation in a Partially Cracked Welded T-Plate, Marine Technology, Vol. 18, No. 4, (1987).

6. P. Rasmussen. Intensity Vector Measurements, ICA Toronto, (1986).

ANALYSIS OF ECHOES FROM A SPHERE IN FOCUSED SOUND FIELD AND ITS APPLICATION TO MATERIAL CHARACTERIZATION

M.Ueda and E.Morimatsu

Tokyo Institute of Technology,Research Laboratory of Precision Machinery and Electronics
Nagatsuta, Midori-ku, Yokohama 227, Japan

INTRODUCTION

Analysis of echoes from a solid elastic sphere is important in acoustical imaging techniques, since it is used as a testing object for evaluating resolving power of ultrasonic pulse echo systems and it is well known that it shows very characteristic pattern such as comet-like pattern on B-mode images. The analysis of the echo from the sphere ,however, has been limited to the case of plane wave illumination.[1-6] Consequently it is rather difficult to apply the results to pulse echo systems directly since the focused ultrasonic beams are used in them.

In this paper a relation , which describes the amplitude and phase of the echo scattered by the sphere in terms of the size and acoustical properties of the sphere as well as the geometry and relative position of ultrasonic transducers which radiate the focused beam to the sphere and receive the scattered waves from it, is derived by modifying the relation which holds for the plane incident wave. This relation is used to estimate echo signals from three kind of spheres, that is, steel, brass, and acryl spheres and these echo signals are compared with experimentally obtained ones. The comparison reveals some deviation between the theoretically and experimentally obtained echoes.

Since the estimation of echoes is carried out using cataloged data of the acoustical properties of the spheres, the optimization of the properties is performed so as to minimize the error between the two signals. Since we have measured the density of the spheres, sound velocity and absorption coefficient of transverse as well as longitudinal waves are taken as the subjects for the optimization. Since it will become too complicated task to optimize the four quantities simultaneously, sound velocity and absorption coefficient are optimized separately. First the sound velocity is optimized using the error function defined in the frequency domain and then the absorption coefficient is optimized using the error function defined in the temporal domain. Very good agreement between the echo signals is obtained after the optimization . Thus the potential of our method for material characterization of the sphere is verified.

1.THEORY OF SCATTERING BY A SPHERE

1.1. Plane incident wave

Scattering from a sphere illuminated by a plane wave has been already analyzed by J.J.Faran[1] and let us review his theory shortly for convenience of later references. A sphere of radius a is placed at an origin of Cartesian coordinates system (x,y,z) (The polar coordinates system (r,θ,ϕ) is also used indiscriminately .) as shown in Fig.1(a) and is illuminated by a plane wave which propagates along the z-direction. Then the incident wave to the sphere cab be written as

$$P_i = P_0 \exp(-jkz) \tag{1}$$
$$= P_0 \exp(-jkr\cos\theta)$$
$$= P_0 \sum_{n=0}^{\infty} (2n+1)(-j)^n j_n(kr)P_n(\cos\theta) \tag{2}$$

where P_0 is the amplitude of the plane wave, k is the wave number of the surrounding medium, j_n is spherical Bessel function of the first kind and P_n is Legendre polynomial. Then the waves scattered by the sphere, P_s, is expressed by

$$P_s = \sum_{n=0}^{\infty} c_n\ h_n^{(2)}(kr)P_n(\cos\theta) \tag{3}$$

$$c_n = -P_0(2n+1)(-j)^{n+1}\sin\eta_n\exp(j\eta_n) \tag{4}$$

where $h_n^{(2)}$ is spherical Hankel function of the second kind and ηn is the phase angle of n-th order diffracted waves and is determined by the diameter and acoustic properties of the sphere as well as the acoustic properties of the surrounding medium.

1.2.Spherical wave illumination

Let us suppose that the sphere is illuminated by a spherical wave which is emitted from a point source located at the point $z=-r_0$ as shown in Fig.1(b). In this case the incident wave, P_i, is expressed by

$$P_i = P_r R_r \exp(-jkR)/R \tag{5}$$
$$R = (r^2 + r_0^2 + 2rr_0\cos\theta)^{1/2} \tag{6}$$

where P_r is the amplitude of the spherical wave at a reference distance R_r from the source and R is the distance between the source and the point of the observation. Then Eq.(5) can be expanded as follows

$$P_i = P_r R_r(-jk)\sum_{n=0}^{\infty} (2n+1)(-1)^n j_n(kr)h_n^{(2)}(kr_o)P_n(\cos\theta) \tag{7}$$

Since Eq.(7) holds for $o\lesssim r\lesssim r_0$, it holds at the surface of the sphere (r=a) and functional form with respect to r and θ for n-th order diffracted wave is just the same as that in Eq.(2). Consequently the scattered waves under the spherical wave illumination can be written as the same form as Eq.(3), but the different coefficient c_n' must be used where it is given by

$$c_n' = P_r R_r k(2n+1)(-1)^n h_n^{(2)}(kr_0)\sin\eta_n\exp(j\eta_n) \tag{8}$$

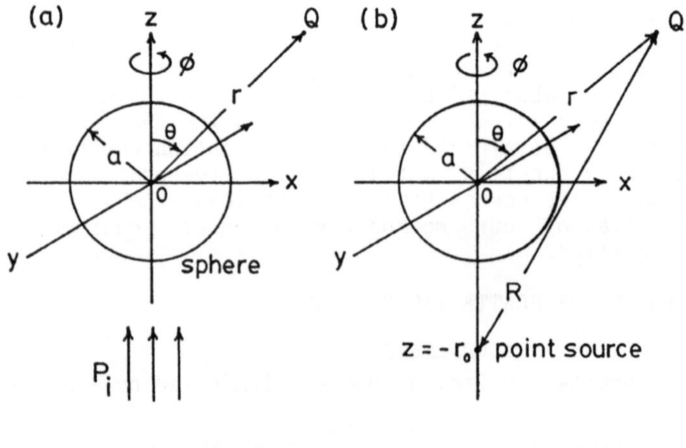

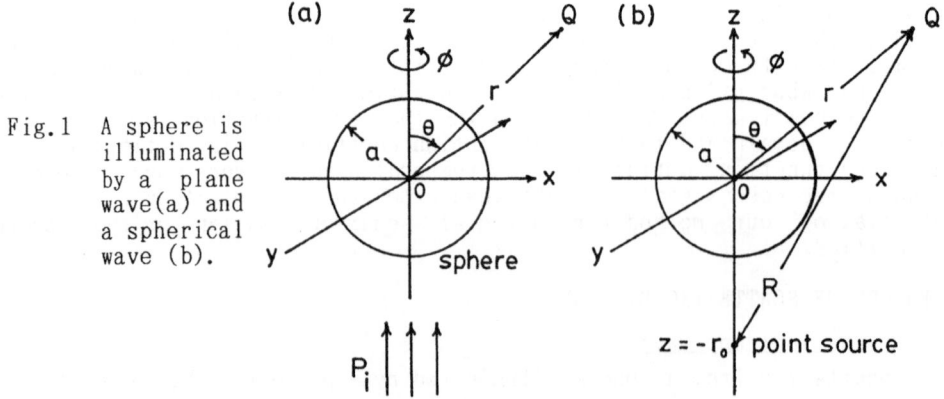

Fig.1 A sphere is illuminated by a plane wave(a) and a spherical wave (b).

Fig.2 An ultrasonic pulse is radiated from a transmitter and waves scattered by a sphere are received by a receiver.

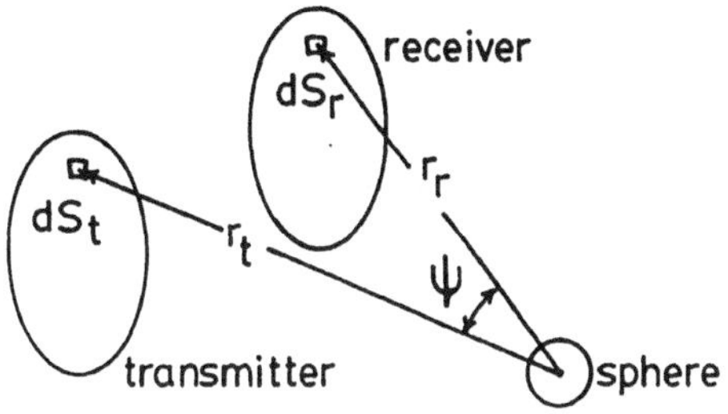

Thus the waves scattered by the sphere illuminated by the spherical wave can be calculated by using Eqs.(3) and (8).

1.3.Echo from a sphere

Let us suppose that the sphere is illuminated by waves insonified by a transmitter and the waves scattered by the sphere are received by a receiver and the echo signal from the sphere is recorded as the output signal of the receiver as shown in Fig.2. If a small surface element on the transmitter is considered as a point source and that on the receiver as a point receiver, then Eqs.(3) and (8) can be applied to calculate transmit-receive characteristics between these elements including the sphere. The spherical wave radiated from the element dSt on the transmitter, dP_i, can be written as

$$\delta P_i = j\omega\rho U(\omega)dS_t \exp(-jkR)/(2\pi R) \tag{9}$$

where ρ is the density of the medium, $U(\omega)$ is the amplitude of vibrating velocity of the surface and an infinite baffle plate is assumed. On the other hand if the scattered waves δP_s is incident to the element dS_r on the receiver, then the output of the element, δE_s, may be written as

$$\delta E_s = R(\omega)\delta P_s dS_r \tag{10}$$

where $R(\omega)$ is the electrical characteristics of the receiver. If the transmitter and receiver are considered as a set of point sources and point receivers respectively, the frequency component of the echo from the sphere, $E_s(\omega)$ can be calculated using Eqs.(3),(8),(9),and (10) and is given by

$$E_s(\omega) = \int_{S_t}\int_{S_r} \delta E_s \, dS_r dS_t \tag{11}$$

$$= (j\omega\rho U(\omega)R(\omega)/2\pi)k \sum_{n=0}^{\infty} (2n+1)\sin\eta_n \exp(j\eta_n)\, I_n$$

$$I_n = \int_{S_t}\int_{S_r} h_n^{(2)}(kr_r)h_n^{(2)}(kr_t)P_n(\cos\psi)dS_r dS_t \tag{12}$$

where S_t and S_r show the surfaces of the transmitter and receiver respectively, r_t and r_r are distances between a center of the sphere and the elements dS_t and dS_r respectively, and ψ is an angle between the lines r_t and r_r as shown in Fig.2. Eqs.(11) and (12) are the basic relations for analyzing the echo signal from the sphere and I_n is the directivity of n-th order diffracted waves and it depends on the geometrical form of the transducers and the position of the sphere relative to them.

2.Analysis of directivity I_n

In the previous chapter simple relations which describe the echo from the sphere are obtained. The directivity I_n in them, however, is expressed by a rather complicated integral as Eq.(12), consequently it is not easy to express I_n in explicit form for the most cases. In this chapter let us suppose that the same circular concave transducer is used as a transmitter and receiver and the sphere is placed at its focal point or the focal plane.

2.1.sphere at the focal point

The sphere is located at the focal point of a transducer (radius b

Fig.3 A sphere is placed at the focal point of a circular concave ultrasonic transducer.

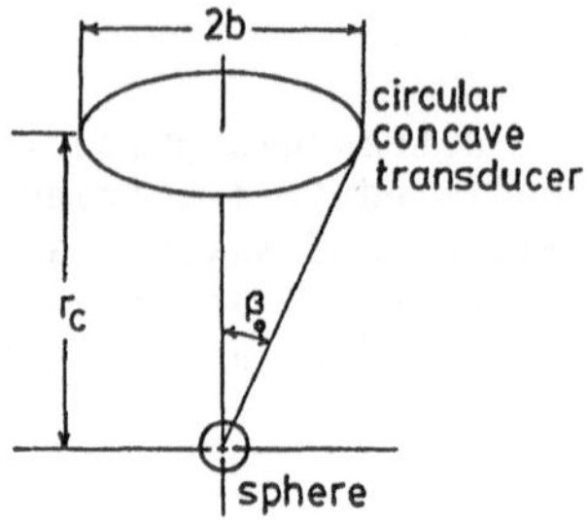

and radius of curvature r_c) as shown in Fig.3. In this case we can make use of the symmetry of the geometry and I_n can be written as

$$I_n = \{ 2\pi r_c^2 h_n^{(2)}(kr_c) \int_0^{\beta_0} P_n(\cos\beta)\sin\beta \, d\beta \}^2$$

$$= \{ 2\pi r_c^2 h_n^{(2)}(kr_c) f_n \}^2 \quad (13)$$

$$f_n = \{ \cos\beta_0 P_n(\cos\beta_0) - P_{n-1}(\cos\beta_0) \}/n \quad (14)$$ [5]

$$\beta_0 = \tan^{-1}(b/r_c) \quad (15)$$

where $2\beta_0$ is an angle spanned by the transducer as shown in Fig.3. Since the electrical characteristics of the transducer, $(j\omega\rho U(\omega)R(\omega)/2\pi)$ in Eq.(11), can be estimated by using an echo signal reflected from a flat end surface of a column placed at the focal point as shown in Appendix, the frequency component of echo, $E_s(\omega)$, can be estimated and the echo signal from the sphere, $e_s(t)$, can be also calculated by

$$e_s(t) = F^{-1}\{ E_s(\omega) \} \quad (16)$$

where F^{-1} shows Fourier inverse operation. Both $E_s(\omega)$ and $e_s(t)$ are used in the inverse problem described later.

2.2. Sphere at the focal plane

The other case where I_n can be calculated analytically is the case in which the sphere is placed at the focal plane of the transducer. And in this case we have succeeded to obtain the approximate analytical expression for I_n and these expressions have been used to study the relation between the side lobe characteristics of echo and acoustic properties of the sphere [7]. These results are not presented in this paper because of the limitation of the space.

3. EXPERIMENTS

The experimental apparatus is shown in Fig.4, where an ultrasonic transducer of diameter 19mmϕ, radius of curvature 100mm, and 2.25MHz broad band characteristics is used. Three kind of spheres, that is, steel, brass, and acryl spheres are used in the experiments . The diameter and acoustic properties of the spheres are listed in Table 1, where the numerical values of density are measured ones and the other values are the cataloged ones. By modifying the wave number to the complex one the effect of absorption in the sphere is taken into consideration and the absorption per wavelength is assumed constant in the sphere. [3] The spheres are placed at the focal point of the transducer and are imbedded into the agar phantom in order to support the spheres and to separate any echoes which are reflected from any intervening boundary planes. A short pulse is radiated from the transducer and an echo from the sphere is digitized by 30 MHz sampling frequency and is sent to a computer for further signal processing of the echo.

The wave form of echo and its spectrum are shown in Fig.5 for a steel, brass, and acryl balls respectively. The echo from the steel ball shows periodic repetition of pulses and the spectrum shows regular repetition of

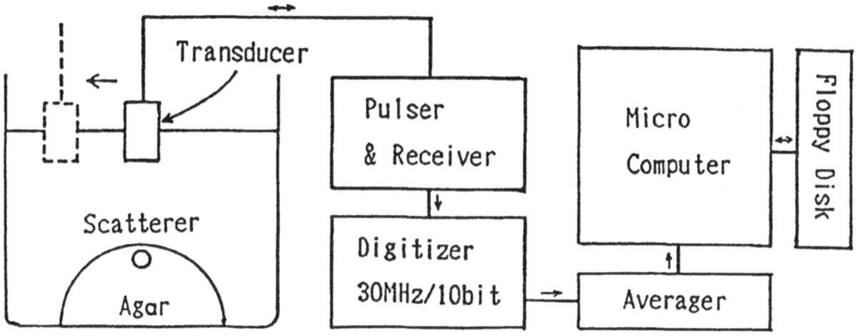

Fig.4 Experimental apparatus used to record the echo signal from the sphere.

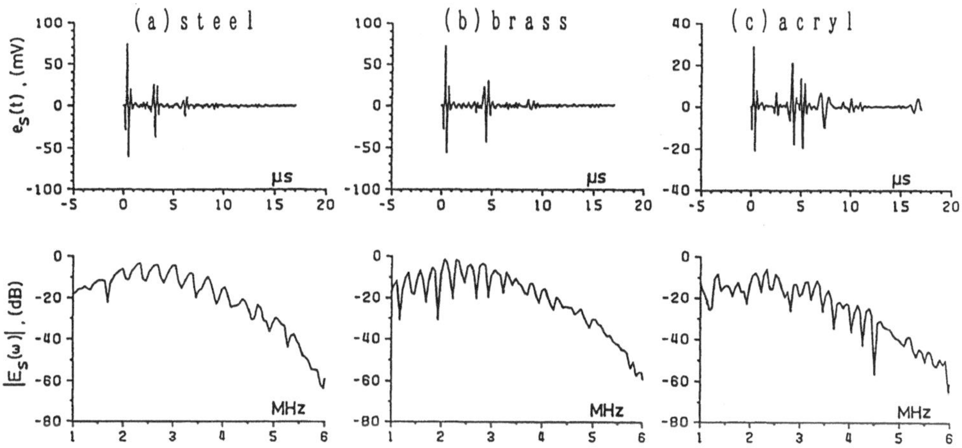

Fig.5 Wave form and spectrum of the echo signal from a steel sphere (a), a brass sphere (b), and a acryl sphere (c).

Table 1 Acoustic properties of the spheres and the surrounding medium. The density of the spheres is determined by the measurement and the other values are the cataloged ones.

	ρ, g/cm³	c_l, m/s	c_t, m/s	$\alpha_l\lambda_l$, dB	$\alpha_t\lambda_t$, dB
steel	7.7	5850	3230	0.0	0.0
brass	8.1	4700	2100	0.0	0.0
acryl[8]	1.2	2720	1460	0.19	0.29
water	1.0	1470	———	———	———

c_l and $\alpha_l\lambda_l$ are velocity and absorption per wavelength of longitudinal waves and ρ is density.
c_t and $\alpha_t\lambda_t$ are velocity and absorption per wavelength of transverse waves.

dips. The diameter of the balls is 3.0 mmϕ (steel), 3.0 mmϕ (brass), and 3.18 mmϕ respectively. The echo from the brass ball shows almost the same appearance as that from the steel ball but the separation of the pulses becomes longer and it reflects slower sound velocity of brass than the steel. The spectrum of the echo from the brass becomes more irregular in terms of the shape and position of the dips. The echo from the acryl ball shows very complicated shape and the regular structure of the dips is almost lost in the spectrum.

4. ESTIMATION OF ECHO AND OPTIMIZATION OF PARAMETERS

First the echo from the steel ball is estimated by following the procedures described in the section 2.1 and using the acoustic properties listed in Table 1. The estimated echo signal and its spectrum are shown by dotted lines in Fig.7(a) and (c) respectively and in these figures those obtained by experiments are shown by solid lines. In these figures the first positive peaks of the echo signal is normalized to 1.0. As seen from these figures there are some deviations between the theoretical and experimental values both in the temporal signal and in the spectrum. This deviation may come from the usage of the cataloged values for the acoustic properties of the ball. Consequently we have tried to optimize the acoustic properties of the ball so as to minimize the deviation between the theory and experiments.

As to the acoustic properties of the ball, since we have measured the density of the ball, the sound velocity and absorption coefficient of longitudinal as well as transverse waves are chosen as parameters to be optimized. But it will become too complicated task to optimize four parameters simultaneously, consequently first the sound velocity is optimized and then the absorption coefficient is optimized.

It is well-known that the shape and position of the dips in the spectrum are closely related to the sound velocity of the ball, an error function defined in the frequency domain and given by Eq.(17) is used in the optimization of sound velocity .

$$ER_F = \int_{f_{min}}^{f_{max}} \{|Es(\omega)| - |E_s(\omega, c_l, c_t, \alpha_l, \alpha_t)|\}^2 \, df \tag{17}$$

On the other hand it may be reasonable to assume that the waveform decay is closely related to the absorption in the ball, consequently an error function defined in the temporal domain and is given by Eq.(18) is used in the optimization of the absorption coefficient .

$$ER_T = \int_0^{T_{max}} \{ e_s(t) - e_s(t, c_l, c_t, \alpha_l, \alpha_t) \}^2 \, dt \tag{18}$$

As to the optimization algorism, a simple gradient search method was used. Optimization process is illustrated in Fig.6, where the variation

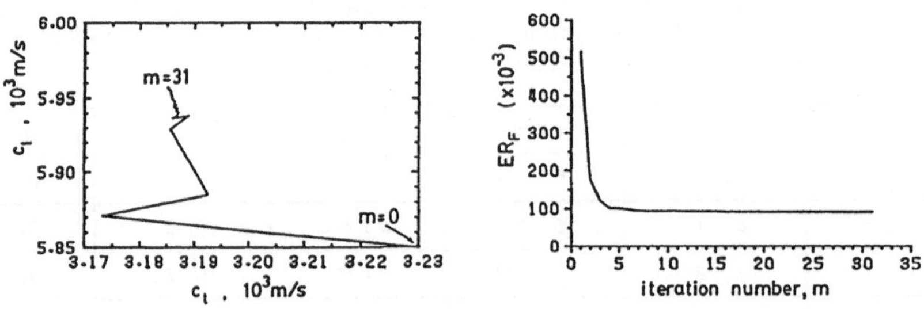

Fig.6 Optimization process of steel ball's sound velocity. (a) The variation of the velocity during the optimization is shown and (b) that of the error function is shown.

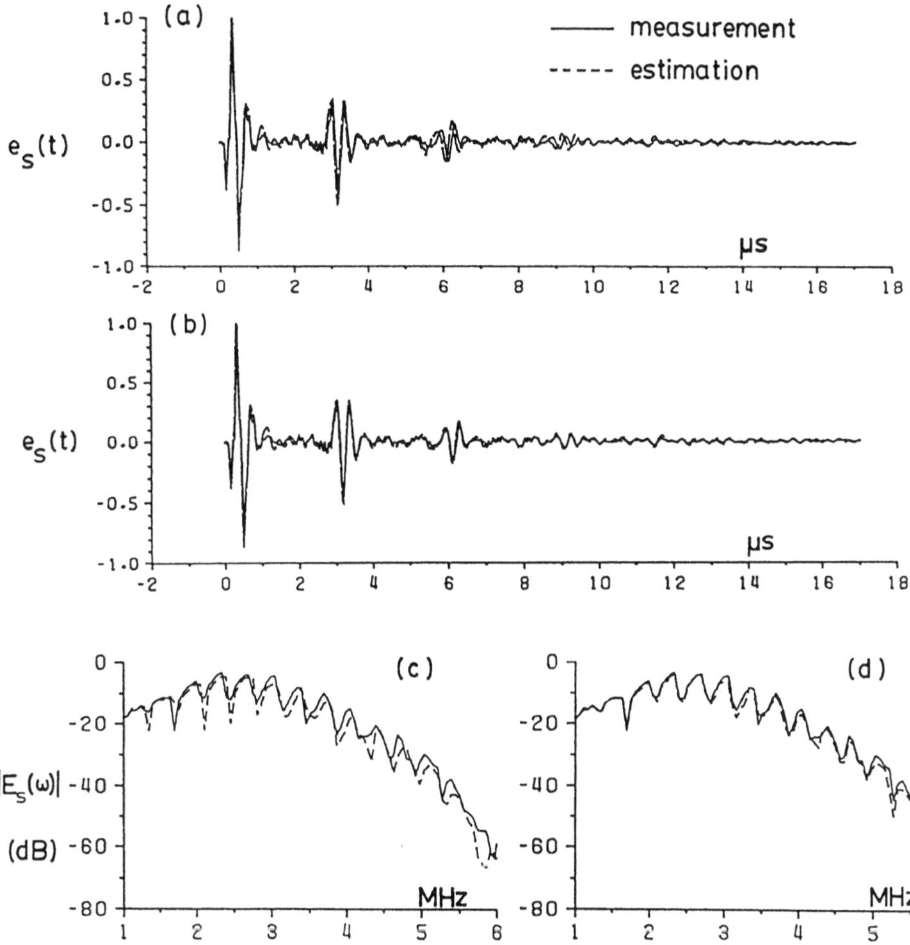

Fig.7 (a) The measured and estimated echo signal from a steel ball of 3mmϕ are plotted. (b) The measured and optimized (velocity) echo signal are plotted. (c) Spectrum of the measured and estimated echo signal are plotted. (d) Spectrum of the measured and optimized echo signal are plotted.

of sound velocity during the optimization process is plotted in Fig.6(a) and that of the error function is shown in Fig.6(b). At the beginning the velocity and the error change rapidly and at the end they settle in the convergent values. In Fig.7(b) and (d) the echo and the spectrum of the steel ball after the velocity optimization are shown. Very good agreements are obtained for both the echo signal and the spectrum. Since the good agreements are obtained by the velocity optimization only, in this case the optimization for the absorption coefficient is not performed and the optimized acoustic parameters are listed in Table 2. The velocity of longitudinal waves is changed as much as 100 m/s.

In Fig.8 the results of the optimization for the brass ball are shown. In this case the velocity optimization is hardly effective and in fact the sound velocity is changed only 5 m/s but the absorption coefficient optimization is effective and rather large absorption coefficients are estimated.

In Fig.9 the results of the optimization for the acryl ball are shown. In this case the initial estimation by the cataloged data gives very poor estimation but after the velocity and absorption optimizations reasonably good agreement is obtained. Thus the potential of our algorism for

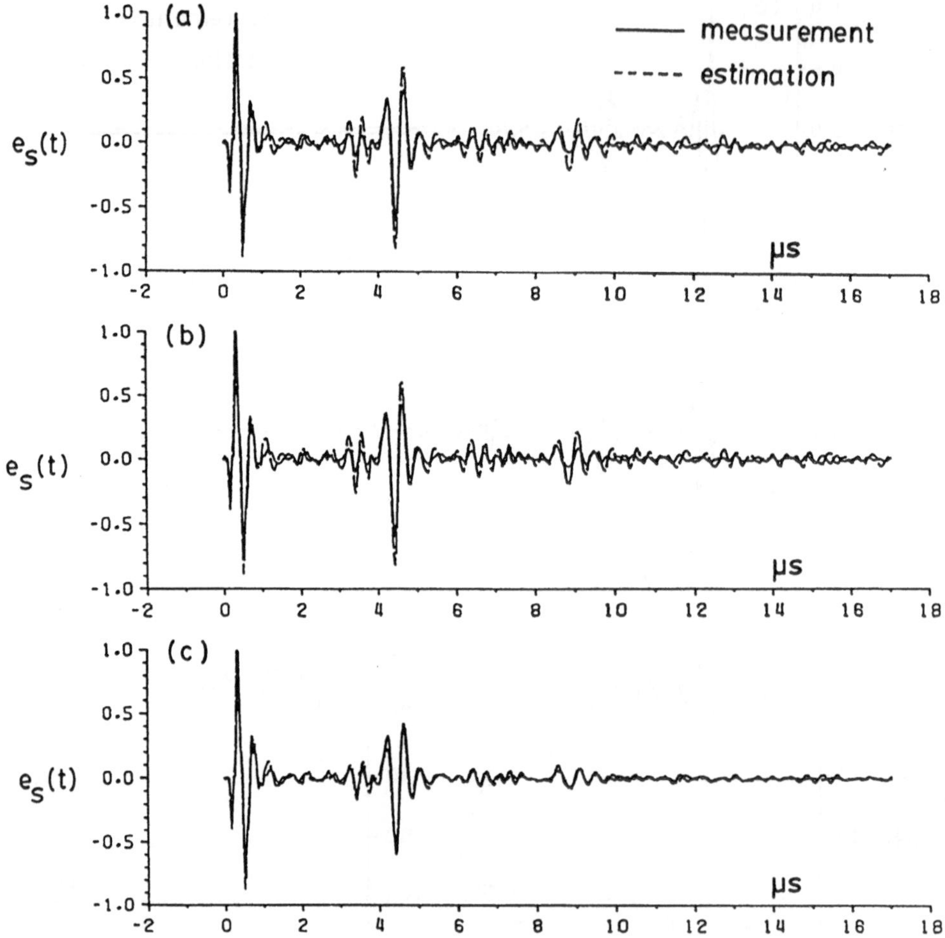

Fig.8 (a) The measured and estimated echo signal from a brass ball of 3mmϕ are plotted. (b) The measured and optimized (velocity) echo signal are plotted. (c) The measured and optimized (velocity and absorption) echo signal are plotted.

echo estimation and optimization of acoustic parameters to material characterization of the sphere is verified. All the echo signals used in the optimization process are normalized ones in which the amplitude of the first positive peak is set to unity.

CONCLUSIONS

In this paper a relation which describes the echo from the sphere obtained by using arbitrary ultrasonic transmitter and receiver is presented and is used to estimate the echo from the sphere that is placed at the focal point of a circular concave transducer. Very good agreement between the theoretically and experimentally obtained echoes after the optimization of acoustic parameters of the sphere. This shows the potential of our algorism for echo estimation and optimization of acoustic parameters to material characterization of the sphere.

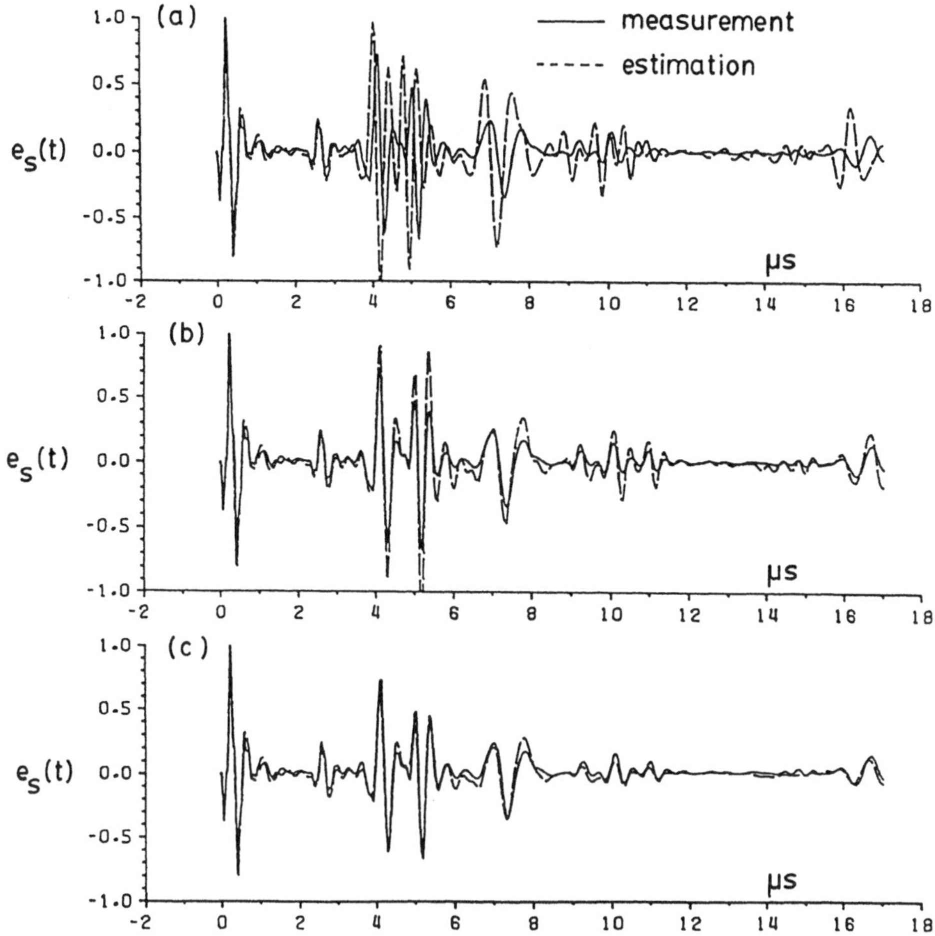

Fig.9 (a) The measured and estimated echo signal from a acryl ball of 3.18 mmϕ are plotted. (b) The measured and optimized (velocity) echo signal are plotted. (c) The measured and optimized (velocity and absorption) echo signal are plotted.

Table 2 Acoustic properties of the spheres after the velocity and absorption optimization. In the case of steel the absorption optimization was not performed since the good agreement has already been obtained after the velocity optimization.

	c_l, m/s	c_t, m/s	$\alpha_l \lambda_l$, dB	$\alpha_t \lambda_t$, dB
steel	5957.2	3185.2	*0.0*	*0.0*
brass	4694.4	2100.0	0.67	0.39
acryl	2736.8	1375.3	0.32	0.82

REFERENCES

1. J.J.Faran,Jr., J.Acoust.Soc.Am.,23, 405 (1951).
2. R.Hickling, J.Acoust.Soc.Am., 34, 1582 (1962).
3. R.H.Vogt, L.Fax, L.R.Dragonette, and W.G.Neubauer, J.Acoust.Soc.Am., 57, 558 (1975).
4. P.L.Edwards and J.Jarzynski, J.Acoust.Soc.Am.,74, 1006 (1983).
5. T.Hasegawa, K.Matsuzawa, and N.Inoue, J.Acoust.Soc.Am.,79,927(1986).
6. V.M.Ayres and G.C.Gaunaurd, J.Acoust.Soc.Am., 82, 1291 (1987).
7. E.Morimatsu and M.Ueda, J.Acoust.Soc.Jpn.,44,16 (1988) (in Japanese).
8. B.Hartmann and J.Jarzynski, J.Appl.Phys.,43, 4304 (1972).
9. M.Ueda and H.Ichikawa, J.Acoust.Soc.Am., 70, 1768 (1981).
10. M.Ueda and H.Ichikawa, Erratum, J.Acoust.Soc.Am., 75, 1012 (1984).
11. Stephen McLaren and John P.Weight, J.Acoust.Soc.Am., 82, 2102(1987).

APPENDIX Estimation of electrical characteristics of transducer

Let us suppose that a flat end surface of a column is placed at the focal point of the circular concave transducer as shown in Fig.A1. The frequency component of echo signal reflected from an object which has uniform acoustic characteristics over its volume can be expressed as [9,10]

$$E_r(\omega) = j\omega\rho U(\omega)R(\omega)R' \int_S \Phi(\omega, r) N\cdot\nabla\Phi(\omega, r)ds \qquad (A1)$$

where R' is the amplitude reflection coefficient of the scatterer, S is the surface of the scatterer, N is outer unit normal to S, and Φ is the velocity potential of the sound field insonified by the transducer and is given by

$$\Phi(\omega, r) = (1/2\pi) \int_T \exp(-jk|r-r'|)/|r-r'| \, ds \qquad (A2)$$

where T is the surface of the transducer and r' is a position vector concerned to the integration. Since the following approximations holds at the end surface

$$N\cdot\nabla\Phi = jk\Phi \qquad (A3)$$

$$\Phi = S_T/(2\pi r_c) \qquad (A4)$$

where S_T is the area of the transducer. Then Eq. (A1) becomes

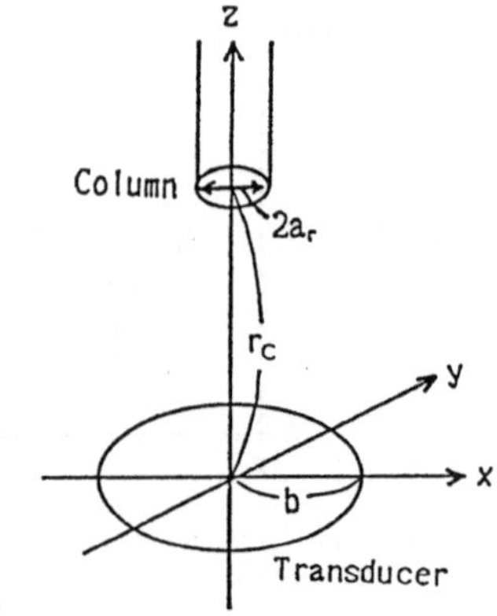

Fig.A1 A column and a transducer

$$E_r(\omega) = j\omega\rho U(\omega)R(\omega)R' jkS_c(S_T/(2\pi r_c))^2 \qquad (A5)$$

where S_c is the area of the flat end surface. Then the electrical characteristics of the transducer, $j\omega\rho U(\omega)R(\omega)$, can be estimated from Eq.(A5). The wave form and spectrum of the echo scattered from a flat end surface of a aluminum column of 1mmϕ which is placed at the focal point of the transducer are shown in Fig.A2. The usefulness of Eq.(A1) for calculating the echo scattered from the column is shown in references 9 and 11.

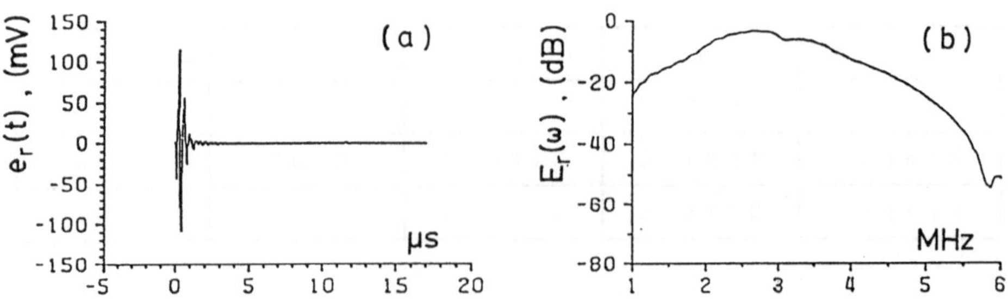

Fig.A2 Wave form and spectrum of echo scattered from flat end of column.

EVALUATION OF POLYMER COATINGS BY ULTRASONIC SPECTROSCOPY

Y. Tsukahara, K. Ohira, M. Saito and G. A. D. Briggs*

Technical Research Institute, Toppan Printing Co. Ltd.
Sugito-machi, Kitakatsushika-gun, Saitama 345, Japan
*Department of Metallurgy and Science of Materials
University of Oxford, Parks Road, Oxford OX1 3PH England

INTRODUCTION

One of the major applications of scanning acoustic microscope (SAM) has been the evaluation of layered structures such as solid substrates overlaid by thin films. For example, the layer thickness was measured by V(z) method (Weglein, 1979) and by impulsive mode (Yamanaka, 1983). SAM has also been used to detect flaws in bonding of thin films to solids (Quate, 1980). These methods in SAM utilized the excitation of leaky Rayleigh waves and the dependence of their dispersion on film thickness and adhesion.

Other than SAM, there have been several methods proposed for the layer thickness measurement: ultrasonic interferometry of normal incident waves (Houze et al., 1984) and a method using pseudo-Sezawa waves (Tsukahara et al., 1984), to name a few. These methods were based on the analysis of the angular and frequency dependence of reflection coefficients on the layer thickness. On the other hand, dependence of reflection coefficients on a boundary condition at an interface between solids has been theoretically analyzed(Schoenberg, 1980), in order to study the adhesion of the interface with ultrasonic waves.

In this paper, we focus on a study of steel substrates with polymer coatings, and analyze the dependence of reflection coefficients on the layer thickness and boundary condition between the coating and substrate. Experimental observations are then made of Fourier spectra of reflected waves of obliquely incident ultrasonic waves, and compared with theoretical analysis.

REFLECTION COEFFICIENT FOR POLYMER COATED STEEL

Calculations of reflection coefficients were made, following a method (Brekhovskikh, 1980), for a steel substrate with a PET film on it. The thickness of the PET film was assumed to be 28 μm. Material constants used in the calculations are listed in Table 1. Measurements of these material constants will be reported elsewhere. As discussed by Pilarski and Rose (1988) for the case of an interface between two solid half spaces, we took into account the effect of the variation of the boundary condition at PET/steel interface by assuming the existence of a very thin (10 nm) intermediate layer between the PET film and the steel substrate(Fig.1). By assuming that the material constants of the intermediate layer had the same values as the PET film, a perfect bonding was simulated. We

Table 1. Material Constants Used in Calcualtions

	ρ	C_l (m/s)	C_t (m/s)	α_l	α_t
water	0.997	1494	-	-	-
PET	1.42	2309	1100	1.7×10^{-5}(s/m)	1×10^{-4}(s/m)
steel	7.86	5915	3251	4.24×10^{-14}(s^2/m)	1.74×10^{-13}(s^2/m)

could simulate a smooth bonding, at which only those components normal to the interface of displacement vectors and stress tensors were continuous, by changing the transverse sound velocity of the intermediate layer to a very small value (1 m/s in the present calculation).

Modulus of the reflection coefficients for a perfect bonding and smooth bonding are shown in Figs. 2 and 3, respectively, as a function of incident angle, θ, and frequency, f. At normal incidence (θ=0°), the dependence of the reflection coefficients on f is identical for both cases. This means that we cannot detect a smooth bonding by the inspection using normal incident ultrasonic waves. At incident angles from 45° to 55°, two figures also look similar. Thus, the layer thickness can be measured regardless of the boundary condition by measuring the spectra of reflected waves with an incident angle set at zero or at an angle between 45° and 55° On the other hand, the two reflection coefficients are quite different at incident angles between 20° and 40°, therefore, smooth bondings can be suitably inspected by using ultrasonic waves with such an incident angle.

EXPERIMENTAL APPARATUS AND METHOD

Fig.4 (a) shows a configuration of experimental apparatus. The apparatus and experimental method were same as used in previous studies (Tsukahara et al., 1986), except that spectra of reflected waves were obtained by FFT of digitized wave forms. For experiments with normal incident waves, a sensor configuration shown in Fig.4(b) was used.

DETECTION OF DEFECTS IN BONDING

Test Specimens

Five blocks (A-E) of steel(20 mm×20 mm×8 mm) were prepared. A surface of each block (20 mm×20 mm) was polished, and 25 µm-thick PET film was bonded on it with an epoxy adhesive. An effort was made to have the thickness of the adhesive as thin as possible. It was found by visual

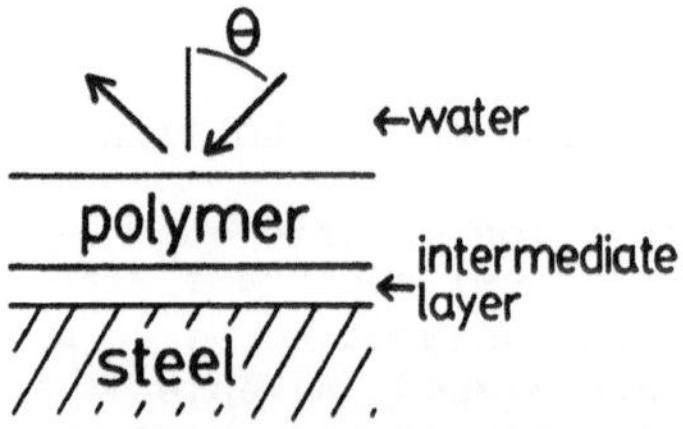

Fig.1 A layered structure with an intermediate layer to simulate a variation in bonding quality

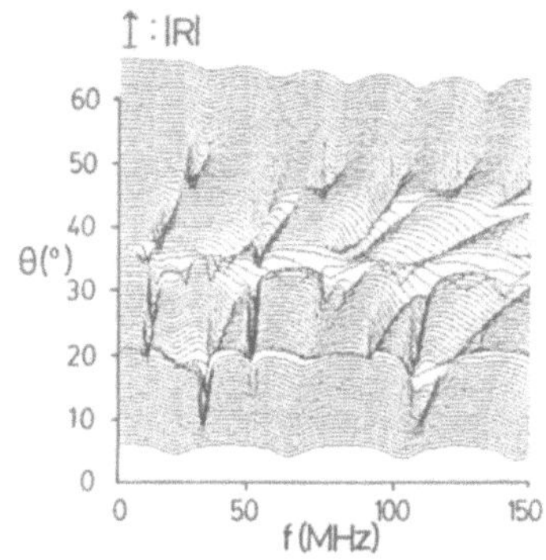

Fig.2 Calculated reflection coefficient for polymer coated steel with a perfect bonding. An arrow indicates the magnitude from 0 to 1.

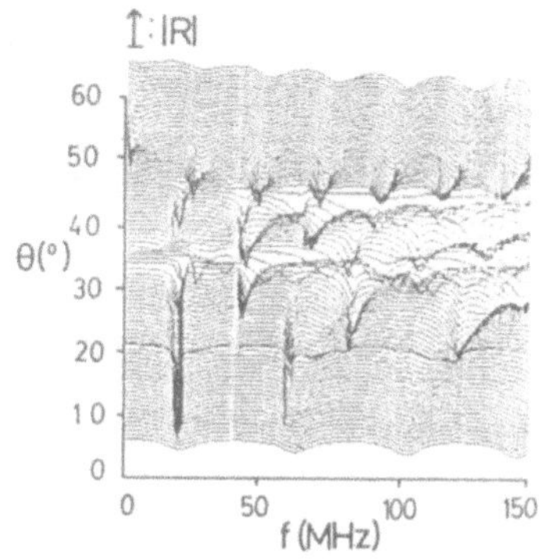

Fig.3 Calculated reflection coefficient for polymer coated steel with a smooth bonding.

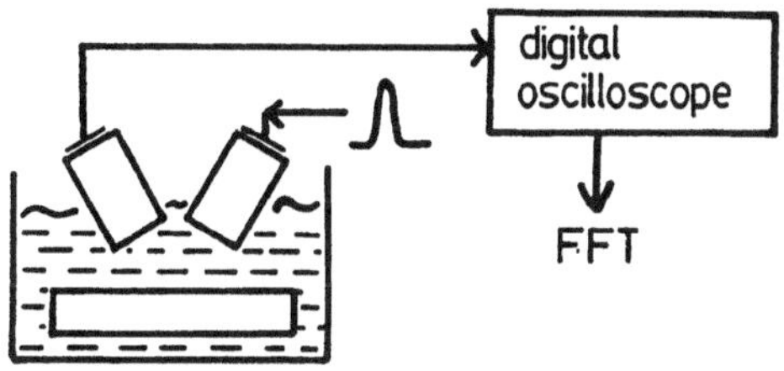

Fig.4(a) Experimental configuration for oblique incidence

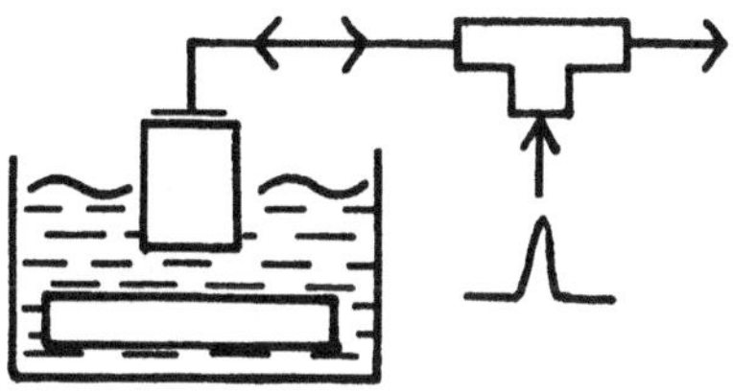

Fig.4(b) Sensor setup for normal incidence

inspection that a part of the PET film, about 2 mm×2 mm in area, of test specimen A was detached. Measurements were made at the center of the surfaces of all specimens except for A, for which measurements were made at the detached area. No defect was found visually on test specimens B-E before experiments. But, after experiments were finished and specimens were taken out of a water bath, periphery of the PET films of test specimens B and C were found to be detached. Faint networks of rust were observed on the surfaces of steel at the detached area, therefore, we inferred that fine chains of defects had been existing between the adhesive layers and the steel substrates of specimens B and C before experiments, and that water had soaked into the interfaces by capillary attraction during experiments in which specimens had been immersed in water, and caused the detachment. Even though the detachment was apparent only at periphery of the surfaces, it might be plausible that a small amount of water penetration might have reached to the central parts of the surfaces. On the other hand, no indication of defects was observed on specimens D and E even after the experiments.

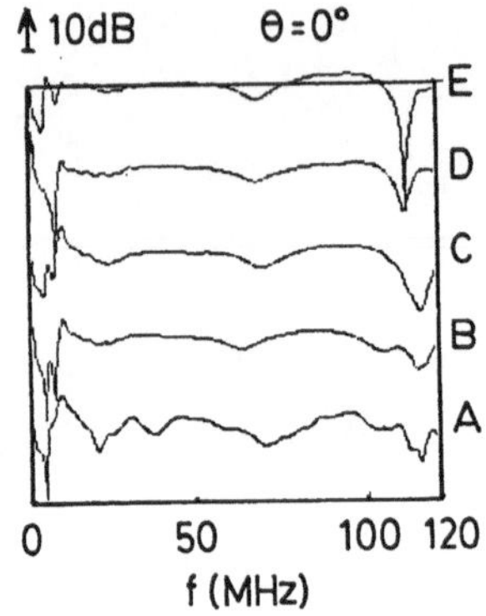

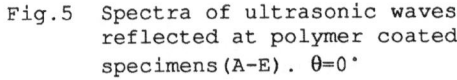

Fig.5 Spectra of ultrasonic waves reflected at polymer coated specimens(A-E). θ=0°

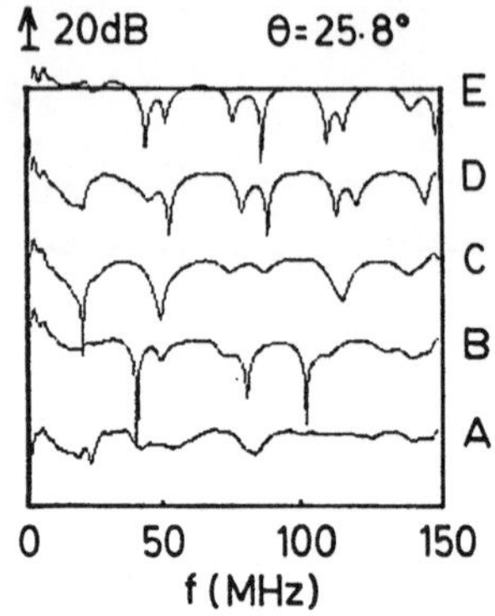

Fig.6 Spectra of ultrasonic waves reflected at polymer coated specimens(A-E). θ=25.8°

Results and Discussions

Fig.5 shows spectra for specimens A-E when θ=0°. The effective frequency range was from 20 MHz to 120 MHz. Spectra for specimen A shows an irregular shape because of the serious defect. Spectra for specimens B-E show similar shapes, except that spectrum for specimen B indicates the thickness of the adhesive layer was thicker. Frequencies of dips, f_{dip}, and peaks, f_{peak}, are related to the effective layer thickness d, under the assumption that the adhesive layer has same material constants as PET film,

$$f_{dip} = (n+1/2)\cdot C_l/(2d) ,$$
$$f_{peak} = (n\cdot C_l)/(2d) ,$$

where C_l is the longitudinal sound velocity of PET film and n is an integer. d was 28.5 μm for specimen B and 26.5 μm for specimens C-E.

Fig.6 shows spectra for all specimens when θ=25.8°. The effective frequency range was from 30 MHz to 140 MHz. Spectrum A shows a severe degradation because of the defect. Spectra D and E have similar shapes showing three pairs of double dips. On the contrary, in spectrum B, double dips are not as clear as in D and E, and spectrum C has two single dips and a pair of shallow double dips. It is helpful here to see the theoretical analysis. In Fig.2, at incident angles nearly equal to 26°, the calculated frequency dependence of the reflection coefficient for a perfect bonding shows three double dips, while in Fig.3 for a smooth bonding there are corresponding single dips. Considering these findings together with the observation of specimens before and after the experiments, we identify the spectra A with a total detachment, B and C with a degradation of bonding by the penetration of water, and D and E with a tight bonding.

LAYER THICKNESS MEASUREMENTS

Test Specimens

Four types of test specimens were used in the following experiments. All specimens consisted of steel plate (230 μm in thickness) with polymer films coated on both sides. Polymer names and their nominal thickness for each type of specimens are listed in Table 2. White PP is PP film

Table 2 Polymer Name, Thickness and Sample Numbers for Each Type of Specimens

	Side 1			Side 2		
		Thickness	No.		Thickness	No.
Type1	PET	15	10-19	PET	15	20-29
Type2	PET	15	30-39	white PP	40	40-49
Type3	PP	20	50-58	PP	20	60-68
Type4	PP	25	70-79	PP	40	80-89

Thickness is nominal in μm.

with suspension of fragments of TiO_2 (0.3 μm in average diameter). Every surface of specimens is identified by numbering as shown also in Table 2.

Results and Discussions

Measurements of spectra were made with an incident angle equal to 30°. Effective frequency range was from 20 MHz to 120 MHz. Spectra for type 2 specimens are shown in Figs.7(a) and (b). Variation in spectral shapes indicates that there were differences in integrity of bonding among specimens. Dip frequencies and corresponding layer thickness are plotted as a function of sample number in Figs.8, 9 and 10 for PET, white PP and PP coatings, respectively.

Measurements of spectra, while changing the incident angle from 19° to 63°, were also made on the following specimens with PP coatings: samples No. 80 (40 μm), No.70 (23.3 μm) and No.52 (21.7 μm). The fd values of dips in spectra are plotted in Fig.11 as a function of an incident angle. At incident angles larger than 29°, fd values for different samples are consistent, although they are scattered at incident angles less than 28°. Because the steel substrates were thin (230 μm), plate waves were excited by ultrasonic waves with an incident angle less than 29° (critical angle of transverse waves in steel) and their dip frequencies could not be normalized by the multiplication with the layer thickness.

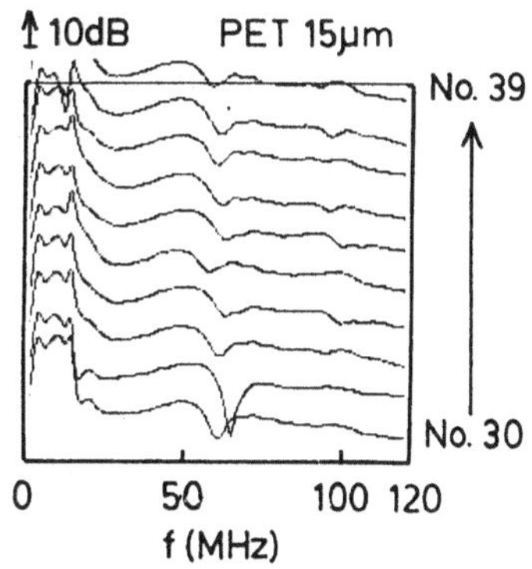

Fig.7(a) Spectra of reflected waves for specimens No.30-39 θ=30°

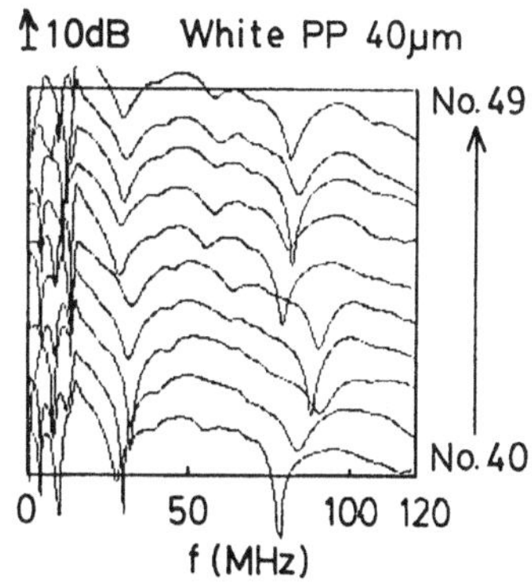

Fig.7(b) Spectra of reflected waves for specimens No.40-49 θ=30°

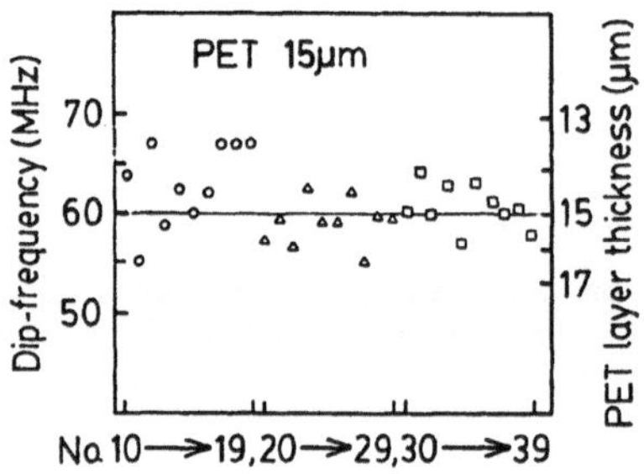

Fig.8 Dip frequencies and corresponding layer thickness for specimens No.10-39

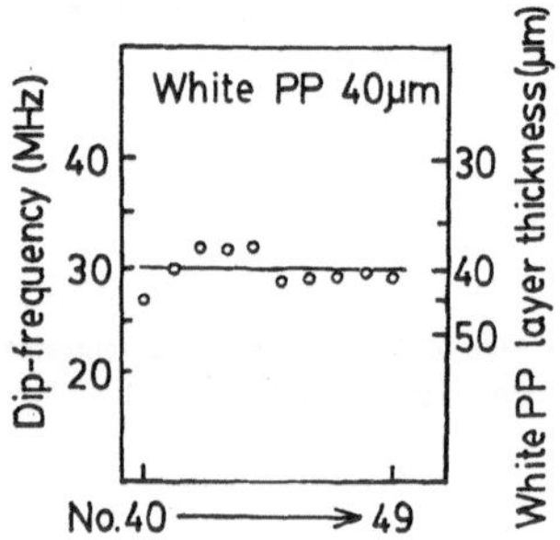

Fig.9 Dip frequencies and corresponding layer thickness for specimens No.40-49

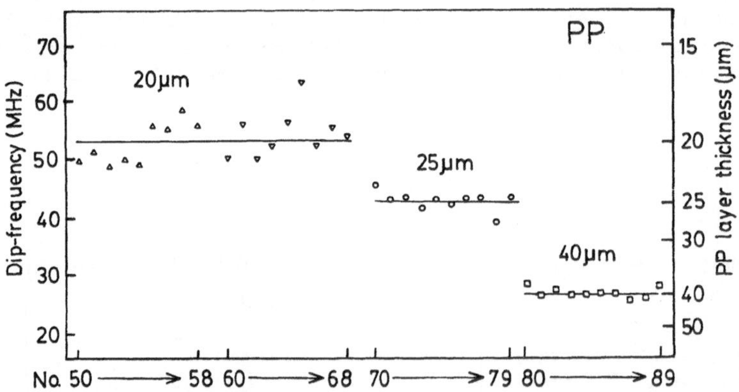

Fig.10 Dip frequencies and corresponding layer thickness for specimens No.50-89

Thus, it was shown that the layer thickness of the samples in the present study could be measured by using ultrasonic waves with an incident angle larger than 29°. Furthermore, an incident angle larger than 45° is desirable, because of the reason discussed in the preceding sections, in order to avoid the effect of bonding quality.

CONCLUSIONS

Calculations of reflection coefficients for a steel substrate with a polymer coating were made with different boundary conditions at an interface between the coating and the substrate (perfect and smooth bondings). At incident angles 20°-40°, frequency dependence of reflection coefficients were quite different for perfect and smooth bondings. At incident angles equal to zero and between 45° and 55°, reflection coefficients were insensitive to the bonding quality. Observations of spectra of ultrasonic waves reflected at polymer coated steel substrates supported the theoretical analysis.

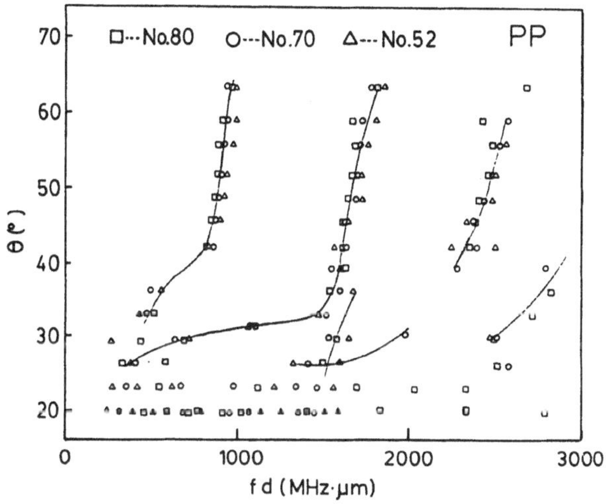

Fig.11 Loci of dips for samples with different layer thickness in fd (dip frequency×layer thickness) and θ (incident angle) plane

ACKNOWLEDGEMENTS

We wish to extend our gratitude to Dr. J. Kushibiki, who made an opportunity for us to have a discussion which has led to the present study. We also wish to thank Professor N. Chubachi for discussions on applications of the present method. Test specimens used in the experiments for bonding quality were prepared by Mr. Hoshino of Japan Probe Co. Ltd. We gratefully acknowledge Mr. H. Masuda for his long standing support of our research.

REFERENCES

Brekhovskikh, L. M., 1980, "Waves in Layered Media," Academic Press, New York.

Houze, M., Nongaillard, B., Gazalet, M., Rouvaen, J. M., and Bruneel, C., 1984, J. Appl. Phys., 55:194-8.

Pilarski, A., and Rose, J. L., 1988, A transverse-wave ultrasonic oblique-incidence technique for interfacial weakness detection in adhesive bonds, J. Appl. Phys., 63:300-7.

Quate, C. F., 1980, Microwaves, acoustics and scanning microscopy, in: Scanned Image Microscopy," E. A. Ash, ed., Academic Press, London.

Schoenberg, M., 1980, Elastic wave behavior across linear slip interfaces, J. Acoust. Soc. Am., 68:1516-21.

Tsukahara, Y., Nakaso, N., Kushibiki, J., and Chubachi, N., An instrument for layer thickness measurement using pseudo-Sezawa waves, 1986, Proc. IEEE Ultrason. Symp. 1986, 1031-5.

Tsukahara, Y., Takeuchi, E., Hayashi, E., and Tani, Y., A new method of measuring surface layer-thickness using dips in angular dependence of reflection coefficients, 1984, Proc. IEEE Ultrason. Symp. 1984, 992-6.

Weglein, R. D., 1979, SAW dispersion and film-thickness measurement by acoustic microscopy, Appl. Phys. Lett., 35:215-7.

Yamanaka, K., 1983, Surface acoustic wave measurements using an impulsive converging beam, J. Appl. Phys., 54:4323-9.

OPTIMAL PARTIAL-DISCRETE ALGORITHM FOR TIME-DELAY ESTIMATION

Hua Lee, John M. Silkaitis, and Douglas P. Sullivan

Department of Electrical and Computer Engineering
University of Illinois at Urbana-Champaign
Urbana, Illinois 61801 USA

ABSTRACT

In this paper, the partial discrete model is applied to the problem of time-delay estimation. Specifically, the cascade form is outlined, although a direct form exists. The direct form [1] is an elegant vector space approach to the problem, but is hindered by the fact that it is difficult to implement because matrix elements are in an integral, rather than closed, form. The cascade form arrives at the same result as the direct form and is easily implemented. This paper will demonstrate why this algorithm is superior to conventional methods, and will also show under what conditions the algorithm degenerates into these other methods. Finally, a step-by-step outline is presented for implementation of the cascade form.

INTRODUCTION

The classical approach to time-delay estimation is to cross-correlate the probing signal with the received signal. This is equivalent to matched-filtering, which results in peaks corresponding to the unknown delays. The amplitudes of these peaks are related to the strength of the returned signal, but are weighted by the cross-correlation function. In general, the resolution is limited since the matched filter is band-limited, due to a lowpass probing signal. If additional information is available, enhancement is feasible, which improves resolution capability and retrieves the amplitudes associated with the corresponding time delays. In particular, when the cutoff frequency of the probing signal and its Fourier transform are known, as well as a bound on the time delays, the estimation problem can be readily implemented with the partial-discrete model outlined in an earlier paper [1].

The partial-discrete model assumes a continuous resultant signal from which samples are obtained and processed to form a continuous profile of the delays within the known bounds. In the optimal algorithm for high-resolution time-delay estimation, the processing is divided into three distinct stages and has become known as the cascade form. Each stage uses a different parameter to provide enhancement in the complete algorithm. The first stage is a spectral estimation which uses the band-limits of the probing signal for enhancement. The second stage is a correction which outputs an initial estimate of the spectrum of the delay profile. In this stage, the Fourier transform of the probing signal is used as the enhancement parameter. In the final stage, the bounds on the time-delay are used for the last enhancement operation resulting in the desired delay profile.

PROBLEM STATEMENT

Let the probing signal, h(t), be known and be band-limited to the frequency range $(-f_c, f_c)$. Furthermore, let the received signal be a sum of delayed versions of the probing signal with amplitudes corresponding to the reflectivity of the target, range to the target, and any number of other factors. This received signal can be written as

$$r(t) = \sum_{k=1}^{K} a_k h(t - t_k) \qquad t \in T, \tag{1}$$

where a_k and t_k are the amplitudes and time delays, respectively. T is the finite observation period, and K is the number of reflecting targets. It will also be assumed that the received signal has been sampled, with uniform spacing Δt which satisfies the Nyquist sampling condition, so that

$$r(n) = \sum_{k=1}^{K} a_k h(n\Delta t - t_k) \qquad t \in T, \tag{2}$$

The unknown delay and amplitude profile is continuous and can be written as

$$d(t) = \sum_{k=1}^{K} a_k \delta(t - t_k) \qquad t_k \in (-\tau, \tau) \tag{3}$$

where τ represents the known bound of the time delays. The received signal can be directly related to the desired signal by a convolution

$$r(t) = d(t) * h(t) \qquad t \in T. \tag{4}$$

This relationship will be used in stage two of the algorithm.

STAGE ONE

As mentioned in the introduction, the first stage involves a spectral estimation of the received signal, which is required for the frequency filtering operation in the second stage. Since only a finite number of samples are available for processing, information outside of the observation interval is lost. Therefore, the discrete Fourier transform (DFT) of the data sequence will be governed by this truncation. The problem of spectrum estimation can be modified into the form of the partial-discrete model. In this case, the objective will be a continuous spectrum which will be derived from the discrete data samples. The problem is outlined as

$$\text{Objective:} \quad R(f) = F\{r(t)\} \qquad |f| \leq f_c$$

$$\text{Subject to:} \quad r(n) = \int_{-f_c}^{f_c} R(f) e^{j2\pi f n \Delta t} df \qquad n=1,2,\ldots,N. \tag{5}$$

The second equation is the measurement constraint, which follows from the definition of the Fourier transform.

To begin the derivation, an inner product is introduced which is defined by

$$< A(f), B(f) > = \int_{-f_c}^{f_c} A(f) B^*(f) df. \tag{6}$$

With this definition, the measurement constraint can be rewritten as

$$r(n) = \int_{-f_c}^{f_c} R(f) e^{j2\pi f n \Delta t} df = < R(f), e^{-j2\pi f n \Delta t} > \qquad n=1,2,\ldots,N. \tag{7}$$

The data sequence is thus a projection of the spectrum upon the N basis functions $e^{-j2\pi f n \Delta t}$, $n=1,2,\ldots,N$. The reconstruction of R(f) from the data samples is then an underdetermined problem of estimating a continuous function from a finite set of data points. As with all underdetermined problems, there are an infinite number of spectra which will satisfy Equation (7). To make the solution unique, the minimum-norm condition will be imposed, and the estimate obtained will be denoted $\hat{R}(f)$. It is known from the *projection theorem* [2] that the minimum-norm solution can be written as a linear combination of the basis function as

$$\hat{R}(f) = \sum_{k=1}^{N} r\hat{\ }(k) e^{-j2\pi f k \Delta t}, \tag{8}$$

where $r\hat{\ }(k)$ is the optimal set of basis coefficients. The problem is now reduced to solving for these coefficients. Since $\hat{R}(f)$ is a solution, it must also satisfy the measurement constraint. Therefore,

$$r(n) = \int_{-f_c}^{f_c} \sum_{k=1}^{N} r\hat{\ }(k) e^{-j2\pi f k \Delta t} e^{j2\pi f n \Delta t} df \qquad n=1,2,\ldots,N. \tag{9}$$

By interchanging the summation and integration

$$r(n) = \sum_{k=1}^{N} r\hat{\ }(k) \int_{-f_c}^{f_c} e^{j2\pi f \Delta t (n-k)} df$$

$$= \sum_{k=1}^{N} r\hat{\ }(k) \frac{\sin[2\pi f_c \Delta t (n-k)]}{\pi \Delta t (n-k)}. \tag{10}$$

Equation (10) can be rewritten in matrix form as

$$[\,r\,] = [\,A\,]\,[\,\hat{r}\,] \tag{11}$$

where

$$A(n,k) = \frac{\sin\,[2\pi f_c \Delta t(n-k)]}{\pi \Delta t(n-k)}. \tag{12}$$

The coefficients can then be evaluated as

$$[\,\hat{r}\,] = [\,A\,]^{-1}\,[\,r\,], \tag{13}$$

and the continuous spectrum is then calculated from Equation (8).

STAGE TWO

From Equation (4), the delay profile is related to the received signal through a convolution with the probing signal. The corresponding transform relationship is

$$R(f) = D(f)H(f). \tag{14}$$

Therefore, by inverse filtering the estimate of R(f) obtained in the last stage by the known H(f), an estimate of D(f) can be obtained. However, this estimate is limited to the same frequency band as H(f), since this spectrum is known to be lowpass. This situation will be addressed in stage three.

Since the processing is to be done digitally, the correction will be done on a sample by sample basis. The frequency sample spacing is controlled by the passband of H(f) and the number of samples, M, desired. To retain the information present in the spectral estimate of stage one, the number of frequency samples must be at least as large as the original data length, i.e. $M \geq N$. The sample spacing is then

$$\Delta f = \frac{2f_c}{M}. \tag{15}$$

Therefore,

$$\hat{R}(m\Delta f) = \sum_{k=1}^{N} \hat{r}(k) e^{-j2\pi m \Delta f k \Delta t}$$

$$= \sum_{k=1}^{N} \hat{r}(k) e^{-j2\pi m k 2 f_c \Delta t / M}. \tag{16}$$

By definition, the sampling frequency, f_s, is the reciprocal of the sample spacing and is required to be greater than the Nyquist frequency, $2f_c$. Define L to be the ratio of these frequencies so that

$$L = \frac{f_s}{2f_c} = \frac{1}{2f_c\Delta t} \geq 1. \tag{17}$$

By letting $M' = ML$, Equation (16) can be rewritten as

$$\hat{R}(m\Delta f) = \sum_{k=1}^{N} \hat{r}(k)e^{-j2\pi mk/M'}. \tag{18}$$

If M' is an integer, then the above can be computed by the DFT. If M' is a power of two, then the spectrum can be computed by an efficient FFT algorithm. After M has been chosen and the frequency samples of $\hat{R}(f)$ have been evaluated, these samples are multiplied to the corresponding frequency samples of the inverse filter to obtain

$$\hat{D}(m\Delta f) = \frac{\hat{R}(m\Delta f)}{H(m\Delta f)}. \tag{19}$$

STAGE THREE

Stage three addresses the dual of the problem encountered in stage one. In this stage, the continuous delay profile within the known time-limit $(-\tau, \tau)$ needs to be estimated from a finite number of samples of its spectrum. As in stage one, the problem can be outlined as

$$\text{Objective:} \quad d(t) = F^{-1}\{\hat{D}(f)\} \qquad |t| \leq \tau$$

$$\text{Subject to:} \quad \hat{D}(m) = \int_{-\tau}^{\tau} d(t)e^{-j2\pi tm\Delta f}dt \qquad m=1,2,...,M. \tag{20}$$

The second equation is the dual of the measurement constraint used in stage one.

A new inner product space is defined in the time domain as

$$<a(t), b(t)> = \int_{-\tau}^{\tau} a(t)b^*(t)dt. \tag{21}$$

Therefore, Equation (20) can be rewritten as

$$\hat{D}(m) = \int_{-\tau}^{\tau} d(t)e^{-j2\pi tm\Delta f}dt = <d(t), e^{j2\pi tm\Delta f}> \qquad m=1,2,...,M. \tag{22}$$

Since the same conditions exist as before, the minimum-norm solution for the delay profile, $\hat{d}(t)$, can be written as

$$\hat{d}(t) = \sum_{b=1}^{M} D'(b)e^{j2\pi tb\Delta f}, \tag{23}$$

where D′(b) are the optimal coefficients for reconstruction. By substituting Equation (23) into Equation (22),

$$\hat{D}(m) = \int_{-\tau}^{\tau} \sum_{b=1}^{M} D'(b) e^{j2\pi tb\Delta f} e^{-j2\pi tm\Delta f} dt$$

$$= \sum_{b=1}^{M} D'(b) \int_{-\tau}^{\tau} e^{j2\pi t\Delta f(b-m)} dt$$

$$= \sum_{b=1}^{M} D'(b) \frac{\sin[2\pi\tau\Delta f(b-m)]}{\pi\Delta f(b-m)}. \tag{24}$$

By converting Equation (24) into matrix form

$$[\hat{D}] = [A'][D'], \tag{25}$$

where the elements of the matrix $[A']$ are defined as

$$A'(m,b) = \frac{\sin[2\pi\tau\Delta f(b-m)]}{\pi\Delta f(b-m)}. \tag{26}$$

Therefore, the optimal coefficients can be generated by

$$[D'] = [A']^{-1}[\hat{D}], \tag{27}$$

and the desired continuous delay profile can be computed from Equation (23).

DISCUSSION

The cascade form uses the cutoff frequency of the probing signal, the transform of this signal, and the bounds on the time delays to increase the resolution over conventional methods. Without knowledge of the cutoff frequency, the first enhancement matrix cannot be formulated. Therefore, the spectrum must be obtained from the DFT of the original data sequence. Since the passband is not known, the inverse filtering of stage two has inherent stability problems. To bypass these problems, the matched filter, $H^*(f)$, is used to filter the spectrum of the received signal. The output now is no longer the spectrum of the delay profile, but rather the transform of the cross-correlation of the received signal and the probing signal. The inverse transform of this spectrum will result in peaks located at the time-delays but with a resolution dependent on the cross-correlation. The output could still be enhanced with knowledge of the bounds on the time delay by the method in stage three. Without any of these enhancement parameters, the cascade form degenerates into the conventional matched filter operation for time-delay estimation.

The cascade algorithm can also degenerate to conventional methods under different auspices. Since both of the enhancement matrices are dependent upon the sample spacing in their respective domains, an unwise choice of this spacing leads to no enhancement. In stage one, if the received signal is sampled at the Nyquist rate, the matrix reduces to the identity matrix and the minimum-norm estimate of the spectrum is simply the DFT of the data sequence. The same reasoning applies to the matrix in stage three with respect to the frequency sample spacing. In each case, the argument of the sine function becomes an integer multiple of pi, and therefore, is always zero. Should both of these matrices become degenerate, the algorithm reduces to the conventional deconvolution method.

A major difficulty encountered when implementing this algorithm is the process of inverting the matrices derived in stages one and three. These matrices are often ill-conditioned and are not associated with an exact inverse. To remedy this problem, singular value decomposition (SVD) is used to generate the Moore-Penrose Pseudoinverse. From the SVD theorem [3], every $M \times N$ matrix has a singular value decomposition from which the unique pseudoinverse can be computed. This procedure is outlined extensively in [4,5,6]. Since the matrices are not usually of full rank, only a portion of the singular values will be nonzero which reduces the dimension of the reconstruction subspace. Two algorithms for improving the reconstruction subspace are the control-point procedure and the iterative weighting procedure, which are discussed in [4,5,6,7].

CONCLUSIONS

In concluding this paper, the algorithm is summarized.

(1) Sample the received signal, r(n).

(2) Generate and invert the first enhancement matrix, [A].

(3) Form the sequence of optimal coefficients, $\hat{r}(k)$.

(4) Decide on frequency spacing, zero pad $\hat{r}(k)$ and take M point DFT.

(5) Inverse filter within the passband obtaining $\hat{D}(m\Delta f)$.

(6) Generate and invert the second enhancement matrix, [A′].

(7) Form the optimal coefficients, D′(b).

(8) Use Equation (23) to compute the continuous delay profile, $\hat{d}(t)$.

The algorithm uses all of the information available for enhancement and divides the processing into three distinct stages. This division allows for sensitivity adjustments when implementing the algorithm, especially when using a threshold for SVD operations. It has also been shown that the algorithm can degenerate into conventional methods if parameters are not known or if sufficient care is not taken when choosing spacings. Additionally, all of the operations involve closed form expressions, which makes it easy to implement them without approximations.

ACKNOWLEDGEMENT

This research is partly supported by the National Science Foundation under grant ECE-8451484, and the Army Research Office under Contracts DAAL 03-86-K-0111 and DAAL 03-87-K-006.

REFERENCES

[1] H. Lee, "High-Resolution Acoustical Image Reconstruction for Finite-Size Objects: The Cascade Form," *Acoustical Imaging*, vol. 15, H. W. Jones Ed. New York: Plenum Press, 1987, pp. 635-646.

[2] D. G. Luenberger, *Optimization by Vector Space Methods*. New York: John Wiley and Sons, 1969, pp. 46-68.

[3] S. Leon, *Linear Algebra with Applications*. New York: Macmillan, 1980, pp. 283-313.

[4] D. Sullivan, "Two New Algorithms for Discrete Time Band-Limited Extrapolation," M.S. thesis, University of Illinois, Urbana-Champaign, 1986.

[5] H. Lee, Z. C. Lin, and T. S. Huang. "Performance and Limitations of Discrete Band-Limited Extrapolation," *Proceedings of the International Conference on Acoustics, Speech, and Signal Processing*, 1986, pp. 1645-1648.

[6] H. Lee and T. S. Huang, "On Discrete Band-Limited Signal Extrapolation," *Proceedings of the International Conference on Acoustics, Speech, and Signal Processing*, 1985, pp. 465-468.

[7] J. M. Silkaitis, "High Resolution Time-Delay Estimation: the Cascade Form," M.S. thesis, University of Illinois, Urbana-Champaign, 1988.

IMAGING THE ACOUSTIC NON-LINEAR PARAMETER

WITH DIFFRACTION TOMOGRAPHY

Anni Cai, Yoshikatsu Nakagawa, Glen Wade and Masahide Yoneyama *

Department of Electrical and Computer Engineering
University of California, Santa Barbara, CA 93106, U.S.A.
* Research and Development Laboratory
RICOH Company, Ltd., Yokohama 223, Japan

ABSTRACT

In previous work, we suggested a way to take into account diffraction when reconstructing tomograms of the acoustic nonlinear parameter in biological specimens. In this type of tomography, two collinear planar waves with different frequencies are employed to insonify the object. The complex-amplitude of the generated secondary wave at the difference frequency is detected for producing the projections. However, the algorithm previously presented is valid only in the case of weakly scattering objects because we neglected scattering effects caused by variations in the refractive index. In this paper, we extend the approach to include compensation for primary-wave scattering and part of the secondary-wave scattering. This makes the approach valid for moderately-scattering objects. Algorithms and results of computer simulations are presented.

INTRODUCTION

Imaging the acoustic nonlinear parameter of biological specimens is of interest because it may provide a sensitive indication of the state of health of tissue [1,2]. In previous work [3], we suggested a way to take diffraction into account when tomographically imaging the acoustic nonlinear parameter in biological specimens. In this approach, as shown in Fig. 1, two sets of planar primary waves, each at a different frequency, ω_1 or ω_2, are projected into the object from the same direction. A variety of secondary waves will be generated due to parametric interaction between the two primary waves. Complex amplitudes of the forward-scattered wave at the difference frequency ω_d are detected with a receiving array perpendicular to the incident waves. By changing the direction of the incident waves and the receiving array, various sets of such diffracted projections can be measured. A tomogram of the spatial variation of the nonlinear parameter is expected to be generated from these projections.

Consider an infinite space filled with a homogeneous loss-free medium with propagation velocity c_o and nonlinear parameter β_o. Imbedded in this medium is a loss-free two-dimensional object of constant density $\rho=\rho_o$, a spatially varying refractive index $n(\vec{r})$ and a nonlinear parameter $\beta(\vec{r})$, where $\vec{r}$ is the position vector. The propagation of the secondary wave at the difference frequency can be approximately described by a nonlinear wave equation [3]

$$\nabla^2 P_{db} + k_{do}^2 P_{db} = k_{do}^2 [n^4(\vec{r})\beta(\vec{r}) \frac{P_1 P_2^*}{P_{1i} P_{2i}^*} - \beta_o] \frac{P_{1i} P_{2i}^*}{\rho_o c_o^2} + k_{do}^2 [1 - n^2(\vec{r})] P_d \tag{1}$$

where P_n and P_{ni} (n=1, *or* 2) are pressures of the total and incident primary waves respectively, k_{do} is the wave number of the difference-frequency wave in the surrounding medium. In this equation, the pressure of the generated secondary wave, P_d, is assumed to consist of two components

$$P_d = P_{dh} + P_{db} \tag{2}$$

where P_{dh} is the secondary wave that would be generated in the homogeneous medium if the object were not present and P_{db} represents the secondary wave produced by the presence of the object. Time dependence is omitted in Eq. (1). Pressures in capital letters represent complex amplitudes, and * indicates taking the complex conjugate.

There are two members in the source term (the right side) of Eq. (1). The first represents the virtual source of the difference-frequency wave generated by parametric interaction between the two primary waves. The second represents the induced source due to the interaction of the generated secondary wave with the object having an inhomogeneity in $n(\vec{r})$.

In order to furnish information about P_1 and P_2 in Eq. (1), a linear equation is used to approximately describe the behavior of the primary waves [3]

$$\nabla^2 P_{ns} + k_{no}^2 P_{ns} = k_{no}^2[1-n^2(\vec{r})]P_n \qquad (n=1,\ or\ 2) \tag{3}$$

where k_{no} is the wave number of the primary wave in the surrounding medium, P_{ns} is pressure of the scattered primary wave and $P_n = P_{ni} + P_{ns}$. The two simultaneous equations, Eqs. (1) and (3), serve as the basis for our imaging method.

Eq. (1) is difficult to solve for $\beta(\vec{r})$. A major reason for the difficulty is that the secondary wave P_d is involved in the source term and is unknown within the object. If we assume that generation of the secondary wave is dominant and that scattering of the secondary wave caused by variations in $n(\vec{r})$ is negligible in comparison, we can simplify Eq. (1) by ignoring the second member of the source term and letting $P_1 P_2^* / P_{1i} P_{2i}^* \approx 1$. $\Delta\beta(\vec{r}) = \beta(\vec{r}) - \beta_o$ can then be reconstructed by employing methods of conventional diffraction tomography [4,5]. However, this approach is valid only in the case of weakly scattering objects [3].

POSSIBILITIES OF COMPENSATING FOR PARTIAL SCATTERING

P_{dh}, a part of P_d, can be determined as long as the surrounding medium and the positions of the transmitter and the receiver are specified. To avoid involving an unknown quantity in the source term of the governing wave equation, only $k_{do}^2(1-n^2)P_{db}$ actually needs to be neglected. In addition, if we detect projection data not only at the difference-frequency but also at ω_1 and ω_2, we can reconstruct $n(\vec{r})$ from the primary-wave data by conventional diffraction tomography. The primary wave distributions P_1 and P_2 can be calculated by using the reconstructed $n(\vec{r})$ and the incident waves P_{1i} and P_{2i}. The assumption that $P_1 P_2^* / P_{1i} P_{2i}^* \approx 1$ would then not be required for reconstructing $\Delta\beta(\vec{r})$. Therefore, it is possible to develop practical algorithms for reconstructing $\Delta\beta(\vec{r})$ in moderately-scattering objects by compensating for the effects of primary-wave scattering and part of the secondary-wave scattering. Fig. 2 shows a schematic diagram of the reconstruction process. The procedure within the double-lined box is the focus of our discussion.

In principle, we can compensate in two ways: (1) we can eliminate the scattering effects from each projection and then combine the compensated projections to obtain an estimate of $\Delta\beta(\vec{r})$; and (2) we can obtain an image in the conventional way [4,5] by using uncorrected data and then correct the image by using a corrector which takes into account the scattering effects. We call the first method pre-compensation and the second, post-compensation.

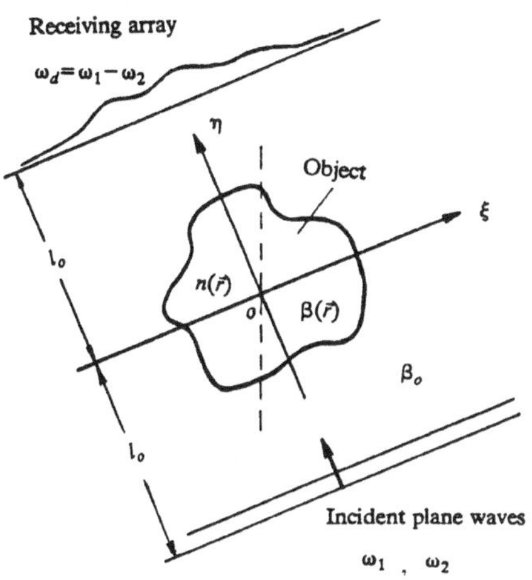

Fig.1 A tomographic configuration for taking diffraction into account in nonlinear parameter imaging.

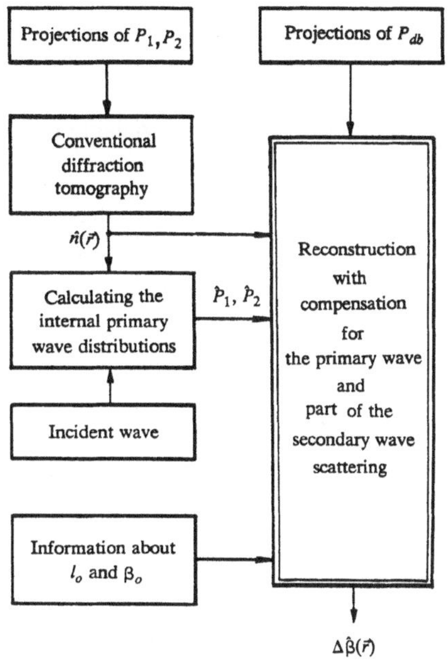

Fig.2 Flow diagram for reconstruction with compensation for partial scattering.

COMPENSATING FOR PART OF THE SECONDARY-WAVE SCATTERING

$k_{do}^2(1-n^2)P_{dh}$, a part of the second member of the source term in Eq. (1), is responsible for the scattering of P_{dh} due to inhomogeneities in $n(\vec{r})$. We now consider how to compensate for this effect by pre-compensation. The secondary-wave component, P_{db}', produced by this source member satisfies

$$\nabla^2 P_{db}' + k_{do}^2 P_{db}' = k_{do}^2(1-n^2)P_{dh} \tag{4}$$

P_{dh} is a plane wave with a linearly-increasing amplitude along the propagation direction and can be expressed as [3]

$$P_{dh} = j\frac{k_{do}\beta_o(l_o+\eta)}{2\rho_o c_o^2} P_{1i}P_{2i}^* \tag{5}$$

where l_o and η are indicated in Fig. 1.

Considering $P_{1i}P_{2i}^*=e^{jk_{do}\eta}$, we can easily show that the relationship between the projection of P_{db}' and the object function is in agreement with the Fourier diffraction projection theorem [4]. Here the object function $o(\vec{r})$ refers to $(1-n^2)(l_o+\eta)$. That is

$$P_{db}' = C\cdot\mathcal{F}^{-1}_{(K_\xi)}\{\frac{e^{j\gamma l_o}}{j2\gamma}\tilde{O}(K_\xi,\gamma-k_{do})\} \tag{6}$$

where $C=jk_{do}^3\beta_o/2\rho_o c_o^2$, $\tilde{O}(\vec{K})$ is the two-dimensional (2-D) Fourier transform of $o(\vec{r})$, $\vec{K}=(K_\xi,K_\eta)$, $\gamma=(k_{do}^2-K_\xi^2)^{1/2}$, and $\mathcal{F}^{-1}_{(K_\xi)}\{.\}$ represents a one-dimensional (1-D) inverse Fourier transform with respect to K_ξ.

By using the properties of the Fourier transform, we can calculate $\tilde{O}(\vec{K})$ from the Fourier transform $\tilde{O}'(\vec{K})$ of $(1-n^2)$

$$\tilde{O}(\bar{K}) = l_o\, \tilde{O}'(\bar{K}) + j\, \frac{d\tilde{O}'}{dK_\eta} \tag{7}$$

Assume that $\tilde{O}'(\bar{K})$ has been available on a square sampling grid in the K_x-K_y plane (see Fig. 3). For a particular incident angle θ, we interpolate $\tilde{O}'(\bar{K})$ from the sample values given on the rectangular grid, first, to the sample values on the semi-circle AB to obtain $\tilde{O}'(K_\xi, \gamma - k_{do})$, and then, to the sample values on the semi-circle $A'B'$ to calculate the following derivative

$$\frac{d\tilde{O}'}{dK_\eta}\Big|_{(K_\xi,\gamma-k_{do})} \approx \frac{\tilde{O}'(K_\xi, \gamma-k_{do}) - \tilde{O}'(K_\xi, \gamma-k_{do}-\Delta K_\eta)}{\Delta K_\eta} \tag{8}$$

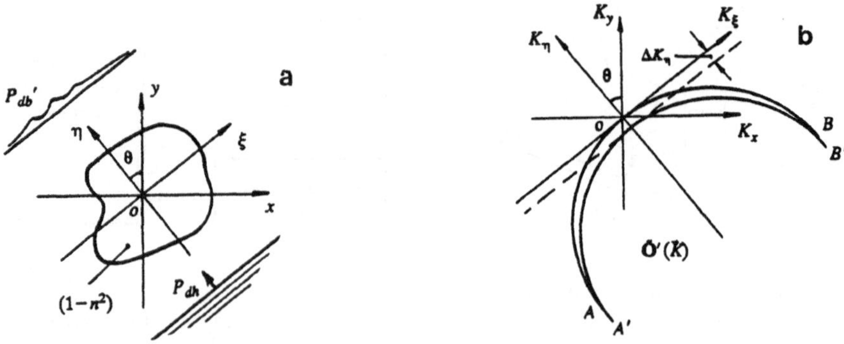

Fig.3 Diagrams illustrating the relationship between the projection of P_{db}' and samples of $(1-n^2)$ in the Fourier transform space.

Thus the value of P_{db}' at the receiver line can be obtained by using Eqs. (6) and (7). This value is then subtracted from the projection of P_{db} at angle θ.

COMPENSATING FOR PRIMARY-WAVE SCATTERING

If $k_{do}^2(1-n^2)P_{db}$, a part of $k_{do}^2(1-n^2)P_d$, in Eq. (1) is neglected, by subtracting both sides of Eq. (4) from Eq. (1), we can obtain

$$\nabla^2 P_{db}'' + k_{do}^2 P_{db}'' = k_{do}^2\left(n^4\beta\frac{P_1P_2^*}{P_{1i}P_{2i}^*} - \beta_o\right)\frac{P_{1i}P_{2i}^*}{\rho_o c_o^2} \tag{9}$$

where $P_{db}''=P_{db}-P_{db}'$. In this section, we will discuss how to compensate for the effects of primary-wave scattering assuming we know $P_1P_2^*/P_{1i}P_{2i}^*$. We call this ratio the normalized wave factor $\psi_n(\bar{r})$.

1. Pre-Compensation

We designate the term in the parentheses on the right side of Eq. (9) as the modified nonlinear parameter $\Delta\beta_m$. Since the internal primary-wave distribution ψ_n, in general, is different for different angles of incidence, $\Delta\beta_m$ is not only a function of positions $\bar{r}$, but also a function of incident angle θ. However, what we are looking for is $\Delta\beta$, which is independent of θ. In order to find the relationship between $\Delta\beta_m$ and $\Delta\beta$, we assume that $\Delta\beta_m$ consists of two terms, one, independent of θ, another, dependent on θ, i.e.

$$\Delta\beta_m = (\beta-\beta_o) + \Delta_\theta \tag{10}$$

Δ_θ can then be derived as

$$\Delta_\theta = \Delta\beta_m - \Delta\beta = \beta(n^4\psi_n - 1) \approx \beta_o(n^4\psi_n - 1) \tag{11}$$

In the last step of Eq. (11), we introduced an approximation in order to make Δ_θ a known function. Errors caused by this approximation exist only within the object region.

The secondary-wave component, P_{db}''', produced by the existence of "object" Δ_θ satisfies

$$\nabla^2 P_{db}''' + k_{do}^2 P_{db}''' = k_{do}^2 \cdot \Delta_\theta \cdot \frac{P_{1i} P_{2i}^*}{\rho_o c_o^2} \tag{12}$$

According to the Fourier diffraction projection theorem, the value of P_{db}''' at the receiver line can be estimated as

$$P_{db}''' = \mathcal{F}_{(K_\xi)}^{-1}\{ \frac{k_{do}^2 e^{j\gamma l_o}}{j2\rho_o c_o^2 \gamma} \tilde{\Delta}_\theta(K_\xi, \gamma - k_{do})\} \tag{13}$$

where $\tilde{\Delta}_\theta(\bar{K})$ is the 2-D Fourier transform of Δ_θ. For each incident angle, the corresponding value of P_{db}''' is subtracted from the projection of P_{db}'' to eliminate the effects caused by Δ_θ.

2. Post-Compensation

Subtracting both sides of Eq. (12) from Eq. (9), we obtain

$$\nabla^2 P_{db}'''' + k_{do}^2 P_{db}'''' = \frac{k_{do}^2 n^4}{\rho_o c_o^2} (\beta - \beta_o) \cdot \psi_n \cdot P_{1i} P_{2i}^* \tag{14}$$

where $P_{db}''''=P_{db}''-P_{db}'''$. For computational efficiency, we will compensate for the remaining effects of primary-wave scattering by post-compensation, i.e., we will make an image in the conventional way by using projections of P_{db}'''', then correct the image by using a proper corrector.

According to the Fourier diffraction projection theorem, the Fourier transform of the projection of P_{db}'''' gives the values of the 2-D Fourier transform of the object along a circular arc as shown in Fig. 4 (a). Here the object function refers to $\Delta\beta\psi_n$. Since ψ_n is dependent

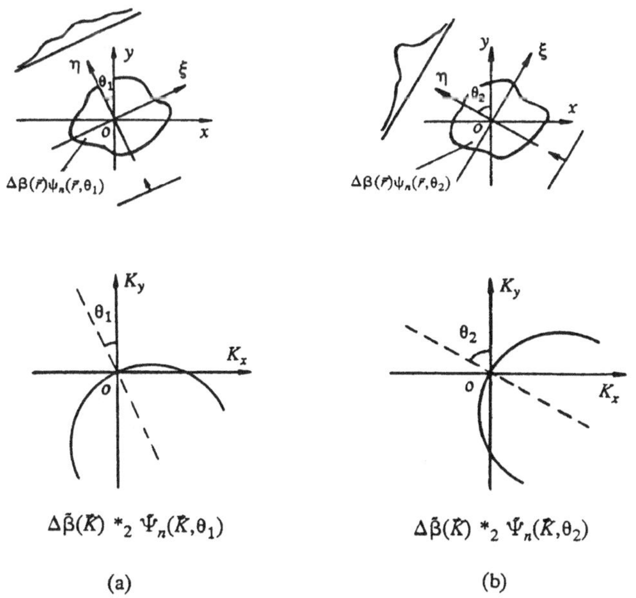

Fig.4 Diagrams illustrating the relationship between the scattered field and samples in the Fourier transform space of $\Delta\beta(\bar{r})\psi_n(\bar{r})$ for two different incident waves.

on the incident angle θ, the arcs obtained at different values of θ do not fall in the same Fourier space (see Fig. 4). In Fig. 4, $\Delta\tilde{\beta}(\vec{K})$ and $\tilde{\Psi}_n(\vec{K},\theta)$ are the 2-D Fourier transforms of $\Delta\beta$ and ψ_n respectively, and $*_2$ represents a 2-D convolution. ψ_n and $\tilde{\Psi}_n$ have been given an argument θ in the figure to indicate their dependence on the incident angle. To obtain a tomogram by using projections of P_{db}'''', we force the data in different Fourier spaces to be put into one space. Because of this, we end up with multiple values for the data in that space. We use the mean value at the intersection of different arcs as the value at that point. By taking the inverse Fourier transform, we can obtain an estimate $\underline{\hat{\beta\psi}}_n(\vec{r})$ of $\Delta\beta(\vec{r})\psi_n(\vec{r})$.

To form a proper method for compensation, we consider the reconstruction of a point source. We assume a hypothetical situation where the nonlinear-parameter distribution is a point inhomogeneity and is expressed by $\Delta\beta\cdot\delta(\xi-\xi_o)\delta(\eta-\eta_o)$, and ψ_n's have the distributions we want to compensate for. The Fourier transform of the projection produced by this source at angle θ is given by

$$\tilde{P}_{db}''''(K_\xi,\eta=l_o) = \frac{e^{j\gamma l_o}}{j2\gamma}\cdot D\Delta\beta\cdot\psi_n(\xi_o,\eta_o,\theta)\cdot e^{-j[K_\xi\xi_o+(\gamma-k_{do})\eta_o]} \tag{15}$$

where D is a constant and equals $k_{do}^2 n^4/\rho_o c_o^2$. The filtered back-propagated image of this object can be expressed by [5]

$$\underline{\hat{\beta\psi}}_n(\vec{r}) = \frac{k_{do}D\Delta\beta}{(2\pi)^2}\int_{-\pi}^{\pi}\psi_n(\xi_o,\eta_o,\theta)d\theta\int_{-k_{do}}^{k_{do}}\frac{|K_\xi|}{2\gamma}\,e^{j[K_\xi(\xi-\xi_o)+(\gamma-k_{do})(\eta-\eta_o)]}dK_\xi \tag{16}$$

The image at point (ξ_o,η_o) can be obtained by integrating the second integral at $\xi=\xi_o$ and $\eta=\eta_o$

$$\underline{\hat{\beta\psi}}_n(\xi_o,\eta_o) = \frac{k_{do}^2 D\Delta\beta}{(2\pi)^2}\int_{-\pi}^{\pi}\psi_n(\xi_o,\eta_o,\theta)d\theta \tag{17}$$

If $\psi_n(\vec{r},\theta)=1$, i.e. if there is no primary-wave scattering, the image at (ξ_o,η_o) should be

$$\underline{\hat{\beta\psi}}_n^{\psi_n=1}(\xi_o,\eta_o) = \frac{k_{do}^2 D\Delta\beta}{2\pi} \tag{18}$$

Obviously, for this nonlinear-parameter distribution, the effects caused by $\psi_n(\vec{r},\theta)$ can be eliminated by multiplying the image obtained by a corrector $c(\xi_o,\eta_o)$. The corrector is given by

$$c(\xi_o,\eta_o) = \frac{\underline{\hat{\beta\psi}}_n^{\psi_n=1}(\xi_o,\eta_o)}{\underline{\hat{\beta\psi}}_n^{\psi_n\neq1}(\xi_o,\eta_o)} = \frac{1}{\frac{1}{2\pi}\int_{-\pi}^{\pi}\psi_n(\xi_o,\eta_o,\theta)d\theta} \tag{19}$$

To generalize this idea for extended distribution of $\Delta\beta(\vec{r})$, we will use the following correction matrix to correct each pixel of the intermediate image $\underline{\hat{\beta\psi}}_n(\vec{r})$

$$c(\vec{r}) = \frac{1}{\frac{1}{2\pi}\int_{-\pi}^{\pi}\psi_n(\vec{r},\theta)d\theta} \tag{20}$$

We note that only values of $c(\vec{r})$ within the object region are significant in the correction since $\Delta\beta$ is zero outside the object.

3. The importance of Pre-Compensation

Fig. 5 (a) shows a plot of a hypothetical variation in $\Delta\beta_m(\vec{r})$. The corresponding object is a cylinder of radius a having variations in $n(\vec{r})$ and $\beta(\vec{r})$ which are shown in (b) and (c) respectively. In this figure, $n_o=1$, $n_1=1.001$, $\beta_o=3.5$, $\beta_1=4$, $\beta_2=5$, $a=16\lambda_1=12\lambda_2=4\lambda_d$, where λ_1, λ_2 and λ_d are wavelengths of the two primary waves and the secondary wave in the surrounding medium respectively. The ripples in Fig. 5 (a) are due to the influence of the nor-

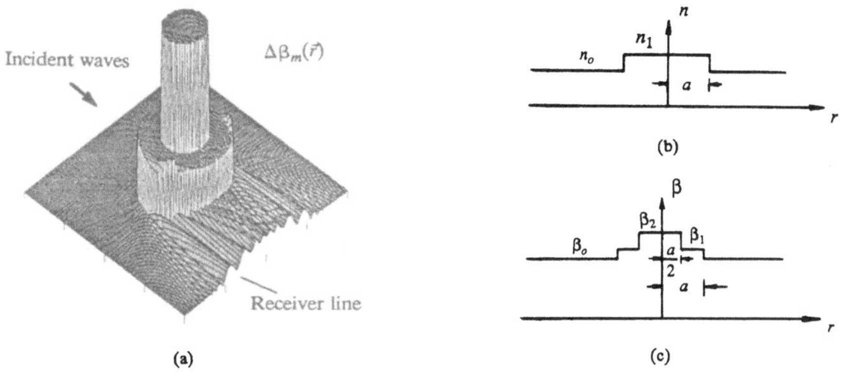

Fig.5 (a) A three-dimensional perspective plot of $\Delta\beta_m(\vec{r})$ for the object depicted in (b) and (c). (b) variation in $n(\vec{r})$ for the object. (c) variation in $\beta(\vec{r})$ for the object.

malized wave factor. The ripples of $\Delta\beta_m$ in the forward direction can be quite large even for an object with a moderate change in refractive index. (For Fig. 5 (a), the change in $n(\vec{r})$ is small.) The secondary wave component generated in the forward region is detrimental to the reconstruction of $\Delta\beta$ because it carries information about primary-wave scattering, not about the nonlinearity of the object.

Pre-compensation eliminates the effects caused by the ripples of $\Delta\beta_m$ outside the object. As expressed in Eq. (14), the remaining uncompensated influence of ψ_n exists only within the object region. From Fig. 5 (a) we see that the scattered primary field within the object is much smaller than that in the forward direction outside the object. Therefore, pre-compensation is more important than post-compensation in compensating for the primary-wave scattering.

Despite the nature of the parametric interaction, $P_1P_2^*/P_{1i}P_{2i}^*$ is similar in numerical value to the ratio of the total field P to the incident field P_i. It is as if a single plane wave at the difference frequency were insonifying the object. Recall that in conventional diffraction tomography, P/P_i is nearly unity when the Born approximation is valid. Therefore, we can see that post-compensation is not necessary if the Born approximation can be invoked at the difference frequency.

To be able to ignore post-compensation, we must have $P_1P_2^*/P_{1i}P_{2i}^* \approx 1$ *inside* the object. To completely neglect the effects of primary-wave scattering requires $P_1P_2^*/P_{1i}P_{2i}^* \approx 1$ both *inside* and *outside* the object. The former condition is easier to satisfy than the latter one.

SIMULATION RESULTS

The geometry of the simulated system is depicted in cross section in Fig. 6. In that figure, $l_o = d$ and $d = 10\lambda_d$. The test object is the same as that shown in Fig. 5 (b) and (c), but with $n_1 = 1.01$. β_o, β_1 and β_2 are chosen to conform to the nonlinearities of water, parenchymal and fatty tissue respectively. The distributions of $n(\vec{r})$ and $\beta(\vec{r})$ are intentionally selected to be different in order to test the fidelity of the $\beta(\vec{r})$ reconstruction.

P_1 and P_2 in Eq. (1) were calculated by using a well-known exact expression for the scattered field from a cylinder with plane and monotonic illumination [6]. An iteration algorithm was used to solve for P_{db} in Eq. (1) [7,8]. The wave fields were simulated over a $4d \times 2l_o$ region. The centrally located part ($2d$) of the calculated secondary wave was used to provide the projection data with 128 sampling points.

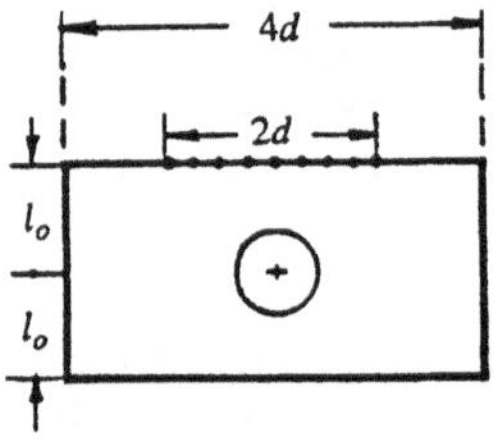

Fig.6 Geometry for the computer simulation.

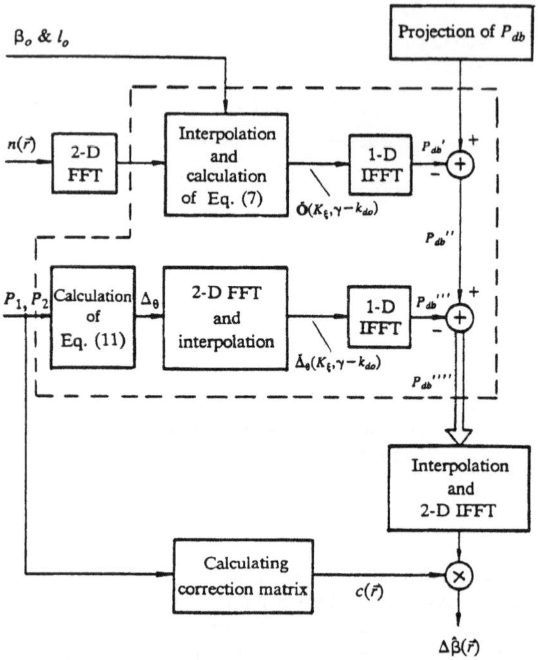

Fig.7 A flow chart showing the compensation for partial scattering.

A flow chart for reconstruction is shown in Fig. 7. Two pre-compensations are performed for each projection of P_{db} by using the procedures indicated in the dashed-line box. All compensated projections are then combined to make an image by Fourier inversion. Finally, this intermediate image is corrected by using $c(\vec{r})$ to yield an estimate of $\Delta\hat{\beta}(\vec{r})$. In order to test the compensation methods themselves, exact (not reconstructed) values of $n(\vec{r})$, P_1 and P_2 were used in the reconstruction.

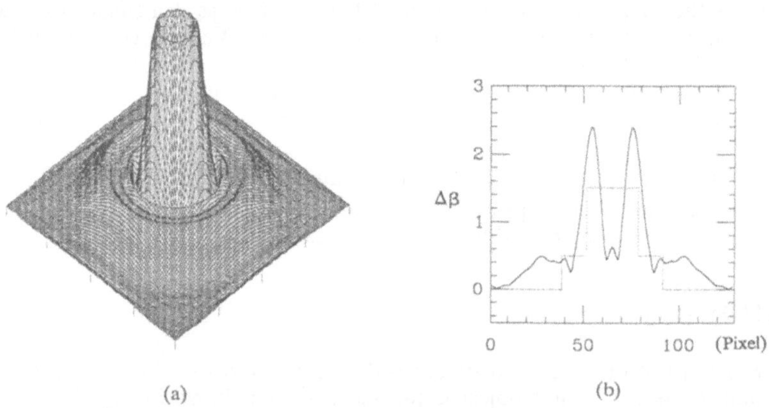

Fig.8 A tomogram generated from the uncompensated projections of an object with 1% change in $n(\vec{r})$. (a) a three-dimensional perspective plot. (b) a central slice of the image.

Fig. 8 gives a tomogram of the object generated from uncompensated projections of P_{db}. The tomogram and the actual object, which is indicated by the dashed line, do not resemble each other. Figs. 9 and 10 show the intermediate image generated from the projections of P_{db}'''' and the tomogram of $\Delta\beta$ after post-compensation, respectively. There is no notable difference between Figs. 9 (b) and 10 (a). This is because the cumulative phase shift for a progressive wave at the difference frequency would be small for this particular object. Post-compensation is not necessary. However, from Figs. 9 (c) and 10 (b), we see that post-compensation does partly correct the "cross-talk" introduced in the idealy zero imaginary part. The remaining imaginary part and the "high shoulder" of the real part of the tomogram reflect the deficiency of our reconstruction method, which is caused by neglecting $k_{do}^2(1-n^2)P_{db}$ in Eq. (1).

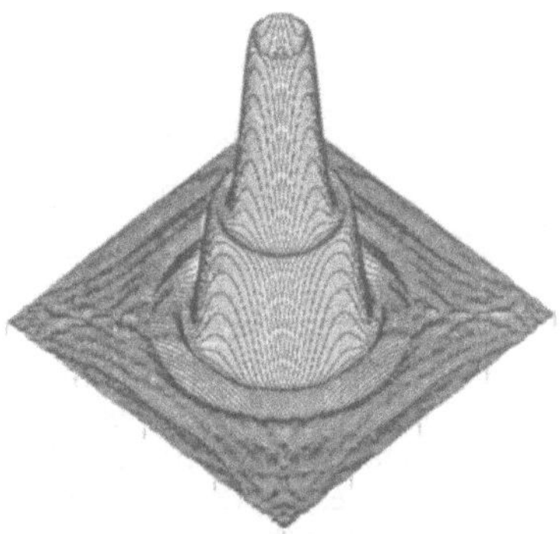

(a)

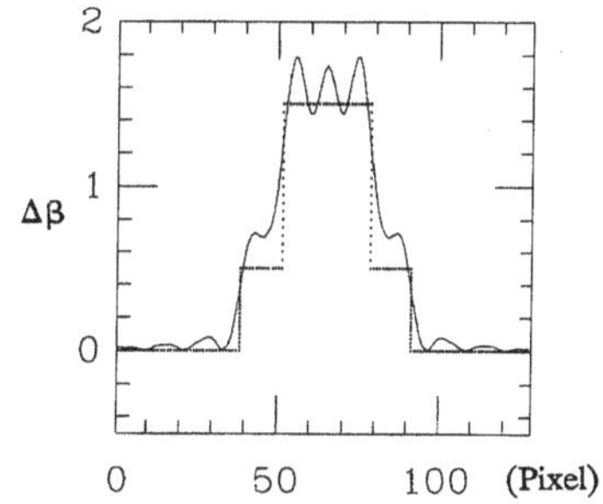

(b)

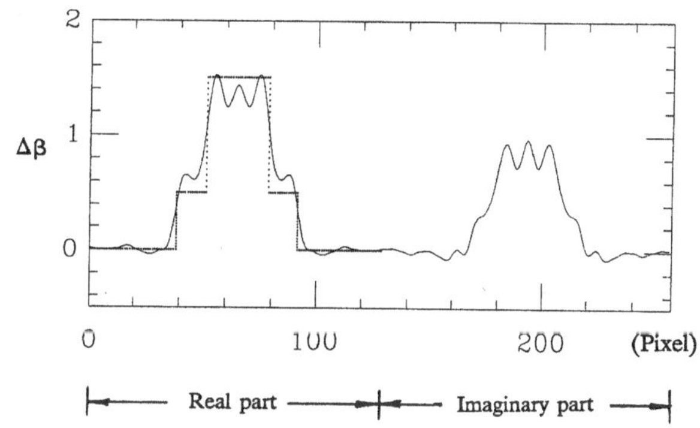

(c)

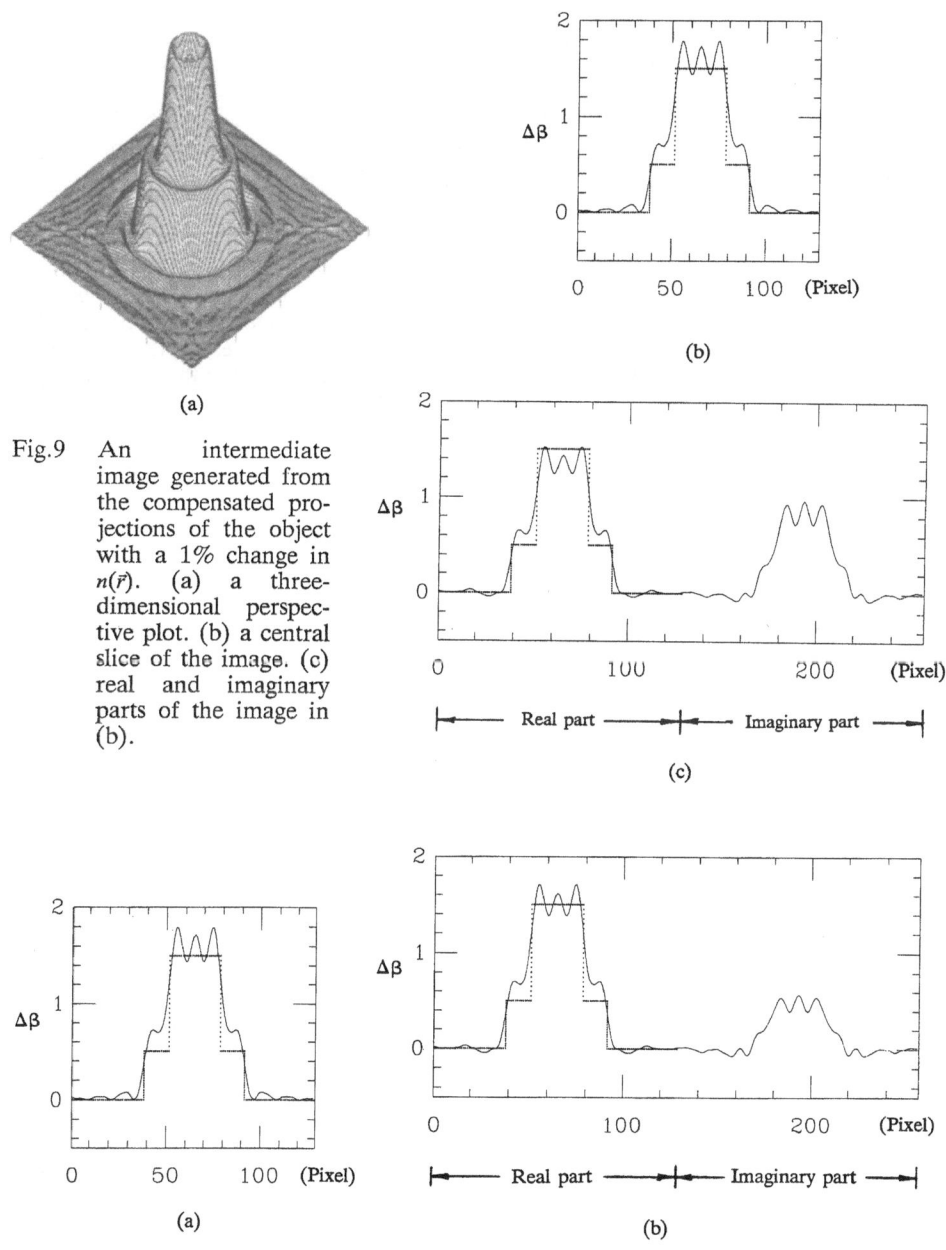

Fig.9 An intermediate image generated from the compensated projections of the object with a 1% change in $n(\vec{r})$. (a) a three-dimensional perspective plot. (b) a central slice of the image. (c) real and imaginary parts of the image in (b).

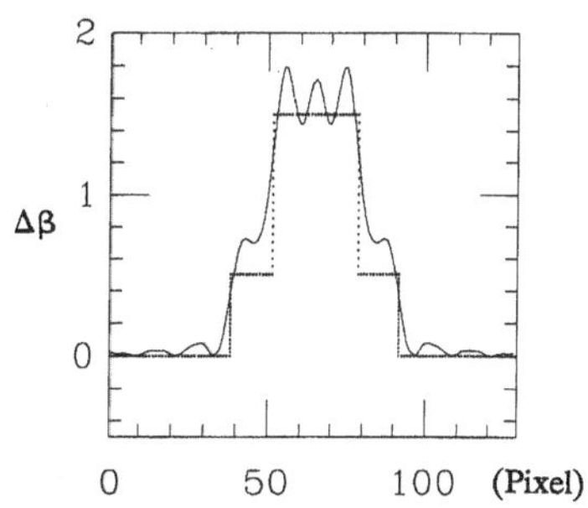

(a)

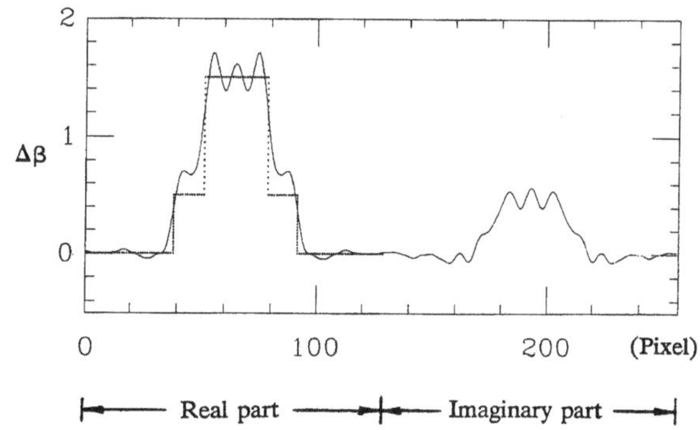

(b)

Fig.10 A tomogram of Δβ obtained from applying post-compensation to the image shown in Fig. 9. (a) a central-slice image of the object. (b) real and imaginary parts of the image in (a).

SIGNIFICANCE OF THE TRANSMITTER POSITION

In conventional diffraction tomography, the position of the transmitter does not matter as long as a plane wave is projected into the object. However, in diffraction tomography of the acoustic nonlinear parameter, the position of the transmitter is significant in the measurements since secondary-wave generation takes place immediately after the primary waves leave the transmitter. The further the transmitter is away from the object, the stronger the secondary wave generated in the surrounding medium. More detrimental to $\Delta\beta$ reconstruction are the scattering effects caused by inhomogeneities in $n(\vec{r})$.

Although we have suggested a method to compensate for the effects due to the scattering of P_{dh}, this method compensates only for the first-order scattering of the secondary wave generated in the surrounding medium. Multiple scattering exists when an object is present in the medium. P_{dh} represents the secondary wave that would be generated in the homogeneous medium if the object were not present. Therefore, $k_{do}^2(1-n^2)P_{dh}$ in Eq. (1) is responsible only for the first-order scattering. The high-order scattering source is included in $k_{do}^2(1-n^2)P_{db}$.

Fig. 11 shows the central slices of two tomograms for the same object which is the same as shown in Fig. 5 (b) and (c), but with $n_1=1.02$. To obtain the image in Fig. 11 (a), the transmitter was assumed to be located at a distance $l_o=10\lambda_d$ from the center of the object. Due to not knowing $k_{do}^2(1-n^2)P_{db}$ in Eq. (1) in our reconstruction model, we did not obtain an accurate tomogram. If the transmitter is placed close to the object, for example if l_o is reduced to $5\lambda_d$, the resultant tomogram shown in (b) bears a better resemblance to the original object.

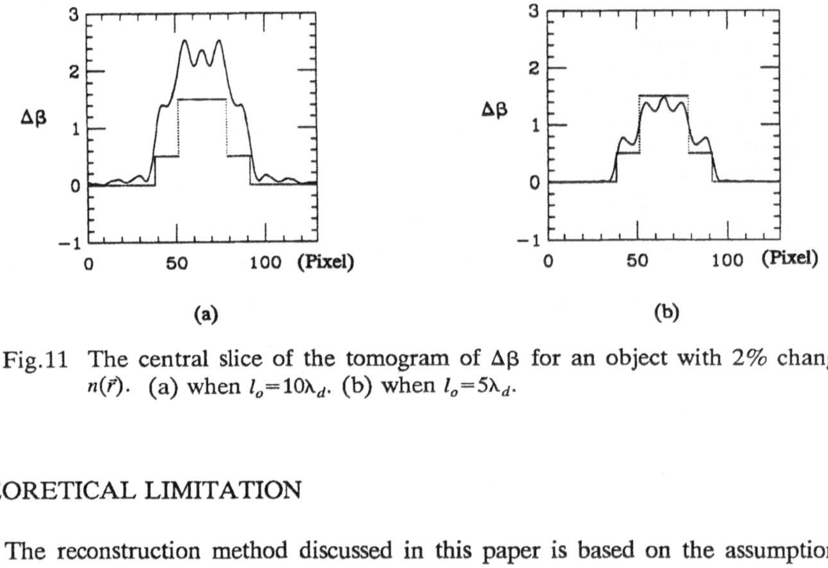

Fig.11 The central slice of the tomogram of $\Delta\beta$ for an object with 2% change in $n(\vec{r})$. (a) when $l_o=10\lambda_d$. (b) when $l_o=5\lambda_d$.

THEORETICAL LIMITATION

The reconstruction method discussed in this paper is based on the assumption that we can neglect $k_{do}^2(1-n^2)P_{db}$ in the governing wave equation. If we take this term into account, Eq. (14) becomes

$$\nabla^2 P_{db}'''' + k_{do}^2 P_{db}'''' = \frac{k_{do}^2 n^4}{\rho_o c_o^2}\Delta\beta \frac{P_1 P_2^*}{P_{1i} P_{2i}^*} P_{1i} P_{2i}^* + k_{do}^2(1-n^2)P_{db} \tag{21}$$

Since $\Delta\beta$ and $(1-n^2)$ have the same region of support, the condition required by neglecting the second term on the right side of Eq. (21) is

$$|P_{db}| \ll \left|\frac{n^4\cdot\Delta\beta}{\rho_o c_o^2\,(1-n^2)}\cdot P_1 P_2\right| \approx \left|\frac{n^4\cdot\Delta\beta}{(1-n^2)}\cdot\frac{v_{1o}}{c_o}\cdot P_2\right| \tag{22}$$

where v_{1_o} is the particle velocity of one of the primary waves and v_{1o}/c_o is recognized as the acoustic Mach number. The Mach number rarely exceeds 10^{-3}, even for very intense waves. For most biological tissue, $\Delta\beta$ is below 3.5. Therefore, the refractive index change is an important factor in Eq. (22). In solving Eq. (21) for $\Delta\beta$, difficulties similar to that in high-order conventional diffraction tomography will be encountered since P_{db} is unknown within the object.

CONCLUSION

We have presented a method to take diffraction into account in nonlinear parameter tomography. Quantitative tomograms of the nonlinear parameter for simulated, moderately-scattering objects are obtained by including compensations for the primary-wave scattering and part of the secondary-wave scattering. The most difficult problem in reconstructing accurate tomograms is to completely compensate for secondary-wave scattering.

ACKNOWLEDGEMENT

This work was performed with support from the RICOH Company, Ltd. of Japan.

REFERENCES

[1] T. Sato, Y. Yamakoshi and T. Nakamura, "Nonlinear Tissue Imaging," *Proc. 1986 IEEE Ultrason. Symp.*, 881-884 (1986).

[2] Y. Nakagawa, W. Hou, A. Cai, N. Arnold, G. Wade, M. Yoneyama and M. Nakagawa, "Nonlinear Parameter Imaging with Finite-Amplitude Sound Waves", *Proc. 1986 IEEE Ultrason. Symp.*, 901-904 (1986).

[3] A. Cai, Y. Nakagawa, W. Hou, N. Arnold and G. Wade, "Diffraction Tomography of the Acoustic Non-Linear Parameter," *Ultrasonics International 87 Conf. Proc.*, 178-183 (1987).

[4] S. X. Pan and A. C. Kak, "A Computational Study of Reconstruction Algorithms for diffraction Tomography: Interpolation Versus Filtered Backpropagation," *IEEE Trans. Acous. Speech & Signal Proc.*, ASSP-31:1262-1275 (1983).

[5] A. J. Devaney, "A Filtered Backpropagation Algorithm for Diffraction Tomography," *Ultrasonic Imaging*, 4:336-350 (1982).

[6] P. M. Morse and K. U. Ingard, *Theoretical Acoustics*, McGraw-Hill, New York (1968).

[7] W. W. Kim, M. J. Berggren, S. A. Johnson, F. Stenger and C. H. Wilcox, "Inverse Scattering Solutions to the Exact Riccati Wave Equations by Iterative Rytov Approximations and Internal Field Calculations," *Proc. 1985 IEEE Ultrason. Symp.*, 878-882 (1985).

[8] M. Slaney and A. C. Kak, *Imaging with Diffraction Tomography*, Tech. Rept. TR-EE 85-5, School of Engineering, Purdue University (1985).

DISPLAY TECHNIQUES OF VOLUME IMAGES OF BURIED OBJECTS IN PILED SNOW BY ACOUSTICAL AND MICROWAVE HOLOGRAPHIC RADAR

Y. Aoki, Y. Takahasi, Y. Sakamoto and M. Ikegami

Dept. of Information Engineering, Hokkaido University
Government Industrial Development Laboratory
N-13 W-8, Sapporo, Japan 060

ABSTRACT

A holographic radar is employed to obtain images of targets in snow, where techniques of holography in two-dimensional plane and frequency sweep with respect to depth axis are combined to collect three-dimensional hologram data with acoustical or microwaves. By numerically processing the hologram data, three-dimensional images, that is volume images are reconstructed and are displayed on a two-dimensional CRT scope as radar images. In this paper we propose display techniques of such volume images in order to diagnose images of a holographic imaging radar. In the proposed display technique we used a CG (Computer Graphics) technique, where the stored image data in voxels of a memory system is searched along a view line connected between a point of view and a voxel. If significant data is found along the view line, the searched image data is projected onto the pixels on the screen. Many types of display methods are proposed and demonstrated by computer simulation. Since the imaging radar discussed here is useful for finding objects buried in snow, the proposed display techniques were examined with actual data collected with microwaves during the snow season. Many objects of various materials, shapes and sizes were used as examples and their reconstructed volume images were displayed.

1.INTRODUCTION

A holographic imaging radar is a promissing radar technique for producing 3D(three-dimensional) images of various objects using long-wavelengths such as acoustical or microwaves. In an imaging radar, not only the locations of targets but also their shapes are imaged and this kind of radar can be utilized in detecting a wide range of objects buried in various depths of snow. However, it is necessary to achieve high resolution of the imaging radar system in order to recognize 3D targets in a 3D environment. Since the principle of the imaging radar proposed here is a multifrequency holographic radar, the resolution of the radar system depends upon the frequency band-width of sweeping frequency and upon the area of hologram aperture corresponding to the scanning area of a receiver. In the case where we cannot expect enough resolution because of a narrow bandwidth of frequency-sweep and limited scanning area, we have to diagnose the displayed radar images in low resolution to recognize the original targets. The display technique proposed in this paper is convenient in diagnosing images in low resolution obtained by holographic imaging radar with waves of long-wavelengths, where 3D volume images are reconstructed from 3D hologram data and images are displayed on a 2D(two-dimensional) CRT scope. Experiments were conducted according to the proposed methods by displaying images of objects buried in snow. Discussions on the advantages and disadvantages of each method are displayed here with the results of the experiments.

2.PRINCIPLES OF RECORDING HOLOGRAPHIC RADAR IMAGES AND SEARCHING IMAGE DATA ROR DISPLAY

Since the detail of the principle of holographic imaging radar has already been presented in the paper of (1), a brief description of the principle is presented here. The process of collecting holographic radar data is as follows; Frequency-swept waves are radiated from the transmitter to the illuminated targets and the reflected waves from the targets are recorded as time series signals. The transmitter and receiver scan a 2D plane of snow covered area, resulting in the recording of 3D data. Image reconstruction is done numerically with the 3D data by using a computer, where pulse compressions with respect to depth axis are done by Fourier transform of the time-series signals. On the other hand 2D pulse compessions, that is the image reconstruction from 2D hologram, are done by Fresnel transform of the 2D hologram data after one-dimensional pulse compressions have been recorded with respect to depth of snow. Thus 3D radar images, that is volume images, are reconstructed and stored in computer memory.

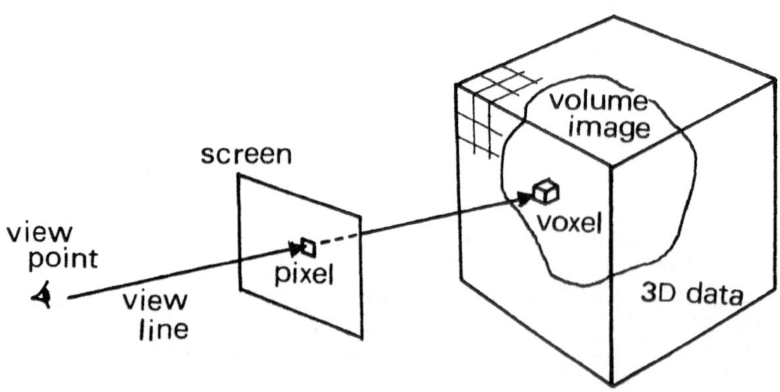

Fig.1 Principle of projecting 3D volume images onto a 2D screen.

Since we use a 2D display equipment, such as CRT to display volume images, it is necessary to project 3D images onto a 2D screen. Here we adopt a technique similar to a ray tracing technique that is used in CG(Computer Graphics), for searching image data and projecting it onto a 2D screen(2).The schematic explanation of the principle is shown in Fig.1, where searching of significant data in the 3D data space is conducted along a view line which originates from a view point through a pixcel on the 2D screen. Here we discuss two methods of searching voxel data in data space, that is, direct searching method and data projection method.

Direct Searching Method

In this method the searching of 3D data is done along a fixed view line through a pixel on the screen as shown in Fig.1, resulting in obtaining one pixel data on the screen after searching data along a certain view line. We change the pixel and repeat the same searching process along a new view line until we scan all the pixels on the 2D screen. Though data on the screen is obtained through the orderly scanning of pixels in this method, the searching order of voxels is not simple, resulting in consumption of access time spent searching volume image data. Therefore different display techniques are necessary to reduce the processing time in searching for data which we will describe later. The advantage of this method is that various techniques of displaying images are possible by processing voxel data along each of the view lines, for example the shading of displayed images as done in CG display. In this paper examples of display are mostly by this searching method.

Data Projection Method

The principle of this method is shown in Fig.2, where 3D data space is divided into 2D parallel data planes. The data of a volume image is assigned to these parallel planes and the data from each of the planes is projected successively onto the screen. In this method the access of voxels in each data plane is orderly, resulting in the reduction of time

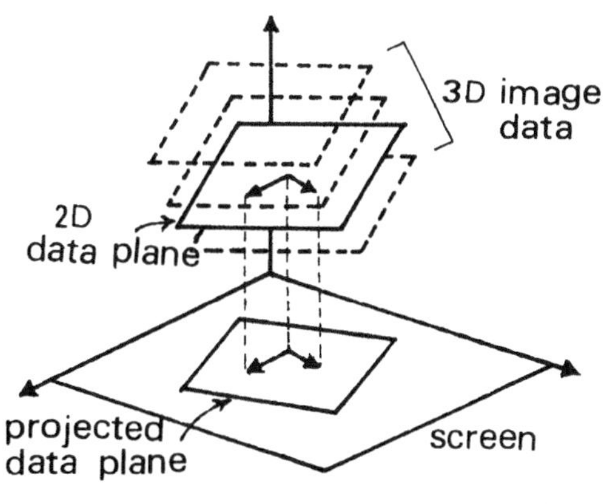

Fig.2 Principle of data projection method.

spent searching for the data of a volume image. The disadvantage of this method is that phenomena related to each of the view lines, such as reflection and refraction, cannot be processed. The proper use of these methods of searching for data and projecting it depends upon what kind images are desired by the users of such imaging radar.

3. DISPLAY METHODS OF VOLUME IMAGES

Various 2D images can be displayed from the same 3D image data by using different projection methods to display volume images onto the screen. In this paper we propose four types of display methods, that is 1)surface shading method, 2)intensity-sum method, 3)intensity-differential-sum method, 4)weighted-differential-sum method.

Surface Shading Method

In this method the surfaces of the 3D targets are displayed with shades corresponding to the distances from the view point to where the surfaces of the targets are located (3). Here the surfaces are determined by the voxels whose intensities exceed the threshold and which appear first in searching for the voxels along the view lines.In the process of searching the surface, the distance from the screen is counted to change the distance into grey level of the displayed image, resulting in displaying depth-shaded image. A 2D depth-shaded image produces the perspective effect. However this method has a disadvantage that the surfaces of targets are often masked by other surfaces of clutter and the selection of threshold is critical to display only the surfaces of the significant targets.

Intensity-Sum Method

In this method all of the intensities of the image data along a view line are summed and the resulting intensity is displayed. A mathematical expression of this method is written as follows,

$$S=\sum_{i=0}^{N-1} v(i) \qquad (1)$$

where S is the intensity of the pixel on the screen and v(i) is the intensity of data of i-th voxel. The special feature of this method is that half-transparent images are displayed. For example, displayed images of targets can be obseved through the faint clutter images. The disadvantage of this method is that all integrated data is displayed and we occasionally cannot clearly distinguish the surfaces of the tagets from the clutter.

Intensity-Differential-Sum Method

In this method the differences between the intensities of neighbouring voxels are calculated along a view line and the absolute values of the differences are summed according to the equation as shown in Eq.(2),

$$S=\sum_{i=0}^{N-1} |v(i)-v(i+1)| \qquad (2)$$

Since the boundaries between targets and surroundings are emphasized by this method, it is possible to display outlines of images regardless of the volume of the targets. The reason why massive target is displayed brighter than a smaller target, is because the summation of intensities of voxels assigned to the image of the massive target is greater than that of the smaller target. Therefore an effect occurs in this method, where the images of the smaller targets are masked by the images of the massive targets. The advantage of this method is that it is possible to display targets with equal weight regardless of the volumes of targets

Weighted-Differential-Sum Method

This is a hybride method of the previously mentioned methods. In this method the intensity S of the pixel is determined by the following equation,

$$S=\sum_{i=1}^{N-1}\left|v(i)-v(i-1)\right|\cdot v(i) \qquad (3)$$

This equation means that the difference of intensities of neighbouring voxels are displayed with weights of the intensities of each of the voxels. The advantage of this method is the elimination of blurry surfaces which appear in intensity-sum method and intensity-differential-sum method. However it is difficult to display perspective in this method. Therefore the following modification of Eq.(3) is proposed,

$$S=\sum_{i=1}^{N-1}\left|v(i)-v(i-1)\right|\cdot v(i)\cdot \exp(-ki) \qquad (4)$$

where ki is a parameter proportional to the distance between the displayed pixel and i-th voxel.

4.EXPERIMENTAL RESULTS

Experiments to display holographic radar images according to the proposed methods were done with computer-simulated data and actual data obtained by microwaves swept

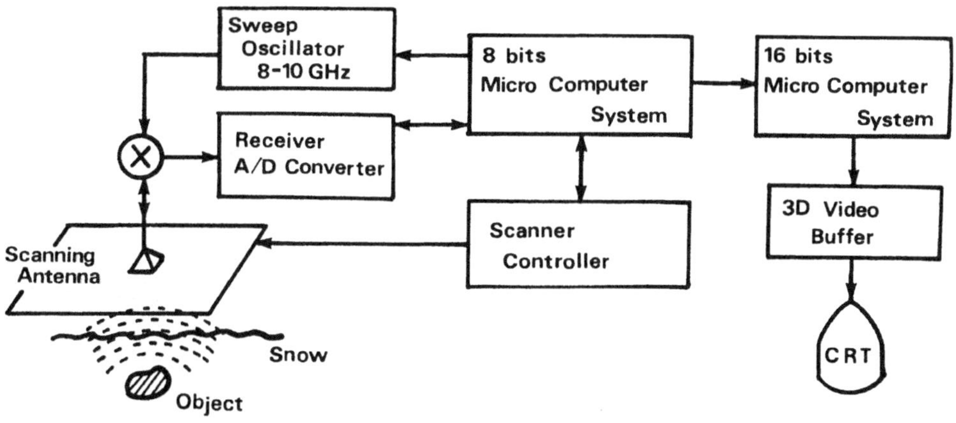

Fig.3 Block diagram of the experimental system of holographic imaging radar.

Fig.4 A scanning antenna of the experimental system of holographic imaging radar assembled on the snow field.

from 8 GHz to 10 GHz.The schematic diagram of the experimental system is shown in Fig.3, where radar data is collected by scanning the 2D snow surface with a transmitting and receiving antenna. The sampling points with respect to 2D scanning plane are 32 x 32, whereas sampling point with respect to frequency axis is also 32. The signals received are processed by 8 bits and 16 bits microcomputer systems and 3Ddata and images are stored in a 3D video buffer as shown in Fig.3.

The experiment was conducted during the snow season when dry snow, that is snow which contains very little moisture, fell and accumulated. Figure 4 shows a photo of the scanning antenna of the experimental system assembled on the campus of Hokkaido University. Many kinds of objects such as metalic pipes, a mannequin covered with silver paper, a container filled with anti-freeze treated water, meat, and concrete blocks were all used as targets buried under snow. Details of the experiment to collect radar data and process it are described in reference (4).

Fig.5
Targets of metalic cans of different lengths.

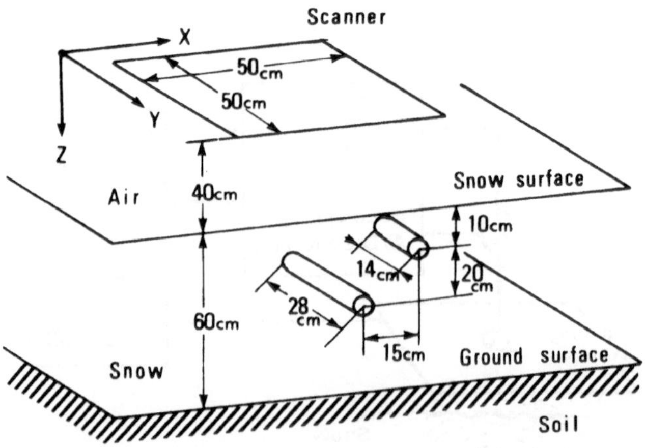

Fig.6 Experimental configuration of locating metalic cans in piled snow.

One example of targets used in the experiment is shown in Fig.5, that is metallic cans 8 cm in diameters and of various lengths. These targets were buried in various depths of snow as shown in the configuration Fig.6. Figure 7(a) and (b) show the processed and displayed images by surface shading method. From these images it is said that the threshold of displaying images is appropriate in Fig.7(a), whereas the images of targets are hidden by the surface of snow and inner boundaries of snow layers in the example of low threshold of Fig.7(b). The results of this experiment suggest that many attempts are necessary to find the appropriate threshold which will display distinct images in surface shading method. The image of Fig.7(c) is computer simulation, where a small sphere, 1/4 disk and 1/8 empty sphere are chosen as targets and images of these targets are displayed by the surface shading method. Experimental results of Fig.7 show the shading effect, by which we can observe the perspective of the displayed images.

Figure 8 shows the images by the intensity-sum method, where (a) is the reconstructed image from the real holographic radar data as previously mentioned, and (b) is the computer simulation. In this method it is not necessary to pay attention to the choice of threshold as was necessary in the surface shading method. However perspective is lost in the image of Fig.8(a). Though the thin shell of a sphere disappears in certain direction as shown in Fig.8(b), the transparent effect is distinct in this image. This suggests that this method is useful in displaying images of targets in piles as a superposition of transparent images.

Figure 9 shows the images produced by the intensity-differential-sum method. In the images of Fig.9(a) changes of intensities of voxel data are emphasized. Figure 9(b) is the smoothed image of (a). Compared to Fig.8(a) and Fig.9(b), we can say that noisy image is reduced in Fig.9(b). Computer simulated image of Fig.9(c) shows the shell of empty sphere is displayed clearly compared with the image of Fig.8(b), however transparent effect is faint in Fig.9(c).

The images of Figs.10 were done by weighted-differential-sum method, where image (a) is displayed according to Eq.(3). Though the image of targets is reconstructed without clutter in Fig.10(a), the outline of the targets is not clear. Therefore, selective local average technique is applied to the image of Fig.10(a), which results in a clearer, more distinct outline of the image in Fig.10(b). However the image of Fig.10(b) looses its perspective and improvement to add perspective is done according to the processing of Eq.(4). The obtained image is shown in Fig.10(c), where we can observe the long metalic can located further away from the short one.

In the display systems proposed here we can easily change the point of view and can diagnose the 3D radar images observed from different view points. But in actuality a high-speed display system is required and we are now developing a special memory system to quickly process volume image data.

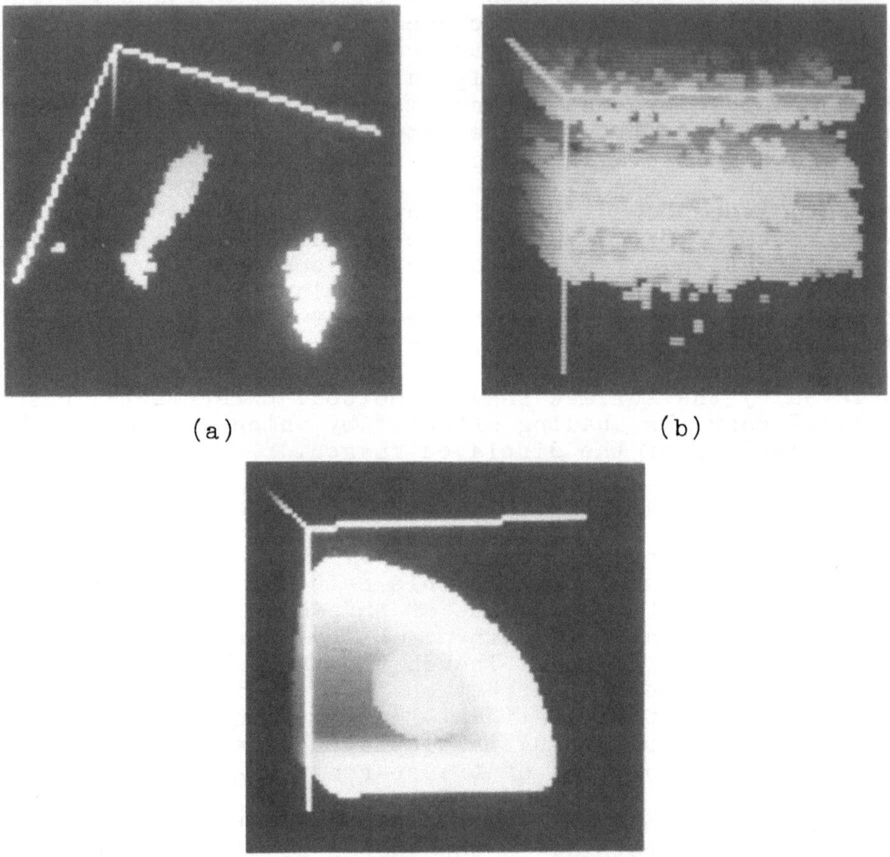

(a) (b)

(c)

Fig.7 Displayed images of metalic cans by surface shading method with appropriate threshold (a), low threshold (b) and computer simulation (c).

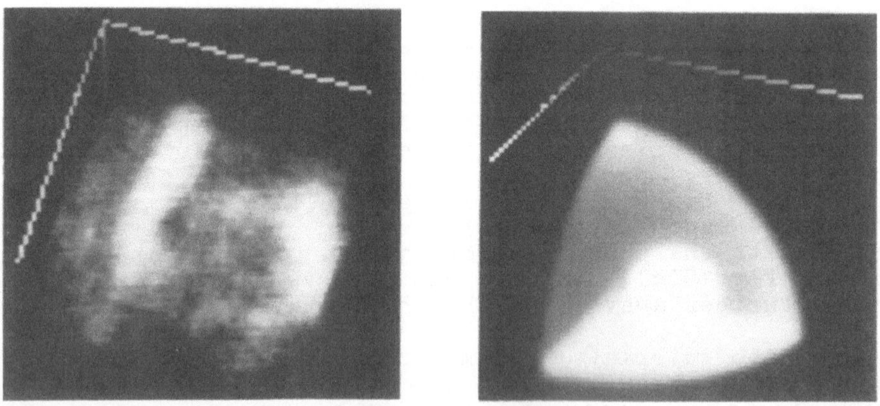

(a) (b)

Fig.8 Displayed image by intensity-sum method of metalic cans (a) and computer simulation (b).

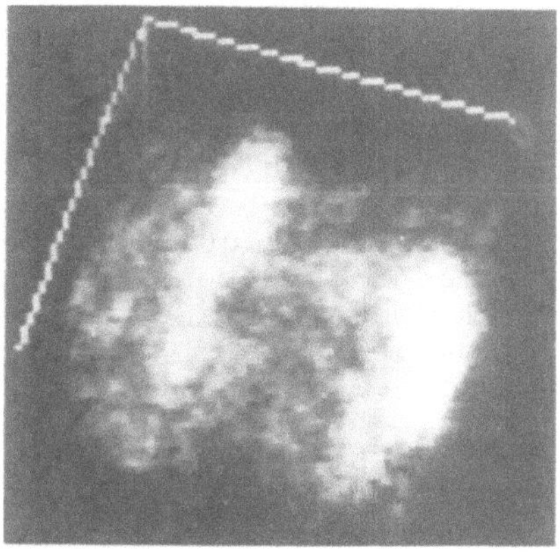

(a)

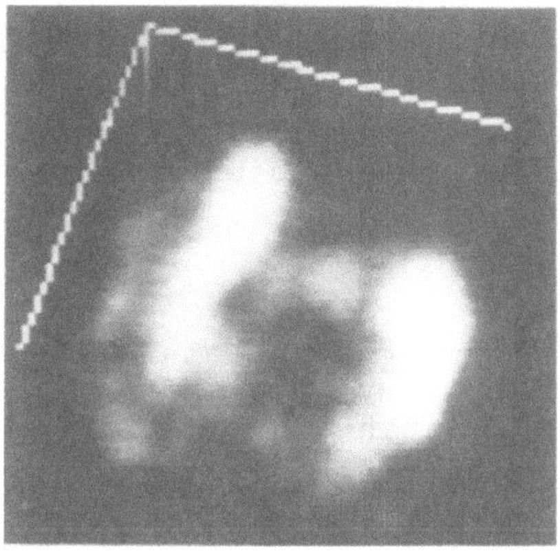

(b)

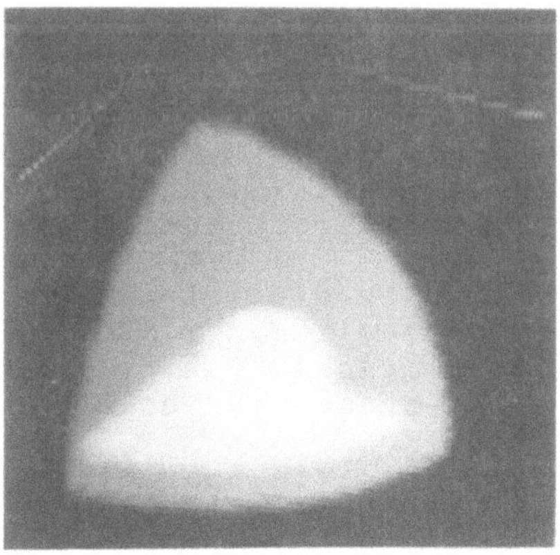

(c)

Fig.9
Displayed images by intensity-differential-sum method (a), its smoothed image (b) and computer simulation (c).

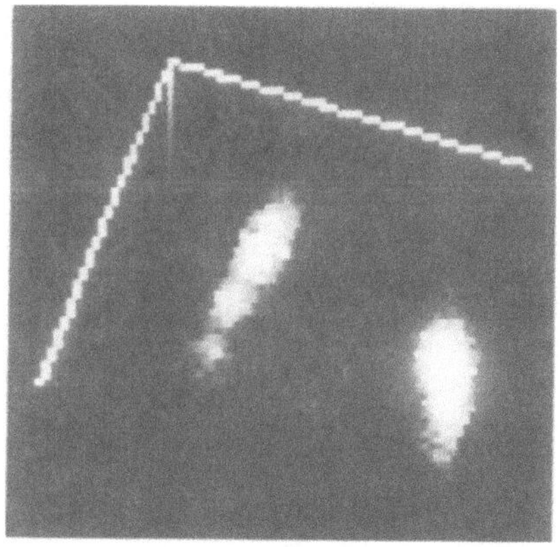

(a)

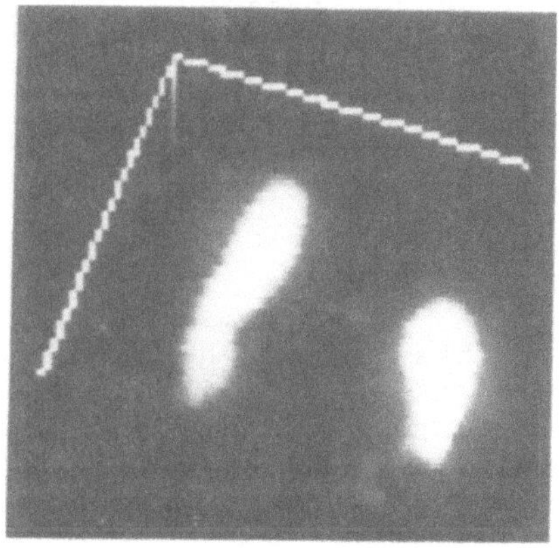

(b)

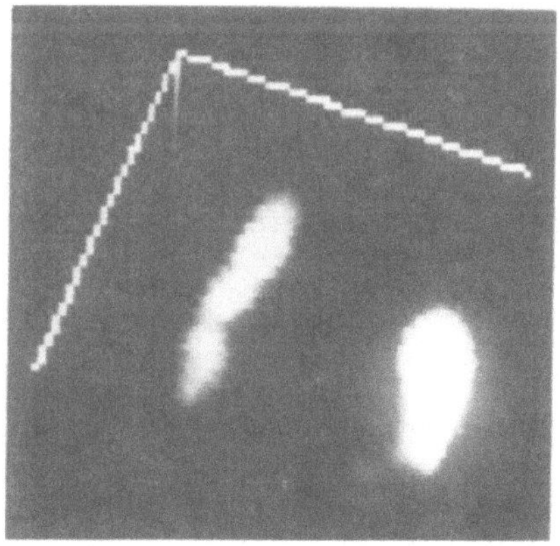

(c)

Fig.10
Displayed images by weighted-differential-sum method (a), modified by average technique (b)and modified with distance parameter (c).

5.CONCLUSION

The proposed techniques of displaying volume images have the potential for diagnosing radar images obtained by using a holographic radar in order to find objects buried in piled snow. Experimental results with X-band microwave confirm the validity of the proposed techniques. However transmittion of microwave through piled snow depends upon the condition of snow, that is transmitted distance becomes shorter as moisture contained in snow increases. Acoustical waves also can be used to collect data according to the same technique of holographic imaging radar, however, we can not expect large transmitted distance in acoustical waves through snow as shown in the experimental result of reference (5).

Few papers of fundamental researches on propagations of long-wavelengths, such as acoustical and microwaves, in piled snow, are reported with the view of holographic imaging radar. Experiments on propagations of such waves in piled snow are necessary to choose suitable frequency bands to obtain three-dimensional data to be processed by the techniques discussed in this paper. The experimental studies recorded here are left for future research projects.

ACKNOWLEDGEMENT

This research was supported by grants, in aid for developmental scientific research, by the Ministry of Education, Culture and Science of Japan and by the Secom Foundation's Research Grant. The authors express their thanks to these financial supporters.

REFERENCES

1. Y.Aoki, Y.Sakamoto and Y.Takahashi, Diagnosis of under-snow radar images by three-dimensional displaying technique in holographic imaging radar, Proc. of IGARSS'87 Symposium, 571 (1987).
2. Y.Takahashi, Y.Sakamoto and Y.Aoki, Display method for 3-D data and its application, Proc. of Sapporo International Computer Graphics Symposium, 172 (1987).
3. K.J.Udupa, Display of 3D information in discrete 3D scenes produced by computerized tomography, Proc.IEEE, 71:420(1983).
4. Y.Sakamoto, K.Tajiri T.Sawai and Y.Aoki, Detection of objects buried in snow using multi-frequency holography, Proc. IEICE, Japan, J70-B:1544 (1987).
5. M.Ikegami, K.Tonooka and Y.Aoki, Acoustic characterstics of snow, Proc. of 17 th International Symposium on Acoustical Imaging, PB-7, (1988).

SPECKLE NOISE REDUCTION USING PM PULSES

Akihisa Ohya, Shin'ichi Yuta*, Iwaki Akiyama**, Takashi Itoh*** and Masato Nakajima

Dept. of Electrical Engineering, Keio Univ., Yokohama Japan and *Inst. of Info. Sci. and Electronics, Univ. of Tsukuba, Tsukuba, Japan and **Dept. of Electrical Engineering, Sagami Inst. of Technology, Fujisawa, Japan and ***Aloka Co., Ltd., Mitaka, Japan

INTRODUCTION

The ultrasonic diagnostic equipment using the pulse echo method has recently come into wide use, and it is used at most hospitals now. It has such advantages as (1)involving no fear of being exposed to X-ray, and (2)making it possible to obtain the tomogram of the living body in real time. But the ultrasonic B-mode image, namely, the tomogram obtained with this equipment is of inferior quality, because it is a random granular pattern called speckle. This speckle arises as a result of the random interference of the ultrasonic waves reflected or scattered at each point of the tissue, and it doesn't coincide with the micro structure of the tissue.

The speckle can be reduced by adding up some images which are independent from one another in terms of the speckle. For the method like this, the frequency compounding[1] and the spatial compounding[2] methods have been proposed. In these conventional methods, the ultrasonic RF pulse waves are used which have nearly constant wave length. But in case the ultrasonic PM(phase modulation) pulse wave is employed, it is considered that the non-stationarity of the wave length hardly brings about interference, resulting in the reduced outbreak of speckle. Further, the interference state of the ultrasonic wave changes according to the way of the modulation and the speckle pattern may change. Consequently, the speckle can be further reduced by adding up the plural images obtained by changing the way of modulation. In this paper, we propose a new technique for the speckle noise reduction using the ultrasonic PM pulse waves generated by modulating the phase of the conventional ultrasonic RF pulse waves with the sinusoidal signals for the purpose of keeping high resolution in the longitudinal direction.

THEORY

Outbreak of Speckle

First, to simplify the case, we examined the case where only two point scatterers (namely, two infinitesimal particles which scatter ultrasonic wave) exist. As shown in figure 1(a), the ultrasonic pulse wave having the wave length L and the pulse duration T is transmitted from the transducer to the scatterers. It is scattered by two scatterers, distance of D apart, and the echo signal is received with the same transducer. (Hereupon, the multiple scatter of the ultrasonics is ignored.) In case of T/2 < D, two echoes scattered by each scatterer are received separately. However, in case of T/2 > D, two echoes interfere with each other. Hence, the amplitude of the received echo signal depends on the relation between the wave length L and distance D between the two scatterers, and it becomes large or small as the result of the interference.

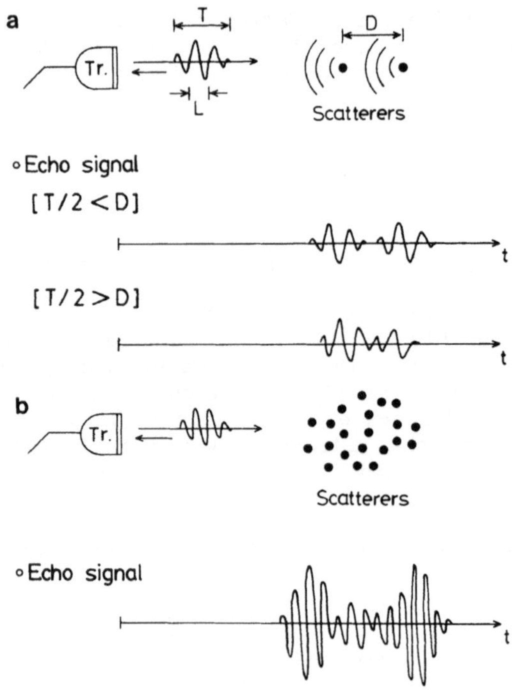

Fig. 1 Outbreak of speckle.
(a) Schema of echo from two scatterers.
(b) Schema of echo from many scatterers.

Next, we examined the case including many scatterers. The tissue (in the living body) can be considered as a group of numerous scatterers gathered at random. Consequently, when the ultrasonic pulse wave is transmitted toward the tissue, the echoes scattered from these many scatterers interfere randomly with one another. As the result, the amplitude of the echo signal received with the transducer has no relation with the distribution of the particles and has only stochastic variance (refer to Figure 1(b)). This variance is observed in the B-mode image as the granular speckle.

Reduction of Speckle

Generally, the speckle can be reduced by summing up the plural images which are independent from one another in terms of the speckle. Based on this theory, the frequency compounding and the spatial compounding methods have been proposed. For the purpose of the acquisition of the independent images, the center frequency or band width of the irradiated ultrasonic pulse wave is changed in the former and the scan position of the transducer in the later. Both these techniques, however, use the ultrasonic RF pulse waves which have nearly constant wave length. As mentioned above, the speckle arises from the coherent interference of the echoes scattered from many particles, the use of such RF pulses results in the generation of the speckle without fail.

Then, the case is considered where the ultrasonic PM pulse waves are employed. In this case, the echo signal isn't the mere summation of the ultrasonic waves whose wave length are fixed, and the interference becomes hard to happen. Consequently, the speckle is considered to appear little here. As the method using the PM pulse like this, the one using the chirp wave has been proposed[3]. In this method, the pulse duration is relatively long and the speckle noise is reduced without any compounding of images. Contrary to this technique, we employ the PM pulses generated by modulating the phases of such ordinary ultrasonic RF pulse waves as are used in the ultrasonic diagnostic equipments, with the sinusoidal signals. Therefore, the pulse length of this PM pulse is almost equal to the ordinary RF pulse's, and the speckle can be reduced without degradation of the resolution in the longitudinal direction (i.e., the direction parallel to the irradiated ultrasonic beam). And, according to the value of the initial phase of the modulation signal, the PM pulse waveform is changed and the speckle pattern may be changed, too. Utilizing this fact, the speckle noise is further reduced by this technique by adding up every image obtained by changing the initial phase.

The ordinary ultrasonic RF pulse waveform f(t), which has the center angular frequency W_0 and the envelope e(t), is expressed in the following equation.

$$f(t) = e(t)\sin(W_0 t). \tag{1}$$

The ultrasonic PM pulse waveform g(t), generated by modulating the phase of this RF pulse f(t) with the sinusoidal signal, is expressed as eq.(2).

$$g(t) = e(t+A\sin(W_1 t+P_0)/W_0)\sin(W_0 t+A\sin(W_1 t+P_0)). \tag{2}$$

In eq.(2), W_1 is the angular frequency of the sinusoidal signal, P_0 is its initial phase, and A is a constant which means the degree of modulation.

Figure 2(a) shows the ordinary ultrasonic RF pulse waveform which has the center frequency 3.5MHz and the Gaussian envelope whose standard deviation is 0.32μsec. (b) and (c) show the ultrasonic PM pulse waveforms generated by modulating the phase of (a) with 1.0MHz sinusoidal waves whose initial phases are 0rad. and πrad., respectively. In case the PM pulses like these are employed, the non-stationarity of the wave length hardly brings about the interference, so the speckle may little appear. As is seen in (b) and (c), when initial phase of the modulation sinusoidal wave is changed, the waveform of the PM pulse is varied and the interference state of the ultrasonic waves reflected or scattered at each point of tissue are changed. As the result, the different distribution of the speckle is obtained in each B-mode image derived from changing the initial phase. Therefore, the speckle noise is further reduced by adding up these images (refer to Figure 3).

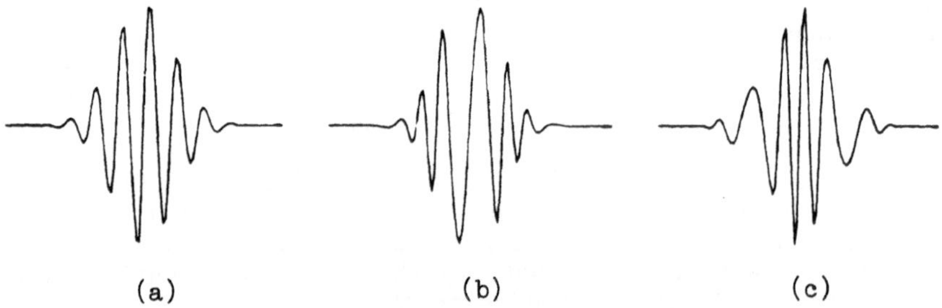

Fig. 2 Ultrasonic pulse waveforms.
(a) Ordinary 3.5MHz RF pulse.
(b) PM pulse(initial phase = 0).
(c) PM pulse(initial phase = π).

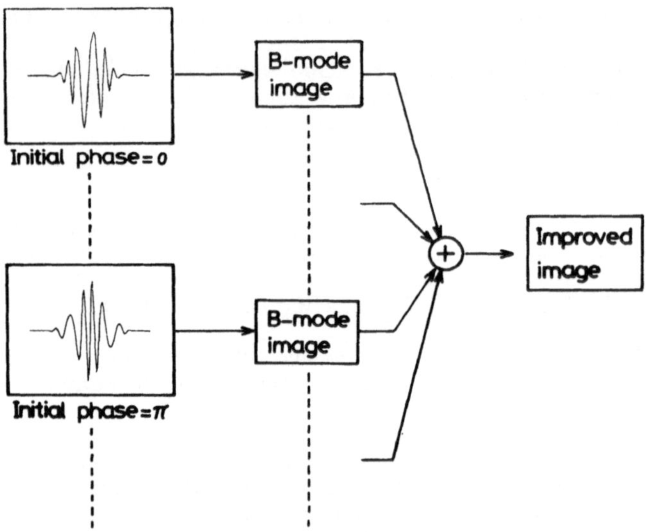

Fig. 3 Concept of speckle reduction.

COMPUTER SIMULATION

We performed the 2-D computer simulations to investigate the availability of the present technique. Figure 4 shows the block diagram of the simulation. First, assuming that the impulse wave is irradiated from the transducer, the impulse response of the phantom received with the same transducer is calculated. Next, this impulse response is convoluted with the ultrasonic pulse waveform transmitted from the transducer, and the result is filtered in consideration of the receiving characteristic of the transducer. Then the echo signal is given and the envelope signal is detected from it. Scanning the transducer linearly, and repeating the calculation mentioned above of the echo signal at every point of the scan, B-mode image is finally obtained.

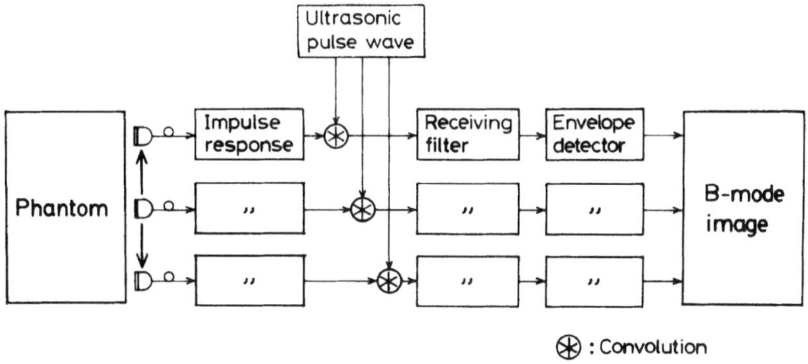

Fig. 4 Block diagram of simulation.

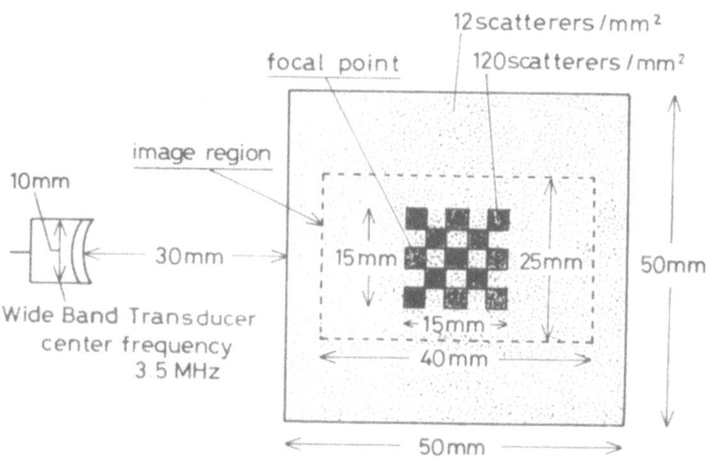

Fig. 5 Simulation phantom.

The simulation phantom is expressed in figure 5. This phantom is constructed from many point scatterers distributed randomly in the region of 50mm × 50mm. The region of 40mm × 25mm is imaged with 50 scanning lines, and the B-mode image is obtained as 80 × 50pixels (1pixel = 0.5mm × 0.5mm). Further, in the center region of 15mm × 15mm, the number of the scatterers is increased by ten times like a checker pattern. Then the scatterer density is 120scatterers/mm^2 in the center checker pattern region, and 12scatterers/mm^2 in the other region. The aperture of the transducer is 10mm and has a concave shape focused on the point of 50mm ahead. The receiving characteristic of the transducer is the Gaussian whose resonance frequency is 3.5MHz and the standard deviation is 1.0MHz. In these computer simulations, the attenuation and the multiple scatter of the ultrasonics is ignored, further, it is assumed that the sound speed is constant.

The results of the computer simulations mentioned above are shown in figure 6. In each image, the ultrasonic beam is irradiated from the upper side of the image. Each image is normalized by its power, and is soft clipped using the threshold. Figure 6(a) shows the distribution of the scatterer density in the phantom. (b) shows the conventional B-mode image obtained by using the ultrasonic RF pulse whose center frequency is 3.5MHz. (c) shows the frequency compounding image obtained by adding up seven images, which are derived from changing the center frequency from 2.0MHz to 5.0MHz every 0.5MHz. (d) and (e) show the images obtained by using the ultrasonic PM pulses shown in figure 2(b) and (c), respectively. Figure 6(f) shows the image improved with this technique, namely, the image obtained by adding up eight images which are derived from changing the initial phases of the sinusoidal signals, from 0rad. to 2πrad. every $\pi/4$rad.

Table 1 R. M. S. error of simulated image.

Image (Fig. 6)	(b)	(c)	(d)	(e)	(f)
R.M.S. error [%]	76.91	60.44	69.35	67.68	60.56

As is seen in figure 6(b), the checker pattern in the phantom is broken by the speckle. But in (d) and (e), by reason of using the PM pulse the speckle doesn't come out as much as in (b) and the arising speckle patterns are different from each other. And in (f) generated by adding up these images and so on, the speckle is reduced to the state where the checker pattern is well understood as much as in (c). In (d), (e), and (f), the use of the ultrasonic PM pulse wave which has nearly the same pulse length as the ordinary RF pulse's, doesn't degrade the resolution in the longitudinal direction.

Table 1 expresses the R. M. S.(Root Mean Square) error of each image referring to the distribution of the scatterer density in the phantom (figure 6(a)). It is found that the images obtained by using the PM pulses, (d) and (e), have less errors than the conventional B-mode image (b), and that the image obtained with the proposed method (f), i.e., the compounding image using the PM pulses, has further less errors and it is nearly equal to the frequency compounding image (c).

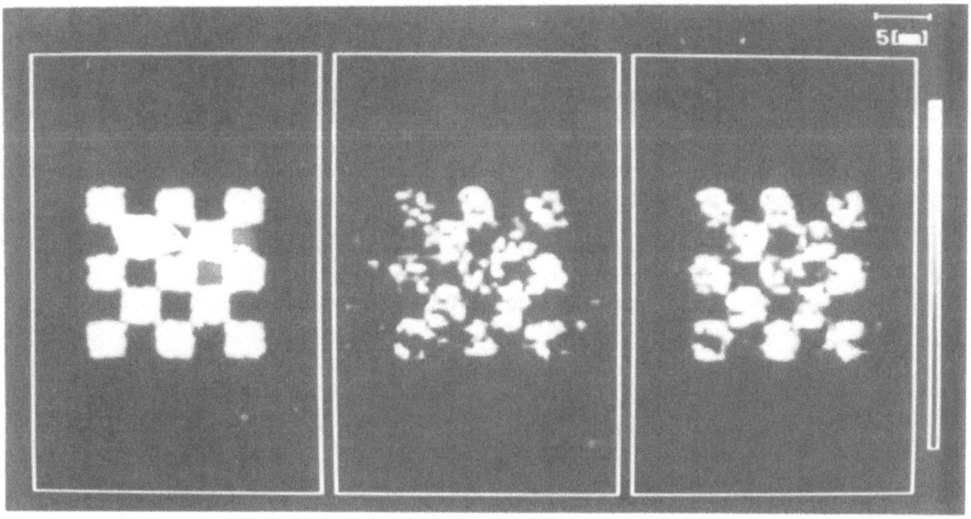

a b c

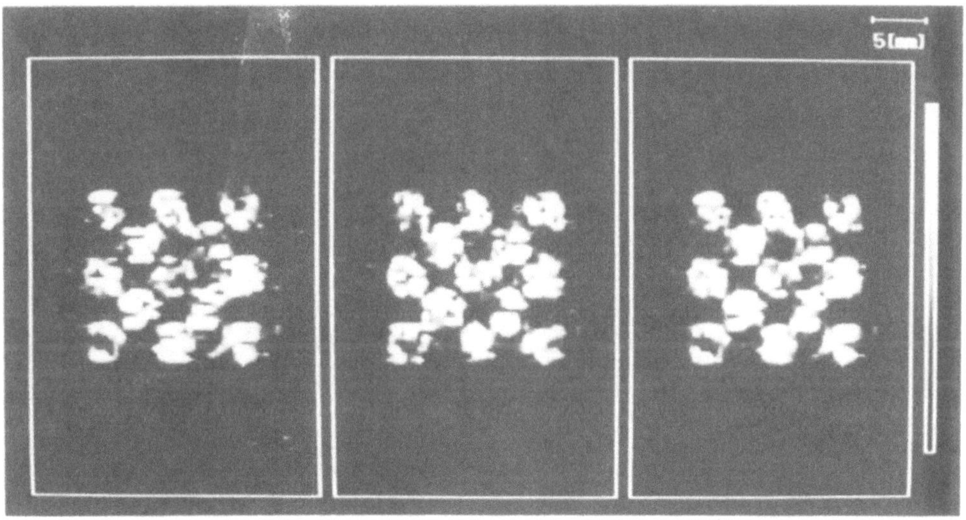

d e f

Fig. 6 Obtained images.
(a) Distribution of scatterer density.
(b) Conventional B-mode image.
(c) Frequency compounding image.
(d) Image generated by using PM pulse(Fig. 2(b)).
(e) Image generated by using PM pulse(Fig. 2(c)).
(f) Proposed method.

CONCLUSION

We proposed a new technique for the speckle noise reduction using the ultrasonic PM pulse waves. We employ the PM pulses generated by modulating the phase of the ordinary ultrasonic RF pulses with sinusoidal signals, so that the resolution in the longitudinal direction doesn't become low. In this technique, the image with little speckle is obtained by using the PM pulses, and the speckle is further reduced by adding up the plural images which are generated by changing the initial phase of the sinusoidal signals and are independent from one another in terms of the speckle. As the result of the 2-D computer simulations, we came to a conclusion that this technique yields nearly the same effect as the frequency compounding method.

The PM pulses can be derived by forcing a transducer with wide band frequency to drive electrically. Consequently, this method can be realized with the simple equipments and can be processed in quasi real time.

REFERENCES

1. P. A. Magnin et al., "Frequency Compounding for the Speckle Contrast Reduction in Phased Array Images", Ultrasonic Imaging 4, 267-281 (1982).
2. C. B. Burckhardt, "Speckle in Ultrasound B-mode Scans", IEEE Trans. Son. Ultrason. 25 (1), 1-6 (1978).
3. S. Singh et al., "Ultrasonic Speckle Reduction Using FM Pulses", IEEE/Eighth Annu. Conf. Eng. Med. Biol. Soc., 1059-1061 (1986).

TEMPORAL PHASE AS AN IMAGING AND CORRECTION TOOL

J.C. Hoddinott, S. Leeman, N. Thomas and D.E. Goss

Dept. of Medical Eng. and Physics, King's College School of Medicine and Dentistry, East Dulwich Grove, London SE22 8PT, U.K.

SUMMARY

The most common conventional approach towards the analysis of ultrasound data is the formation of an image from the envelope of the received signals. In so doing, any information which may be encoded in the phase of the signals is disregarded. More recently, applications such as tissue characterisation and doppler techniques have begun to make indirect use of this additional information, but its full potential remains unrealised.

In this paper we illustrate how phase can be used to provide vital information, of benefit in tissue characterisation and also concerning image artefacts, such as speckle, leading to novel techniques for speckle reduction which can realistically be implemented in real-time data analysis. We also show how these techniques can be extended to include the processing of doppler data and images.

INTRODUCTION

The usefulness of ultrasound as a diagnostic tool in medicine has, in the past, been severely hampered by the presence of a number of artefects. Much previous work has concentrated on the development of correction factors to take into account diffraction effects caused by typical transducers[1]. However, our work has lead us to suggest that such artefacts may, in certain cases, be overcome by the use of appropriate measurement techniques[2]. Moreover, we have found that it is not neccesarily diffraction which is the most problematic artefact, but rather that the emphasis should be placed on the correction of interference effects, which appear to pose the greater problem both in attenuation estimation and imaging. The following simple demonstration underlines this point.

A number of simulated A-lines were created, by the convolution of a gaussian pulse with a series of randomly placed reflectors of varying magnitude, so that interference effects between the 'echoes' occurred, but no attenuation or diffraction effects were included. Figure 1 shows a plot of mean frequency obtained from short-time Fourier analysis versus depth, for one such A-line chosen at random. It might be assumed, on inspection, that a diffraction correction was appropriate, the shape of the curve being typical of such an effect. However, in the context of the simulation it is clear that

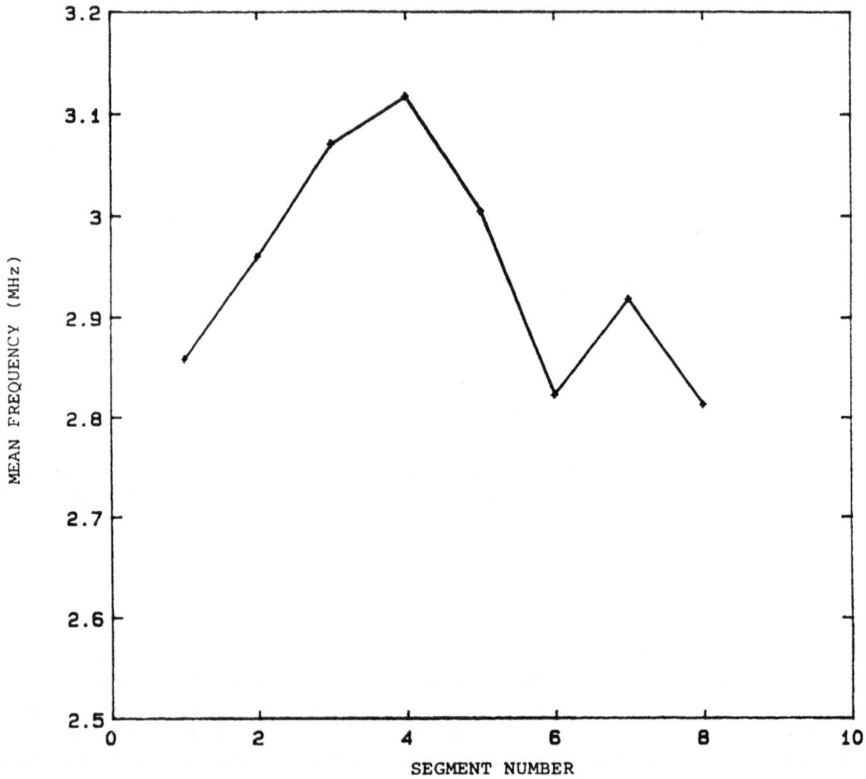

Fig. 1. Mean frequency vs depth for simulated A-line, showing how interference effects may give rise to a shape typically ascribed to diffraction.

it is interference and not diffraction that is responsible for the appearance of the results.

While it is not suggested that such behaviour is typical of all results obtained from such simulations, it does illustrate that it is possible that some effects which might unwittingly be ascribed to diffraction, may in fact be due to interference effects.

Since it is the coherent nature of the ultrasound waves which directly results in those interference effects which cause such difficulties when analysing the data, it seems logical to investigate such artefacts in the phase domain, where they ultimately originate. Our technique utilises the instantaneous frequency (time derivative of the temporal phase) to locate the interference artefact in the time domain of a band-limited signal such as a typical R.F. A-line.

Signal envelope, phase and instantaneous frequency may be defined in terms of the analytic signal. Given a real, time-varying, band-limited function, f(t), the associated analytic signal a(t), is defined as

$$a(t) = f(t) + jH[f(t)]$$

where $j=\sqrt{-1}$, and H[f(t)] denotes the Hilbert transform of f(t). Signal envelope, e(t), is given by

$$e(t) = |a(t)|$$

and instantaneous phase, $\Phi(t)$, is given by

$$\Phi(t) = \arg[a(t)]$$

Signal instantaneous frequency, $\Phi'(t)$, is then given by

$$\Phi'(t) = (1/2\pi)d\Phi/dt = (1/2\pi)d/dt(\arctan(H[f(t)]/f(t)))$$

Note that the phase, $\Phi(t)$, is limited to the range $\pm\pi$ and must therefore be 'unwrapped' prior to differentiation to avoid discontinuities in $\Phi'(t)$ whenever $|\Phi(t)|$ extends beyond π.

INTERFERENCE AND INSTANTANEOUS FREQUENCY

Interference occurs when we have two or more overlapping ultrasound pulses. Figure 2 shows a simulation of the simple case involving only two, identical gaussian modulated sine waves separated by varying distances. The resultant instantaneous frequencies are illustrated for two different separations and the disruption caused by interference effects may instantly be recognised by the occurrence of a spike in the instantaneous frequency at the same location as the interference dip in the combined amplitudes, the size of the spike indicating the severity of the interference.

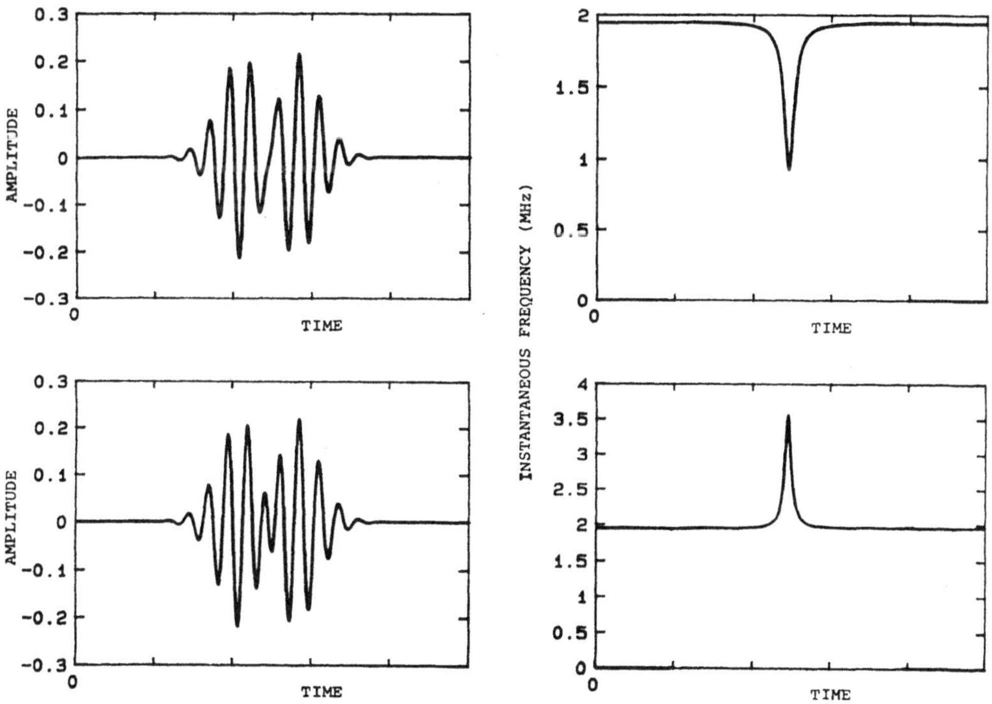

Fig. 2. Two interfering gaussian pulses (left), together with their corresponding instantaneous frequencies (right). Mean frequency values are 1.80 MHz (top) and 2.11 MHz (bottom), compared to an undeviated value of 1.95 MHz for a single pulse.

Interference also has a marked influence on mean frequency estimates obtained via short-time Fourier analysis. Consider two identical pulses, f(t),
separated by a short distance, a, the resultant waveform,F(t), is given by

$$F(t) = f(t) + f(t-a)$$

The corresponding Fourier transform is then

$$G(w) = g(w) + e^{iwa}g(w)$$

Or, in terms of their power spectra

$$|G| = 4|g|[coswa/2]$$

So that the resultant power spectrum for our two overlapping gaussians is also gaussian, but modulated by a cos term, which will produce gashes in the spectrum at varying intervals depeding on the separation of the pulses, a. In this way certain frequencies are suppressed, depending on the degree of interference and it can easily be seen that such an effect can cause mean (spectral) frequency values to be biased.

We might expect the deviations in the instantaneous frequency to act as an indicator to tell us when mean frequency values have been corrupted by interference effects and an attempt was made to verify this hypothesis experimentally.

The medium investigated was a tissue equivalent phantom with an attenuation coefficient claimed by the manufacturers to increase linearly with frequency. The material was linearly scanned to aquire 128 pulse-echo A-lines, lying in a plane through the phantom and separated by a lateral distance of 1mm. A commercially available, 3.5 MHz centre frequency, 13mm diameter, focused (4-10 cm) transducer was used and the data processed as follows.

A portion of A-line, approximately 2cm long, was randomly chosen from the data set. This was then divided into eight segments and the mean Fourier domain frequency calculated for each. When these spectral means are plotted against depth into the medium, the slope of the best linear fit to the points may be directly related to the frequency dependence of the attenuation coefficient, so that the corruption of mean frequency values within segments would give rise to noisy attenuation estimates[9]. The instantaneous frequency of the signal was then plotted as the time (or depth into the medium) dependence of the deviation from the mean. We attempted to see whether instantaneous frequency deviations could be used to identify and thus exclude corrupted data segments in order to improve slope estimates. However, it was found that while the deviations in the instantaneous frequency, are related to the severity of the interference phenomenon, they may not be used directly to correct for the effects on the mean frequency estimates. The reason for this may most simply be demonstrated by plotting deviation in instantaneous frequency against deviation in mean frequency, for the simulation previously described, viz. two interfering gaussian pulses (Figure 3). It may be seen that large deviations in the instantaneous frequency are not neccesarily associated with large deviations in mean frequency - in fact the most extreme spikes are an indication that the mean frequency is close to its true value. The region of instantaneous frequency which indicates the worst corruption of mean frequency values is not easily defined, especially in the more general case where the parameters of the pulse may be different from those involved in such a simplified example.

An alternative approach to this problem was then investigated, in which

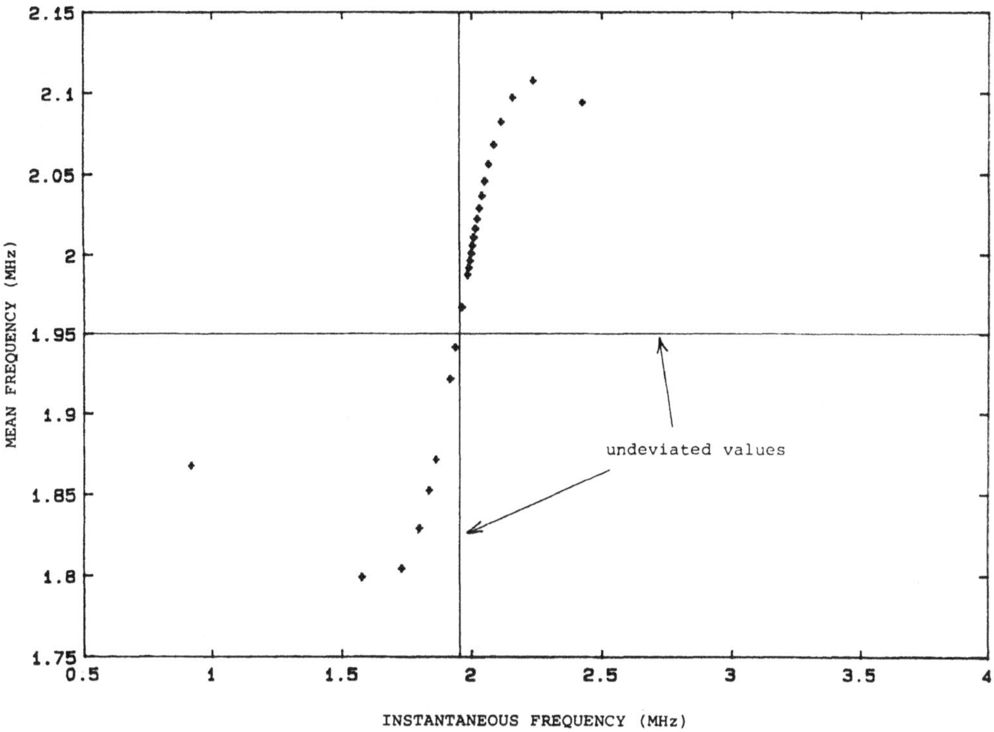

Fig. 3. Variation of mean frequency with instantaneous frequeny for two interfering gaussian pulses at varying separations.

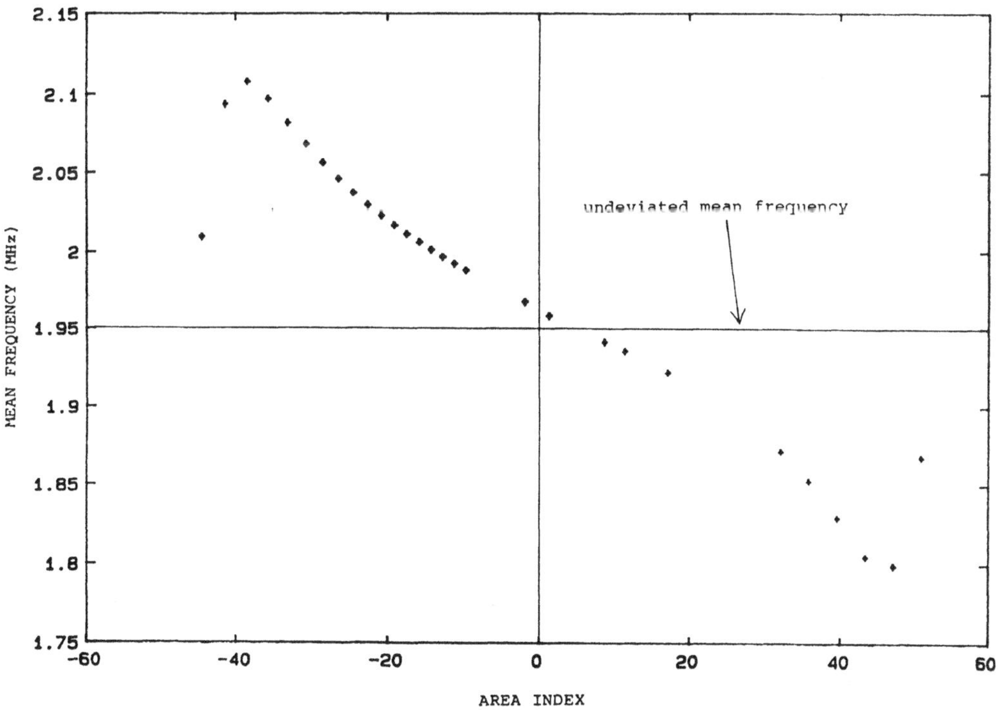

Fig. 4. Variation of mean frequency with calculated 'area index' for two interfering gaussian pulses at varying separations.

the total of the deviations from the mean instantaneous frequency were summed within each segment, to produce an 'area' index. The simulation was repeated for this new parameter, and the results plotted against mean frequency as before. Figure 4 shows the resultant graph and it may be seen that the region of interest is far easier to define, being the shallowest part of the slope on either side of zero 'area'. This led to the proposal that if the 'area' of the instantaneous frequency spikes is small, then the deviation from the mean frequency will be small. It should be noted that data which give the largest area index will be classed as 'bad' even though the mean frequency may not be greatly deviated. On the other hand, no corrupted data segment would be classed as 'good'.

This technique was tested on the simulated A-line used earlier, consisting of a pulse convoluted with a reflector sequence. Again, no diffraction or attenuation were included in the model, so that the mean frequency should ideally have been perfectly constant and undeviated. The area index and mean frequency were then calculated for each of eight data segments.

A typical plot of mean frequency versus 'depth' is shown in figure 5, figure 6 is the corresponding plot of area index. Three data points, indicated by arrows, lie close to zero on the area index and the corresponding mean frequencies are very close to the correct value (dotted line). The remaining segments of data were classed as bad even though one of them (ringed) lies close to the true mean frequency. This technique appears, then, to correctly identify all corrupted mean frequency values, although it may also lead to the rejection of some good data. Therefore, when correctly

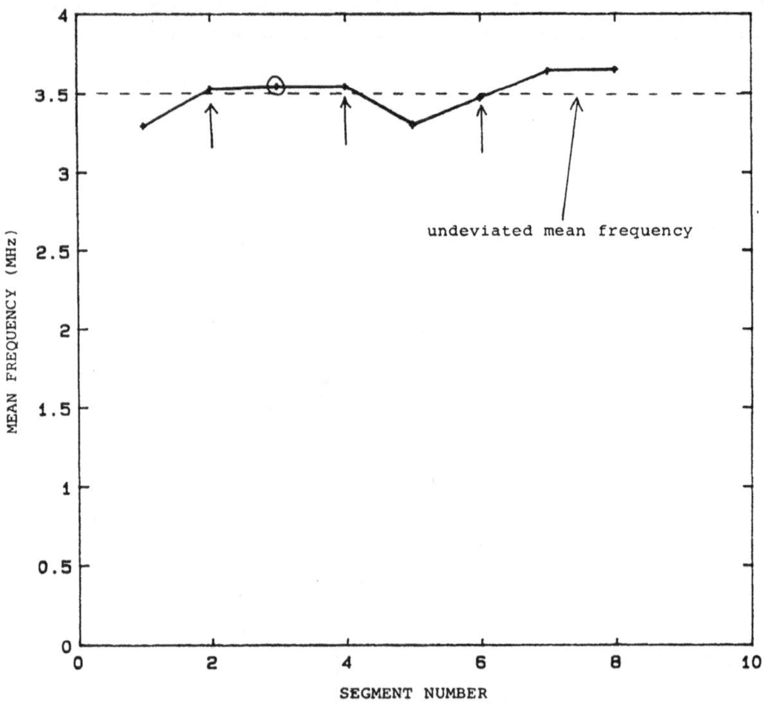

Fig. 5. Mean frequency versus depth (segment) for a simulated A-line. 'Good' values of mean frequency as identified by the area index (Fig. 6) are indicated by arrows. The one good point which was rejected is circled.

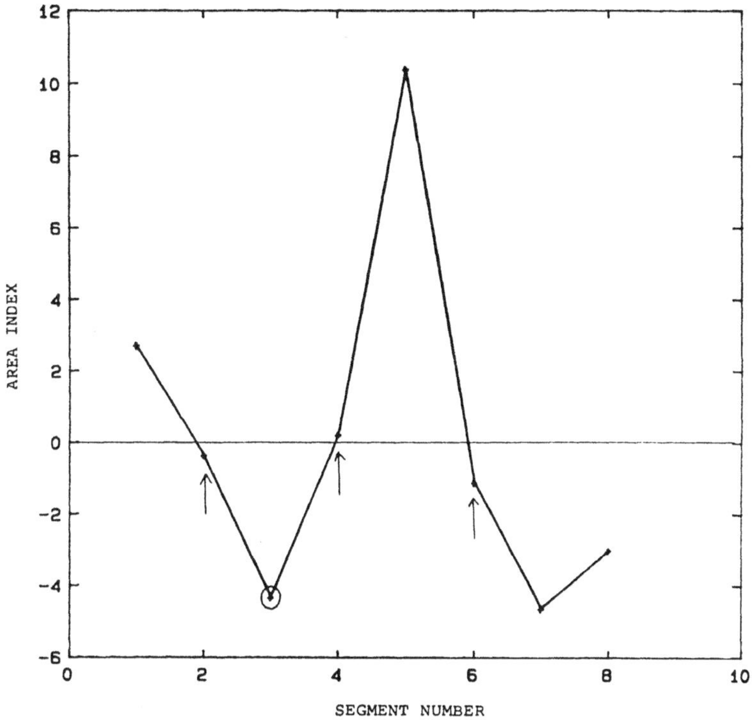

Fig. 6. 'Area index' versus depth for the A-line in Fig. 5. Good and bad data points are identified in the same way.

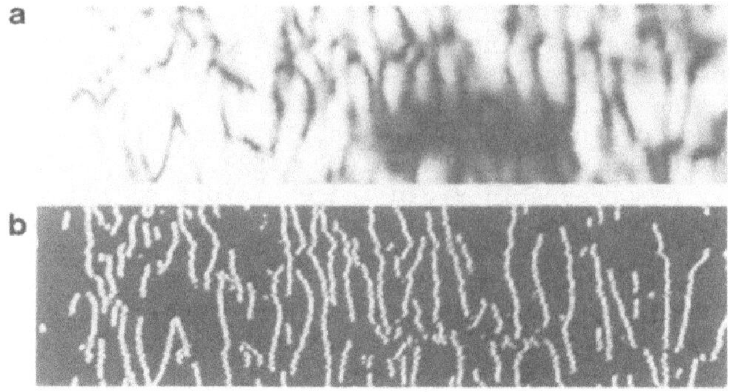

Fig. 7. a. Envelope image of low scatter region in homogeneous segment of tissue phantom.
b. Map of interference effects for image in a.

applied, this technique is capable of giving an accurate attenuation slope estimate on the basis of a <u>single</u> A-line.

SPECKLE REDUCTION

The next stage in the analysis was the extension of this technique to the processing of ultrasound images[4,5]. Figure 7a shows the conventional

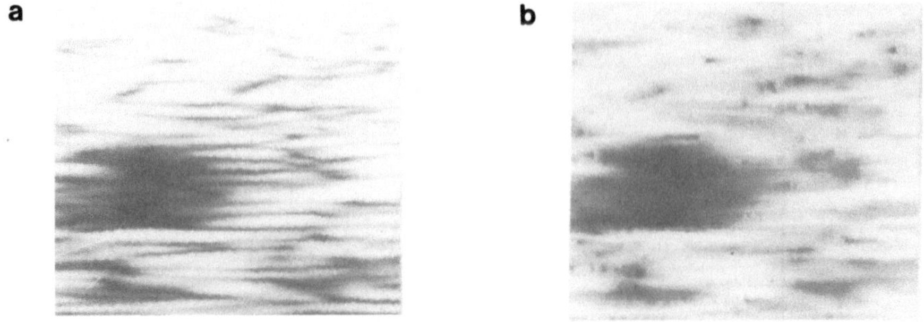

Fig. 8. a. Axially compressed version of envelope image as in Fig. 6a.
b. Speckle reduced envelope image from above.

envelope image obtained from a tissue phantom containing a region of low scattering. Speckle, due to coherent interference effects is shown to be well developed throughout the phantom. Figure 7b shows the corresponding image of destructive interference generated by displaying the locations of large instantaneous frequency excursions, as calculated from each echo waveform individually. In this case only fluctuations of greater than 0.65MHz from the centre frequency (3.5MHz) were displayed. Note that the dark boundaries of the individual speckle elements are extremely well mapped to the outlines in the interference image.

The final stage is the re-assignation of the image values which are coincident with large instantaneous frequency excursions, as specified by the interference map. Preliminary results were obtained using the following technique. Each image point to be adjusted was replaced by the average of the two nearest maxima on either side, on a line-by-line basis. This new image then undergoes local one-dimensional averaging only at those points specified by the interference map. Further speckle reduction is achieved by effectively broadening the interference map and repeating the above process. Figure 8a shows the result of employing this speckle reduction technique on the envelope image depicted in Figure 7a. In this case the length of the averaging window was 0.6mm (approximately 1.5λ). The speckle reduced image has been axially compressed to avoid image distortion and an axially compressed version of the original envelope image is provided for comparison (figure 8b).

Figures 8a and 8b show that the effect of the speckle reduction procedure is to de-emphasise or 'fill-in' image minima arising from destructive interference, while the clarity of the low scatter zone boundaries remains relatively unaffected. Although the technique in its present form does not remove the effect of speckle on the image, it can be seen that the impression of artefactual fine structure is diminished. It is encouraging to note that even at this preliminary stage, the technique appears to perform at least as favourably as other established speckle reduction methods, without the need for the time-consuming averaging of many images, or two-dimensional filters. Also, since the A-lines are treated individually, this opens the way towards the real-time implementation of speckle reduction.

PROCESSING DOPPLER DATA

In many estimates of the velocity of fluid flow using the doppler technique the mean frequency shift of the echoes arising from the flow is

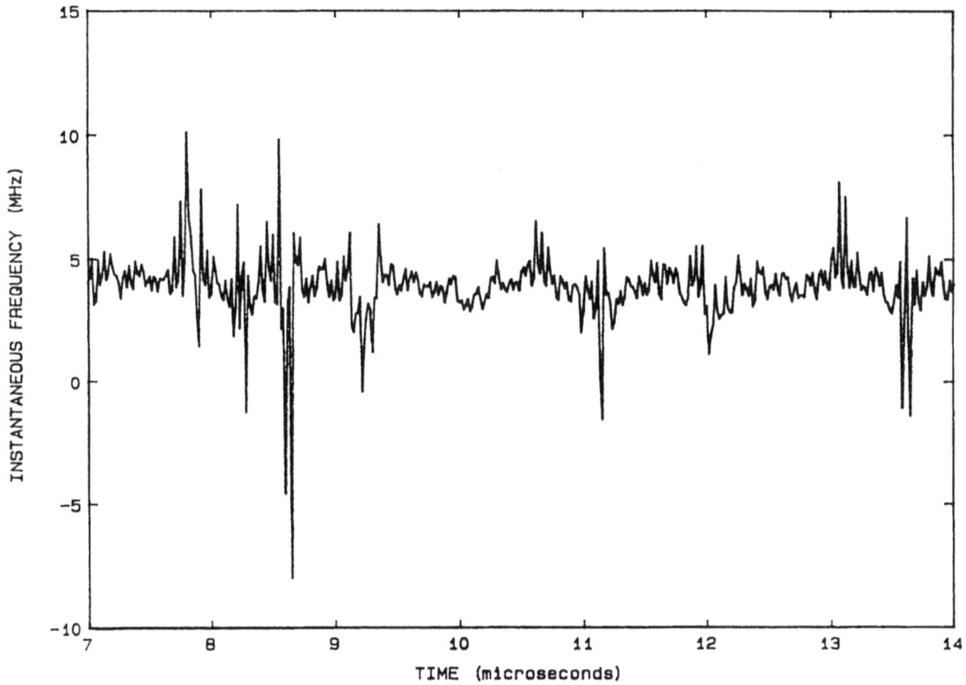

Fig. 9. Instantaneous frequency versus depth for a segment of A-line from a flow phantom, showing spikes indicative of interference effects.

related directly to the mean velocity of the moving scatterers. However, pulsed doppler is also plagued by artefacts [6] which, although they cause changes in mean frequency, are not associated with flow. Of particular interest within the context of this paper are the effects of interference.

We have already shown how interference corrupts mean frequency values and that this is reflected in the instantaneous frequency by the occurrence of a large deviation from the mean. The zero crossing density for an A-line will also be noisy, as would be expected from its close relationship to the mean frequency.density at the indicated locations. This may explain the lack of success which researchers have experienced when attempting to use zero crossings to calculate flow velocities [7]. However, our work also shows that care should be taken when using Fourier techniques to calculate flow from mean frequency estimates, since these results are also open to corruption from the same source. Figure 9 shows the instantaneous frequency derived from an A-line obtained from a flow phantom, with the ultrasound transducer at right angles to the flow, so that no doppler shift was recorded. Excursions from the mean of several MHz may be observed, indicating the presence of severe interference effects and indeed, mean frequency estimates obtained from phantom data showed significant (though not systematic) variation, even when no frequency shift should have been observed.

CONCLUSIONS

While interference has long been recognised as the cause of speckle in imaging, its importance as a major artefact in other ultrasound pulse-echo work has long been overlooked. Our work illustrates how far-reaching the effects of this neglect are particularly in tissue characterisation (attenuation estimation) and doppler flow. Making use of the additional information encoded in the phase of ultrasound signals, allows this obtrusive artefact to be accurately located in the signals and we have gone some way towards establishing a technique which allows us to correct the individual A-lines accordingly. Improvement in attenuation estimates and image quality has been shown to be possible by this means and the fact that the data may be processed on a line-by-line basis opens the way towards attenuation and flow estimates from single A-lines and real-time image processing.

REFERENCES

1 Robinson, D. E., Wilson, W. S. and Bianchi, T., Beam pattern (diffraction) correction for ultrasonic attenuation, Ultrasonic Imaging 6. 2993-286 (1984).

2 Costa, E. T., Leeman, S., Richardson, P.A.C., and Seggie, D. A., Measurement and calibration of transducer fields, Proc. Inst. Acoustics 8. pt. 2. 113-118 (1986).

3 Kuc, R., Estimating acoustic attenuation from reflected ultrasound signals: comparison of spectral shift and spectral-difference approaches, IEEE Trans. Acoustics, Speech and Signal Processing ASSP-32 No. 1. 1-6 (1984).

4 Seggie, D. A., Digital processing of acoustic pulse-echo data, PhD Thesis, University of London (1986).

5 Leeman, S. and Seggie, D. A., Speckle reduction via phase, Proc. SPIE, 768 (1987).

6 Leeman, S., Roberts, V. C. and Wilson, K., Quantitative doppler with ultrasound pulses, chapter 17, report 47, Physics in medical ultrasound (Inst. physical sciences in medicine), (1986).

7 Flax, S. W., Webster, J. G. and Updike, S. J., Statistical evaluation of the doppler ultrasonic blood flowmeter, ISA Trans., 10(1), 1-20, (1971).

PHASE RECONSTRUCTION FROM AMPLITUDE BASED ON THE RYTOV TRANSFORMATION OF THE WAVE EQUATION

Mehrdad Soumekh

Department of Electrical and Computer Engineering
State University of New York at Buffalo
Amherst, NY 14260

ABSTRACT

This paper addresses the problem of phase reconstruction from amplitude for a diffracted wave. We utilize the Rytov transformation of the Helmholtz wave equation to relate the phase and amplitude functions; this results in two partial differential equations. We show that each of these PDEs yields a perturbation scheme to reconstruct the phase from amplitude. We also examine issues involved with the implementation of these methods in practice.

INTRODUCTION

In this paper, we examine the problem of phase reconstruction from amplitude for a scattered wave via a new framework that is based on the Rytov transformation of the Helmholtz wave equation [2]. The phase reconstruction problem arises in asynchronous and phase-unreliable imaging systems where the only available or reliable information is the amplitude of the scattered waves. Examples include holography [3], crystallography [4], and electron/acoustic microscopy [5,6].

In a recent paper [1], we develop a practical algorithm for phase reconstruction from amplitude. This algorithm is one of the two perturbation solutions for the phase function obtained from the Rytov's Riccati wave equation. Our purpose in this paper is to present both of these two perturbation solutions. We then addresses the problems associated with the practical use of these algorithms. In particular, we show that one of the perturbation solutions, cannot be realized when the wavenumber distribution of the medium of propagation is unknown.

The other perturbation solution, however, can be implemented in practice if the medium of propagation is not absorptive. Moreover, this algorithm can also be utilized for phase unwrapping problems of diffraction tomography [1]. We begin with a brief derivation of the Rytov's wave equation. Then, by separating the real and imaginary parts of the resultant PDE, we develop the two perturbation solutions. Finally, a description of the phase reconstruction algorithm is presented. Practical issues involved in the digital implementation of and determining the zero-th order perturbation for this method are discussed in [1].

WAVE EQUATION MODEL

Consider a single frequency diffracted wave function, $\psi(x,y)$, propagating in an inhomogeneous medium on the two-dimensional (x,y) plane. The governing Helmholtz wave equation for this wave function is

$$\nabla^2\psi(x,y) + k^2(x,y)\psi(x,y) = 0, \tag{1}$$

where $\nabla^2 \equiv \frac{\partial^2}{\partial x^2} + \frac{\partial^2}{\partial y^2}$ is the Laplacian operator and $k(x,y)$, complex wavenumber, is related to the variations of index of refraction (or speed of propagation) and attenuation coefficient on the (x,y) plane. We denote the complex phase of $\psi(x,y)$ by

$$\begin{aligned}\Omega(x,y) &\equiv ln[\psi(x,y)] \\ &\equiv \alpha(x,y) + j\phi(x,y),\end{aligned} \tag{2}$$

where $\alpha(x,y)$, log-amplitude, is a known function. We are interested in determining the phase function, i.e., $\phi(x,y)$.

We choose $\psi_0(x,y)$ and $k_0(x,y)$, respectively, to be known estimates of $\psi(x,y)$ and $k(x,y)$ such that

$$\nabla^2\psi_0(x,y) + k_0^2(x,y)\psi_0(x,y) = 0. \tag{3}$$

Moreover, we denote the complex phase of $\psi_0(x,y)$ by

$$\begin{aligned}\Omega_0(x,y) &\equiv ln[\psi_0(x,y)] \\ &\equiv \alpha_0(x,y) + j\phi_0(x,y),\end{aligned} \tag{4}$$

that is assumed to be a known complex function.

We define the scattered complex phase (Rytov function [2]) by the following:

$$\begin{aligned}\Omega_s(x,y) &\equiv \Omega(x,y) - \Omega_0(x,y) \\ &\equiv \alpha_s(x,y) + j\phi_s(x,y).\end{aligned} \tag{5}$$

The phase function ϕ_0 is known. Hence, it is sufficient to estimate $\phi_s = \phi - \phi_0$ to reconstruct ϕ. It can be shown from (1)-(5) that [1,2]

$$\nabla^2\Omega_s(x,y) + 2\nabla\Omega_0(x,y)\cdot\nabla\Omega_s(x,y) + \nabla\Omega_s(x,y)\cdot\nabla\Omega_s(x,y) + k^2(x,y) - k_0^2(x,y) = 0. \tag{6}$$

We then separate the real and imaginary parts of (6). The real part of (6) results in a *nonlinear* spatially-varying differential equation of the form

$$\begin{aligned}\nabla^2\alpha_s + 2\nabla\alpha_0\cdot\nabla\alpha_s - 2\nabla\phi_0\cdot\nabla\phi_s + \nabla\alpha_s\cdot\nabla\alpha_s - \nabla\phi_s\cdot\nabla\phi_s + \\ Re[k^2 - k_0^2] = 0,\end{aligned} \tag{7}$$

where $Re[C]$ denotes the real part of the complex number C. The imaginary part of (6) yields a *linear* spatially-varying differential equation as follows:

$$\begin{aligned}\nabla^2\phi_s(x,y) + 2\nabla\alpha(x,y)\cdot\nabla\phi_s(x,y) + 2\nabla\alpha_s(x,y)\cdot\nabla\phi_0(x,y) + \\ Im[k^2 - k_0^2] = 0,\end{aligned} \tag{8}$$

where $Im[C]$ denotes the imaginary part of the complex number C.

Note that both (7) and (8) are PDEs that relate the unknown phase function to the known amplitude function. We next present two perturbation techniques for the reconstruction of the phase function from the amplitude data based on (7) and (8).

PHASE RECONSTRUCTION BASED ON (7)

The left side of (7) contains a term related to the real part of $k^2(x,y)$. Hence, (7) is not a suitable model in our problem when $k(x,y)$ is unknown. Suppose $k(x,y)$ is known. We let $\phi_n(x,y)$ be the response of a linear shift-varying system when the input source function is $s_n(x,y)$ such that

$$2\nabla\phi_0(x,y)\cdot\nabla\phi_n(x,y) = -s_n(x,y) \qquad for\ n = 1,2,...,\infty. \tag{9}$$

We define the source functions by the following:

$$s_1(x,y) \equiv -\nabla^2\alpha_s - 2\nabla\alpha_0\cdot\nabla\alpha_s - \nabla\alpha_s\cdot\nabla\alpha_s - Re[k^2 - k_0^2], \tag{10a}$$

and

$$s_n(x,y) \equiv \sum_{i=1}^{n-1} \nabla\phi_i(x,y)\cdot\nabla\phi_{n-i}(x,y) \qquad for\ n = 2,3,...,\infty. \tag{10b}$$

In this case, it can be shown from (7), (9) and (10) that the unknown phase is the sum of all the ϕ_n responses; i.e.,

$$\phi_s(x,y) = \sum_{n=1}^{\infty} \phi_n(x,y). \tag{11}$$

It is clear that $s_n(x,y)$ can be determined from the knowledge of $\phi_i(x,y)$, $i = 1,...,n-1$ [see (10)]. Moreover, $\phi_n(x,y)$ is uniquely defined by (9) if $s_n(x,y)$ is known. In this case, equations (9)-(11) are the basis of a perturbation solution for ϕ_s. However, due to the fact that (9) is a shift-varying PDE, it might not be possible to implement this scheme in practice for any choice of ψ_0.

A possible choice for the initial wave estimate that yields a numerically manageable solution is a decaying plane wave; i.e.,

$$\psi_0(x,y) = exp[j(k_x x + k_y y)],$$

where $k_x^2 + k_y^2 = k_0^2$. It is also required that (k_x, k_y) is a complex pair such that the linear equation $k_x\mu + k_y\nu = 0$ does not have a solution in the *real* two-dimensional domain defined by (μ,ν) other than $(\mu,\nu) = (0,0)$ (we will discuss why this constraint is essential.)

In this case, the left side of (9) becomes

$$2\nabla\phi_0(x,y)\cdot\nabla\phi_n(x,y) = 2jk_x\frac{\partial\phi_n}{\partial x} + 2jk_y\frac{\partial\phi_n}{\partial y}. \tag{12}$$

Using (12) in (9) and taking the spatial Fourier transform of both sides of (9) yields

$$\Phi_n(\mu,\nu) = \frac{S_n(\mu,\nu)}{2k_x\mu + 2k_y\nu}, \tag{13}$$

where $\Phi_n(\mu,\nu)$ and $S_n(\mu,\nu)$, respectively, are the two-dimensional spatial Fourier transforms of $\phi_n(x,y)$ and $s_n(x,y)$.

We chose the complex numbers (k_x, k_y) such that the denominator of the right side of (13) never becomes zero on the real plane of (μ,ν) except for the dc point, i.e., $(\mu,\nu) = (0,0)$. Thus, with the knowledge of $S_n(\mu,\nu)$, $\Phi_n(\mu,\nu)$ can be reconstructed using (13) at all spatial frequencies except the dc point. Now consider (10b) for the next step of the iteration. We observe that $s_{n+1}(x,y)$ is a function of the gradients (partial derivatives) of $\phi_i(x,y)$, $i = 1, ..., n$. In other words, the dc values of $\phi_i(x,y)$, $i = 1, ..., n$, are not required to be known for finding the source function for the next step of the iteration.

Thus, the iteration can be continued to find, except for the dc value, all spatial frequency components of the series on the right side of (11). In this case, $\phi_s(x,y)$ can be determined within a dc value. If $\phi_s(x,y)$ is known at a single point in the spatial domain, then we can uniquely determine the unknown scattered phase (see [1] for the details).

PHASE RECONSTRUCTION BASED ON (8)

Equation (8) that relates the unknown phase to the amplitude function depends on the imaginary component of $k(x,y)$; this component is a representative of absorptive properties of the medium of propagation. Hence, one is capable of developing a practical algorithm for the phase reconstruction from amplitude when the wavenumber distribution of the propagating medium is real (though unknown) or the absorption coefficient of the medium is known; this the case in many imaging problems of microscopy and holography. We next present the basic equations used in obtaining this algorithm when $Im[k^2 - k_0^2] = 0$; practical issues involved in implementing this scheme are discussed in [1].

We let $\phi_n(x,y)$ be the response of a linear shift-invariant system when the input source function is $s_n(x,y)$ such that

$$\nabla^2 \phi_n(x,y) = -s_n(x,y) \qquad for\ n = 1, 2, ..., \infty. \tag{14}$$

We define the source functions by the following:

$$s_1(x,y) \equiv 2\nabla \alpha_s(x,y) \cdot \nabla \phi_0(x,y), \tag{15a}$$

and

$$s_n(x,y) \equiv 2\nabla \alpha(x,y) \cdot \nabla \phi_{n-1}(x,y) \qquad for\ n = 2, 3, ..., \infty. \tag{15b}$$

In this case, it can be shown from (8), (14) and (15) that the unknown phase is the sum of all the ϕ_n responses; i.e.,

$$\phi_s(x,y) = \sum_{n=1}^{\infty} \phi_n(x,y). \tag{16}$$

Note that the series solution in (16) is not the same as the one obtained in (11).

Taking the spatial Fourier transform of both sides of (14), one obtains

$$\Phi_n(\mu,\nu) = \frac{S_n(\mu,\nu)}{\mu^2 + \nu^2}. \tag{17}$$

Hence, with the knowledge of $S_n(\mu,\nu)$, $\Phi_n(\mu,\nu)$ can be reconstructed using (17) at all spatial frequencies except the dc point. Note that $s_{n+1}(x,y)$ is a function of the gradient of $\phi_n(x,y)$ [see (15b)]. In this case, the knowledge of the dc value of $\phi_n(x,y)$ is not essential in determining the source function $s_{n+1}(x,y)$; i.e., the source function for the next step of the iteration can be uniquely determined.

Similar to the procedure described in the previous section, the above perturbation scheme results in reconstruction of all spatial frequency contents of the phase function except for its dc value. The dc value can be determined if the phase function is known at a single point in space [1].

ALGORITHM

Based on the results of the previous section, we can now summarized the phase reconstruction method that utilizes the amplitude-phase relationship in (8). The algorithm for the phase reconstruction based on (7) can also be outlined in a similar fashion.

Step 1: Start. Evaluate the partial derivatives of α, α_s and ϕ_0.

Step 2: Evaluate s_1 using (13a).

Step 3: Set $n = 1$. Set $\phi_s = 0$ at all pixel points.

Step 4: Evaluate ϕ_n from s_n using (12).

Step 5: Add the result to ϕ_s. Check if there has been no significant change in ϕ_s in the last two iterations. If so, go to Step 8.

Step 6: Evaluate the partial derivatives of ϕ_n. Increment n by one.

Step 7: Evaluate s_n using (13b). Go to Step 4.

Step 8: Stop.

Note that Step 5 specifies a criterion for stopping the iteration. One may use other means to identify the minimum iteration number required for obtaining an *acceptable* solution.

CONCLUSIONS

The Rytov transformation of the wave equation was utilized to develop two perturbation solutions for phase from amplitude. However, only one of these schemes can be utilized in practice when the properties of the propagating medium are unknown. This scheme has been successfully applied in phase reconstruction/unwrapping problems of imaging systems [1]. A similar perturbation solution exists for determining the amplitude function of a diffracted wave from its phase data [1]. Various studies have shown a correlation between the phase and amplitude of scattering in a random medium [7,8,9]. Rytov's wave equation exhibits the amplitude-phase coupling via a deterministic PDE [see (6)-(8)]. The success of these deterministic principles in phase recovery from amplitude, as indicated by our numerical results in [1], underscores the importance of diffraction and refraction phenomena in inverse problems of the wave equation.

REFERENCES

1. M. Soumekh, "Phase reconstruction/unwrapping from amplitude for diffracted waves using a perturbation solution of the wave equation," *IEEE Transactions on Acoustics, Speech and Signal Processing*, vol. 36, no. 7, July 1988.

2. S.M. Rytov, "Diffraction of light by ultrasonic waves," Izv. Akad. Nauk SSSR, Ser. Fiz. 2, p. 223, 1937.

3. E. Wolf, "Is a complete determination of the energy spectrum of light possible from measurements of the degree of coherence," Proc. Phys. Soc. 80, 1269-1272, 1962.

4. G. Ramachandran and R. Srinivasan, *Fourier Methods in Crystallography*, Wiley, New York, 1970.

5. H.A. Ferwerda, "The phase reconstruction problem for wave amplitudes and coherence functions," in *Inverse Source Problems in Optics*, H.P. Baltes, Ed., Spring-Verlag, Berlin, 1978.

6. R. Lemons and C. Quate, "Acoustic microscope: Scanning version," Appl. Phys. Lett. 24 (4), 163-165, 1974.

7. V.I. Tatarski, *Wave Propagation in a Turbulent Medium*, McGraw Hill, New York, 1961.

8. L.A. Chernov, *Wave Propagation in a Random Medium*, McGraw Hill, New York, 1968.

9. A. Ishimaru, *Wave Propagation and Scattering in Random Media*, Academic Press, 1978.

ULTRASONIC IMAGING USING CORRELATION TECHNIQUES COMPUTER SIMULATIONS AND EXPERIMENTAL RESULTS

M. Gindre, C. Lebeault, J. Perrin, F. Rieuneau, and W. Urbach

C.H.U. Cochin-Laboratoire de Biophysique (CNRS-UA 593)
24, rue du Fb St. Jacques, 75674, Paris France

INTRODUCTION

Correlation overcomes the problem of conventional pulse-echo systems (average transmitted power and resolution) by transmitting a continuous pseudorandom signal and then compressing it into a short high resolution pulse at the receiver. The total Signal-to-Noise Ratio (SNR) is improved by band-compression and allows one to retrieve echo signals buried in the receiver noise.

Although some theoretical approaches [1,2] to the optimization of echographic systems were made in the early 1980s, there are few experimental attempts to assemble an experimental echographic system [3]. We report here the results of comparative experimental studies between pulse-echo and correlation imaging systems. The latter is optimized by computer simulation in order to reduce the Signal-to-Clutter Ratio.

After a brief presentation of the theoretical background, we shall describe schematicaly our simulation program in section 2. The third section will be devoted to the presentation of simulated results followed in the fourth section by experimental results

1-THEORETICAL BACKGROUND

A B-scan imaging device is schematicaly illustrated in Fig 1. The system generates periodically a short duration high voltage pulse $e_B(t)$, which is sent to the transducer and causes a pressure impulse to be launched into the medium. Part of the emitted energy is reflected (or backscattered) by various discontinuities of the medium. In the linear regime of operation, the received signal, s(t), can be written as

$$s(t) = \int_0^{\infty} h(\tau)\; e_B(t-\tau)\, d\tau = e_B(t) * h(t) \qquad (1)$$

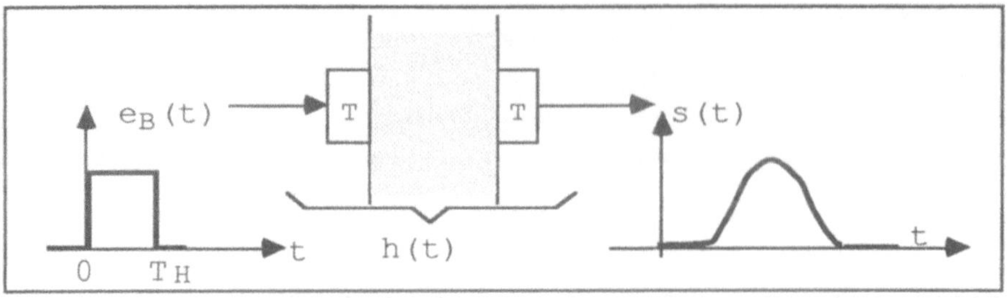

Fig. 1. Schematic representation of an echographic system. $e_B(t)$ and s(t) are, respectively, the emitted impulse and the signal received by the same transducer, T. Formally this is equivalent to the two identical transducers represented in the figure. h(t) is the impulse response of both the medium and the transducers working in their emission and reception modes respectively.

where h(t) is the combined impulse response of both the propagating medium and of the transducer working alternatively in its transmission and reception mode. If $e_B(t)$ is short enough ($e_B(t) \approx \delta(t)$) the received signal, s(t), provides, to a good approximation, the impulse response of the system.

As in the pulsed echo mode, correlation provides the impulse response of the system. It can be shown [4] that for linear systems the crosscorrelation:

$$R(\tau) = \int e(t).s(t-\tau)\,dt \qquad (2)$$

is related to the impulse response h(t) by the following equation:

$$R(t) = C_{ee}(t) * h(t)\ , \qquad (3)$$

where $$C_{ee}(\tau) = \int e(t).e(t-\tau)\,dt\ . \qquad (4)$$

Again, if $C_{ee}(t)$ is short (i.e. $C_{ee}(t) \approx \delta(t)$), equation (3) is equivalent to equation (1).

A relatively brief $C_{ee}(t)$ can be obtained when e(t) is a pseudorandom binary code made of a set of rectangular elements a_i of duration T_H. Each a_i is equal to -1 or +1. The length of such a code is equal to $P = 2^n - 1$, where n is the degree of the generator polynom [5,6]. The corresponding autocorrelation function, $C_{ee}(t)$, is represented in fig 2. If the duration, T_H, of each binary element is short, the autocorrelation function can be considered as a Dirac impulse to a good approximation.

Although both pulsed echo and correlation systems appear equivalent, it should be pointed out that correlation technique provides an enhanced Signal-to-Noise Ratio. Let us consider, for example, a classical, B-type echographic system.

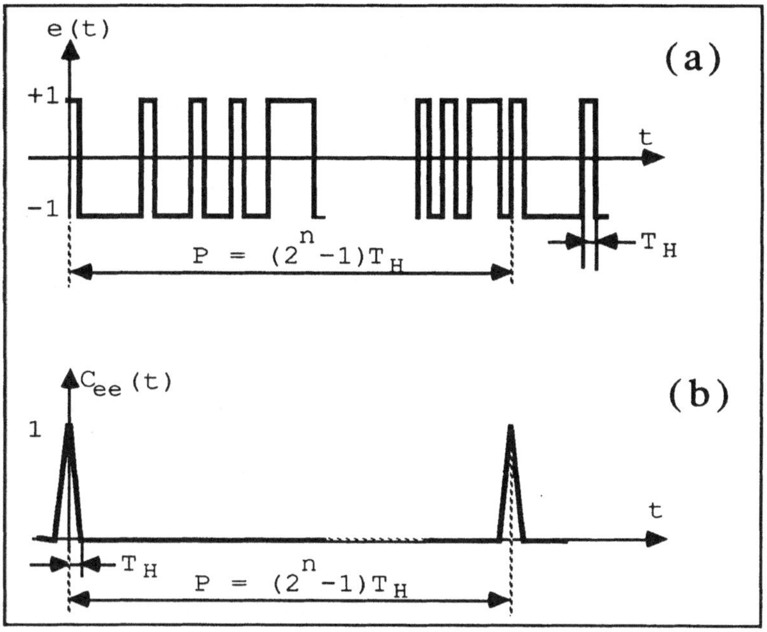

Fig. 2. (a) Pseudorandom binary code e(t)
(b) and its autocorrelation function $C_{ee}(t)$

For the sake of simplicity we choose for $e_B(t)$ a rectangular impulse of duration T_H and of constant amplitude = 1. The corresponding output bandwidth is $(BW) = \frac{1}{T_H}$. In that case the emitted energy W_B is simply proportional to the pulse duration: $W_B = kT_H$. As far as the correlation technique is concerned the emitted energy is proportional to the duration of the pseudorandom code: $W_C = PkT_H$ and the output bandwidth is still $(BW) = \frac{1}{T_H}$. Thus the emitted signal energy is increased by a factor of P, and due to the bandwidth compression, the noise is reduced by the same factor P. The SNR is therefore increased by a factor of P^2.

To explore a biological tissue of thickness 15 cm, a code of 1023 elements is needed. The corresponding gain in SNR is therefore $(1023)^2$, i.e. roughly 60dB.

2-COMPUTER SIMULATIONS

PRINCIPLES

to demonstrate the applicability of the correlation technique to ultrasonic imaging, we performed computer simulations of an echographic A-mode line. Propagation mechanisms such as diffraction, attenuation and diffusion were considered separately first, and then all together in order to provide a comparison between the correlation and the conventional echo system. Propagation medium absorption is considered weak enough to separate attenuation and diffraction effects. Furthermore, we neglect multiple scattering (Born's approximation).

Our simulation program permits to choose the excitation signal shape and transducer characteristics. The medium is described by a set of small scatterers. The scatterer density, i.e. the number/λ^3 (λ=wavelength), and their spatial distribution are chosen prior to the simulation. Most of the calculation was performed in the frequency domain, and the temporal behaviour was obtained by inverse fast Fourier transform of the computed results.

The constitutive parts of the system were modelized as follows:

The transducers are modelized by Gaussian linear phase bandpass filters centered on the ultrasonic frequency f_0 . The corresponding amplitude transfer function is :

$$|H_T(f)| = \exp\left(-\frac{(f-f_0)^2}{2B_T^2}\right) \tag{5}$$

where B_T is the transducer's bandwidth and f_0 the central frequency of the transducer. The transfer function of the transducer may eventually be set to a Dirac impulse if one is only interested in the echographic signal modifications induced by the propagating medium.

The interaction of the ultrasonic wave with the medium is characterized by three mechanisms:

1-The attenuation is represented by a causal minimum phase filter [7]. The corresponding transfer function is :

$$H_a(f) = |H_a(f)|\exp(j\phi(f)), \tag{6}$$

where $20\log|H_a(f)| = \beta f d$. β and d are, respectively, the attenuation coeficient (in dB/(cm.MHz)) and the penetration length (in cm). The Hilbert's transform of the modulus provides the phase:

$$\phi(\omega) = \frac{1}{2\pi} VP \int_{-\pi}^{+\pi} Ln(|H_a(\psi)|) .\cot(\frac{\psi-\omega}{2})\, d\psi \tag{7}$$

where VP is the Cauchy's Principal Value and $\omega = 2\pi f T_H$

2-The diffusion transfer function is described by using a second degree polynomial approximation [8]:

$$H_a(f) = a_o + a_1 f + a_2 f^2 \tag{8}$$

where a_o, a_1 and a_2 are determined by the size of each target.

3-The diffraction is described by a space-time filtering procedure which takes into account the distance variation from the target to the different locations of the transducer's front plane. We use the Stepanishsen method [9] to calculate the corresponding transfer function for each target.

The simulated pulse echo system is shown in the Fig. 1. $e_B(t)$ is a rectangular pulse which is modulated at the central frequency of the transducer in order to optimize the energy transfer into the propagating medium. The pulse duration is T_H. For the correlation system, e(t) is the pseudorandom binary sequence generated by the primitive polynom [5,6]

$$X^{10} + X^3 + 1 = 0 \qquad (9)$$

The duration of each binary element was set to be equal to the duration, T_H, of the pulsed system. As previously, e(t) is modulated at the central frequency of the transducer. As the emitting and the receiving transducers are Gaussian filters, they induce ghost lobes in the received radiofrequency signal, s(t) (Fig. 3).

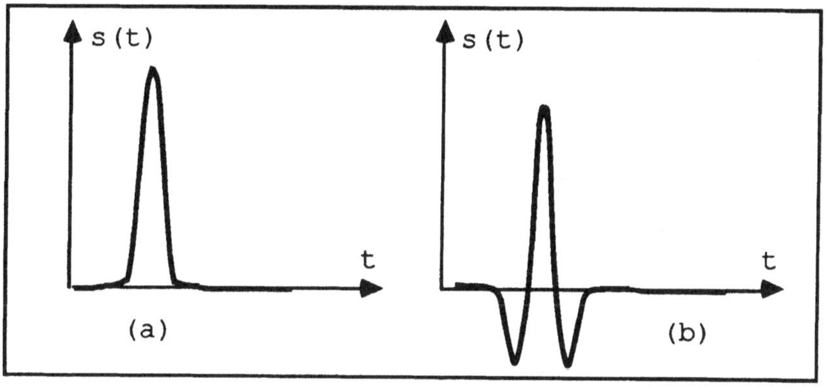

Fig. 3. a) s(t) obtained for infinite band transducers
b) s(t) obtained with finite band transducers

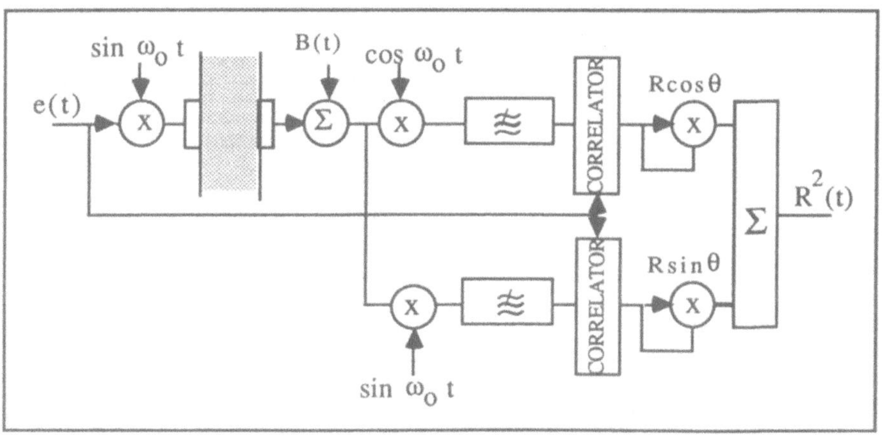

Fig. 4. Schematic representation of correlation setup. See text for more details

In order to preserve the axial resolution of our correlation system, we choose to suppress the ghost lobes by a double demodulation with two carriers in quadrature (Fig. 4). As a result, the demodulated spectrum contains a supplementary lobe centered at $2f_o$. This harmonic lobe is eliminated by a lowpass filter (6th order Butterworth) and the filtered signal is then correlated with e(t).

Due to the demodulation the output of one channel is $R\cos\theta$ where $\theta = \frac{4\pi d}{\lambda}$ contains the phase information. The second channel yields $R\sin\theta$. Thus R^2 is obtained by squaring and adding the two outputs. It should be emphasized at this point that, when needed, the θ value is easily determined and can be used for further signal processing.

In some simulations a white noise, B(t), within the band of the transducer was added to the recieved signal, s(t). This noise is generated by an independant pseudorandom sequence in the computer. Finally, in order to allow an easy comparison with $R^2(t)$, in most figures $s^2(t)$ is represented instead of s(t).

SIMULATION RESULTS

It should be remembered that in the absence of noise, the correlation technique, with e(t) as described previously, is formally equivalent to a pulsed echo system but with $e_B(t) = C_{ee}(t)$, i.e. triangular shape pulse (cf. equations (1) and (3)).

In order to illustrate briefly how the attenuation and the diffraction of the propagating medium interact with the echographic signal shape, we performed simulations with a single target located at different distances, d, from the transducer center.

For these simulations the attenuation coefficient was set to 0.5dB/(cm.MHz), and the duration of the normalized echographic signal was T_H = 160ns. The diameter of the planar transducer is 0.8 cm and its impulse response was a Dirac function. For low values of d, the effect of attenuation was small and the pulse broadening was due mostly to diffraction.

The next step was to simulate the echoes arising from a medium composed of three distinct zones (Fig. 5). On a background of randomly disposed targets we superimposed two network zones 1 and 2 cainting targets located on a square lattice. The detailed description of the medium is given in table 1. The 0.8 cm diameter transducers were limited to a bandwidth of 6MHz and their central frequency was set to f_o = 6.25 MHz.

The simulation results are presented in Fig. 6 for both the classical and correlation echographic systems. In Fig. 6 we also present the signals issued from the two in the quadrature demodulation channels (Fig. 6a and 6b).

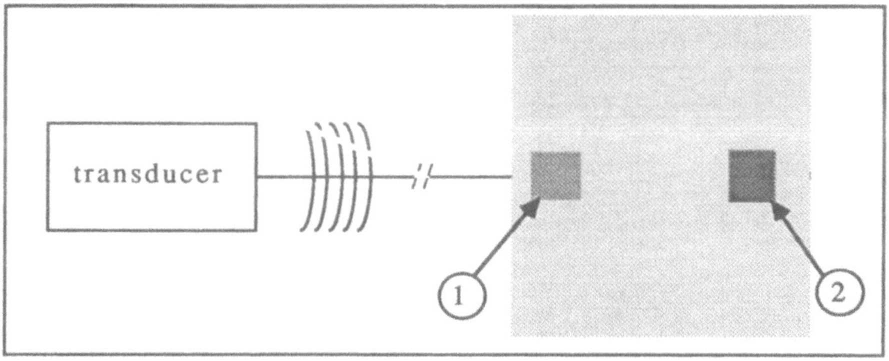

Fig. 5. Simulated medium for computer processing of an echo A-type line in pulsed and correlation mode respectively. See text and table 1 for further details.

As expected the two imaging systems lead, in the absence of noise, to equivalent information. In both cases zone 2 is hardly visible. This is due to the fact that, although the targets are relatively dense, their backscattered force (nearly 0.1±10%) is randomly distributed. Thus the resulting backscattered signal vanishes by destructive interference. Zone 1 is easily visible. It must be emphasized, however, that the extra lobes observed in Fig. 6d are not informative. They are due to the fact that we did not apply the demodulation procedure to this particular simulation.

When white noise (SNR ≈ 0 dB) is added to the signal, the advantage of the correlation technique is clearly demonstrated in Fig. 7. The correlation system still provides practically unmodified information (Fig. 7a) whereas the signal provided by echo system is burried in noise (Fig. 7b).

Table 1 Propagating medium characteristics. λ is ultrasonic wavelength

		background	zone 1	zone 2
Attenuation		0.5 dB/(cm.MHz)		
Target disposition		Random	Network	
Number of targets $/\lambda^3$		1	2	4
Diffusion	sign	Random		
	Amplitude	$0<\lvert a_0\rvert<0.1$	$0.1<\lvert a_0\rvert<0.9$	$0.09<\lvert a_0\rvert<0.11$

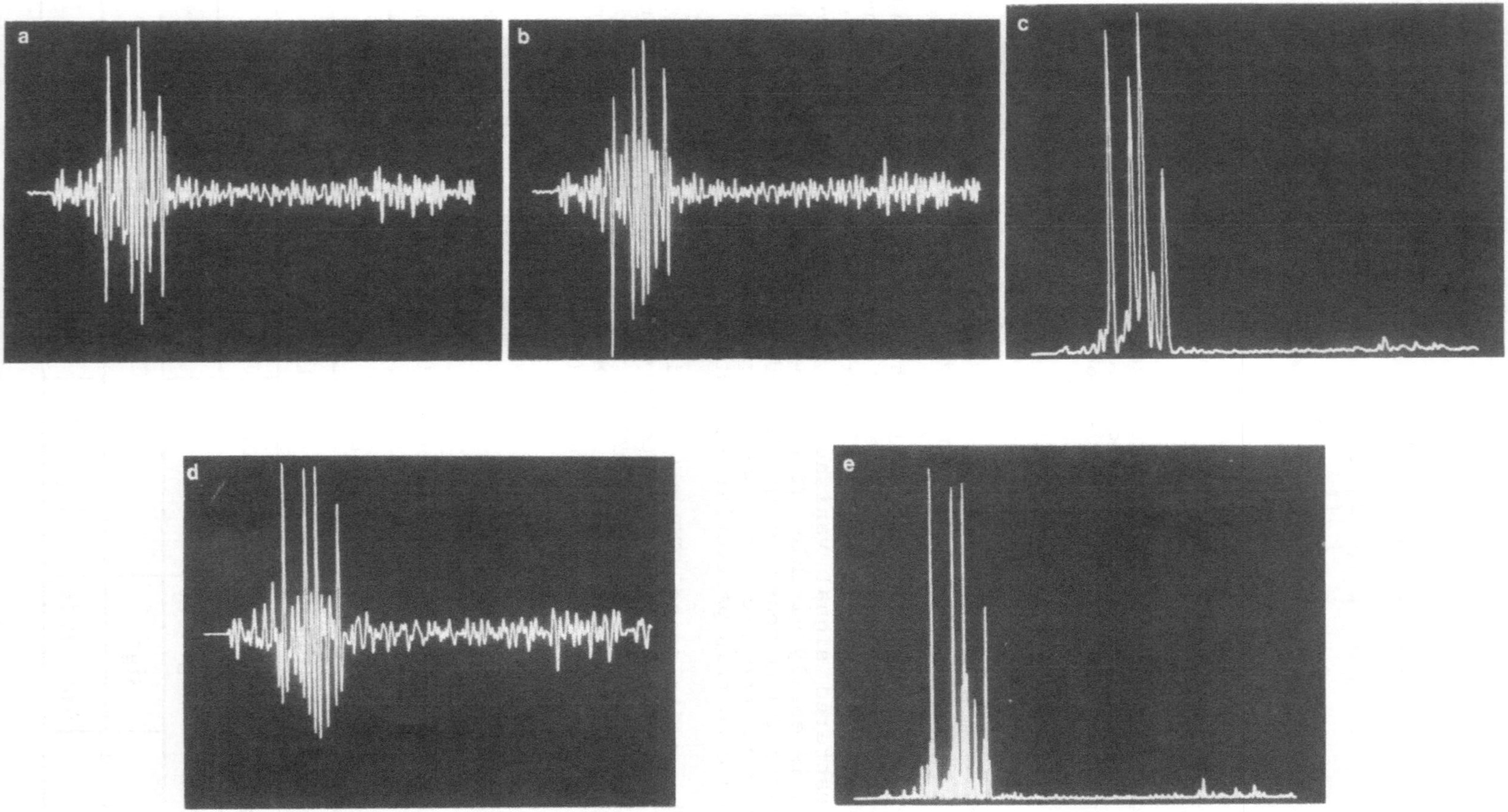

Fig. 6. Simulated A-type line. Correlation [(a): $R\sin\theta$ (b): $R\cos\theta$; (c): $R^2(t)$] and pulsed echo [(d): $s(t)$ and (e): $s^2(t)$]

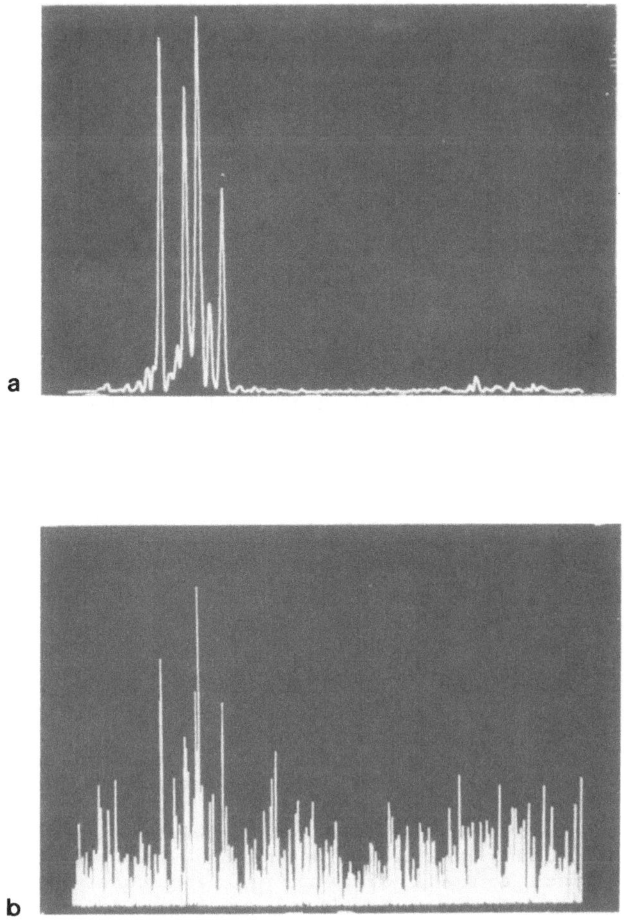

Fig. 7. When, in order to simulate a real medium, white noise is added to the received signal, only the correlation technique (a) allows the extraction of useful information. This is not the case for the classical system (b).

3-EXPERIMENTAL RESULTS

The simulation experiments allowed us to design a signal processing apparatus that uses the correlation procedure.

The experimental system is schematically illustrated in Fig. 8. Two transducers are mounted on the top of a water tank on two independent movable stages allowing easy independent, spatial orientation. The target is a 1.5 mm diameter silicon tube embedded in a synthetic sponge. The sponge was used because the high reflectivity of its internal structures yields a noisy background in ultrasonic images. This background mimics the speckle echoes of biological tissues. We used a pair of disc shaped transducers having similar transfer functions. Their working frequency, f_o, was 2.5 MHz, and their corresponding bandwidth, BW, was 0.5 MHz . The electric signal was a pseudorandom code of 511 bits. The duration of each bit was T_H = 1 ms, and its amplitude was randomly set to +1 or -1 volt. The correlation between the emitted and received signals was made by using a Data Precision model Data 6000 signal processing unit, which samples the input signals at a 4 MHz rate. Because only one demodulation-correlation channel was used, the phase information was preserved. The correlation image was made point-by-point, and the intermediate results were stored in the computer. Fig. 9 shows an A-mode echographic line obtained with the correlation technique. The significant peaks correspond to (1) the front plane of the sponge, (2) the front wall of the silicon tube, (3) the back of the silicon tube. At (4) the rear plane of the sponge is seen and at (5) the wall of the water tank.

Using the same experimental apparatus, we performed a similar experiment in pulsed mode. A 130 volt modulated pulse was necessary to extract unambiguously the tube echo signal from ambient noise

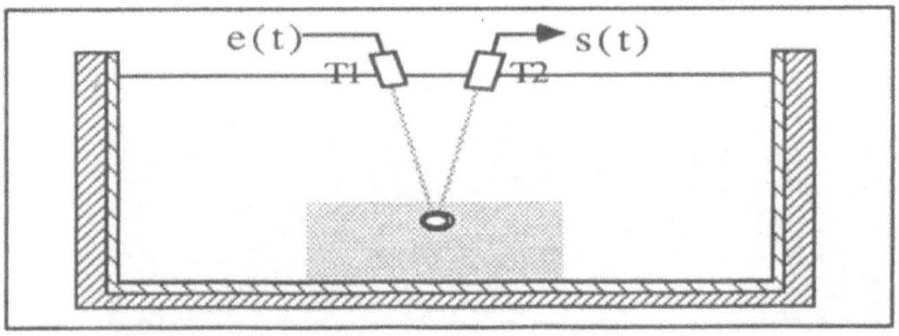

Fig. 8. Schematic representation of the experimental apparatus used in the correlation and classical pulsed echo systems. T_1 and T_2 are the emitting and receiving transducers. The diffusing medium was a sponge in which was placed a small diameter silicon tube. The walls of the water tank are covered with ultrasound absorbing material.

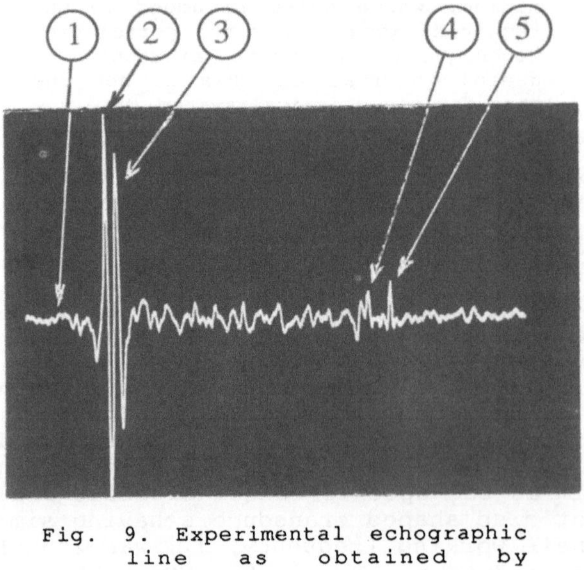

Fig. 9. Experimental echographic line as obtained by correlation

CONCLUSION

This preliminary study shows the advantages and the feasibility of the correlation technique in ultrasonic imaging. In the correlation experiments, the emitted energy was 10^4 times lower than the energy needed in the classical ultrasonic imaging technique. The increase in the SNR obtained by correlation technique partially overcomes the dependence of the attenuation on the frequency and on the depth. This allows us :

1) to explore biological tissues of increased thickness without increasing emitted energy

2) for a given tissue thickness, to significantly increase the signal frequency, leading to enhanced image resolution.

Futhermore, because the correlation technique preserves phase information, further signal processing of received echoes could be done in order to provide quantitative information about the medium.

ACKNOWLEDGEMENTS

This work was supported by ANVAR. It is a pleasure to thank Mr. L. Bachner, Mrs. G. Berger, Mr. R. Coursant, Mr. M. Fink and Mr. P. Laugier for valuable discussions and Mr. P. Kahn for critical reading of the manuscript.

REFERENCES

1. W. K. Kim, S. B. Park, S. A. Johnson, Signal to noise ratio and bandwidth for pseudorandom codes in an ultrasonic imaging system, Ultrasonics. 3:313(1984)
2. D. Nahamoo, A. C. Kak, Ultrasonic echo imaging with pseudorandom and pulsed sources: A comparative study, Ultrasonics. 3:1 (1981)
3. D. Cathignol, J. Y. Chapelon, J. L. Mestras, Critical review on ultrasonic imaging and doppler measurement by correlation technique, ITBM. 3:169 (1982)
4. J. Max, "Méthodes et techniques de traitements de signal et applications aux mesures physique," Mason, Paris FRANCE (1985)
5. W. W. Peterson, "Error correcting codes," J. Wiley and sons, N.Y. USA (1961)
6. G. Cullman, "codes détecteurs et correcteurs d'erreurs," Dunod, Paris FRANCE (1967)
7. R. Kuc, Generating a minimum phase digital filter model for the acoustic attenuation of soft tissue, presented at the IEEE ultrasonics symposium (1983)
8. K. Gurumurphy, report on the meeting in Ferrara nov 1975, Subtopic group A-scan simulation
9. M. Fink, J. F. Cardoso, P. Laugier, Diffraction effect analysis in medical echography, Acta electronica,26:59 (1984)

RECONSTRUCTION OF LOCAL REFLECTIVITY

WITH 3-D DECONVOLUTION

H. Schwetlick, M. Ueda, and M. Tabei

Tokyo Institute of Technology
Research Laboratory of Precision
Machinery and Electronics
Nagatsuta, Midori-ku, Yokohama 227, Japan

INTRODUCTION

The determination of local reflectivity characterizing the interior of a unknown object from acoustic impulse echo data might be considered as a standard problem of acoustical imaging. In many cases this inverse problem can be formulated as a deconvolution problem. By one-dimensional axial deconvolution of a single time trace, assuming plane wave propagation, an estimate of the acoustic impulse response of the scatterer is obtained within the usable band-width of the transducer. This one dimensional scattering function enables an improved discrimination of echoes of one or more reflectors [1]. For special cases like the analysis of a discrete layered medium, a priori information can be used in order to extend the usable frequency range [2]. In [3] the lateral resolution is increased by deconvolution with the spatial characteristics of the transducer. In these cases only one part of the complex transducer characteristics is considered.

In the approach given here, all characteristics of the transducer are included in the reconstruction model. The echoes of a weak scattering object, obtained on a plane aperture are considered as the convolution of the local reflectivity of a scattering volume with the spatial and temporal characteristics of a transducer in pulse-echo mode. Hence the deconvolution represents the inverse problem of finding the local reflection coefficients in the target region from the measured echoes.

RECONSTRUCTION MODEL

The geometry of a transducer insonifying and receiving the reflections from a scattering volume is shown in Fig.1. In Ueda [4,5] an expression for the Echo $\tilde{E}(\omega)$ in the frequency domain

$$\tilde{E}(\omega) = j2\pi\rho_0 G(\omega) \iiint_V R(x,y,z) \left(jk\tilde{\Phi}(x,y,z,\omega) \right)^2 dxdydz \qquad (1)$$

is derived. The transducer is located in the origin of the coordinate system. The impedances of the scattering volume z(x,y,z) and the

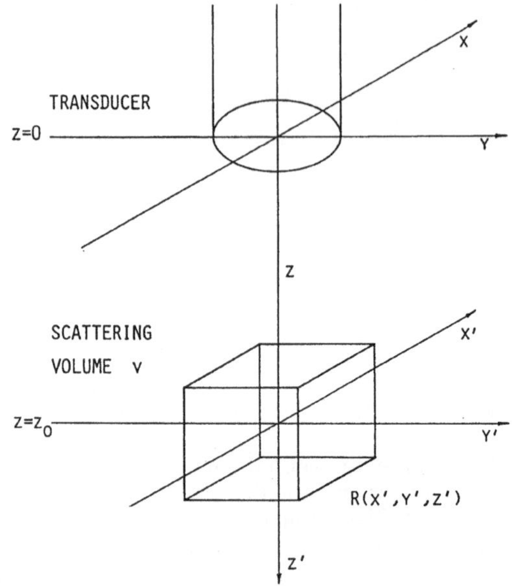

Fig.1 Geometry of a transducer in pulse-echo mode and a reflecting volume.

surrounding impedance z_0 determine the local reflection coefficient

$$R(x,y,z) = \frac{z(x,y,z) - z_0}{z(x,y,z) + z_0} .$$

in the volume v, outside the reflection coefficient is zero. $\tilde{\Phi}(x,y,z,\omega)$ is the velocity potential generated by the transducer with unity excitation at the point (x,y,z). $G(\omega)$ represents the transfer function of the electro-mechanical conversion, ρ_0 is the density, $k=\omega/c$ the wave number, and c the velocity of sound. Then the echo in the Plane z=0 is given by the convolution

$$\tilde{E}(x,y,\omega) = j2\pi\rho_0 G(\omega) \iiint_{-\infty}^{\infty} R(x',y',z') \, (jk\tilde{\Phi}(x-x',y-y',z,\omega))^2 \, dx'dy'dz \qquad (2)$$

Near the focal point in the volume of interest, the velocity potential is besides a phase term independent of the distance z and can be approximated by

$$\tilde{\Phi}(x,y,z,\omega) = \tilde{\Phi}(x,y,z_0,\omega) \exp(-jk(z-z_0)). \qquad (3)$$

The function

$$\tilde{D}(x,y,\omega) = j2\pi\rho_0 G(\omega) \, (jk\tilde{\Phi}(x,y,z_0,\omega) \exp(jkz_0))^2 \qquad (4)$$

describes the characteristics of the transducer and can be interpreted as the echo of an ideal point reflector in the focal plane. The first term $j2\pi\rho_0 G(\omega)$ as a function of ω only, will subsequently be referred to as the temporal characteristics of the transducer, the second term $(jk\tilde{\Phi}(x,y,z_0,\omega)$ describes the response as a function of space and will be referred to as spatial characteristics. By inserting eq.3 and eq.4 into eq.2,

$$\tilde{E}(x,y,\omega) = \iiint_{-\infty}^{\infty} R(x',y',z') \exp(-j2kz') \, \tilde{D}(x-x',y-y',\omega) \, dx'dy'dz'. \qquad (5)$$

is obtained. The two-dimensional Fourier transform with respect to the spatial coordinates

$$\approx{E}(k_x,k_y,\omega) = \iint_{-\infty}^{\infty} \tilde{E}(x,y,\omega) \exp(-jk_x x - jk_y y)\, dxdy \quad (6)$$

leads to an expression for the echo in the wavenumber-frequency domain

$$\approx{E}(k_x,k_y,\omega) = \approx{R}'(k_x,k_y,\omega)\, \approx{D}(k_x,k_y,\omega) \quad (7)$$

where

$$\approx{D}(k_x,k_y,\omega) = \iint_{-\infty}^{\infty} \tilde{D}(x,y,\omega) \exp(-jk_x x - jk_y y)\, dxdy \quad (8)$$

and with the travel time coordinate z=ct/2

$$\approx{R}'(k_x,k_y,2\omega/c) = 2/c\ \approx{R}(k_x,k_y,2\omega/c)$$

$$= \iiint_{-\infty}^{\infty} R'(x',y',ct/2) \exp(-ik_x x - jk_y y - j\omega t)\, dx'dy'dt. \quad (9)$$

Note that $\approx{R}'(k_x,k_y,2\omega/c)$ is the Fourier transform of $R(x',y',z')$ multiplied with 2/c. Eq.7 shows the convolution as a product in the wavenumber-frequency domain. Hence the inversion for the unknown reflection coefficient becomes

$$\approx{R}'(k_x,k_y,2\omega/c) = \approx{E}(k_x,k_y,\omega)\ \approx{D}^{-1}(k_x,k_y,\omega), \quad (10)$$

where $\approx{D}^{-1}$ is the inverse of $\approx{D}(k_x,k_y,\omega)$, its determination is discussed below. By the inverse Fourier transform

$$R(x,y,ct/2) = 1/(8\pi^3) \iiint \approx{R}'(k_x,k_y,2\omega/c) \exp(jk_x x + jk_y y + j\omega t)\, dk_x dk_y d\omega. \quad (11)$$

the reflectivity function is determined.

NUMERICAL PROCEDURE

The derivation in the previous chapter shows the solution of the reconstruction in terms of Fourier transforms. For the numerical treatment these Fourier transforms are approximated by discrete Fourier transforms. The functions E and D at discrete points $(l\Delta x, m\Delta y, n\Delta t)$ are designated by the indices l,n,m in the time domain, given at L and M sampling points in space and N sampling points in time. Correspondingly in the wavenumber-frequency domain the values are given at $(p\Delta k_x, q\Delta k_y, r\Delta\omega)$ and are designated by p,q,r. The distances between sampling points in the wavenumber-frequency domain are given by

$$\Delta k_x = 1/(L\Delta x),$$
$$\Delta k_y = 1/(m\Delta y),$$
$$\omega = 1/(N\Delta t).$$

The reflectivity R(x,y,z) is given at the same discrete points in x and y direction, in z-direction at $z=z_0+n\Delta tc/2$ in the space-time domain and at $2\omega/c=r\Delta\omega 2/c$ in the wavenumber-frequency domain. The transform pair

$$\approx{E}_{p,q,r} = DFT^3\{E_{l,m,n}\}$$

$$= \sum_{l=-L/2}^{L/2-1} \sum_{m=-M/2}^{M/2-1} \sum_{n=-N/2}^{N/2-1} E_{l,m,n} \exp(-j2\pi lp/L - j2\pi mq/M - j2\pi nr/N) \quad (12)$$

$$E_{l,m,n} = DFT^{-3}\{ \tilde{\tilde{E}}_{p,q,r}\}$$

$$= 1/LMN \sum_{p=-L/2}^{L/2-1} \sum_{q=-M/2}^{M/2-1} \sum_{r=-N/2}^{N/2-1} \tilde{\tilde{E}}_{p,q,r} \exp(j2\pi lp/L + j2\pi mq/M + j2\pi nr/N) \quad (13)$$

defines the relation between the discrete space-time domain and wavenumber-frequency domain and is computed by an FFT algorithm.

In the process of deconvolution the measured or simulated echoes are first transformed from the space-time domain into the wavenumber-frequency domain by eq.(12). The next step involves the multiplication as shown in eq.(10) of this data by the inverse of the transducer function. From this product, the reflectivity is obtained by the inverse transform eq.(13).

Because of the limited band-width of the probing transducer, the resulting reflectivity function is expected also to be band-limited. Moreover, since the absolute value of the reflectivity might be more suitable for the interpretation of the image, the envelope

$$A(x,y,t) = \sqrt{(R^2(x,y,ct/2) + (H\{R(x,y,ct/2)\})^2)} \quad (14)$$

of the resulting reflectivity was chosen for display. $H\{R(.)\}$ is the Hilbert transform of of the function $R(x,y,ct/2)$, calculated by FFT from $j\tilde{R}(x,y,2\omega/c)$.

For the spatial sampling L=M=32 points were chosen, for the temporal sampling N=128 points. With a spatial sampling width of $\Delta x=\Delta y=0.6$mm an $d\Delta t=0.1$us, a target region of $19.2*19.2*19.2 mm^3$ can be reconstructed. The computation time, mainly dependent on the two 3-D Fourier transforms, amounts to about 20 minutes for one reconstruction.

SIMULATION

Even this method might find application for general transducers, in this paper the application is restricted to a focus transducer, i.e. a focus transducer with a diameter of 0.75 inch and a focal distance of 4 inch. Since it is intended to apply this reconstruction technique to all types of reflectors for general imaging, the numerical simulation includes the basic types: plane, line, and point reflectors.

The numerical computation of echo data is based on an expression for the sound field of a focus transducer given in [6]. This Method assumes a uniform piston motion and expresses the velocity potential 0 as a function of space and frequency by a series of Bessel functions. The response of a point reflector is then obtained by eq.(4). The multiplication with $(jk)^2$ in eq.(4) in the frequency domain corresponds to double differentiation in the time domain. For reasons of the discretization, the differentiation was carried out in the time domain and approximated by first order differences. Fig.2 show the time functions of the expression $(jk\Phi(x,y=0,t))^2$ at different positions off center and might be considered as spatial impulse response of a transducer with uniform piston motion and a point reflector. The temporal characteristics of the transducer, $j\omega\rho_0 G(\omega)$, are not considered in this graph, and the wavelets represent only the spatial characteristics.

The echo of an ideal line reflector was determined by integrating the expression for the point reflector along one line. In this case the length of the numerical integration was taken about twice the length of the diameter of the transducer and the computations were at about 200 sampling points.

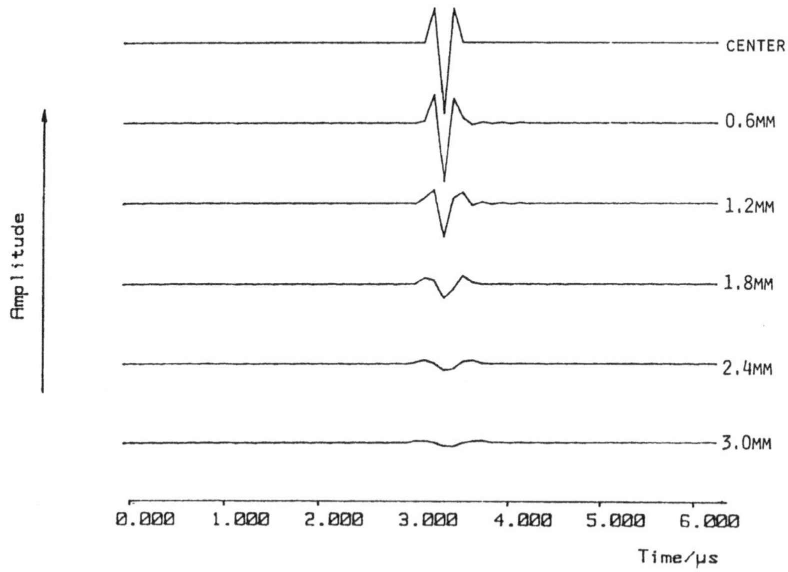

Fig.2 Spatial response of a transducer and a point reflector in the focal plane in different distances from the center line.

The echo of a plane was obtained using the mirror image principle, consequently the data where obtained by integrating numerically the sound field in a distance of $2z_0$. Cylindrical coordinates were used and the integral was approximated by summation over 20 points.

By time shifting and appropriate amplitude weighting and summing these responses a resulting response is obtained. Multiplication with the temporal transducer characteristics $j\omega\rho_0 G(\omega)$ issues a realistic simulation of the experimental echoes.

A comparison of simulated and measured echoes of a point and a plane reflector at normal incidence is shown in fig.3. The Signal was synthesized having about the same signal amplitudes as the corresponding measured data, the echoes of a plane agar surface and a 0.5mm diameter glass sphere. The frequency characteristics of the transducer are computed by deconvolving a measured echo of a plane metal reflector by its computed impulse response. The signals at x=y=0 are almost identical, the deviation in the last part of the wavelet of the measured point reflector might arise from the finite dimension of the sphere. However, at distances more far away from the center, measured and simulated data deviate. The general behavior remains the same, but the focusing of the simulated transducer appears to be stronger than the one of the real transducer. The deviation might be due to the assumption of a uniform piston motion, which is only approximately valid.

THE INVERSE FUNCTION

The determination of the inverse function $\tilde{\tilde{D}}^{-1}$ is a crucial point in obtaining a accurate reconstruction. In fig.4 the spatial characteristic. $(jk\tilde{\tilde{\Phi}}(k_x,k_y,\omega))$ from simulation and the experimentally obtained spatial and temporal characteristics $\tilde{\tilde{D}}(k_x,k_y,\omega)$ of a focus transducer are shown in the wavenumber-frequency domain. It can be seen, that both functions assume non zero values in the center part of the graphs. In regions outside the center part, values become small and the inverse is consequently not well defined. Particular with experimental data with a limited signal to noise

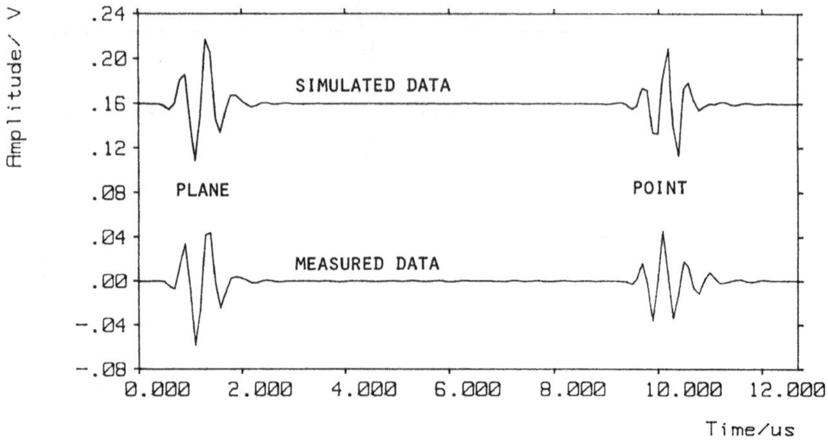

Fig.3 Comparison of echo data from a plane and a sphere.

ratio, the error in small values will be amplified by inversion, and can lead to instabilities. In an easy approach, by thresholding, the inverse of small values is set to zero. In another approach, equivalent to the Wiener filter, the inverse is determined by

$$\tilde{\tilde{D}}^{-1} = \tilde{\tilde{D}}^{*} / (\tilde{\tilde{D}}\tilde{\tilde{D}}^{*} + \varepsilon). \qquad (15)$$

For a very small values of the parameter ε, $\tilde{\tilde{D}}^{-1}$ approaches $1/\tilde{\tilde{D}}$, for very large values of the parameter $\tilde{\tilde{D}}^{-1}$ approaches $\tilde{\tilde{D}}^{*}$. But within a wide range of values, this inverse provides a smooth transition between valid and invalid data in the wavenumber-frequency domain. As in one-dimensional deconvolution, towards smaller value of the parameter ε the background noise will increase, towards larger values the effect of deconvolution will be reduced.

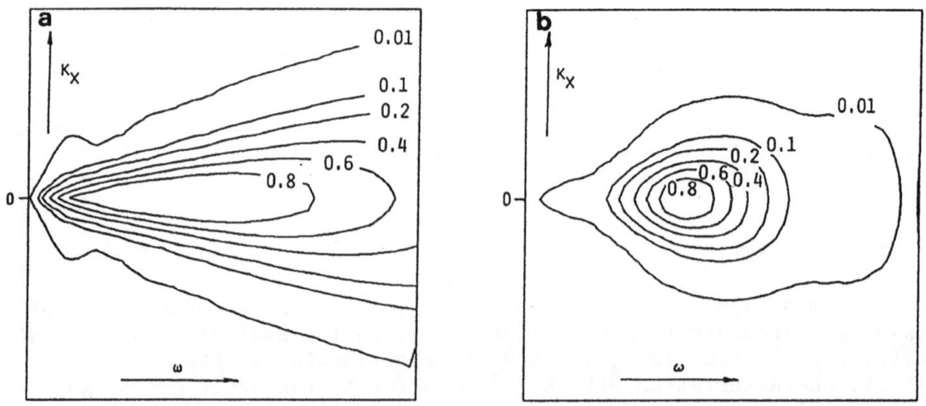

Fig.4 Simulated and measured transducer characteristics in the wavenumber -frequency domain, (a) simulated spatial characteristics. (b) measured temporal and spatial characteristics. Contour lines relate to max-value.

Since ultrasonic transducers with wide bandwidth and high axial resolution are available, also the case was considered, to deconvolve with the spatial characteristics only. The inverse $\tilde{\tilde{D}}^{-1}$ for this deconvolution is obtained by inverting the spatial characteristics of the transducer, i.e. the function $(jk\tilde{\tilde{\Phi}}(k_x, k_y, \omega))^2$, only. This would enable, to calculate the inverse within a wider area in the wavenumber-frequency domain with higher accuracy. The result is a reflectivity function convolved with the temporal transducer characteristics. Up to now, this deconvolution is restricted to simulation data. Since the computation ofan inverse requires very accurate data, it was found to be necessary to deconvolve experimental data with an inverse from experimental data. But with a more elaborated model for the spatial characteristics of thetransducer, it is expected, to obtain numerically a more accurate inverse.

RECONSTRUCTION RESULTS

The result of a reconstruction experiment with simulated data is shown in fig.5. The echo data of a point, a line and a plane were computed as described above, the temporal characteristics of the transducer in the space-freaquency domain were assumed to follow a cosine-bell curve.

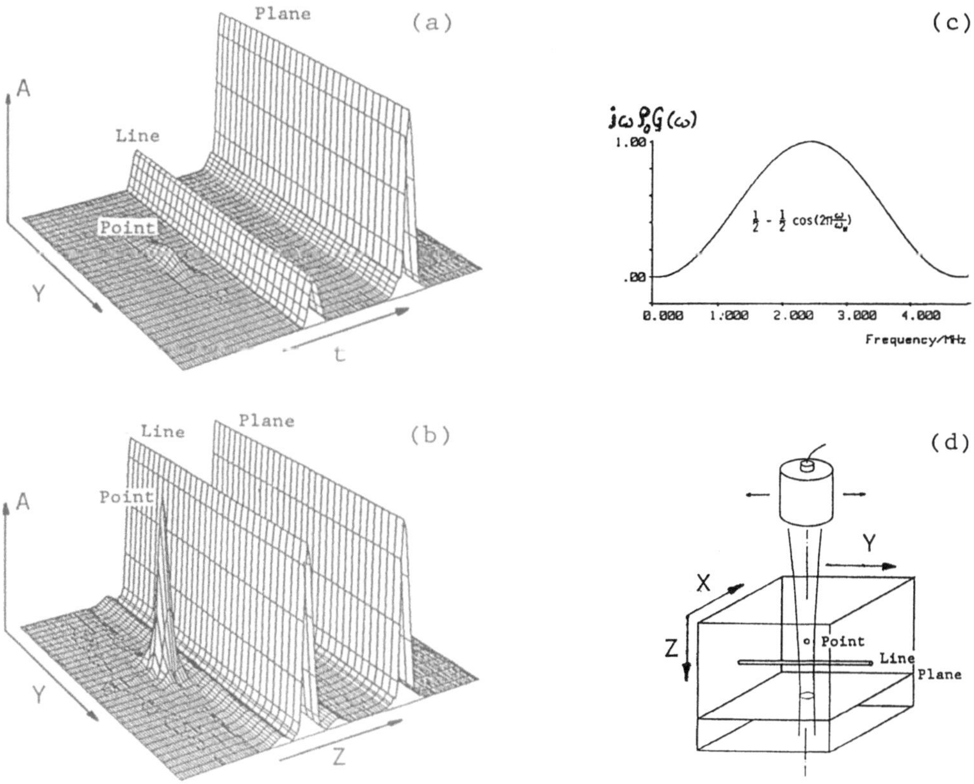

Fig.5 Reconstruction from simulated echo data from a point, a line and a plane reflector, (a) envelope of the echo signal, (b) temporal frequency characteristics of the transducer, (c) envelope of the resulting reflectivity, (d) geometry of the simulated specimen.

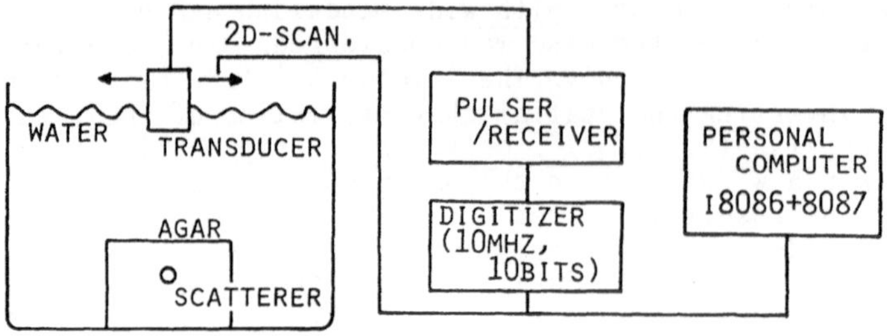

Fig.6 Experimental system

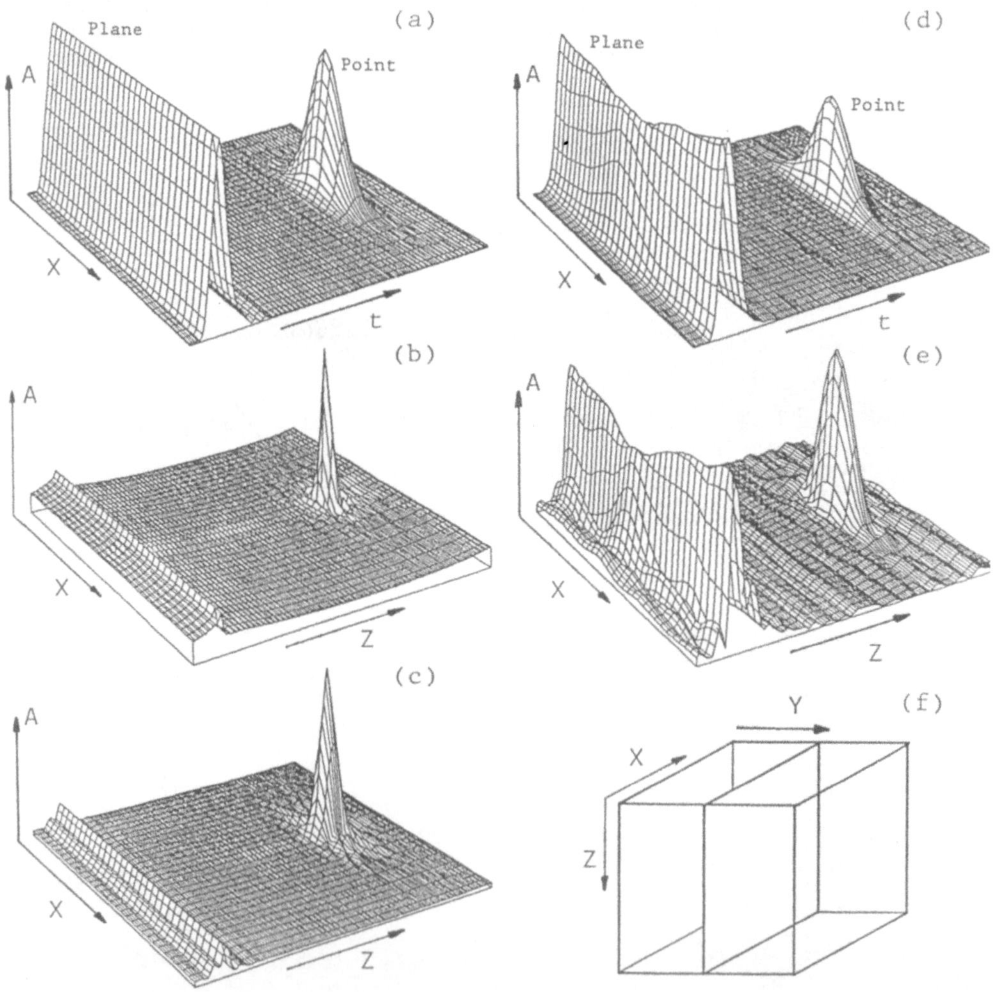

Fig.7 Reconstruction from simulated and measured echo data, (a) simulated echo data, (b) deconvolution with temporal and spatial transducer characteristics, (c) deconvolution with spatial characteristics only, (d) measured echo data, (e) deconvolution with experimentally obtained transducer characteristics, (f) displayed section of the target volume.

The inverse function D was computed from the spatial transducer characteristics only. Assuming equal reflectivity, the echoes of a plane reflector have a higher amplitude compared to the echoes of a line or point reflector. Since the temporal characteristics of the transducer were not included in the inversion, the reconstructed reflectivity is also a band-limited function, but still possesses a high resolution. The reconstruction then displays an estimate of the local reflectivity, which is more independent of the actual geometry of the reflectors.

Experimental echoes were acquired by the system shown in fig.6. An ultrasonic transducer in pulse echo mode scans on a plane aperture. Averaging the signal 256 times increased the signal to noise ratio. The reflecting target was a glass sphere of 0.5mm diameter, which was buried in agar. It was placed in the focal region of the transducer. Fig.7(d) shows the envelope of the reflection data. The spatial and temporal characteristics of the transducer for the computation of the inverse function were obtained by measuring the echo signal D(x,y,t) of a small reflector, also a 0.5mm glass sphere. The result of the deconvolution is shown in fig.7(e), fig.7.(f) indicates the displayed section of the reconstructed volume.

Parallel to this experiment, the echoes from same specimen were computed by numerical simulation (fig.7(a)). Different from the simulation experiment described above, the actual temporal transducer characteristics from experimental data were included as in fig.3. This echo data

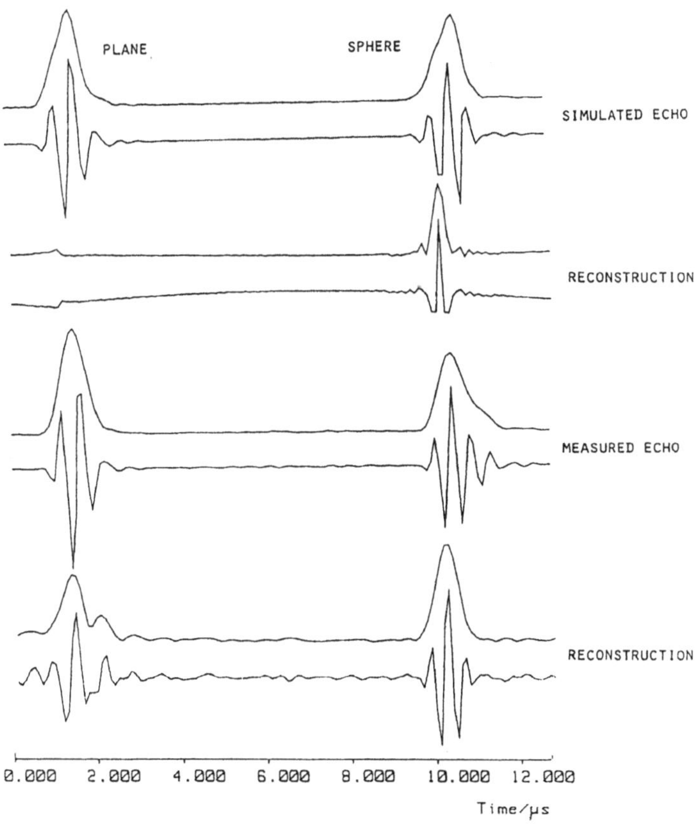

Fig.8 Echoes and reconstructed reflectivity in the center line.

were deconvolved with two different inverse functions. In the first reconstruction (fig. 7(b)), the inverse was calculated from the simulated function D(x,y,t), where spatial and temporal characteristics of the transducer are included. In the second reconstruction (fig. 7(c)) the inverse was computed from spatial characteristics only. In the later case, low frequency components in the reconstruction, which might be erroneous, are suppressed.

The reconstructions from both, simulated and measured echoes, indicate, that the reconstructed reflectivity is more independent of the actual geometry of the reflector and more related to the local reflectivity than the unprocessed echoes are. The reconstruction can be thought of as an additional focusing. The graphs in fig. 8 of echoes and reconstructions demonstrate this behavior at the center line at x=y=0.

Experiments with different values of ε were performed. The results shown are obtained with a ε of 0.1 of the maximum value, which was optimal for the experimental data.

CONCLUSION

Three-dimensional deconvolution of pulse-echo data provides an estimate of the local reflectivity with an increased resolution. The signal processing consists mainly on the computation of two three-dimensional discrete Fourier transforms. The first experiences with data obtained by scanning indicate, that this method can be extended to larger target volumes or applied to data from an ultrasonic sensor array. The critical point, the determination of an appropriate inverse function, which could be possibly improved by an accurate transducer model, is subject of our current investigation.

ACKNOWLEDGMENT

The support by the Japan Society for the Promotion of Science and the German Alexander von Humboldt-Stiftung is gratefully acknowledged.

REFERENCES

[1] Nabel, E., Die Impulsantwort als Ergebnis der Ultraschallspektrometrie Versuche zur exakten Bewertung von Ultraschallechos, Materialpruefung 19:496 (1977)

[2] Schwetlick, H., Miyashita, T., Schickert, M., Acoustical Imaging with incomplete Data, Proceedings "Ultrasonics International 87" London (1987)

[3] Hundt, E. E., Trautenberg, E. A., Digital Processing of Ultrasonic Data by Deconvolution, IEEE Trans. Sonics and Ultrason. 27:249 (1980)

[4] Ueda, M., Ichikawa, H., Analysis of an Echo Signal Reflected from a Weakly Scattering Volume by a discrete Model of the Medium, J. Acoust. Soc. Am. 70:1768 (1981)

[5] Ueda, M., Spectrum of Echo Scattered by Simple Object Placed at the Focal Plane of Ultrasonic Transducer, "Acoustical Imaging" Vol. 13, Kaveh, M. et al. editors, Plenum Press (1984)

[6] Ueda, M., Fast Converging Series Expansion for Velocity Potential of Circular Piston Source, J. Acoust. Soc. Jpn. 6:35 (1985)

OPTIMAL TIME DOMAIN DECONVOLUTION FOR THE HIGH-RESOLUTION COMPUTERIZED ULTRASOUND ECHO TOMOGRAPHY

A.Yamada and H.Yamabe

Faculty of Engineering, Toyohashi University of Technology
Toyohashi-shi, 440 Japan

ABSTRACT

A time domain deconvolution filtering technique is proposed for reconstructing high-resolution cross-sectional images of ultrasonic reflectivity from the reflection data collected surrounding an object. The deconvolution filtering for compensating the time domain distortion of the system, defined by the transducer characteristics etc., is incorporated into the conventional filtered-back-projection image reconstruction process. Specifically, the deconvolution filter is optimized so as to maximize the resolution capability subject to the constraint of the small processing noise. Theoretical formula for the resolution achievable with this techniques is presented. Experimental results are also shown to demonstrate that high-resolution images comparable to the theoretical limit is obtained.

INTRODUCTION

Ultrasound imaging methods are widely used in medical diagnostics and industrial inspection. Among them, B-scan method is most popular. However,the resolution in lateral direction is much less than in axial direction due to the nature of the sound wave such as diffraction phenomena. In contrast, there have been made a number of reports in recent years on the application of CT scanning approach to the ultrasound pulse-echo mode tomography [1]-[8]. In these methods, reflectivity of the object is imaged by a computer based reconstruction procedure from the pulse-echo data collected surrounding an object. The wavefront of sound wave is considered under the assumptions of the object being weakly reflecting, speed and absorption of sound being constant. The difficulties in B-scan method can, therefore, be circumvented to give high resolution in all directions.

In the method discussed above, the resolution of the image at last is limited by the temporal response of the pulse-echo system, defined by such as the transducer property and sound wave propagation, etc. If these properties can be compensated in the reconstruction procedure, further improvement of the resolution will be expected. From such a point of view, some authors [1],[3] have suggested the basic idea based on the concept of the inverse filtering technique. However, the mitigation of the noise amplification compatible with high-resolution capability has been the remaining problem hindering the practical use of this technique[8].

In this paper, a time domain deconvolution filtering technique is proposed for reconstructing high-resolution cross-sectional images of ultrasonic reflectivity from the reflection data collected surrounding an object. The deconvolution filtering [9], for compensating the time domain distortion of the system, is incorporated into the conventional Filtered-Back-Projection (FBP) image reconstruction process. Specifically, the deconvolution filter is optimized so as to maximize the resolution capability subject to the constraint of the small processing noise. Theoretical formula for the resolution achievable with this technique is presented. Experimental results are also shown to demonstrate that high-resolution images comparable to the theoretical limit are obtained.

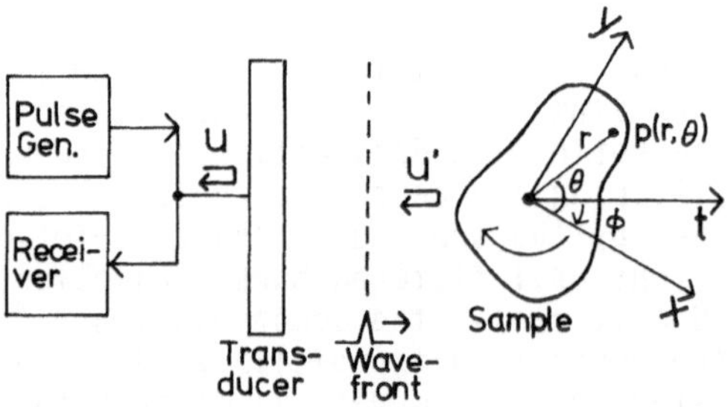

Fig.1. Schematic of a pulse-echo system for the reconstruction tomography.

MODEL OF THE PULSE-ECHO SIGNAL

Consider the pulse-echo mode measurement system as shown in Fig.1, where the transducer is excited by short pulse, and the echo waveform reflected from an object is received through the same transducer. By rotating the object, pulse-echo data $u(t,\phi)$ are collected, as a function of time t and rotation angle ϕ. We shall consider the problem for reconstructing the cross-sectional distribution of a reflectivity parameter $p(x,y)$ in two-dimensional(2D) xy plane.

Let the temporal response of the system be given by $f(t)$, summarizing the properties such as those of the transducer and the sound wave propagation, etc. Further, we consider the original acoustic reflection $u'(t,\phi)$, which can be observed with an idealized distortion-less measurement system. Then, the convolution relation can be assumed between u, f and u' as

$$u(t,\phi) = \int_{-\infty}^{+\infty} u'(t-t',\phi)f(t')dt' = u'(t,\phi) * f(t), \qquad (1)$$

where $*$ denotes one-dimensional(1D) convolution. We maintain that the following band-limited property should be assumed with respect to the Fourier spectrum $F(\omega)$ of f.

$$|F(\omega)| \fallingdotseq 0, \qquad \text{at } |\omega|<\omega_l, \omega_h<|\omega|, \tag{2}$$

where ω is angular frequency, and ω_l and ω_h are the lower and higher cut-off frequency, respectively.

In this paper, we assume the object being weakly reflecting such as the biological soft tissues, and the wavefront of the insonifying wave being a straight line with uniform amplitude. Under these condition, reflection u' is represented by the integration of the reflectivity p along a straight line wavefront of the insonifying wave. That is,

$$u'(t,\phi)=\int\int_{-\infty}^{+\infty} p(x,y)\delta(2(x\cos\phi+y\sin\phi)/c-t)dxdy, \tag{3}$$

where δ is 1D Dirac delta function, c is mean acoustic velocity in the medium. Thus, the relation between u and p are obtained with the substitution of eq.(3) into eq.(1) as

$$u(t,\phi) = f(t) * u'(t,\phi)$$
$$= f(t) * \int\int_{-\infty}^{+\infty} p(x,y)\delta(2(x\cos\phi+y\sin\phi)/c-t)\, dxdy. \tag{4}$$

BASIC METHOD

The FBP method was adopted to the case where reconstruction of a two-dimensional image p is made from a set of one-dimensional projections u based on the model described in eq.(4). That is, each projection data $u(t,\phi)$ for different rotation angle ϕ is convolved with filter function $h(t)$. Then, the image for a given point is reconstructed by summing these pre-filtered data that pass through it. Thus, by relating the image point at (r,θ) and the projection point at (t,ϕ) as

$$t = (2r/c)\cos(\phi-\theta), \tag{5}$$

the mathematical equation describing the reconstruction procedure is

$$p(r,\theta)=\int_0^{2\pi} u(t,\phi) * h(t)\, d\phi, \tag{6}$$

where the polar coordinate (r,θ) is related with the cartesian coordinate(x,y) as

$$r = \sqrt{x^2+y^2}, \quad \theta = \tan^{-1}(y/x). \tag{7}$$

Note that we consider here the problem for reconstructing an image based on the model described in eq.(4), in contrast to the conventional FBP method based on the model in eq.(3). In the present case, the image can be recovered only in an approximate form on account of the band-limited nature of f. Hence, we consider the problem for maximizing the resolution capability under the constraint of the small processing noise for the compensation of f. In the following, we will develop the method for determining the optimal filter h_{opt} based on the concept mentioned above.

DERIVATION OF THE POINT SPREAD FUNCTION

Consider the point reflector at (r_0, θ_0)

$$p(x,y) = \delta(r-r_0, \theta-\theta_0), \tag{8}$$

where $\delta(\cdot,\cdot)$ denotes 2D Dirac delta function.

Substituting eq.(8) into eq.(4), the observed echo signal is

$$u^r(t, \phi) = f(t - t_0), \tag{9}$$

where

$$t_0 = (2r_0/c) \cos(\phi - \theta_0). \tag{10}$$

The equation (9) describes the temporal impulse response of the system. Furthermore, the backprojection point spread function (spatial impulse response) g is evaluated by substituting eq.(9) into eq.(6). It follows that

$$g(r, \theta) = \int_0^{2\pi} f^t(t - t_0)\, d\phi, \quad \text{for} \quad t = (2r/c) \cos(\phi - \theta), \tag{11}$$

where f^t represents the total temporal response of the system including the convolution filtering process as

$$f^t(t) = h(t) * f(t). \tag{12}$$

Equation (11) can be represented in terms of the Fourier spectrum $F^t(\omega)$ of $f^t(t)$. The result is

$$g(r') = \int_{-\infty}^{+\infty} F^t(\omega)\, J_0(2|\omega| r'/c)\, d\omega, \tag{13}$$

where $J_0()$ is the zero-order Bessel function, and

$$r' = \sqrt{r^2 + r_0^2 - 2rr_0 \cos(\theta - \theta_0)} \tag{14}$$

is the distance between the observation point at (r, θ) and the point reflector at (r_0, θ_0).

Equation(13) implies that contribution to g(r') is only the even part of $F^t(\omega)$. In addition, when $F^t(\omega)$ is an even function, the back-projection can be done over a 180 degree aperture in stead of a full 360 degree aperture. Hence, it is convenient to set F^t such that

$$F^t(\omega) = F^t(-\omega) = |k|\, G(|k|)/c, \tag{15}$$

where

$$k = 2\omega/c \tag{16}$$

is the spatial frequency. Substituting eq.(15) into eq.(13), results in

$$g(r') = \int_0^{+\infty} G(k)\, J_0(kr')\, k\, dk. \tag{17}$$

The above shows that g(r') and G(k) are related by the Hankel transform.

DETERMINATION OF OPTIMAL FILTER

It is desirable that g concentrates the energy around the peak as much as possible in order to realize a high resolution image. As a measure of 2D signal concentration, the 3-rd order moment M of g around the origin is introduced as

$$M[g]=\int_0^{+\infty}\int r^3\, |g|^2\, dr d\theta \;/\int_0^{+\infty}\int r\, |g|^2\, dr d\theta, \tag{18}$$

where the peak of g is chosen at the origin for the sake of simplicity.

In this discussion, g may be chosen to be an impulse from a viewpoint of the minimization of M. However, owing to the band-limited property of f described in eq.(2), spectrum F^t of f^t must also be restricted to the same band-limited region in order to suppress the noise sensitivity. On the other hand, the Hankel transform G of g is related with F^t by the relation in eq.(15). Thus, G also need be band-limited in the same region. The optimum value of g is therefore be obtained by the solution of the problem:

$$\begin{aligned}&\text{Minimize } M[g(r)]\\ &\text{subject to } G(k) = 0, \quad \text{at } |k|<k_l, k_h<|k|,\end{aligned} \tag{19}$$

where k_l and k_h is related with ω_l and ω_h by the relation:

$$k_l = 2\omega_l/c,\; k_h = 2\omega_h/c. \tag{20}$$

The problem described in eq.(19) is solved as

$$\begin{aligned}G_{opt}(k) &= Z_0(k), && \text{at } k_l<|k|< k_h\\ &= 0, && \text{elsewhere}\end{aligned} \tag{21}$$

$$M[g_{opt}(r)] = \lambda^2 \tag{22}$$

where

$$Z_0(k) = \cos\alpha\; J_0(\lambda k) - \sin\alpha\; N_0(\lambda k) \tag{23}$$

is the linear sums of zero-order Bessel function of first kind $J_0()$ and second kind $N_0()$. α and λ are determined so that k_l and k_h coincide with the first and second zeros of $Z_0(K)$, respectively. The optimal moment $M[g_{opt}]$ in eq.(22) can be evaluated by the approximate expression

$$M[g_{opt}] \fallingdotseq (\pi/\Delta k)^2. \tag{24}$$

The desired temporal impulse response f^t_{opt} should therefore be

$$f^t_{opt}(t) = F^{-1}\{|k|G_{opt}(|k|)/c\}, \tag{25}$$

where $F^{-1}\{\ \}$ denotes inverse Fourier transform. Finally, based on the relation in eq.(12), the optimal filter h_{opt} is determined so that the output through the filter for reference input signal should be as close as possible to the desired response f^t_{opt} developed above.

RESOLUTION CAPABILITY

Consider the circular function circ() with diameter a as

$$\begin{aligned} circ(r;a) &= 1, \quad && \text{at } |r| < a/2 \\ &= 0, && \text{elsewhere.} \end{aligned} \tag{26}$$

Substituting eq.(26) into eq.(18), the moment of circ() is

$$M[circ(r;a)] = a^2 / 8. \tag{27}$$

We shall define the resolution Δl of response g with the diameter of function circ having same moment with that of g. By replacing M[circ] and a in eq.(27) by M[g] and Δl, respectively, it is found to be

$$\Delta l = 2\sqrt{2}M[g]. \tag{28}$$

Substituting the approximate value of $M[g_{opt}]$ in eq.(24) into the above, it follows that

$$\begin{aligned} \Delta l_{opt} &\fallingdotseq \sqrt{2}(2\pi/\Delta k) \\ &= c/(\sqrt{2}\Delta f), \end{aligned} \tag{29}$$

where

$$\Delta f = (\omega_h - \omega_l)/2\pi . \tag{30}$$

Equation (29) implies that resolution is limited by only the temporal band-width of the pulse-echo measurement system, to give high-resolution images independent of the direction.

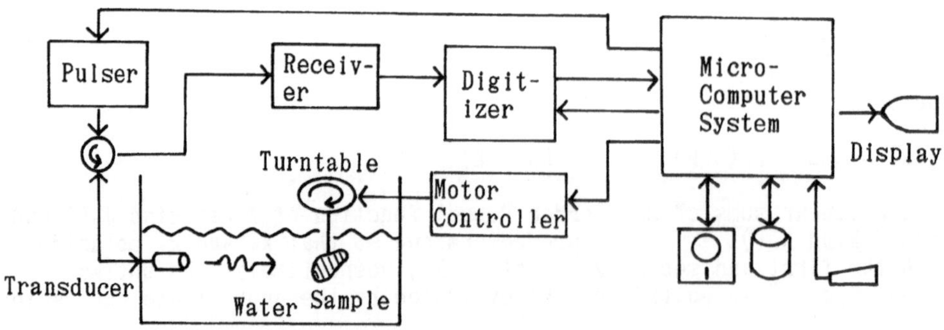

Fig.2. Configuration of the system.

EXPERIMENTAL EVALUATION

An experimental image reconstruction system is shown in Fig.2. The transducer and the object were immersed in a water tank. The object was mounted on a turntable with a stepping motor. We used a narrowband circular disk transducer of 10mm diameter and 4MHz center frequency. In order to estimate the resolution, tungsten wires of 0.1mm diameter, placed parallel to the transducer face at a distance about 50mm, were used as an object. The observed echo signals were digitized with the sampling frequency of 50MHz, and transferred to a microcomputer system (Intel

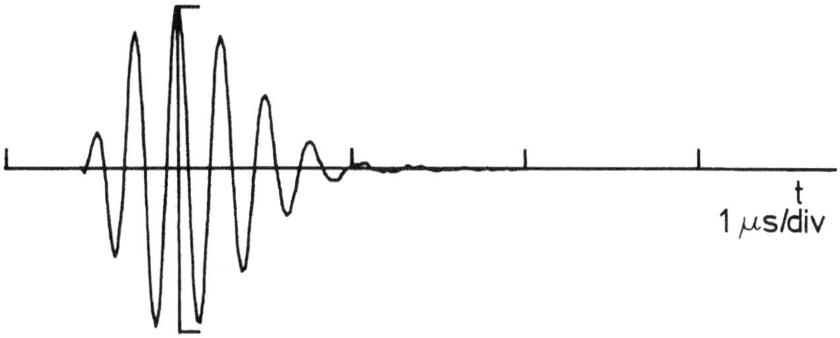

Fig.3. Observed reference signal.

(a) Temporal impulse response.

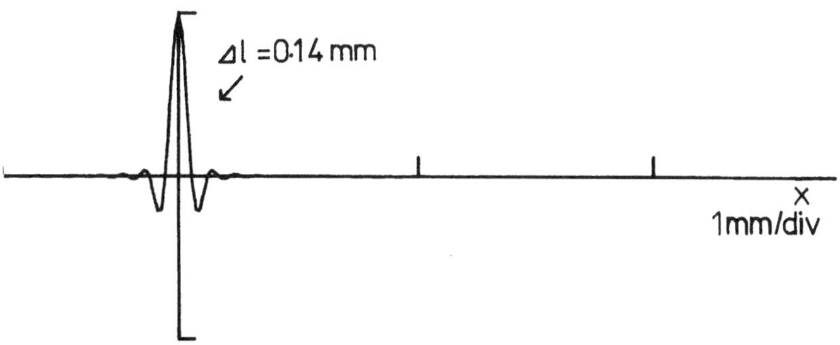

(b) Point spread function.

Fig.4. Theoretical responses.

8086 CPU)in order to implement the image reconstruction.

A single wire target was prepared as the object for reference. Based on the observed reference signal as shown in Fig.3, the filter coefficient h_{opt}, with $\omega_l/2\pi$=1MHz and $\omega_h/2\pi$=8MHz, was calculated. While, the theoretical temporal impulse response f^t_{opt} and point spread function g_{opt} were calculated as shown in Fig.4(a) and (b), respectively. The theoretical resolution capability Δl was estimated as 0.14mm with the use of eq.(28).

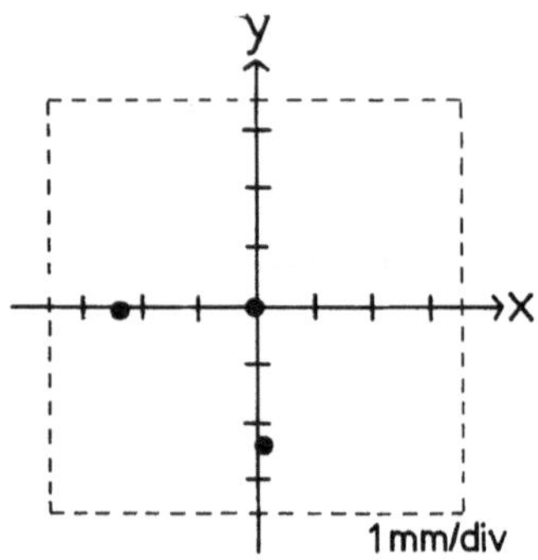

Fig.5. Arrangement of tungsten wire targets.

A test target consisting of tungsten wires as shown in Fig.5 was imaged. A total of 100 projections over a 180 degree aperture with 481 samples per projection were used. Images were reconstructed on a 321× 321 pixel grid for 7.1×7.1mm square region. The results are shown in Fig.6, where (a) was calculated based on the synthesized data, while (b) was based on the experimental data. As expected, three wires appear sharply exhibiting a uniform resolution in all directions. The resolution capability was estimated as 0.15mm for the result in (a) and 0.24mm for (b), respectively. Fairly good agreement is obtained between the result based on the synthesized data in Fig.6(a) and the theoretical value in Fig.4(b). Whereas, the resolution of the experimental result in Fig.6(b) is a little bit worse than these values, due to the lack of the accuracy in parallel alignment between three wires.

CONCLUSION

The FBP method has been adopted to the case where the reconstruction of a two-dimensional ultrasonic reflectivity image is made from a set of pulse-echo data surrounding an object. The deconvolution filter for the compensation of the temporal response of the system has been incorporated, where compensations are made only in the band-limited spectrum region so that the noise amplification can be mitigated. In particular, the moment of the point spread function has been minimized subject to the band-limited condition. As a consequence, the deconvolution filter has been optimized so as to maximize the resolution capability under the constraint of the small processing noise. Based on this result, the theoretical formula for the resolution capability by the proposed method

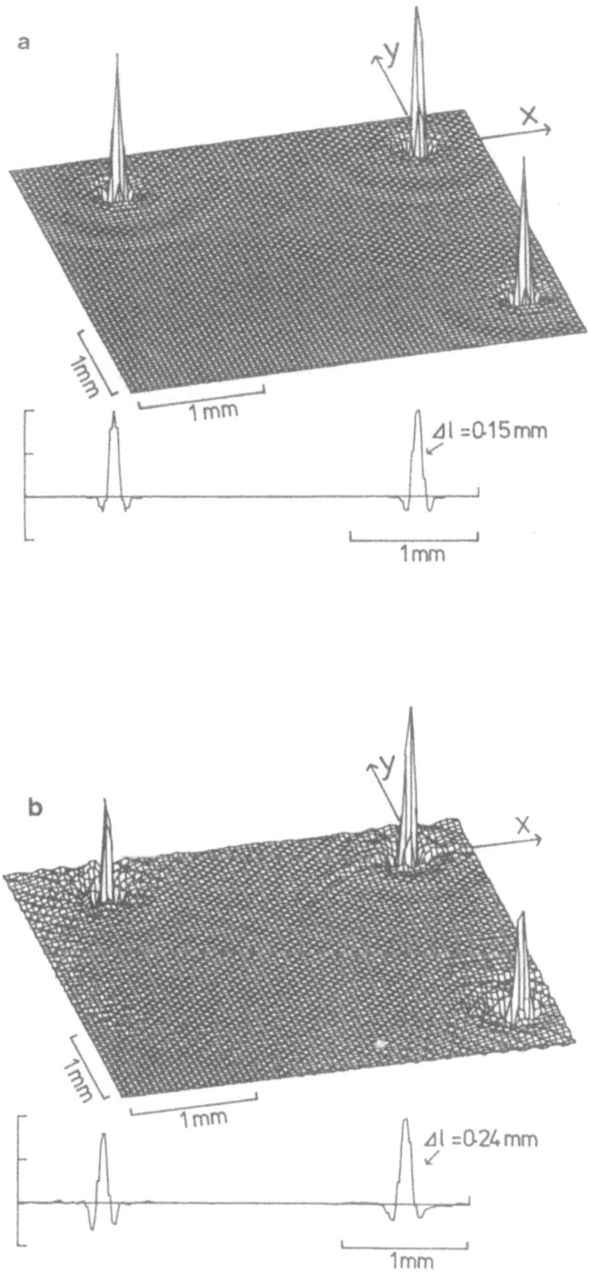

Fig.6. Reconstructed images. (a)Image based on the synthesized data. (b)Image based on the experimental data. 161×161 pixel portions of the entire 321×321 pixel images are displayed. The lower insets are the cross-sectional plot through the two peaks near x axis.

has been presented. It has been demonstrated that isotropic high-resolution image in proportional to the bandwidth of the temporal response is achievable. At last, experimental evaluations have been done to verify that high-resolution images comparable to the theoretical limit can be obtained.

In an practical application, it is frequently impossible to collect the data over a 180 degree angle of view. In such cases, image has to be reconstructed with the projection data over a limited view angle. Authors intend to carry out the further investigations towards these directions.

ACKNOWLEDGMENTS

This work was partially supported by Scientific Grant-in Aid from the Ministry of Education of Japan, and the Mechanical Industry Development & Assistance Foundation.

REFERENCES

[1] S.J.Norton and M.Linzer: "Ultrasonic reflectivity tomography: reconstruction with circular transducer arrays", Ultrasonic Imaging, 1, pp.154-184 (1979).

[2] G.Wade, S.Elliott, I.Khogeer, G.Flesher,J.Eisler, D.Mensa, N.S.Ramesh and G.Heidbreder: "Acoustic echo computer tomography", Acoustical Imaging, 8, pp.565-576(1980).

[3] S.A.Johnson, J.F.Greenleaf, B.Rajagopaln and M.Tanaka: "Algebraic and analytic inversion of acoustic data from partially or fully enclosing apertures", Acoustical Imaging, 8,pp.577-598 (1980).

[4] S.J.Norton: "Reconstruction of a two-dimensional reflecting medium over a circular domain: Exact solution", J.Acoust.Soc.Am., 67, 4, pp.1266-1273 (1980).

[5] G.Maderlechner, E.Hundt, E.Kronmuller and E.Trautenberg: "Experimental results of computerized ultrasound echo tomography", Acoustical Imaging, 10, pp.415-425 (1982).

[6] M.Moshfeghi:"Ultrasound reflection-mode tomography using fan-shaped-beam insonification", IEEE Trans.Ultrason.Ferroelect. & Freq.Cont., UFFC-33,3, pp.299-314 (1986).

[7] K.A.Dines, S.A.Goss: "Computed ultrasonic reflection tomography", IEEE Trans.Ultrason.Ferroelect. & Freq.Cont., UFFC-37,3, pp.309-318 (1987).

[8] H.Yamabe,A.Yamada: "Ultrasound reflection mode reconstruction tomography by incorporating the deconvolution signal processing", Paper of the Technical Group on Ultrasonics, IECE Japan, US87-61, pp.53-60 (1988).

[9] A.Yamada :"On-line deconvolution for the high resolution ultrasonic pulse-echo measurement with narrow-band transducer", Proc. 1987 IEEE Ultrasonics Symposium, pp.1027-1030(1987).

AN APPLICATION OF AR MODEL TO MULTIFREQUENCY HOLOGRAM

Toyokatsu Miyashita

Department of Electrical Engineering
Kyoto Institute of Technology
Kyoto 606, Japan

INTRODUCTION

The extrapolation of short available data or incomplete data with the aid of the prediction theory has been vigorously investigated to obtain a better resolution of impulse response, especially in the field of seismology [1], and to get a fine resolution of point-targets in the reconstructive holographic imaging[2-5].

One method of the extrapolation used in seismology is by L_1-norm minimization[1], where the band-limited original data is extrapolated in the frequency domain in both directions, i.e., down to zero frequency and up to Nyquist's frequency, so that the impulse response in the time domain may have the minimum L_1-norm keeping the given band-limited frequency component unchanged. One merit of this method is the possibility of more complicated objective functions than the simple L_1-norm. The biggest demerit is its computer-consuming calculation.

Another means of the extrapolation used in seismology is application of an autoregressive (AR) model to the band-limited spectrum of the data[1]. AR coefficients of the model are usually determined by the Burg method using Levinson's recursive formula, and the given band-limited frequency component is actually extrapolated using the AR coefficients. The principle is involved in maximum entropy method (MEM). Although this method has a difficulty of having additional constraints, it requires only short computation and, further, it is almost insensitive to the order of the AR model because the actual finite data-extrapolation is adopted, and the optimum order is easily determined by the L_1-norm of the reconstructed impulse response[6].

We proposed an application of the AR model to the direct extrapolation of the multifrequency hologram matrix (MFHM)[7] of few frequency elements, and demonstrated improvement of the reconstructed images, especially, resolution of unresolved images of point-targets by this method[2-4]. This report discusses, in detail, how the frequency-array component of the multi-frequency hologram matrix can be extrapolated by application of the AR model and how the extrapolated MFHM with a widened frequency aperture reconstructs images of fine resolution and good image quality.

APPLICATION OF AR MODEL TO HOLOGRAPHIC IMAGING

Exactly speaking, there are two different ways of applying the AR model to the reconstructive holographic imaging. The difference exists in the way of data-extrapolation. One makes an actual finite extrapolation of the hologram data according to the calculated AR prediction coefficients in the first stage, and images are then reconstructed by an appropriate transformation of the extrapolated hologram[2-4]. The other executes a direct imaging using the prediction coefficients from the well-known prediction formula of power-spectrum density which corresponds to the intensity-image of the holographic imaging[5].

In the latter case, the extrapolation is made equivalently infinitely; therefore, the optimum prediction order should be found. Otherwise, as known well in the MEM theory, the reconstructed point-like images, which correspond to sharp narrow line-spectra in the usual application of MEM theory, have not only absolutely but also relatively incorrect intensities depending on the prediction order, and, in the worst case, one point splits into two points with too high prediction orders. Another demerit is the long computation time necessary to calculate the prediction formula of power-spectrum density.

In the former case, the actual finite data-extrapolation is almost insensitive to the prediction order above a critical value and also it is stable. In addition, the imaging can be performed quickly by FFT. In this report, the actual data-extrapolation is applied to the MFHM.

AUTOREGRESSIVE PROPERTY OF THE MULTIFREQUENCY HOLOGRAM MATRIX

The multifrequency hologram matrix is shown to have an autoregressive property in its frequency components. A wide-aperture antenna transmits a two-dimensional plane wave of a frequency f_k $(k=1,2,\ldots,K)$ and illuminates objects composed of point-targets. The lth point-target has a reflection coefficient S_l $(l=1,2,\ldots,L)$. The complex amplitude $H_k(r)$ detected by each receiver element at r at each frequency element f_k of a frequency array is recorded, and an operation on $H_k(r)$, which is called MFHM[7], reconstructs an image of the objects.

The MFHM is described as[7]

$$H_k^L(r) = \sum_{l=1}^{L} S_l \exp[-2\pi i Y_l(r)\bar{f}_k] , \tag{1}$$

where $Y_l(r)$ is the two-way distance from the transmitter aperture to the lth point-target and back to a receiver element at r, which is measured in the wavelength of the central frequency f_c of the frequency array, and $\bar{f}_k$ is the normalized frequency, i.e., $\bar{f}_k = f_k/f_c$.

It is shown below by mathematical induction that this series of quantities $H_k{}^L(r)$ $(k=1,2,\ldots,K)$ fulfills an AR equation. The following abbreviation is used for simplicity:

$$B_l \equiv S_l \exp[-2\pi i Y_l(r)\bar{f}_k], \tag{2}$$

$$\phi_l \equiv 2\pi Y_l(r)\Delta f , \tag{3}$$

where Δf is the normalized element spacing of the frequency array.
1) When $L=1$, i.e., only one point-target exists,

$$H_k^1(r) = B_1 , \qquad H_{k-1}^1(r) = B_1 \exp[i\phi_1] ,$$

therefore, obviously,

$$H_k^1(r) = a_1^1(r) H_{k-1}^1(r) , \quad (4)$$

where

$$a_1^1(r) = \exp[- i\phi_1] .$$

This shows that adjacent frequency components are related by a unit complex coefficient $a_1^{\,1}(r)$, i.e., by a constant-angle rotation in the complex plane. 2) Now let the following formula hold for $L-1$,

$$H_k^{L-1}(r) = \sum_{l=1}^{L-1} a_l^{L-1}(r) H_{k-l}^{L-1}(r) , \quad (5)$$

then, by a simple calculation, the following formula is derived.

$$H_k^L(r) = \sum_{l=1}^{L} a_l^L(r) H_{k-l}^L(r) , \quad (6)$$

where

$$a_l^L(r) = (-)^{l+1} \sum_{m=1}^{L} \prod^{l} \exp[-i\phi_m] . \quad (7)$$

Here, definition of the notation $\sum_{m=1}^{L} \Pi^l b_m$ is as follows. We take l different elements b_m's out of $\{b_m | m=1,2,\ldots,L\}$, and multiply them to get a term. We make the term for every case of taking l different elements out of L elements, and sum up all the terms. It has been proved that AR equation (6) holds for all positive integer of L.

The equation (6) is deterministic. However, information about the object, especially the number of point-targets, is a priori unknown. Therefore, the following AR equation (8) with an unknown order N should be considered.

$$H_k^N(r) = \sum_{l=1}^{N} a_l^N(r) H_{k-l}^N(r) . \quad (8)$$

If $N > L$, Eq.(8) is overdetermined, and if $N < L$, it is underdetermined. Therefore, it is considered that the optimum order of the AR model L should be found by some appropriate criterion. However, when the coefficients $a_l^{\,N}$'s are determined by the Burg method, the final result, i.e., the reconstructed images from the extrapolated hologram is almost insensitive to the order N above a critical value and the extrapolation is also very stable without divergence. These properties are investigated, somewhat in detail, in the following sections.

COMPUTER-SIMULATION STUDY OF THE EXTRAPOLATION OF MFHM

The following multifrequency holographic imaging system is simulated. The aperture width of the transmitter is virtually infinite. The receiver array, which is placed on the same line as the transmitter, has 32 elements with an equal element-spacing of $(4/3)\lambda_c$, where λ_c is the wavelength of the central frequency of the frequency array, which will have 32 frequency elements and a total aperture of $0.15 f_c$ when it is extrapolated. These arrays are so designed that the point-spread of the reconstructed image at $140\lambda_c$ in front of the antenna is $3.74\lambda_c$ with the weight function of the modified Taylor distribution of a side-lobe level of -30dB ($\bar{n} = 5$).

Here a comparison is made between images reconstructed from the MFHM H_{nk} ($n=1,2,\ldots,32$; $k=1,2,\ldots,32$) extrapolated from the incomplete MFHM of $K' = 16$ up to the full size of $K = 32$. The central part of the frequency array is assumed to be known.

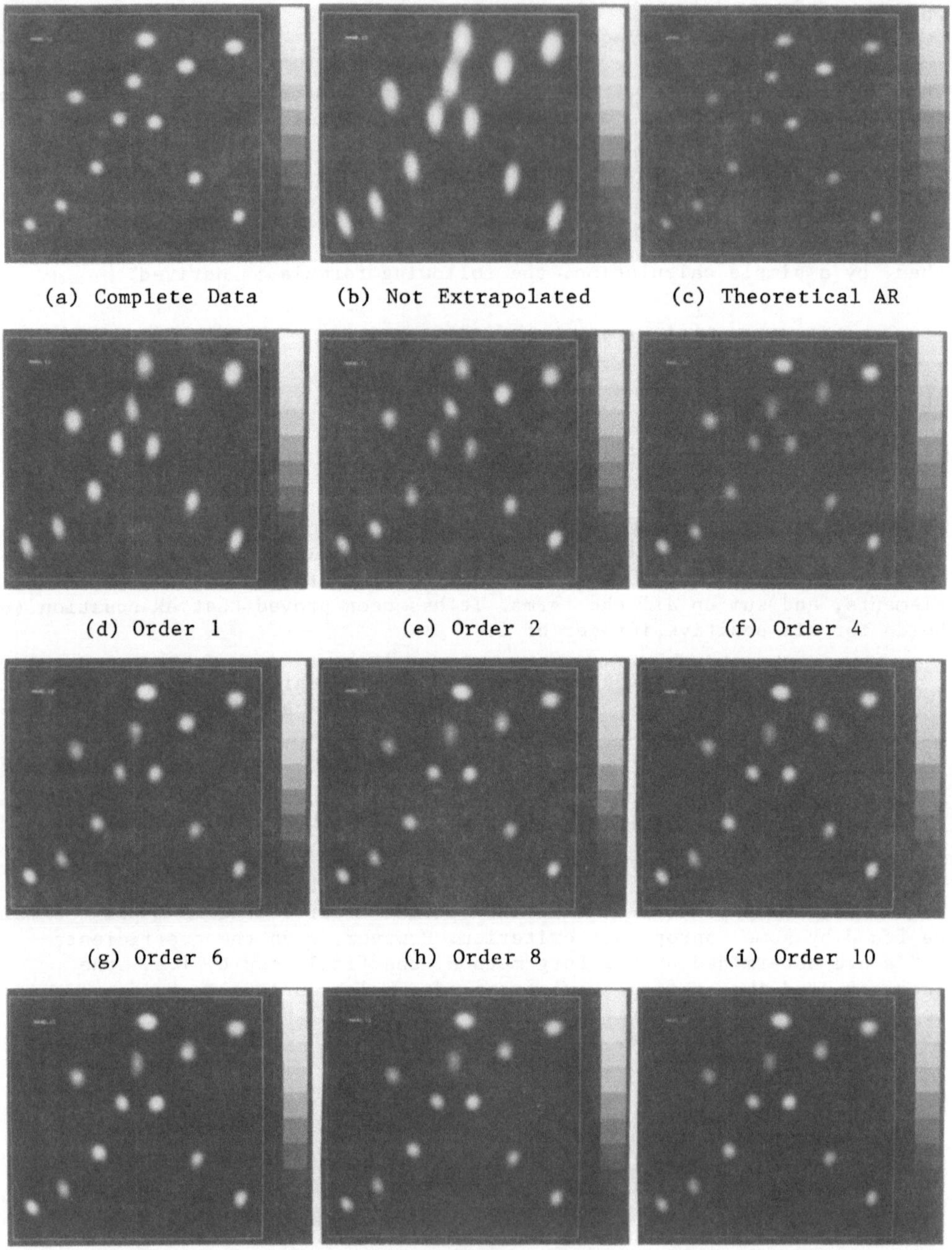

(a) Complete Data (b) Not Extrapolated (c) Theoretical AR

(d) Order 1 (e) Order 2 (f) Order 4

(g) Order 6 (h) Order 8 (i) Order 10

(j) Order 12 (k) Order 14 (l) Order 16

Fig.1. Reconstructed images of 12 point-targets with 22 effective bits for the frequency elements. (a)Reconstructed image from the original complete MFHM (N=32, K=32). (b)Reconstructed image from the non-extrapolated MFHM (N=32, K=16). (c)Reconstructed image from the MFHM extrapolated by Eq.(7) from K'=16 to K=32. (d)-(l)Reconstructed images from the MFHM extrapolated by Burg method with different orders from K'=16 to K=32.

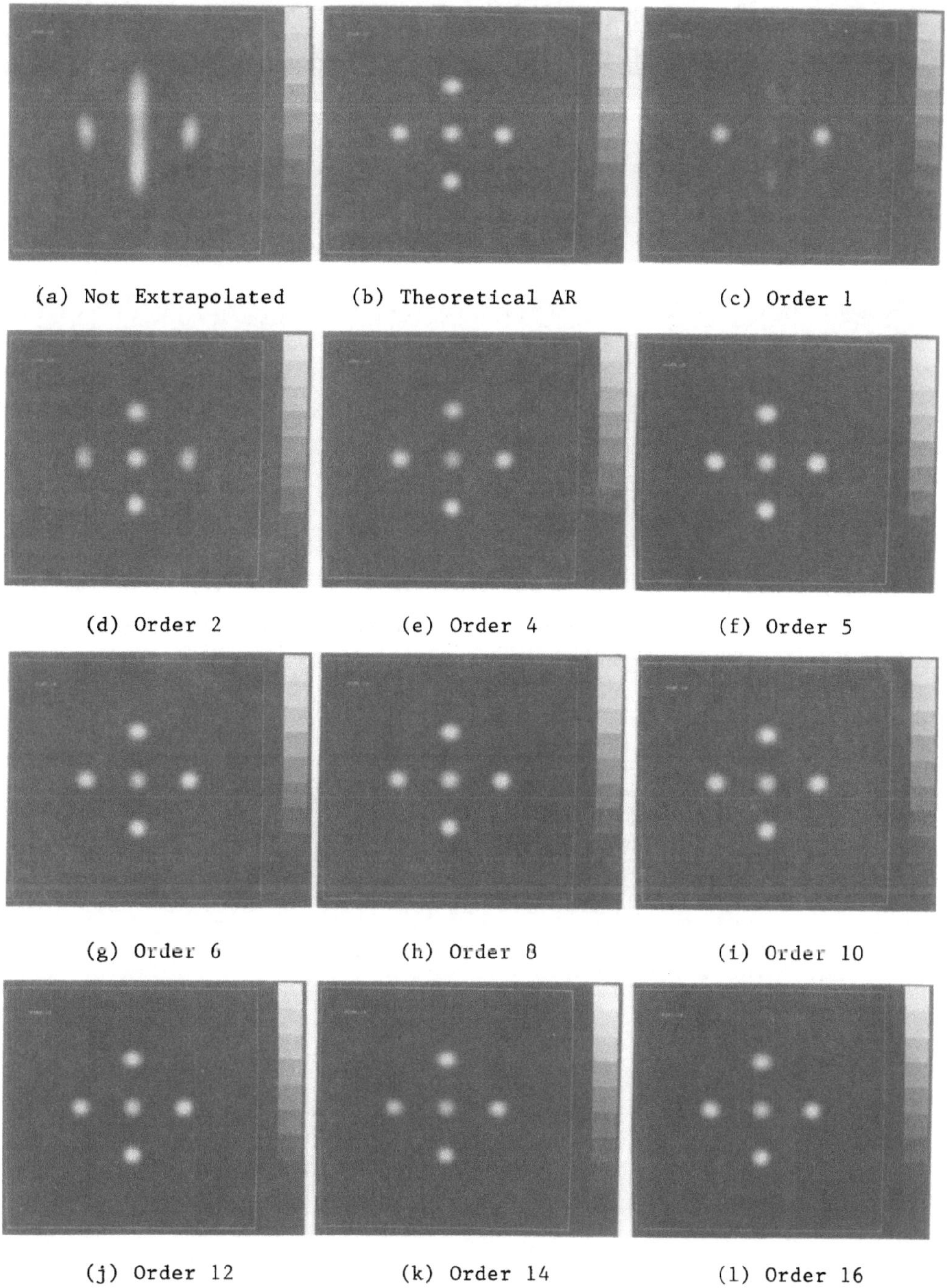

Fig.2. Reconstructed images of 5 point-targets with 22 effective bits for the frequency elements. (a)Reconstructed image from the non-extrapolated MFHM (N=32, K=16). (b)Reconstructed image from the MFHM extrapolated by Eq.(7) from K'=16 to K=32. (c)-(l)Reconstructed images from the MFHM extrapolated by Burg method with different orders from K'=16 to K=32.

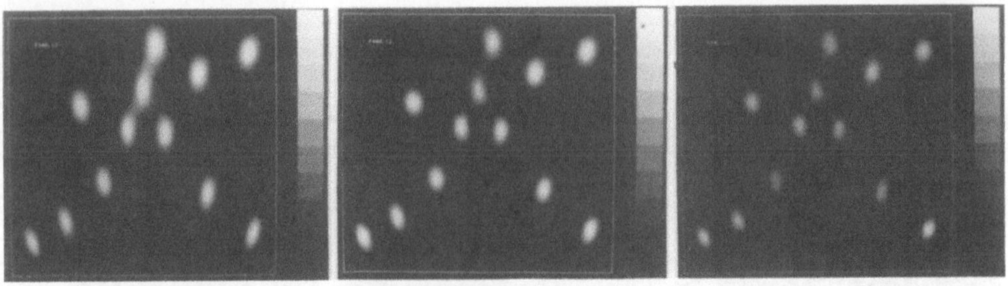

(a) Not Extrapolated (b) Order 1 (c) Order 4

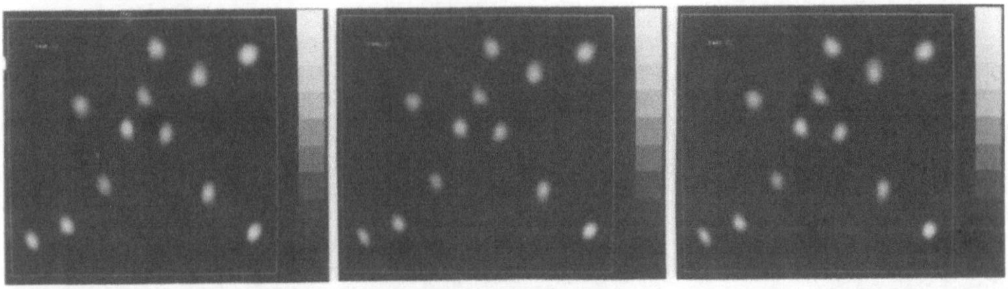

(d) Order 8 (e) Order 12 (f) Order 16

Fig.3. Images of 12 point-targets with 10 effective bits for frequencies. (a)Image from nonextrapolated MFHM. (b)-(f)Images from MFHM extrapolated by Burg method with different orders from $K'=16$ to $K=32$.

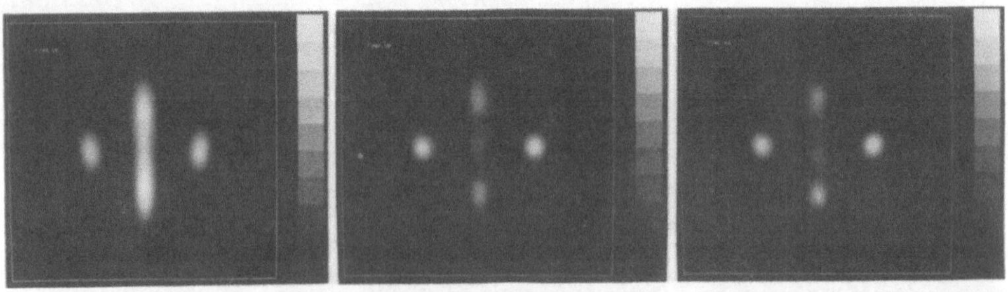

(a) Not Extrapolated (b) Order 6 (c) Order 8

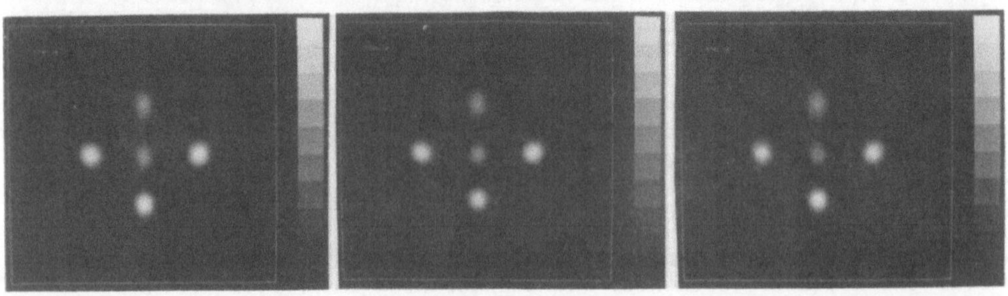

(d) Order 10 (e) Order 12 (f) Order 16

Fig.4. Images of 5 point-targets with 10 effective bits for frequencies. (a)Image from nonextrapolated MFHM. (b)-(f)Images from MFHM extrapolated by Burg method with different orders from $K'=16$ to $K=32$.

Table 1. AR Coefficients and Magnitudes of MFHM of 5 Point-Targets with 22 Effective Bits for Frequency Elements (at $n = 16$)

AR Coefficients by Eq.(7) in Order			AR Coefficients by Burg Method		
(1.953, 4.231), (-5.829, 6.831)			(1.739,-3.555), (-4.204,-5.457)		
(-8.644,-2.435), (-.7304,-4.602)			(-6.392, 1.254), (-.8971, 3.234)		
(.8310,-.5562)			(.5747, .5162)		

Magnitude of Holograms			Magnitude of Holograms		
k	True Value	Extrapolated	k	True Value	Extrapolated
1	257.0276	260.7502	1	257.0276	255.7246
2	214.8667	217.5028	2	214.8667	215.5872
3	171.2644	171.8731	3	171.2644	162.9728
4	192.2026	190.9531	4	192.2026	175.9371
5	244.1192	242.6671	5	244.1192	231.4203
6	262.0559	261.0120	6	262.0559	255.0957
7	228.5249	227.9620	7	228.5249	225.7922
8	174.0228	173.8517	8	174.0228	173.4677
9	174.1462	=	9	174.1462	=
10	230.4130	=	10	230.4130	=
11	266.9075	=	11	266.9075	=
12	250.3105	=	12	250.3105	=
13	193.0662	=	13	193.0662	=
14	156.3206	=	14	156.3206	=
15	191.6229	=	15	191.6229	=
16	234.9807	=	16	234.9807	=
17	232.9899	=	17	232.9899	=
18	185.7494	=	18	185.7494	=
19	149.5759	=	19	149.5759	=
20	192.0104	=	20	192.0104	=
21	253.9706	=	21	253.9706	=
22	272.6423	=	22	272.6423	=
23	234.2695	=	23	234.2695	=
24	166.3384	=	24	166.3384	=
25	144.4939	146.7762	25	144.4939	146.4356
26	189.5020	198.7536	26	189.5020	195.0501
27	218.8689	240.5348	27	218.8689	227.9646
28	197.0022	239.4500	28	197.0022	210.6515
29	145.2549	216.2999	29	145.2549	167.5224
30	148.4299	224.6598	30	148.4299	174.7138
31	219.2348	276.0627	31	219.2348	237.3240
32	272.2991	324.8325	32	272.2991	279.5437

Figures 1-4 show the effect of the extrapolation of the hologram which appears in the reconstructed images. The upward direction of the figures is the range direction from the antennas, which is resolved by the frequency array. In Figs.1 and 3, 12 point-targets are distributed in a square region of $64\lambda_c \times 64\lambda_c$. In Figs.2 and 4, five point-targets are placed at (0, 0), (0, $\pm 10\lambda_c$), and ($\pm 10\lambda_c$, 0). Intensity-images in Figs.1 and 2 are reconstructed from the holograms of the frequency array with 22 effective bits, and those in Figs.3 and 4 are with 10 effective bits. The images reconstructed from the nonextrapolated original holograms have an overlapping on the center line and those points are not resolved. The extrapolated holograms resolve clearly these points.

1) Robustness of Multifrequency Holographic Imaging

In the examples shown here, exact hologram-extrapolation is obtained only for MFHM of five point-targets with 22 effective bits for frequency as shown in Table 1. All the others have at least 10%, usually about 50% error in magnitude. Non the less, reconstructed images from those holograms

Table 2. AR Coefficients and Magnitudes of MFHM of 5 Point-Targets with 10 Effective Bits for Frequency Elements (at $n = 16$)

AR Coefficients by Eq.(7) in Order	AR Coefficients by Burg Method
(2.084, 4.166), (-5.385, 7.177)	(.3667,-.5402), (.1643,-.0632)
(-8.827,-1.611), (-1.300,-4.472)	(.0564, .1199), (-.0205,-.0045)
(.7340,-.6792)	(-.0433, .1170)

Magnitude of Holograms			Magnitude of Holograms		
k	True Value	Extrapolated	k	True Value	Extrapolated
1	258.2368	166403.7	1	258.2368	10.38683
2	216.0382	132042.8	2	216.0382	10.74946
3	170.9972	98484.56	3	170.9972	15.76506
4	194.9227	67645.50	4	194.9227	24.91109
5	247.7361	41617.45	5	247.7361	47.01135
6	262.5754	21951.81	6	262.5754	39.84712
7	230.7831	9128.879	7	230.7831	70.22818
8	174.2415	2395.660	8	174.2415	117.1423
9	175.4901	=	9	175.4901	=
10	234.2427	=	10	234.2427	=
11	265.4609	=	11	265.4609	=
12	252.9442	=	12	252.9442	=
13	194.7083	=	13	194.7083	=
14	156.3421	=	14	156.3421	=
15	194.1754	=	15	194.1754	=
16	236.9041	=	16	236.9041	=
17	235.2517	=	17	235.2517	=
18	188.4665	=	18	188.4665	=
19	149.5639	=	19	149.5639	=
20	193.8220	=	20	193.8220	=
21	256.8230	=	21	256.8230	=
22	271.2319	=	22	271.2319	=
23	238.4372	=	23	238.4372	=
24	168.2943	=	24	168.2943	=
25	144.6249	2045.315	25	144.6249	155.6969
26	191.5551	8435.480	26	191.5551	91.20824
27	219.0753	20723.97	27	219.0753	62.22340
28	200.9507	39638.16	28	200.9507	50.02625
29	147.5753	64696.29	29	147.5753	29.58859
30	147.6184	94371.94	30	147.6184	9.010986
31	220.7911	126616.7	31	220.7911	2.844512
32	273.8040	159561.4	32	273.8040	2.805549

have exactly essential features of the targets.

2) Effect of Extrapolation of the Frequency-Array Components of MFHM

As shown in Figs.1-4, the extrapolated holograms reconstruct images of good resolution. With Burg method, in general, and also by the theoretical AR coefficients (7), the magnitude of the hologram components increases about 50% generally as they are extrapolated towards both high-frequency and low-frequency sides. This causes a little bit sharper point-spread of the reconstructed images than that of the original images from the complete data.

In the practical application, the possibility of the hologram-extrapolation is meaningful, especially in the following sense. The phase characteristics of the practical measuring system is not satisfactorily uniform all over the entire frequency aperture. Usually, only within a narrow bandwidth, uniform phase characteristics is obtained. Then the extrapolation from the part of uniform frequency dependence is very meaningful.

3) Robustness of Extrapolation with the Aid of Burg Method

All examples shown here assume that the central part of the frequency array, 16 elements, has hologram data, and that the whole part of the transducer array has data. Extrapolation of the hologram is made for each transducer element from the 16 known hologram data for the frequency array up to 32 hologram data.

In this direct extrapolation of the hologram, the reconstructed images are always stable, and their dependence on the order of Burg method is almost ideal, i.e., for example in Fig.1, the reconstructed image is already almost satisfactory with the order of 4, and with the order higher than 8, it remains unchanged, and the same behavior can be seen also in Figs.2-4.

4) Dependence of the Extrapolated Hologram on the Accuracy of the Incomplete Original Hologram

We considered two kinds of holograms to investigate the dependence of the data-extrapolation on the accuracy of the original data. One is the hologram data calculated using the frequency array data of 22 effective bits (about 6 effective digits), and the other is of 10 effective bits (about 3 effective digits).

The extrapolation by the theoretical AR coefficients (7) is very sensitive to the accuracy of the given data. With the frequency accuracy of 22 effective bits, the extrapolation recovers the hologram very exactly but with a small increment at the higher frequencies (k=25-32) as shown in Table 1. With the frequency accuracy of 10 effective bits, however, the extrapolation is completely divergent and destructive.

On the other hand, the extrapolation with Burg method is insensitive to the accuracy of the given data as shown in Figs.1-4. Even with the frequency accuracy of 10 effective bits, the extrapolation recovers the hologram very stably, and the reconstructed image becomes better as the prediction order increases until the image stops changing.

5) Relation between the Theoretical AR Coefficients (7) and Prediction Coefficients given by Burg Method

It should be noticed that the AR coefficients calculated from the theoretical formula (7) and those estimated by Burg method are completely different not only theoretically but also numerically as shown in Tables 1 and 2. However, Burg method gives very good and stable data-extrapolation as shown in Table 2.

CONCLUDING REMARKS

It was derived that the frequency-array components of the multi-frequency hologram matrix fulfill an autoregressive relation. Therefore, extrapolation of the hologram of few frequency elements can widen the frequency aperture of the hologram and improve the resolution of the reconstructed image or recover the original image.

Burg method was shown to be a robust one which gives the coefficients of the autoregressive relation. Calculation of the coefficients are performed effectively using Levinson's recursive formula and the hologram extrapolation itself is also very rapid by numerical calculation by computer. Burg method gives very stable nondivergent extrapolation even if the accuracy of the hologram data is poor and even if the prediction order is too high. Reflecting the stable behavior of the extrapolated hologram data, reconstructed image is also stable and has a good image-quality.

In the course of the derivation of the autoregressive relation of the multifrequency hologram matrix elements, we obtained the theoretical formula of autoregressive coefficients. The coefficients, which can be obtained only if all the relevant information about the object is completely known, can not give any stable extrapolation of the hologram of an insufficient accuracy of the frequency elements.

We had expected that L_1-norm minimum of the reconstructed image will give us a good criterion for the determination of the optimum prediction order also in this multifrequency holographic imaging as in the case of seismic data[6]. However, it proved not true, because we look for the impulse response in the seismic-data processing, and we allow a finite point-spread in the holographic imaging. L_1-norm measure should be applied on the data which have no finite point-spread.

Fortunately, the extrapolation of the hologram data by Burg method needs not such an additional criterion. Increasing the prediction order, the reconstructed image has better resolution and it stops changing practically.

REFERENCES

[1] D.W.Oldenburg, T.Scheuer, and S.Levy: Geophysics 48 (1983) 1318.
[2] T.Miyashita: Paper of Technical Group of IECE Japan (1986) MW86-1 [in Japanese].
[3] T.Miyashita: Japan. J. Appl. Phys. 26 Suppl. 26-1 (1987) 70.
[4] T.Miyashita: IEEE Proc. Letters (to be published).
[5] T.Yamamoto and Y.Aoki: IECE Japan J65-C (1982) 390 [in Japanese].
[6] T.Miyashita, H.Schwetlick, and W.Kessel: Acoustical Imaging (Plenum Press, New York, 1985) ed., A.J.Berkhout et al., vol. 14, p. 247.
[7] T.Miyashita, J.Nakayama, and H.Ogura: Acoustical Imaging (Plenum Press, New York, 1980) ed., K.Wang, vol. 9, p. 23.

ULTRASONIC PHASE CONJUGATOR USING MICRO PARTICLE SUSPENDED CELL AND ITS APPLICATION

Takuso Sato, Hiromi Kataoka, Takeaki Nakayama and Yoshiki Yamakoshi

Graduate School at Nagatsuta, Tokyo Institute of Technology, 4259 Nagatusta, Midori-ku, Yokohama-shi 227 Japan

1. INTRODUCTION

The phase conjugated wave for a given wave is defined as the wave which has the time-reversed wavefront of the original wave, hence it has many promising applications such as a real-time compensation of phase turbulence or an imaging without lens. To obtain the ultrasonic phase conjugated wave, several methods have been proposed.[1,3] For instance, N.P.Andreeva et.al. proposed a method which is based on the use of small deformations of a liquid surface for the applied ultrasonic waves. This method, however, uses three waves simultaneously, that is signal, reference and phase conjugated waves, hence, the scattered waves of high intensity signal and reference waves are necessarily superposed as the extra noises over the desired weak phase conjugated waves. Moreover, the phase conjugated wave's direction is fixed to the downward direction from the surface because the liquid surface is used. The later may give severe restriction when it is applied to practical system.

In this paper, a new ultrasonic phase conjugator which is based on the use of real-time hologram generated in a micro particles suspended liquid cell is proposed. When both the signal and reference waves are applied to the cell, the micro particles with the refractive index different from that of the liquid are subjected to redistribution by the acoustic radiation force due to the interference field of the two waves resulting in the generation of the corresponding phase hologram. Then for the read out wave applied to the cell from just the opposite direction of the reference wave, the desired phase conjugated wave of the signal wave is generated. In this method, the read out wave can be applied at any time after stopping high intensity signal and the reference waves, hence the phase conjugated wave with high signal to noise ratio can be obtained easily at the desired time, moreover, there is no restriction for the orientation of the phase conjugator.

2. PRINCIPLE

2.1 Definition of Phase Conjugated Wave

The phase conjugated wave for a given wave is defined as the wave which has the time reversed wavefront for the original wave, that is, for a given wave $\mathbf{E}(\mathbf{r},t)$;

$$\mathbf{E}(\mathbf{r},t) = E \exp[j(\mathbf{kr} + \phi - \omega t)] \quad (1)$$

the phase conjugated wave $\mathbf{E}^*(\mathbf{r},t)$ is given by

$$\mathbf{E}^*(\mathbf{r},t) = E \exp[j(-\mathbf{kr} - \phi - \omega t)] \quad (2)$$

where $\mathbf{r}$ is the position vector, $\mathbf{k}$ is the wave number vector and ϕ is a phase shift.

2.2 Basic Principle of Generation of Phase Conjugated Wave by using Micro Particles Suspended Liquid Cell

Figure 1 shows a schematic construction of the micro particles suspended cell. When the signal wave $\mathbf{E}_1$ and the reference wave $\mathbf{E}_2$ are applied to the cell with the angle θ, the interference field generated in x direction is given by

$$\begin{aligned} I_x &= \mathbf{E}_1\mathbf{E}_2^* + \mathbf{E}_1^*\mathbf{E}_2 \\ &= 2\,E^2 \cos k_x x \end{aligned} \quad (3)$$

where $k_x = (2\pi/\lambda)\sin\theta$ and λ is the wavelength.

Now, if radius of the micro particles a is small enough so that

$$k_x\, a \ll 1 \quad (4)$$

is satisfied, then the radiation force which is applied to the particles is given by[4]

$$\begin{aligned} F(x) = \pi a^2\, 4(k_x a)\cos(2k_x x) \\ [(2E)^2/(\rho_0 c_0{}^2)]\, f(\varepsilon,\sigma) \end{aligned} \quad (5)$$

where

$$f(\varepsilon,\sigma) = \frac{\varepsilon + \{2(\varepsilon-1)/3\}}{1 + 2\varepsilon} - \frac{1}{3\varepsilon\sigma^2} \quad (6)$$

and $\varepsilon = \rho_p/\rho_1$, $\sigma = c_p/c_1$, ρ and c are the density and the sound velocity with the subscripts 1 and p representing the liquid and the particle, respectively. Hence, the equation of motion of the particle in x direction is derived as follows

$$\frac{4}{3}\pi a^3 \rho_1 \frac{d^2x}{dt^2} + 6\pi a \eta_1 \frac{dx}{dt} = F(x) \quad (7)$$

where η_1 is the viscosity of the liquid. In the derivation of eq.(7), the effect of the particle's collisions is neglected.

The redistribution of the particles as the result of their movements generates the hologram in the cell. Its transmission coefficient is given by

$$T = KI_x \quad (8)$$

Hence, if the read out wave $\mathbf{E}_3 = \mathbf{E}_2^*$ is applied to the hologram from the opposite direction, then the phase conjugated wave of the signal wave is generated as follows

$$\mathbf{E}_c = KE^2\mathbf{E}_1^*$$

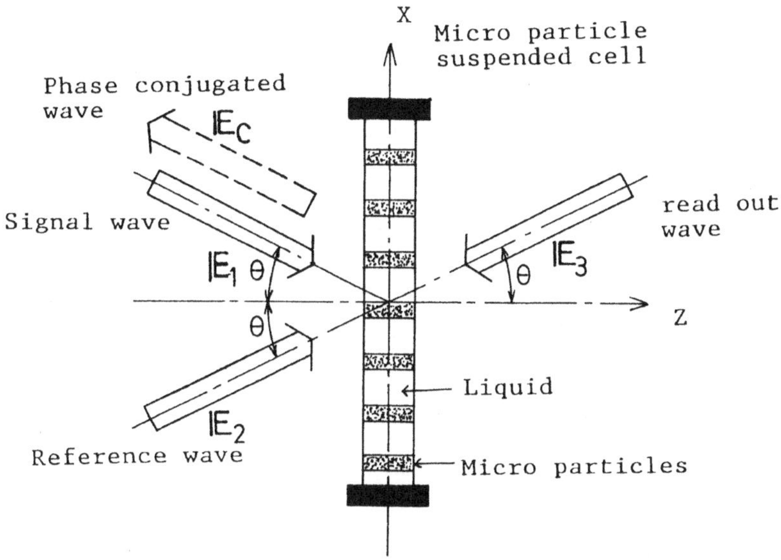

Fig.1 Schematic explanation of generation of ultrasonic phase conjugated waves by using micro particles suspended liquid cell.

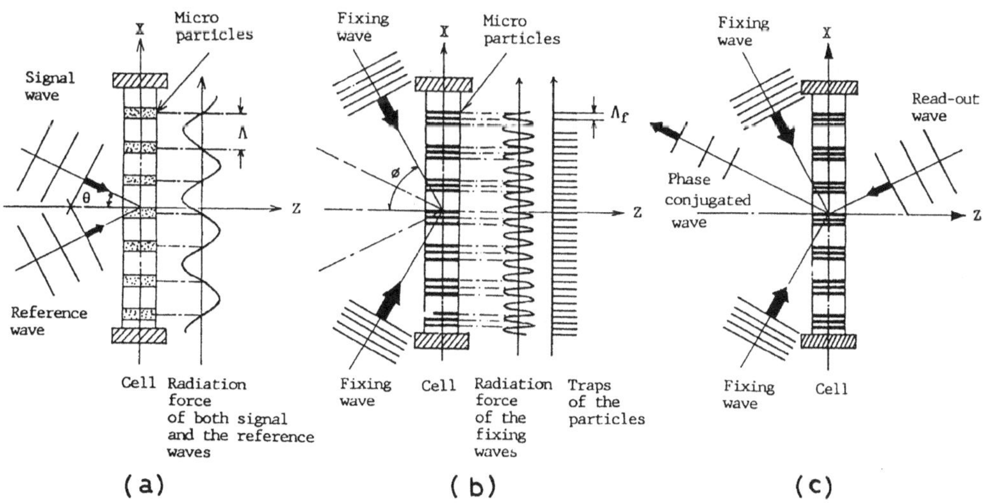

Fig.2 Mechanism of hologram fixing. (a) : write in state, (b) : hologram fixing state and (c) : read out state.

2.3 A Method of Fixing the Hologram

In the proposed method, when the read out wave is applied to the cell to get the phase conjugated waves, the hologram will be destructed by the radiation force of the applied read out waves resulting in the decrease of the amplitude of the generated phase conjugated waves.

To avoid the destruction of the hologram in read out state, a method of fixing the hologram is considered. It is based on the use of additional two ultrasonic waves (hologram fixing waves). The mechanism of the hologram fixing is shown schematically in fig.2. After generation of hologram (fig.2(a)), the hologram fixing waves with frequency much higher than that of the signal and reference waves are applied with the angle ϕ which is chosen so that $\phi>\theta$ (fig.2(b)). Then the acoustic radiation force due to the hologram fixing waves gives further redistribution of the micro particles resulting in the periodic distribution of the particles with the period Λ_f which is much smaller than that of the original hologram Λ. In this case, the ratio of the force due to the hologram fixing waves F_f and the force due to the read out wave F_r can be written as follows

$$r = \frac{F_{f,max}}{F_r} = \frac{6}{\sigma^2-1}\frac{k_{f,x}}{k^4 a^3 \sin\theta}\left(\frac{E_f}{E_r}\right)^2 \tag{10}$$

where $F_{f,max}$ is the maximum amplitude of the F_f; $k_{f,x}$ is the wave number of the fixing waves in x direction; E_f and E_r are the amplitudes of the fixing waves and read out wave, respectively. From eq.(10), we find that the force due to the hologram fixing waves is much larger than that due to the read out wave provided that the particle's diameter a is so small that $a^3 \ll (k_{f,x}/k^4)$ is satisfied. Hence, in this case, the hologram is fixed even when the read out wave is applied. In other words, it may be said that the particles in the original hologram are caught by many traps with very small intervals generated by the hologram fixing waves.

3. EXPERIMENTS

Experimental set-up for observing the phase conjugated wave is shown in fig.3. T_{r1}, T_{r2}, T_{r3} are the ultrasonic transducers for signal, reference and read out waves, respectively. Additional transducers T_{r4} and T_{r5} are used for the hologram fixing. The acrylic spheres of 60 μm diameter are used as the particles and they are suspended in the saline solution whose specific gravity is adjusted so that it is equal to that of the particles.

The processes of phase conjugated wave generation are as follows; first the signal and reference waves are applied to the cell from T_{r1} and T_{r2} to produce the hologram in the cell (write in state), then after stopping these waves, the hologram fixing waves are applied from T_{r4} and T_{r5} to fix the hologram (hologram fixing state). Next if the read out wave is applied from T_{r3}, then the phase conjugated wave is generated and it is observed by using T_{r1} (read-out state). In the preliminary experiments, the following parameters are used; volume ratio of the particles is 10 %; frequency of the signal, reference and read out waves is 1.0 MHz; frequency of hologram fixing waves is 3.5 MHz.

Figure 4 shows the photographs of the holograms generated in the cell. Fig.4(a) is the initial state, (b) shows the state at 60 sec after starting the write in process and (c) shows the state at 40 sec after starting the read out process. In this experiment, hologram fixing waves

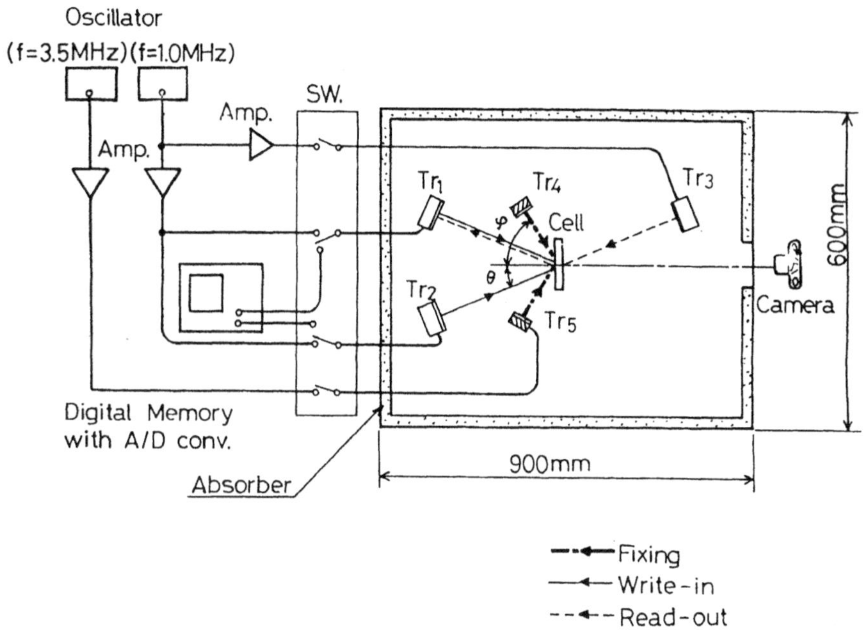

Fig.3 Experimental set up for observing the phase conjugated wave. T_{r1}, T_{r2} and T_{r3} are the ultrasonic transducers for signal, reference and read out waves, respectively. Additional transducers T_{r4} and T_{r5} are used for hologram fixing.

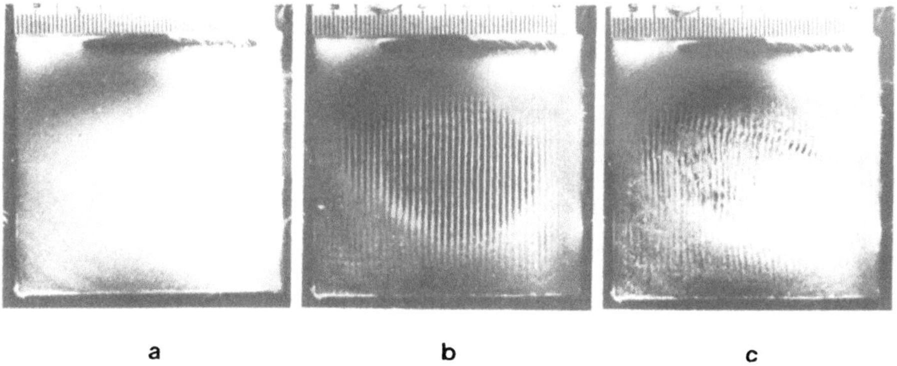

Fig.4 Photographs of the holograms produced in the cell. (a) : initial state, (b) : write in state (after 60 sec) and (c) : read out state (after 40 sec).

are not applied. These results show clearly the processes of formation and destruction of the hologram.

Figure 5 shows the efficiency of the generation of the phase conjugated wave which is given by

$$\eta = (P_c / P_t) \tag{11}$$

where P_c is the amplitude of the phase conjugated wave and P_t is the amplitude of the directly transmitted wave which is measured by using T_{r2}. Fig.5(a) and (b) show the increase of the amplitude of the phase conjugated wave in write in state and its decrease in read out state, respectively. From the results, it is clear that the hologram formation is completed within about 30 sec and the hologram is destructed rapidly due to the radiation force of the read out wave.

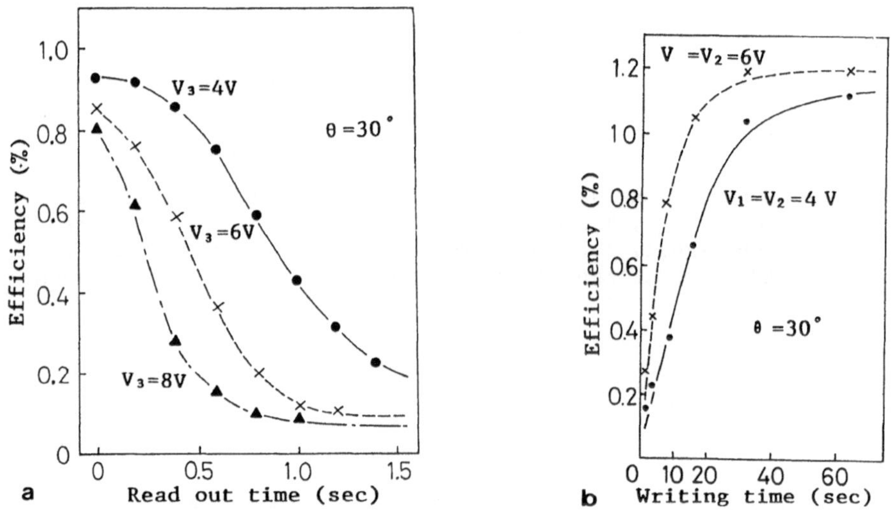

Fig.5 Efficiency of generation of the phase conjugated wave. (a) : write in state and (b) : read out state.

Next, to demonstrate the usefulness of the hologram fixing method, the efficiency of the generated phase conjugated wave in read out state is measured by adding the hologram fixing waves. Figure 6 shows the experimental results for several values of applied voltage to the T_{r4} and T_{r5}. From the results, it is clear that the hologram is not destructed by the radiation force of the read out wave and the continuous phase conjugated wave can be obtained provided the applied voltage to T_{r4} and T_{r5} is higher than 2V.

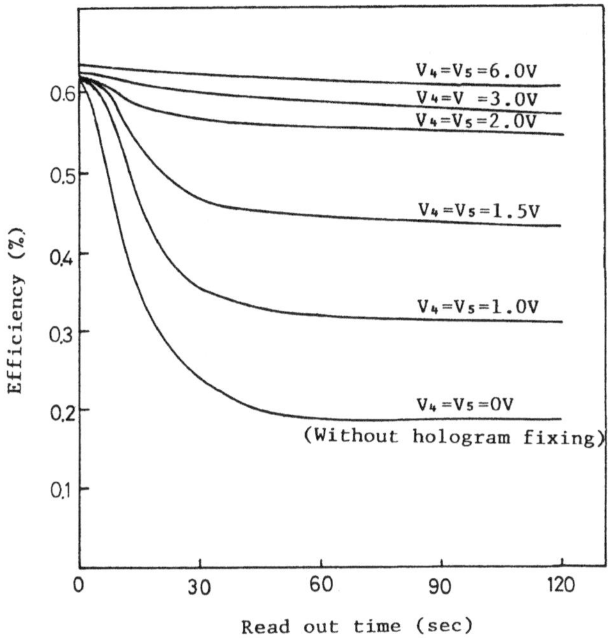

Fig.6 Efficiency of generation of the phase conjugated wave for different applied voltages to the hologram fixing transducers T_{r4} and T_{r5}.

4. APPLICATIONS

4.1 Lens Less Imaging using Phase Conjugated Wave

Figure 7 shows the experimental set-up to observe images formed by the phase conjugated wave. In the experiment, first, the object is placed at the plane 0 and the hologram is stored in the cell by applying both signal and the reference waves. Then, after stopping these waves, the hologram fixing waves are applied from T_{r4} and T_{r5} to fix the hologram. Next the object is removed and ultrasonic transducer array is placed instead of it to get the object's image formed by the phase conjugated wave. In the experiments, the frequency of the signal, reference and read out waves is 3.0 MHz and their incident angle is 25 deg; the frequency of the hologram fixing waves is 4.5 MHz and their incident angle is 60 deg.

Figure 8 shows an example of the experimental results. Figure 8(a) shows the object (rectangular aperture of 12mmx13mm), fig.8(b) is the image which is observed by setting ultrasonic transducer array at the cell plane in write in state. From the figure, the effect of degradation of the image due to the diffraction effect of the waves is seen. Fig.8(c) is the image formed by the phase conjugated wave observed at the plane 0 in read out state. We can see fairly clear image. Fig.8(d) shows the numerical results of the image obtained by the ideal phase conjugated wave by taking into account the aperture size of the read out wave at the cell plane. From the results, we may say that almost perfect phase conjugated wave is generated by the constructed phase conjugator.

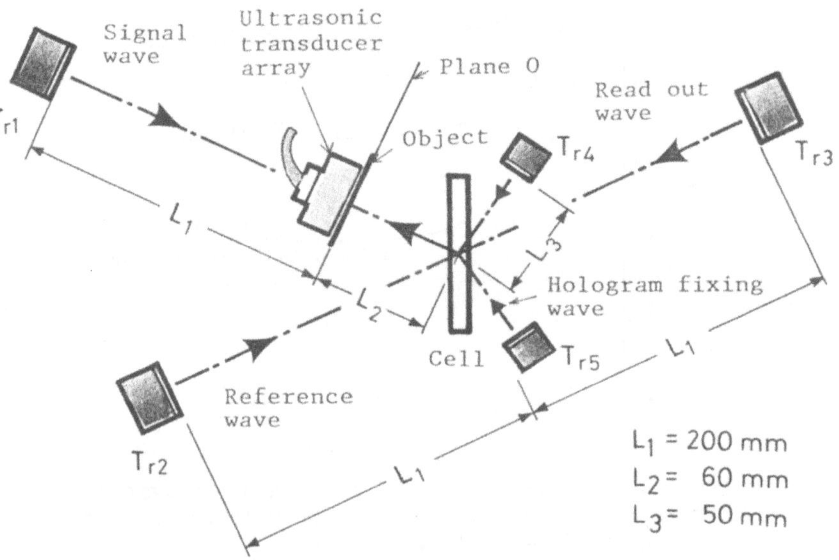

Fig.7 Experimental set up for lens less imaging system.

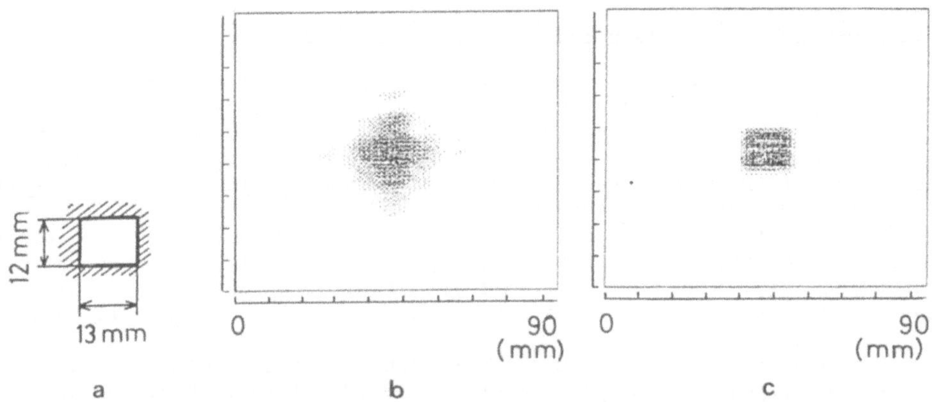

Fig.8 Image formed by the phase conjugated wave. (a) : object, (b) : image at the cell plane and (c) : image formed by the phase conjugated wave.

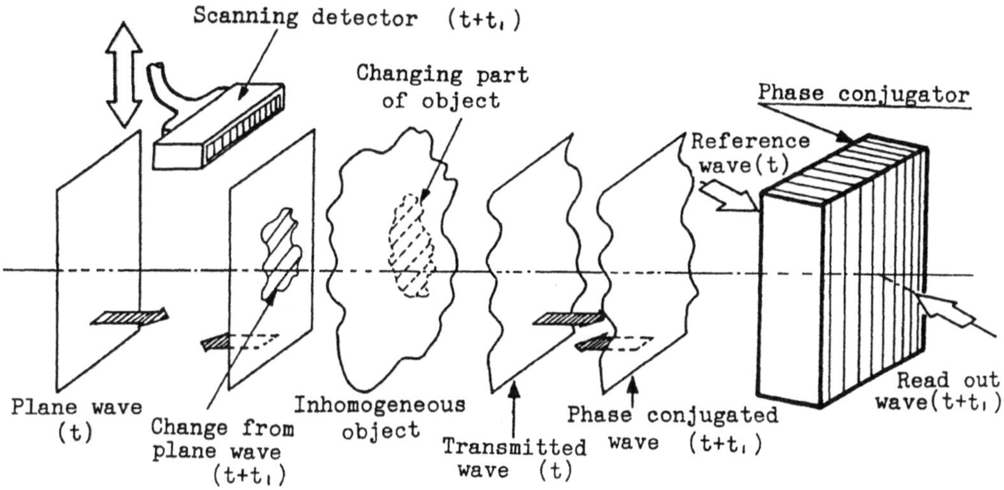

Fig.9 Schematic diagram of dynamic interferometric novelty imaging system.

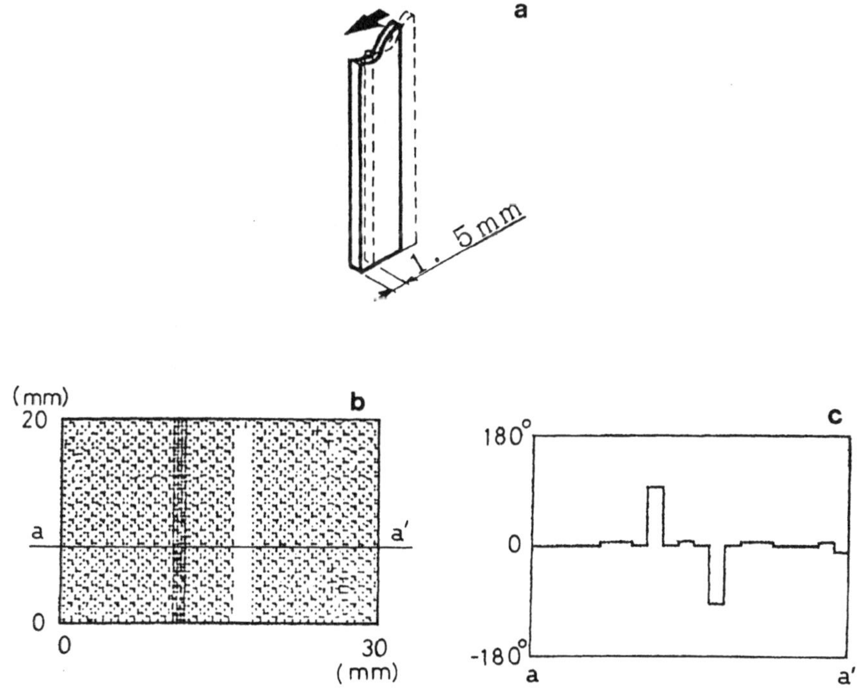

Fig.10 Experimental result of the novelty imaging system. (a) : object (object is an acrylic plate bar (phase object) and it is moved between time t and $t+t_1$), (b) : observed phase image and (c) : cross sectional image along a-a' line in fig.(b).

4.2 Dynamic Interferometric Novelty Imaging

Figure.9 shows a schematic diagram of dynamic interferometric novelty imaging system which can extract only the changing parts of an inhomogeneous object between any given times t and $t+t_1$. The principle of the method is as follows; when the uniform plane wave (signal wave) is passed through the inhomogeneous medium (object) at time t=t, the complex amplitude at the output plane is given by

$$S(x,y:t) = a(x,y:t)\exp(j\phi(x,y:t)) \tag{12}$$

where a(x,y:t) and $\phi(x,y:t)$ are the object's amplitude transmission coefficient and the phase transmission coefficient, respectively. This wavefront is stored as the hologram in the cell in write in state. Then the read out wave is applied at time $t=t+t_1$, the phase conjugated wave passed through the object from the reverse direction gives the following wavefront at the detecting plane;

$$R(x,y:t) = a(x,y:t)a(x,y:t+t_1) \times \exp[j\phi(x,y:t+t_1) - j\phi(x,y:t)] \tag{13}$$

Hence, if the phase of the received wave is detected, for instance, only the changing parts of the object's phase can be extracted regardless the stationary complex inhomogeneous structures of the object. In other words, only the novel parts of the object are imaged by cancelling its complex stationary parts.

Figure 10 shows an example of the phase image when a bar of acrylic resin is used as the object and is moved a little between time t and $t+t_1$ as is shown in fig.10(a). Figure 10(b) is the phase image observed at plane 0 in read out state, and fig.10(c) is the cross sectional image along a-a' line in fig.10(b). In this case, the image corresponds to the object's phase change, that is, $\phi(x,y:t+t_1) - \phi(x,y:t)$. From the result, we can clearly see the desired characteristics of the system as a novelty imaging.

5. CONCLUSION

A new ultrasonic phase conjugator which is based on the use of the real-time hologram produced in a micro particle suspended liquid cell was proposed. This method has a special feature that the phase conjugated wave with high signal to noise ratio can be obtained easily. A method of hologram fixing by using additional ultrasonic waves were also introduced to avoid the destruction of the hologram in the read out state. As the applications of the phase conjugator, a lens less ultrasonic imaging system and a dynamic interferometric novelty imaging system were shown.

REFERENCES

1. B.Ya.Zel'dovich, "Principles of phase conjugation," Springer-Verlag, N.Y. (1984).
2. B.Ya.Zel'dovich, N.F.Pilipetsky, A.N.Sudarkin and V.V.Shkunov, Wave-front reversal by an interface, Dokl.Akad.Nauk USSR, 252:92 (1980).
3. N.P.Andreeva, F.V.Bunkin, D.V.Vlasov and K.Karshiev, Experimental observation of acoustic phase conjugation at a liquid surface, Pis'ma Zh.Tekh.Fiz., 8:104 (1982).
4. K.Yoshioka and Y.Kawashima, Acoustic radiation pressure on a complete sphere, Acoustica, 5:167 (1955).

SONIC VELOCITIES IN BONE AT 10GHZ FREQUENCIES

S.Lees*, N-J. Tao** and S.M. Lindsay**

*Bioengineering Department
Forsyth Dental Center
140 Fenway, Boston MA 02115,USA

**Physics Department, Arizona State University
Tempe AZ 85287 USA

ABSTRACT

Brillouin light scattering is a noncontacting means for determining the sonic velocity with wavelengths of 100-1000 nm, comparable to the length of a collagen molecule. It was used previously to measure sonic velocity in soft tissues. Mineral crystallites embedded in mineralized tissue induce Raleigh and Mie scattering, which can obscure the weaker inelastically scattered light in the Brillouin spectrum. In addition the water line in wet tissues can dominate the spectrum, making it difficult to assign the spectral lines associated with the collagen.

A high performance nine pass tandem Fabry-Perot interferometer provided sufficient contrast to yield results in most instances for 50 to 100 um thick mineralized and demineralized specimens. A description of the equipment is presented, together with results for rat tail tendon collagen, mineralized turkey leg tendon, deer antler and cow tibia. The specimens were tested on two axes, along the collagen fibrils and perpendicularly to the fibril axis. The tests were performed on wet and air dried specimens, mineralized and after demineralization.

INTRODUCTION

Brillouin light scattering represents the inelastic scattering of light by thermally excited elastic waves propagating in the medium, an effect that was predicted by Leon Brillouin in 1922. Brillouin scattering is a technique for detecting a very narrow band of all these vibrations in the lowest order modes, the acoustic branch. It is sensitive to the direction of wave propagation and can distinguish between longitudinal and transverse acoustic propagation modes by the polarization of the scattered light. Since it is a very weak effect, it requires an instrument with high resolution, high sensitivity and high contrast.

While Brillouin light scattering is used primarily in the study of liquids and inorganic crystals there have been some successful applications on biological materials. Rat tail tendon fibers collagen was investigated by at least four groups (1,2,3,4). Cornea collagen and eye lens crystallins were studied by Randall and Vaughan (5,6). Lindsay and his group(7,8) have been employing the technique to investigate DNA molecular dynamics and configurational transformations of DNA. It is of particular interest in biological systems because the elastic wavelength is comparable to the length of large linear proteins. Some investigators have hoped to determine the intra- and intermolecular force constants in collagen but this goal has not yet been attained (1,3).

PERTINENT EQUATIONS AND EXPERIMENTAL CONSIDERATIONS

Light interacts with high frequency elastic waves because periodic fluctuations in the density of the medium modulates the dielectric constant. The details of the interaction can be found in several ways, but it may be regarded as a Doppler shift as the light goes through a moving modulation of the medium. The inelastically scattered light has components with frequencies

$$f_s = f_i \pm f_e \qquad (1)$$

f_s = inelastic scattered frequency
f_i = incident light frequency
f_e = elastic wave frequency

The velocity of the elastic wave can be determined from the expression

$$c = \frac{f_e \lambda}{2n \sin(\theta/2)} \qquad (2)$$

where θ = internal scattering angle (the angle between the incident and scattered light beams)

λ = wavelength of the incident light falling on the specimen
n = index of refraction of the specimen

When as in Fig 1 the optical path is symmetrical, Snell's Law is applicable and Eq(2) becomes

$$c = \frac{f_e \lambda}{2 \sin \psi/2} \qquad (3)$$

where ψ = angle of incidence . For the 45° - 45° geometrical arrangement

$$c = \frac{f_e \lambda}{\sqrt{2}} \qquad (4)$$

$\lambda = \lambda_i n_m$, λ_i = laser wavelength,

n_m = index of refraction of the medium, which is 1.0 for air and 1.33 for water. In this experiment a single mode argon-ion laser was used with wavelength 514.5 nm.

Referring to Fig 1 again, four possible sources for inelastic scattering are displayed schematically. The "A" path produces the primary line for which Eq (4) is applicable. Since the specimen has a finite thickness, there are reflections at each of its two surfaces which may scatter light collected by the spectrometer. The reflected light can, in turn, induce inelastic scattering elsewhere. The first reflection from the incident plane can generate a line from the medium surrounding the specimen, designated as "B". When the medium is water this line can be very strong. A second reflection from the excident plane can produce a secondary inelastic line from the specimen, marked "C". A fourth possible line has its source in the enveloping medium from the light beam that is transmitted through the specimen ("D"). The several expressions shown in the figure are appropriate for data reduction.

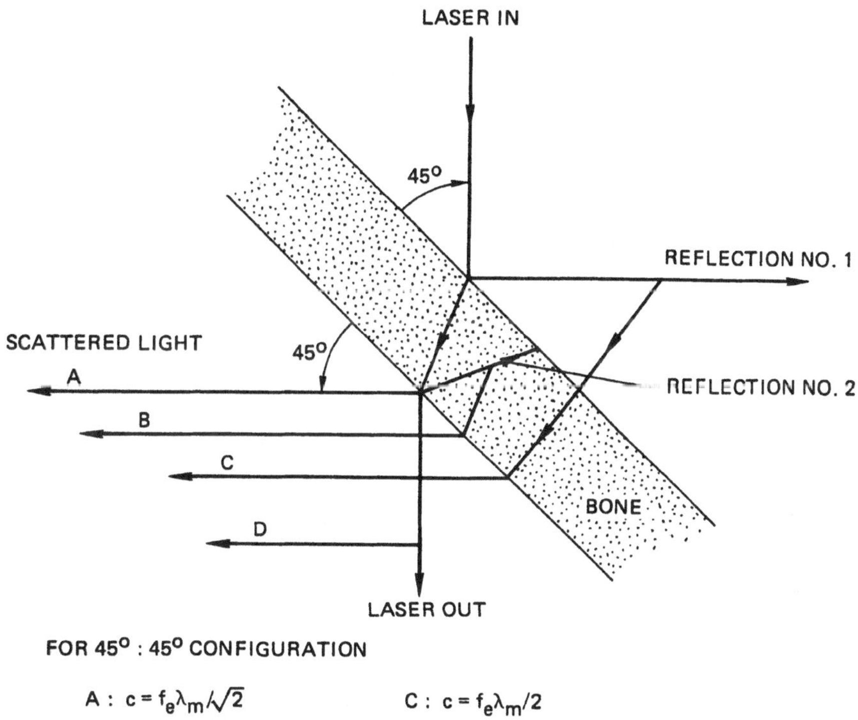

FOR 45° : 45° CONFIGURATION

A : $c = f_e\lambda_m/\sqrt{2}$ C : $c = f_e\lambda_m/2$

B : $c \cong f_e\lambda_m/2n$ D : $c = f_e\lambda_m/\sqrt{2}$

Fig 1. Geometry of the sample for Brillouin scattering experiment. Note the four possible sources of light scattering. The expressions show how to calculate the sonic velocity, c, from the Brillouin shift, f_e, the laser wavelength in the external medium, λ, and the index of refraction of the sample, n.

THE HIGH PERFORMANCE FABRY-PEROT INTERFEROMETER

While 10 GHz represents a very high acoustic frequency it is not so optically. It is necessary to employ a high resolution spectrometer to distinguish the inelastic scattering line from the very bright elastic central line in the spectrum. The previously cited workers used a three pass Fabry-Perot interferometer, but the resolution and contrast capability of that system was inadequate for the Brillouin spectral line emitted by bone and demineralized bone. The nine pass tandem Fabry-Perot interferometer was used successfully for most of the conditions in which these materials were examined (9). More complete details of the instrument are given in the reference. The arrangement of the instrument seen in Fig 2, is provided to give the reader an appreciation of instrumental design. Light from the argon laser (upper right corner) is focussed onto the sample in the lower right corner. The scattered light is collected by a colimating lens system and passed through a pair of Fabry-Perot interferometers which share a common scanning stage. The inclination of the interferometers results in different spectral ranges of that the tandem combination act as a pair of incommensurate comb filters in frequency. Thus only a unique combination of orders is simultaneously transmitted by both. This eliminates the overlappin-orders problem of the conventional Fabry-perot, allowing the full resolution of the instrument to be used. The common scanning stage maintains the

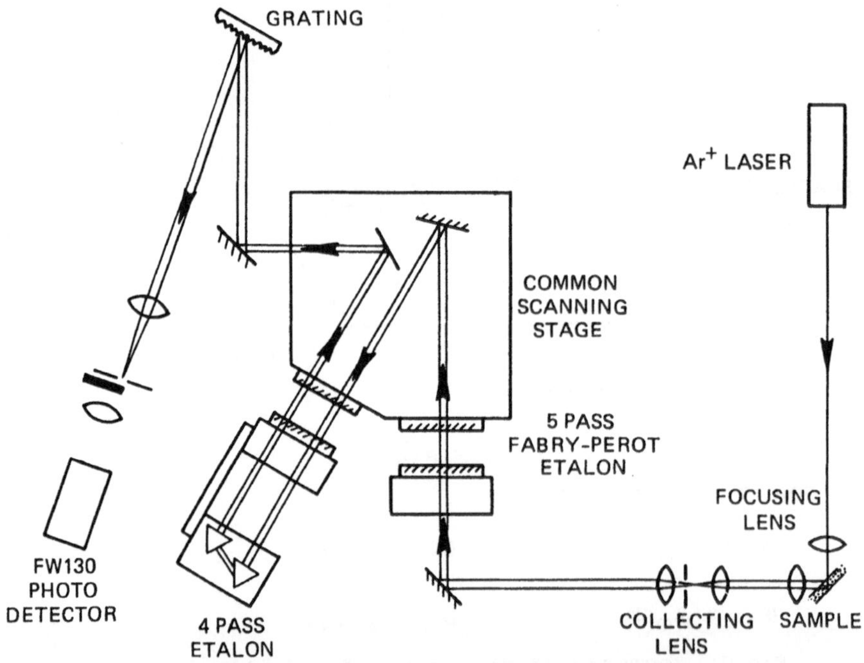

Fig 2. Schematic layout of the tandem nine pass Fabry-Perot interferometer. The first etalon is a five pass, the second etalon is a four pass Fabry-Perot interferometer. The common scanning stage, activated by a piezoelectric bimorph motor, surveys the available spectrum divided into 512 bins. More complete details are given in the reference (9).

coincidence of the overlapped orders during a scan as described in the reference (9). Each Fabry-Perot is, in addition, multipassed, which yields a large increase in the contrast of the instrument. Light from the interferometer is further filtered by a diffraction grating and then focussed onto a photon counting photomultiplier. Photon count rate is recorded as a function of scan position on a multichannel scaler.

The high performance instrument was shown in the reference to have an ultimate resolution of 10^{-3} cm^{-1}, a maximum scanning range free of overlapping orders of 50 cm^{-1} and a contrast greater than 10^{17}. For this application where the inelastic scattering line is so broad, the inaccuracy was estimated to be about 5% of the observed sonic velocity.

MATERIALS

Three kinds of material were examined: mineralized turkey leg tendon (density 1.50 g/cc), deer antler (1.77 g/cc) and cow tibia (2.05 g/cc). The tissues were examined wet and after air drying, along the specimen axis and perpendicular to the axis. Some of the tissue was demineralized in 0.5M EDTA solution at pH 7.5 (to preserve the proteins). The tests were repeated with the demineralized tissues. The specimens were cut to expose the radial orientation. They were less than 100um thick, and they were sufficiently transparent to permit printing to be read when placed in contact with the page.

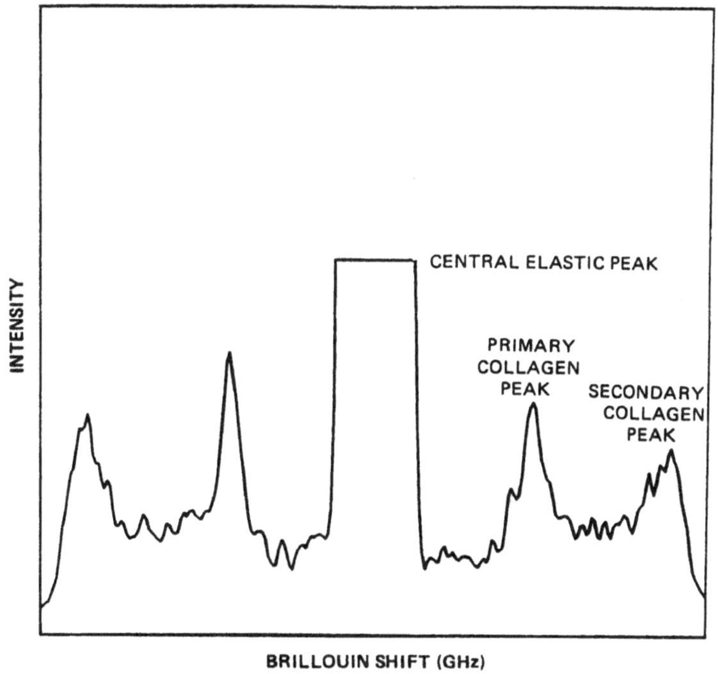

Fig 3. Brillouin spectrum of airdried rat tail tendon fiber parallel to the axis. Note both the primary and secondary Brillouin peaks were recorded. Since the specimen was mounted in air there are no peaks from the embedding medium.

EXPERIMENTAL RESULTS

Measurements were performed at room temperature using the 45°-45° geometry (14). The maximum power incident upon the sample chamber window or the air dried sample was 40mW, which was reduced when necessary to minimize damage to the sample.The results were reported with the electric vector either parallel or perpendicular to the specimen axis.

One specimen of airdried rat tail tendon fiber collagen was tested to provide comparison with the literature.

Typical spectra are shown in Figs 3 - 6. The one in Fig 3 corresponds to airdried rat tail tendon fiber collagen and may be considered to be a model for the other spectra. Rat tail tendon collagen is much more highly ordered than bone collagen and the Brillouin spectrum is quickly and readily displayed. The central peak is the elastically scattered line. The two inner peaks are the primary inelastically scattered lines, while the two outer peaks are the secondary lines. There are no water lines because the specimen was in air. The first requirement is to determine the value for f_e by utilizing Eq (1).

The sonic velocity is calculated from Eq (4). The index of refraction of the specimen when it is desired to determine the sonic speed corresponding to the secondary peaks. Instead, we found the index of refraction by assuming the sonic velocity is the same in "A" and "B" in Fig 1.

Fig 4 represents the spectrum of wet cow tibia, fully mineralized. The Brillouin spectral line is extremely broad and requires curve fitting to determine the locus of the peak. By contrast, Fig 5 is the spectrum for the same tissue when fully dried in air. The primary spectral line is very well defined and much narrower. The other part of the spectrum closer in to the central peak may be due to other modes of acoustical wave propagation like shear waves. The contrast between the two figures shows the strong influence of water on the properties of the mineralized tissue.

Fig 6 is the spectrum obtain from demineralized cow tibia that has been air dried. Again, there are several well defined peaks symmetrically distributed around the central elastically scattered line. This spectrum closely resembles that in Fig 4 for rat tail tendon collagen. The inner peaks are the primary inelastically scattered lines and the outer peaks are the secondary lines. The same equations that were used for rat tail tendon collagen were applied here and the index of refraction was obtained.

These spectra show that water and mineral strongly influence the dynamic characteristics of bone. It is not clear whether the effects are due to the variation of the optical properties of the tissue or that it is solely an indication of different elastic properties. These data will be discussed in greater detail in a forthcoming paper (14). The results are presented in tabular form, together with values obtained for similar tissues at lower frequencies. The 10 MHz data were obtained by pulse echo methods which yield the group velocity. All other data at higher frequencies report the phase velocity.

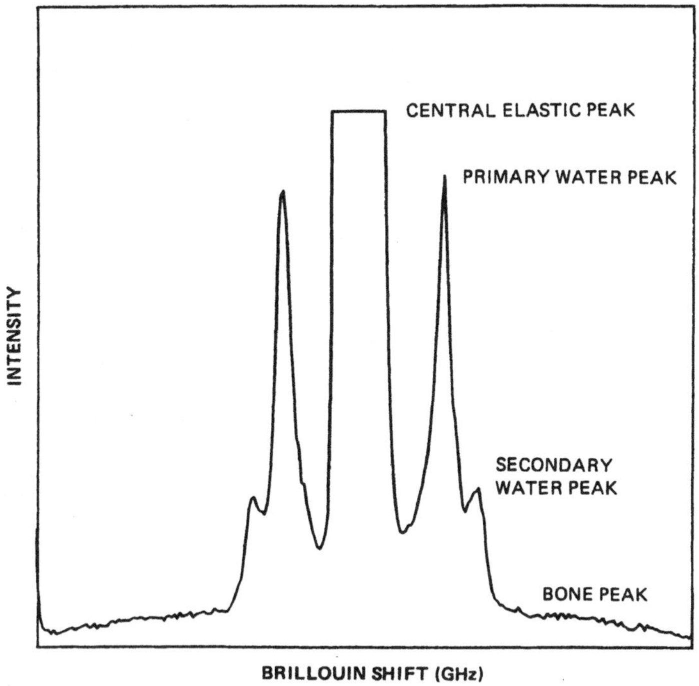

Fig 4. Brillouin spectrum of wet cow tibia along the bone axis. The water peaks dominate. The single bone peak is a broad spectral line outside the secondary water peak, identified because it modifies the shoulder of the water lines. The location of the bone peak and the Brillouin shift were found by fitting Lorentzians to the spectrum.

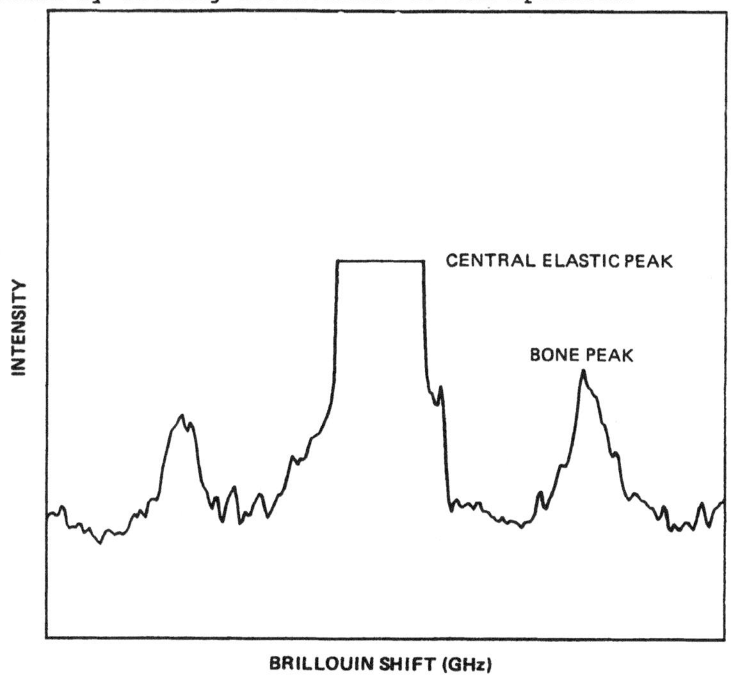

Fig 5. Brillouin spectrum of airdried cow tibia along the bone axis. There are no lines from the embedding medium. The bone peak is much narrower and better defined than in Fig 4.

The wave propagation mode can be distinguished in most cases by the orientation of the electric vector of the light beam. The laser beam is polarized which establishes the orientation of the incident beam. When the scattered light has the same polarization as the incident light (the V-V mode), the sonic wave is in the longitudinal mode. When the polarizer in the scattered light path is crossed (the V-H mode) it is the shear wave that is detected. We were unable to make use of this technique because there was no difference whether the polarizers were parallel or crossed. The dimensions of the mineral crystallites are comparable to the wavelength of light, as are the structures of the bone matrix, such as the collagen fibrils. Light is scattered by particles with optical wavelength dimensions. Mie scattering is dominant when the particles are larger than a light wave length, where Rayleigh scattering dominates for much smaller particles. Presumably the internal structure of the bone matrix scatters the light by a combination of Mie and Rayleigh processes so as to destroy the polarization of the existing light in demineralized tissue. The mineral crystallites apparently act in a similar fashion in the fully mineralized state.

It is assumed that the sonic velocities in the table refer to longitudinal wave propagation because the values are equal to or greater than the value for rat tail tendon collagen which are known to be longitudinal (3). The technique of Brillouin light scattering may be selective in that it detects the motion of the collagen moiety within the more complex structure of the mineralized tissue. The contributions of the other components of bone may be indirect. It is planned to obtain sonic velocity

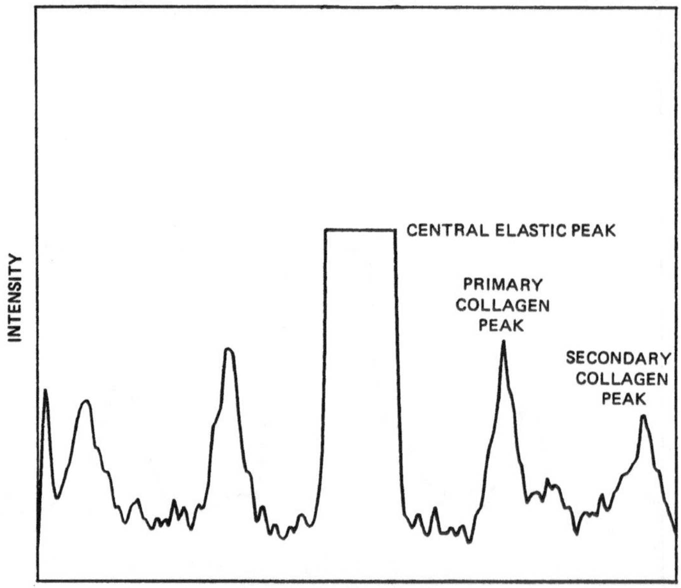

Fig 6. Brillouin spectrum of airdried demineralized cow tibia along the bone axis. Both primary and secondary collagen peaks were recorded. The spectral lines are as sharp and well defined as the corresponding lines in Fig 3 for rat tail tendon collagen.

measurements of mineralized tissues at ultrahigh frequencies by more conventional methods, where all of the components of the tissue share in the wave propagation. A comparison of the two kinds of results should provide information about the interaction of the mineral component with the organic structure.

TABLE 1

MINERALIZED TISSUES

Tissue		freq	ax	rad	dens	U^
Turkey leg tendon	Wet	11GHz	3.07	2.86km/s	1.50g/cc	5%
	Dry	11GHz	3.88	3.60*	1.55	10%
Deer antler	Wet	11GHz	3.17	2.81	1.77	5
	Dry	11GHz	3.82	3.46	1.77	2
Cow tibia	Wet	16GHz	4.06	3.64	2.06	5
	Dry	13GHz	4.86	3.88	2.08	2
Deer antler+	Wet	10MHz	3.08	2.40	1.70	1%
	Dry	10MHz	3.72	2.81	1.70	1
Cow tibia+	Wet	10MHz	4.18	3.32	2.06	1
	Dry	10MHz	4.47	3.49	2.08	1

* tissue burned during test U^ = uncertainty
+ from Lees et al. (10)

TABLE 2

SOFT & DEMINERALIZED TISSUES

Rat tail tendon	Wet*	7GHz	2.64	1.89	1.12	1%
	Dry	10GHz	3.70	2.94	1.25	1%
Turkey leg tendon	Wet	-----	----	----	----	---
	Dry	11GHz	4.0	3.8	1.3	2
Deer antler	Wet	8GHz	2.17	2.09	1.18	5
	Dry	10GHz	3.83	3.74	1.25	2
Cow tibia	Wet	-----	----	----	----	---
	Dry	10GHz	3.62	3.49	1.25	2
Rat tail tendon	Wet	100MHz	2.1+++	1.73++	1.12	1
Deer antler+	Wet	10MHz	1.88	1.69	1.18	1
Cow tibia +	Wet	10MHz	1.82	1.68	1.18	1

* Cusack and Miller (3)
+ from unpublished data
++ Goss & O'Brien (11)
+++ Lees et al. (12)

Acknowledgements - The work was supported in part by National Institute on Aging Grant 02325 (S.L.) and by ONR Grant N0001487K-0487 and NSF Grant BBS 8615653 (N.J.T & S.M.L.).

REFERENCES

1. Harley,R., James,D., Miller,A., & White,J.W. 1977
Phonons and the elastic moduli of collagen and muscle
Nature 267:285-287

2. Randall,J.T. & Vaughan,J.M. 1979
Brillouin scattering in systems of biological significance
Phil. Trans. R. Soc. Lond. A293:341-348

3. Cusack,S. & Miller, A. 1979
Determination of the elastic constants of collagen by Brillouin light scattering
J. Mol.Biol. 135:39-51

4. Cusack,S. & Lees,S. 1984
Variation of longitudinal acoustic velocity at gigaherz frequencies with water content in rat-tail tendon fibers
Biopolymers 23:337-351

5. Vaughan,J.M. & Randall,J.T. 1980
Brillouin scattering, density and elastic properties of the lens and cornea of the eye.
Nature 284:489-491

6. Randall,J.T & Vaughan,J.M. 1982
The measurement and intepretation of Brillouin scattering in the lens of the eye.
Proc. R. Soc. Lond. B214:449-470

7. Lee,S.A., Lindsay,S.M., Powell,J.W., Weidlich,T., Tao,N.J., Lewen, G,D, & Rupprecht,A. 1987
A Brillouin scattering study of the hydration of Li- and Na-DNA films
Biopolymers 26:1637-1665

8. Tao,N.J., Lindsay,S.M. & Rupprecht,A. 1987
The dynamics of the DNA hydration shell at gigaherz frequencies
Biopolymers 26:171-188

9. Lindsay,S.M., Anderson,M.W. & Sandercock,J.H. 1981
Construction and alignment of a high performance multipass vernier tandem Fabry-Perot interferometer.
Rev. Sci. Instrum. 52:1478-1486

10. Lees,S., Ahern,J.M. & Leonard,M. 1983
Parameters influencing the sonic velocity in compact calcified tissues of various species
J. Acous. Soc. Amer. 74:28-33

11. Goss,S.A. & O'Brien,W.D. Jr. 1979
Direct velocity measurements of mammalian collagen threads
J.Acous. Soc. Amer.85:507-511

12. Lees,S., Heeley,J.D., Ahern,J.M. & Oravecz,M.G. 1983
Axial phase velocity in rat tail tendon fibers at 100MHz by ultrasonic microscopy
IEEE Trans. Sonics & Ultrasonics 30:85-90

13. Torchia,D.A., Hiyama,Y., Sarkar,S.K., Sullivan,C.E. & Young,P.E. 1985
Multinuclear magnetic resonance studies of collagen molecular structure and dynamics
Biopolymers 24:65-75

14. Lees,S., Tao,N.J. & Lindsay,S.M. 1988
Sonic velocities in compact hard tissues at gigaherz frequencies by means of Brillouin light scattering
(in preparation)

EDUCATIONAL FILM FOR ULTRASONIC ENGINEERS (PART II)

K. Harumi

Tokyo University of Information Sciences
1200 Yatou-cho Chiba, Chiba, Japan

M. Uchida

College of Industrial Technology, Nihon University
Narashino, Chiba, Japan

T. Saitou and T. Fujimori

Research Laboratory, Simizu Construction Co. Ltd.
Kootooku, Tokyo, Japan

ABSTRACT

Educational film showing the behavior of ultrasonics in a solid are presented by using computer simulation, and this film is the revised and complete film, and it will takes about 27 minutes long. It consists of fourteen parts, including vector or lattice representation, and they are 1) Longitudinal wave, 2)Transverse wave, 3) Radiation of ultrasonics from a normal transducer, 4) Reflection of longitudinal wave on a free surface, 5) Reflection and mode conversion of the longitudinal wave on a free surface for 20° incidence, 6) Reflection and mode conversion of the transverse wave on a free surface for 20° incidence, 7) Reflection of the transverse wave on a free surface for 50° incidence, 8) Reflection and refraction of the longitudinal wave on a steel-water interface, 9) Reflection and refraction of the longitudinal wave on a water-steel interface, 10) Reflection and refraction of the longitudinal wave on a plastic-steel interface for 45° incidence (angular probe), 11) Reflection of the longitudinal wave by a ribbon crack for normal incidence, 12) Reflection of the longitudinal wave by a crack for 60° incidence, 13) Reflection of the transverse wave by a crack for 60° incidence, 14) Reflection of the transverse wave by a crack on a free surface.

1. INTRODUCTION

Since 1978 we have produced four films of the computer simulation of elastic waves for research or educational purposes. In Japan we are now using an educational film in educational meetings. These film are useful in understanding the behavior of elastic wave in a solid, for instance, we can see ultrasonic in steel. Longitudinal wave and transverse or surface wave can especially be distinguished, if the vector representation is understood, and that will not be difficult for ultrasonic engineers. The films also have another advantage in that they promote understanding of the mechanisms of elastic waves, such as why we have mode conversion on a free surface, or on a crack or corner reflections.

2. HOW TO USE THESE FILMS

From our experience, it is best to explain the outline or detail of the films, and meaning of vector representation through the use of slides or an overhead projector. For this purpose we have prepared slides of an abstract of the films, and in some cases you may wish to use only these without the film in explaining, for instance, the outline of elastic waves. This is a more economical way. We have also prepared cassette tapes for use in explanation after having been fully explained by the slides.

3. LONGITUDINAL WAVE

When no stress is applied on the surface, the medium is represented by a lattice with square lattices.

Continuous sinusoidal normal stress is applied by a transducer, and one period is divided into 12 equal parts. Thus the time-distance k becomes $\pi/6$.

When a positive pressure is given to the SURFACE, the displacements with positive y-components appear near the surface up to a half-cycle, the steel is compressed and shrunk to a half cycle. When negative pressure is then given after a half cycle, the part in front of the transducer is stretched and expanded. After one cycle, by repeating this process, longitudinal wave may be seen with both those compressed and expanded parts, propagating to downwards. In this deformation the rectangular shape does not change. However, the area of the rectangle does change and therefore the longitudinal wave become a compressional wave in Fig. 1.

As steel is solid, if the rectangle is compressed in the y-direction, it expands to the x-direction. The ratio of compression $\Delta x/x$ to expansion $\Delta y/y$ is called Poisson's ratio, and it take 0.283 for steel and 0.33 for Aluminium. The wavelength in this case is shown by the length between two compressed portions and is about 11 spacings.

In observing longitudinal waves as vector representation, as shown in Fig. 2, the displacement vectors are given from points in the undeformed medium to points in the deformed medium.

In the first place, when a positive pressure is applied on a surface the displacements with positive y-components appear near the surface up to a half cycle, after half cycle a negative pressure is applied up to one cycle, vectors with negative y-components are observed, and they propagate to downwards. Thus, by repeating this process, we can see that the longitudinal wave with vectors parallel to the propagation direction propagates downwards.

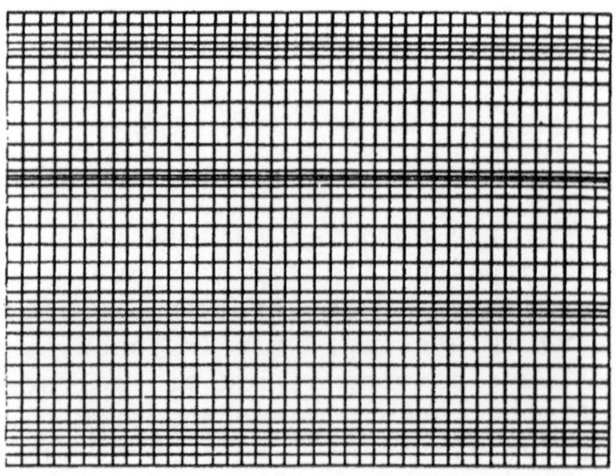

Fig. 1 Lattice representation of longitudinal wave.

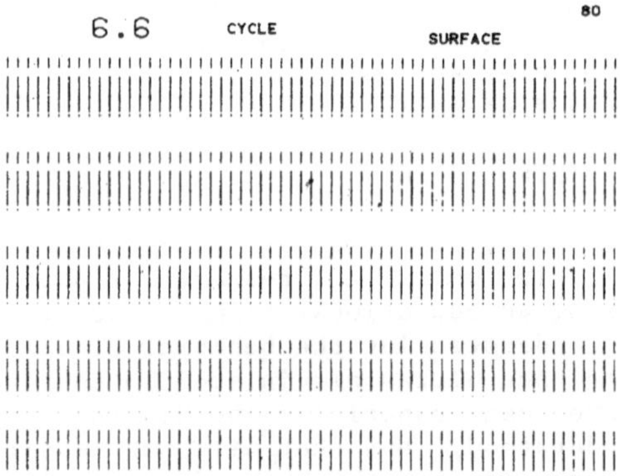

Fig. 2 Vector representation of longitudinal wave.

4. TRANSVERSE WAVE

Part two of this film presents transverse wave. In this case continuous sinusoidal tangential stress is applied by the transducer, and the rectangular shape changes its shape to a diamond up to a half cycle. Between a half cycle and one cycle, then negative tangential stress is applied, another diamond shape is observed. By repeating this process, simple shearing deformation propagates downward, This is the transverse wave, and it has shearing deformation with no change in area of the rectangule as shown in Fig. 3. In the same process, in vector representation, displacement vectors with positive x-components appear as the positive stress is applied. Between a half cycle and one cycle, negative tangential stress is applied, and we find vectors with negative x-components. By repeating this process, we can see that the transverse wave with vectors perpendicular to the propagation direction propagates downwards as shown in Fig. 4. While the longitudinal wave change their areas without changing shape, as they did in Part 1, the transverse waves in contrast, change their shape without changing their area. Therefore the longitudinal wave is the compressional wave and the transverse wave is the shearing wave.

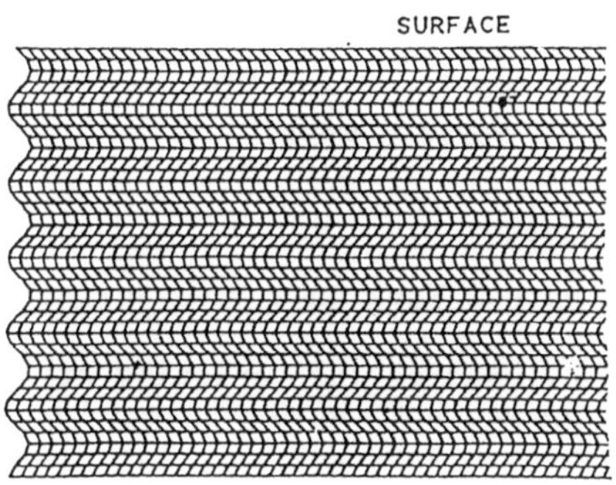

Fig. 3 Lattice representation of transverse wave.

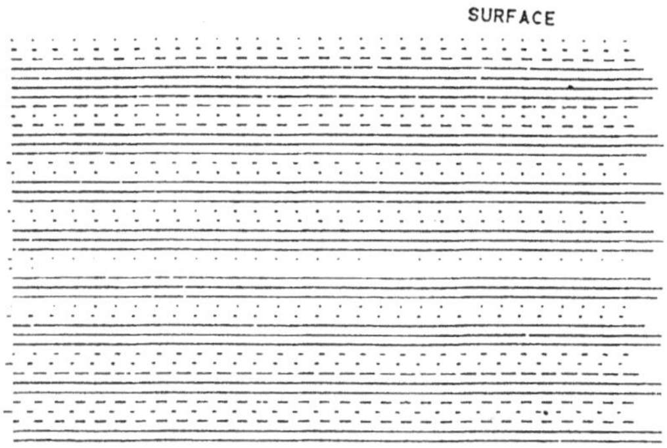

Fig. 4 Vector representation of transverse wave.

5.SURFACE WAVES AND THE GENERATION OF ELASTIC WAVE FROM A NORMAL TRANSDUCER

In order to see the radiation of the transverse wave and surface waves from a normal transducer, we use a rather small normal transducer and this width becomes 17 spacings. The width of the transducer is assumed to be 10 mm in the case of one megaherz, if the medium is steel.

As the picture is symmetric to the center line, only the right quarter is indicated in Fig. 5. In front of the SOURCE we observe the longitudinal wave propagates downward with compressed or expanded parts, but outside the SOURCE we have no deformation. We can observe a very strong deformation of shearing near the edge of the transducer because this part is situated between the deformed area of the longitudinal waves and the undeformed area. This strong shearing deformation extends to the transverse waves and propagates obliquely downwards. On or near the surface, the deformed part of shearing near the edge of the transducer extends to the surface waves. These are quite similar to those of water waves, and they are confined within only about seven spacings, one surface wavelength, from the surface. They propagate along the surface. These are the surface waves.

Let's observe the surface waves in a vector representation in Fig. 6. You will observe that the vectors with negative y-components appear near the edge of the transducer due to the negative normal stress from a half cycle to one cycle. These vectors draw the arch vectors "A" and extend to the transverse wave "T" and the surface wave vectors "S". They correspond to the strong shearing deformation in the lattice representation near the edge of the transducer. On and near the surface, we observe an elliptic motion of the vectors, and you will find an elliptic, anticlockwise motion. The ratio of the length of the major-axis of the ellipse to the minor-axis is about 1.7 : 1.0 as shown in clockwise motion. You will find only vertical vectors in the solid more than two spacings below the surface.

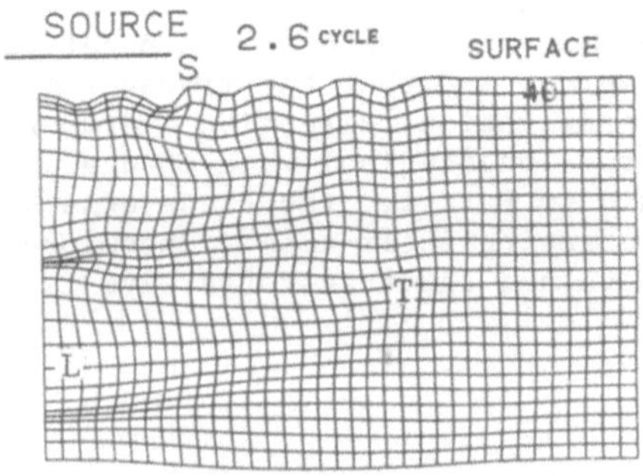

Fig. 5 Lattice representation of surface wave.

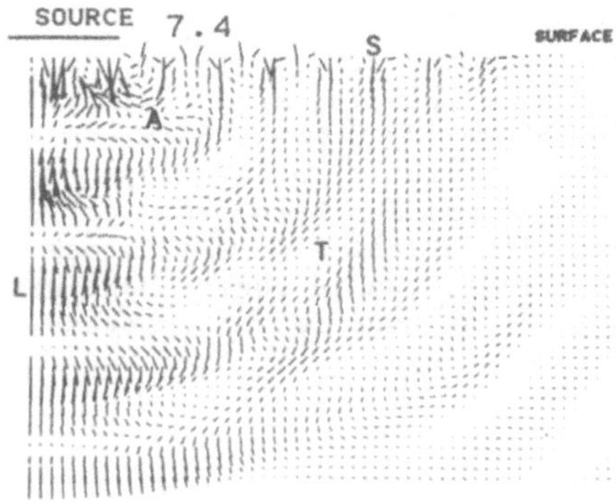

Fig. 6 Vector representation of surface wave.

6. REFLECTION AND MODE CONVERSION AT FREE SURFACE

6-1 Reflection of the Normal Incident Longitudinal Wave on a Free Surface

In this case, we have no mode conversion, and as the vanishing of the stress on a free surface, we observe the displacement vectors grow large, and they become twice as large as that of the incident wave. This large vectors are the surface vectors, and half of the surface vector leave the surface, and come back upward as shown in Fig. 7.

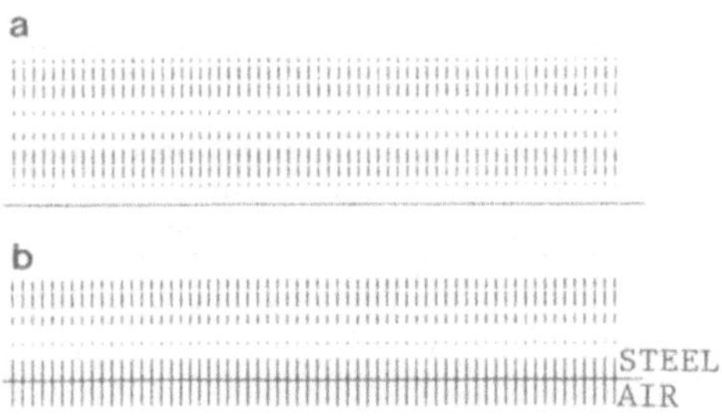

Fig. 7 Reflection of longitudinal wave on a free surface.
(a) Vectors in a steel,
(b) large surface vector.

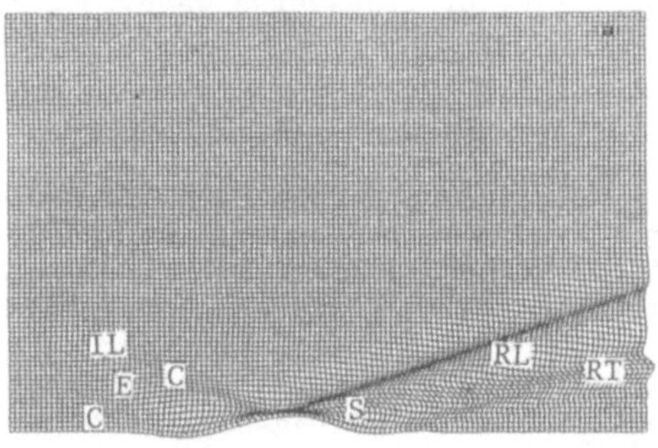

Fig. 8 Reflection of longitudinal wave on a free surface.

6-2 Reflection of Longitudinal Wave of 20 Degree Incidence

6-2-1 Lattice Representation

Figure 8 shows the lattice representation of the reflection of the longitudinal wave of 20 degree incidence on a free surface, in which IL is the incident longitudinal wave with compressional part C and expanded part E, and RL or RT is the reflected longitudinal or transverse wave, respectively. Mode converted transverse wave RT with S shape shear deformation is caused by the strong S-shape deformation S near the surface. This means the surface deformation is the causes of the generation of the mode converted transverse wave.

6-2-2 Vector Representation

Figure 9 shown the vector representation of the same case. The vector S is the surface deformation vector, about twice as large as the incident wave vector. We can decompose the surface deformation vector AD (IL) into the incident longitudinal wave vector AB (IL), reflected longitudinal wave vector BC (RL), and the reflected transverse wave vector CD (RT) in Fig. 10. The ratio of these vectors AB : BC : CD : AD is given as 1.0 : 0.71 : 0.8 : 1.9, and this ratio agrees with the theoretical results in Krautkramer (6) except for the conversion ratio 0.55 in BC to stress ratio.

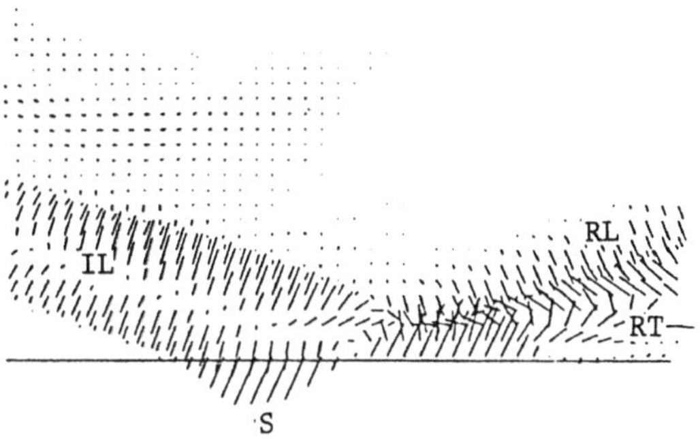

Fig. 9 Reflection of longitudinal wave of 20° incidence on a surface.

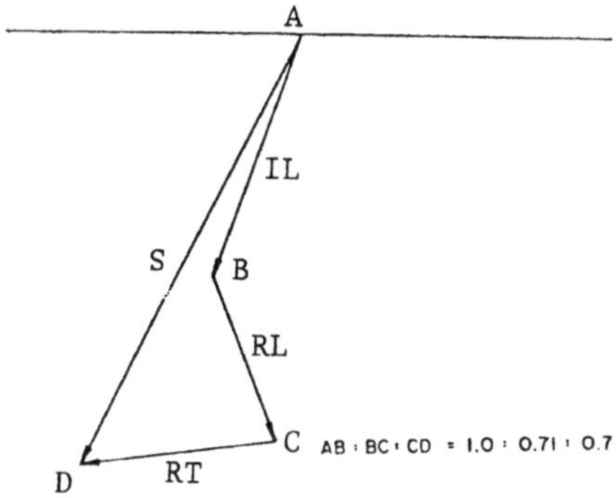

Fig. 10 Dicomposition of surface wave S into IL, RL, RT, of Fig. 9.

6-3 Reflection of the Longitudinal Wave of 50° Incidence

6-3-1 Lattice Representation

In this case, we can see the strong S-shape deformation S near the surface as shown in Fig. 11, though the incident wave is the longitudinal wave having expanded part E and compressed part C. This S-shape deformation extends to the transverse wave.

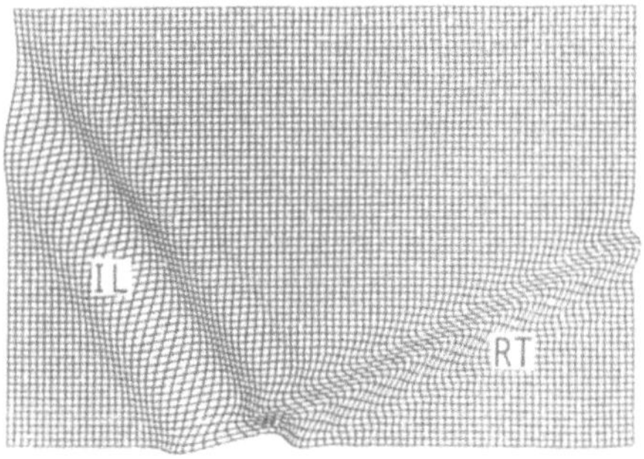

Fig. 11 Reflection of longitudinal wave of 60° incidence on a free surface.

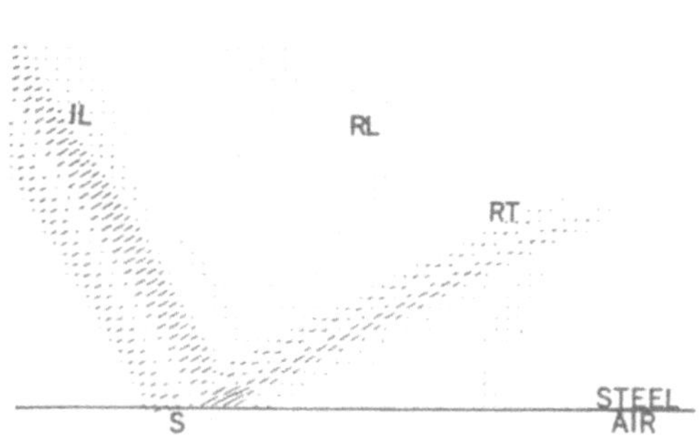

Fig. 12 Vector representation of Fig. 11.

6-3-2 Vector Representation

In Fig. 12, you will see large surface vector S on the surface, and have a very small reflected longitudinal wave. We can see the reflected transverse wave propagating to the upper left side with a propagation direction of 28.4°. The incident longitudinal wave propagates with a velocity of $v_L/\sin\alpha_L$ along the surface, and the reflected transverse wave propagates with a transverse wave velocity to the upper leftside. Thus we find the sine of the reflected angle α_T becomes $\sin(\alpha_T) = [(v_L/\sin \alpha_L)^{-1} v_T]$ as shown in Fig. 13 and the reflected angle α_T results in $\sin^{-1}(v_T\sin\alpha_L/v_L)$, about 28.4°. This is the explanation of Snell's law.

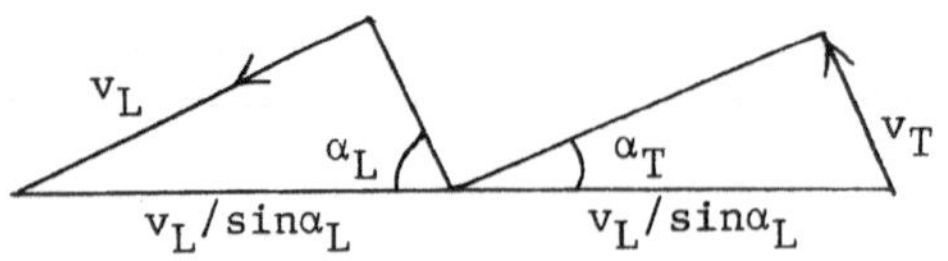

Fig. 13 Illustration of Snell's law.

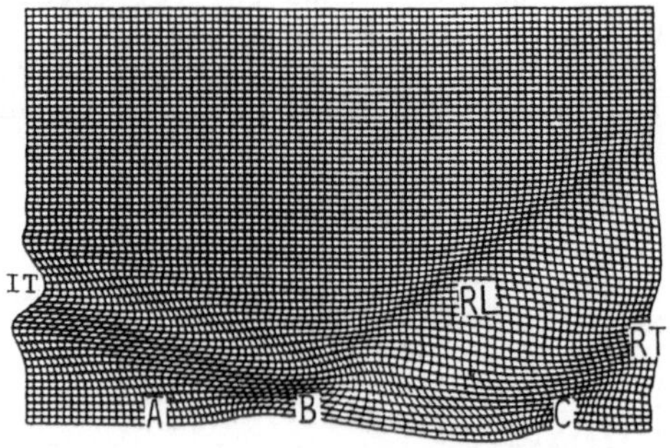

Fig. 14 Lattice representation of transverse wave on a free surface.

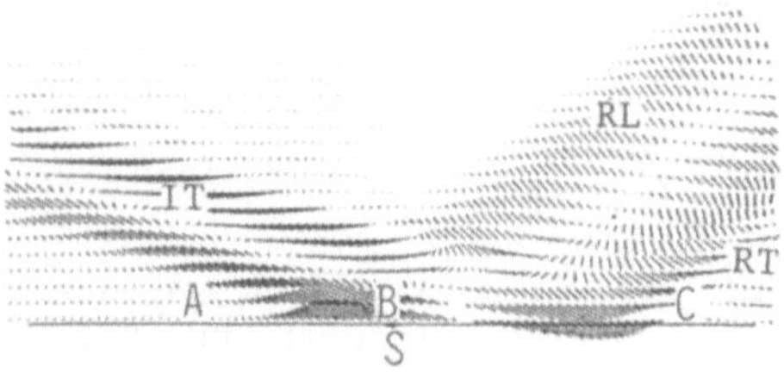

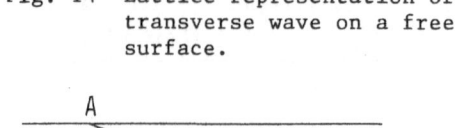

Fig. 15 Vector representation of Fig. 14.

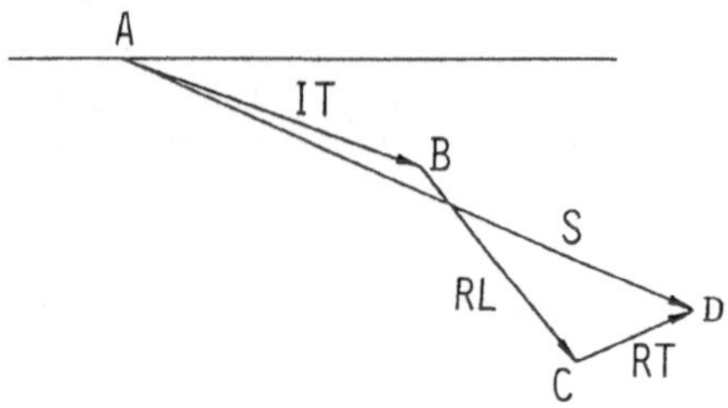

Fig. 16 Vector decomposition of surface wave S for Fig. 15

6-4 Reflection of the Transverse Wave of 20° Incidence on a Free Surface

6-4-1 Lattice Representation

Figure 14 shows the reflection of the transverse wave having 20 degree incidence on a free surface. IT is the incident transverse wave with S-shape deformation, and RT or RL is the reflected transverse or longitudinal wave. We observe two compressed parts, A and C, expanded part, B, on the surface, while the incident transverse wave has no such compressional parts. These compressed and expanded parts form the longitudinal wave and propagate upwards as the reflected longitudinal wave.

6-4-2 Vector Representation

When the incident transverse wave with 20° incident angle making an angle with a normal of the free plane surface, and having upper-left vectors at its front part, reaches a free surface, the vectors of the surface wave are generated. These surface wave vectors grow about twice making 23° with the surface (rather than 20°) in Fig. 15.

We shall illustrate the mechanism of the generation of mode converted longitudinal wave from these vectors. In Fig. 16, AB is the vector of incident T wave of unity length, having upper-left vector,and making a 20° angle with the surface. The reflected T wave vector CD is lower-left vector of making 20° angle with the surface, and the reflected L wave vector BC is upper-left vector, making 38.5° [= $\sin^{-1}(\sin 20°/0.55)$] angle with the surface. We observe the surface vector has about twice as large as that of the incident T wave, and making about 23° angle with the surface. We cannot compose the surface vector AD only by the use of the transverse wave vectors AB and CD, therefore we have to use such longitudinal wave vector as BC in order to obtain the surface vector, and from the length in Fig. 15, we have the length of AD as about 1.8 times as large as AB, then we have the decomposition of AD = 1.86 into three vectors AB : BC : CD = 1.0 : 0.72 : 0.4. Thus we know that mechanism of the generation of the mode converted longitudinal wave can be explained from the surface deformation.

6-5 Reflection of the Transverse Wave of 50° Incidence on a Surface

6-5-1 Lattice Representation

In this case we have no mode converted longitudinal wave, but only a reflected transverse wave, and we observe only the surface deformation in Fig. 3 without compressional deformation, as shown in Fig.17, on a surface.

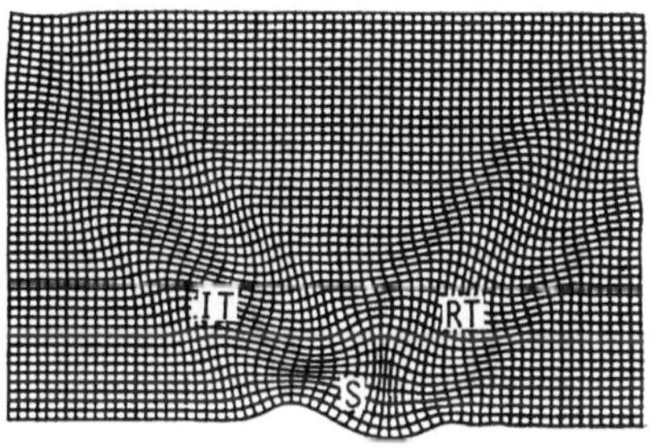

Fig. 17 Reflection of transverse wave of 50° incidence on a free surface.

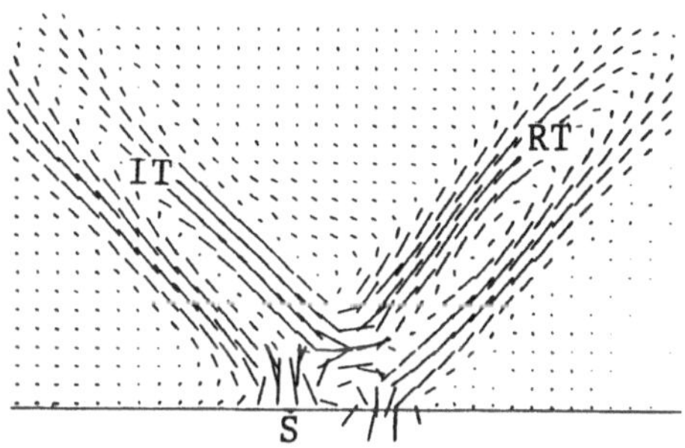

Fig. 18 Vector representation of Fig. 17.

6-5-2 Vector Representation

The front part of an incident T wave has upper left vectors, and the surface wave with upward vector S is generated as it reaches a free surface. Another surface wave vectors S2 with downward vectors is generated at the back part of the reflected wave RT in Fig.18.The ratio of these vectors IT, RT, and S is given as 1.31 : 1.31 : 2.0 and we can decompose the vector AC of the surface wave into AB and BC of the incident and reflected wave in Fig. 19, therefore having no need of a reflected longitudinal wave.

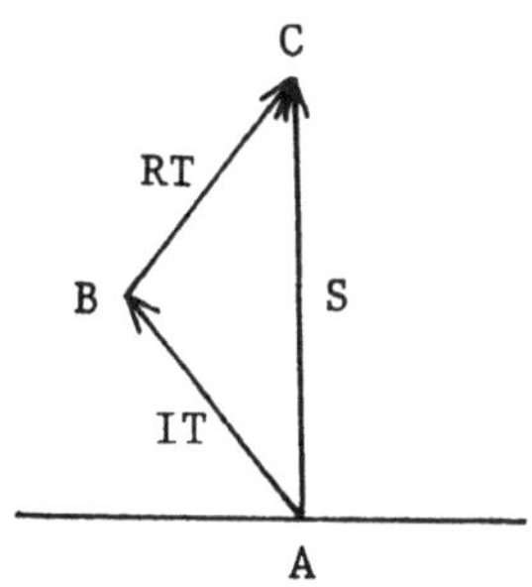

Fig. 19 Vector decomposition of surface wave for Fig. 18.

7. REFLECTION AND REFRACTION ON A STEEL-WATER, OR WATER-STEEL INTERFACE

7-1 Normal Incident of the Longitudinal Wave into Water

We shall observe the longitudinal wave propagates from a steel into water for normal incidence in Fig. 20. As the impedance of water is about one-33rd of that of a steel, that is much softer than a steel, the strength of the reflected wave is almost equal to that of the incident wave. As the same as the free surface, the reflected L wave with almost same intensity to that of the incident wave come back to upwards. As the continuity of the normal component of displacement, the compressional wave of positive or negative y-components is generated in a water, when the L wave of positive or negative y-components reaches the surface. the length of the displacement vector is the same as that of the incident wave. However, as the impedance of water is about 1/33 of that of steel, the pressure in water is about 1/15 of that of steel. The wavelength is steel is about 23 spacings, but as the velocity ratio 5900 : 1480, and the wavelength in water becomes about 9 spacings, and the water-wave proceeds downwards with 1/4 velocity of that in a steel.

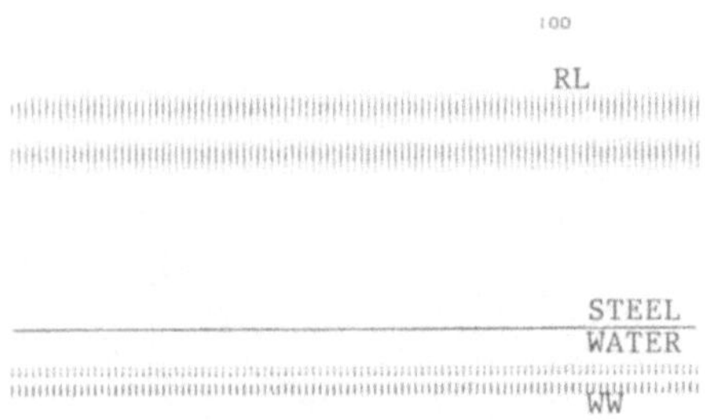

Fig. 20 Reflection of longitudinal wave at steel-water interface.

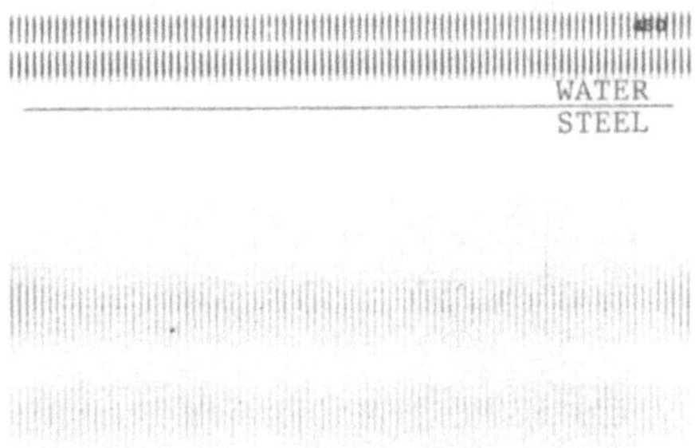

Fig. 21 Reflection of longitudinal wave at water-steel interface.

7-2 Reflection of Longitudinal Wave on a Water-Steel Interface

Let's observe the reflection and penetration of the longitudinal wave on a water-steel interface. In this case, the reflected wave has 94% displacement of the incident wave, and the penetrating wave into steel has 6% displacement and 193% stress of that of the incident wave. At first we shall look at in displacement representation in Fig. 21. When the wave reaches the water-steel interface, 94% of the incident wave is reflected and comes back to upwards. Penetration wave into steel has only 6% displacement of the incident wave, and we observe very weak displacement in a steel. As the velocity in steel is about 5900 m/sec, and 1430 m/sec in water, the wavelength in steel becomes about 4 times as long as that of in water. As you see in the film reflected wave has 94% stress as large as that of the incident wave, and the penetration wave has 193% stress in a water. The ratio of acoustic impedance of a water to of a steel is given as 1.5 : 46.5, the stress is very strong in a steel, 193%, and we observe very strong stress in a steel, compared with that in water.

8. REFRACTION OF LONGITUDINAL WAVE ON A PERSPEX-STEEL INTERFACE

When the incident longitudinal wave IL having 45° incident angle reaches the interface, part of the wave reflected, and we can see the reflected longitudinal wave RL and mode converted reflected transverse wave RT in a perspex, as shown in Fig. 22. When the incident wave penetrates steel, the front part of it pushes the interface downwards, and the rear part pulls the interface upwards. From the displacement vectors A having downwards vectors to the rear part B having upwards vectors, the arch-shaped vectors C connecting the two vectors A and B are formed as shown in Fig. 23. The arch-shaped vectors C have lower right vectors in front of C, and upper left vectors in the rear part of C, and it propagate to leftside. Since, we have the refracted transverse wave having vectors perpendicular to the propagation direction, and from the Shell's law the propagation angle is given by about 56.7°, as the velocity of the longitudinal wave in perspex is about 2730 m/sec and the velocity of the transverse wave in a steel is about 3230 m/sec. We observe very weak reflected longitudinal wave in a perspex, and we have no refracted longitudinal wave in a steel. Because, when the incident angle exceeds 27.6°, we have no refracted longitudinal wave in a steel.

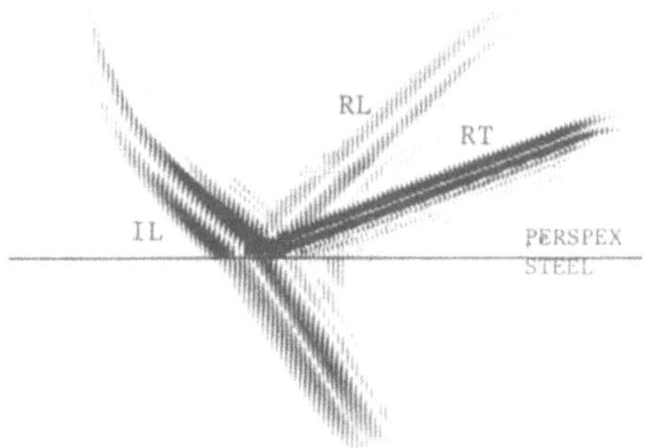

Fig. 22 Reflection of longitudinal wave on a at perspex-steel interface.

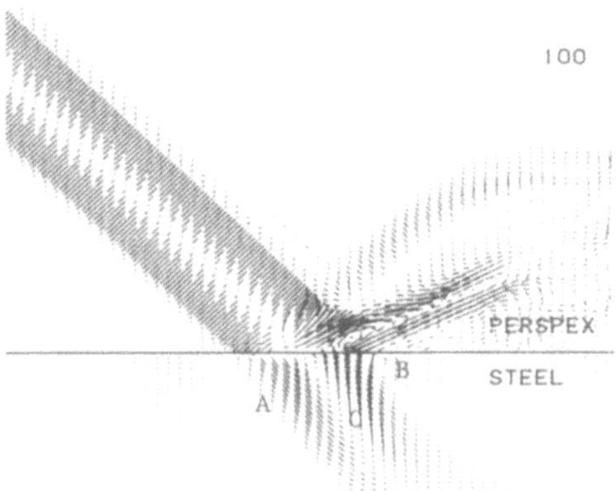

Fig. 23 Arch-shaped vector C in a steel.

9. REFLECTION OF LONGITUDINAL WAVE NORMAL INCIDENT ON A CRACK

Let's observe the reflection of the longitudinal wave by a crack. The incident angle is taken as 90°, and the length d of the crack is given about twice as large as the wave length, and the value kd become about 6.5. When the incident wave reaches the crack, reflected longitudinal wave RL1 is generated in front of the crack, and comes back to downwards. After the reflected longitudinal wave RL1, two arch shaped vectors RT1 and RT2 are observed, each having a propagating center X near the two edges of the crack as shown in Fig. 24. As the arch vectors propagate along the surface with a surface velocity slightly slower than the transverse wave velocity, but they propagate with the transverse wave velocity in an other direction. Hence, we have the propagating center X slightly outside of the two edges. We observe also two reflected longitudinal wave RL1 and RL2 having their propagating center near the edges of the crack, and we have a strong reflected longitudinal wave at the overlapping point of these two longitudinal edge waves.

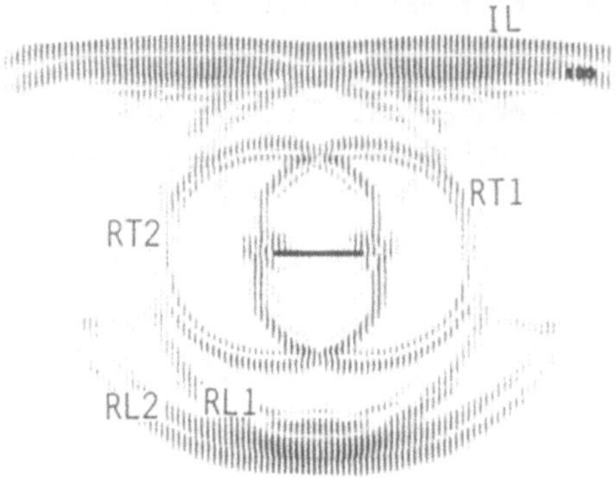

Fig. 24 Reflection of normal incident longitudinal wave by a crack.

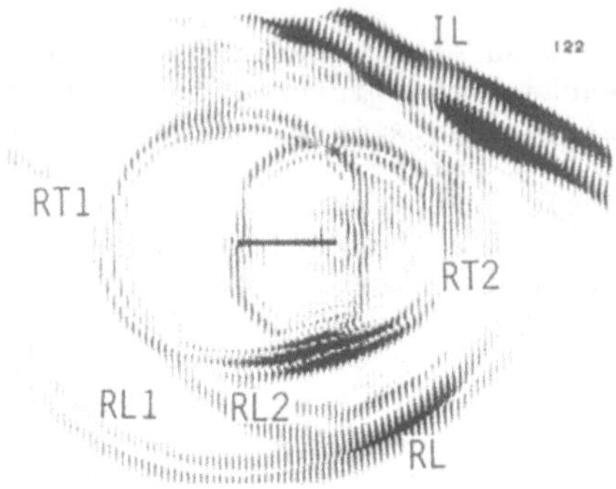

Fig. 25 Reflection of 60 incident longitudinal wave by a crack.

10. REFLECTION OF LONGITUDINAL WAVE OF 60° INCIDENCE ON A CRACK

We will see the reflection of a longitudinal wave by a crack. The incident angle is taken as 60° and the length of the crack is the same as in previous case. When the incident wave reaches the crack, the vectors grow large on the crack surface, and the reflected longitudinal wave propagates to the lower right side. Then, strong vectors having a component almost parallel to the crack follow and propagate to the lower right side, becoming the reflected transverse wave. This reflected transverse wave is formed by two edge waves, RT1 and RT2, having their propagating center near the two edges of the crack shown in Fig. 25. We also observe that the reflected longitudinal wave, RL is formed by overlapping of two longitudinal edge waves, RL1 and RL2, also having their propagating center near the two edges of the crack. When the transverse edge waves pass the opposite edges, other transverse edge waves having their propagating centers near the opposite edges are observed.

11. REFLECTION OF TRANSVERSE WAVE OF 60° INCIDENCE ON A CRACK

We shall observe the reflection of the transverse wave having the incident 60° angle by a crack. The length of the crack is the same as in the previous case. When the wave reaches the crack, the vectors of the front part of the wave having lower right vectors grow large, these vectors propagate to the lower right and extend the reflected longitudinal wave. When the incident wave passes the left edge of the crack, the transverse edge wave RT1 is generated, and propagates as if it has its propagating center near this edge. We also observe another transverse edge wave RT2,having its propagating center near the right edge as shown in Fig. 26. As the front part of the incident wave, IT, has lower right vectors, the front part of the reflected edge wave RT1 has upper right vectors because it is comprised mainly of the reflected transverse wave; on the other hand, the front part of the reflected edge wave RT2 has lower left vectors. Therefore the polarity of the displacement vectors of the two edge wave is opposite. We observe a strong reflected transverse wave RT at the overlapping part of the two edge waves.

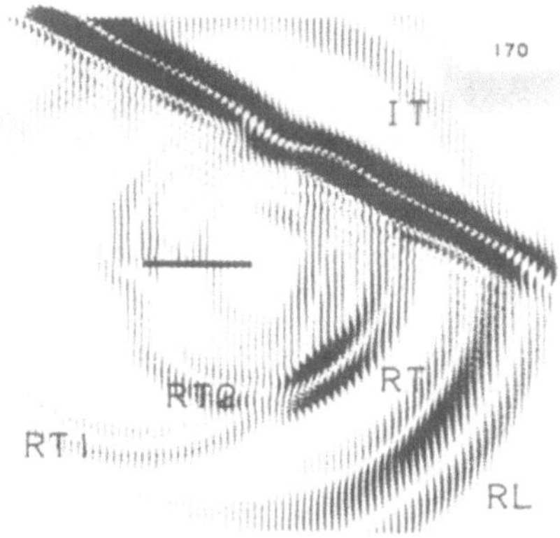

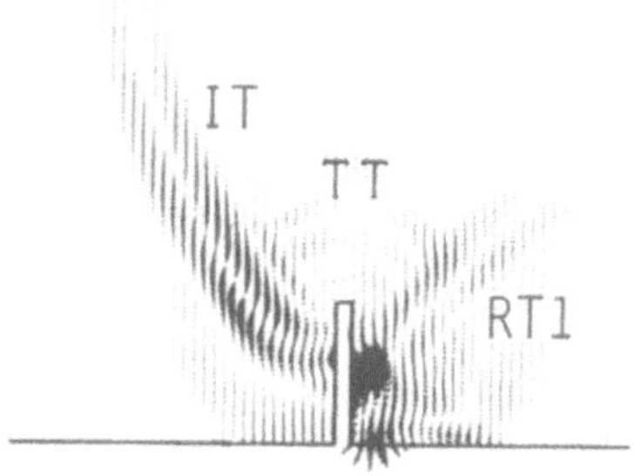

Fig. 26 Reflection of 60° incident transverse wave by a crack.

Fig. 27 Reflection of transverse wave by A60J30

12.REFLECTION OF TRANSVERSE WAVE BY A CRACK ON THE SURFACE OF 60° INCIDENCE

The reflection of this case is similar to that of the corner reflection in the 5th film. The incident angle is given by 60 degree and as the T wave reaches the crack the reflected T wave RT1 is generated at the bottom plane. When the incident wave reaches the tip of the crack, the reflected transverse tip wave TT, though very weak,is generated as shown in Fig. 27. We also find that the reflected longitudinal wave RL2 propagates from the vertical plane of the crack, as if it has its propagating center near the tip of the crack. The reflected RL2 and RT3 are generated at the bottom, and they proceed upwards as shown in Fig. 28. The reflected transverse wave RT4 is also generated at the bottom plane when the RT2 is reflected on the bottom. While these RL2,RT3,and RT4 overlap each other until they leave the tip, they separate after the 200th step, and RL2 propagates as if it has its propagating center at the tip of the crack; RT3 propagates as a head wave with the surface wave S1 on the crack.

We observe that two transverse tip waves TT2 and TT3 are generated as shown in Fig. 29, as RT4 and RT3 pass the tip of the crack. The maximum of reflected longitudinal wave RL2 has propagation direction of about 70° or 80° with the bottom plane, and this is confirmed by experiments.

The reflected RT4 propagates as a head wave connecting the tip wave TT2 with the surface wave S2 on the bottom plane.

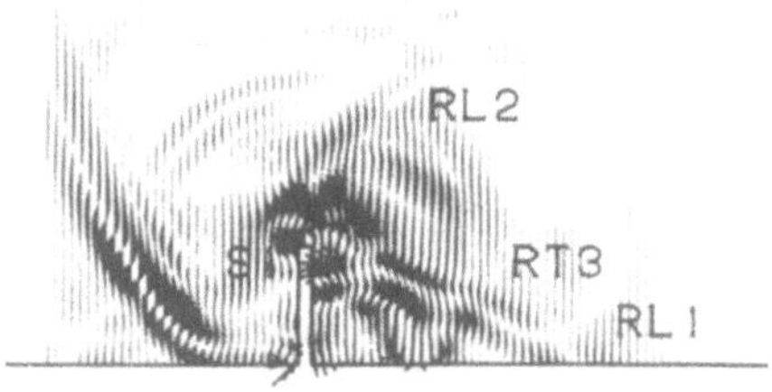

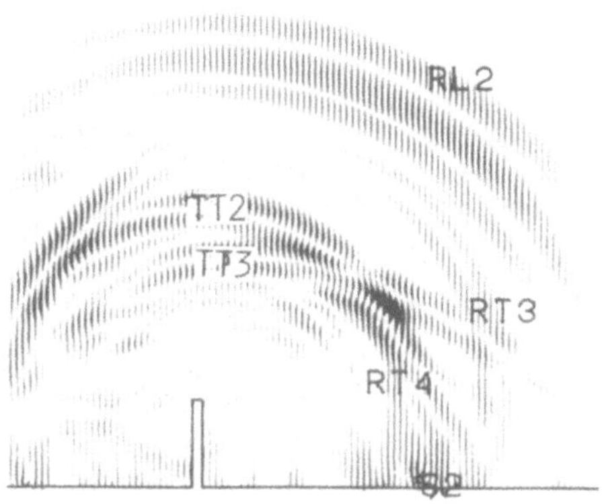

Fig. 28 RL2:Reflected longitudinal wave.
RT3:Reflected transverse wave.

Fig. 29 TT2, TT3:Transverse tip wave
RT3, RT4:Reflected transverse wave.

13. CONCLUSION

By using lattice or vector representation of the results of the computer simulation, the behavior of elastic wave in a solid was visualized, and the mechanism of ultrasonics was clarified in this film for the purpose of the education of ultrasonic engineers. This film will be effective for the understanding of elastic waves in a solid, otherwise no one can realize the behavior of ultrasonics in a solid.

In the first and second parts, the significance of the longitudinal or the transverse wave have been illustrated. In the third part, mechanism of the generation of the transverse waves from a normal transducer have been clarified. In the next parts, the mechanism of mode conversion of elastic waves on a free surface was explained, and the clarification of these mechanism will be possible only by the use of graphical representation of computer simulation.

Reflection and refraction on a steel-water or water-steel interface are also visualized and explained in the following parts. Then, the mechanism of refraction and reflection of the longitudinal wave on a perspex-steel interface of an angle-probe have been explained by using vectors representation. Reflection of the longitudinal or the transverse wave by a smooth ribbon crack is also visualized, and the importance of the edge wave have been recognized. In the last part of this film, the reflection of the transverse wave having 60° incidence by a crack on a free surface have been visualized, and very complicated reflections have been explained.

From these visualizations of elastic waves in a solid, the behavior of ultrasonics will be easily understood. Therefore, this film will be effective in understanding of elastic waves for the educational purpose of ultrasonics engineers. This film, or in video-type from with slides, are now being used in many advanced countries, and we hope this film will be used in other countries.

ACKNOWLEDGEMENTS

The computation of this work is was done by the use of HITAC-280H in Tokyo University. We express our thanks to many people who advised us of this film.

REFERENCES

1. Harumi, K., Saitou, T. & Fujimori, T. "Motion Picture of the Computer Simulation of Elastic Waves Generated by Transducer",Proc. 1st Symp. on Ultra. Charact., 1987, pp. 521-532.
2. Harumi, K., Saitou, T. & Fujimori, T. "Motion Picture of the Computer Simulation of Elastic Wave in a Solid (Part 2)" Proc. 9th world Cong. of NDT in Melbourne. 4H8, 1979.
3. Harumi, K., Saitou, T. & Fujimori, T. "Motion Picture of the Computer Simulation of Elastic Waves in a Solid (Part 3)" Proc. Ultrasonic Symp. in Graz., 1979, pp. 194-199.
4. Harumi, K., Saitou, T., Fujimori, T. & Ootsuki, T., "Motion Picture of the Computer Simulation of Elastic Waves (Part 4) Reflection of Elastic Waves by several Kinds of Defects" Proc. IEEE Ultra. Symp., vol.2, 1982, pp. 1079-1083.
5. Harumi, K., Okada, H., Saitou, T. & Fujimori, T., "Motion Picture of the Computer Simulation of Elastic Waves (Part 5) Mode Conversion on a Free Surface and Reflection by a Corner" Proc. IEEE Ultra. Symp., vol. 2, 1982, pp. 1084-1087.
6. Krautkramer J., Krautkramer H. "Ultrasonic Testing of Materials" Springer Verlag, Berlin, (1977).

THE PERFORMANCE OF A LEAKY SURFACE ACOUSTIC WAVE TRANSDUCER CONSISTING OF PVDF AND FUSED QUARTZ

Kohji Toda and Akihiro Sawaguchi

Department of Electrical Engineering
The National Defense Academy
Hashirimizu, Yokosuka 239, Japan

ABSTRACT

A leaky surface acoustic wave transducer operating at a liquid-solid boundary radiates a longitudinal wave into water. The structure of a PVDF/interdigital transducer/fused quartz plate is analyzed numerically to understand the transducer performance including the phase velocity and efficiency for the acoustic beam radiation into water in comparison with the characteristics in a structure incorporating a piezoelectric ceramic plate instead of PVDF. The zeroth-order mode has a large transducer efficiency, although the electromechanical coupling constant is small. The results are useful in designing an ultrasonic transducer for acoustical imaging and sensor technology.

INTRODUCTION

A polyvinylidene fluoride (PVDF) film is a popular material in various fields, such as underwater acoustics, medical ultrasound imaging and nondestructive testing.[1,2] PVDF transducers are expected to get sharp impulse responses, and accordingly, to give high resolution images, because its acoustic impedance is closer to that of water or human body tissues than is the case for piezoelectric ceramic.[3] The film is mainly used in the thickness mode.[4,5] An interdigital transducer (IDT) can operate by radiating a longitudinal wave into water.[6] The introduction of this type of IDT has some merits including the use of well-establised surface acoustic wave (SAW) technology. The potential for using the leaky SAW transducer is developing in the field of acoustic imaging and sensing technology.[7,8]

In this paper, we describe the results of the numerical analyses of the transducer performance in a structure of PVDF/IDT/fused quartz. The real and imaginary velocities of leaky SAWs propagating along the liquid-solid boundary, group velocity, electromechanical coupling constant, and transducer efficiency for acoustic beam radiation into water are given as a function of the product of the wave number and film thickness. Another structure incorporating this piezoelectric ceramic plate as an alternative of using a PVDF film is discussed for comparing the difference of the characteristics of two kinds of piezoelectric materials.

ANALYTICAL PROCEDURE

A coordinate system to explain the leaky SAW transducer performance is shown in Fig. 1, in a sagittal plane. The leaky SAW travel in the x_1 direction along the water-solid boundary. In the piezoelectric film with a crystal symmetry of 2mm, the mechanical displacement U_j along the x_j direction and the electric potential ϕ are given as follows:

$$U_j = A_j \exp(ikbx_3)\exp\{ik(x_1 - vt)\}, \qquad j=1,2,3 \tag{1}$$

$$\phi = A_4 \exp(ikbx_3)\exp\{ik(x_1 - vt)\}, \tag{2}$$

where A_j and A_4 are the amplitude of U_j and ϕ, respectively. v is the leaky SAW velocity, k is the wave number, and b is decay constant.

The mechanical displacement component U_{sj} in the fused quartz plate of infinite thickness and the displacement U_{wj} in water are expressed by the following forms:

$$U_{sj} = A_{sj} \exp\{ikb_s(x_3 + d/2)\}\exp\{ik(x_1 - vt)\}, \qquad j=1,3 \tag{3}$$

$$U_{wj} = \frac{\partial}{\partial x_j} A_w \exp\{ik_w b_w(x_3 - d/2)\}\exp\{ik(x_1 - vt)\}, \tag{4}$$

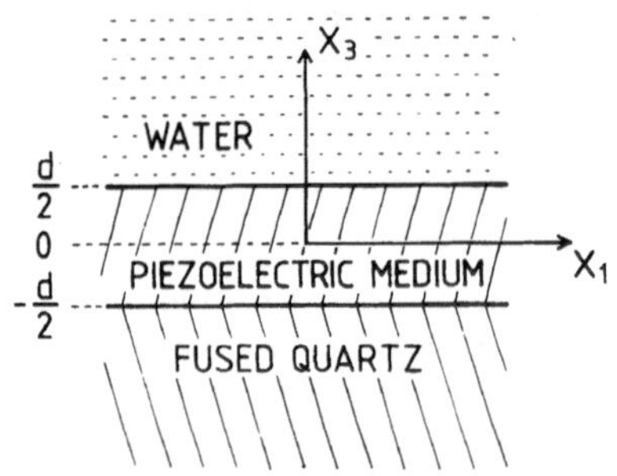

Fig. 1 Coordinate system.

where k_w $(=\omega/v_w)$ is the wave number of the longitudinal mode in the liquid, v_w is the longitudinal wave velocity in the liquid, and b_w is the decay constant defined by $\{1-(v_w/v)^2\}^{1/2}$.

In a layer structure incorporating IDT, four kinds of electrical boundary conditions as illustrated in Fig. 2 must be considered. The numerical calculation of leaky SAW velocity is the main frame of the present study, from which the electromechanical coupling factor k^2 and the transducer efficiency η for the operating of a SAW transducer are derived.

NUMERICAL CALCULATION RESULTS AND DISCUSSION

The numerical analysis was carried out by developing the method proposed by Farnell.[9] The material constants used for the analysis was the same as those in Ref. 10. Velocity curves for the leaky SAWs as a function of the product of the wave number and the film thickness are shown in Fig. 3. In this case, two boundary surfaces were electrically opened. The leaky SAWs propagate along the boundary surface of water-PVDF, while connected into water in a longitudinal mode. Thus, a large value of the imaginary part v_i of the velocity is desirable for the leaky SAW transducer. It is noticeable from Fig. 3 that the zeroth-order mode has the largest v_i value around kd of 2.0. The distribution of the longitudinal and vertical displacement components are shown in Fig. 4, for in a case of kd=2.2. The results are valid for designing the thickness of fused quartz.

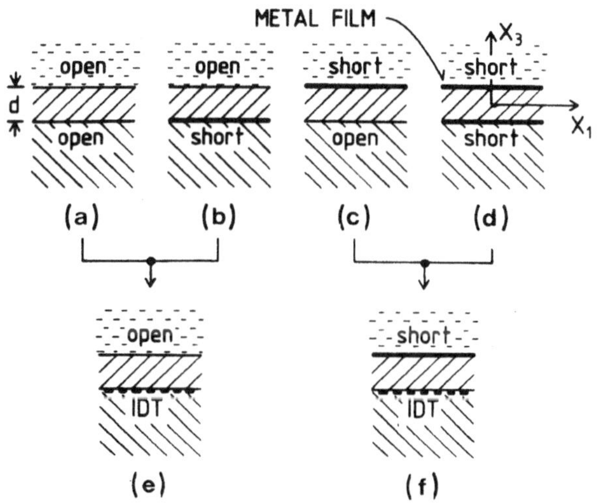

Fig. 2 Four kinds of electrical boundary conditions and two types of transducer configurations.

Values of the calculated electromechanical coupling constant k^2 as a function of kd are shown in Fig. 5, for the transducer conjunction without a counter electrode on the boundary between the water and the PVDF. The leaky SAWs excited on the substrate by applying an ac voltage are mode-converted to the longitudinal wave in water, obeying the following relation:

$$C = 1 - \exp(4\pi v_i/v_r). \qquad (5)$$

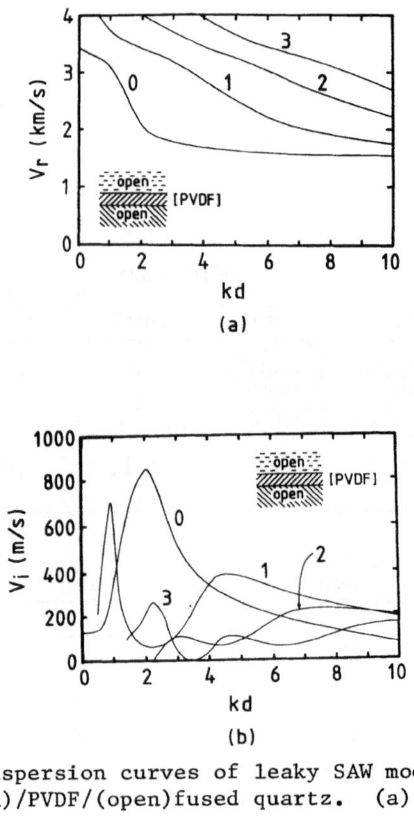

Fig. 3 Calculated dispersion curves of leaky SAW modes under condition of water(open)/PVDF/(open)fused quartz. (a) for real component, (b) for imaginary component.

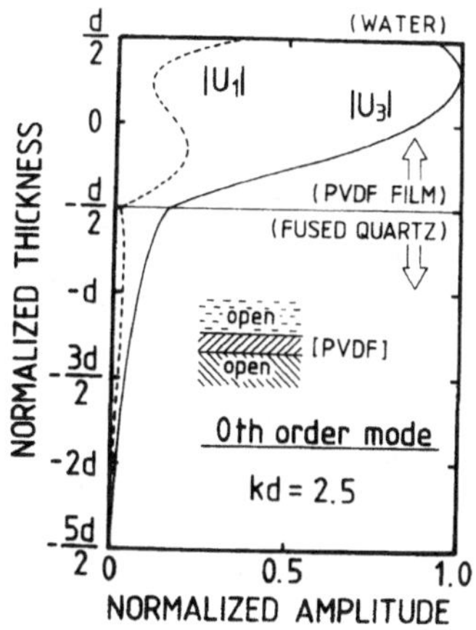

Fig. 4 Distribution of horizontal and vertical displacement of zeroth-order mode, at water(open)/PVDF/(open)fused quartz state.

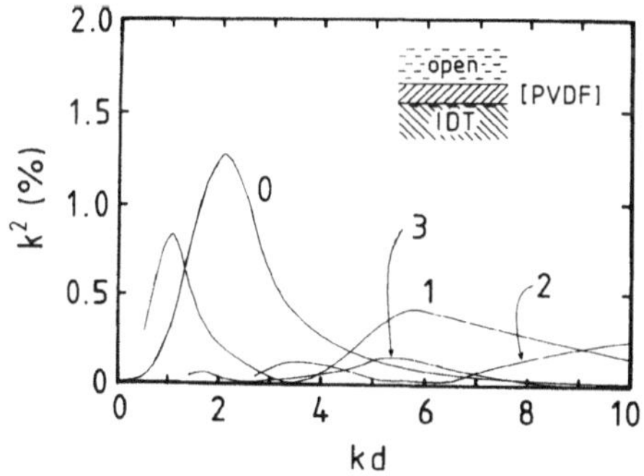

Fig. 5 Calculated relations between k^2 and kd for leaky SAW transducer consisting of PVDF and fused quartz, without counter electrode.

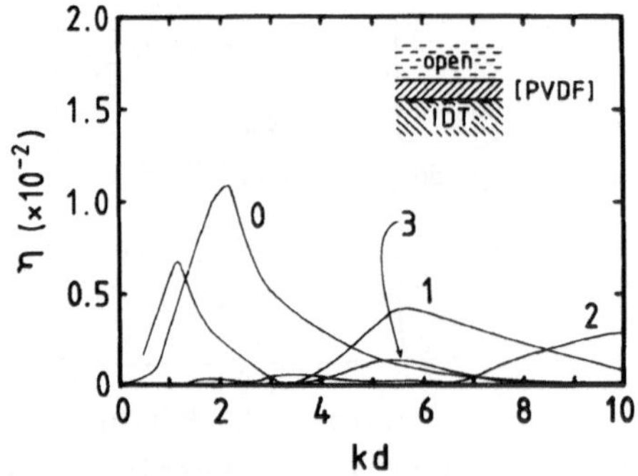

Fig. 6 Calculated η-*kd* relations for PVDF and fused quartz transducer without counter electrode.

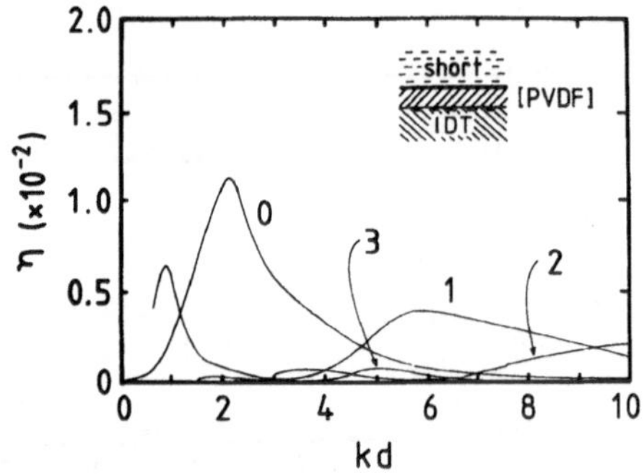

Fig. 7 Calculated η-*kd* relations for PVDF and fused quartz transducer with counter electrode.

The transducer efficiency η for underwater sound is obtained from the product of k^2 and C. The calculated relations between η and kd are shown in Figs. 6 and 7 for two types of transducer conjunctions, with and without the counter electrode, respectively. It is obvious that use of the zeroth-order mode around kd=2.2 is significant. The difference arisen from the transducer configurations is very small.

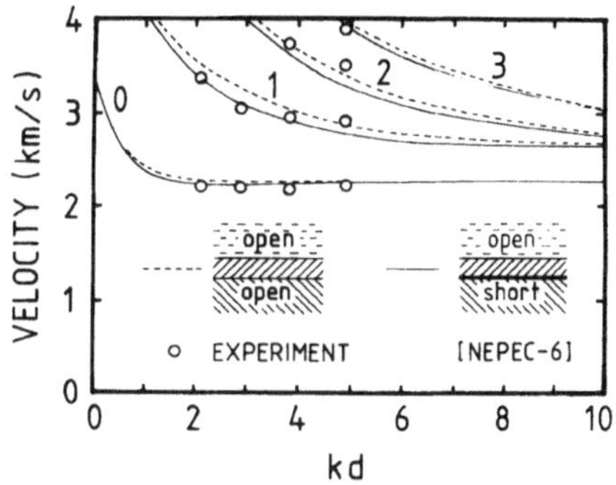

Fig. 8 Calculated dispersion curves of leaky SAW modes at water(open)/ NEPEC-6/(open/short)fused quartz state.

The same analytical procedure was applied to a layer structure composed of piezoelectric ceramic thin plate, IDT and fused quartz. Calculated and experimental velocity curves as a function of kd are shown in Fig. 8. The material used in this case was a piezoelectric ceramic NEPEC-6 produced by Tokin Corp. The material constants necessary for numerical calculation are listed in Ref. 11. The velocity curves have a similar tendency with those in Fig. 3. Experimental results were obtained from the product of measured center frequency and the interdigital periodicity of the IDT for four kinds of device constructions, which are in agreement with the calculated curves. Calculated results of k^2 and η as a function of kd for the structure of NEPEC-6/IDT/fused quartz are shown in Figs. 9 and 10, respectively.

The difference indicated in the two kinds of piezoelectric media is not remarkable, although the coupling constant is weak in the case of the use of PVDF. The kd value having the largest transducer efficiency is almost twice the value in the case of the ceramic substrate. This means that the use of PVDF is favorable for higher frequency operation. The structure of PVDF/IDT/fused quartz has also some advantages in terms of the flexibility and thinner film state of PVDF.

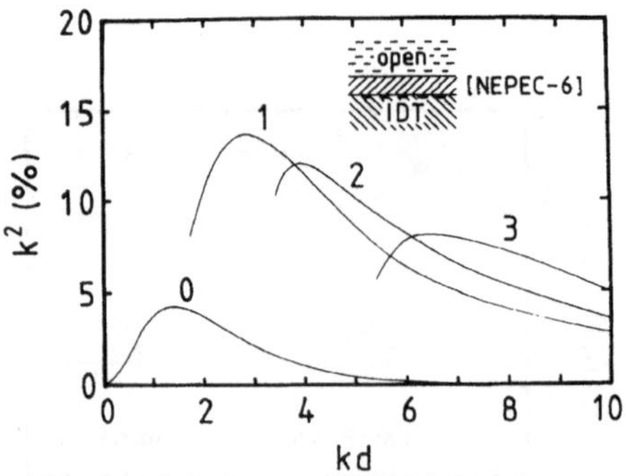

Fig. 9 Calculated results of k^2 as function of kd for transducer structure of water(open)/NEPEC-6/(IDT)fused quartz.

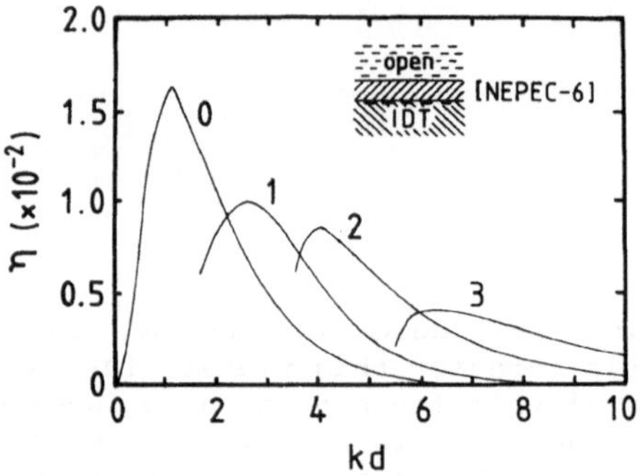

Fig. 10 Calculated results of η as function of kd for transducer structure of water(open)/NEPEC-6/(IDT)fused quartz.

CONCLUSION

The performance of a leaky SAW transducer composed of a PVDF film/IDT /fused quartz structure was analyzed numerically. The zeroth-order mode has a peak efficiency around kd=2.2 from the large mode conversion efficiency, although it has a small k^2. This type of leaky SAW transducer is favorable for high frequency operation and for concave structure, compared with the use of a thin piezoelectric ceramic plate. The calculation results on this type of transducer using a PVDF will be useful for sensor technology and acoustic imaging.

ACKNOWLEDGEMENTS

The authors thanks M. Hayama for his useful assistance. This research was partially supported by a Grant-in-Aid from the Ministry of Education, Sience and Culture, Japan.

REFERENCES

1. N. Chubachi and T. Sannomiya, "Composite Resonator Using PVDF Film and its Application to Concave Transducer for Focusing Radiation of VHF Ultrasonic Waves", Proc. IEEE Ultrason. Symp., pp.119-123 (1977).
2. W. H. Chen, H. J. Shaw, D. G. Weinstein, and L. T. Zitelli, "PVDF Transducers for NDE", Proc. IEEE Ultrason. Symp., pp.780-783 (1978).
3. H. J. Nguen, P. Hartemann, and D. Broussoux, "Single Element and Array PVDF Transducers for Acoustic Imaging", Proc. IEEE Ultrason. Symp., pp.832-836 (1982).
4. D. R. Bacon, "Characteristics of a PVDF Membrane Hydrophone for Use in the Range 1-100 MHz", IEEE Trans. Sonics Ultrason., SU-29, pp.18-24 (1982).
5. W. R. Leung and K. K. Yung, "Internal Losses in Polyvinylidene Fluoride (PVDF) Ultrasonic Transducer", J. Appl. Phys., 50, pp.8031-8033 (1979).
6. K. Toda and Y. Murata, "Acoustic Focusing Device with an Interdigital Transducer", J. Acoustic. Soc. Am., 62, pp.1033-1036 (1977).
7. G. W. Farnell and C. K. Jen, "Microscope Lens Using Conversion of Rayleigh to Compressional Waves", Proc. IEEE Ultrason. Symp., pp.673-676 (1980).
8. K. Kobayashi, T. Moriizumi and K. Toda, "Longitudinal Acoustic Wave Radiated from an Arched Interdigital Transducer", Appl. Phys. Lett., 52, pp.5386-5388 (1981).
9. G. W. Farnell, "Symmetry Considerations for Elastic Layer Modes Propagation in Anisotropic Piezoelectric Crystals", IEEE Trans. Sonics Ultrason., SU-17, pp.229-238 (1970).
10. R. S. Wagner, "PVF_2 Elastic Constants Evaluation", Proc. IEEE Ultrason. Symp., pp.464-468 (1980).
11. Tokin Products Catalogue (Piezoelectric ceramics).

DIFFRACTION ARTEFACTS AND THEIR REMOVAL

E.T. Costa[1,2] and S. Leeman[1]

[1]King's College School of Medicine and Dentistry, Department of Medical Engineering and Physics, Dulwich Hospital, East Dulwich Grove, London SE22 8PT, UK and
[2]Department of Biomedical Engineering, Biomedical Engineering Center, UNICAMP, Caixa Postal 6040, 13100 Campinas, SP, Brazil

INTRODUCTION

Gore and Leeman (1977) verified experimentally for the medical ultrasound case, in the context of the 'replica pulse' formulation of Freedman (1961), that the transient pressure field produced by a piston-like transducer can be described as a sum of two major components: 'direct' and 'edge' waves (Weight and Hayman, 1978). The edge waves are produced by the periphery of the transducer and are the cause of the diffraction effects observed in many measurement and calibration schemes based on techniques using small hydrophones to probe pulsed ultrasound fields.

The diffraction effects, as can easily be seen, are an inevitable consequence of the wavelengths and transducer aperture sizes used in virtually all practical applications. They are very troublesome in many quantitative techniques, where they manifest themselves as measurement artefacts which compromise the accuracy of the method. As yet, no truly effective 'diffraction correction' technique has been proposed, but much effort has been expended on devising special transducers, or, for the pulse-echo case, in devising digital filters to correct for the effect.

We propose here a rather different approach towards the problem. We analyse the concept of diffraction, and come to the conclusion that the difficulty may be resolved by resorting to a novel means of field measurement. A specially designed PVDF hydrophone (which we term 'diffraction-insensitive', or 'DI', hydrophone) has been constructed for this purpose, and experimental data are presented to show that the device is indeed insensitive to the diffracted component of the emitted field. Some applications of the purpose-built hydrophone are also shown below.

THE DI HYDROPHONE

The purpose-built hydrophone consists of a thin (25 μm), uniformly poled and electroded PVDF membrane, stretched over a large supporting annulus, so that a 75 mm diameter active planar surface is obtained (which is sufficient to intercept effectively the entire ultrasound field produced by the transducers available in our laboratory). The theory that supports the working of our purpose-built hydrophone has been presented elsewhere (Costa et al, 1986) and will not be repeated here. What is important to notice is that, in fact, such a hydrophone acts as a filter which responds only to those components of the ultrasound field's (three-dimensional) spectrum whose wave vectors are orthogonal to the plane of the membrane. Thus, field measurement with such a hydrophone gives the same output as would be obtained with an equivalent 'one-dimensional' field: in this way, diffraction effects are not seen at all, or even the focusing of fields (Costa et al, 1987), thus indeed allowing us to called it a 'diffraction-insensitive' (DI) hydrophone.

Response to direct and edge waves

A pulsed ultrasound field was produced by the the electric excitation of a planar 15 mm diameter commercial PZT transducer. This field was probed using a 0.5 mm diameter PVDF hydrophone (Marconi bilaminar shielded), located on axis very near to the transmitter in order to obtain separation between the direct and edge waves (Fig. 1.a). The far field pulse on axis was also measured (Fig. 1.b). The transducer's field was probed also with our DI hydrophone, arranged with its face perpendicular to the pulse propagation direction, at the same distances from the transducer face (Fig. 1.c). The DI hydrophone shows a single, location-invariant pulse, very similar to the direct wave component observed with the point hydrophone. We consider diffraction to arise entirely from the presence of the edge wave and it is clear that our hydrophone is totally insensitive to that component of the transient field when employed in this measurement configuration. This finding is in accordance with formal theoretical arguments, which are not developed here. Fig. 1.d shows the calculated derivative of the waveform measured with the DI hydrophone, and it is seen to be remarkably similar in shape to the waveform obtained with the point hydrophone in the far field. This accords with theoretical predictions that the far field, on axis, has the same form as the derivative of the direct wave. As further evidence of the close similarity of the waveforms, we show their normalised amplitude spectra in Fig. 2.

Attenuation measurements

A substitution method was used to estimate the transmission loss of a glycerine sample relative to that of water, using a 19 mm diameter, 3.5 MHz (nominal), focused transducer as a transmitter, and either the DI hydrophone or the point hydrophone (0.3 mm diameter ceramic PZT transducer) as receiver. Measurements were performed in the 1 to 5 MHz frequency range, using sinusoidal tone-bursts of at least 20 cycles in duration. The transmitter was fixed and both the sample and the receiver could be translated axially. Experiments were performed in the following configurations: 1) sample at 3 cm and receiver at 9 cm from the transmitter; 2) sample at 6 cm and receiver at 19 cm from the transmitter.

The results are shown in Fig. 3. For the point hydrophone, there are quite significant differences between the results obtained in the two experimental configurations, due to diffraction artefacts that are still present, even in the substitution method, as predicted by Hutchins and Leeman (1981). However, the results obtained with the DI hydrophone show a good overlap between measurements in the two experimental configurations, and it is appreciated that, contrary to the demands of measuring with conventional transducers, our technique does not present setting-up difficulties, as the relative locations of the transmitter, sample and receiver did not significantly affect the results (as theoretically predicted). Measurements can be carried out in the near or the far field, with focused or unfocused transmitters, as the circumstances dictate.

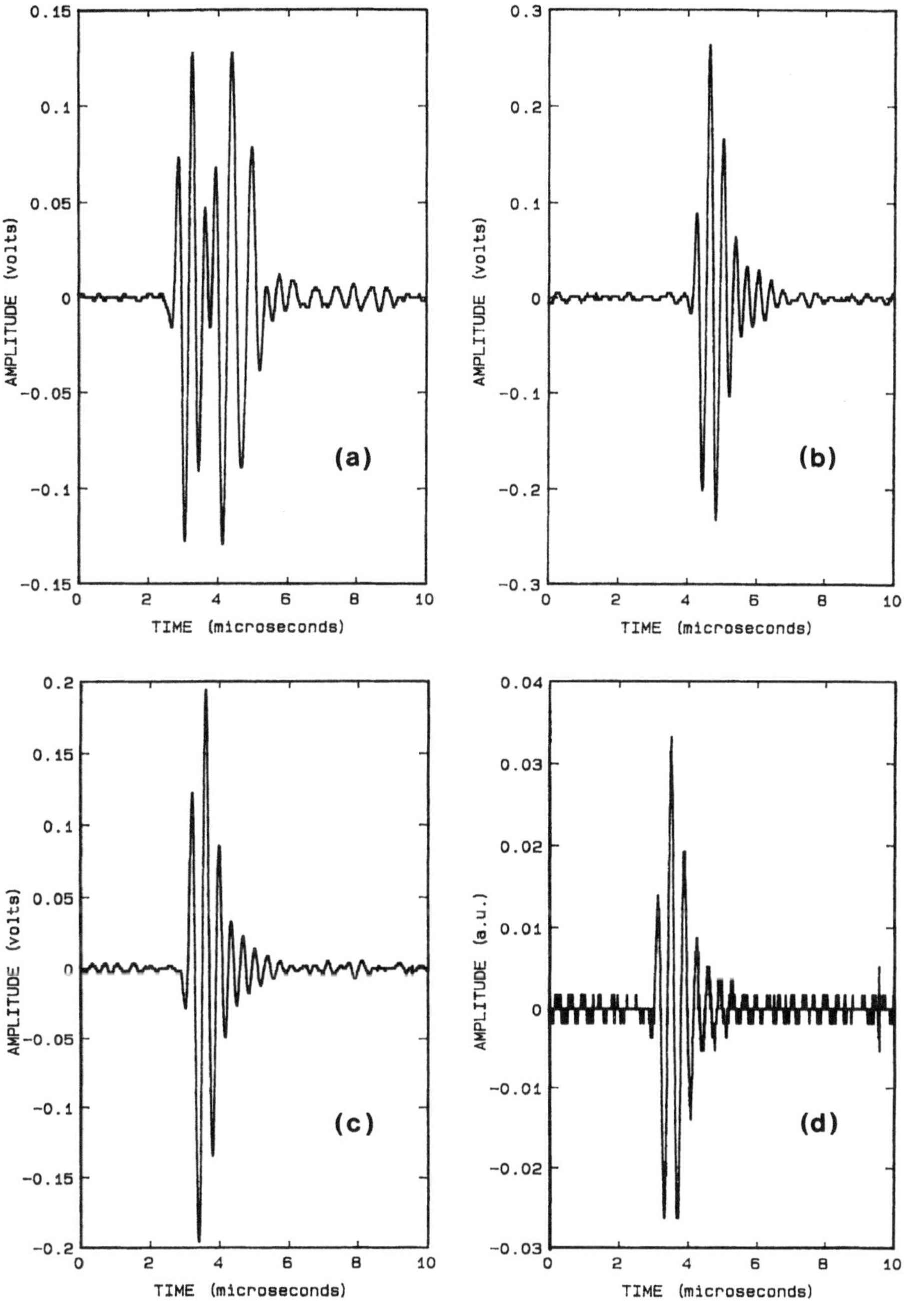

Fig.1. (a) Output of a point hydrophone showing the direct and edge waves, on axis, of a pulsed ultrasound field as explained in the text; (b) Output of the same hydrophone, in the far field, on the acoustic axis of the transducer; (c) Output of a DI hydrophone placed in the same field and at each location detailed in (a) and (b). It shows a location invariant output; (d) The derivative of the pulse shown in (c).

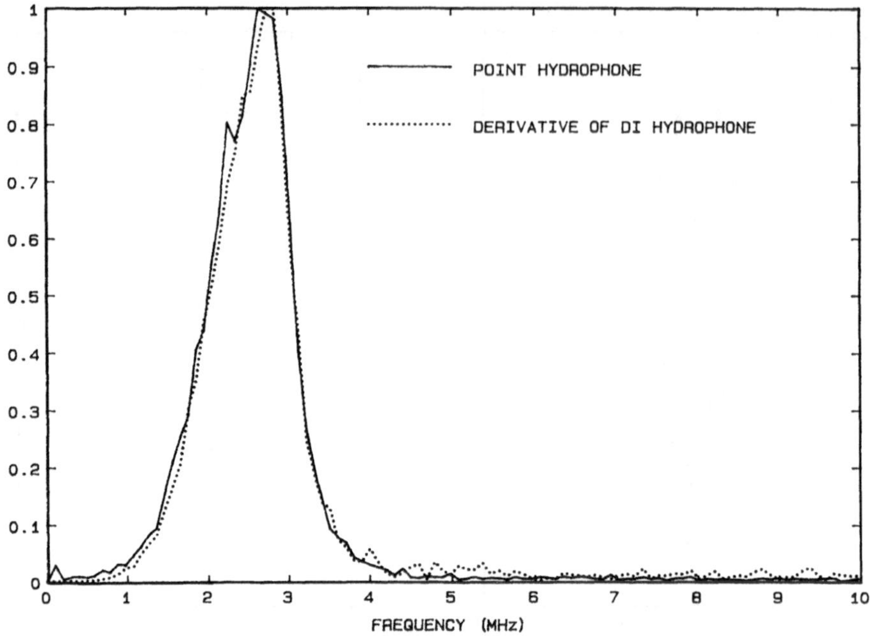

Fig. 2. The normalised amplitude spectrum of the pulse measured with the point hydrophone (shown in Fig. 1.b) and the normalised amplitude spectrum of the derivative (shown in Fig. 1.d) of the pulse measured with the DI hydrophone.

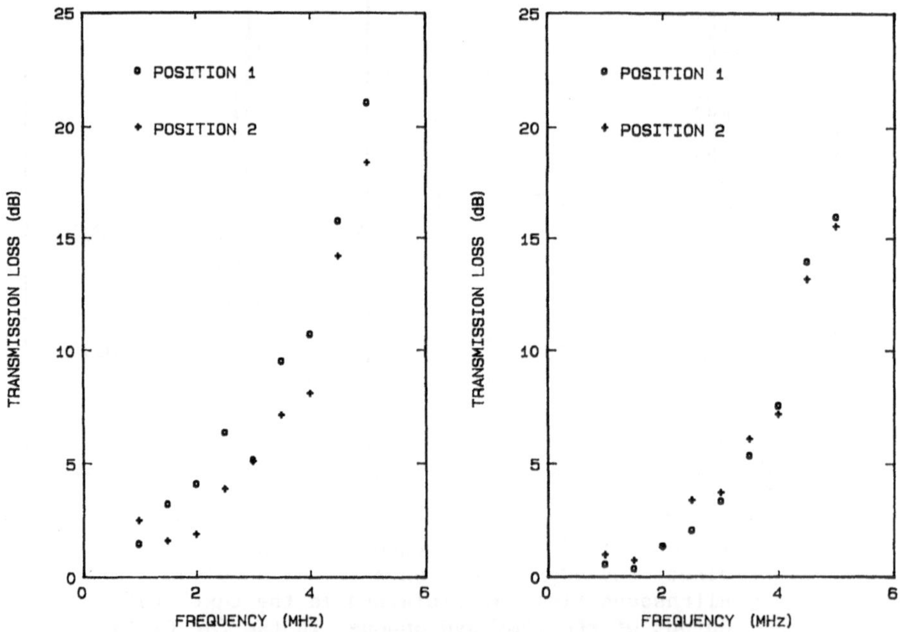

Fig. 3. Frequence dependence of the transmission loss of a glycerine sample, relative to that of an identical water sample, for two different experimental configurations as described in the text; (left) as measured with a point hydrophone and (rigth) as measured with a DI hydrophone.

Reflectivity Measurements

We have devised a novel technique for measuring characteristic parameters of different materials via the ultrasound reflectivity function obtained for planar boundaries between different media. Consider the well known expression for the amplitude reflectivity function:

$$R = (Z_2 - Z_1)/(Z_2 + Z_1)$$

where Z_1 and Z_2 are the characteristic acoustic impedance of medium 1 and medium 2 respectively. Thus, reflectivity measurements encode information about the impedance ratio Z_2/Z_1. Since Z is the product of the medium's density (ρ) and acoustic velocity (c), and since ρ is frequency-independent, we see that it is possible to extract information about velocity dispersion, in principle, by measuring the frequency dependence of the reflectivity function!

The following equation shows the influence of absorption on the reflectivity function:

$$R = (Z_2\beta_1 - Z_1\beta_2)/(Z_2\beta_1 + Z_1\beta_2)$$

where $\beta_j = \alpha_j c + i\omega$, with α denoting the absorption and ω the frequency.

From the above expression we see that absorption processes modify the reflectivity function, even in the presence of an impedance matched boundary, i.e. $Z_1=Z_2$, by the parameter αc, dubbed 'absorbance' by Leeman et al (1979). When there is no absorbance mismatch across the boundary, the reflectivity function reverts to its conventional (impedance model) value. Moreover, in general, a frequency dependence of the reflectivity function results, even when the velocity dispersion is negligible.

The experimental verification of the above ideas requires a plane wave, a plane interface, and a means of independently measuring both the incident and the reflected wave amplitudes, without disturbance. Conventional techniques make use of finite aperture transducers (that do not produce purely plane waves), and great care is necessary in experiments in order to recognise and remove diffraction artefacts that are normally present in these situations. Moreover, the data have to be calibrated against the reflectivity of a 'perfect reflector' (as a polished stainless steel plate).

The technique we devised relies on the following features of the DI hydrophone: it is insensitive to the diffraction artefacts and, moreover, by permitting the ultrasound field to pass through unhindered, it allows both the incident and reflected waves to be measured by the same hydrophone. The method is thus, in principle, a self-calibrating procedure, obviating the need for calibrating results against those obtained with a perfect reflector. Measurements can be made with planar or focused transducers, and the only requirement is that the ultrasound field be entirely and orthogonally intercepted by the DI hydrophone.

A number of machined samples of different materials were maintained with their flat faces parallel to the plane of the DI hydrophone in a water bath at 17° C. Pulsed ultrasound fields were generated with a number of different focused transducers in the 1 to 8 MHz frequency range. Accurate pulse envelopes of the measured incident and reflected waveforms were obtained via digital computation and the analytic signal approach, and reflectivities assessed from a comparison of the incident and reflected pulse peak amplitudes. The frequency of the measurement was taken to be that of the peak spectral amplitude component.

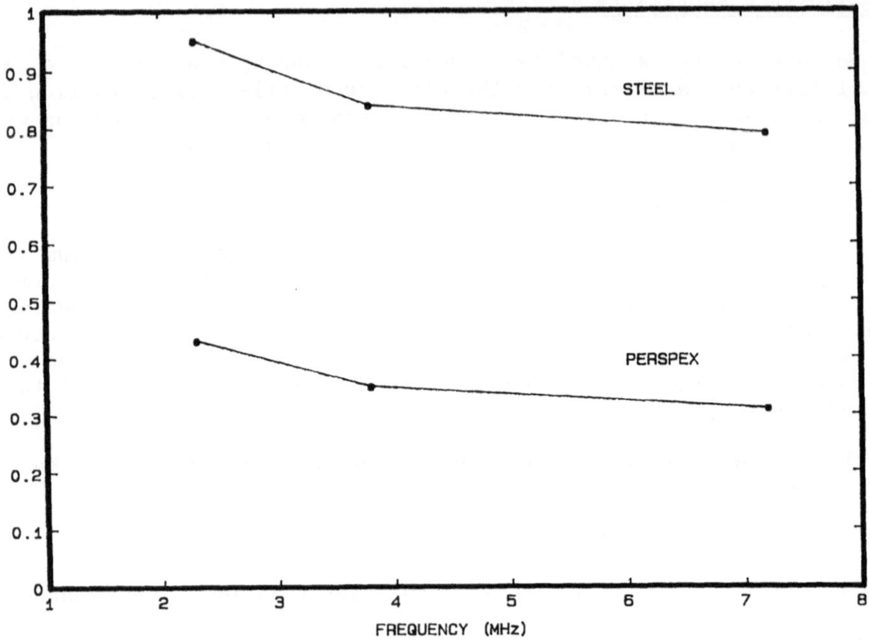

Fig. 5. Measured reflectivities of steel and perspex. Results are not corrected for hydrophone transmissivity.

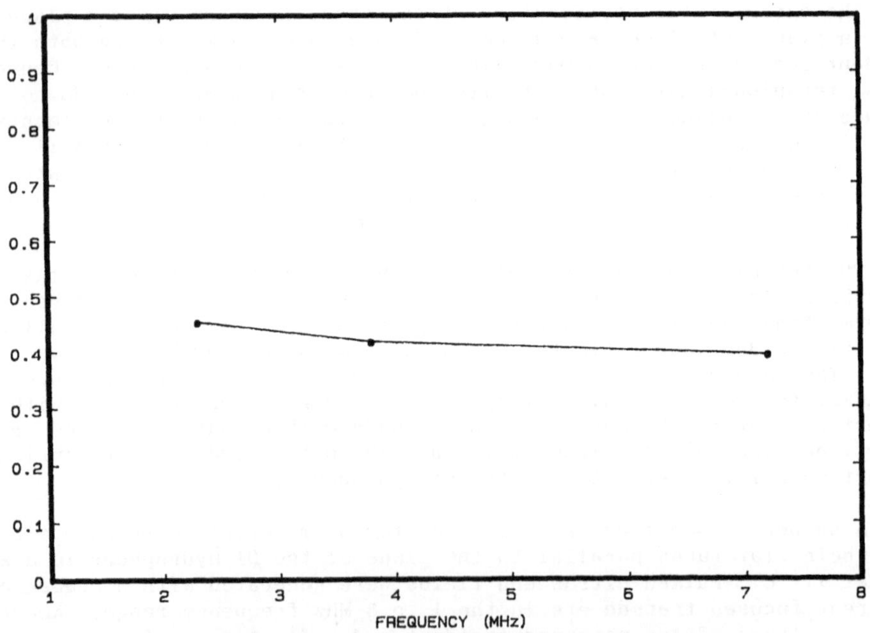

Fig. 6. Ratio of measured reflectivities of perspex and steel. The frequence dependence of this ratio indicates that absorption processes may influence reflection properties.

The results for steel and perspex samples can be seen in Fig. 5. It is clearly seen a decline of the measured values of reflectivity with frequency, for both samples. These results are influenced by the frequency dependence of the hydrophone transmissivity, but the influence of absorption effects on the reflectivity function may be seen by taking the ratio of the perspex to steel results as shown in Fig. 6. If the trends in Fig. 5 were entirely due to the transmission properties of the PVDF membrane of our hydrophone, there should be no trend in the data in Fig. 6. In fact, there was an approximately 10% decline over about 5 MHz, thus confirming a possible frequency dependence of the reflectivity function due to the absorption processes in the different media. It is felt that velocity dispersion effects are, in this case, small compared to the influence of the different frequency dependences of the absorption processes in steel and perspex.

The DI hydrophone and non-linear propagation

We have already shown the insensitivity of the DI hydrophone to the diffractive nature of (linear) ultrasound fields. In fact, such a hydrophone is of use when measuring non-linear fields, as well (Costa et al, 1987). We present here results obtained with both the point and the DI hydrophone at the same locations in a non-linear pulsed ultrasound field produced with a 2.25 MHz, 19 mm diameter, 9cm focused transducer. The measurements were performed at three distances, on axis, from the face of the transmitter: 5 cm (before focus); 9cm (at focus); and 21 cm (beyond focus - far field).

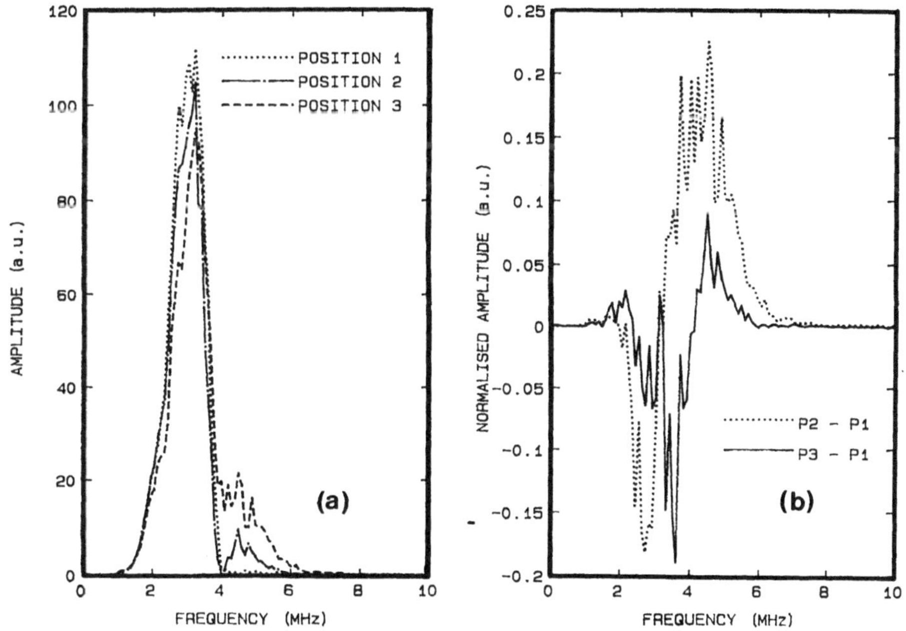

Fig. 7. (a) The non-normalised amplitude spectra in a non-linear pulsed field, as measured by a point hydrophone in the locations described in text; (b) the spectral differences of the normalised amplitude spectra (as described in text).

In Fig.7 we show: (a) the non-normalised amplitude spectra at each position, as measured by the point hydrophone, and (b) the spectral differences between the normalised amplitude spectrum at position 2 and at position 1, and between the normalised spectrum at position 3 and at position 1. The stronger presence of higher frequency components at larger ranges are clearly seen, in accord with theoretical predictions. The spectral differences, on the other hand, show a very complicated pattern, making it very difficult to separate the non-linear effects from the diffraction artefacts and focusing effects embedded in the measurements.

Fig. 8 shows the results obtained with the DI hydrophone and, as for the point hydrophone, the non-normalised amplitude spectra (a) and the spectral differences (b) were taken at the same locations in the same

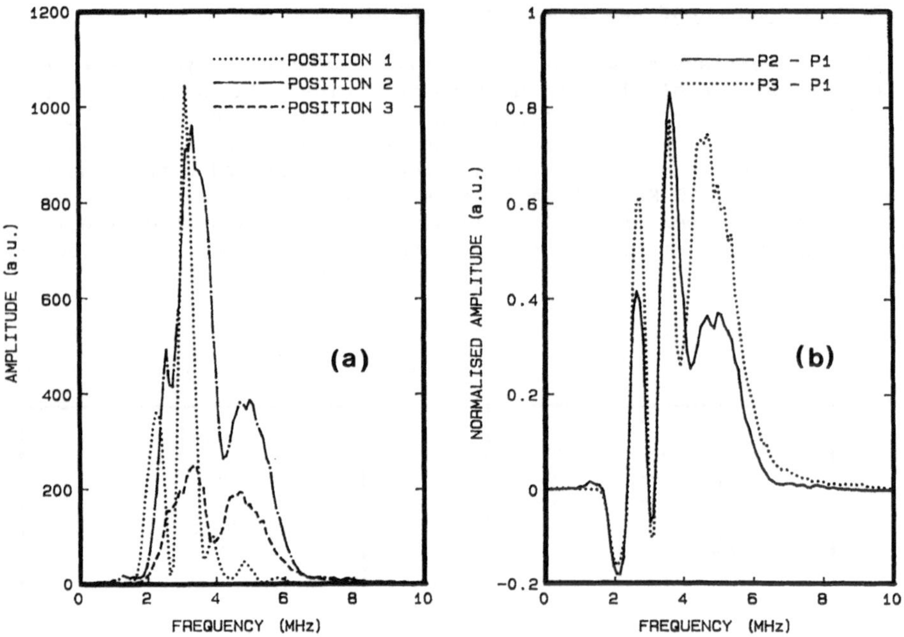

Fig. 8. (a) The non-normalised amplitude spectra at the same locations in the same non-linear field as for Fig. 7 as measured with a DI hydrophone; (b) the spectral differences of the normalised amplitude spectra.

field. As the measurement locations were placed further away from the transducer, the higher frequency harmonics become more evident. The essential features of the phenomena are not masked by the focusing and diffraction effects revealed by the point hydrophone measurements. Nearer the transducer, the low frequency components have not yet shown appreciable non-linear effects, and the higher frequency components of the original pulse start to show some depletion due to the non-linear pumping to higher harmonics. Further away in the field, the presence of frequency components outside the original band is very marked, and even the lower frequency components of the original pulse show depletion. These results show clearly the frequency and range dependence of the non-linear propagation effects, phenomena which are not so obvious in the point hydrophone measurements.

CONCLUSION

We have presented experimental evidence to justify the theoretical predictions of a novel technique to produce diffraction-free measurements with a special hydrophone developed in our laboratory, and have shown its wide applicability and better performance compared to using conventional hydrophones and transducers when measuring attenuation and reflectivity of different materials. It was shown that it indeed measures the direct wave of the ultrasound field, being completely insensitive to diffraction and focusing effects normally seen in conventional methods of assessing the ultrasound field. Furthermore, we gave evidence that its role in analysing the non-linear ultrasound propagation is of great importance, since it separates this effect from the diffraction artefacts and give much more insight into the phenomenon.

REFERENCES

Costa, E.T., Leeman, S. Richardson, P.A.C., and Seggie, D.A., 1986, Measurement and calibration of transducer fields, in: "Proceed ings of The Institute of Acoustics", vol. 8, part 2, R. Lawrence (ed.), The Institute of Acoustics, Edimburgh.

Costa, E.T., Leeman, S., and Hoddinott, J.C., 1987, Overcoming the ultrasound diffraction artefact. II. Attenuation estimation, characteristic impedance and nonlinear propagation, Revista Brasileira de Engenharia, Caderno de Engenharia Biomedica, 4(1):47.

Freedman, A., 1961, PhD Thesis (University of London), released as an Establishment Report entitled 'The formation of acoustic echoes in fluids' by the Admiralty Underwater Weapons Establishment in 1962, Portland, UK.

Gore, J.C., and Leeman, S., 1977, Echo structure in medical ultrasound pulse-echo scanning, Phys. Med. Biol., 22(3):431.

Hutchins, L. and Leeman, S., 1981, Tissue parameter measurement and imaging, in: "Acoustical Imaging 11", J.P. Powers (ed), Plenun, New York.

Leeman, S., Leeks, R., and Sutton, P., 1979, Analysis of pulse-echo ultrasonic images, Les Colloques de l'INSERM, 88:35.

Weight, J.P., and Hayman, J., 1978, Observations of the propagation of very short pulses and their reflection by small targets, J. Acoust. Soc. Am., 62(2):396.

TESTING SURFACE LAYER OF PAVEMENT

WITH AN ULTRASONIC METHOD

Romuald Sztukiewicz

Institut of Civil Engineering
Technical University
Poznań, Poland

INTRODUCTION

In the structural arrangement of flexible pavement the surface layer is based on foundations. It is normally made of asphaltic concrete which has viscoelastic qualities. The definition of the surface layer suggests that it is made of the layer surface and surface layer area, the latter being situated under the real surface. Thus the structure of the surface layer embraces the geometrical structure of the surface and the physico-chemical structure of the area situated under the surface.

Standard methods of testing pavement that have been used so far, are based mostly on measuring quantities characterizing the geometrical surface of the layer, Wojdanowicz[1]. Radiational methods of testing the surface layer area are an example of non-destructive methods that can be used, Christory[2]. Geometrical deformations of the layer surface are secondary changes and they are caused by structural changes that take place within the surface layer area. Thus, methods describing the structure of the surface layer area should be used for testing the surface layer of flexible pavement.

It is possible to use ultrasonic methods, limited to small powers of the ultrasound, where the viscoelastic qualities of medium can be treated as independent from the amplitude of ultrasonic waves, to describe the structure of the surface layer. By applying the ultrasonic method of echo we can establish the velocity of the longitudinal ultrasonic wave which can further be used to describe the structure of the surface layer area of flexible pavement. In this "surface" method two transducers, the sending and the receiving, are placed on the surface layer. Assuming a constant distance between the transducers l' it is possible to establish the longitudinal ultrasonic wave velocity within the surface layer area. Thus we can obtain a relationship between the propagation velocity of the longitudinal ultrasonic wave and bulk specific gravity of the surface layer area. While conducting experiments in the same spots of the surface layer area in time intervals we can observe the changes of bulk specific gravity that take place during the exploitation of pavement[3].

Changes in the surface layer which are due to traffic and external agents can be divided into two stages. The first is concerned with changes of bulk specific gravity that take place during the initial period of exploitation, and these changes are reversible. The second stage includes the creation and expansion of micro-scratches and micro-deffects that cause irreversible changes within internal structure of the surface layer.

The purpose of this paper is to present a surface ultrasonic method that enables to obtain an acoustic image of the surface layer of flexible pavement during the initial period of exploitation.

THE TECHNOLOGY OF TAKING MEASUREMENTS

The one - side access to the surface of the layer forced us to use the method of echo. This "surface" method can be used to mark the basic parameter of ultrasonic waves propagation, i.e. the wave velocity. The surface layer made of asphaltic concrete about 0.05 m thick can be treated as a thin slab. In such slab, where the wave length is bigger that slab's thickness, the first wave to reach the receiving transducer is the longitudinal wave, Wehr[4]. If we register the time of wave propagation and assume a constant distance between the transducers, we can establish the longitudinal ultrasonic wave velocity. The measurements were taken with an ultrasonic material probe that registers the time of the longitudinal ultrasonic wave propagation. The applied transducers frequency of 40 kHz enabled the emission of wave's length bigger than the thickness of the surface layer.

Permanent places for taking meassurements were set on an especially prepared, newly - made section of Serbska street, Poznań, Poland. This section fulfilled all the requirements for experimental road sections, Birula[5]. The ultrasonic measurements were taken in 6 cross - sections on 54 measuring fields. The measuring fields were rectangles 3.0x0.3 m marked on areas of both heavy and sporadic traffic (the axis of a lane). It was assumed that each measuring field was to be an representative volume element (RVE). The surface layer of flexible pavement was made of moderate granular asphaltic concrete of closed structure, 0.05 m thic, according to technological requirements. On each measuring field, six results of longitudinal wave propagation were registered with a constant measuring base 1' = 0.20 m. Knowing the distance between the transducers 1' and the propagation time, the velocity of longitudinal ultrasonic wave was established. By calculating velocities of propagation for the longitudinal ultrasonic waves several times, it was possible to find a velocity distribution for a measuring field.

The time of the longitudinal ultrasonic wave propagation was taken four times a year and we simultaneously registrered the temperature of the surface layer area at the depth of 0.02 m. An temperature probe V.40.33 was used to measure the temperature, applying an multimeter V 640. The ultrasonic measurements taken in different seasons were then brought down to one comarative temperature. The correction of average velocities of the ultrasonic waves was done according to the following formula:

$$c'_{Lr} = c'_{Lp} + \triangle\,(t_r - t_p) \qquad (1)$$

where: c'_{Lr} - the ultrasonic wave velocity for comparative temperature,

c'_{Lp} - the ultrasonic wave velocity calculated in the temperature of a measurement,

t_r - comparative temperature (10° or 20^oC)

t_p - temperature of asphaltic concrete during measurements,

$\triangle$ - the ultrasonic wave velocity variation gradient.

Except for the ultrasonic tests, normal observations were also made. Data concerning the type and structure of traffic were registered. We also gathered meteorological data such as average monthly temperature of air, rainfall, insolation and the number of times when temperature passed the point 0^oC.

THE APPLICATION OF THE SURFACE ULTRASONIC METHOD "IN SITU" ON THE EXPERIMENTAL SECTION

The surface ultrasonic method tests were made in eight terms from October 1984 to September 1986. For each measuring field an average velocity of ultrasonic waves c_L', standard deviation and variation coefficient ν_{cL} were calculated. The tests results were brought to one temperature. They are presented on an exemplary diagram that shows the changes of velocity of the ultrasonic wave and of the variation coefficient during the period of testing. Thus we obtained an acoustic image of the surface layer area as a change of the ultrasonic longitudinal wave propagation velocity (Fig. 1) and a change of the variation coefficient (Fig. 2).

These diagrams show that during the first 6 months of pavement exploitation the velocity of the longitudinal ultrasonic waves went down. It was caused by the process of destruction during winter time (temperature passed the point 0^oC 72 times). Only on lane one the velocities went up and this was caused by heavy trucks that normally take that lane. In the second half of 1985 the velocities of the waves went rapidly up. During that period the traffic became much heavier because a parallel international highway No 2 had been closed. At the same time the insolation of the surface was very heavy. An increased traffic and weather conditions influenced the compaction process of asphaltic concrete within the surface layer area. After winter time which was the period of destructive weather influence and low temperatures the velocities of the waves went rapidly down.

Knowing the velocity of the longitudinal ultrasonic wave we can establish the bulk specific gravity of the surface layer area made of asphaltic concrete. The following formula shows this relationship:

$$c_L' = 1344.0\ \rho_{pm} + 39.692 \qquad (2)$$

where: c_L' - the longitudinal ultrasonic wave velocity (surface method),

ρ_{pm} - bulk specific gravity of the surface layer area.

Similarly, it is possible to analyse the changes of the variation coefficient of the ultrasonic wave velocity in time function.

CONCLUSIONS

Experiments that had been carried out proved that changes of the surface layer area structure in the initial stage of exploitation lead to changes of bulk specific gravity and can be registered by an ultrasonic material probe. Observations of variations in ultrasonic waves velocities c_L' and the relationship between bulk specific gravity ρ_{pm} and velocity c_L' enable to determine changes of bulk specific gravities within the surface layer area made of asphaltic concrete.

The above remarks suggest that the ultrasonic method may be very helpful in describing the structure of the surface layer area in the initial stage of exploitation. The results of the ultrasonic investigations can be presented as diagrams creating an acoustic image of the surface layer area of flexible pavement.

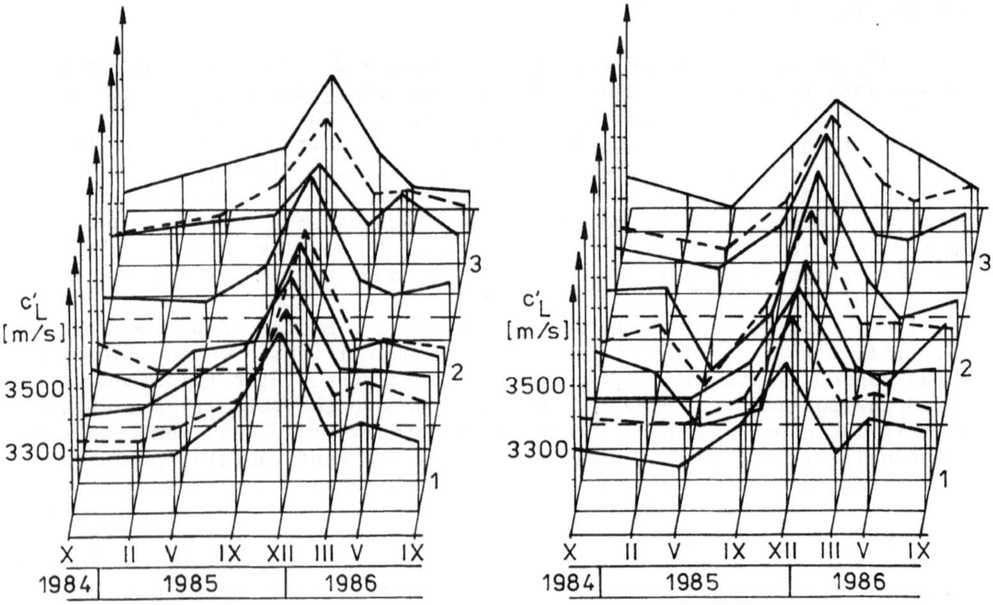

Fig. 1. Diagram showing ultrasonic wave velocities changes in time function for cross - sections 3 and 4

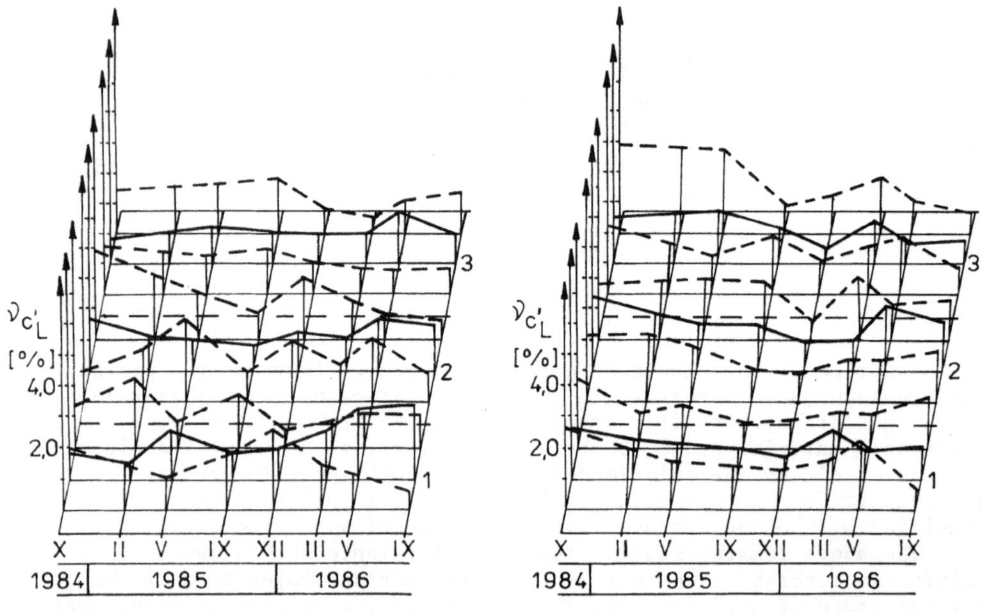

Fig. 2. Diagram of variation coefficient changes in time function for cross - sections 1 and 2

REFERENCES

1. S. Wojdanowicz, Zagadnienia oceny stanu technicznego nawierzchni drogowych, Prace IBDiM, 4 (1985) 183:196.
2. I. P. Chistory, G. Laurent, Methodologie l'utilisation du GDM 45, Ministere de l'Equipement, Paris (1974).
3. R. J. Sztukiewicz, Testing surface pavement with an ultrasonic method, Ultrasonics International 87 Conf. Proc. London (1987) 366:371.
4. J. Wehr, Pomiary prędkości i tłumienia fal ultradźwiękowych, PWN, Warszawa (1972).
5. A. K. Birula, Przydatność eksploatacyjna nawierzchni drogowych, WKiŁ, Warszawa (1971).

ACOUSTIC CHARACTERISTICS OF SNOW

M. Ikegami[+], K. Tonooka[+], and
Y. Aoki[++]

[+] Government Industrial Development Laboratory
Hokkaido, Sapporo, 004, Japan
[++]Department of Information Engineering, Hokkaido
Sapporo, 060, Japan

ABSTRACT

This paper is concerned with the reconstruction of ultrasonic imaging under fallen snow. We are improving the radar technique under fallen snow. After the attenuation constant of snow is measured, one and two dimensional imaging method are proposed using a noise reduction technique. This technique is used because the acoustic reflected signals from objects under fallen snow are so weak, that the reflected signals acquired are mixed with noises.

INTRODUCTION

In our snowy and cold area, snow removing work is a serious subject. 4500Km of roads must be kept clean and a few people die from the snow which has fallen from the roof. The technique of imaging under fallen snow is important in order to mechanize the work and to reduce the labor of snow removing workers.

Microwave is suitable for imaging objects under fallen snow, since the attenuation of microwave is relatively small in snow and ice. However there are some defects because of the large attenuation in moist snow and regulations under the wireless telegraphy act.

Acoustic wave is not dangerous to life and easy to use in public. There are many types of snow: dry snow, moist snow, and almost ice removed by a snow plow. There are various characteristics of snow which affect acoustic propagations: the size of particles, the percentage of moisture content, specific gravity, etc. It is difficult to exactly measure these characteristics, because the conditions of measuring vary easily. To image some objects under fallen snow, the most important acoustic characteristics are the relationship between the attenuation and the frequency, and the relationship between the attenuation and characteristic of the snow.

We improved the measurement technique of the acoustic attenuation of fallen snow, the noise reduction technique of acoustic images and the two-dimensional imaging technique as explained below.

ATTENUATION MEASUREMENT OF SNOW

Many kinds of acoustic characteristics effect acoustic propagation. How long acoustic waves propagate under fallen snow depends on the attenuation constant of the snow and the reflection at the boundary between air and snow principally. So we tried to measure the attenuation constant of the snow.

The difficulty in measuring the acoustic characteristics of snow exists in identifying the kind of snow which has various kinds of structures. Although it is possible to measure the attenuation of snow, we can not know what kind of snow is measured. In this experiment, the temperature and density of the snow and the air temperature were measured. However they seem to not be sufficient in order to determine the characteristics of snow. Simple measurement methods of characteristics of snow need further investigation.

The attenuation(Zl) between transmitter and receiver includes three factors. These are the defused reflection(Ri) on the surface of snow, the attenuation(Ai) of snow and the defused reflection(Ro) on the surface of air as follows:

$$Z(l_n) = Ri + Ai(l_n) + Ro \tag{1}$$

where l_n is the thickness of the snow wall. $Z(l_n)$ can be measured, and Ri and Ro must be eliminated. To seek the relationship between Z and l_n, $Z(l_n)$ is measured many times varying the l_n. Figure 2 shows this relationship. Ai equals the attenuation of the snow divided by the distance of the acoustic propagating path. Ai is obtained by calculating the gradient of the line of these graphs.

After assuming that the snow is not melted, the experiment was carried out at an air temperature below 0 degrees centigrade. Figure 1 shows the experimental set-up which was used to measure Ai and which was used to reconstruct one and two dimensional images also. The frequencies of 25 or 40KHz were used. The distance between the transmitter and receiver was about 15cm. The transmitting wave was 10 cycle burst wave. When the strength of the signal was measured, the receiver samples the signal directly in order to monitor the signal and to make a distinction between the signal directly propagated in air and the signal propagated through snow.

The strength of the signal was measured about 10 times because it varied widely at the same thickness, l_n. Figure 2 shows these recorded results while changing the l_n. From this figure, the attenuation at 25KHz and 40KHz are 2.0 dB and 2.7 dB. Moreover, it seems that the recorded signal levels($Z(l_n)$) at 25KHz are higher than the level at 40KHz. It is

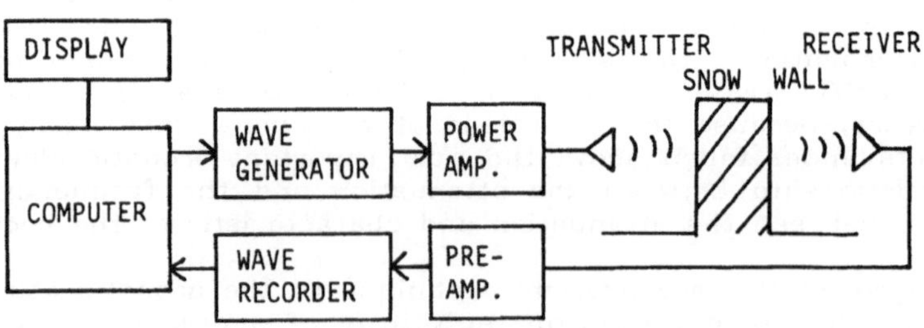

Fig. 1. Experimental configuration.

because the reflection from the surface of the snow at 40KHz is stronger than the level at 25KHz. Table 1 presents other results. From these results we can not find the relationship between the attenuation and the density. This is because the density is not measured precisely or there is another parameter which effects the acoustic attenuation.

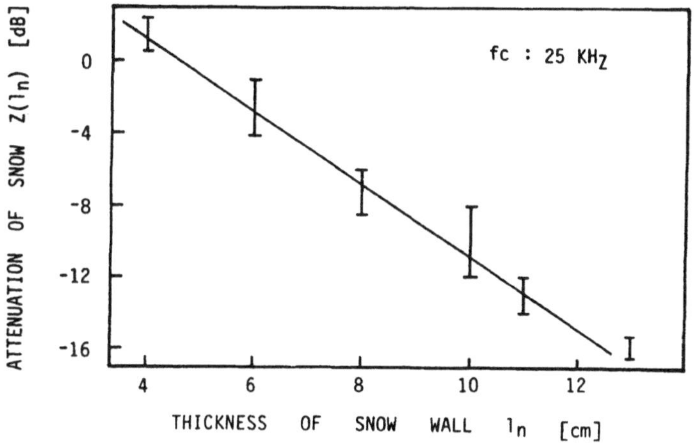

Frequency is 25KHz.

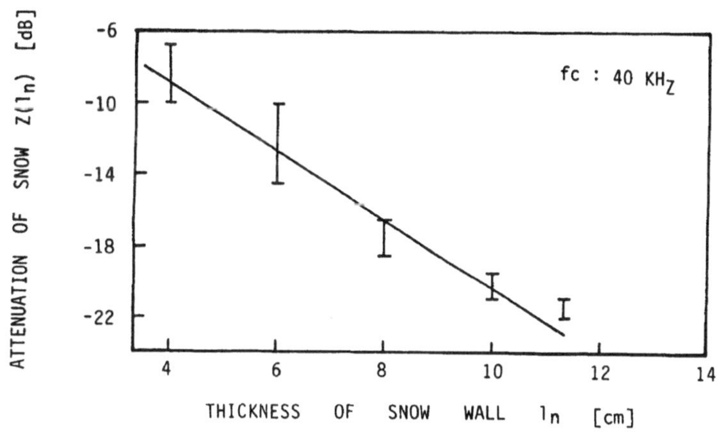

Frequency is 40KHz.

Fig. 2. Relation between thickness of snow wall: l_n, and attenuation of snow. The density of snow is $0.19 g/cm^3$. The gradient of graph denotes the attenuation of snow.

Table 1. Values of attenuation and density.

Density (g/cm^3)	0.19	0.18	0.29
Attenuation at 25KHz(dB)	2.0	1.7	1.9
Attenuation at 40KHz(dB)	2.7	1.9	2.9

NOISE REDUCTION OF ONE DIMENSIONAL IMAGE

Many types of noise signals are recorded with reflected signals in the analog circuit of the imaging system. One is generated by analog amplifiers which amplify the small received signals and the other is generated by digital circuit which controls the whole imaging system. These noises cover the weak signals which are reflected from objects at a far distance and decrease the special resolution. In general, these signals have wide band unknown spectrum, or narrow band known spectrum. Because the transmitting waves are short burst signals which have mono frequency spectrum in our system, it is possible to separate reflected waves from noises.

In general, one_dimensional images are obtained by recording the strength of the reflected signals. If the noises are recorded with the reflection, then the noises overlap with the reflection. The burst transmitting wave T(t) is given by:

$$T(t) = \begin{cases} \sin(w_0 t) & 0 < t < T_0 \\ 0 & T_0 < t \end{cases}, \tag{2}$$

where w_0 is carrier frequency and T is burst duration. The reflected signals R(t) from the objects O(l) which are located l far from the transducer can be expressed as:

$$R(t) = \int T(t - 2*l/v) * O(l)\, dl + N(t) \tag{3}$$

where N(t) is the noise signals.

To reconstruct the one_dimensional image, R(t) must be detected. The analog detection circuit can be substituted for the moving average D(t):

$$D(t) = \frac{1}{T} \int_{t-T/2}^{t+T/2} | R(t') |\, dt' \tag{4}$$

where integration period is T. By this method, D(t) is a reconstructed image which includes noise. There are no suppressive effects of noises.

The transmitting waves have mono spectrum: w_0, and noises have widely spread spectrum. So when the received signal is processed by band pass filter, the center frequency of w_0, it is possible to reduce the noises. We made such a filter using the short period frequency analysis technique. The reconstructed image which is filtered D'(t) can be expressed as:

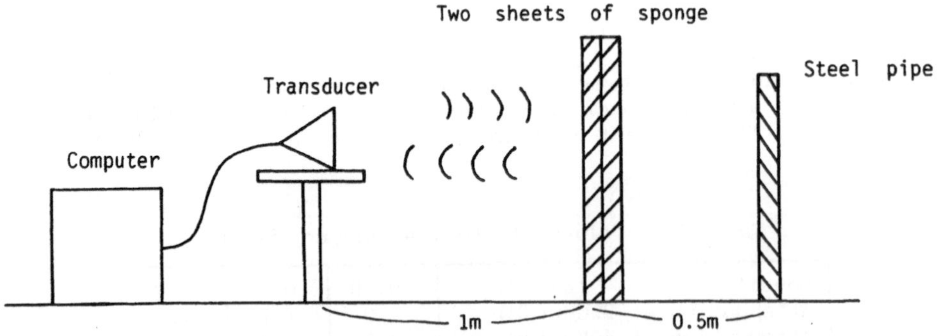

Fig. 3 Experiment set up.

$$D'(t) = \int R(t') * e^{-jw_0 t'} * U(t'-t)\, dt' \quad (5)$$

$$U(t) = \begin{cases} 1 & 0 < t < T_0 \\ 0 & \text{otherwise} \end{cases}, \quad (6)$$

where U(t) is window function and T_0 is window size.

The experiment was carried out using 25KHz frequency in the room. The transmitter and receiver are made by piezoelectric ceramics. The band width is 750Hz in -3dB points. Transmitted waves form 5 cycle burst waves. In the receiver section, the reflected base band signal is sampled directly using an A/D converter. The sampling rate is 2μS. The burst duration time is equal to the 100 times the sampling interval. The detector is constructed using a digital filter by software in the computing processes. Test objects consist of: a steel pipe, which is 6cm in diameter, and two sheets of sponge, which are 8cm in thickness. Transmitting signals become weak through these sponges. The attenuation of the sponge is about -17dB and strength of reflection from the steel pipe is about -19dB compared with an aluminum plate. Figure 3 shows the experimental arrangement.

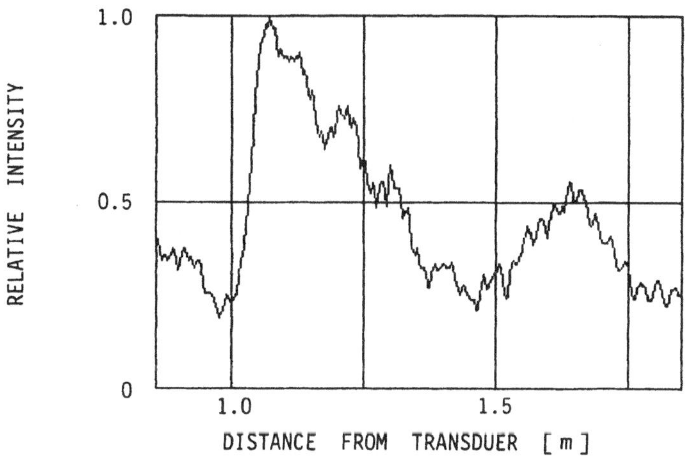

Fig. 4. Reflected signal filtered by moving average.

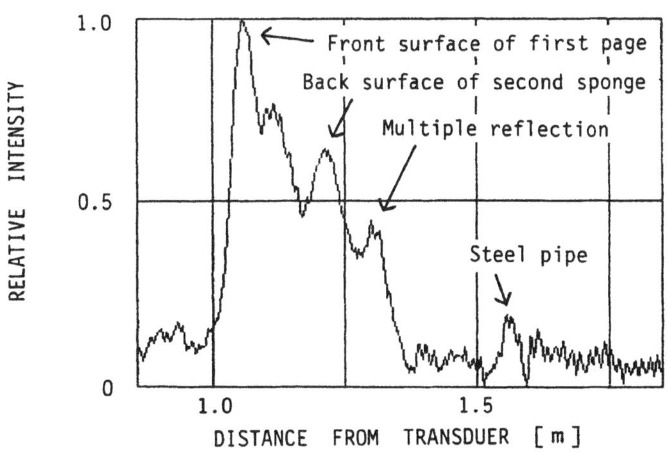

Fig. 5. Reflected signal filtered by short period spectra analysis.

Figures 4 and 5 show one_dimensional reconstructed images. Fig. 4 is obtained by the detection of the moving average. The window size is 128 points(256μs). It is equal to the width of the transmitting burst signal. The detection is confirmed by moving the window by 10 sampling points steps. Fig. 5 is obtained by the short period spectrum analysis technique. The transmitting signal is 25KHz. So the spectrum of the window of 128 points is analyzed and the strength of the 25KHz spectrum is rearranged according to the distance factor.

We can observe three big peaks on reconstructed images. The first peak is the reflection from the front surface of the first sponge. The second peak is the reflection from the front surface of the second sponge. The third peak is the back surface of the second sponge. Another peak is corresponding to multiple reflections of these three surfaces. We should recognize the reflection from the pipe, 50cm behind the front surface of the first sponge. The noise level shown in Fig. 4 is bigger than that of Fig. 5. In Fig. 5 the reflection from the pipe can be observed clearly. But the reflection can not be obtained clearly in Fig. 4, because the noise signals are so strong compared with the reflection from the steel pipe.

TWO DIMENSIONAL IMAGE IN AIR

We tried to make two dimensional images, one axis is the time domain and another axis is corresponding to the rotating angle of the pair of transmitter and receiver. The directivity of the transmitter and receiver are obtained using an acoustic horn. To optimize the length of the horn, the relationship between length and beam profile is measured. Figure 6 shows this relationship. The beam profile is measured at the angel where the signal is -10dB weak compared with the maximum level. When the length of the horn increases, the beam profile becomes narrower up to 15cm. However, the beam profile is not significantly improved over 15cm. It seems that the accuracy of the horn structure drops as the length of the horn increases.

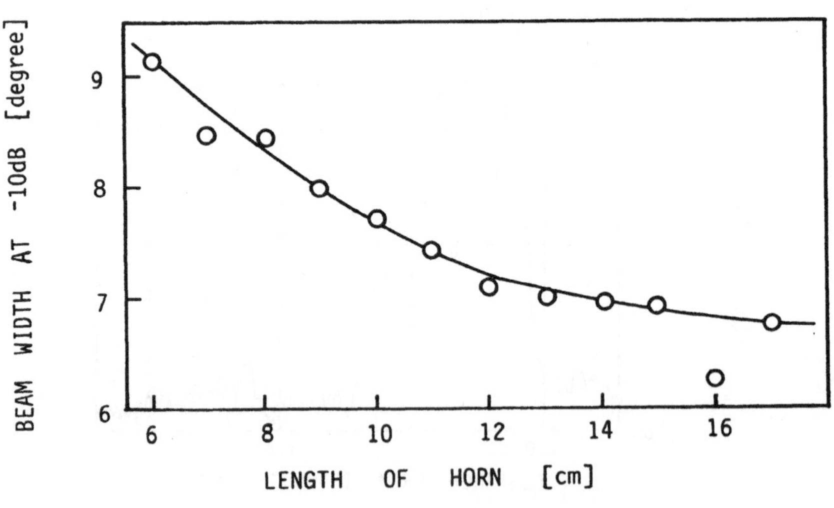

Fig. 6. Relation between beam profile of -3dB point and horn length.

Figure 7 shows the beam profile of the horn which were used in this experiment, where the beam width at -3dB angle is 8 degrees. Fig. 8 shows the two dimensional images when rotating the transducer from 0 to 45 degrees. The two objects which are steel pipes of 15cm diameter were placed 50cm the from transducer. The distance between the two objects is 3cm(in the upper figure) and 4cm(in the low figure). In the lower figure the two objects are recognized separately.

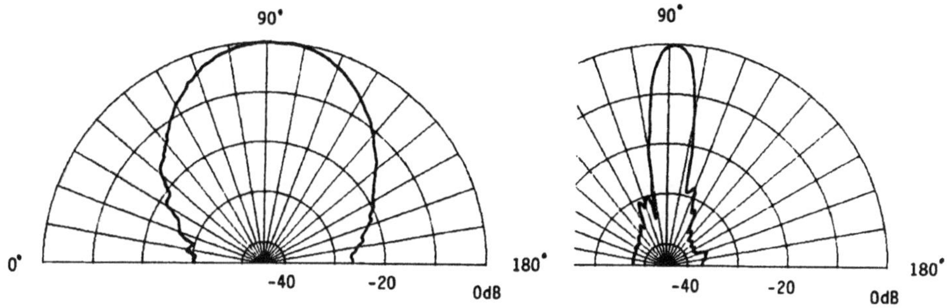

Fig. 7. Beam profile of transducer used in this experiment.

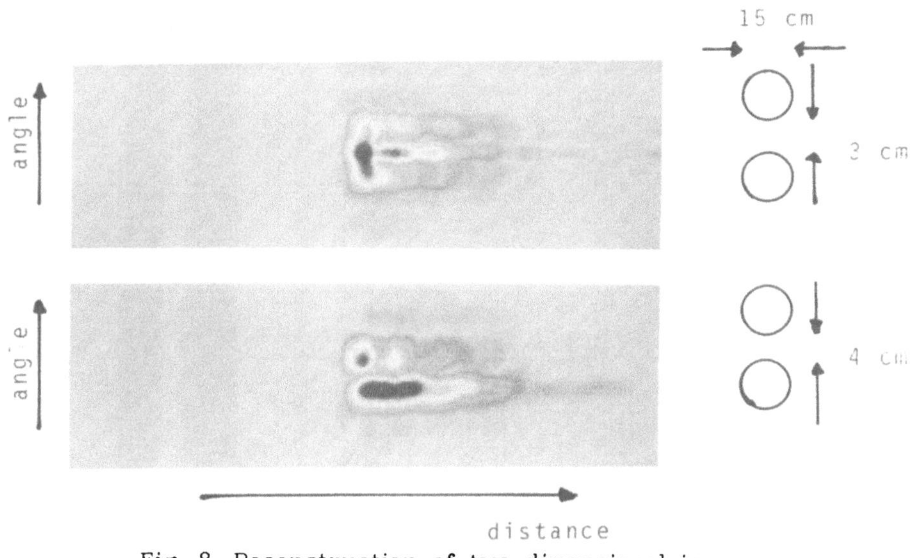

Fig. 8. Reconstruction of two dimensional image.

CONCLUSIONS

The attenuation constants of several kinds of fallen snow have been measured. However these value were too big to propagate at a long distance under fallen snow, it seems to be possible to image in near field, within 1m, using a lower frequency acoustic wave than 25KHz or 40KHz. Because the reflected signals were estimated to be so weak and noisy, the noise reduction technique using short period frequency analysis was proposed. We were able to obtain improved spatial resolution and noiseless image. Furthermore, by adapting this technique, we obtained two dimensional images. After conducting this experiment, we realized that an easy characterization technique of snow is important.

REFERENCES

A. Ishimaru: *Wave Propagation and Scattering in Random Media*, Academic Press, pp.50-51, 1978.

H. R. Reed, C. M. Russell: *Ultra High Frequency Propagation*, John Wiley & Sons, Inc., 1953.

Stiles, W. H. and F. T. Ulaby: *Microwave Remote Sensing of Snowpacks*, NASA Contractor Report 3263, (1980).

M. Ikegami, T. Yamamoto, Y. Aoki: *Ultrasonic Reflection Mode CT with Projection Data by A-mode Image*, Acoustical Imaging Vol.15, Plenum, New York, 1986

E. E. Hundt, E. A. Trautengerg: *Digital Processing of Ultrasonic Data by Deconvolution*, Vol.SU-27, No.5, pp.249-252, 1980.

ACKNOWLEDGMENT

The author would like to thank N. Hiraike and K.Okada for their assistance in arranging the experimental set-up. Also the valuable advice and support of Drs. S. Suzuki, S. Sayama, K. Suzuki is gratefully acknowledged.

A STATISTICAL APPROACH TO SOUND

SCATTERING IN RANDOM INHOMOGENEOUS MEDIUM

W. S. Gan

Acoustical Services Pte Ltd
29 Telok Ayer Street
Singapore 0104, Republic of Singapore

INTRODUCTION

In this paper, the problem of strong scattering of sound wave in a highly inhomogeneous medium for acoustical imaging problem is being considered from a statistical approach. Although the inverse scattering of sound wave has been considered from a statistical approach [1,2] before, it is based on Born approximation and Rytov's approximation. Here the above approximations are avoided. The medium is considered as a generalised communication channel[3]. First the Langevin equation for an inhomogeneous stochastic medium is used. To solve this equation, conventional techniques of solution, such as standard perturbational and variational techniques fail, because the random inhomogeneity operator which describes the interaction of the incident field with the inhomogeneous medium is random in nature and these media are space-time variable in their inhomogeneities. Hence the statistical approach is used.

DERIVATION OF THE LANGEVIN EQUATION

Using the statistical communication theory[3], the general situation is embodied in the operational relation:

$$\{v\} = \vec{T}_{R_0} (\vec{T}_{AR} \vec{T}_M^{(N)} \vec{T}_{AT}) \vec{T}_{T_0} \{u\} \tag{1}$$

where $\{u\}$ is a set of "messages" to be transmitted and $\{v\}$ is the ensemble of received messages which are consequent upon the set $\{u\}$. The $\vec{T}_{AT}$, $\vec{T}_M^{(N)}$, $\vec{T}_{AR}$ are general operators which describe, respectively the process of coupling to the medium by the transmitting aperture, the effects of the medium itself on the resulting injected space-time signals, and the receiving aperture, by which the received field is converted into a (temporal) wavem for further signal-processing by the receiver itself. The operators

$\vec{T}_{T_0}$, $\vec{T}_{R_0}$ are temporal processes only, representing the overall "encoding" and "decoding" of the original and final "message" sets $\{u\}$, $\{v\}$, in (1). This $\vec{T}_{T_0}$ includes the actual encoding process whereby "messages" are converted into signals, and then are suitably modulated to drive the transmitting aperture $\vec{T}_{AR}$. Conversely, $\vec{T}_{R_0}$ includes the corresponding decoding process, as well as any appropriate usually nonlinear signal processing. In this paper, we are concerned essentially with the "signal" portion of Eqn.(1), namely the injected signal $S_{in}(t)$, the received wave $\vec{X}(t)$, following the receiver aperture $\vec{T}_{AB}$, and with the detected wave $Y = T_{Det}\{\vec{X}\}$ (before any decoding):

$$Y(t) = T_{Det}\left\{X(t)\right\} = \vec{T}_{Det}\left\{\vec{T}_{AR}\vec{T}_M^{(N)}\vec{T}_{AT}\left\{S_{in}\right\}\right\} = \vec{T}_{Det}\vec{T}_{AR} \tag{2}$$

where S_{in} is an injected signal and where

$$\alpha(\vec{R},t) = \vec{T}_M^{(N)}\vec{T}_{AT}\left\{S_{in}\right\} = \vec{T}_M^{(N)}\left\{-G_T\right\} \tag{3}$$

is the field at some point $(\vec{R},t)$ in the medium in question, and G_T is a source function. Here G_T and X(t) are explicitly given by

$$G_T = \hat{G}_T S_{in} = G_T(\xi,\ t/S_{in})$$

$$= \int_{-\infty}^{\infty} h_T(t-\tau,\ t/\xi)\, S_{in}(\tau,\xi)\, d\tau$$

$$= h_T * S_{in} \tag{4}$$

and $X(t) = R\,\alpha$

$$= \int_R dV_R(\vec{R}) \int_{-\infty}^{\infty} h_R(t-\tau,\ t/\vec{R}\in V_R)\,\alpha(\vec{R},\tau)\, d\tau$$

$$= \int_R h_R * \alpha\, dV_R \tag{4b}$$

where the source and array operators $\hat{G}_T$, $\hat{R}$ are respectively,

$$\hat{G}_T = \hat{G}_T\,(t,\xi/\tau) \int_{-\infty}^{\infty} h_T(\tau,t/\xi)(\ \ )_{R,t-\tau}\, d\tau \tag{5}$$

$$\hat{R} = \hat{R}(t/\vec{R},\ \tau)$$

$$= \int_R dV_R \int_{-\infty}^{\infty} h_R(\tau, t/\vec{R} \in V_R)(\ \ )_{R,t-\tau}\, d\tau$$

$$(= T_{AR}) \tag{6}$$

For teh linear media assumed in this paper, the field obeys a partial integro-differential equation of the form:

$$(\hat{L}^{(o)} - \hat{Q})\alpha = -G_T + \text{b.c.'s} + \text{i.c.'s} \tag{7}$$

where $\hat{L}^{(o)}$ is a linear (scalar) partial differential operator with $\hat{L}^{(o)}$ associated with the homogeneous portion of the medium. Here $\hat{Q}$ is, in general, a scalar, linear integro-indifferential scattering operator, which describes the interaction of the incident, homogeneous field with the (differential, or local) scattering(i.e., reradiating elements) of the inhomogeneous portion of the medium. Boundary conditions (n.c.'s) and initial conditions (i.c.'s) are a necessary part of Eqn.(7), as indicated.

Since we are dealing generally with random media, inasmuch as Q contains random as well as deterministic components, we are concerned with the appropriate ensembles, $\{\alpha\}$, $\{X\}$, etc. The ensemble of equations(7) governing propagation of the field is now the Langevin equation. Since, formally,

$$\{\alpha = (\hat{L}^{(o)} - \hat{Q})^{-1}(-G_T)\} \tag{8}$$

In comparison with Eqn.(3), shows that

$$\{T_M^{(N)} = (\hat{L}^{(o)} - \hat{Q})^{-1}\} \tag{9}$$

SOLUTION OF THE LANGEVIN EQUATION

For the purely deterministic special cases we seek solutions of Eqn. (7) directly. However, for the general situation of random media, the "solutions" of the corresponding Langevin equation (8) are the various statistics of the field $\{\alpha\}$, e.g. the various moments $\langle\alpha\rangle, \langle\alpha_1\, \alpha_2\rangle$, etc. and the moments of the (linear functionals) of the field, $\{X(t)\}$, e.g., $\langle x\rangle$, $\langle X_1 X_2\rangle$, etc and more comprehensively, the various probability densities (and distributions) of α and X. Thus we are interested here primary in the means $\langle\alpha\rangle, \langle X\rangle$, the intensities $\langle\alpha^2\rangle$, $\langle X^2\rangle$, and the covariances ($\sim\langle\alpha_1\, \alpha_2\rangle$, $\sim\langle X_1 X_2\rangle$). When X is a Gaussian process, these first- and second-order moments are sufficient for subsequent signal processing, to yield $\{v\}$. But, as is often the case, X here is not Gaussian, so that a least first-order distributions of X must be developed for effective processing.

Starting with the appropriate equations of state, mass, and momentum conservation, one can obtain by suitable approximations of the resulting nonlinear propagation equations, linearized "wave equations" which are special variants of Eqn.(7). However, in this paper we will deal with the nonlinear wave equations instead. It is more convenient to proceed with the canonical form Eqn.(7) directly. Here $\hat{L}^{(o)}$ a linear differential operator, with coefficients independent of (R,t), while Q also linear, may be an integro-differential (or global) operator and is dependent on position (R,t). For the moment, we regard Q and hence Eqn.(7) as deterministic. For our case of sound propagation in highly inhomogeneous medium, the medium in inhomogeneous absorptive. Then Equ.(7) reduces to

$$\left\{\left(1 + \tau_x(\vec{R},t)\frac{\partial}{\partial t}\right)\nabla^2 - \frac{1}{c_o^2}\left[1 + \epsilon(\vec{R},t)\frac{\partial^2}{\partial t^2}\right\}\alpha = - G_T \tag{10a}$$

where $\hat{L}^{(o)} = \left(1 + \tau_{ox}\frac{\partial}{\partial t}\right)\nabla^2 - \frac{1}{c_o^2}\frac{\partial^2}{\partial t^2}$

$$\hat{Q} = \frac{\epsilon}{c_o^2}\frac{\partial^2}{\partial t^2} - \tau_{ox}\gamma_x\frac{\partial}{\partial t}\nabla^2 \tag{10b}$$

with $\tau_x = \tau_{ox}'(1 + \gamma_x[\vec{R},t])$ generally. When $\tau_{ox} = 0$, Eqn.(10a) reduces to the "extended" inhomogeneous Helmholtz equation

$$\left\{\nabla^2 - \frac{1}{c_o^2}\left[1 + \epsilon(\vec{R},t)\right]\frac{\partial^2}{\partial t^2}\right\}\alpha = - G_T \tag{11a}$$

$$\hat{Q} = \frac{\epsilon}{c_o^2}\frac{\partial^2}{\partial t^2}$$

$$\hat{L}^{(o)} = \nabla^2 - \frac{1}{c_o^2}\frac{\partial^2}{\partial t^2} \tag{11b}$$

Here $\vec{R} = \hat{i}_x x + \hat{i}_y y + \hat{i}_z z$ (12)

and rectangular coordinates are assumed throughout, so that $\vec{\nabla}\cdot\vec{\nabla}\cdot \quad = \nabla^2$, etc. Specifically, c_o = constant wavefront speed, embodies the effects of velocity gradients, internal wave phenomena and /or (weak) local turbulence, while τ_x represents the effects of relaxation absorption.

With inhomogeneous (but still deterministic) media conventional techniques of solution fail, largely because the medium is not reciprocal and there is no general way of applying and evaluating the initial conditions over the various volume integrals which appear in the development of the generalized Huygen's principle(GHP) when $\hat{Q} \neq 0$. Moreover, since these media are space-time variable in their inhomogeneities, e.g., $\hat{Q} = \hat{Q}(\vec{R},t/\ldots)$ either locally, or because of local Doppler, such media do not support space- and time-harmonic solutions nor are the standard perturbational and variational techniques of the "classical" approaches generally applicable, particularly when $\hat{Q} = \hat{Q}(\vec{R}, t/\ldots)$ and is a random operator and is the case here. Finally, in addition to all these difficulties is the usually insuperable one of bounded media when the boundaries themselves are random and moving.

Accordingly, entirely new approaches[3] are required, for manageable solutions of Eqn.(7) generally and Eqn.(10a) in particular which embody the physical realities of the application and which in particular, include both the random character of $\hat{Q}$, i.e. ϵ (R,t) in Eqn.(10a) etc. and the random Dooppler effects, as well.

For linear media, with $\hat{L}^{(o)\,-1} = \hat{M}$ (which also includes boundary and initial conditions), it can be shown that, in operator form, Eqn.(7) becomes, for each member of the ensemble of propagation equations

$$
\begin{aligned}
(\hat{1} - \hat{M}\hat{Q})\,\alpha &= \hat{M}(-G_T)\\
\alpha &= \alpha_H + \alpha_I\\
\alpha_H &= \hat{M}\,(-G_T)
\end{aligned}
\tag{13a}
$$

where $\hat{1}\alpha = \alpha$, etc. We note that α_H, Eqn.(13a) is the unscattered field, of the purely deterministic, i.e. homogeneous portion of the medium here. Moreover, because any boundaries are included in $\hat{Q}$, $\hat{M}$ is replaced by M_∞, so that

$$
\alpha_H = \hat{M}_\infty \quad (-G_T)
\tag{13b}
$$

where the infinite domain operator for the homogeneous component of the medium becomes

$$\hat{M}_\infty = \hat{M}_\infty(\vec{R}, t/\vec{R}', t')$$

$$
= - \int_{-\infty}^{\infty} dt' \cdot \int_{(\infty)} d\vec{R}'\, g(\vec{R}, t/\vec{R}', t')_\infty \;(\quad)_{\vec{R}', t'}
\tag{14a}
$$

$$
= - \mathcal{F}_S^{-1}\,(\hat{Y}_{0,\infty})
\tag{14b}
$$

$$= - \mathcal{F}_K\, \mathcal{F}_S^{-1} \left\{ \mathcal{Y}_{0,\infty} \right\}$$

Here

$$
\begin{aligned}
\hat{\mathcal{Y}}_{0,\infty} &= \hat{\mathcal{Y}}_0(\vec{k}, S) \int dt' \int_{(\infty)} e^{-i\vec{k}\cdot\vec{R}' - st'}\,(\quad)_{\vec{R}', t'}\, d\vec{R}'\\
\hat{Y}_{0,\infty} &= \hat{Y}_{0,\infty}(S/\rho) \int dt' e^{-st'}\,(\quad)_{t'}
\end{aligned}
\tag{14c}
$$

and $\vec{k} = \hat{i}_x k_x + \hat{i}_y k_y + \hat{i}_z k_z = 2\pi \nu$

(14d)

is a vector warenumber, with ν a vector spatial frequency. Here g_∞ is the Green's function of the corresponding infinite, homogeneous medium with $\hat{Q} = 0$.

For the inhomogeneous Helmholtz equation (11a), (11b), Eqn.(14c) for the homogenous part reduce to

$$Y_{0,\infty} = \frac{e^{-\rho S/c_o - st'}}{4\pi\rho}$$

$$\mathcal{Y}_{0\infty} = \left(k^2 + \frac{s^2}{c_o^2}\right)^{-1}$$

$$M(\vec{R}, t/\vec{R}', t')_\infty = \frac{\delta(t - t' - \rho/c_o)}{4\pi\rho} \tag{15}$$

REQUIREMENT OF STATISTICAL APPROACH TO LANGEVIN EQUATION

Usually, the inhomogeneous Helmholtz equation is solved by perturbation technique[4]. But these involve random operator and divergence problem of the diffraction integral. Hence, to achieve useful results, the Langevin equation must be converted into various deterministic forms[3], which represent the various moments of the scattered field. This leads to suitable replacement of the random $\hat{Q}$ by an appropriate deterministic operator $\hat{Q}_1^{(d)}$, $\hat{Q}_{12}^{(d)}$, etc. for the first-, second- and higher order moments of the field.

To use the statistical approach, the (random) channel operator $T_M^{(N)} = (\hat{1} - \hat{\eta})^{-1}\hat{M}$ is written as

$$T_M^{(N)} = \left\{\hat{1} + \frac{\hat{\eta}}{\hat{1} - \hat{\eta}}\right\}\hat{M} = \hat{M} + \hat{I} = \hat{\mathcal{M}}$$

$$\therefore \hat{I} = \frac{\hat{\eta}}{1 - \hat{\eta}}\hat{M} \tag{16}$$

where we separate the homogeneous ($\hat{M} = \hat{M}_\infty$ here) from the scatter operator $\hat{I}$. We note that for ambient fields, in addition to the desired signal source (S_{in}), we replace the source function $-G_T$ in the Langevin equation (Eqn.(7)) by $-G_T - G_A$, where now G_A is a localized or a distributed, ambient "noise" source with an associated field $\alpha_A = \hat{M}_\infty \quad (-G_A)$.

Mostly for reasons of analytical complexity, channel characterization is usually formulated in terms of the lower moments, e.g. the mean and co-variance operators associated with I. The moment operators are, directly from Eqn.(15),

$$\hat{m} = \hat{M} + \langle \hat{I} \rangle$$

$$\hat{K}_I = \langle \hat{I}_1 \hat{I}_2 \rangle - \langle \hat{I}_1 \rangle \langle \hat{I}_2 \rangle$$

$$\langle \hat{m}_1 \hat{m}_2 \rangle = \hat{M}_1 \hat{M}_2 + \hat{M}_1 \langle \hat{I}_2 \rangle + \hat{M}_2 \langle \hat{I}_1 \rangle + \langle \hat{I}_1 \hat{I}_2 \rangle$$

and with higher moments similarly determined. These moment operators are deterministic, as a result averaging.

We should like to point out here, the advantage of the statistical approach over the perturbation technique, is that in the perturbation technique, it involves the random operator which is not deterministic. But in the statistical approach, the random operators become deterministic as a result of averaging. For example, I, the scatter operator is not deterministic. But it becomes deterministic after averaging:

$\hat{I}_{1,2} = \hat{I}(\vec{R}_{1,2}, t_{1,2} / \vec{R}', t')$ where $(\vec{R}_1, t_1)$ and $(\vec{R}_2, t_2)$ are different points in the field.

Expressions for the corresponding moments of the received wave X(t) are found from

$$X(t) = \hat{R}\,\alpha = \hat{R}(\hat{M} + \hat{I})(-G_T - G_A) \tag{17}$$

Thus the mean received waveform and various second moments are respectively,

$$\langle X \rangle = \hat{R}\langle \alpha \rangle = \hat{R}(\hat{M}_\infty + \langle \hat{I} \rangle)(-G_T - G_A) \tag{18}$$

$$\langle X_1 X_2 \rangle = \hat{R}_1 \hat{R}_2 \langle \alpha_1 \alpha_2 \rangle = \hat{R}_1 \hat{R}_2 \hat{m}_1 \hat{m}_2 G_1 G_2 \tag{19}$$

where $G = G_T + G_A$ with

$$K_X(t_1, t_2) = \langle X_1 X_2 \rangle - \langle X_1 \rangle \langle X_2 \rangle = \hat{R}_1 \hat{R}_2 (\langle \alpha_1 \alpha_2 \rangle - \langle \alpha_1 \rangle \langle \alpha_2 \rangle)$$

$$= \hat{R}_1 \hat{R}_2 \hat{K}_I (G_1 G_2) \tag{20}$$

The array operators in Eqns.(4b) and (6) which appear in Eqns.(18) to (20) for the received wave, can be rewritten alternatively as

Here V_R is the region occupied by the physical array itself (in the field α) and η locates the array element $d\eta$ with respecto to O_R, the coordinate origin of the receiving array. The array is space-time linear filer. In practice, most physical arrays consist of an assemblage of m = 1,2, ..., M discrete, nut distributed elements, each of which samples the radiation field α. Thus the array operator in Eqn.(21) may be expressed in more detail formally as a vector operator

$$\hat{R} = \int_{-\infty}^{\infty} d\tau \int_{R_m} (\ ) h_m(\eta, t-\tau)_R d\eta = \{\hat{R}_m\} \tag{22}$$

which defines each "component", $\hat{R}_m$.

Finally, we can see that the solution of the Langevin equation in Eqns.(18) to (20) can be expressed in explicit forms with substitutions for G, R, M, I from Eqns (11a), (22), (16) and (17) respectively.

REFERENCES

1. J.M. Richardson and K. A. Marsh, "Probabilistic Approach to the Inverse Problem in the Scattering of Elastic and Electromagnetic Waves", preprint.
2. L. A. Chernov, "Wave Propagation in a Random Medium", Dover Publications, New York (1960).
3. D. Middleton, "Channel Modeling and Threshold Signal Processing in Underwater Acoustics: an Analytical Overview", IEEE Journal of Oceanic Engineering, OE-12:4 (1987).
4. M. Slaney and A. C. Kak, "Imaging with Higher Order Diffraction Tomography, paper presented at the 1985 IEE Ultrasonics Symposium, San Francisco, USA, October 1985.

COMPARISON OF NUMERICAL ULTRASONIC IMAGING WITH PHOTOELASTIC IMAGING

K. Harumi
Tokyo University of Information Sciences
1200 Yatou-chou Chiba, Chiba, Japan

K. Date
Miyagi National College of Techology
Natori, Miyagi, Japan

M. Uchida
College of Industrial Techology, Nihon University
Narashino, Chiba, Japan

H. Shimada
Akita National College of Technology
Akita, Akita, Japan

ABSTRACT

A new photoelastic visualization of ultrasonics proportional to the intensity of ultrasonics has been done by Date. The results obtained by this means and by numerical visualization are in very good agreement, thus verifying the reliability of the two methods.

1. INTRODUCTION

Several earlier reports on photoelastic (P-E) visualization were made by Date[1) 2)] and be also proposed a new method[3)] in which the light intensity is proportional to the intensity of ultrasonics. Harumi performed visualization by numerical experiments, and the results agree with experimental results within 2 or 3 dB error, his results are now used in verification and prediction of experiment.[4)] Numerical experiments have been performed in the same manner as the experiments by Date, and the two results show very good agreement even in the details of many pictures. This is true even though the materials are assumed to be steel in the numerical experiments and glass in the photoelastic experiments. We believe these results verify the reliability of the two methods qualitatively.

2. REFLECTION OF TRANSVERSE WAVE OF 45 DEGREE INCIDENCE BY A CYLINDRICAL HOLE

Figure 1 shows a photoelastic picture of reflection of a transverse wave by a cylindrical hole, and Fig. 2 is the vector diagram of displacement vectors corresponding to Fig. 1. Two minima in the reflected transverse wave RT are clearly observed in both figures, as are the two reflected longitudinal waves RL1 and RL2. In Fig. 1 the intensity of the longitudinal wave is illustrated rather weakly (about half the intensity of the transverse wave), because the wavelength of the longitudinal wave is about twice that of the transverse wave and the light intensity is proportional to the stress.

Figure 3 shows a P-E figure in which the incident transverse wave IT has passed a hole, and the two rotating surface waves on the hole exactly come together on the surface; three parts of the reflected transverse waves and two minimums are observed. Figure 4 shows numerical results of the same instance, and again, the two roating surface waves exactly come together in the incident direction (shown by an arrow) on the hole.

Three parts, RT1, RT2, and RT3 of the reflected transverse waves and the two reflected longitudinal waves RL1 and RL2 are also been observed. Figure 5 represents the P-E picture slightly after the incident wave IT has passed, and Fig. 6 is a numerical figure of the displacement vectors. These figures clearly prove the very good agreement between P-E and numerical results.

3. REFLECTIONS BY AN UPPER TIP OF A CRACK FOR 45 DEGREE INCIDENCE

Figure 7 shows a P-E picture of the reflection of the transverse wave of 45 incidence by the upper edge of a crack with rectangular tip shape, and Fig. 8 is the corresponding numerical picture when the incident wave IT has passed the edge. RL is the reflected longitudinal wave and RT is two reflected transverse waves having their propagating center at the two corners of the upper rectangular edge of the crack in Fig. 8; these are also observed in Fig. 7. Figure 9 shows the P-E figure when the incident wave has passed the lower edge, and Fig. 10 is the corresponding numerical picture. RT is the transverse wave reflected on the right surface of the crack, and H denotes the head wave connecting the incident wave IT with the reflected longitudinal wave RL. Two edge waves RT1 and RT2 from the upper edge are also observed in Fig. 10, and again in Fig. 9. Very complicated edge waves from the lower edge are seen in the P-E picture of Fig. 11, and the same waves, S1 and S2 are also visible in Fig. 12.

4. REFLECTION OF LONGITUDINAL WAVE BY A CRACK PLACED PARALLEL WITH THE PROPAGATION DIRECTION

Figure 13 shows the reflection of the longitudinal wave produced from a normal transducer placed on a free surface, and IL is the incident longitudinal wave propagating downwards. Two transverse waves from the two edges of the normal transducer are observed, as well as transverse and longitudinal edge waves RL and RT. These latter waves are also found in Fig. 14, and the two transverse waves from the transducer are denoted as EW in this figure.

Figure 15 is a photoelastic picture in which the incident wave IL has just passed the lower edge of the crack, and Fig. 16 is the corresponding numerical picture. RT2 in Fig. 16 is a head wave connecting the transverse edge wave RT1 with the incident longitudinal wave; the longitudinal edge wave RL reaches and is reflected on an upper free surface of medium, and in both figures two transverse waves EW arrive at the upper edge of the crack.

Figure 17 is a photoelastic picture and Fig. 18 is the numerical picture in which longitudinal edge wave RL in Fig. 16 is reflected on an upper free surface; it becomes RLL and mode converted transverse wave RLT in both figures. Edge wave EW from the lower edge is also observed in both figures.

5. REFLECTION OF A LONGITUDINAL WAVE BY A CRACK PLACED PERPENDICULAR TO THE PROPAGATION DIRECRION

Figure 19 is a photoelastic picture of reflection of the longitudinal wave from a normal transducer by a crack placed perpendicular to the propagation direction of the incident longitudinal wave. Fig. 20 is the numerical picture of Fig. 19. IL is the incident wave from the normal transducer and is reflected by a rectangular shape crack. The reflected transverse and longitudinal waves RT1 and RL1 are observed in both figures. Transverse waves EW shown in the previous section are also observed. Figure 20 and 21 show the reflected longitudinal wave RL and the mode converted transverse wave RT; transverse edge wave TE with its propagating center near the right edge of the crack is clearly observed in both figures. S is a surface wave at the end of TE propagating on the two free surfaces of the crack, and H is the head wave connecting TE with the reflected longitudinal wave RL on the lower side of the crack.

The very good agreement of the two results verifies the validity of the two methods, even in the details of corresponding pictures.

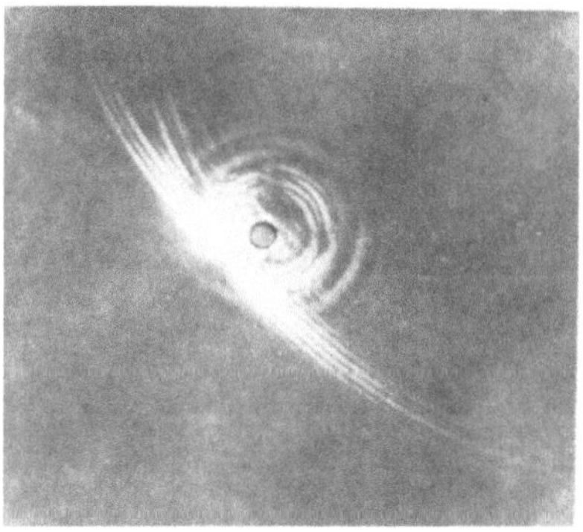

Fig. 1 Reflection of transverse wave by a cylindrical hole.

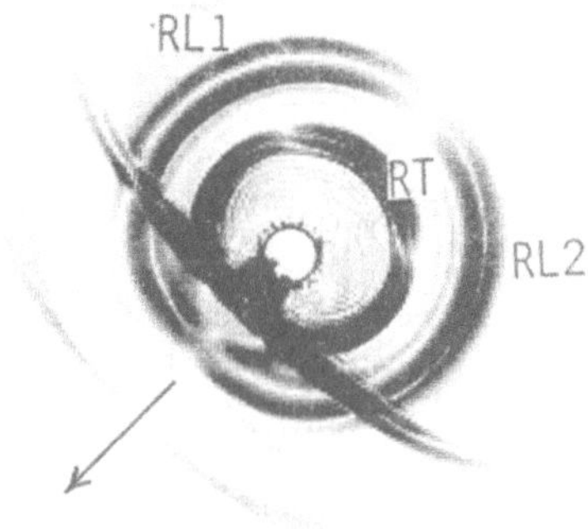

Fig. 2 Displacement vector diagram for Fig. 1. RT; reflected transverse wave; RL1,RL2,reflected longitudinal waves.

CONCLUSION

We thus can see in these four cases that the two very different methods of photoelastic and numerical visualization exhibit remarkable satisfactory agreement, at least qualitatively. This is true despite differences in materials that is steel for numerical experiments and glass for photoelastic experiments, and the difference of the directivities of incident waves.

Fig.3 Reflection by a cylindrical hole.

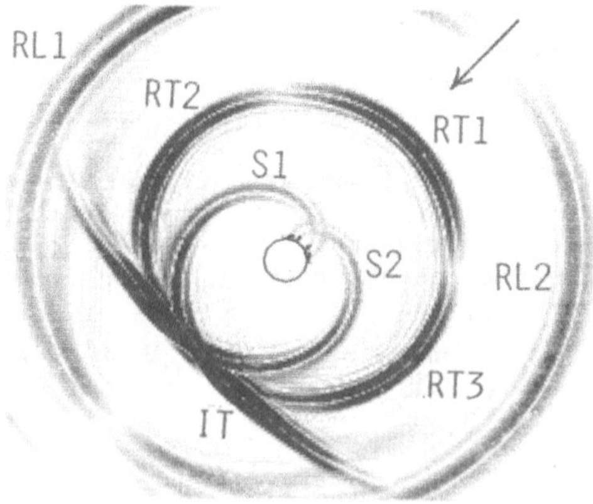

Fig. 4 Numerical result for Fig. 3; RT1, RT2, RT3, reflected transverse waves; RL1, RL2, reflected longitudinal waves; IT, incident wave; S1,S2,rotating surface waves.

Fig. 5 Reflection by a cylindrical hole.

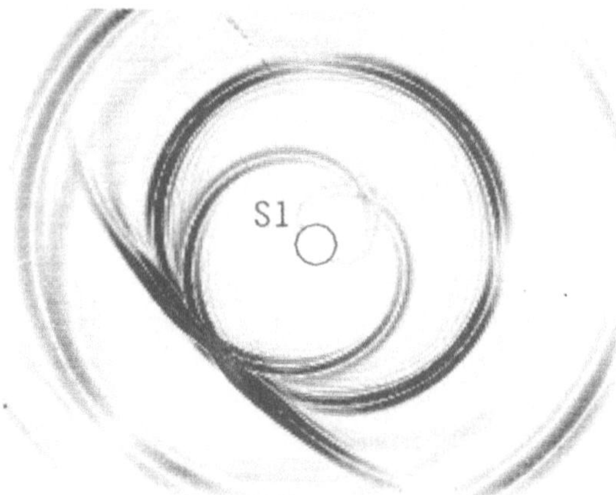

Fig. 6 Numerical result correspond to Fig. 5.

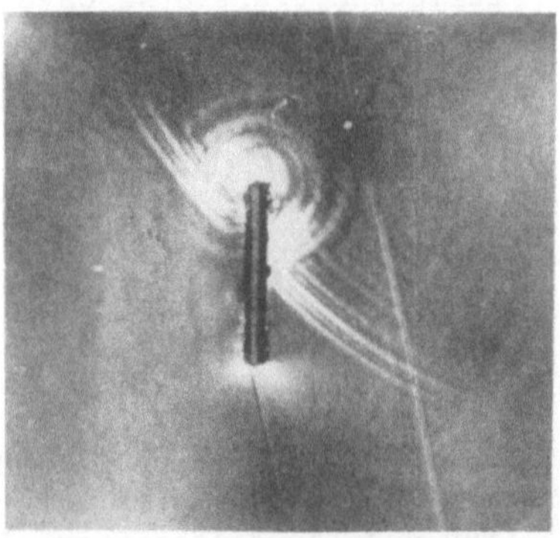

Fig. 7 Reflection by an upper edge of a crack.

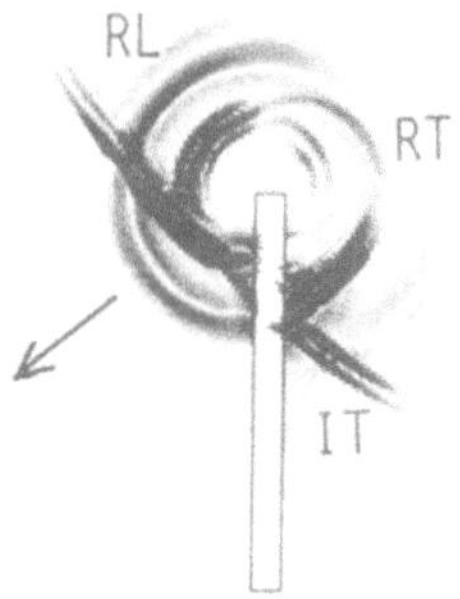

Fig. 8 Numerical result for Fig. 7.
IT, incident transverse wave;
RL, reflected longitudinal waves;
RT, reflected transverse wave.

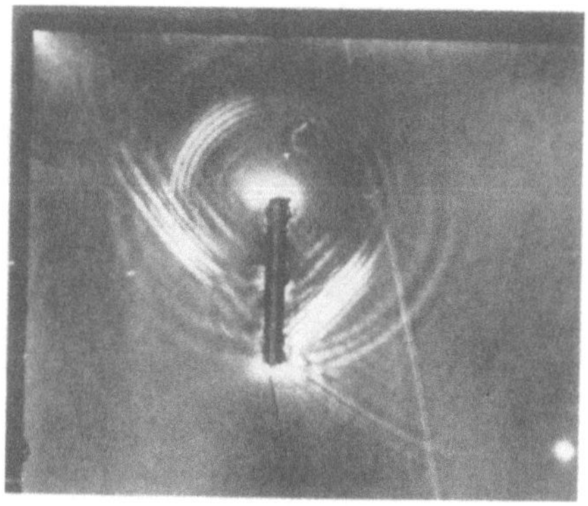

Fig. 9 Reflection by a crack

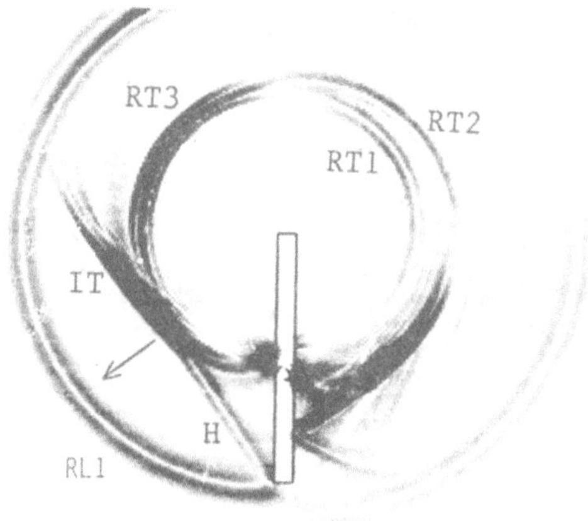

Fig. 10 Numerical result for Fig. 9.
RT1, RT2 edge waves from
two corners of an upper edge.

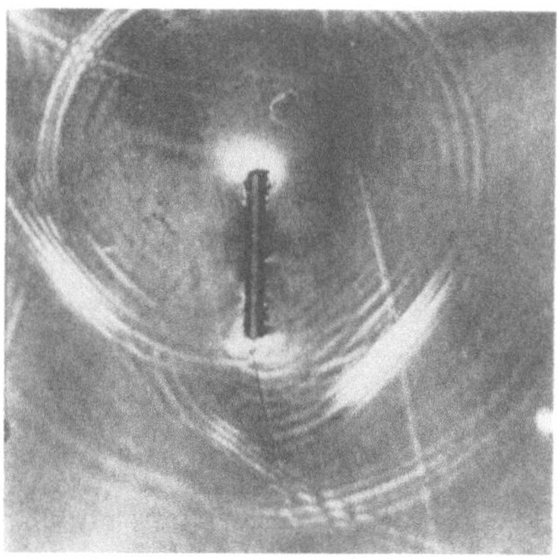

Fig. 11 Reflected waves near the lower edge.

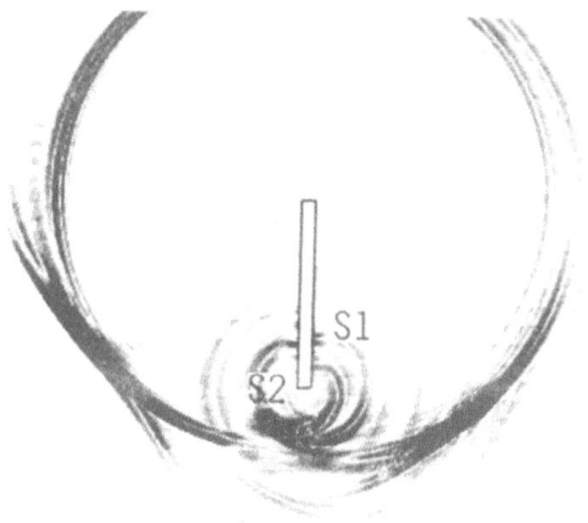

Fig. 12 Numerical result for Fig. 11.
S1, S2, edge waves.

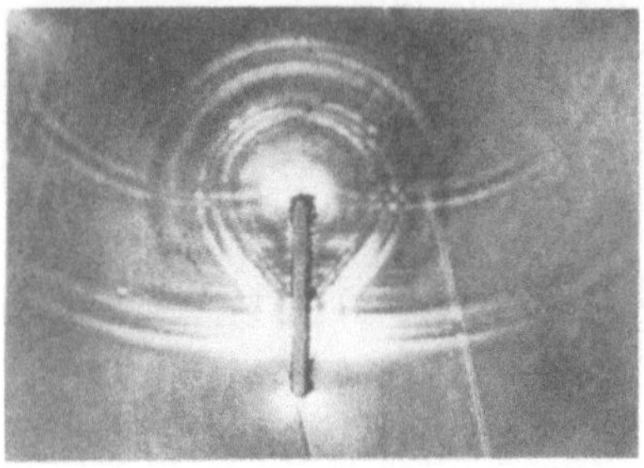

Fig. 13 Reflection of the longitudinal wave by a crack from a transducer.

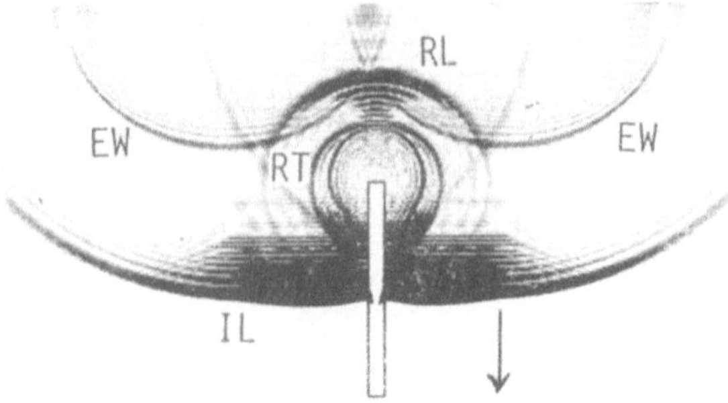

Fig. 14 Numerical results for Fig. 13.

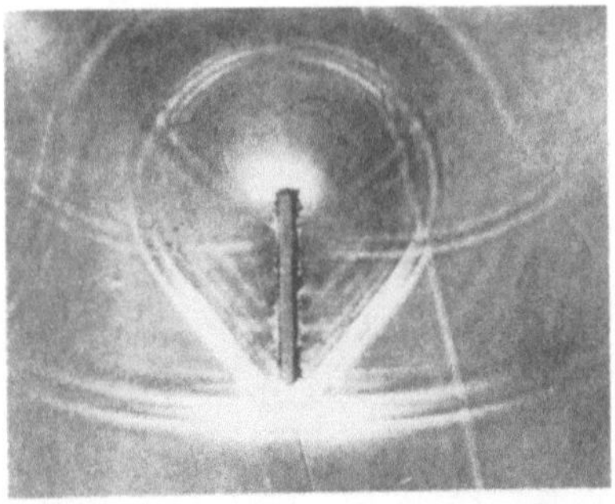

Fig. 15 Incident longitudinal wave pass the lower edge of the crack.

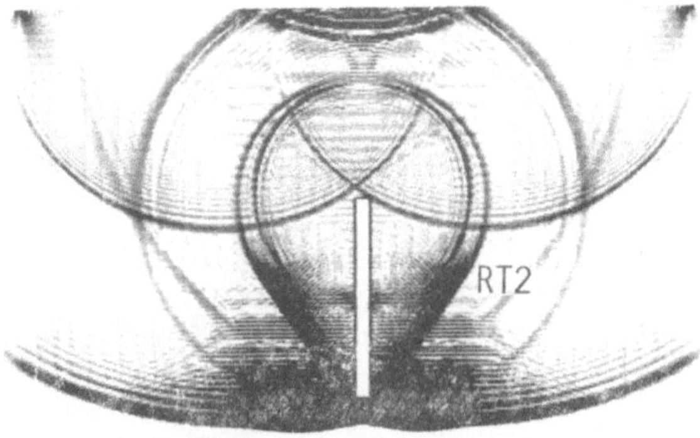

Fig. 16 Numerical result for Fig. 15. RT1 transverse edge wave; RT2, head wave.

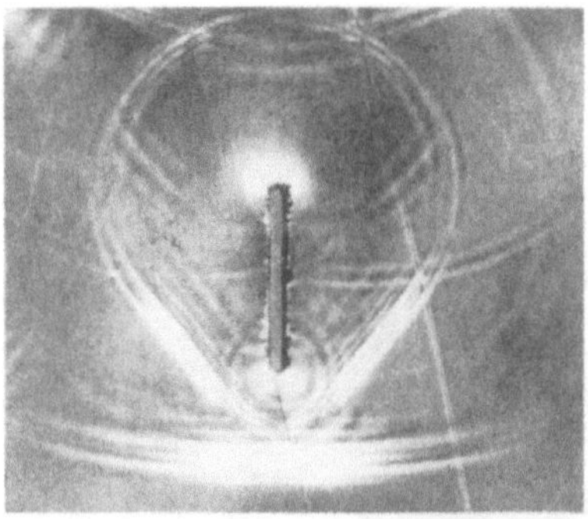

Fig. 17 Edge waves EW from the lower edge.

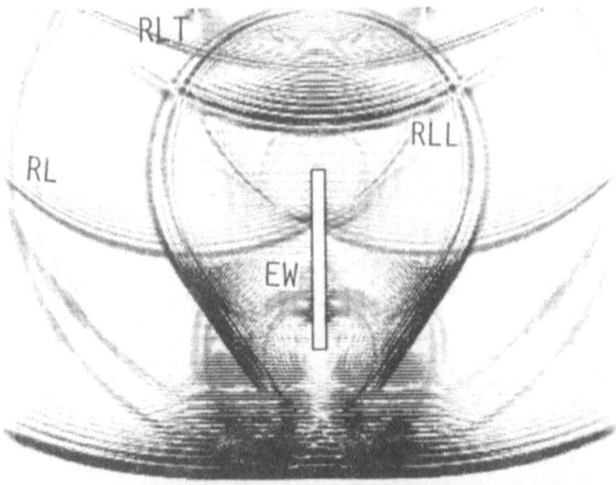

Fig. 18 Numerical result for Fig. 17. RL reflected longitudinal wave;RLL, RLT, reflected longitudinal and transverse waves by an upper free surface.

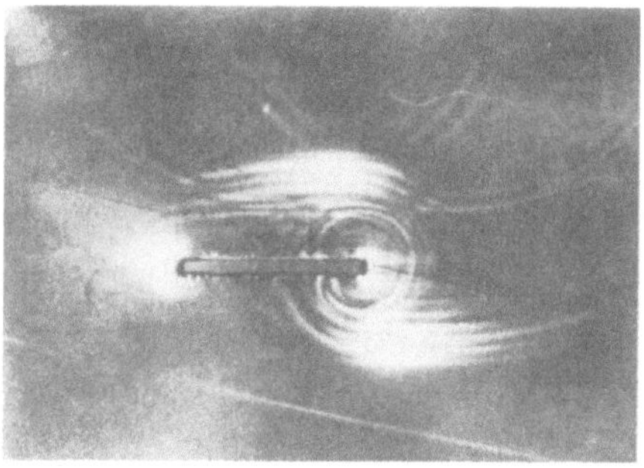

Fig. 19 Reflection of the longitudinal wave by a crack placed perpendicularly.

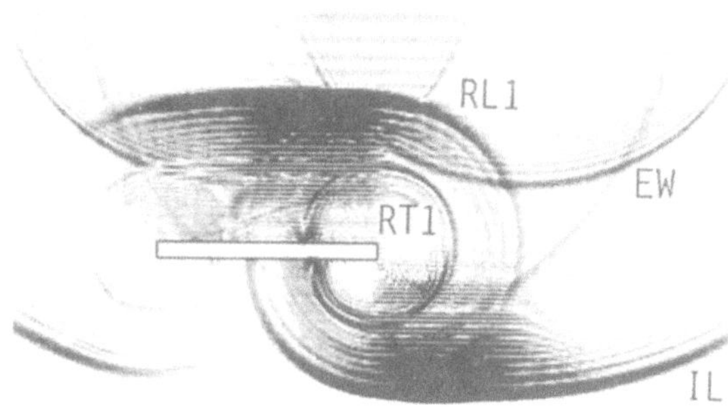

Fig. 20 Numerical result for Fig. 19. EW turansverse wave from the transducer; RL reflected longitudinal wave; RT reflected transverse wave.

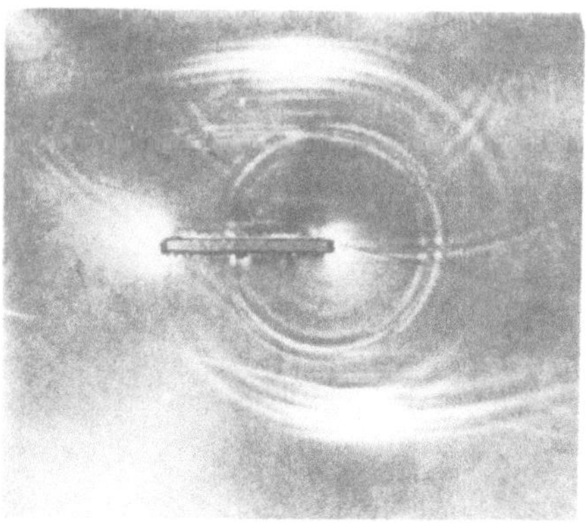

Fig. 21 Reflection by a crack placed perpendicularly.

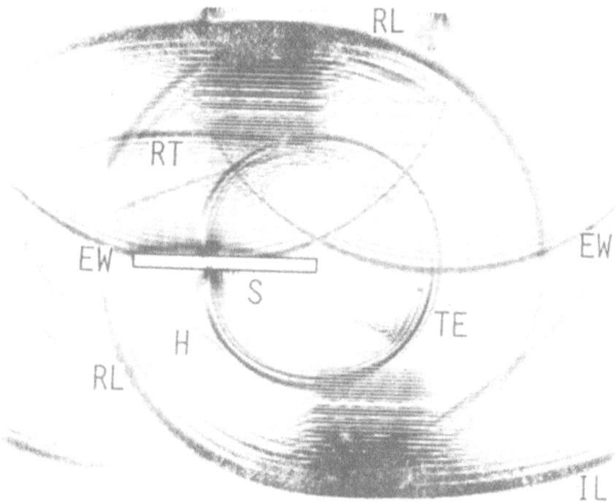

Fig. 22 Numerical result for Fig. 21. TE transverse edge wave from the right edge of the crack; H head wave; S surface wave.

ACKNOWLEDGEMENT

The computation of this work was done by the use of HITAC-S810-20 in Tokyo University.

REFERENCES

1. K. Date, Y. Ito, H. Shimada, "Observation of Ultrasonic Wave in Solid Using Stoboscopic Photoelasticity", Jn. Japan Soc. NDI vol. 33, pp. 513(1984).
2. K. Date, H. Shimada, "Measurement of Ultrasonic Wave Stress by Stroboscopic Photoelasticity", Jn. Japan Soc. NDI vol.35, pp.148(1986).
3. K. Date, S. Kikuchi, H. Shimada, "Visualization of Ultrasonic Wave Interaction with Defect by Stroboscopic Photoelasticity", Jn. Japan Soc. NDI vol. 35, pp.150 (1986).
4. K. Harumi, "Computer simulation of ultrasonics in a solid", NDT International vol. 19. pp. 315 (1986).

MEASUREMENT OF SAW VELOCITY BY FOCUSSING SYSTEM USING FRESNEL PHASE PLATE IDT

T. Nomura, S. Mizuno, and T. Yasuda

Department of Electrical Communication
Shibaura Institute of Technology
Tokyo 108, Japan

ABSTRACT

A method of measuring the surface acoustic wave (SAW) velocity using an interdigital transducer (IDT) is presented. The system to measure the variation of the SAW velocity on a sample substrate has been constructed using the Fresnel Phase Plate type IDT (FPP-IDT). In the measurement system, the FPP-IDT is used to excite a focussing Leaky SAW(LSAW) on the surface of sample materials. The measurement of the LSAW velocity variation is achieved by recording the phase difference between the two output of the close IDT's. The system is applied to measure the LSAW velocity variation on a SAW device material. The measurement are made for anisotropic materials such as 128° rotated Y-cut $LiNbO_3$ and X-cut $LiTaO_3$ at a frequency of 45 MHz. The results of the initial experiments show that the FPP-IDT system is suitable for mapping the two dimensional variation of LSAW velocity on a SAW device wafer.

INTRODUCTION

Ultrasonic techniques provide a powerful means of characterization and imaging in material science, biology and thin film technology. In the application of the ultrasonic techniques, the surface acoustic waves (SAW's) play important role. For example, the enhanced contrast of the acoustic image in the reflection acoustic microscope is due to the interference between the specularly reflected signal from the sample surface and a delayed SAW, which reradiates to the lens while propagating along the surface of the sample.

On the other hand, the piezoelectic single crystals, such as $LiNbO_3$ and $LiTaO_3$, have been widely used for SAW device. The SAW filter and resonator have been used in the industrial fields. It is expected that the characterization of these SAW device requires more sensitive because the operating speed of the various electronic equipments becomes higher. SAW propagation directly depends on the elastic property of the sample. Especially SAW phase and attenuation constant is considerably influenced by the change of the material construction and the defect of the crystalline structure. Therefore the SAW velocity is one of the most important parameters which determine the elastic property. Recently the evaluation of inhomogenity on a piezoelectric substrate by scanning acoustic microscope (SAM) has been reported[1]. However, the SAM requires a sophisticated electronic signal processing and a elaborated mechanical apparatus. Therefore, the simple techniques for measuring the variation on a SAW device wafer are needed for evaluating SAW devices such as filter and resonator.

An interdigital transducer (IDT) developed to excite a SAW, radiates acoustic wave in oblique directions with respect to the substrate, and so the IDT is used as the acoustic wave source for nondestructive evaluation system[2]. We had periously used the IDT system for acoustic imaging and for elastic characterization of materials.

In this paper, we present a new method to map the elastic anisotropy using a focussed beam. The focussed SAW beam is obtained by a one-dimensional Fresnel Phase Plate type IDT. Experimental verification of the two dimensional mapping using a FPP-IDT have been carried out.

PRINCIPLE OF MAPPING

If an acoustic beam is incident on a solid-water interface at or near the Rayleigh angle, a SAW is generated on the solid surface in addition to the normal specular reflection. The SAW reradiates an acoustic wave into the water as it propagates along the solid surface. Hence the SAW is called a leaky SAW (LSAW). The reflected distribution consists of the superposition of the specular reflection component and the component generated by the LSAW. The trailing field of the reradiated acoustic wave depends on the velocity and the attenuation constant of the LSAW[3].

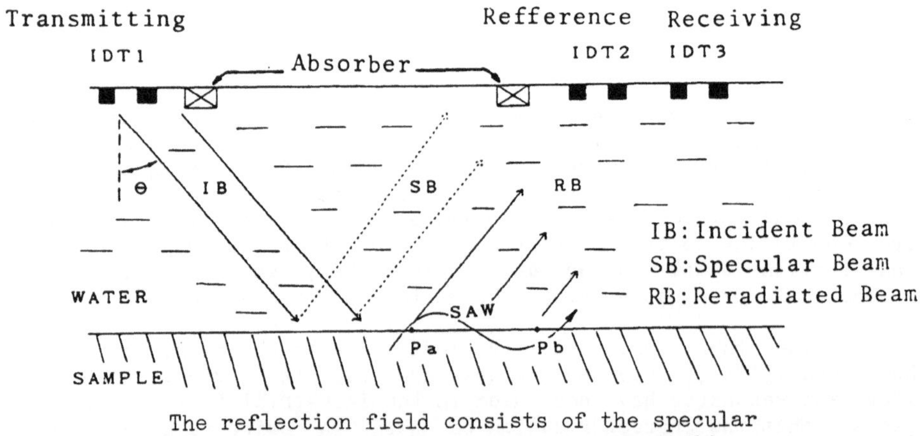

The reflection field consists of the specular reflection field and the reradiation field.

Fig. 1. Schematic representation of the present measurement.

Figure 1 shows a schematic representation of the measurement using IDT. The IDT's, transmitting, reference, and recieving IDT are deposited on the same piezoelectric substrate, positioned parallel to the sample surface. The IDT, which excites SAW's, radiate acoustic waves in oblique directions with respect to the substrate when it is placed in water. The radiation angle Θ is expressed by the equation, $\Theta=\sin^{-1}(Va/Vs)$, where Va is the acoustic wave velocity in the water and Vs is the SAW velocity on the piezoelectric substrate [2]. If the SAW velocity of a sample in water is nearly equal to that of the IDT substrate, the acoustic beam radiated from IDT1 is incident on the sample at or near the Rayleigh angle. As mentioned above, in addition to the normal specular reflection, an acoustic beam generates a LSAW on the sample, which reradiates an acoustic wave into water as it propagates along the surface. Therefore, the reflected field consists of these field components. The reflected wave distribution is detected by the receiving and reference IDT, which are constructed with close two IDT's, IDT2 and IDT3, deposited on the same piezoelectric substrate. If the incident beam width is limited by the absorber as shown in figure, the trailing field of the reradiated field distribution depends on only the velocity and the attenuation constant of the LSAW on the sample materials. In Fig.1 the gap height between the substrate and the sample surface is chosen so that the receiving IDT can detect only the trailing field. The two signals, which are intercepted by the receiving and reference IDT's separated by a small amount L, are proportional to the amplitude and the phase of LSAW on the sample surface. In this case these components differ only in path length on the sample surface but otherwise experience much the same enviromental disturbance such as thermal drift. Therefore, the velocity and the attenuation constant of the LSAW can be obtained from the phase and amplitude characteristics of the trailing field.

The SAW velocity measurment is achieved by comparing the phase difference of two signals. One signal is reradiated from the point Pa on the sample, the other signal undergoes a lateral shift before reradiating from the point Pb. The amount of shift is equal to the small distance between the IDT2 and the IDT3, as shown in Fig.1. Except for the lateral phase shift, two signals in this model have identical propagation delays. The differential phase delay between the two signals is due to the lateral shift L. Therefore, the additional phase lag of the shifted wave relative to the first reradiated one is given by

$$\varphi = 2\pi fL/Vs \qquad (1)$$

where f is frequency of the acoustic waves and L the distance between the receiving IDT and the reference IDT and Vs the SAW velocity of the sample substrate. The phase difference of the two reradiated component is proportional to the applied frequency and the distance L. Mapping of LSAW velocity is achieved by recording the phase difference in a phase bridge. If there is no local variations of LSAW velocity, the resultant phase difference will be constant. Velocity variations produces a differential phase shift. The phase shift $\Delta\phi$ is given by

$$\Delta\phi = 2\pi fL\,\Delta Vs/Vs^2 \qquad (2)$$

where ΔVs is velocity variation of the sample surface. Thus the phase difference between the two signal at the fixed frequency is proportional to the local velocity change ΔVs.

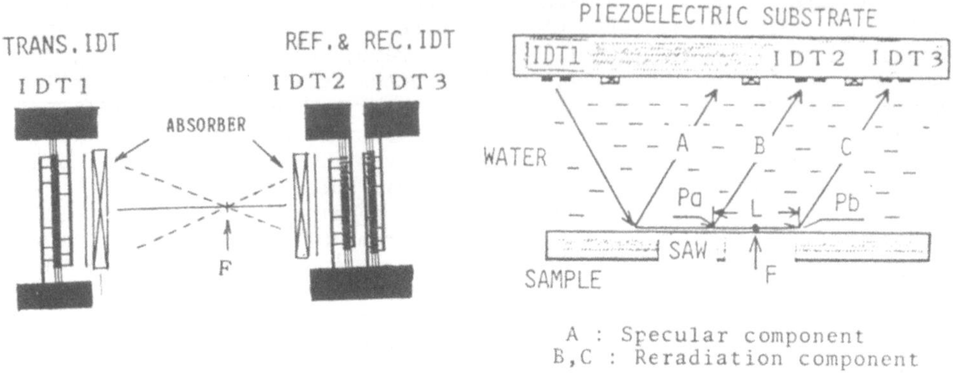

(a) IDT configuration (b) Schematic representation

Fig. 2. The configuration of the Fresnel Phase Plate IDT and schematic representation of the measurement.

A. Fresnel Phase Plate IDT

The purpose of the present study is to display the distribution of the LSAW velocity on the sample surface. To obtain a high spatial resolution, it is necessary to use a focussed acoustic beam on the sample surface. For this purpose, we have used the focussing IDT, which is constructed with a Fresnel Phase Plate type IDT (FPP-IDT) shown in Fig.2(a)[4]. An interdigital electrode of each IDT is divided into several section, each of which is placed to form one dimentional Fresnel zone plate (FZP) on a piezoelectric substrate. In the present study, we place the electrodes with a reversed polarity in the nonexciting regions of the FZP, and thereby construct one dimensional Fresnel phase plate (FPP) using the interdigital electrodes. Three FPP IDT, 1,2, and 3 are prepared on the piezoelectric substrate and used for transmitting, receiving, and reference IDT. Figure 2(b) illustrates the focussed SAW beam formed on the sample surface. The SAW generated from each electrode segment of the transmitting IDT is converted into the LSAW through the water couplant, and the contributions from all segments are brought to focus at the point F on the sample surface by diffraction effects. The LSAW beam from the focal point then diverges and two close FPP-IDT's receive the acoustic beam reradiated from the neighborhood of the focal point, F.

B. Electronics

Figure 3 shows the block diagram for mapping the LSAW velocity variation on the sample. Mapping is at constant frequency. Two receiving signals are first made the same amplitude by a limitter. The receiving signal is delayed by 90° using a phase shifter and fed into the phase bridge. The reference signal is fed into the other port of the phase bridge. The phase difference signal is proportional to the sine of the phase difference. This increases the sensitivity of the

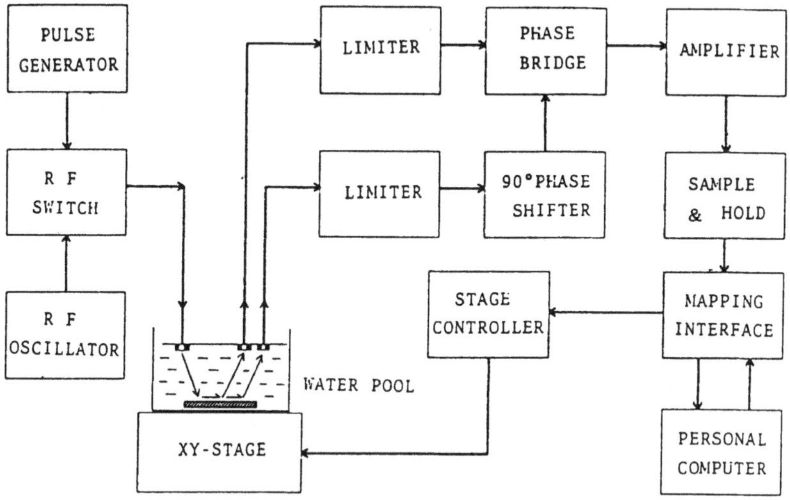

Mapping is at constant frequency and in RF tone burst mode.

Fig. 3. Block diagram of the measurement system for the LSAW velocity variation.

detector in the mapping of LSAW velocity variations. In the measurement, we fixed the frequency so that the output of the phase bridge is related to the LSAW velocity alone. The present system detects the phase difference between two outputs by receiving FPP-IDT, and records the variation of this phase difference for scanning X axis and Y axis direction over the sample. Both XY stage mechanically scanning and the phase difference output recording are automatically controlled by the personal computer.

EXPERIMENTS

FPP-IDT for the measurement is designed by the following parameter : the SAW center frequency is 45 MHz, the number of the electrode pairs is 3, the distance between the receiving and reference is 0.8 mm, the number of the Fresnel zone is 6. Measurement of the phase difference is achieved to record the phase output. Figures 4,5 show experimental results of two dimensional velocity variation measured by the present system.

A. Velocity Perturbation Due to Thin Film Overlayer

The velocity distribution was examined for a 128° YX $LiNbO_3$ substrate, on which an aluminum electrode of about 3000 Å thickness was deposited by vacuum evaporation. Our purpose was to measure the perturbation of the LSAW velocity caused by the aluminum film. The

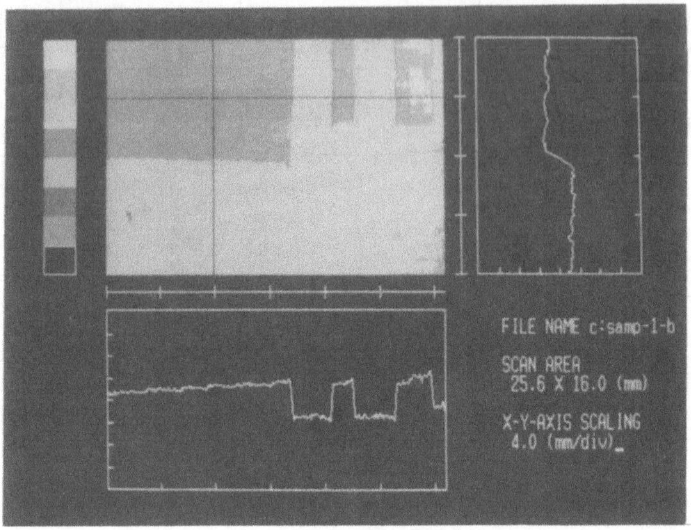

(a) Al strip on 128°Y cut $LiNbO_3$.

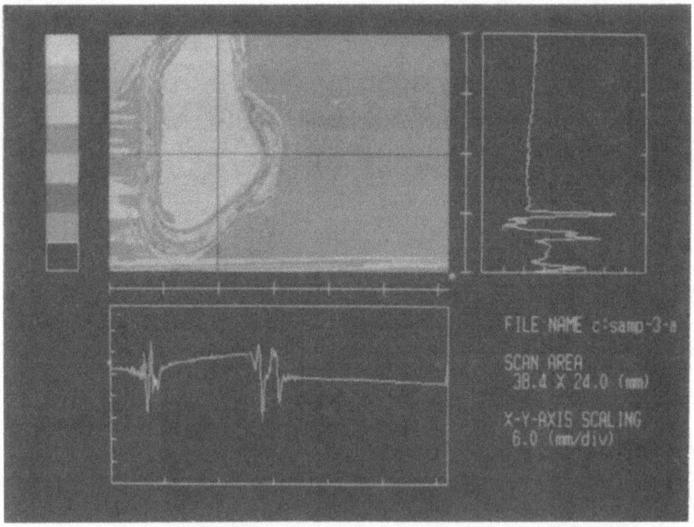

(b) 128°Y cut $LiNbO_3$ locally coated with photo resist.

Fig. 4. Velocity perturbation due to thin film on 128°Y cut $LiNbO_3$. Horizontal and vertical line represent scan line; lower trace is velocity variation along 90°X-direction; side trace is velocity variation along X-direction. LSAW propagates along the X-direction.

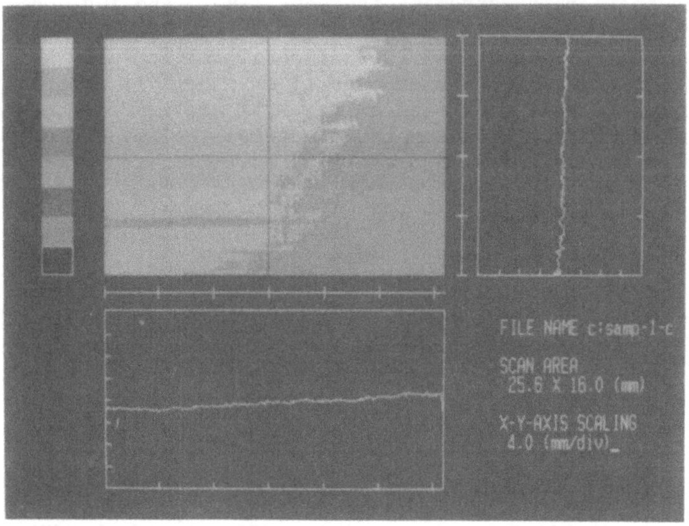

(a) 128°Y cut $LiNbO_3$.
Lower trace is velocity variation along 90° X-direction; side trace is velocity variation along X-direction.

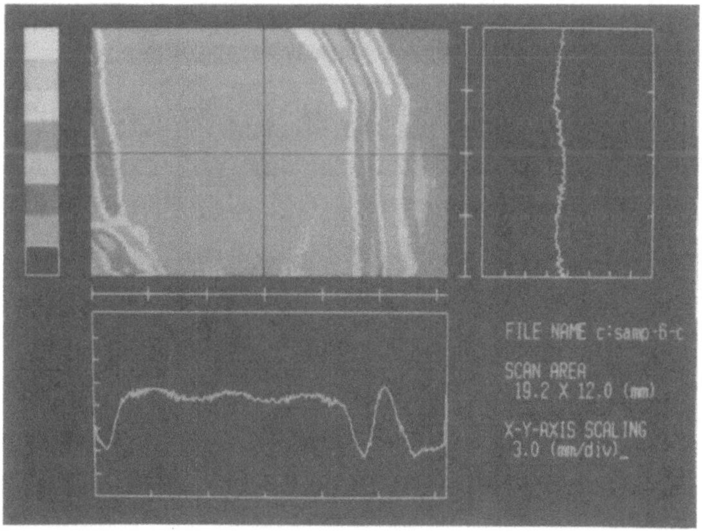

(b) X cut $LiTaO_3$.
Lower trace is velocity variation along 22° Y-direction; side trace is velocity variation along 112° Y-direction.

Fig. 5. Velocity variation of sigle crystal, $LiNbO_3$ and $LiTaO_3$. Horizontal and vertical line represent scan line; LSAW propagates along the vertical line.

change of velocity due to the existance of the aluminum film was mapped in two-dimension as shown in Fig.4(a). Vertical and horizontal line in the photograph represent the scan line, and two white line show the velocity variation measured along each line. The measured phase difference between the surface of metallization and the surface of 128° YX $LiNbO_3$, which corresponds to a velocity perturbation of 1.02 percent. The fluctuation on the film are due to the surface flaws of the evaporated film and the electrical noise. The velocity difference between the bare and aluminum coated is found to be about 40 m/s from the phase difference. The computed velocity difference is 47 m/s between electricaly open and shorted surfaces.

The second sample was photo resist electrode pattern deposited on a 128°Y cut $LiNbO_3$ substrate. Figure 4(b) shows two dimensional mapping of the measured velocity perturbation due to the photo resist film. It is found that the velocity variation due to the photo resist is about 17 m/s.

B. Velocity Variation of the Single Crystals

Simple techniques for measuring the variation on a SAW device wafer are needed for evaluating SAW devices such as resonators and filters. Recently, SAM with a line focus beam have been used to map elastic anisotropy [1]. As mentioned above, the IDT with a focussed beam is suitable for measuring the uniformity of the LSAW velocity. We have used the system to investigate velocity variations on SAW device wafers. Experiments have been made for 128°Y cut $LiNbO_3$ and X cut $LiTaO_3$. Figure 5 show two dimensional mapping of the measured results of velocity variation. Figure 5(a) shows velocity variation measured for 128°YX $LiNbO_3$. The line scan in the photograph show the results of velocity variation for scanning along the X- and 90° to the X-direction. Figure 5(b) shows velocity variation measured for X 112° Y $LiTaO_3$. In this case, the line scan represent the velocity variation along 112° to the Y- and 22° to the Y-direction. In both measurement, the maximum difference is 0.1 % in each direction. It is thereby found that the velocity variation for 128° YX $LiNbO_3$ and X 112° Y $LiTaO_3$ wafer are within 0.3 %.

In the measurement, the instrumental defects such as thermal, electrical, and mechanical variations, are nearly perfectly suppressed, so that the results show the variations due to the sample alone.

CONCLUSION

Method of measuring LSAW velocity variation on the SAW wafer has been presented. The Fresnel phase plate IDT for focussed SAW beam has been used to map the LSAW velocity variation on the sample surface. It is found that the FPP-IDT system is suitable to map the local variation of the leaky SAW velocity with high spatial resolution. In experimental results measured by the present system, it is successed that the small velocity variation can be detected and displayed on the TV monitor. The system will be applied for the evaluation system of SAW device wafer in the mass production line.

ACKNOWLEDGMENTS

The authors wish to thank Prof.T.Moriizumi of the Tokyo Institute of Technology and Dr.S.Shiokawa of Shizuoka university for helpful discussions.

REFERENCES

1. J.Kushibiki, H.Asano, T.Ueda, and N.Chubachi, "Application of line-focus-beam acoustic microscope to inhomogeneity detection on SAW device materials", in Proc. 1986 IEEE Ultrason. Symp., pp.749-753
2. T.Nomura, S.Shiokawa, and T.Moriizumi, "Measurement and mapping of elastic anisotropy of solid using a leaky SAW excited by interdigital trasducer", IEEE Trans. Sonics Ultrason., vol.SU-32, no.2, pp.235-240, Mar. 1985
3. H.L.Bertoni and T.Tamir, "Unified theory of Rayleigh angle phenomena for acoustic beams at liquid-solid interface", Appl. Phys., vol.2, pp.157-172, 1973
4. T.Nomura, S.Shiokawa, T.Moriizumi, and T.Yasuda, "Two-dimensional mapping of SAW propagation constants by using Fresnel-phase-plate interdigital transducer", in Proc. 1983 IEEE Ultrason. Symp., pp.621-626

USEFULNESS OF ULTRASONIC IMAGING IN THE MEDICAL FIELD

Motonao Tanaka

Department of Medical Engineering and Cardiology
The Research Institute for Chest Diseases and Cancer
Tohoku University, Sendai 980 Japan

INTRODUCTION

Three dimensional texture of the biological tissue and organs are very complex. And also in diseased state, pathological changes observed at the initial stage are limited to a small area. They gradually expand with a lapse of time and not distribute uniformly through out the organ when examined at a certain stage.

Therefore, for the accurate clinical diagnosis and evaluation of the structure and function of the tissue or organ, it is necessary for the method of evaluation to meet the following requirements:
1) The site and size of, and the structure within a target organ or tissue can be identified on three dimension and distinguished clearly from the adjacent organ and tissue with accuracy prior to evaluation.
2) In disease status, the size, extent and location of abnormal changes in structure and the functions of the target tissue and organs can be evaluated quantitatively.
Thus, the two dimensional display of the tissue and organ, the so-called cross-sectional imaging, meets these requirements. In this regard, ultrasound is one of the most suitable media. Consequently a various technique for measurement by using ultrasound have been developed.

Two imaging techniques of two dimensional display have been developed. One is the direct two-dimensional display on CRT screen of the data obtained by ultrasound. The other is the display of data after processing by a computer.
And also, from the clinical view points, following two subjects has been attracted clinicians attention.
1) Evaluation of changes in characters of the heart tissue.
2) Non-invasive evaluation of cardiac function:
The ultrasonic imaging technique plays an important role for approaching these subjects.

In this paper, the new two dimensional imaging techniques in the field of cardiology, which have been developed in our laboratory, and their usefulness will be presented.

1) TWO-DIMENSIONAL DISPLAY OF THE STRUCTURE

(a) Direct display of the macroscopic structure of the heart tissue

The heart is semi-ellipsoid and consists mainly of two tissue

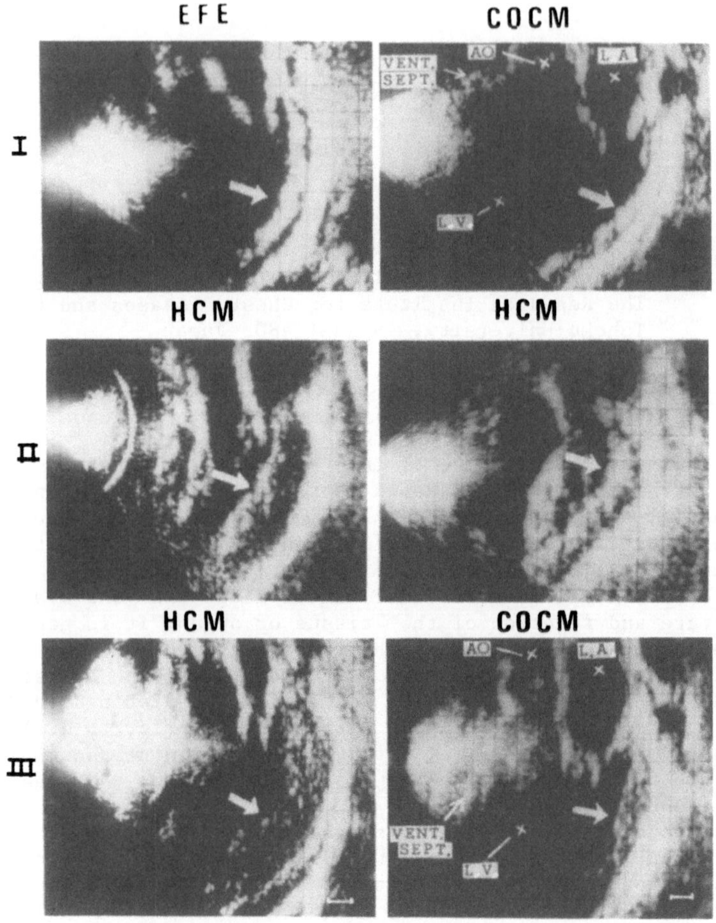

Fig.1. Two-dimensional echocardiograms in cases with cardiomyopathy. Typical abnormal intensified echoes (white thick arrows) are observed in the diseased left ventricular wall.
I : band-shape echo
II : mixed echo with type I and III
III: freckle-like echo
EFE: Endocardial Fibroelastosis
COCM: Congestive Cardiomyopathy
HCM: Hypertrophic Cardiomyopathy
AO : Aorta, LA : leftatrium, LV : left ventricle
VENT SEPT: ventricular septum

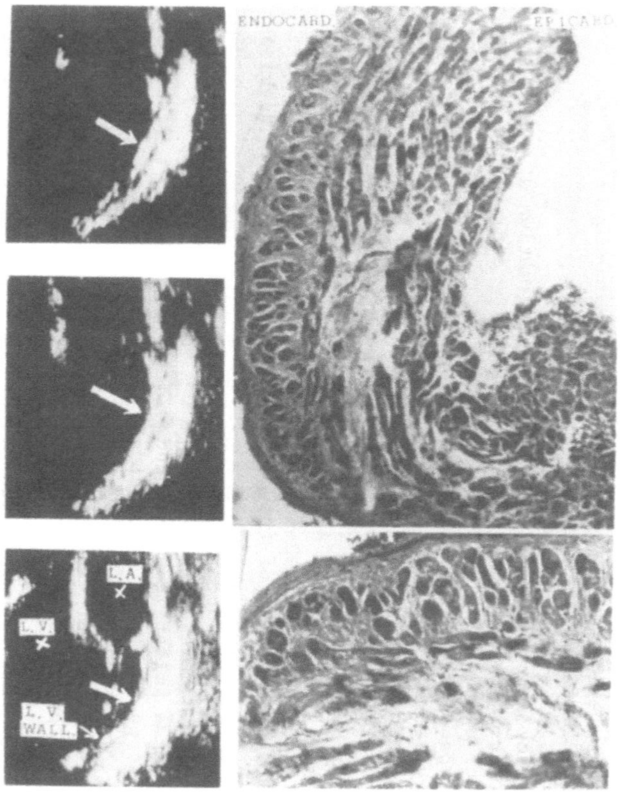

Fig.2. Comparative findings between two-dimensional echocardiograms and actual histological spacimen of the ventricular wall in the case of dilated cardiomyopathy with type I echo (thick white arrows).

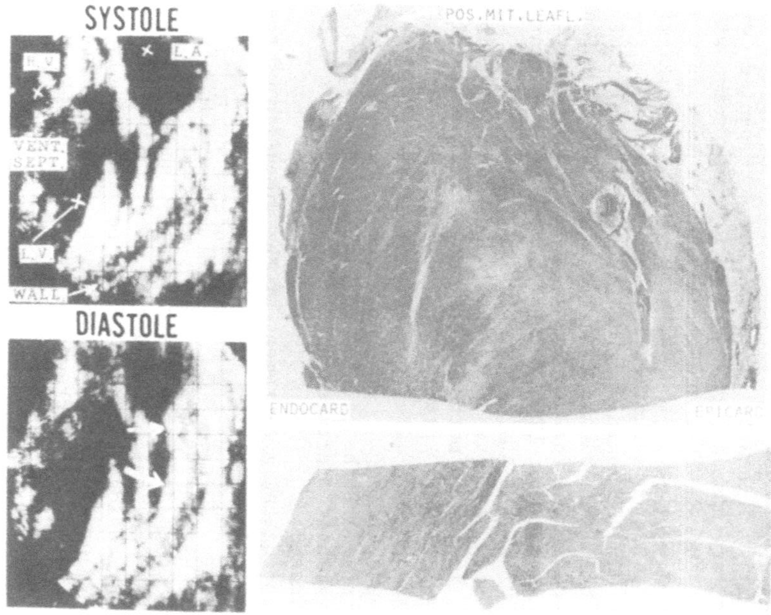

Fig.3. Comparative findings between two-dimensional echocardiograms and actual histological specimen of the ventricular wall in the case of HCM with type III echo. The area of blue color show the fibrotic tissue.

elements: namely, muscle tissue and membraneous or fibrous tissue such as valve, pericardium and endocardium.
Furthermore, many of heart diseases are accompanied with structural abnormalities or tissue degeneration, and some with functional abnormalities without patho-anatomical changes such as aryhthmias.
Therefore, in cardiology, the two dimensional display of pathological changes in relation to structure is in strong demand.
The technique of ultrasonic imaging for the evaluation of morphological structure of the heart has been developed since 1963 in our laboratory. An original equipment for obtaining a two dimensional picture of the heart was constructed in 1964[1] and a cross section picture of the heart was taken in 1965[4] for the first time in the world by the use of new devices such as a concave transducer, ultrasonic sector scan in which the proximity immersion method and ECG synchronized display method were applied. The name given to this method was ultrasonic Cardio-tomography. It is now very familiar to clinicians as an important diagnostic tool:

The advantage of the cardiotomographic method is as follow.

1. Anatomical structure and its abnormalities can be estimated.
2. Morphological measurement of the dimension of the heart can be performed.
3. Dynamic events of the heart structure during a cardiac cycle can be analyzed.
4. Characters and patho-histological changes of the heart tissue are possible to detect.

The two dimensional display of the heart shows clearly intensified abnormal echoes of the damaged myocardium as demonstrated in the Fig.1. As shown by white thick arrows, the band-shape echoes, freckle-like echoes and smog-like echoes are observed.
As represented in Fig 2 and 3, fibrous tissue or tissue degeneration is detected always in the area where the abnormal intensified echoes are demonstrated. These results strongly suggest that it is possible to evaluate the tissue character by ultrasonic method.

(b) <u>Quantitative two-dimensional display of the microscopic structure of the heart tissue</u>

In order for the method of non-invasive estimation of changes in tissue character to be complete, it is essential to analyze the acoustic characteristics of the heart tissue and tissue elements.
Thus, a new type of acoustic microscope has been developed since 1980 in our laboratory for the purpose of quantitative evaluation of the acoustic characteristics of the tissue elements.

By the use of this method, the microscopic tissue structure is displayed as a two dimensional image, and the velocity and attenuation of ultrasound in target tissue element can be determined with the tissue element identified on two dimensional image.
Fig 4 shows a block diagram of the acoustic microscopic system developed in our laboratory in 1985[3]).
The diameter of the transducer is about 1.6mm. The ultrasonic beam is converged by a concave sapphire lens.
The frequency of a pulsatile ultrasound is 100 to 500 MHz. Resolution of this microscope is of the order of about 8 micron.
A 10-micron thick specimen is placed at the point of focusing and examined in a small amount of water or saline used as a coupling medium.

The transducer is driven mechanically at a speed of about 125 mm/s to 250 mm/s along x-axis and a specimen placed on stage moves with the speed of about 0.5 mm/s along y axis. Thus the specimen is scanned with an ultrasonic beam on two dimension and the so called C-mode scan is performed.

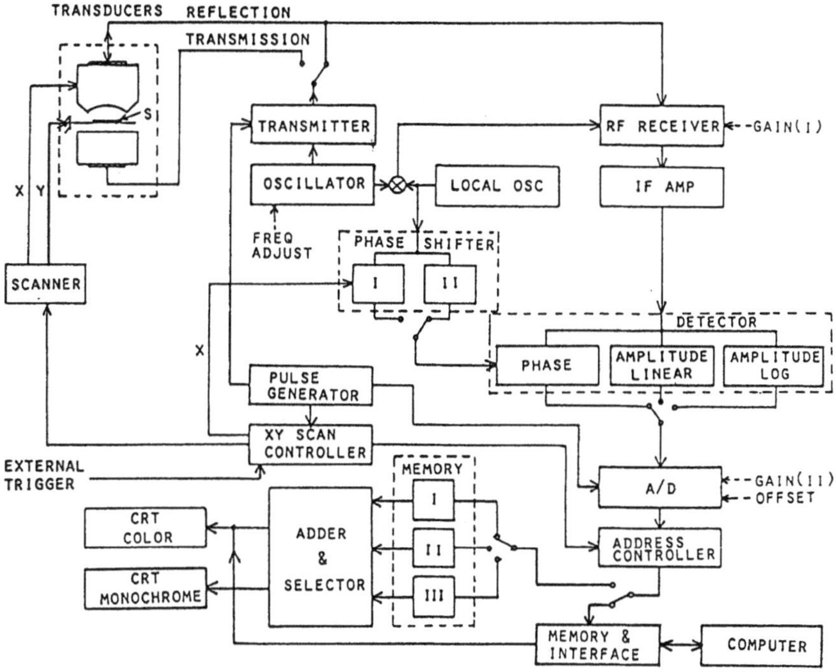

Fig.4. Blockdiagram of the acoustic microscopic system developed in our laboratory.

The rate of attenuation and the velocity of ultrasound traveling through the tissue specimen is calculated from the amplitude of the signal received and from the difference in phase between signals transmitted and received in comparison with that of the coupling medium (water or saline).

The rate of attenuation and velocity are converted into color signals and displayed on color CRT as a two dimensional C mode pattern.

It is essential to determine the thickness of a specimen as accurately as possible for the quantitative evaluation of the rate of attenuation and velocity of ultrasound in tissue. Therefore, a new method of compensation for thickness was introduced.

The frequency dependency of amplitude and phase shift of the waves reflected from a thin specimen was found to produce a characteristic pattern as a function of the thickness of specimen. Providing that the frequency dependency pattern is calculated theoretically and the amplitude and phase shift are measured by changing the frequency of transmission ultrasound, the exact thickness of the specimen can be determined at any point-area by comparing the theoretical pattern, as demonstrated in Fig 5, with an actual one.

In our microscope, the data of 16 frames of amplitude and phase mode which is different in frequency were taken by changing the frequency of ultrasound in the range between 100 and 200 MHz, and stored in the memory of the computer.

The data of amplitude or phase shift which are different in frequency were read out at the same point-area on 16 frames, and thus an actual frequency dependency pattern was determined at one point-area of a specimen by using these 16 data.

The same procedure was performed at every point-area of the specimen, the corrected thickness of the specimen was determined and the attenuation and velocity of the tissue through the specimen were obtained as an absolute value.

In this way, the data of attenuation and velocity are shown in color.

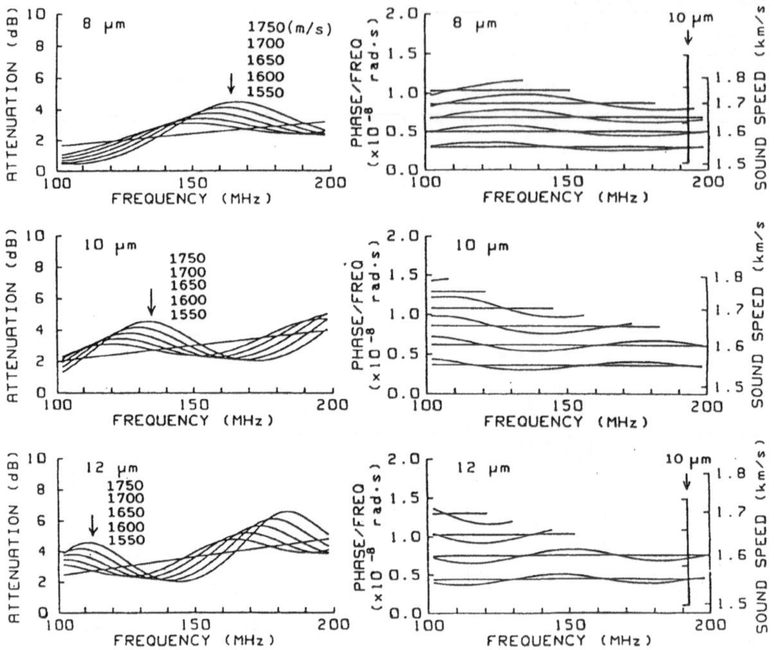

Fig.5. Calculated frequency dependency curves of the attenuation and of the phase ratio of the received signal.

As demonstrated in Fig. 6, 11 color steps are divided at intervals of 0.25 dB/mm/MHz for the attenuation rate and at intervals of 30 m/s for the sound speed. Red color shows the area where attenuation is largest and the velocity is fastest. Black color shows the area where attenuation is smallest and the velocity slowest.

Fig.7 and 8 show the optical and the acoustic microscopic images of the myocardium from a case of myocardial infarction. The area where the ultrasound is propagated very fast is colored red. The area which is colored green, the so-called scar tissue, in optical image corresponds to the area which is displayed in red or orange on the acoustic image of velocity.

Introduction of the quantitative acoustic microscopic method has enabled us to measure the acoustic properties of the tissue element selectively, which is very significant for tissue characterization.

2) TWO-DIMENSIONAL DISPLAY OF THE FUNCTION OF THE HEART BY USING THE DATA PROCESSING TECHNIQUE BY A COMPUTER

The force generated in the myocardium is transmitted to the intracardiac blood and thus the intra-ventricular blood is ejected. Therefore, information concerning the movement of the intraventricular blood, in other word, the flow dynamic parameters, such as flow velocity, acceleration, velocity vector, dynamic pressure and so on, are under direct influence of the cardiac function.

Providing that an adequate technique of velocity data processing is utilized and that hydrodynamic data are obtained, the cardiac functions will be estimated quantitatively and non-invasively. For this purpose, the ultrasonic Doppler method is the most useful.

The two-dimensional color flow mapping technique has been used in practice recently. However, this technique is inadequate for the quantitation of hydrodynamic parameters, because the mean value of Doppler shift in the beam direction is used in this quantitation.

In this paper, a new method which has been developed in our

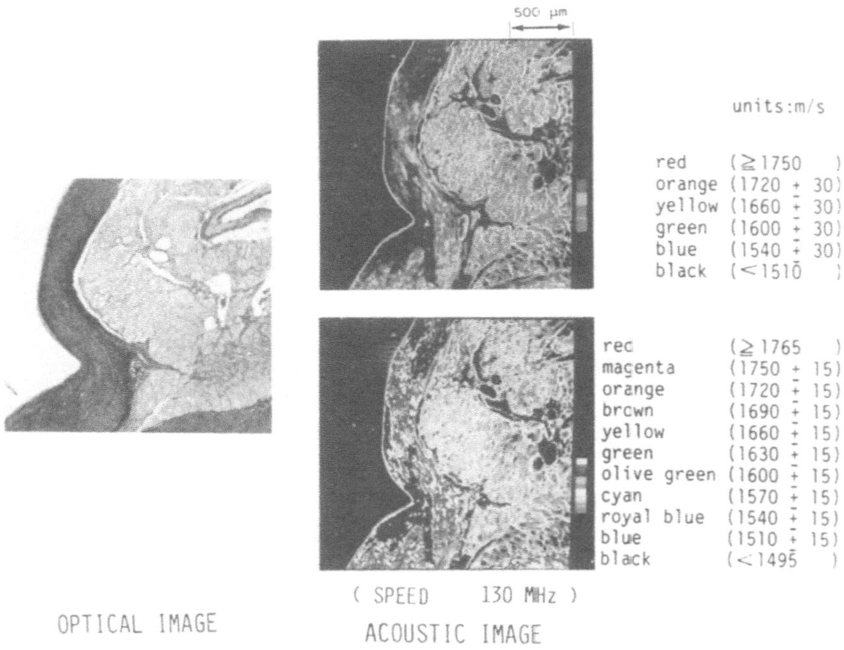

Fig.6. Color scale display of the velocity data of damaged myocardium. Color step is divided at intervals of 60m/s (right uper picture) and of 30m/s (right lower picture).
Left side picture is optical microscopic image in the same portion of the ventricular wall.

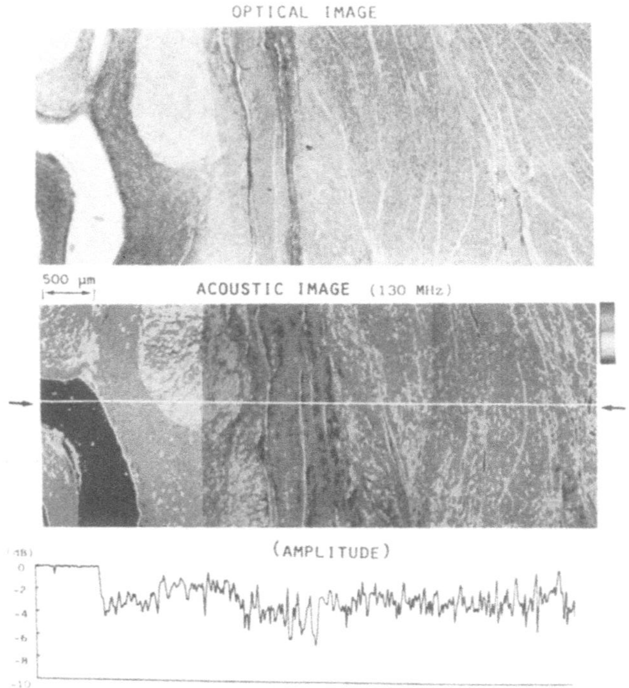

Fig.7. Optical (uper) and acoustic (middle) microscopic images of the damaged myocardium in the case with myocardial infarction. Lower curve show the A-mode pattern of the anplitude along the white line showed in the middle picture.

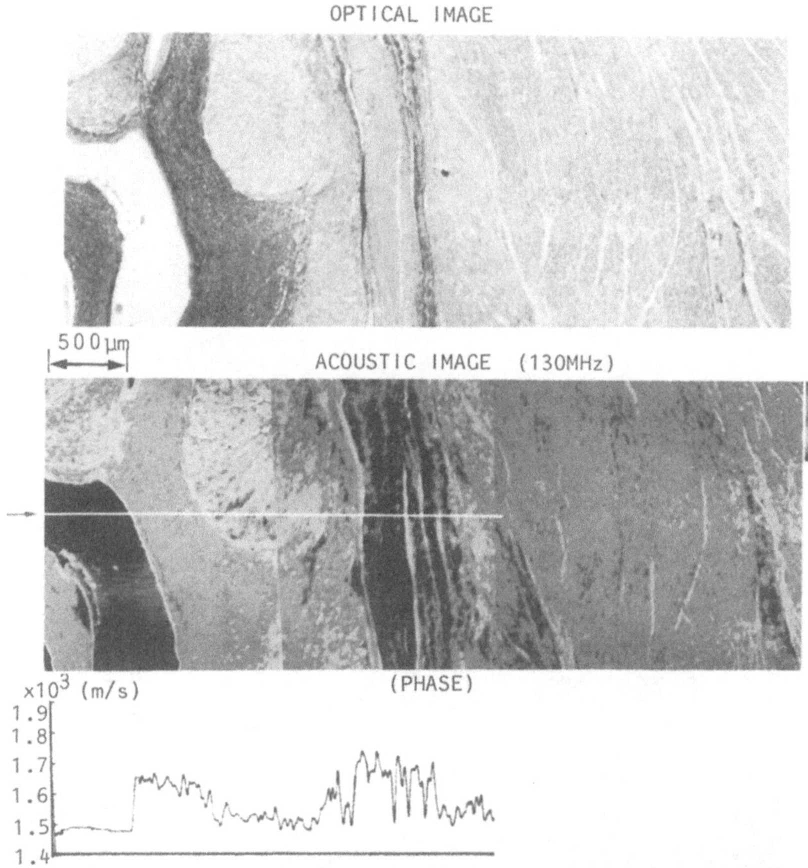

Fig.8. Optical (upper) and acoustic (middle) microscopic images of the damaged myocardium.

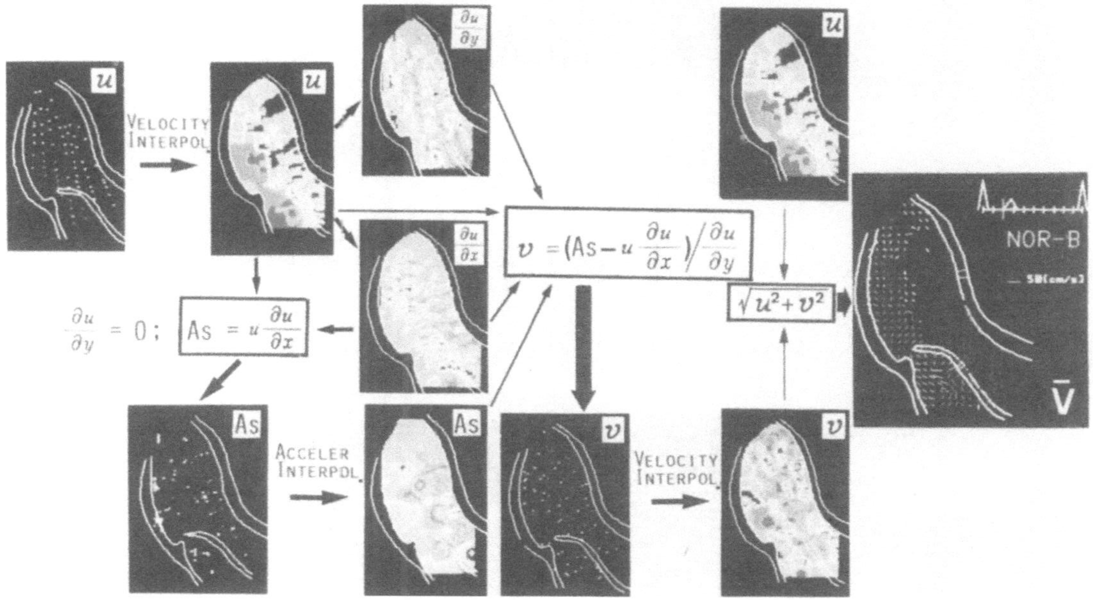

Fig.9. Processing chart of the Doppler velocity data for obtaining the rectangular velocity component (V) from the velocity component along the beam direction (u) and the velocity vector ($\overline{V}$).

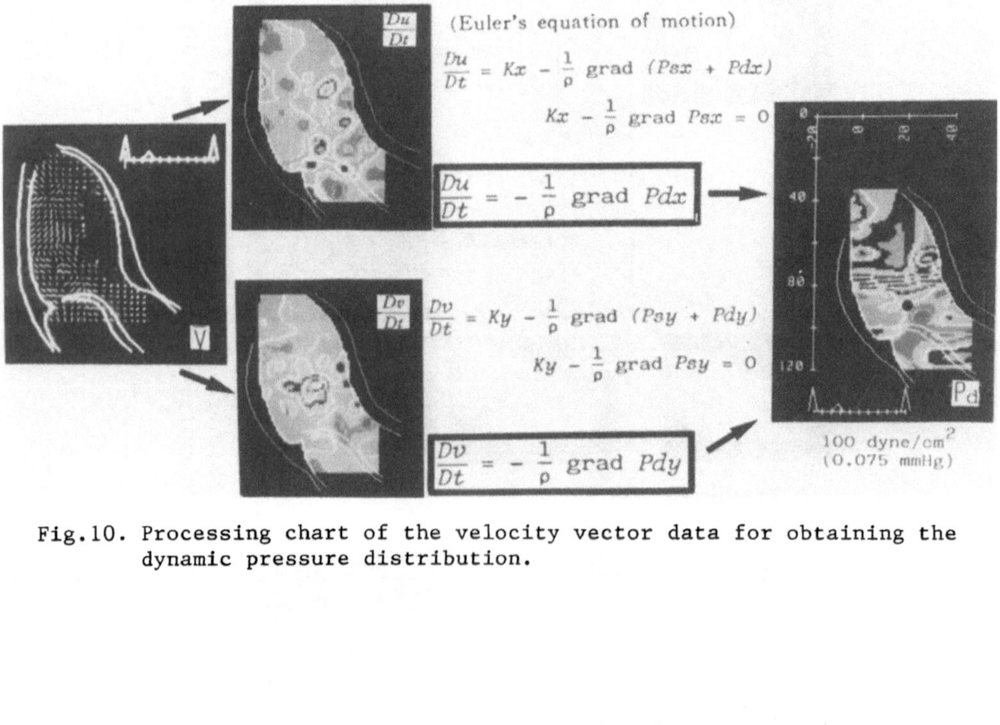

Fig.10. Processing chart of the velocity vector data for obtaining the dynamic pressure distribution.

laboratory for extracting flow dynamic parameters of the intracardiac blood flow from Doppler velocity data will be demonstrated and two-dimensional imaging of these parameters and its medical significance will be shown.

The flow parameters obtained by the present method are velocity, acceleration, velocity vector, dynamic pressure and their two dimensional distribution.

A pulsed ultrasound of 3 MHz in frequency were used. Doppler signals were recorded at 70 to 150 sampling points on the apical long-axis section plane which covered both the whole length of the in- and out-flow tracts, and were analyzed with FFT method in real time. Envelope velocity data of FFT output signals in one cardiac cycle at each sampling point were put into a computer after manual revision of frequency aliasing.

The left side picture of Fig.9 shows the velocity data thus obtained and plotted at a systolic phase (picture u in Fig.9). The cardiac phase is shown by a vertical yellow line on ECG. Warm color shows an outward flow and cold color an inward flow. From the data thus obtained, an equi-velocity pattern (picture (u) of Fig.9) is drawn by the application of an interpolation technique which has been developed in our laboratory.

The velocity data are represented by a color step arranged in the order of yellow, pink and red for the outward flow, and in the order of blue, light blue and green for the inward flow. One step is 5 cm/sec. Accordingly, the highest velocity area in this phase is at the subaortic area and the lowest velocity area at the apex.

From the equi-velocity data thus obtained, the velocity components tangential to the beam axis are calculated by applying the law of conservation of momentum of the flow. (formula 1)

$$\frac{Du}{Dt}=\frac{\partial u}{\partial t}+u\frac{\partial u}{\partial x}+v\frac{\partial u}{\partial y}+w\frac{\partial u}{\partial z} \cdots\cdots (1)$$

if $(w)=0$

$$\frac{Du}{Dt}=\frac{\partial u}{\partial t}+u\frac{\partial u}{\partial x}+v\frac{\partial u}{\partial y}$$

$$=\frac{\partial u}{\partial t}+As \cdots\cdots (2)$$

Providing that the flow is symmetric to the cross section plane, the velocity component (W) which is tangential to the plane is theoretically 0.

Accordingly, acceleration is given by the formula (2). The velocity component (V) which is tangential to the beam axis is calculated by the equation AS.

$$As=u\frac{\partial u}{\partial x}+v\frac{\partial u}{\partial y} \cdots\cdots (3)$$

In the equation AS, at the point of which the equi-velocity line is rectangular to the beam axis, the values of are 0.

Therefore, value of AS is calculated by the formula (4)

$$As=u\frac{\partial u}{\partial x} \cdots\cdots (4)$$

and the distribution of AS is obtained by using the acceleration interpolation technique as shown in picture AS in Fig.9.

The velocity component (V) is calculated from the value of

$\frac{\partial u}{\partial y}$ (picture $\frac{\partial u}{\partial y}$ in Fig. 9), $\frac{\partial u}{\partial x}$ (picture $\frac{\partial u}{\partial x}$ in Fig. 9) and AS thus obtained.

In this way, the velocity vector and its distribution on the longitudinal section plane by using velocity data V and U can be calculated easily.

Furthermore, local acceleration of can be calculated without difficulty. In addition, the total acceleration of can be obtained.

Therefore, as shown in the Fig.10, by utilizing the Euler's equation, the pressure gradient at each point on the same section plane is calculated and the pressure distribution which is normalized for the pressure data at the standard point set up in the ventricle can be obtained.

The distribution pattern of the flow parameter during systole of a normal case are demonstrated in Fig.11.

The equi-velocity lines show a relatively smooth pattern which is convex towards the apex. The interval between lines becomes shorter in the vicinity of the aorta. These results show that the maximum velocity area is near the outlet of the ventricle, and that the velocity increases gradually from the apex to the outlet of the ventricle. No marked turbulence is observed. In the acceleration pattern, one color step in 20 cm/s^2, the area of acceleration in cold color is along the ventricular septum. The area of deceleration in warm color is along the ventricular free wall from the mitral valve to the apex. These findings suggest that the downward movement of the mitral ring and valve leaflets, and the anterior displacement of the free wall are of larger magnitude than the posterior displacement of the ventricular septum.

The difference in flow direction can be shown by utilizing the velocity vector distribution picture. The majority of the out-flow moves toward the aorta.
However, the minority of the outward flow runs backward under the mitral leaflet and changes itself into the flow toward the apex. The cold color at the postero-basal portion on the equi-velocity pattern coincides with this backward flow.

Also a small eddy is observed at the area of valsalva sinus. It is considered that these findings suggest clinical evidence substantiating Bellhans's hypothesis.
A difference in pressure within the ventricle can be detected without difficulty based on the pressure distribution pattern. In this phase, the high pressure area in warm color is located apicoposteriorly and the low pressure area in cold color at the basis. The lowest pressure area at the outflow tract coincides with the area where the velocity vector is largest and acceleration is highest.
The area of the highest pressure is at the center of the apex. The equi-pressure line is almost parallel to the short axis of the ventricle. The pressure is relatively high just below the mitral leaflet.

These findings suggest that the force generated by contraction of the ventricular free wall and the septum participate to the same extent to the flow along the short axis. However, along the long axis, the force generated by the downward movement of the mitral complex is of larger magnitude than that by the upward movement of the apex.

On the contrary, all of these patterns change markedly in the diastolic phase. The equi-velocity lines are convex towards the apex and the intervals between two lines are almost equal. The oblique inflow runs from the postero-basal to the antero-apical portion. A large eddy appears at the outflow tract just behind the anterior mitral leaflet (Fig.12).
As regards the distribution of pressure, the high pressure appears in a wide area at the apical, atrial and subaortic portions of the ventricle. The low pressure areas are at the basal portion as shown by the blue band which is parallel to the short axis of the ventricle.

By using the new technique mentioned above, various kinds of flow parameters can be quantitated.

Therefore, when these flow parameters and their distribution thus

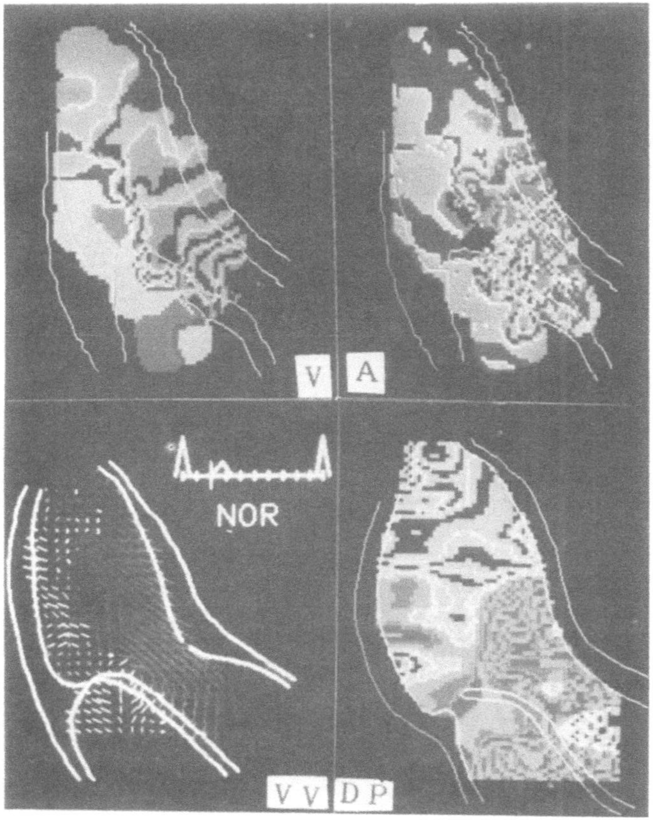

Fig.11. Two-dimensional distribution patterns of the flow parameters of the intra ventricular blood flow in the ejection phase.
V : equi-velocity pattern
A : equi-acceleration pattern
VV: Velocity vector distribution pattern
Dp: Dynamic pressure distribution pattern

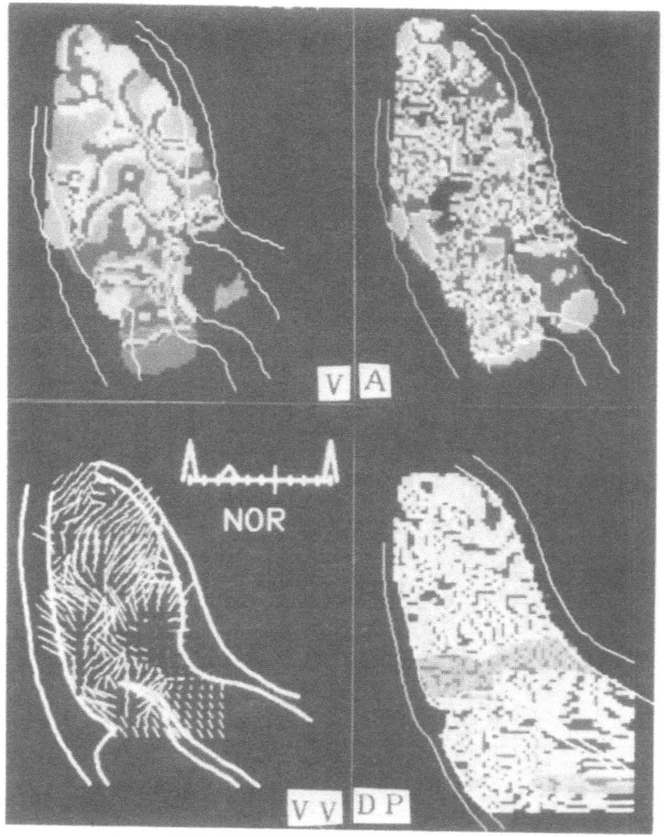

Fig.12. Two-dimensional distribution patterns of the flow parameters in the rapid filling phase.

obtained are synthesized, the blood flow in the cardiac chamber, the local myocardial function, and pump function can be analyzed qualitatively and quantitatively.

CONCLUSION

The usefulness and significance of ultrasonic two dimensional imaging are concluded as follows:

1) The macroscopic and microscopic structure of the tissue and organ can be characterized as a two dimensional picture. The size of a target organ and tissue can be identified and localized on two dimensional image.

2) A comparative study of the structural and functional abnormalities of the tissue and organ can be done simultaneously in more than two areas on two dimensional plane.

3) Informations which are out of reach of one point or one-dimensional imaging can be extracted by expansion of dimension from one to two.

REFERENCES

1) T.Ebina, M.Tanaka etal: The diagnostic application of ultrasound to the diseases in mediational organs.
----- Ultrasono-tomography for the heart and great vessels (the first report). Sci. Rep. Res. Inst. Tohoku Univ. -C. 12: 58 1965
Kokenshi, 17: 1 1964 (in Japanese)

2) M.Tanaka etal: Ultrasono-tomography of the heart and great vessels in living human subject.
Med. Ultrason. 4: 47 1966
Sci. Rep. Res. Inst. Tohoku Univ. -C. 14: 1 1967

3) M.Tanaka etal: Development of the acoustic microscope for the medical and biological use and its medical application.
Kokenshi, 37: 377 1985 (in Japanese)

4) M.Tanaka etal: Flow vector distribution patterns constructed from the informations obtained from the unidirectional tracing.
Med. Ultrason. 13(suppl.I): 230 1986

5) S.Ohtsuki etal: A method of flow vector mapping deduced from Doppler information.
J.A.S.J. 43: 764 1987

A METHOD OF FLOW VECTOR MAPPING DEDUCED DOPPLER DATA ON SECTOR SCANNED PLANE

Shigeo Ohtsuki, Motonao Tanaka, and Motoyoshi Okujima

Tokyo Institute of Technology
Research Laboratory of Precision Machinery and Electronics
4259 Nagatsuta, Midori-ku, Yokohama 227, Japan

INTRODUCTION

Ultrasonic equipment has been used in popular as a clinical diagnostic tool. There are two kinds of machine which are different in principle. One is ultrasonic tomographic system and the other ultrasonic pulsed Doppler system. The former is used for imaging of anatomical structure of a body by the pulse reflection technique, and the later for detection of the information on the motion of the structure.

Combining these two techniques, two-dimensional flow velocity images of the heart or large vessels are possible to obtain in real time by using a sector-scan or linear-scan ultrasonic Doppler system. Doppler velocity obtained by this method means the velocity component in the beam direction detected by Doppler effect of reflected ultrasound from a moving target. By coding the magnitude of Doppler velocity to a color, its distribution of the flow velocity is displayed in color image overlapped with ultrasonic cross section picture of the structure. On the other hand, the two-dimensional distribution of flow vectors proposed in this paper give us more useful information than the Doppler color imaging mentioned above.

In this paper, we propose a method to get two-dimensional flow velocity vector components from Doppler data on an observing plane.

THE ACCELERATION OF TRANSPORT IN FLUID AND THE INFORMATION BY DOPPLER EFFECT IN REFLECTED WAVES

There are two kinds of viewpoints for analyzing the characteristics of a flow. One is the tracing method of a "particle" which is a element of fluid. This is so-called Lagrangian viewpoint. The other the analyzing method of flow velocity, acceleration and so on at one fixed point. This is so-called Eulerian viewpoint. Doppler technique with range resolution such as pulsed Doppler technique is belong to Eulerian viewpoint. And substantial acceleration can be expressed as

$$\frac{DV}{Dt} = \frac{\partial V}{\partial t} + (V \cdot \text{grad})V \tag{1}$$

The first term of the right side of the equation (1) is local acceleration and the second is the acceleration of transport or convective acceleration.

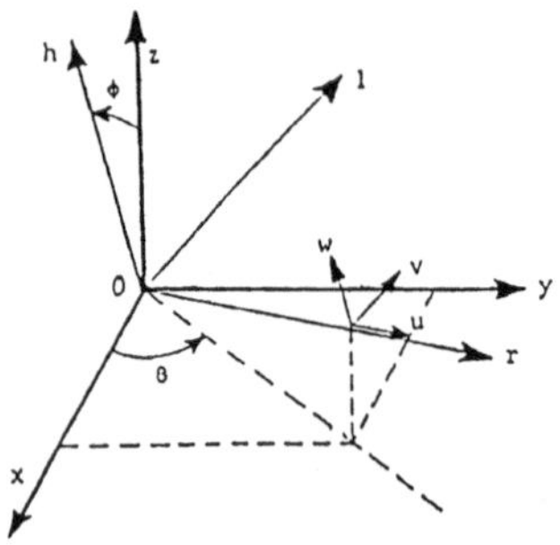

Fig.1 Absolute and rotated right angle coordinates.

Let us fix a set of Cartesian coordinates (x, y, z) the origin of which is the position of a ultrasonic probe for sector scan as shown in Fig.1. When the set of the coordinates (x, y, z) is rotated by the angle θ around z-axis and by the angle ϕ in the x-z plane, the orientation of the x-axis agrees with that of ultrasonic beam. In this case, the set of coordinates is denoted by (r, l, h) as shown in Fig.1. Let us call the set of the coordinates (r, l, h) a rotated set of Cartesian coordinates.

Using the components u, v and w of the velocity V on the axes r, l and h respectively, the r-component of the acceleration expressed by equation (1) can be described as following:

$$\frac{Du}{Dt} = \frac{\partial u}{\partial t} + u\frac{\partial u}{\partial r} + v\frac{\partial u}{\partial l} + w\frac{\partial u}{\partial h} \tag{2}$$

Then acceleration component A_{cr} of transport is

$$A_{cr} = u\frac{\partial u}{\partial r} + v\frac{\partial u}{\partial l} + w\frac{\partial u}{\partial h} \tag{3}$$

By the way, Doppler information obtained by pulsed Doppler technique with a single transducer is velocity component v_d (Doppler velocity) in the direction of ultrasonic beam as shown in Fig.2. This Doppler velocity is proportional to the amount f_d (Doppler frequency) of frequency shift of reflected wave by Doppler effect. Then,

$$v_d = c\frac{f_d}{2f_o} \tag{4}$$

where c is propagating speed and f_o is carrier frequency. This velocity component v_d is positive when the target moves toward the probe. So the velocity component u is negative of v_d. That is,

$$u = -v_d \tag{5}$$

The unknown variables included in the right side of equation are v and w. The other variables such as u and $\partial u/\partial l$ are measurable with shift sector scanning as shown in fig.3.

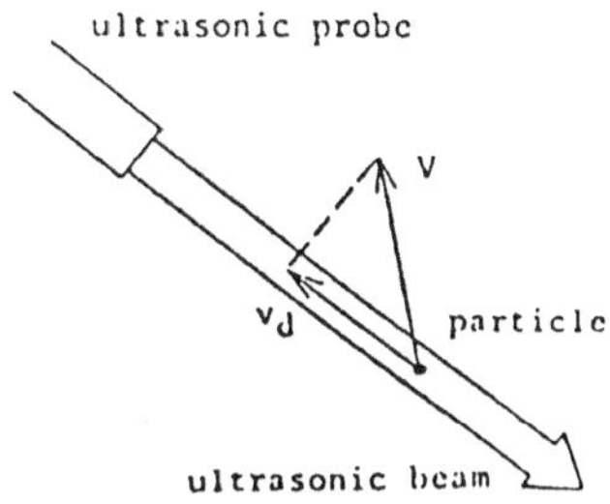

Fig.2 Doppler velocity v_d and ultrasonic beam.

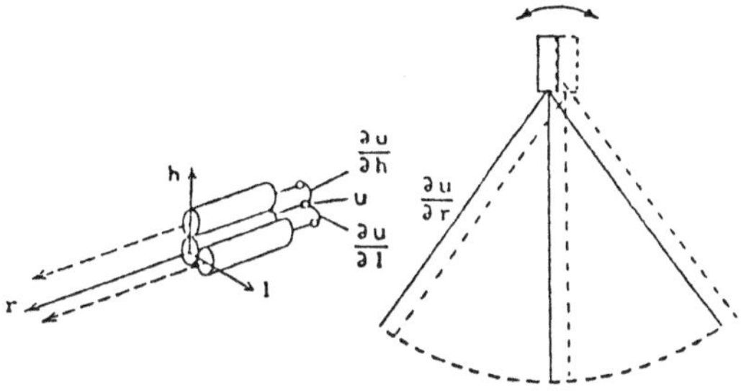

Fig.3 Shift sector scan.

ESTIMATION OF VELOCITY COMPONENT

Generally, the acceleration A_{cr} cannot be evaluated because of unknown variables v and w. However, the value A_{cr} of acceleration of transport is independent of v and w, when the condition is satisfied.

$$\frac{\partial u}{\partial l} = \frac{\partial u}{\partial h} = 0 \tag{6}$$

Then the expression (3) is

$$A_{cr} = u\frac{\partial u}{\partial r} \tag{7}$$

The right side of above expression can be evaluated with measured Doppler velocity distribution on an observing plane.

Next step is interpolation or exterpolation of acceleration A_{cr} of transport on the plane. For its accurate estimation, following conditions should be considered. (1) Non-uniformity of data density and (2) range weight. It will be a topic in future to develop an excellent estimation technique. For the points on the conditions:

$$\frac{\partial u}{\partial l} \neq 0, \quad \frac{\partial u}{\partial h} = 0 \tag{8}$$

acceleration A_{cr} of transport can be expressed as

$$A_{cr} = u\frac{\partial u}{\partial r} + v\frac{\partial u}{\partial l} \tag{9}$$

From (9), velocity component v is derived as

$$v = \frac{A_{cr} - u\frac{\partial u}{\partial r}}{\frac{\partial u}{\partial l}} \tag{10}$$

For other points, the value of v may be estimated from the values of adjacent points.

For the case under the condition

$$\frac{\partial u}{\partial h} \neq 0 \tag{11}$$

the velocity component w is estimated from (3) as

$$w = \frac{A_{cr} - u\frac{\partial u}{\partial r} - v\frac{\partial u}{\partial l}}{\frac{\partial u}{\partial h}} \tag{12}$$

For the case that the condition (11) does not meet, the value w may be estimated with adjacent values.

Above process generates flow vector (u, v, w) distribution on the region of interest.

EXAMPLE OF ESTIMATION OF FLOW VECTOR DISTRIBUTION

There is a laminar flow in a tube to use above mentioned technique theoretically. Let us start to get the data which will be measured by Doppler technique. There is a laminar flow of fluid suspending ultrasonic scatters in a tube immersed in water. The observing plane sector-scanned by ultrasonic beam contains the axis of the tube as shown in Fig. 4(a). The velocity distribution is shown in Fig. 4(b). The magnitude of velocity along the radial axis of the tube is expressed as

$$M = V_{max}(1-(s/a)^2) \tag{13}$$

where V_{max} is the speed on the axis of the tube, a is the radius and s is the distance from the axis.

The velocity component in the direction of ultrasonic beam is measurable by Doppler technique as shown in Fig. 4(c). So we get the distributions of Doppler velocity (Fig. 4(d)), u (Fig. 4(e)), $\partial u/\partial r$ (Fig. 4(f)) and $\partial u/\partial l$ (Fig. 4(g)) as basic data.

We can get the aggregation of points in Fig. 4(h) where A_{cr} can be fixed by above measurable values. In this case, the distribution of A_{cr} can be easily estimated as Fig. 4(i) where

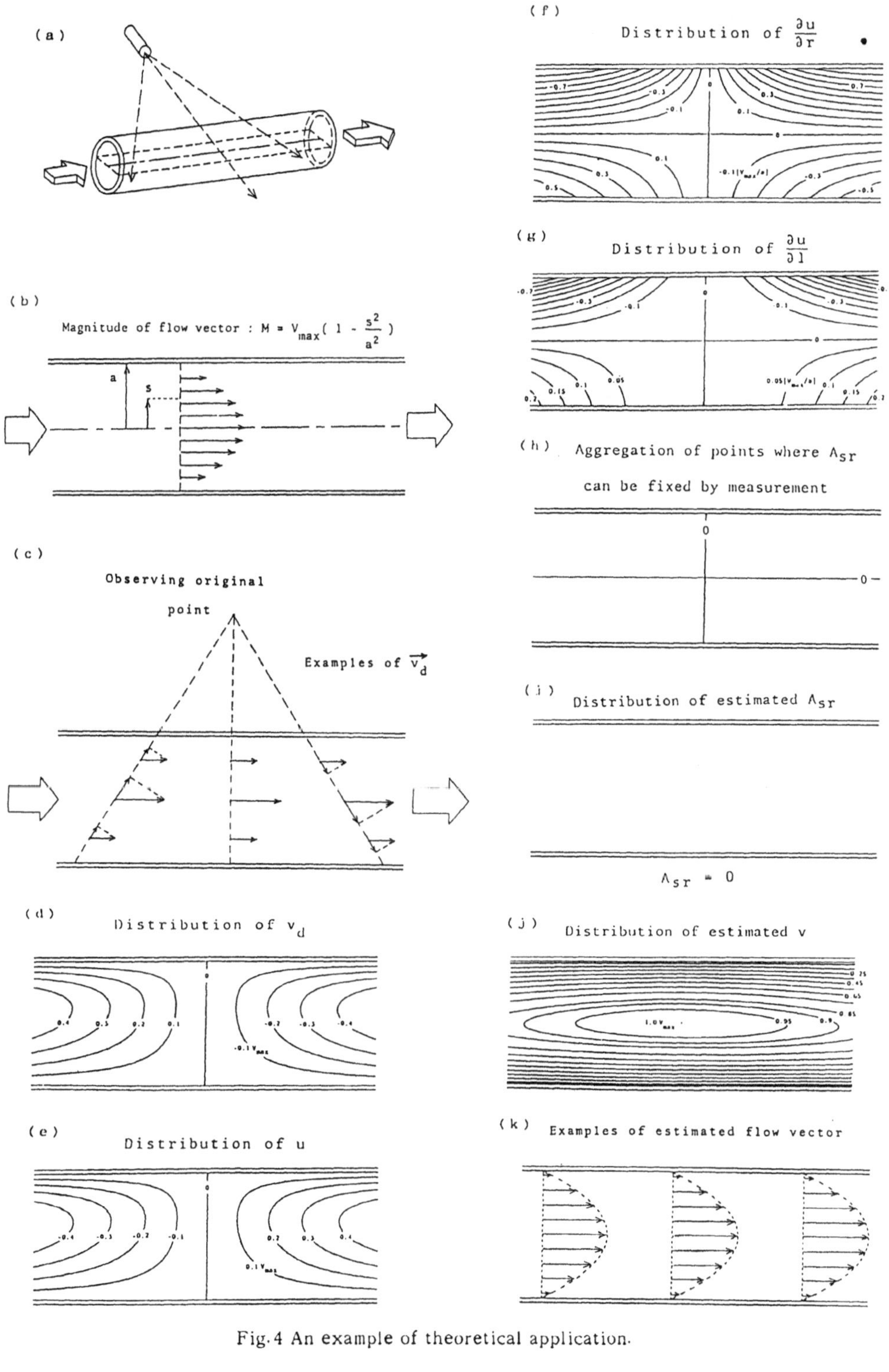
(a)
(b)
Magnitude of flow vector : M = $V_{max}(1 - \frac{s^2}{a^2})$
a
s
(c)
Observing original
point
Examples of $\vec{v_d}$
(d)
Distribution of v_d
(e)
Distribution of u
(f)
Distribution of $\frac{\partial u}{\partial r}$
(g)
Distribution of $\frac{\partial u}{\partial l}$
(h) Aggregation of points where Λ_{sr}
can be fixed by measurement
0
(i) Distribution of estimated Λ_{sr}
$\Lambda_{sr} = 0$
(j) Distribution of estimated v
(k) Examples of estimated flow vector

Fig.4 An example of theoretical application.

A_{cr} equals zero. So we can get the distribution (Fig. 4(j)) of v by the equation (10). Then we can estimate velocity distribution as Fig. 4(k) and this distribution agrees with the original distribution in Fig. 4(b).

CONCLUSION

We introduced the new technique to estimate flow velocity vector distribution deduced from ultrasonic Doppler method.

REFERENCES

1. S. Ohtsuki, M. Okujima, and M. Tanaka, "Blood Flow Mapping Based on Doppler Information," Journal of Ultrasound in medicine (Abstract of 31st AIUM Annual Convention), Vol. 5, No. 9, p. 86, 1986
2. S. Ohtsuki, M. Okujima, and M. Tanaka, "A Method of Flow Vector Mapping Deduced from Doppler Information," The Journal of the Acoustical Society of Japan, Vol. 43, No. 10, pp. 764-767, 1987 (in Japanese)

TWO-DIMENSIONAL MAPPING OF THE VELOCITY VECTOR DISTRIBUTION OF THE INTRAVENTRICULAR BLOOD FLOW

A. Yamamoto, M. Tanaka, N. Endoh, K. Takahashi, S. Ohtsuki*, and M. Okujima*

Department of Medical Engineering and Cardiology
Institute for Chest and Cancer, Tohoku University, Japan
*Tokyo Institute of Technology, Japan

INTRODUCTION

In order to quantify the cardiac flow dynamics in a non-invasive manner, the method for two-dimensional mappings of the equi-velocity and equi-acceleration distribution has been developed in our laboratory during the past several years[1,2]. In this report, 2-dimensional mapping of the flow velocity vector in the left ventricle and its usefulness in medical diagnosis is described.

METHODS

The pulsed Doppler equipment employs a 3.5 MHz carrier frequency and a 4.4 kHz in repetition rate.

By the apex approach (Fig 1), an echo-tomograph through the left ventricle was chosen for which the intraluminal dimension was maximal. The Doppler velocity pattern was recorded at 50-100 sampling points on the plane. The pulsed Doppler signals of one cardiac cycle obtained from each point were manually traced, i.e., the envelope of the velocity pattern, which was analyzed with the FFT technique, was input into the computer. The velocity data were obtained at each cardiac phase when one cardiac cycle was equally divided into 20 portions.

Equi-velocity mapping was then drawn by the application of a special interpolation technique developed[1,2].

From the equi-velocity distribution data , the velocity components tangential to the beam axis (V) were calculated by applying the law of conservation of momentum, as shown in Fig. 2. Assuming that the flow is symmetric to the cross section plane, the velocity component(W), which is tangential to the section plane, is theoretically 0. Accordingly, the acceleration (Du/Dt) consists of local and convective acceleration.

The velocity component(V), which is tangential to the beam axis, was calculated by the equation (Asr, as shown in Fig. 2). In the equation for Asr, at the points where the equi-velocity line is perpendicular to the beam axis, the values of $\partial u / \partial l$ becomes 0. Accordingly by the values of

Asr calculated with the formula ($Asr = u\frac{\partial u}{\partial r}$), the distribution of Asr on section plane was calculated from the dispersed data obtained by using the acceleration interpolation technique. The velocity component(V) was calculated from the data of $\partial u / \partial l$, $\partial u / \partial r$ and thus Asr obtained. The velocity vector and its distribution were obtained on the longitudinal section plane by the velocity data of both V and U.

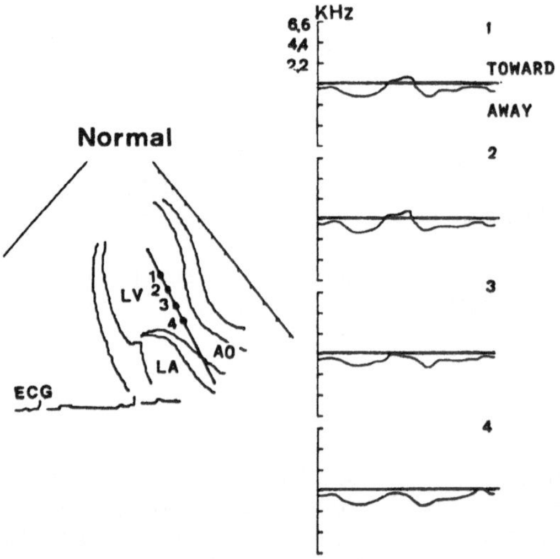

Fig 1
Manual tracing of the Doppler flow velocity during one cardiac cycle at the points of the left ventricle (for example ,1,2......) illustrated on the figure (B mode). Normal case.

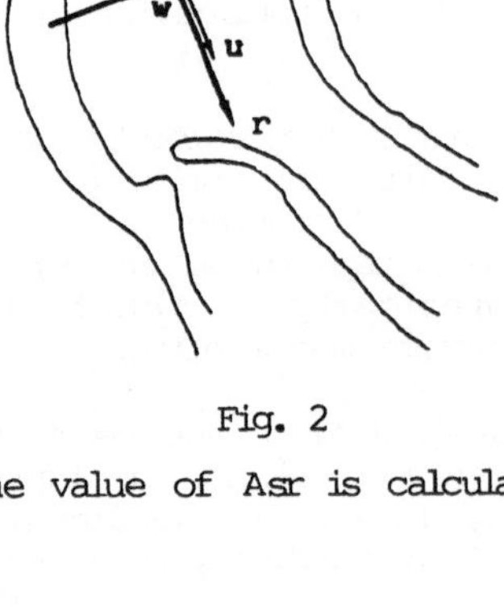

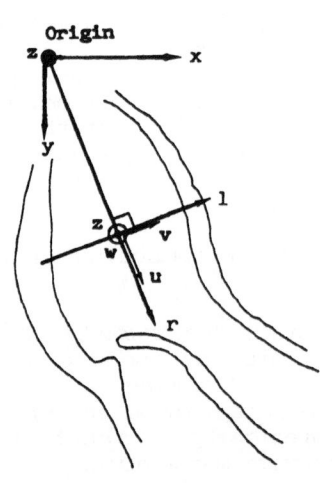

$$\frac{Du}{Dt} = \underbrace{\frac{\partial u}{\partial t}}_{(1)} + \underbrace{u\frac{\partial u}{\partial r} + v\frac{\partial u}{\partial l} + w\frac{\partial u}{\partial z}}_{(2)}$$

(1): local acceleration
(2): convective acceleration(Asr)

if (w) =0

$$\frac{Du}{Dt} = \frac{\partial u}{\partial t} + u\frac{\partial u}{\partial r} + v\frac{\partial u}{\partial l}$$

where

$$Asr = u\frac{\partial u}{\partial r} + v\frac{\partial u}{\partial l}$$

from u and v

$$V = |\sqrt{u^2+v^2}|$$

Fig. 2
The value of Asr is calculated by the formula given here.

RESULTS

Figure 3 is an example of 2-dimensional mapping at the isovolumic contraction phase where the equi-velocity distribution is shown on the left, the equi-acceleration distribution in the middle and the velocity vector on the right. These maps were obtained at the cardiac phase indicated by the long bar on the ECG in the figure.

On the equi-velocity distribution maps, the flow away from the apex, i.e., the "out flow", was indicated by repetitive sequence of red, magenta and yellow colors at an interval of 5cm/second in flow velocity. Repetitive sequences of blue, light blue and green indicate the direction of the flow toward the apex, i.e., the "in flow".

The equi-acceleration distribution is also indicated by the cold color sequence for acceleration and warm color sequence for decceleration at the interval of the $10cm/sec^2$.

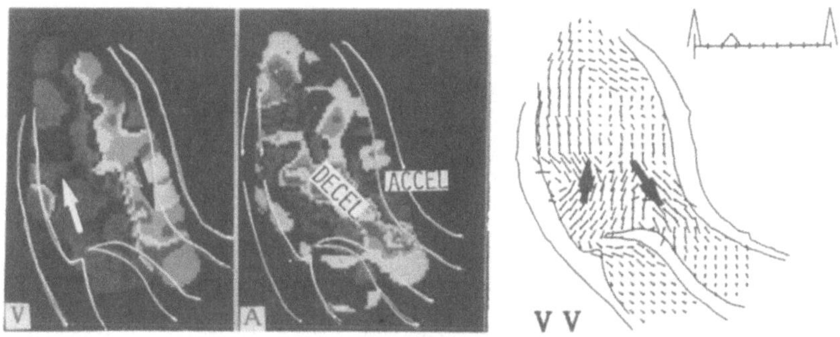

Fig. 3
Isovolumic contraction phase
V: equi-velocity
A: equi-acceleration
VV: velocity vector

During the isovolumic contraction phase, the equi-velocity map revealed that the "flow away" stream was observed chiefly over the septal side of the ventricle and was faster behind the anterior leaflet of the mitral valve. At the same time, there was "flow toward" the apex along the left ventricle wall and which was produced by the downward movement of the mitral annulus.

The equi-acceleration velocity at this phase was apparent at the area near the septum, reflecting the forward movement of the posterior wall toward the septum.

The flow vector distribution at this phase is shown on Fig. 3 (the right figure). The warm colors indicate upward flow and the cold colors, downward flow. The faster components of the flow vector were observed at the area near the outflow tract and the eddy was seen at the basal central area of the ventricle.

During the ejection phase (Fig 4), the equi-velocity mapping revealed parabola-shaped distributions in the outflow tract of the ventricle. Thus, the equi-velocity patterns get closer and red zones narrower, indicating an increase in flow-away velocity and the extent of the flow-away stream at the central part of the outflow tract.

The equi-acceleration pattern is shown on the middle frame. The area of acceleration indicated by cold colors is found in the area along the ventricular septum and posterior wall. The area of decceleration shown by warm colors, is observed from the center of the ventricle to the apex.

On the right frame (Fig. 4) which shows the velocity vector distribution in the ejection phase, a bulk of outflow moved toward the aorta and the faster components of the flow vector were observed at the area in the outflow tract. However, some of outward flow turns backward under the mitral leaflet.

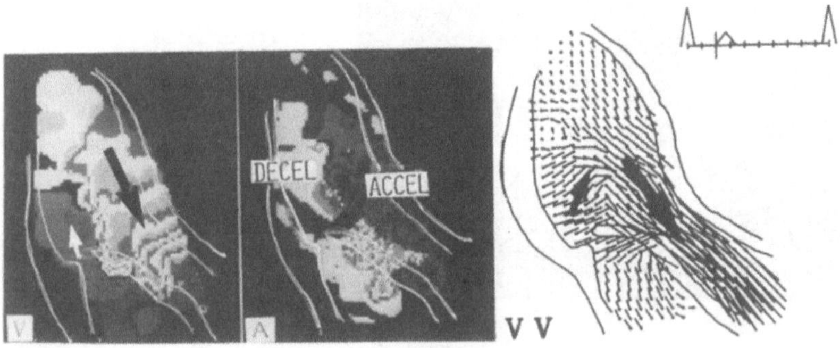

Fig. 4
Ejection phase
V: equi-velocity
A: equi-acceleration
VV: velocity vector

In order to explain these findings, the wall motion was studied by A-mode and B-mode echocardiography. The movements of the septum and ventricular wall were evaluated by 2 channel M-mode recordings at the apical, central and basal parts on the long-axis view (Fig 5).

The onset and peak of the posterior wall contraction were measured and the time-sequence evaluated by 2-channel M-mode recording performed along the long-axis view. The onset and peak of the inward movement of the posterior wall and septum were measured at the cardiac base, apex and center, and the time-sequence was evaluated by 2-channel M-mode recording performed along the long-axis view. The lower frame illustrates these relationships. The inward movement of the ventricular wall appeared very early at the apical portion, but those of the central and basal part were slightly delayed.

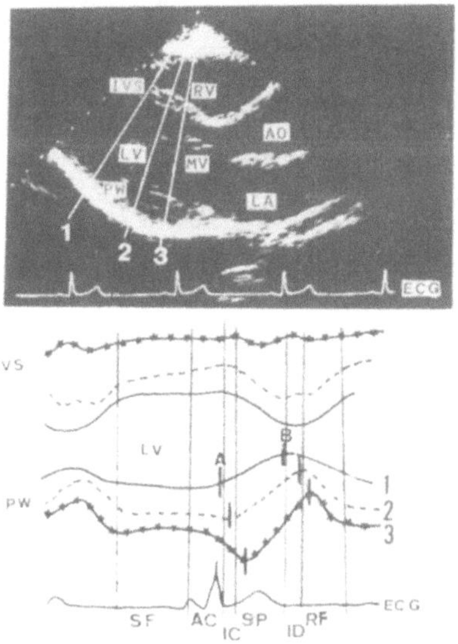

Fig. 5
The ventricular wall movement evaluated by 2-channel M-mode recording
A:The onset of the contraction
B:The peak time

IC:isovolumic contraction
SP:systolic phase
ID:isovolumic dilatation
RF:rapid filling phase
SF:slow filling phase
AC:atrial contraction

1: Apex
2: Center
3: Base

In Figure 6, the left side frame, shows the B-mode echography in systolic phase. The right figure shows a schematic representation of the 2-dimensional movement of the left ventricle analyzed with M-mode and B-mode echograms.

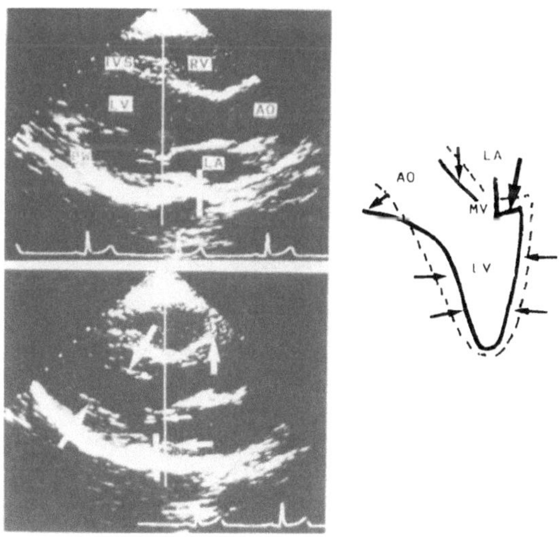

Fig. 6
Schematic representation of the movement of the left ventricle during ejection phase.
White bars on echogram show the positions of the posterior mitral ring in early systole (upper picture) and late systole (bottom picture).
White and black arrows indicate the direction and extent of the movement.

..... isovolumic contraction
—— systolic phase

When the systolic phase was evaluated by M-mode and B-mode echography (Fig 7), the inward movement of the posterior wall and the septum appeared very early at the apex and continued up to the mid-systolic phase. The most delayed contraction was observed in the basal part (Fig.5). It seemed that the major influence of the force for ejection was generated by the shortening of the long axis, produced mainly by the antero-downward movement of the mitral complex and by the peristaltic movement of the ventricle occurring from the apical to basal parts, during this phase.

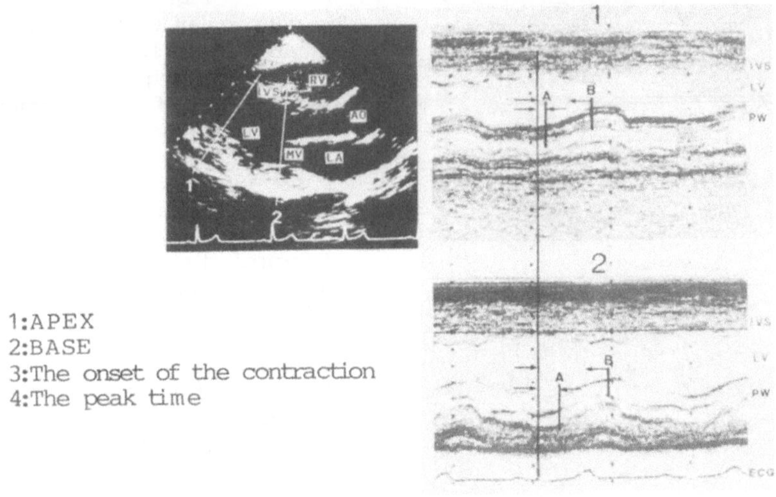

Fig. 7
B-mode and M-mode echocardiograms in the normal case.
(1) and (2) of the M-mode echography were taken along the direction of the ultrasonic beam.

During the rapid filling phase (Fig. 8), the equi-velocity lines were convex toward the apex and the intervals between each line were almost equally separated. At the area of outflow tract, a flow reversal was observed.

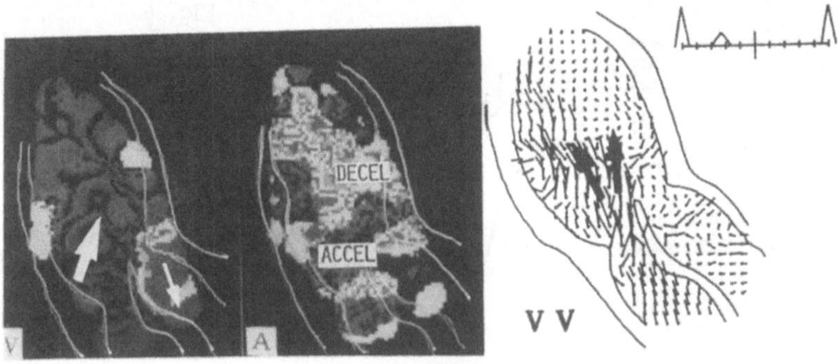

Fig.8
Rapid filling phase
V: equi-velocity
A: equi-acceleration
VV: velocity vector

Acceleration of flow was seen at the basal, apical and posterior portions, and the deccelerated flow was seen in the central to posterior area of the ventricle.

The flow-vector distribution indicated that there were two downward flows, one toward the posterior wall and the other toward the septum. Also, a remarkable clockwise eddy appeared behind the anterior mitral leaflet. The eddy might have given rise to a closure of the mitral leaflet.

When the echography was observed during the filling phase(Fig. 9), the mitral complex moved upward rapidly while the mitral valve was open.

In this stage, the ventricle elongated along the long axis and widened along the short axis. These suggested that in the rapid filling phase, the elongation of the long axis was more influential in forming this particular flow pattern. In contrast, the elongation along the short axis appeared to play a greater role in hemodynamics of the slow filling phase.

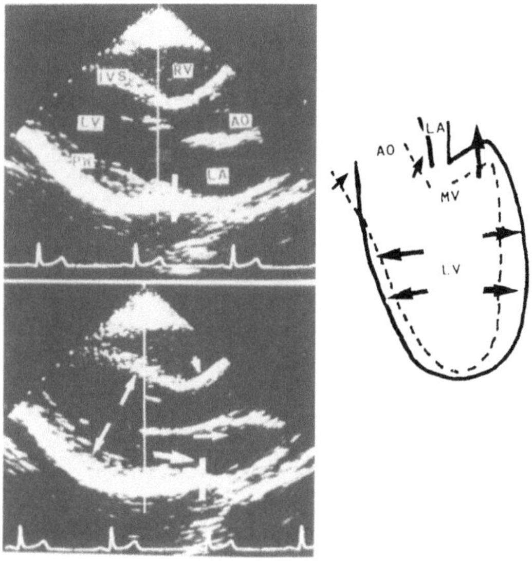

Fig. 9
Schematic representation of the diastolic movement of the ventricle.
White bars on the echogram show the position of the posterior mitral ring in early diastole (upper picture) and late diastole (bottom picture).
White and black arrows show the direction and extent of the movement.

---- isovolumic dilatation
—— rapid filling phase

DISCUSSION

The 2-dimensional display of the left ventricular hemodynamics, i.e., the equi-velocity, equi-acceleration and flow-vector distributions, described above, enables us to visualize and understand the flow dynamics occurring in site in the heart, irrespective of the direction of the ultrasonic beam.

The flow vector distribution obtained by this approach clearly reflects the complex effects of local pressure, movements of the blood and dynamic changes of the acceleration.

When the dynamics of the heart structure, that is, the ventricular walls and valves, are considered, flow-vector distributions reflect clearly the movements of the ventricular walls, papillary muscles, and the valvular complex. Thus, 2-dimensional flow-vector distributions should serve as an important addition to the understanding of cardiac hemodynamics in a non-invasive manner.

REFERENCES

1. A. Yamamoto, M. Tanaka, N. Endoh, K. Takahashi, H. Ohkawai, S. Ohtsuki, and M. Okujima, Two-dimensional mapping of the dynamic pressure distribution of the left ventricle, 2nd Internat. Congr. Cardiac Doppler, Kyoto, Japan (1986).

2. A. Yamamoto, M. Tanaka, S. Ohtsuki, N. Endoh and K. Takahashi, Two-dimensional mapping of the flow velocity vector and dynamic pressure distributions as functions of the normal left ventricle, 2nd Asian-Pacific Conference on Doppler and Echocardiography, San Diego, USA (1987).

CONSIDERATIONS ON SIGNAL-TO-NOISE RATIO IN A CONTINUOUS WAVE DOPPLER SYSTEM USING AN ARRAY TRANSDUCER

T. Mochizuki and C. Kasai

Aloka Co., Ltd.
6-22-1 Mure, MItaka-shi, Tokyo 181, Japan

INTRODUCTION

Phased array systems using ultrasound waves have been widely used to produce cross-sectional images of the heart in recent years. Ultrasonic Doppler techniques also supply useful diagnostic information on blood flow in living organs. In the Doppler technique, continuous wave Doppler (CWD) is used to measure high blood flow velocity quantitatively, and it is very desirable that this CWD function be implemented in phased array systems. There has been some commercial equipment in which the CWD is available, however, it lacked measurement sensitivity.

When a phased array transducer is used in a CW Doppler system, it is expected that the ratio of piezo-elements used for transmission and reception in the array transducer will alter the signal-to-noise ratio (i.e., SNR) of the received signal, even if the total number of elements is constant. This paper describes how this ratio influences CWD sensitivity.

PRINCIPLE

Ultrasonic transmission/reception model

We think of a model consisting of arrayed sound sources and arrayed receiving elements arranged in a circular arc, as a two-dimensional model shown in Fig. 1. There are N_T transmitting elements and N_R receiving elements. The ultrasonic waves are emitted from the transmitting elements and reflected at a target located at the focal point F. The reflected waves are received by N_R receiving elements, added to each other in phase and become received signal Es.

Equations of the receiving signal and noise

Generally, Rayleigh's equation is used to express the sound field generated by a circular concave-surface transducer.[1] When this equation is applied to our two-dimensional array transducer model described earlier, the sound pressure P_T at an arbitrary point q on the plane is inversely proportional to the distance r_i between i-th transducer element and the point q, and proportional to the transducer-exciting voltage V_T. When the sound propagation coefficient is expressed as $\gamma = \alpha + j\beta$, we obtain P_T as Eq. (1).

$$P_T = AV_T \sum_{i=1}^{N_T} \frac{d}{r_i} exp\{-\gamma r_i\} \tag{1}$$

where, α is the attenuation coefficient, β is the phase constant, A is the proportional constant, and d is the width of transmitting element on an array transducer. The d is assumed to be much smaller than the wavelength so that the sound waves emitted from the transducer can propagate concentrically from each element. When the sound waves are reflected at the point q where the sound pressure is P_T, the received signal E_S can be written as Eq. (2).

$$E_S = BP_T \sum_{j=1}^{N_R} \frac{1}{r_j} exp\{-\gamma r_j\} \tag{2}$$

Where B is the proportional constant, and N_R is the number of receiving elements. Accordingly, the received signal E_S is obtained by substituting Eq. (1) into Eq. (2), as shown in Eq. (3).

$$E_S = CV_T \sum_{i=1}^{N_T} \sum_{j=1}^{N_R} \frac{d}{r_i r_j} exp\{-\gamma(r_i + r_j)\} \tag{3}$$

Where, C=AB.

In our transmission/reception model shown in Fig. 1, the distances between r_i and r_j coincide at the focal point F so that the equation can be simplified for easier analysis. Therefore, we will discuss SNR only for the case in which sound waves are reflected entirely at point F. First, for reasons of convenience, we will assume that sound waves do not attenuate during propagation. Later, we will consider the influence of attenuation. In this condition, Eq. (3) is simplified to Eq. (4) by abbreviating the phase term.

$$E_S = CV_T \frac{d}{R^2} N_T N_R \tag{4}$$

Where R is the distance between the transducers and point F.

In contrast, when the average noise voltage generated in one receiver is assumed n_0, the summed noise voltage En becomes as given by Eq. (5).

$$E_n = n_0 N_R^{1/2} \tag{5}$$

As receiver noise is random, the total noise signal En is proportional to the square root of N_R, the number of receivers.

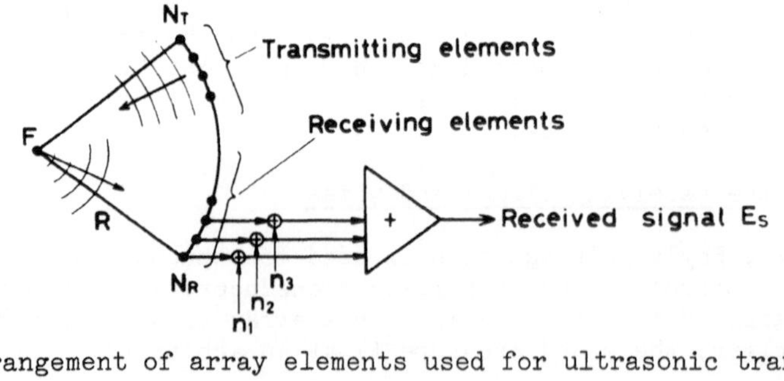

Fig. 1. Arrangement of array elements used for ultrasonic transmission and reception.

Transmission voltage

As a common way to express the degree of convergence of sound waves emitted from a circular concave-surface transducer, factor D defined by Torikai is widely used.[2,3]

$$D = \frac{a^2}{\lambda R} \tag{6}$$

Where λ is a wavelength, a is the radius of a transducer and R is the radius of the concave transducer. We applied the above equation to the array transducer and defined the D-factor of the transmission transducer D_T as Eq. (7). Here, radius a in Eq. (6) is replaced by $N_T d/2$.

$$D_T = \frac{1}{\lambda R}\left(\frac{N_T d}{2}\right)^2 \tag{7}$$

As noted earlier, we assume that sound waves are emitted into the two-dimensional sound field. Accordingly, the maximum sound pressure Pmax produced by the array elements is believed to be nearly proportional to the square root of the maximum pressure produced by the circular concave transducer which has the same aperture length as the array transducer.[4] By using this relation, relative sound pressure Pmax/Po curve for an array transducer is derived from the values for a circular concave vibrator obtained by using Torikai's formula [2]. The broken line in Fig. 2 shows the calculated result. Where Po is the sound pressure on the surface of the transducer. When the D factor is 2 or less, the broken line can be approximated as Eq. (8) and it is shown as the solid line in Fig. 2.

$$Pmax/Po = 1.4+0.8\ D_T \tag{8}$$

Since Po is proportional to the transmission voltage V_T, Eq. (8) can be rewritten as Eq. (9).

$$Pmax = (1.4+0.8\ D_T)\kappa V_T \tag{9}$$

Where κ is the proportional constant. The variation of D_T changes the maximum sound pressure Pmax for a constant driving voltage V_T. In our model, however, the driving voltage was controlled such that Pmax is always held to a constant value K. In this condition, the driving voltage for the transmission transducer elements becomes:

$$V_T = \frac{K}{\kappa(1.4+0.8\ D_T)} \tag{10}$$

Substitution of Eq. (7) in Eq. (10) makes Eq. (11).

$$V_T = \frac{K'}{1.4+0.8\mu N_T^2} \tag{11}$$

where, $K' = K/\kappa$ and $\mu = d^2/4\lambda R$.

Discussion so far has been based on the assumption that the sound waves were propagated through an attenuation-free medium. However, the actual tissue of an organism has propagation attenuation. Figure 3 shows qualitatively how the sound pressure of the transmitted sound waves along the central axis of the array transducer decreaces from the constant acoustic power K through their propagation. As the transmitting element number N_T increases, the peak point of sound pressure moves farther away from the sound sources, and hence the maximum sound pressure attenuates more. This attenuation must be compensated to keep the maximum sound pressure constant. To do this, we have to know the relation between Pmax corresponding to each D-factor and the position Zm at which Pmax is obtained. However, no graphs to show this relation are generally prepared. Therefore we first graphed the relationship between Pmax and Zm when the attenuation is zero, and then we estimated Zm when attenuation exists. The broken line in Fig. 4 indicates the Zm position for D_T-factor in the attenuation-free medium. This curve can be approximated by Eq. (12) within 5% error when the D_T-factor is 2 or less. The solid curve in Fig. 4 shows the approximate curve.

$$Zm = (1-10^{-0.55 D_T})R \tag{12}$$

In the attenuation medium however, Zm becomes approximately 10% shorter (i.e., closer to the transducer) than the above value when the attenuation coefficient α is between 0.3 and 0.6 dB/MHz/cm. By taking these factors into account, the driving voltage V_T is obtained as Eq. (13).

$$V_T(N_T) = \frac{K'}{(1.4+0.8\mu N_T^2)\exp\{-0.9\alpha f_o Zm\}} \tag{13}$$

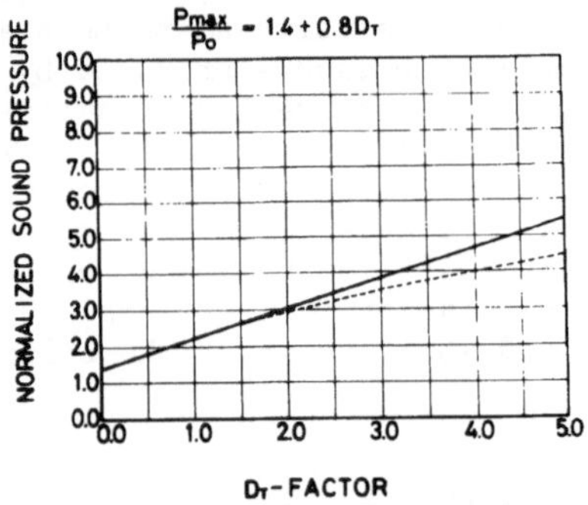

Fig. 2. Normalized maximum sound pressure Pmax/Po for D_T-factor.

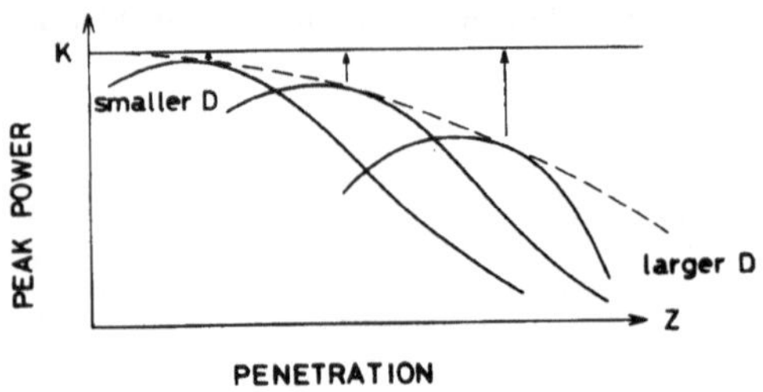

Fig. 3. Influence of the propagation and its compensation.

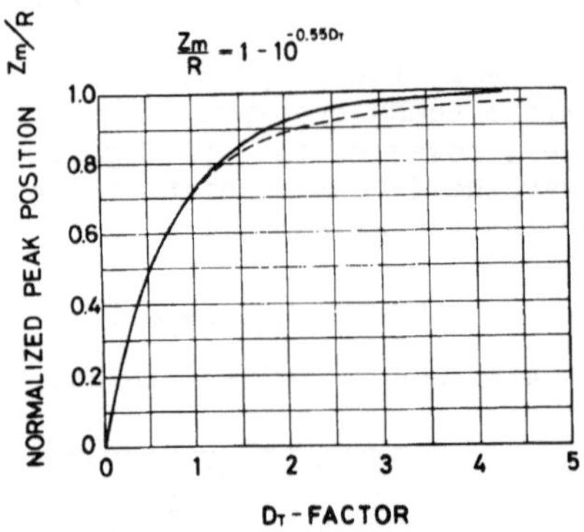

Fig. 4. Normalized peak position for D_T-factor.

Signal to Noise Ratio (SNR)

Es signal received by the transducers is obtained by substituting Eq. (7), Eq. (12) and Eq. (13) into Eq. (4), and multiplying the result by $\exp\{-2\alpha foR\}$, which indicates sound wave attenuation in the round-trip propagation distance 2R. Thus, we obtain Eq. (14).

$$Es(N_T) = \frac{d\exp\{-2\alpha f_o R\}}{R^2} N_T N_R \frac{K'}{(1.4+0.8\mu N_T^2)\exp\{-0.9\alpha f_o R(1-10^{-0.55\mu N_T^2})\}} \tag{14}$$

Accordingly, SNR is defined as the power ratio of the signal Es and En.

This is shown in Eq. (15).

$$SNR = \frac{\left\{ \frac{d\exp\{-2\alpha f_o R\}}{R^2} N_R N_T \frac{K'}{(1.4+0.8\mu N_T^2)\exp\{-0.9\alpha f_o R(1-10^{-0.55\mu N_T^2})\}} \right\}^2}{\left\{ n_o N_R^{1/2} \right\}^2} \quad (15)$$

CALCULATION EXAMPLES

In this section, actual calculation is made using Eq. (15). Parameters were set as follows: d=0.31mm, R=100mm and fo=2MHz. The calculation was carried out with three attenuation coefficients: =0, 0.3, and 0.6dB/MHz/cm. The calculation results are plotted in Fig. 5, where the number of transmitting elements N_T is on the X axis, and the relative SNR value on the Y axis. The total number of elements in the array transducer ($N_T + N_R$) is 64. Therefore, when N_T is 43, the number of receiving elements N_R is $64-N_T = 21$.

As shown in Fig. 5, the peak value of SNR shifts towards the right depending on the attenuation coefficients. When there is non-attenuation, the optimum number of transmitting elements N_T which produces the maximum SNR is 35. In this case, the optimum ratio between N_T and N_R (i.e., N_T/N_R) is 1.2. When the attenuation is 0.3dB/MHz/cm, N_T is 43 ($N_T/N_R \fallingdotseq 2.0$), and it is 47 ($N_T/N_R \fallingdotseq 2.8$) when the attenuation is 0.6dB/MHz.cm. That is, the optimum ratio N_T/N_R which maximizes SNR is larger than 1, and this ratio increases as the attenuation coefficient increases. However the decrease of SNR from the maximum value when N_T/N_R is varied from the optimum value, is mild. For example, when α is 0.3dB/MHz/cm which is close to the attenuation coefficient of living tissue, the decrease of SNR value is only 0.5dB from the peak, even if N_T is set at 49($N_T/N_R \fallingdotseq 3.3$).

These results provide very useful information for actual system design because it is generally more difficult and expensive to manufacture receiving circuits than transmitting circuits. That is, the smaller the number of receiving circuits, the easier system hardware design becomes.

EXPERIMENTAL METHOD

We made an experimental system to confirm the theory described above. To simply the experiment, we employed the following method. Generally, ultrasound signals reflected by blood cells in the living body are extremely small and random, so it is considerably difficult to quantitatively measure the SNR of such signals with an accuracy of one or two decibels. It is also difficult to make a propagation medium with the same attenuation coefficient as the living body. To avoid such difficulties, we carried out the experiments in water and then estimated the relative SNR of the system based upon experimental data.

First, a hydrophone (1mm diameter, band width of 1-10MHz) was placed in water on the main beam axis of the transmitting array transducer, as shown in Fig. 6(a). Then, the sound pressure along the central axis of the transmission beam was measured. At this time, driving voltage V_T was controlled such that the maximum pressure Pmax was held constant. Second, as shown in Fig. 6(b), the hydrophone was placed at the focal point F. The sound waves of which the intensity is proportional to the sound pressure measured at this focal point are emitted from the same hydrophone.

Es signal was measured at the output terminal of the adding amplifier by making sound signal intensity much greater than that of noise. In this condition, the output signal is considered to be almost completely sound wave signal, while receiver noise En was measured after stopping sound wave generation. From these values, the relative SNR was calculated as the ratio of the two (i.e., Es/En).

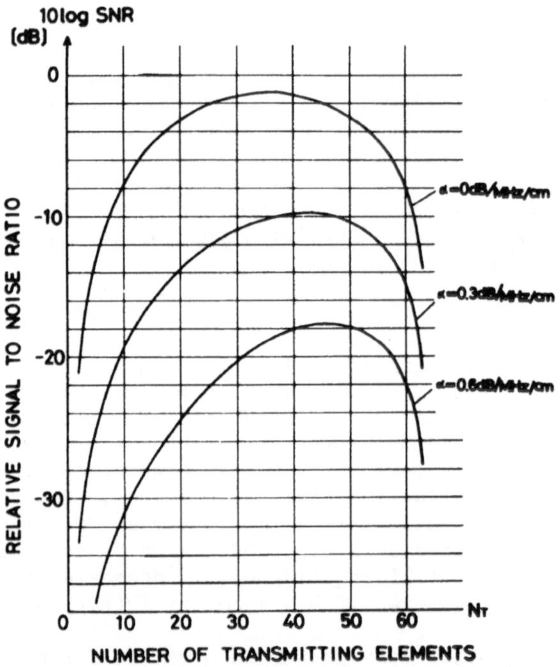

Fig. 5. Theoretical curves of signal-to-noise ratio.
Conditions: The total number of elements ($N_T + N_R$) is 64, radius R=100mm, width of transmitting element d is 0.31mm, frequency fo is 2MHz. Ultrasonic propagation attenuation coefficients are α =0, 0.3 and 0.6 dB/MHz/cm.

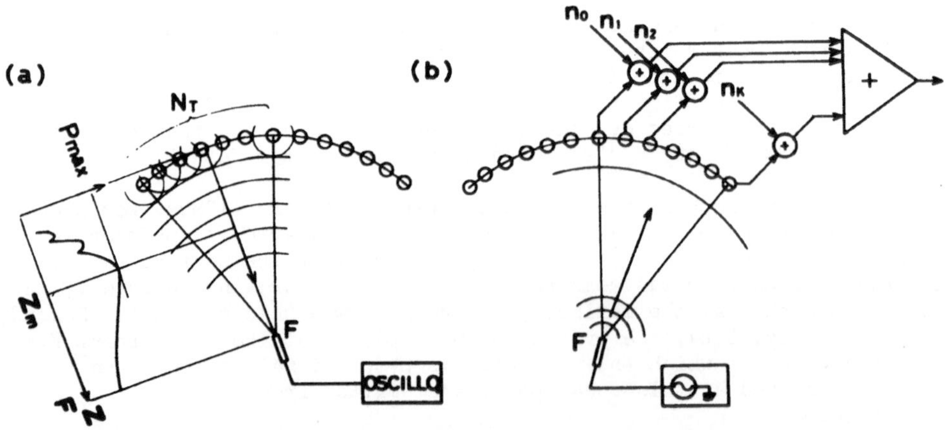

Fig. 6. SNR measurement. (a) Measurement of transmitted sound. (b) Measurements of received signal and noise.

EXPERIMENTAL RESULTS

The probe used in the experiment has an aperture length of 20mm along the array direction and 10mm across the array direction. Its ultrasonic frequency is 2MHz. The total number of array elements was 64 and the focal distance R was set as 100mm, the same as the calculation examples. The number of transmitting elements N was varied in four ways ---31, 39, 43, and 47. Figure 7 shows the measured sound pressure curves on the central axis of the transmission sound beam by the means illustrated in Fig. 6(a). In this Fig. 7 the horizontal axis indicates the distance Z between the transducer surface and the hydrophone, and the vertical axis the sound wave amplitude detected at the hydrophone. Since the curves in Fig. 7 are the ones obtained in water where the propagation attenuation is negligibly small, the attenuation in living tissue must be taken into account.

Figure 8 shows the replotted data where the attenuation coefficient is assumed to be 0.3 dB/MHz/cm. In the Fig. the maximum sound pressures were made to become a constant value, K'. As shown in Fig. 8, when N_T is increased from 31 to 47, the sound pressure peak point moves away from the transducers, and the sound pressure value at the focal point Z=100mm increases. Next, sound waves in proportion to the values at Z=100mm for each N_T value in the figure were emitted from the hydrophone, then the received signal Es at the summing amplifier was measured. After that, noise power was measured by stopping the sound generation. Thus the relative SNR was calculated based upon the aforementioned procedure. The same operation was performed for the cases in which α=0 and α=0.6 dB/MHz/cm. These results were plotted in black dots in Fig. 9 together with the theoretical curves obtained in Fig. 5.

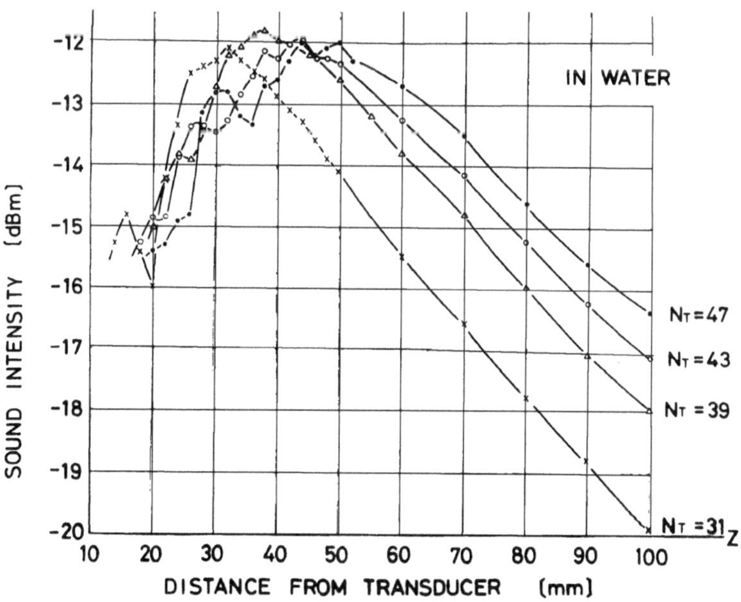

Fig. 7. Measured data of sound pressure on the central axis of transmitting transducers for four values of N_T. Conditions: Radius R=100mm, width aperture length is 10mm and sound frequency is 2MHz.

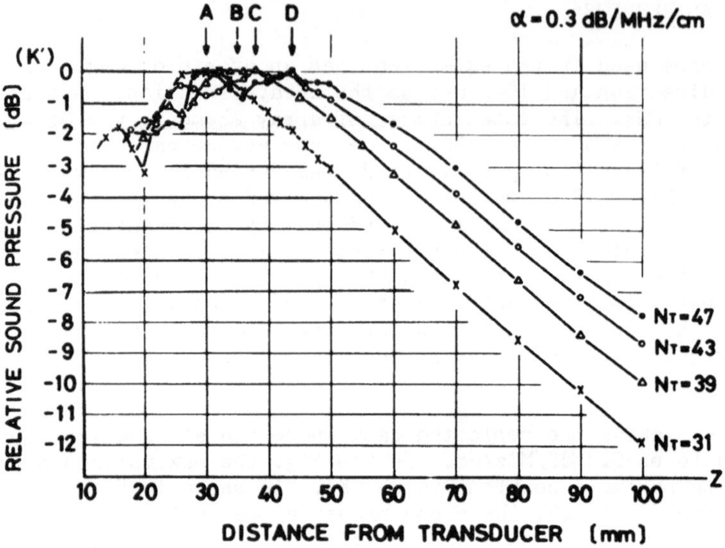

Fig. 8. The replotted curves of sound pressure in Fig. 7.
The propagation attenuation coefficient is assumed to be 0.3 dB/MHz/cm and the maximum acoustic intensity is adjusted to a constant value K'.

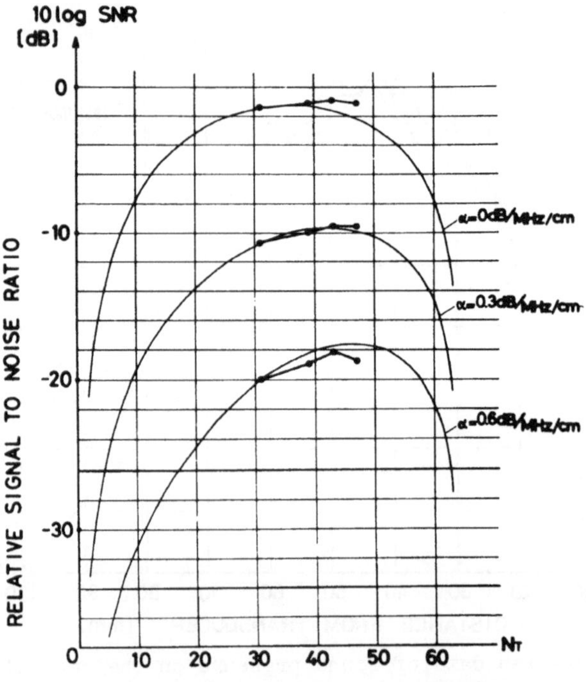

Fig. 9. Experimental results together with theoretical curves.

In Fig. 9, the experimental data are not sufficiently coincident with the theoretical curves. However, we can note the tendency that as the attenuation coefficient increases, the optimum number N_T which maximizes the SNR, increases. The reason for this result is considered that the width aperture length (perpendicular to the array direction of the transducer) most likely influenced the sound field. In theory, the width aperture length of the probe was assumed to be infinite. However, the transducer used has a 10mm width length, and should have influenced the sound field to some extent.

DEVIATION OF THE ULTRASONIC BEAM

As described so far, SNR can be improved by making the ratio of elements used for transmission and reception uneven. However, it is feared that this may cause the composite ultrasonic beam axis of transmission and reception to deviate from the central axis of the array transducer. To learn the overall directivity of the composite transducer, we carried out a simulation calculation using a computer. We used the same two-dimensional model as shown in Fig. 1. However, a point reflector was placed at an arbitray position in the field instead of the focal point. The overall directivity pattern was calculated using Eq. (3). Figure 10 shows the result when $N_T=43$ ($N_T/N_R \doteqdot 2$), R=100mm and $\alpha=0.3$ dB/MHz/cm.

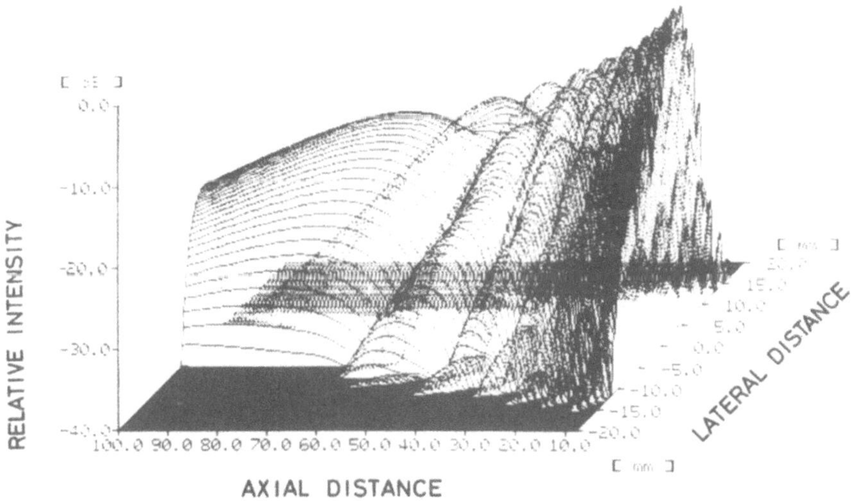

Fig. 10. Overall directivity pattern of transmission and reception. Conditions: $N_T=43$($N_T/N_R \doteqdot 2.0$), $\alpha=0.3$ dB/MHz/cm.

In order to observe the deviation of the sound beam in detail, a beam profile at a given distance Z from the transducer surface can be extracted using the directivity pattern. Figure 11 shows the beam profile at Z=60mm. It can be seen that the intensity peak point deviates from the central axis by approx. 0.5mm. When this value is converted to an angle, the deviation is within 0.5 degrees so that it would not be likely to cause any problem in practical use.

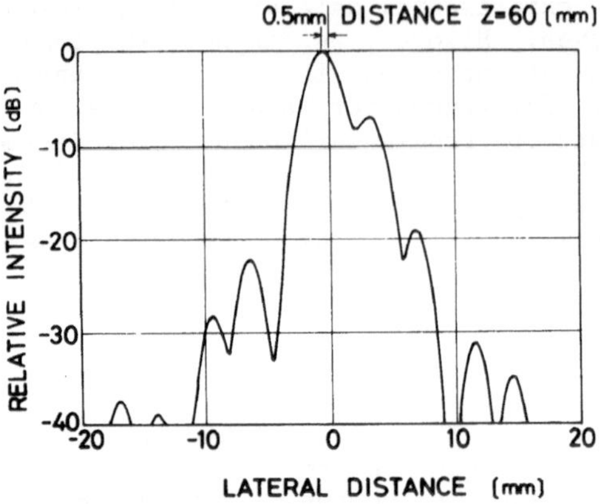

Fig. 11. The beam profile along array direction at Z=60mm.

CONCLUSIONS

We carried out theoretical calculations concerning signal-to-noise ratio in a continuous Doppler system using an array transducer and obtained the following results:

(1) For a continuous wave Doppler system using an array transducer, the optimum N_T/N_R ratio which makes the signal-to-noise ratio (SNR) maximum varies with the attenuation coefficient. This optimum ratio increases as the attenuation coefficient increases.

(2) The optimum N_T/N_R ratio is approximately 2 in theory when the attenuation coefficient is 0.3 dB/MHz/cm.

(3) Even when the N_T/N_R ratio is increased to 3, the decrease in the SNR from the peak value is only about 0.5 dB. This increase in the N_T/N_R ratio helps facilitate the design of system hardware.

(4) When N_T/N_R =2, the sound beam deflection from the central axis of an array transducer is within about 0.5 degrees. This value is considered tolerable in practical use.

Experiments were conducted and a similar tendency obtained. Deviation from theory is considered in that the cross direction of the array transducer influenced the sound field.

ACKNOWLEGEMENT

We wish to express our deep gratitude to Mr. M. Hirose, Mr. T. Nakano, Mr. T. Yoda, and Mr. Y. Tabuchi, who kindly cooperated with us in making the experimental equipment and the measurements.

REFERENCES

1. Lord Rayleigh ," Theory of Sound "(Dover Publ., New York, 1945) Vol. 2, P. 107

2. Y. Torikai, IECE Technical report on Ultrasound (in Japanese). Nov. 16, 1962

3. J. Saneyoshi, Y. Kikuchi, O. Nomoto , "Technical Handbook on Ultrasound" (Nikkan Kougyo Shinbun), 1968, P. 1415

4. M. Okujima, S. Ohtsuki, Approximation evaluation of beam width of a concave-surface transducer and variable aperture transducer, J.J.M.U., Vol. 5, No. 2, P. 9

A FUNDAMENTAL EVALUATION OF IN VIVO SOUND SPEED MAPPING TECHNIQUE BY CROSSED BEAM METHOD

Masafumi Kondo[1], Kinya Takamizawa[1], Makoto Hirama[2], Kiyoshi Okazaki[2], Kazuhiro Iinuma[2], and Yasuaki Takehara[3]

[1]Toshiba Research and Development Center, Kawasaki, Japan 210
[2]Toshiba Medical Engineering Laboratory, Tochigi, Japan 329-26
[3]Kanto Central Hospital, Setagaya-ku, Tokyo, Japan 158

INTRODUCTION

Recently, several kinds of in vivo sound speed measurement techniques, using a pulse echo method, have been developed for the purpose of ultrasound tissue characterization.[1,2,3] The crossed beam method was proposed as a simple method by Haumschild and Greenleaf[4] and Nishimura et al.[5] This method uses two single probes. One probe is used for transmitting the ultrasound pulsed wave and the other for receiving the wave scattered from the region where the beams from the two probes cross. From the propagation time of the pulsed wave, the sound speed value is calculated. This method can also be realized using a linear array probe.[6]

The above techniques, including the crossed beam method, all relate to the measurement of average sound speed value. Thus, it is difficult to obtain the sound speed information of local pathological regions (e.g. liver tumors). From the clinical point of view, it is necessary to judge whether or not a liver tumor is malignant. There are significant histological differences between several kinds of liver tumors. It is possible that these tumors may have different acoustic properties (e.g. sound speed). Thus, in vivo sound speed mappings are earnestly required. However, these have not been studied, sufficiently.

The main purpose of our research is to propose the in vivo sound speed mapping technique using the crossed beam method and evaluate its clinical effectiveness. This paper discusses the potential of this technique by a mapping simulation based on the ray tracing technique and an experiment with cylindrical agar phantoms that mimic liver tumors. In the above experiment, a modified electronic linear scanner is used as the in vivo sound speed mapping system.

PRINCIPLES OF LOCAL SOUND SPEED ESTIMATION TECHNIQUE

Fig. 1 shows principles of the local sound speed estimation technique using the crossed beam method, employing a linear array probe. Transmitting regions, T_1 and T_2, and receiving regions, R_1 and R_2 , are arranged on the human body surface as shown. The local sound speed value C_0 is estimated in the region surrounded by center axes of transmitting and

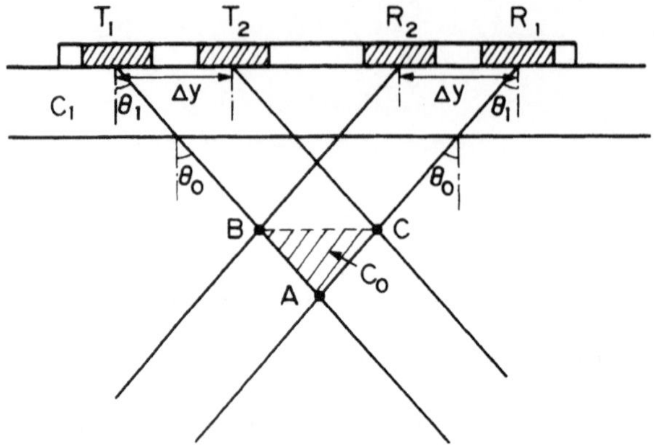

Fig. 1 Principles of local sound speed estimation technique using crossed beam method

receiving ultrasound beams. The pulsed wave, transmitted from T_1 or T_2, is reflected from scatterers around beam crossed points A, B or C and then is received by R_1 or R2. Propagation times along paths T_1AR_1, T_1BR_2 and T_2CR_1 are t_{11}, t_{12} and t_{21}, respectively. Each propagation time tij is determined by detecting the peak address of the received pulsed wave. The propagation time difference Δt is defined as follows.

$$\Delta t = t_{11} - (t_{12}+t_{21})/2 \tag{1}$$

From this simple geometry, the local sound speed value C_0 is given by

$$C_0 = \frac{\Delta y}{\Delta t.\sin\theta_0} \tag{2}$$

where Δy is the interval between centers of transmitting or receiving regions and θ_0 is the beam steering angle. By application of Snell's law, at the boundary of the surface layer, we obtain the following :

$$\frac{\sin\theta_0}{C_0} = \frac{\sin\theta_1}{C_1} = \frac{\Delta\tau}{a} = \alpha \tag{3}$$

where C_1 is the sound speed of the surface layer, θ_1 is the beam steering angle in the surface layer, a is the element pitch length of the linear array probe and $\Delta\tau$ is the delay time difference of each channel. By combining equations (2) and (3), the compensated sound speed C_0 is given by

$$C_0 = \frac{\Delta y}{\alpha.\Delta t} \tag{4}$$

In the equation(4), refracting influences at the surface layer are significantly reduced.

SOUND SPEED MAPPING SIMULATION

For estimating sound speed images in the existence of the local region that has a different sound speed value from that of the surrounding region, A mapping simulation based on the ray tracing technique is adopted. Fig. 2 shows the mapping simulation model. The model consists of three regions. For example, in the liver, these regions are respectively the suface layer(I), the normal liver tissue region(II) and the liver tumor region(III), whose sound speed values are C_1, C_2 and C_3. The shape of region III is circular. The local sound speed value C_0 is estimated in the oblique line region. In the calculation, individual ultrasound beams are regarded as sound rays and are independently traced. Refraction angles containing the total reflection are calculated at each intersection between sound rays and boundaries. Coordinates for points A, B and C are given as sound rays intersections. The propagation time difference Δt and the local sound speed value C_0 is obtained by calculating individual propagation times tij along each path. The sound speed image can be obtained from application of equation(4) at each point in the plane.

Fig. 3 shows a simulated image in the existence of a circular region whose sound speed value is set 1550 m/s. This region is 20 mm in diameter and the depth of the circle center is 90 mm. The sound speed value of the surrounding region is 50 m/s lower than that of the circular region. The beam steering angle θ_0 is 13.5° , the lateral width of ROI (region of interest) Δy is 14.4 mm and the measurement pitch length in lateral and depth direction are 1.44 mm and 1.5 mm, respectively. The display range of the sound speed value is from 1450 m/s to 1650 m/s. White parts in the image corredpond to high sound speed values. Behind the circular region (black parts in the image), estimated values of sound speed are smaller than true values. This reason is that estimated values of propagation time difference Δt decrease with smaller ROI size due to refractifng influences and this leads to an under-estimation of the sound speed values. In the region that starts from both sides of the circular region in the streering beam angle direction, sound speed values are overestimated. This is because the ROI size becomes greater due to refracting influences. Thus, the exact estimation of sound speed values is quite difficult especially behind the circular region. However, the sound speed information for the circular region (i.e. its relative

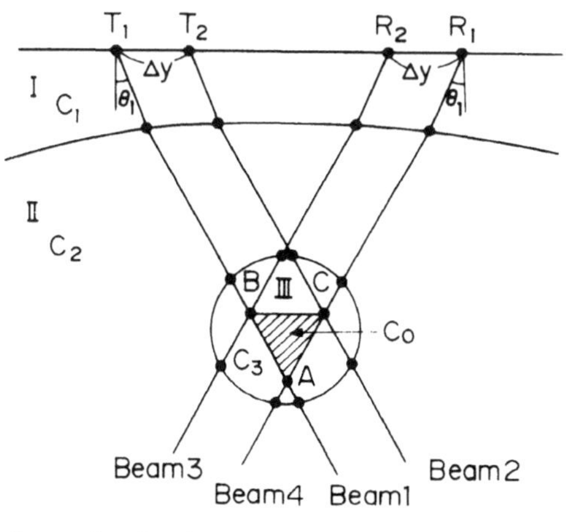

Fig. 2 Mapping simulation model; I : surface layer region; II : normal liver region; III : liver tumor region

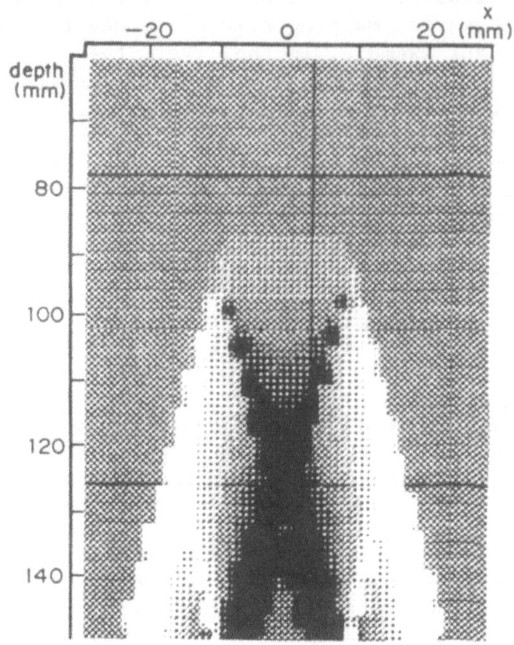

Fig. 3 Sound speed image obtained by mapping simulation

magnitude for the surroundidng region) is emphasized behind the circular region as a refraction mapping pattern. This result should be checked in the phantom experiment.

IN VIVO SOUND SPEED MAPPING SYSTEM

The electronic linear scanner is modified for clinical application of sound speed mapping. Fig. 4 shows the data acquisition method using the linear array probe. The mapping system has following functions.

Optimum Scanning As the received pulsed wave becomes wide, the estimation error increases because the peak address of the received wave is likely to be measured at an incorrect time. Thus, optimum conditions (e.g. aperture size, beam steering angle etc.) should be determined from minimizing the width of the received signal from each beam-crossed point. These conditions are that the beam steering angle is constant and that the aperture size increases not monotonically as the depth of beam-crossed point increases. The above conditions are determined under several restrictions concerning the linear scanner.

Fast data Acquisition One of the serious problems in in vivo sound speed mapping is the influence of the tissue movement arising from respiratory motion. In order to reduce this influence on the sound speed image, data acquisition should be completed within a time such that the movement effect is negligible. This system has fast static memories of large capacities. Received wave data (max. 1 Mbytes) and peak data (max. 1 Kbytes) from beam crossed points (max. 1 K) can be acquired within 2 seconds.

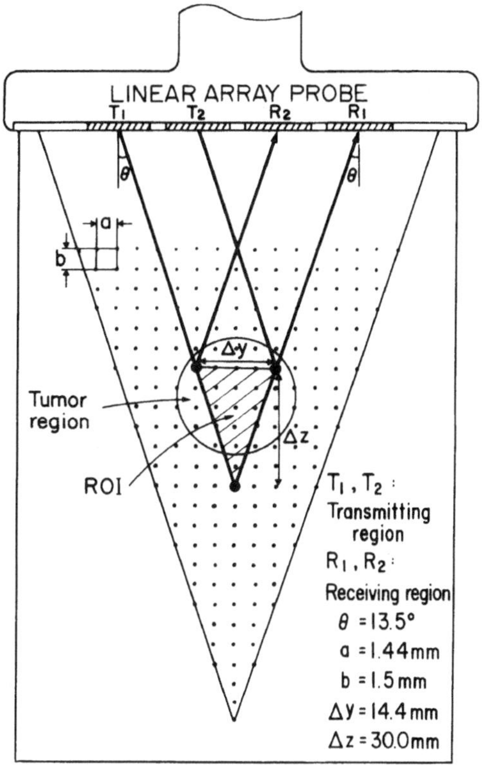

Fig. 4 Data acquisition method using linear array probe

Automatic gain control Received signals from various beam crossed points have different values. Consequently, it is probable that the received signal is not within the limits of the dynamic range and that the estimation error of the propagation time increases. In this system, the gain value set for each beam crossed point can be controlled around the optimum level using the automatic gain control system.

Fig. 5 shows a block diagram of the mapping system. The received signal is amplified and the envelope is detected by the receiver. The amplification ratio is controlled by the automatic gain control system. In the data acquisition system (DAS), the digitized wave data are stored and the peak data are detected. These data are transfered from the DAS to the mini-computer system and are transformed into propagation times. Using the propagation times, the sound speed image is constructed and is displayed on the CRT as half-tone image. Simultaneously, a B mode image is displayed.

Sound speed mapping is carried out under following measurement conditions partially shown in fig. 4 :

* Probe frequency : 3.75 MHz, Total number of elements: 128
* Beam steering angle : 13.5°(θ), Dynamic range : 20 dB
* Region of interest(ROI) : 14.4 mm in lateral width(Δy)
* Measurement pitch length : 1.44 mm in leteral direction (a)
1.50 mm in depth direction (b)
* Number of gray scales : 16, Display range : 1450 m/s ∼ 1650 m/s
* Data acquisition time : 2 s, Automatic gain control time : 5 s

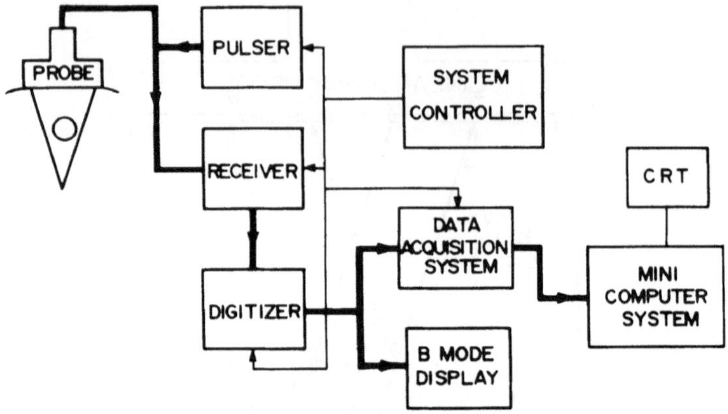

Fig. 5 System block diagram of in vivo sound speed mapping system

CYLINDRICAL PHANTOM EXPERIMENT

Fig. 6 shows a sound speed image of a homogeneous phantom. This phantom consists of agar and graphite. The sound speed value of the phantom is 1550 m/s(25°C) and the attenuation coefficient value is 0.5 dB/MHz/cm. As shown in Fig. 6, a homogeneous image is obtained. The average value and the standard deviation value of sound speed in the measurement plane are 1550 m/s and 41.1 m/s, respectively. Thus, there are few influences of attenuation on the sound speed image.

Fig. 7 shows a sound speed image of an attenuating cylindrical phantom surrounded by an attenuating medium. This phantom consists of agar and graphite. The sound speed values of the cylindrical region and the surrounding medium are, respectively, 1580 m/s and 1510 m/s. The attenuation coefficient is 0.5 dB/MHz/cm in both regions. The cylindrical region is 19 mm in diameter and the depth of the center is 70 mm. Fig.7(b) shows

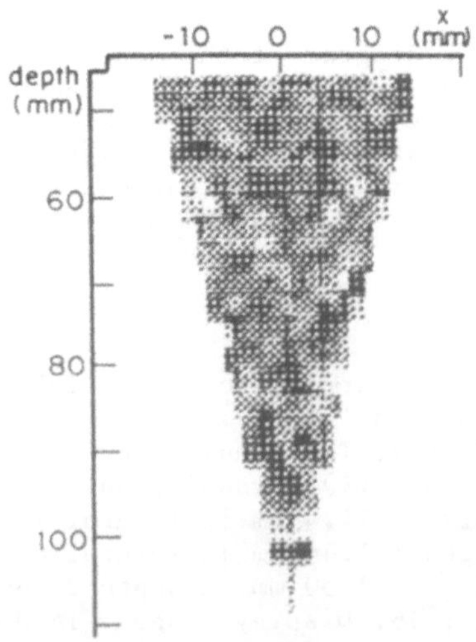

Fig. 6 Sound speed image of homogeneous graphite agar phantom; C1=1550 m/s(sound speed); α=0.5 dB/MHz/cm(attenuation coefficient)

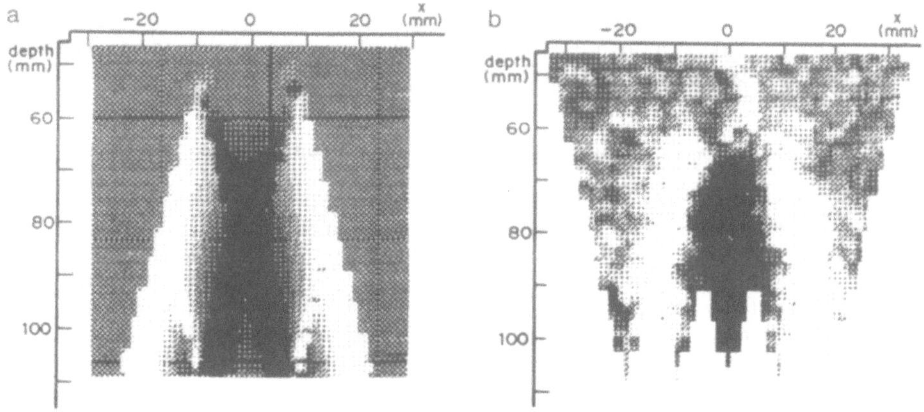

Fig. 7 Sound speed image of attenuating cylindrical phntom surrounded by attenuating medium, when C2 is higer than C1;
(a) mapping simulation image (b) experimental image;
C1=1510 m/s(surrounding region) C2=1580 m/s(cylindrical region)
α =0.5 dB/MHz/cm(both regions)
ϕ =19 mm(diameter of cylinder) d=70 mm(depth of cylinder center)

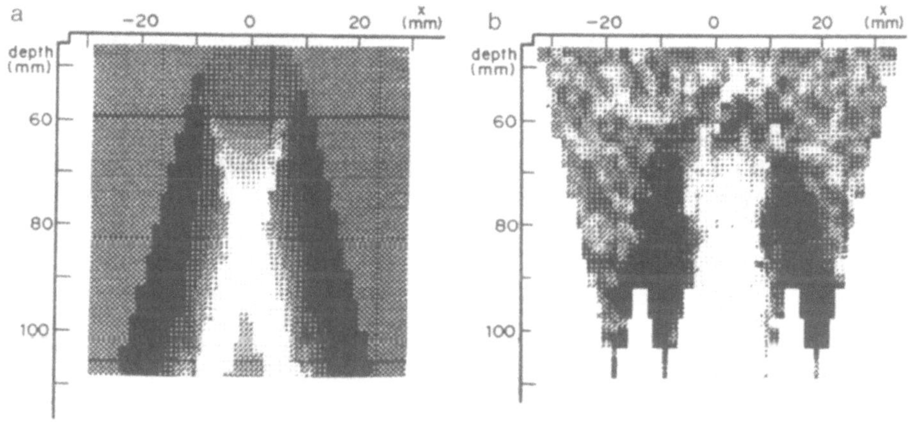

Fig. 8 Sound speed image of attenuating cylindrical phantom surrounded by attenuating medium, when C2 is lower than C1;
(a) mapping simulation image (b) experimental imge;
C1=1580 m/s C2=1510 m/s α=0.5 dB/MHz/cm ϕ=19 mm d=70 mm

the experimental image. This image agrees well with the simulated image (Fig. 7(a)). The mapping simulation was performed on a non-attenuating model. Thus, the attenuation effect has few influences on the sound speed image. Further, it is shown that the ray tracing technique is effective in estimating the effect of ultrasound beam refraction and that the mapping system functions accurately.

Fig. 8 shows a sound speed image of another attenuating cylindrical phantom. The sound speed values of the cylindrical region and the surrouding medium are 1510 m/s and 1580 m/s respectively. Other parameter values are same as those of the above experiment. The experimental image agrees well with the simulated image (Fig. 8(a)). The sound speed image has a reverse mapping pattern as compared with that of the above image (Fig. 7) because refraction angles are all of opposite sign to those of the above. Thus, the sound speed information for the local region (whether or not the sound speed value of this region is higer than that of the surrounding region) is emphasized in the image as a refraction mapping pattern.

It is probable that the boundary shape of liver tumors is not always spherical. Therefore, the influence of the boundary shape degradation on the sound speed image should be evaluated. From the experiment of the cylindrical phantom that has a degraded shape, it is shown that this influence is not so great. However, the shape degradation degree is not quantified. Thus, this influence needs to be investigated.

CONCLUSION

The in vivo sound speed mapping technique, using the crossed beam method, was proposed. The potential of this technique was discussed. From the mapping simulation based on the ray tracing technique and experiment with cylindrical phantom mimicking liver tumors, following results are obtained.

* It is difficult to estimate exactly values of local sound speed especially behind the cylindrical region.
* the sound speed information for the cylindrical region (its relative magnitude to the surrounding medium) is quite emphasized behind the cylindrical region as a refraction mapping pattern.
* Attenuation has few influences on the refraction mapping pattern.

Thus, it is probable that the refraction mapping pattern is of clinical usefulness. Further, clinical applications will be performed on several kinds of liver tumors to evaluate the clinical effectiveness of this sound speed mapping technique.

REFERENCES

1. Robinson D. E., Chen F. and Wilson L. S.: Measurement of velocity of propagation from ultrasonic pulse echo data, Ultrasound in Med. & Biol, 8: 413-420, 1982.
2. Akamatsu K., Miyauchi S., Nishimura N., et al.: Asimple new method for in vivo measurement of ultrasound velocity in human liver and its clinical usefulness, Jpn. J. Med. Ultrasonics, 12: 338-347, 1985 (in Japanese)
3. Ogawa T., Katakura K., Umemura Y. et al.: In vivo measurement of mean sound speed usisng focus adjustment method (FAM) in ultrasonotomograph, Jpn. J. Med. Ultrasonics, 12: 31-36, 1985 (in Japanese).
4. Haumschild D. J. and Greenleaf J. F.: A crossed beam method for ultrasonic speed measurement in tissue, Abstracts in Eighth Int. Symp. Ultrasonic Imaging and Tissue Characterization: Ultrasonic Imaging, 5: 168, 1983.
5. Nishimura N., Akamatsu K., Ohkubo H. et al.: Measurement of ultrasound velocity as a diagnostic tool for diffuse liver diseases, Proc. Jpn. Soc. Ultrasonics Med. 45: 21-22, 1984 (in Japanese).
6. Sumino Y., Hirama M., Okazaki K., et al.: A proposal of crossed beam method using a linear array probe for in vivo measurement of sound velocity of tissue, Proc. in Jpn. Soc. Ultrasonics Med. 46: 7-8, 1985. (in Japanese).

MOVEMENT ANALYSIS USING B-MODE IMAGES

I. Akiyama, N. Nakajima*, and S. Yuta**

Dept. of Electrical Engineering, Sagami Inst. of Techonology Fujisawa, Japan, *Dept. of Electrical Engineering, Keio Univ. Yokohama, Japan and **Inst. of Information Science and Electronics, Univ. of Tsukuba, Tsukuba, Japan

INTRODUCTION

One of the major abilities of ultrasonic diagnositic equipment is to present an image in a real time format. Due to this ability the motion of various kinds of organs and fetus is easily recognized by viewers of the B-mode images. However, the method for measurement of the motion using a time series of B-mode images has not been established yet.

This paper describes a study on the measurement of the motion of various kinds of organs and fetus by processing a time series of B-mode images. Since the B-mode image presents a noisy pattern of a granular structure called speckle, it is important to discuss the variation of the speckle. We will first construct a model of behavior of the speckle with tissues' movement and then we will confirm this model by the phantom experiments. Results of this study reveal that the speckle pattern moves along the direction of tissues' movement without changing its pattern when the movement length is short. Therefore the tissues' movement are estimated by computing the speckle's movement on the image.

For the image from which contours are easily recognized such as cardiac valves and a fetus, the motion of contours is rather important. In this case to reduce the speckle is significant, since the resultant displacement is not properly computed due to the presence of the speckle. In this paper zero-crossing method was used for contours detection and then optical flow method was used for movement estimation along the resultant contours. The results of the experiments employing a commercially available ultrasonic equipment are successfully obtained for the motion of the cardiac valve and the fetus.

SPECKLE PATTERN

A Model of Speckle Variation

A B-mode image presents a granular pattern called speckle arising from the high coherence of ultrasound wave. We first explain a model of the speckle variation in basis of a motion of a mass of small particles.

Now we assume that a biological tissue is constructed of a numerous number of small particles which are randomly distributed. A received waveform at a transducer obtained by pulse echo is considered to be phase randomly interfered by a number of wavelets reflected from each particle. Resultant amplitude of the received echo varies stochasticly as a local change in the position of each particles, because a way of interference of reflected wavelets is complicated. Therefore the speckle pattern is determined by the spatial relationship among the particles and the transducer.

Then a variation of the speckle will be considered when the tissue moves. Three direction of movement is defined as shown in Fig.1; x is a lateral direction, y is a vertical direction against the image plane and z is a range direction.

When the tissues move without change in spatial relationship of the particles toward the x-direction as shown in Fig.2, a mass of particles located in the region of A as shown in Fig.2(a) shifts to the position of A as shown in Fig.2(b). In this situation, if acoustical and electrical characteristics of the each transducer constructed by the array probe is completely same, the echo from the region of A in Fig.2(a) is same as that in Fig.2(b). Thus the image speckle does a translation movement when the tissues move toward x direction.

Then the movement toward z direction is considered by using Fig.3. An area of A is corresponding to that of the square area performed by pulse duration length and beam width. Ultrasonic fields at the position of A before the movement are different from that after the movement. However, if dimension of the region A is much smaller than of the length between A and the transducer, both ultrasonic fields are almost same.

Finally we will consider the movement toward the y direction. In this situation the resultant speckle pattern is obviously different before and after movement, because the imaging plane is different. Consequently the motion of tissues will be determined by shift of the speckle appeared on B-mode images.

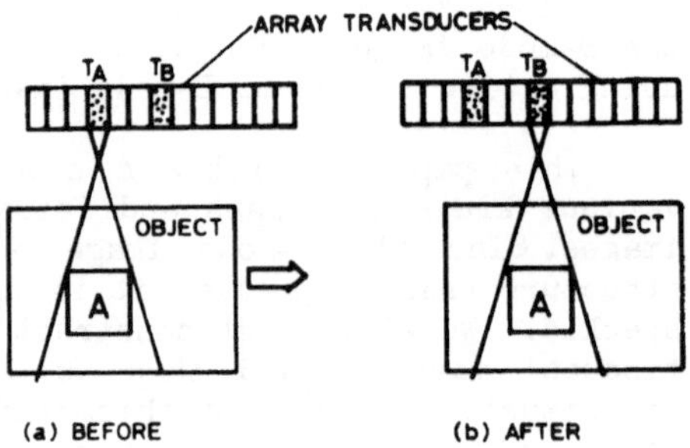

Fig.2 Schematic of movement of x-direction.

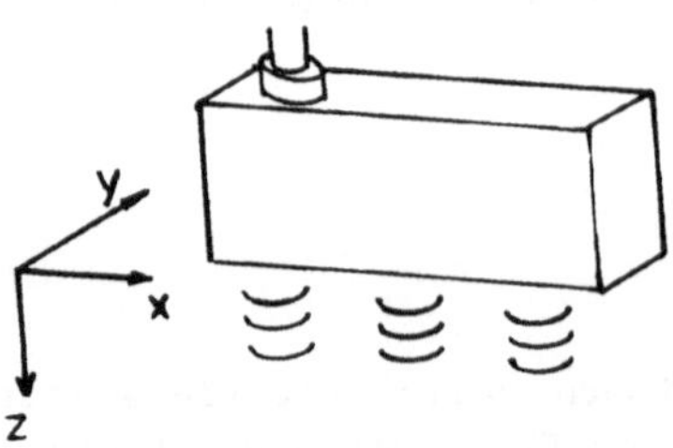

Fig.1 Definition of xyz direction.

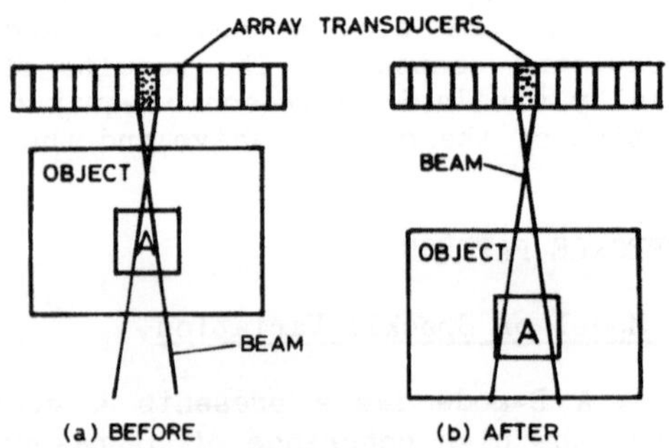

Fig.3 Schematic of movement of z-direction.

Phantom Experiments

In order to confirm the afore mentioned model we made phantom experiments by using a commercially available ultrasonic diagnostic equipment in which a 3.5MHz linear phased array probe was implemented. The phantom used was made of graphite powder suspended in agar-gel. The probe was fixed at xyz scanner to be moved along x, y, and z direction with an interval of 0.1mm, 0.5mm and 0.4mm respectively. B-mode images were computed from received echo digitalized at 25MHz sampling frequency. The resultant image is 64-by-64 in pixel, which was corresponding to an area of 3cmx1.5cm. Then the cross correlations were computed by FFT as a function of moving distance of the probe. The peak value of the cross correlation was used to investigate the variation in the speckle.

Fig.4 shows the variation of the correlation coefficient with the distance of the movement. The results for the movement along x and z direction say that the speckle shifts toward the direction of movement on the B-mode image without variation of their patterns, since the correlation coefficient is higher than 0.6 as shown in Fig.4(a) and (c). On the other hand, the results of the movement along y direction show that the rapid change was occurred as shown in Fig.4(b). Therefore the image speckle shift toward the direction of practical movement without variation of its pattern when the movement is limited on the x-z plane.

ESTIMATION OF DISPLACEMENT VECTORS

Calculation of Speckle Shift

In the fields of motion image processing, correlation method[1] is available to evaluate the displacement of a target presented on two images before and after movement. This method estimates the target's movement from the shift of peak of cross correlation between those two images. However, since the practical movement of biological tissues is complicated, it is necessary to get a small area for calculation of

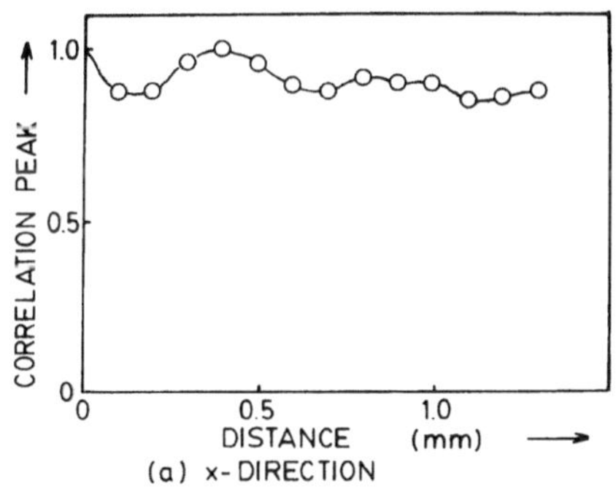

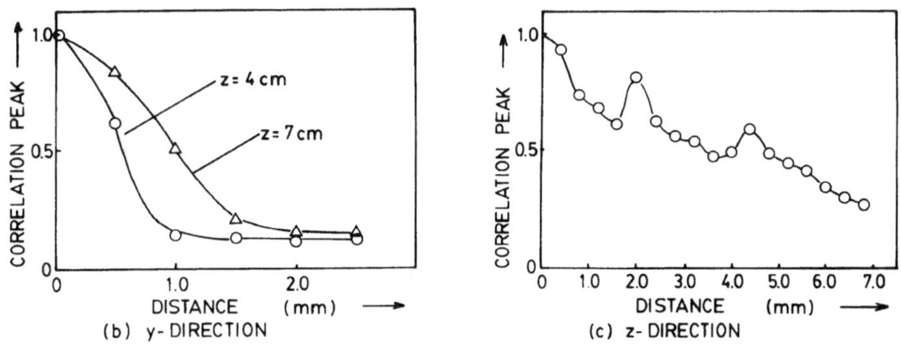

Fig.4 Relationship between correlation peak and movement in length. (a), (b) and (c) are x, y and z direction, respectively.

the cross correlation by a spatial window. An aperture of window used in the experiments was of 6x6mm. We used the layer phantoms which were constructed and allowed to be moved as shown in Fig.5. The resultant displacement vectors were displayed on the B-mode images as shown in Fig.6. All vectors computed were successfully obtained in both movement along x and z direction.

Estimation of Displacement Vectors of Contours

Detection of contours. One of methods for detection of contours is zero-crossing method[2]. This method provides the contours as follows. First a blurred image is computed by Gaussian filter and then Laplacian of the filtered image is computed. The resultant image provides the contours by connecting zero-crossing points.

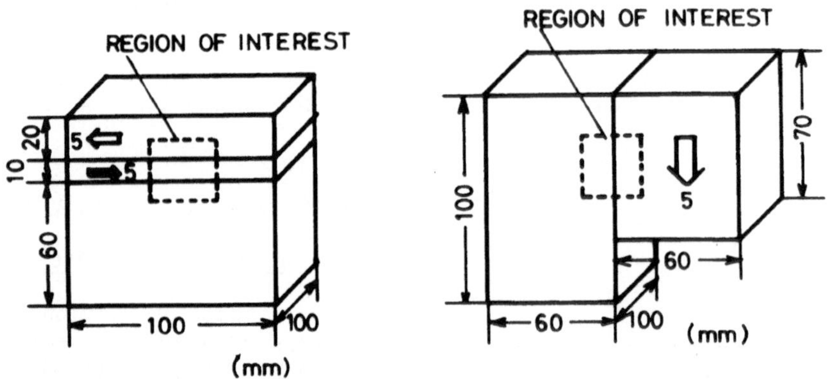

Fig.5 Schematics of layer phantoms made from graphite powder suspended in agar-gel.

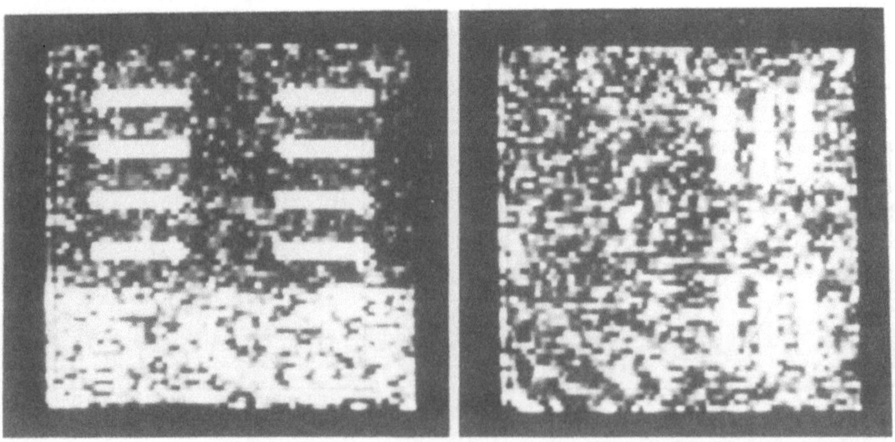

(a) (b)

Fig.6 Calculated speckle's shifts by correlation method.

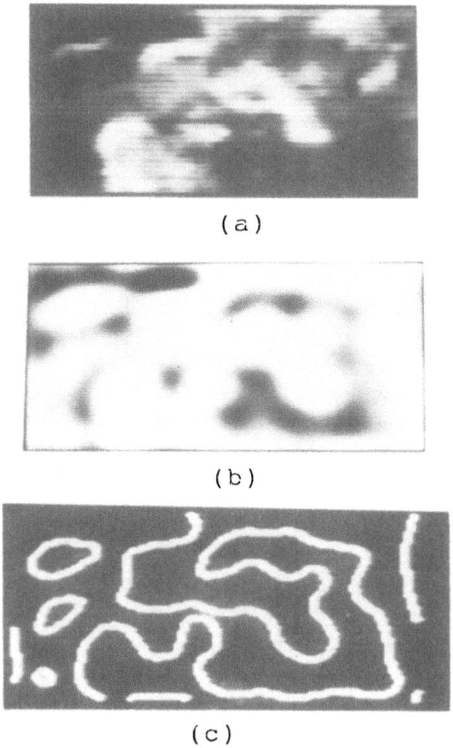

Fig.7 Results of contours detection. using fetus' face image. (a) is B-mode image, (b) is filtered image and (c) is contours.

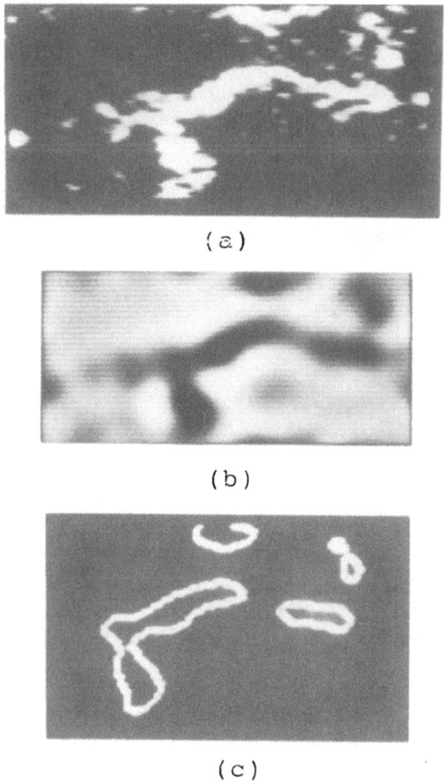

Fig.8 Results of contours detection using cardiac valve image. (a) is B-mode image, (b) is filtered image and (c) is contours.

Fig.7 and Fig.8 show the B-mode images (a), filtered and Laplacian images (b), and contours (c) for cardiac valve and fetus' face, respectively. It is found that contours are successfully determined for the both images.

Calculation of displacement vectors along the contours. Displacement vectors are calculated from the image data of before and after movement in terms of optical flow method[3]. This method access the displacement vectors from the slope of gray level with neighboring position and time. In the experiments we computed the displacement vectors along the contours which is obtained by zero-crossing method. The results of this procedure, however, does not presents proper vectors as shown in Fig.9. These results can be explained in terms of the speckle which exerts a wrong influence on the calculation of the slope. In order to improve the vector estimation we used the filtered images, mentioned above, in substituting of B-mode images. We obtained the results shown in Fig.10 and Fig.11. It can be seen that the displacement vectors are successfully obtained.

CONCLUSION

Motion analysis of biological tissues by processing a time series of B-mode images which was obtained in a real time format was studied. It

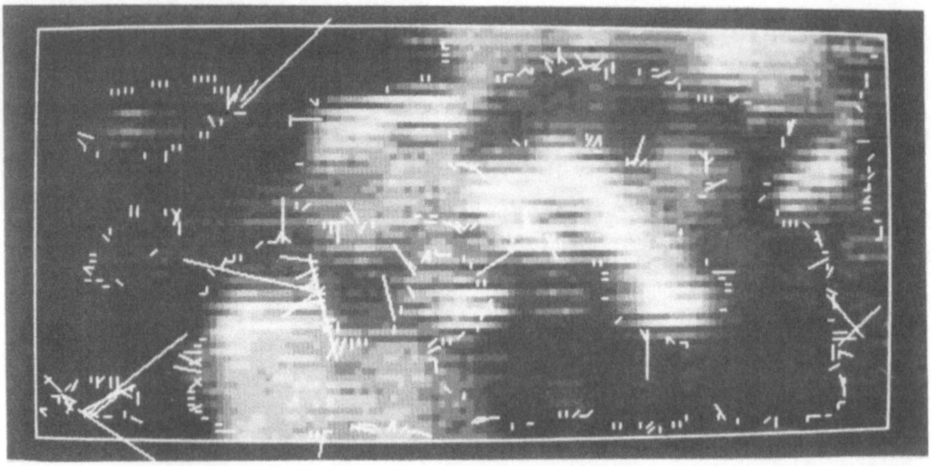

Fig.9 Experimental results of motion detection.

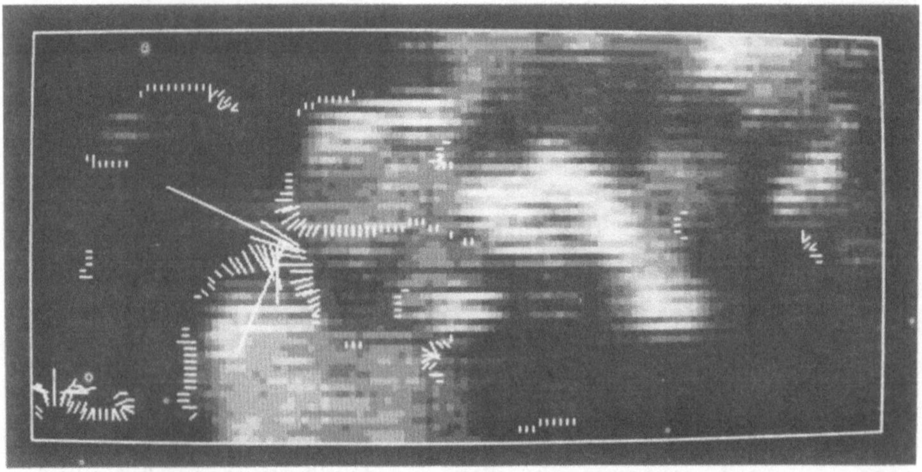

Fig.10 Experimental results of motion detection. Laplacian of Gaussian filtered images were used.

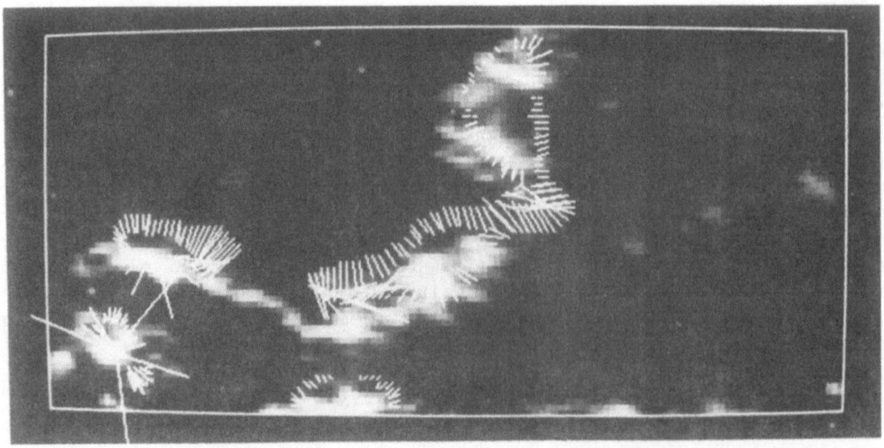

Fig.11 Experimental results of motion detection. Laplacian of Gaussian filtered images were used.

is significant to accomplish the motion analysis how the speckle appeared on the B-mode image behaves with motion of the biological tissues. Due to the discussion and the experiments of the speckle's behavior the speckle shifts along the direction of the tissues' movement without change of speckle pattern and thus the motion analysis is accomplished by the image speckle.

However, for the images from which the contours are easily detected the contours' motion is significant. Then we made the experiments by using the B-mode images of cardiac valve and fetus. As a result we successfully obtained the movement of contours.

REFERENCES

1. Takai N. and Asakura T. "Vectorial measurements of speckle displace ment by the 2-D electronic correlation method.", Applied Optics, 24:660 (1985).
2. Marr D., Ullman S. and Poggio T. "Bandpass channels, zero-crossings and early visual information processing.", J.Opt.Soc.Am., 69:914 (1979).
3. Hildreth E.C. "Computations underlying the measurement of visual motion.", Artificial Intelligence, 23:309 (1984).

ULTRASONIC SPECKLE VELOCIMETRY

T. Kanazawa, M. Shingyouuchi, I. Akiyama*, S. Yuta**, T. Itoh***, and M. Nakajima

Department of Electrical Engineering, Keio University Yokohama, Japan and *Department of Electrical Engineering Sagami Institute of Technology, Fujisawa, Japan and **Institute of Information Science and Electronics University of Tsukuba, Ibaraki, Japan and ***Aloka Co., Ltd. Mitaka, Japan

INTRODUCTION

It is needless to say that the blood flow velocity measurement is of great importance in the clinical diagnosis. The Ultrasonic Pulse Doppler method was proposed and has widely been used. We may say that this method reached its peak with 2-D color mapping technique. The remaining subject for further study concerning the blood flow velocity measurement is to establish a new method to measure the blood flow of either extremely high or low velocity. The ULTRASONIC SPECKLE VELOCIMETRY proposed in this paper may rather be the method for the velocity of the slow blood flow. This method is based on the statistic dynamic properties of the speckle pattern in the B-mode image generated according to the object movement. Thus, the principle of this method is quite different from that of the Doppler method. When this method is put into practical use, it makes it possible to measure the velocity of such extremely slow blood flows as in the capillaies or around the heart wall, which has been hard to measure by the Doppler method. Therefore, the method will enlarge the measurable range of the blood flow velocimetry and it is expected to obtain some new clinical diagnostic information.

We define the tissue parenchyma as a group of many point scatterers densely distributed without order. When the ultrasound is transmitted toward it, the echo signal is formed by a random phase interference of the wavelets reflected from each scatterer and its magnitude is attended with a stochastic variance which is unrelated to the spatial distribution of the scatterers. Such a magnitude variance, which dose not coincide with the microstructure of the tissue, is called a speckle. This speckle is a main factor to degrade the quality of the B-mode image. When the object is moving with a constant velocity toward one direction, the speckle fluctuates temporally corresponding to the velocity of the object. Consequently, the velocity can be determined by measuring the fluctuation number of the speckle on the received echo signal.

In this paper, we propose a new simple method to measure the blood flow velocity in real time and without any invasions by means of the signal processing of the speckle which appears on the echo signal with statistical properties. Firstly, we describe the properties of the speckle, particularly the relation with the scatterer density. Secondly, we explain the theory of the way to measure the flow velocity. Then, we describe the procedures, the results and the discussion on the fundamental experiments carried out to verify the effectiveness of this method.

PRINCIPLE

Concept

We firstly explain the concept of this method with a phantom consisting of many point scatterers lined on one dimension as shown in Fig.1. These point scatterers are distributed without order and are moving with a constant velocity. The ultrasonic continuous wave (CW) is let off by the transmitting transducer and the echo reflected at each

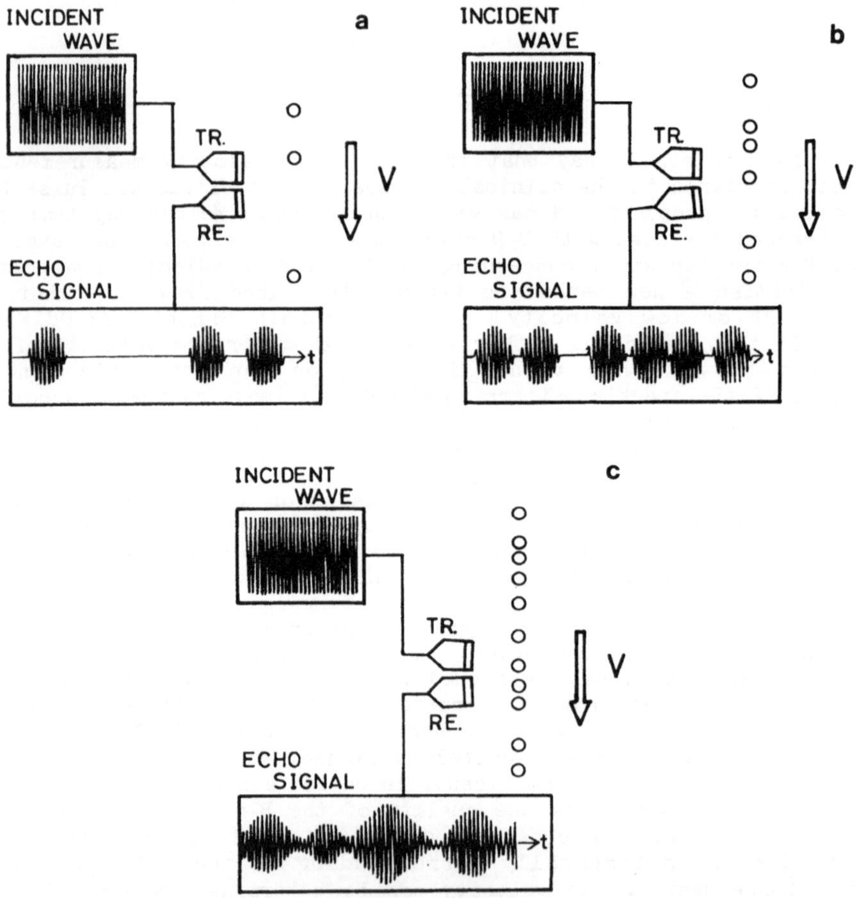

Fig.1. Schema of echo from scatterers.
(a) low density.
(b) medium density.
(c) high density.

scatterer is detected by the receiving transducer. Fig.1(a) shows that when the scatterer density is low (the average distance among scatterers is long), the envelope of the echo signal corresponds to the distribution of the scatterers since the echoes reflected from scatterers do not interfere with one another and are detected separately. Secondly, as shown in Fig.1(b), when the velocity is same and the scatterer density is double as in (a), the magnitude fluctuation number becomes twice the one in (a) since the average distance among the scatterers becomes half. In short, the fluctuation number of the envelope signal increases in proportion to the scatterer density. However, as shown in Fig.1(c), in case the density is very high, the average distance becomes so short that the echoes returned from scatterers pile up and interfere with one another. And the envelope of the echo signal begins to fluctuate statistically. Such a magnitude, which dose not correspond to the spatial distribution of the scatterers, is called a "Speckle". Once this speckle appears on the echo signal, the fluctuation number of the envelope signal does not increase even though the scatterer density becomes high.

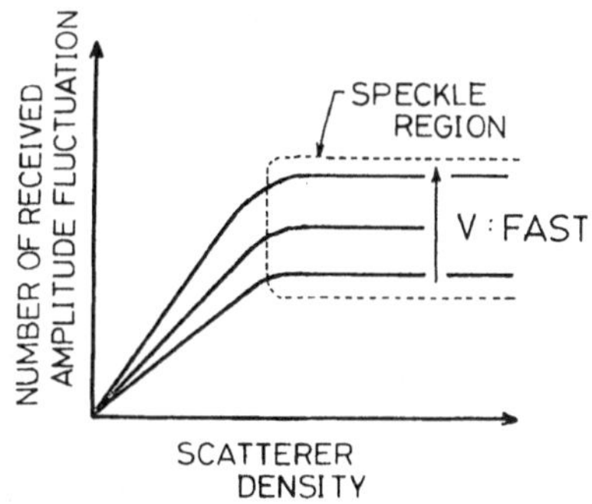

Fig.2. Relation between number of received amplitude fluctuation and scatterer density.

Fig.2 shows the conceptual figure of the relationship between the fluctuation number of the envelope signal and the scatterer density, when the object is moving with a constant velocity. In Fig.2, there is a region where the speckle dominates in the echo signal and the fluctuation number is independent of the scatterer density and depends only on the flow velocity of the scatterers. Hereinafter we will call this region "Speckle region". In the blood flow velocimetry in the clinical field, a scatterer is a red blood corpuscle in blood, and so on. The density of the red corpuscles is high enough to allow the wavelets scattered by them to interfere with one another. So the density belongs to the "speckle region" without question.

When the flowing object which consists of a large number of the point scatterers distributing without order and high density is irradiated by CW having a constant amplitude as shown in Fig.3, the speckle appearing on the echo signal fluctuates corresponding only to the flow velocity. In other words, the fluctuation is quick in case of high velocity, while it is loose in case of low velocity. We attempt to estimate the flow velocity of the object through quantitative measurement of the degree of the speckle magnitude fluctuation. The zerocrossing technique was adopted for its quantitative measurement. That is to say,

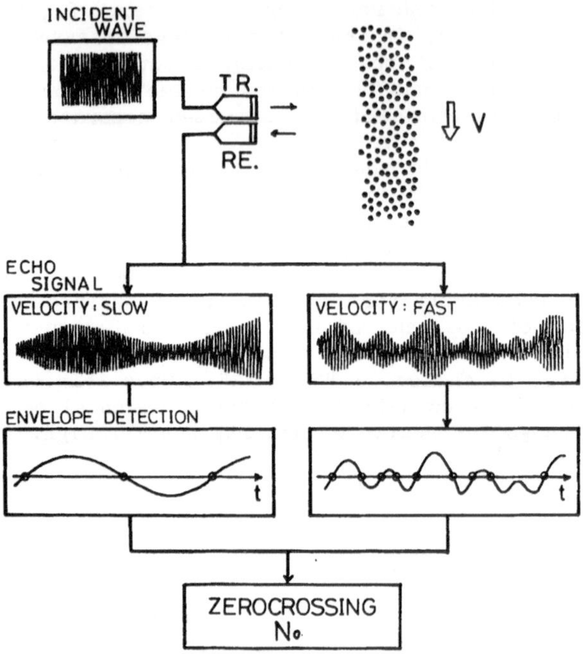

Fig.3. Concept of this method.

the envelope of the echo signal is detected at first in this method, then a fluctuation characteristics of the envelope is obtained by subtracting the mean from it, and finally the object flow velocity is estimated by counting the number of the curve crossing zero axis in a unit time.

Theory

In this section, we consider theoretically the relationship between the fluctuation number of the speckle received by the transducer and the moving velocity of the object, using a simple model as shown in Fig.4 [1]. In obtaining the equation to show this relation, we make use of the following statistical property. In case the autocorrelation function of the speckle intensity can be approximated to normal distribution, the zerocrossing number of the speckle fluctuation in a unit time is in inverse proportion to the standard deviation of this function.

In the first place, the plane (called the object plane) is supposed to be a model of the object, where a large number of tiny point scatterers are distributed without order and finely enough. And, the observation plane is in parallel with the object plane and the distance between these plane is R. Here, the Cartesian coordinates system of the object plane and observation plane are defined by $\mathbf{X} = (x, y)$ and $\mathbf{L} = (l, m)$ respectively in the vector notation. The z-axis is through each origin on the $\mathbf{X}$ plane and the $\mathbf{L}$ plane vertically, and the moving velocity of the scatterers on the object plane is denoted by $\mathbf{V} = (Vx, Vy)$.

The scatterer distribution is written as the complex reflective coefficient on the object plane, namely,

$$\mathbf{r}(\mathbf{X}) = r_0(\mathbf{X}) \exp[j\theta(\mathbf{X})], \tag{1}$$

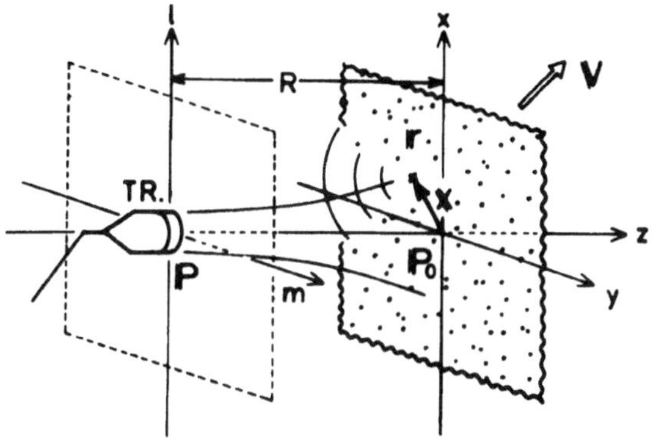

Fig.4. Coordinate system.

where $r_0(\mathbf{X})$ is the magnitude term, and $\theta(\mathbf{X})$ is the phase term replacing the distribution in the z direction. In case a fancied transducer located on the origin of L coordinate irradiates the object plane with CW, the echo signal that is a ultrasound pressure received by the same transducer is written as

$$\mathbf{P}(t) = \int P_0(\mathbf{X}) \exp(-j\omega t)\, \mathbf{r}(\mathbf{X})\, \mathbf{K}(\mathbf{X})\, d\mathbf{X}, \tag{2}$$

where $P_0(\mathbf{X})$ is the magnitude distribution of the irradiating ultrasound on the object plane, ω is its angular frequency, and $\mathbf{K}(\mathbf{X})[= \mathbf{K}(\mathbf{X},\mathbf{L})|_{\mathbf{L}=0}]$ is the propagation function of the ultrasound pressure from the object to the transducer.

In the next place, we investigate the autocorrelation function of $\mathbf{P}(t)$, which is defined as

$$\Gamma_P(\tau) = \overline{\mathbf{P}(t)\, \mathbf{P}^*(t+\tau)}, \tag{3}$$

where τ is a time interval, the superscript * and $\overline{\quad}$ stand for the complex conjugation and the expected value, respectively. By using (1) and (2) in (3), the autocorrelation function $\Gamma_P(\tau)$ is given by

$$\Gamma_P(\tau) = \iint P_0(\mathbf{X}_1)\, P_0(\mathbf{X}_2) \cdot \overline{\exp[j\theta(\mathbf{X}_1-\mathbf{V}t)]\, \exp^*[j\theta(\mathbf{X}_2-\mathbf{V}(t+\tau))]} \cdot \mathbf{K}(\mathbf{X}_1)\, \mathbf{K}^*(\mathbf{X}_2)\, d\mathbf{X}_1 d\mathbf{X}_2, \tag{4}$$

where we consider that the speckle is caused by a random phase interference of the echoes reflected from the scatterers, and is independent of the magnitude randomness of the complex reflective coefficient, then we assume $r_0(\mathbf{X}) = 1$. And, a form of the expected value is applied

only to the phase term θ. Assuming that $\overline{\theta^2} = \overline{\theta(\mathbf{X}_1)^2} = \overline{\theta(\mathbf{X}_2)^2}$, the exponential term in (4) is written as

$$\overline{\exp[j[\theta(\mathbf{X}_1-\mathbf{V}t)-\theta(\mathbf{X}_2-\mathbf{V}(t+\tau))]]} = \exp[-\overline{\theta^2}[1-\Gamma_\theta(\mathbf{X}_1-\mathbf{X}_2+\mathbf{V}\tau)]], \tag{5}$$

where $\Gamma_\theta(\mathbf{X})$ is the autocorrelation function of $\theta(\mathbf{X})$. When the phase distribution $\theta(\mathbf{X})$ is fine enough, its autocorrelation function $\Gamma_\theta(\mathbf{X})$ is approximated to

$$\Gamma_\theta(\mathbf{X}) = \begin{cases} 1: & |\mathbf{X}| \leqq \Delta S \\ 0: & |\mathbf{X}| > \Delta S, \end{cases} \tag{6}$$

where ΔS is the correlation area of $\theta(\mathbf{X})$. Therefore, use of (5) and (6) in (4) yields the autocorrelation function $\Gamma_P(\tau)$ as

$$\Gamma_P(\tau) = \Delta S \int P_0(\mathbf{X}-\mathbf{V}\tau/2)\ P_0(\mathbf{X}+\mathbf{V}\tau/2)\ \mathbf{K}(\mathbf{X}-\mathbf{V}\tau/2)\ \mathbf{K}^*(\mathbf{X}+\mathbf{V}\tau/2)\ d\mathbf{X}. \tag{7}$$

In Eq.(7), we assume that $\overline{\theta^2} \gg 1$.

We approximate the magnitude distribution of the incident ultrasound to the normal distribution, then the magnitude distribution $P_0(\mathbf{X})$ is given by

$$P_0(\mathbf{X}) = \exp(j2\pi R/\lambda)\ \exp(-|\mathbf{X}|^2/W^2), \tag{8}$$

where W and λ are the beam width and the wavelength, respectively. And assuming that the transducer is located in the Fresnel diffraction field, the propagation function $\mathbf{K}(\mathbf{X})$ is approximated to

$$\mathbf{K}(\mathbf{X}) = \exp(j\pi|\mathbf{X}|^2/\lambda R). \tag{9}$$

Substituting (8) and (9) in (7), the absolute value Γ_P become

$$|\Gamma_P(\tau)| = C_1\ \exp(-|\mathbf{V}|^2\tau^2/2W^2)\ \exp(-\pi^2W^2|\mathbf{V}|^2\tau^2/2\lambda^2R^2), \tag{10}$$

where C_1 is a constant.

On the other hand, the intensity I(t) of the received ultrasound $\mathbf{P}(t)$ is given by $I(t) = |\mathbf{P}(t)|^2$, and $\Delta I(t) = I(t) - \overline{I(t)}$ is the fluctuating element of it. Then, the autocorrelation function of $\Gamma_{\Delta I}(\tau)$ is written as [2]

$$\begin{aligned} \Gamma_{\Delta I}(\tau) &= |\Gamma_P(\tau)|^2. \\ &= C_2\ \exp[-[1/W^2+1/S^2]|\mathbf{V}|^2\tau^2], \end{aligned} \tag{11}$$

where C_2 is a constant and the definition $S = \lambda R/\pi W$ is used.

Therefore, the autocorrelation function of ΔI is obtained as the normal distribution, then the zerocrossing number of ΔI in a unit time is written, by using the standard deviation of $\Gamma_{\Delta I}$, as[3]

$$N_0 = (1/\pi)\ [2[1/W^2+1/S^2]]^{1/2}\ |\mathbf{V}|, \tag{12}$$

where it shows that the zerocrossing number N_0 has no connection with the scatterer density in the object plane. And in (12), the coefficient of proportionality concerning the zerocrossing number N_0 and the velocity $\mathbf{V}$ is given by

$$\beta = (1/\pi)\ [2[1/W^2+1/S^2]]^{1/2}. \tag{13}$$

As is stated above, it is clarified theoretically that the temporal fluctuation number of the speckle has a linear relation to the velocity of the object. The coefficient of proportionality β is the function of the beam width W, the wavelength λ, and the distance R, as is denoted in (13).

EXPERIMENTAL VERIFICATION

In this section, we describe the procedures, the results and the discussions on the fundamental experiment, separately, which was carried out to verify the effectiveness of the ultrasonic speckle velocimetry.

Procedure

In this experiment, we adopted the pulse echo technique using one transducer which is expected to have large applicability in the future. The use of this technique simplifies its measuring system and enables us to choose sample volume freely. Fig.5 shows the block diagram of the experimental system.

The phantom used in this experiment was a water flow through a acylic tube (the inside diameter is 20mm and the outside diameter is 24 (mm) fixed in the water tank, which included a organic matter, dextran particle size: 20 ~ 80 μm) as the scatterer. We set up the transducer both to transmit and to receive ultrasonics, 50mm away from the acylic tube and toward the vertical direction where the flow velocity could not be measured by the Doppler method. This transducer has a resonance frequency of 5 MHz and a concave circular aperture, and its diameter is 17mm. And the focal distance of this transducer is 50mm. It transmitted the radio frequency pulse (RF pulse) to the object at a recurrence frequency of 1.6 kHz.

First, the A-mode signal was obtained by the envelope detection of the echo signal received by the same transducer. The speckle was observed in the oscilloscope, which existed between two large magnitudes resulting from the front and rear of the acylic tube (as is seen in Fig.6). Corresponding to the flow velocity, its magnitude fluctuates. Next, for the purpose of obtaining the magnitude of the echo which was scattered only by the water flow, the value of the A-mode signal was read at the time (t = 66 μsec.) corresponding to the round-trip distance (10cm) from the transducer to the center of the tube. And this instantaneous value was held until the next sample time. In short, by putting the A-mode signal in the Sample and Hold circuit to carry out this processing, we could obtain the continuous waveform of the temporal speckle fluctuation signal. Fig.7 shows the concept of the signal processings mentioned above.

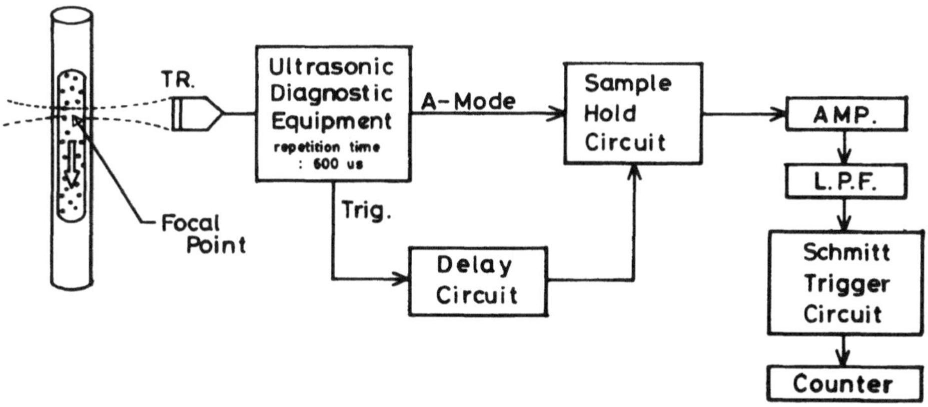

Fig.5. Block diagram of experimental system.

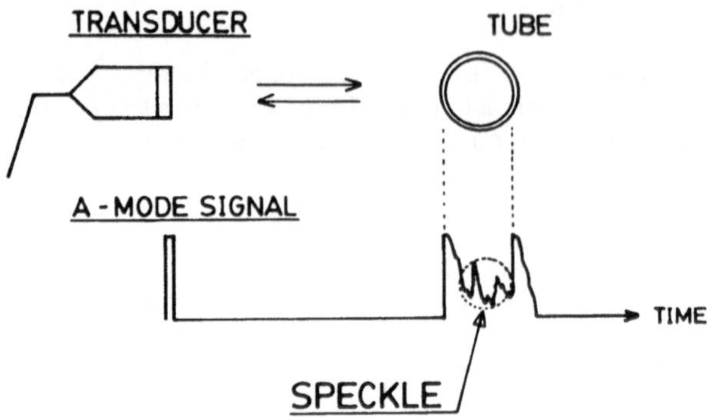

Fig.6. Schema of A-mode signal.

Then, the obtained signal passes through AMP., Low Pass Filter and Schmitt Trigger Circuit in order, and finally the zerocrossing number of signal was counted in one second by the counter.

Result

Fig.8 shows the example waveform of the speckle fluctuation obtained by this experiment, where the velocity of interest is slow. In Fig.8 (a), (b), and (c), the flow velocity are 5.3cm/s, 15.9cm/s, and 26.5cm/s, respectively. It is evident that the fluctuation number of the speckle magnitude increases gradually in accordance with the increase of the flow velocity. The relationship between the zerocrossing number and the flow velocity is shown in Fig.9 as the result of this experiment,

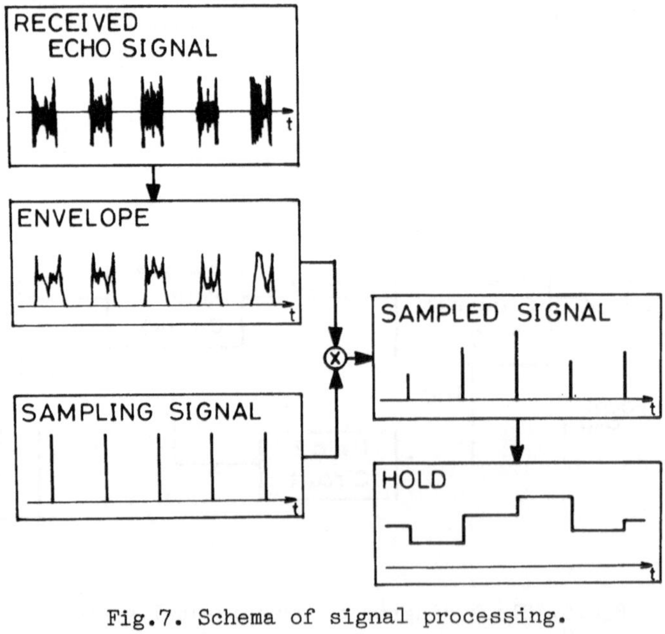

Fig.7. Schema of signal processing.

where the sign expresses the mean and the standard deviation of 50 times measurement at each velocity. The measuring time to obtain one datum is 1 sec. The result indicates the good linear relationship between the flow velocity and the zerocrossing number.

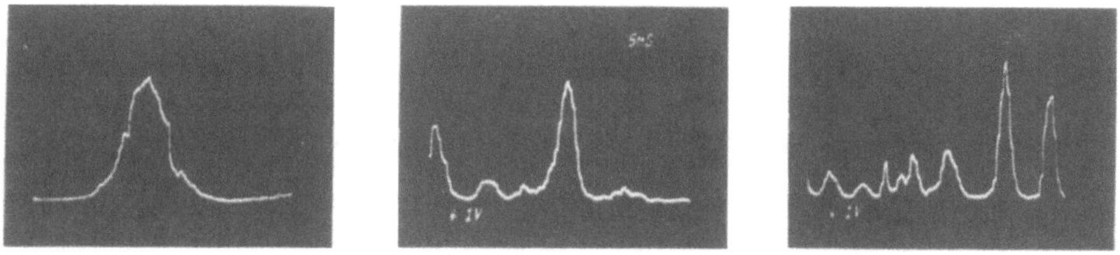

(a) V = 5.3 (cm/s) (b) V = 15.9 (cm/s) (c) V = 26.5 (cm/s)

Fig.8. Speckle fluctuation.

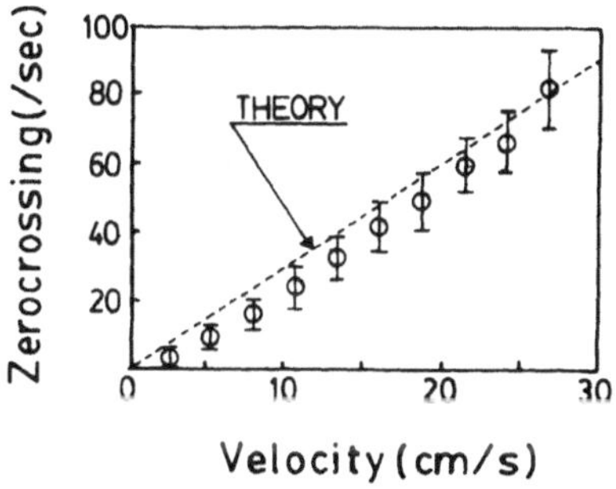

MEASURING TIME = 1.0 sec.

Fig.9. Relation between zerocrossing and flow velocity.

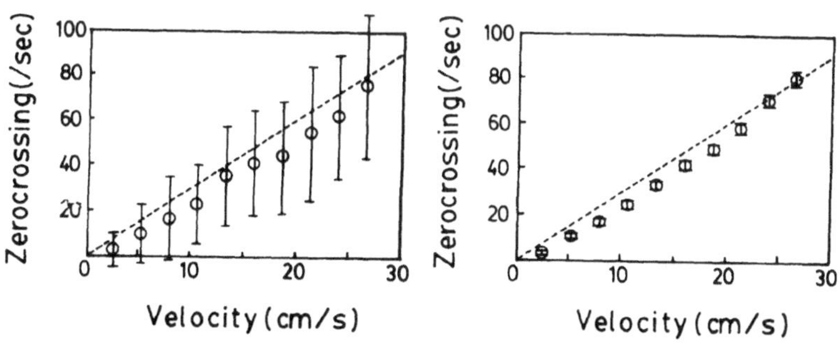

(a) MEASURING TIME = 0.1sec. (b) MEASURING TIME = 10.0sec.

Fig.10. Relation between accuracy data and measuring time.

Discussion

We compare the theoretical value with the experimental results. In this experiment, the incident beam width W at the transducer's focal point is about 2mm, the wavelength λ of the indicate ultrasonics is 0.3mm and the distance R between the object and the transducer is 50mm. Therefore, when the object is located on the focal point, the coefficient of the proportionality is written, by using (13), as $2.94cm^{-1}$. The theoretical straight line obtained by using it is denoted in Fig.9 as a dotted line. The cause of the error between the theoretical value and the experimental result may be that the beam width in the object was not really constant in the experiment, the fancied receiver is assumed as a point detector in the theory, and so on. Considering the accuracy of present measuring system, however, the result of this experiment can be said to coincide fairly well with its theoretical value.

MEASURING TIME AND ACCURACY

In this method, the degree of the speckle fluctuation corresponding to the object flow velocity is measured quantitatively by its zerocrossing number. It is anticipated that the longer measuring time brings about the smaller dispersion of the measured data because the magnitude fluctuation of the speckle is the stochastic phenomenon. Then, we investigate quantitatively that the relation between the measuring time and the dispersion of the measured data. The results are shown in Fig.10 (a) and (b), where the measuring time to obtain one datum is 0.1sec. and 10 sec. respectively. The measuring time in Fig.9 is 1 sec. There is a tendency that the dispersion decreases and the precision improves as the measuring time becomes long. But, when the measuring time is so long, the real time advantage of this method may be lost.

CONCLUSION

As a new method for the blood flow velocity measurement, we proposed the ultrasonic speckle velocimetry, gave theoretical explanation on its principle, and described the result of the fundamental experiment. Both the theoretical expression using the simple model and the experimental results indicated that the object flow velocity and the fluctuation number of the speckle are in good linear relationship. This method is different from the Doppler method in principle and enables the measurement of extreme slow blood flow and the flow streaming in vertical direction.

The subjects for future study are construction of the measuring system for the clinical experiment, and comprehensive discussions on its practical use.

REFERENCES

1. T. Asakura, N. Takai, "Dynamic Laser Speckles and Their application to velocity measurements of the diffuse object," Appl. Phys. 25, 179-194 (1981).
2. J. W. Goodman, "Statistical Optics," Jhon Wiley & Sons, New York,109 (1985).
3. J. S. Bendat, "Principles and Applications of random noise theory," Jhon Wiley & Sons, New York, 370-385 (1958).

ULTRASOUND ATTENUATION ESTIMATION IN SOFT TISSUE USING THE ENTROPY DIFFERENCE OF PULSED ECHOES BETWEEN TWO ADJACENT ENVELOPE SEGMENTS

H.S. Jang, T.K. Song and S.B. Park

Department of Electrical Engineering, KAIST

P.O. Box 150, Cheongryang, Seoul, Korea

ABSTRACT

Among several parameters in ultrasonic tissue characterization(UTC), attenuation has gained the most attention since it provides valuable information in clinical diagnosis and in quantitative B-scan imaging. Although attenuation is intrinsically defined by way of transmission measurement, it can be estimated from analysis of the backscattered echo waveform. But, because of the random spatial variation in the backscattered signal amplitude, the estimator variance becomes very large, making it difficult to estimate the attenuation slope from the backscattered ultrasound signal.

In this paper, we describe a new method of estimating the ultrasound attenuation coefficient of soft tissue based on the entropy difference between two adjacent envelope segments of narrowband ultrasound pulse echoes. Assuming uniform attenuation in the region under investigation, the attenuation is estimated by minimizing the difference of entropies in the two segments as the attenuation is continuously compensated for. The simulation and experimental results with a tissue equivalent (TE) phantom show that the proposed estimator is quite robust to the receiver noise and requires less data length as compared with the conventional methods. At present an experimental result with a TE phantom shows that a resolution of 10mm x 10mm can be attained with echoes from a depth of 3 cm using a 2.25 MHz transducer.

1. INTRODUCTION

Pulse echo ultrasonics has become, in recent years, an important and widely accepted method for non-invasive imaging of the human body, especially for fetal, cardiac, and abdominal applications, and appears to offer great potential. Results in varied clinical settings have shown medical ultrasound to be an accurate, versatile and inexpensive technique yielding cross-sectional tomographic images with negligible discomfort [1] and with reasonable resolution.

The present success of ultrasound has been realized through the utilization of only a small portion of the information actually available in the echo waveform. All ultrasound systems now in routine clinical service provide data on only two tissue properties : backscattered echo strength and the average attenuation in the medium, which can not be separated experimentally, nor is measured quantitatively. The paper by Ophir et al. [2] addressed the difficulties of using pulse echo techniques in vivo. The fundamental difficulty is closely related to the scattering process. In order to obtain a statistically stable result, it is often required to have a large tissue sample volume with a global statistical property.

The statistical characteristics of sample volumes of the envelope of ultrasound echo from soft tissue have been studied by several investigators [3]. The time domain method of estimating the attenuation coefficient from the mean value of the samples of echo envelope was recently proposed. Greenleaf et al. [4] showed that the noise to signal ratio (NSR) of the echo "envelope peaks" can be successfully used in estimating the attenuation. The attenuation coefficient of the tissue may then be obtained by adjusting the gain function so as to minimize the NSR of the echo envelope peaks. Also, the second order maximum entropy method (MEM) to estimate the central frequency shift between two adjacent regions of ultrasound signals was proposed [5].

In this paper, we present a new time-domain approach to estimating the ultrasound attenuation in soft tissue, which is based on the entropy of the "envelope sample values", not the envelope peak values, of the echo signal. The attenuation is estimated by finding the minimum difference of the entropies for two adjacent regions as the attenuation is compensated for. To explore the robustness off the estimator, measurements were conducted with a TE phantom which was designed to mimic the acoustical properties of soft tissues.

2. STATISTICAL SIGNAL MODELLING OF THE ECHO ENVELOPE

In order to extract information about the attenuation of soft tissue from ultrasound pulse echoes, one needs not only an accurate physical model to describe the interaction between the sound wave and the tissue scatterer, but also a careful statistical analysis of the signal contained in ultrasound pulsed echoes. Since coherent interference diminishes the independence of echoes from individual scatterers, statistical analysis is necessary in associating the tissue parameters. Consider first the distribution of the echo amplitudes received from a statistically homogeneous region of randomly positioned scatterers. If the transmitted ultrasound pulse is narrowbanded and the number of scatterers within one resolution element is large [6], the phase of the scattered waves is uniformly distributed between 0 and 2π. The echo waveform can then be described as follows :

$$r(t) = x(t)\cos 2\pi f_0 t - y(t)\sin 2\pi f_0 t, \tag{1}$$

where f_0 is the central frequency of the echoes, x(t) and y(t) are lowpass, random functions called the in-phase and the quadrature-phase modulations, respectively. The envelope R(t) is then defined as

$$R(t) = (x^2(t) + y^2(t))^{1/2}. \tag{2}$$

It is known that the echo envelope R(t) follows a Rayleigh distribution with the probability density function(pdf) :

$$p(R) = \frac{R}{\psi^2} \exp(-R^2/2\psi^2) , \qquad R > 0 \tag{3}$$

where ψ is the expectation of the echo envelope R divided by $(\pi/2)^{1/2}$. For different tissues having different backscattering properties, the Rayleigh constant ψ may differ. In the time domain analysis, the statistics such as mean and variance are useful as a measure of the behavior of the tissue state. The investigators were able to distinguish different tissue characteristics based on an analysis of the probability density function of the echo amplitudes. For the the attenuation medium, the statistical property changes and the histogram shape can not be predicted. Therefore, the pdf characteristics of the echo amplitude must be determined to estimate the attenuation. In order to examine the pdf characteristic of the ultrasonic pulse echo which propagates through an attenuating medium, we need the histogram of the echo. We assume a simple exponential model for the backscattered echo envelope to account for the attenuation [7] :

$$R(d) = \frac{R_0}{d}\sigma(d)\exp[-2 \int_0^d \alpha(r)dr], \tag{4}$$

where σ is the backscattering coefficient, d the nonzero distance to the transducer, α the attenuation coefficient, and R_0 a constant of proportionality. The factor of two in the exponent in Eq. (4) arises from the round trip to the reflector and back to the transducer. In our model, the effect of finite transducer beam width and finite pulse length, which also contribute to the amplitude fluctuations, are included to the assumption that the backscatter echo envelope is a random variable. The echo, otherwise having stationary statistics, will change in its envelope statistics, as it propagates through an attenuating medium, which is fully dominated by the backscattering term in Eq. (4). Therefore, the attenuated random echo envelope R(d) can be considered to be generated by multiplying the depth-dependent monotonic gain function with the input driving random signal.

In the following we assume that the input driving random process is X and output random process Y, and answer the question: what is the conditional probability of the output random variable ? The distribution function of random variable Y, $F_Y(y)$, can be expressed by

$$F_Y(y) = P[\,Y < y\,] = \lim_{T\to\infty} \frac{1}{T}\int_0^T P[\,a(t)\,X < y\,]\,dt \tag{5}$$
$$= \lim_{T\to\infty} \frac{1}{T}\int_0^T P(\,X < \frac{y}{a(t)}\,)\,dt$$
$$= \lim_{T\to\infty} \frac{1}{T}\int_0^T F_X(\,\frac{y}{a(t)}\,)\,dt,$$

where a(t) represents the envelope variation and T is the measured time interval. By differentiating Eq. (5), we get the probability density function of Y as

$$f_Y(y) = \frac{\partial F_y(y)}{\partial y} = \lim_{T\to\infty} \frac{1}{T}\int_0^T \frac{1}{a(t)} f_X(\frac{y}{a(t)})dt \tag{6}$$

To see that $f_Y(y)$ is actually the probability density function, we integrate it from 0 to infinity :

$$\int_0^\infty f_Y(y) = \lim_{T\to\infty} \frac{1}{T}\int_0^T\int_0^\infty \frac{1}{T} f_X(\,\frac{y}{a(t)}\,)dydt. \tag{7}$$

By changing the variable,

$$\int_0^\infty f_Y(y) = \lim_{T\to\infty} \frac{1}{T}\int_0^T\int_0^\infty f_X(y)dydt = \lim_{T\to\infty} \frac{1}{T}\int_0^T dt = 1. \tag{8}$$

The expected value of a random variable Y is

$$E(Y) = \int_0^\infty y f_Y(y)\,dy = \lim_{T\to\infty} \frac{1}{T}\int_0^T\int_0^\infty \frac{y}{a(t)} f_X(\frac{y}{a(t)})\,dy\,dt$$
$$= \lim_{T\to\infty} \frac{E(X)}{T}\int_0^T a(t)\,dt. \tag{9}$$

The second moment of Y is defined by

$$E(\,Y^2\,) = \int_0^\infty y^2 f_Y(y)\,dy$$
$$= \lim_{T\to\infty} \frac{1}{T}\int_0^T\int_0^\infty \frac{y^2}{a(t)} f_X(\frac{y}{a(t)})\,dy\,dt$$
$$= \frac{E(X^2)}{T}\int_0^T a^2(t)\,dt. \tag{10}$$

These are general results. For a given distribution, we can calculate several statistic measures such as variance, skewness and kurtosis using Eqs. (9) and (10). As the variance of the envelope decreases, the skewness increases. The variance of the random variable Y for the measurement time T is calculated as

$$\sigma_Y^2 = E(Y^2) - E^2(Y)$$
$$= E(X^2)\frac{1}{T}\int_0^T a^2(t)dt - \frac{E^2(X)}{T^2}\left[\int_0^T a(t)dt\right]^2 \tag{11}$$

The pdf characteristic of the echo envelope through the attenuating medium will now be examined through theoretical formulation. We assume the the input random variable is a Rayleigh distribution with a constant ψ. In our problem, tissue attenuation can be considered as a monotonic decreasing gain function of time. The finite time interval T is the observation time which is determined from the dimension of the tissue equivalent phantom. For the particular case of our interest, that is, $a(t) = e^{-2\beta t}$, the variance of Eq. (11) is calculated as follows :

$$\frac{1}{T}\int_0^T a(t)\,dt = \frac{1}{T}\int_0^T e^{-C\beta t}\,dt$$
$$= \frac{1}{C\beta T}(1 - e^{-C\beta T})$$
$$= \frac{2}{C\beta T}e^{\frac{-C\beta T}{2}}\sinh\frac{C\beta T}{2}. \tag{12}$$

Also,

$$\frac{1}{T}\int_0^T a^2(t)\,dt = \frac{1}{T}\int_0^T e^{-2C\beta t}\,dt$$
$$= \frac{-1}{2C\beta T}(e^{-2C\beta T} - 1)$$
$$= \frac{1}{C\beta T}e^{-C\beta T}\sinh C\beta T \tag{13}$$

Therefore, the variance σ_Y^2 is calculated as

$$\sigma_Y^2 = 2\psi^2\frac{1}{T}\int_0^T a^2(t)dt - \frac{\pi}{2}\psi^2\left[\frac{1}{T}\int_0^T a(t)dt\right]^2 \tag{14}$$

The observed output distribution through the attenuating medium shows deviation from the Rayleigh distribution as shown in Fig. 1. The Rayleigh distribution has a single peak located exactly at the square root of the corresponding Rayleigh coefficient ψ. The skewed modified Rayleigh distribution due to attenuation is compared to the fitted Rayleigh distribution which has a peak at the same grey level. Since the pdf characteristic is changed, the statistical measure has been changed. The variance of the attenuated signal changes with the observation interval T. Eq. (14) is plotted in Fig. 2, where we see that the variance becomes minimum at $\beta = 0$. This implies that, as we gradually increase the amount of compensation, β will attain a minimum point at the correct value of β. The variance difference between two adjacent regions can also be used as an ultrasound attenuation estimator. In general, the variance estimator as a statistical measure is not robust to the signal fluctuation. Therefore we will consider another statistical measure, namely entropy difference to be described below.

3. ENTROPY DIFFERENCE ESTIMATOR BASED ON THE ENVELOPE HISTOGRAM

The entropy is closely related to other statistical measures such as variance and skewness. As the variance of the envelope statistics decreases, the skewness increases, which in turn decreases the entropy value. Since the histogram of the echo envelope shows fluctuations different from the theoretical Rayleigh distribution [10], the entropy, which is less sensitive to the fluctuation of the histogram than the variance or skewness, is adopted to estimate the attenuation.

As the attenuation is gradually compensated, the entropy values of adjacent two regions increase. But the rate at which the entropy increases is faster in the far region than in the near region, and finally the difference of two entropy values reaches a minimum point at a certain amount of attenuation compensation, which is used as the estimated attenuation.

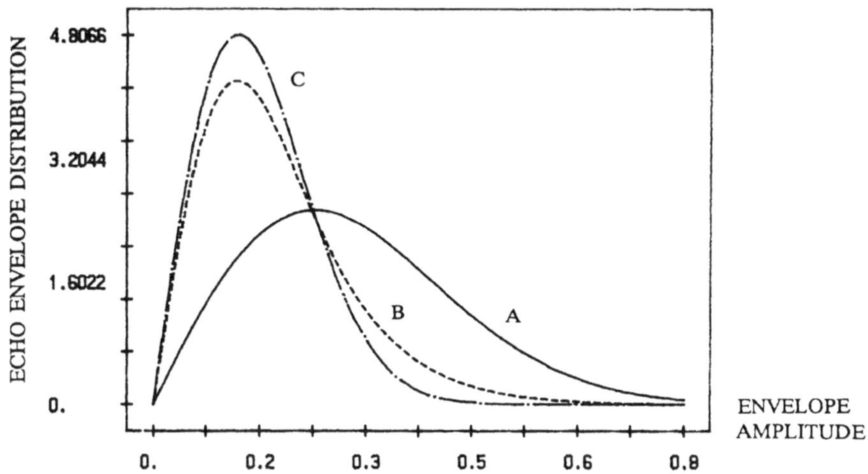

Fig. 1. Effect of attenuation on the echo envelope distribution. A : original Rayleigh distribution. B : after attenuation. C : Rayleigh distribution with the same peak as B.

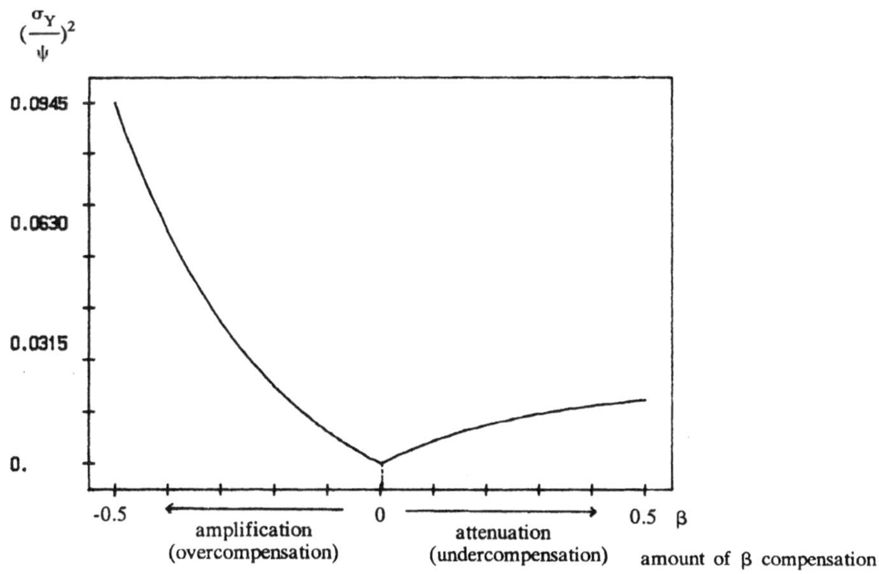

Fig. 2. Variance as a function of β (or as a function of the amount of β compensation).

Fig. 3b shows that the histogram shape from the far region has its peak skewed toward the low grey level since more attenuation is incurred in this region. Therefore, if we have a priori knowledge of the pdf characteristic in the attenuation free medium, attenuation can be perfectly compensated by maximizing the extent of similarity between the two pdf's. Unfortunately, the pdf characteristic of the echo envelope with no attenuation is unknown. Here, we adopt the compensation methodology so that the resultant pdf characteristics of the two segmented regions can be made as similar as possible. As the effect of attenuation is gradually compensated, the pdf similarity gradually increases to reach a peak and then decreases.

Consider next the entropy of the envelopes of two adjacent regions whose attenuation coefficient is to be estimated. The histogram on the bins is a function h(j), for j=1, 2, 3, ; · · , B, where B is the number of bins. Then h(j) is defined as

$$h(j) = \sum_{x=1}^{N} \chi_j g(x) \tag{15}$$

where χ_j is the characteristic function of the j th bin ; i.e.,

$$\begin{aligned} \chi_j(g) &= 1 \quad , \text{if } b_{j-1} < g < b_j \\ &= 0 \quad , \text{otherwise,} \end{aligned} \tag{16}$$

where b_{j-1} and b_j are the lower and upper boundaries of the j th bin. The total number of counts in the histogram is

$$N = \sum_{j=1}^{B} h(j) \tag{17}$$

The normalized histogram is

$$P_h(j) = \frac{h(j)}{N} \tag{18}$$

and $P_h(j)$ is an approximation to the pdf of the original echo signal. Each element $P_h(j)$ can be taken as an estimate of the probability of an echo sample lying within the interval [b_{j-1}, b_j]. The entropy stochastic based on the normalized histogram $P_h(j)$ is

$$H(P_h) = -\sum_{j=1}^{B} P_h(j)\log P_h(j) \tag{19}$$

The entropy can be seen as the average under the assumed pdf $P_h(j)$ of the information capacity while it indicates the randomness of the data samples in the image processing. The shape of an echo envelope histogram is closely related to the amount of entropy. For example, a narrowly distributed histogram indicates a low entropy value while a uniformly distributed one indicates the maximum entropy value. The entropy difference (ED) is defined as follows :

$$ED = |(\text{entropy of near region}) - (\text{entropy of far region})|$$

$$= \sum_{i=1}^{B} P_N(i)\log P_N(i) - \sum_{i=1}^{B} P_F(i)\log P_F(i), \tag{20}$$

where B is the number of grey levels of the histogram of the echo envelope of each of two adjacent regions of equal length, and the subscripts N and F stand for "near" and "far", respectively. The envelope changes with depth due to frequency-dependent attenuation, which results in the downshift of the center frequency of the pulse spectrum. This fact must be taken into account in the attenuation compensation. Thus, for the Gaussian waveform the envelope must be corrected by multiplying the envelope by a depth-dependent gain function of the following form [11] :

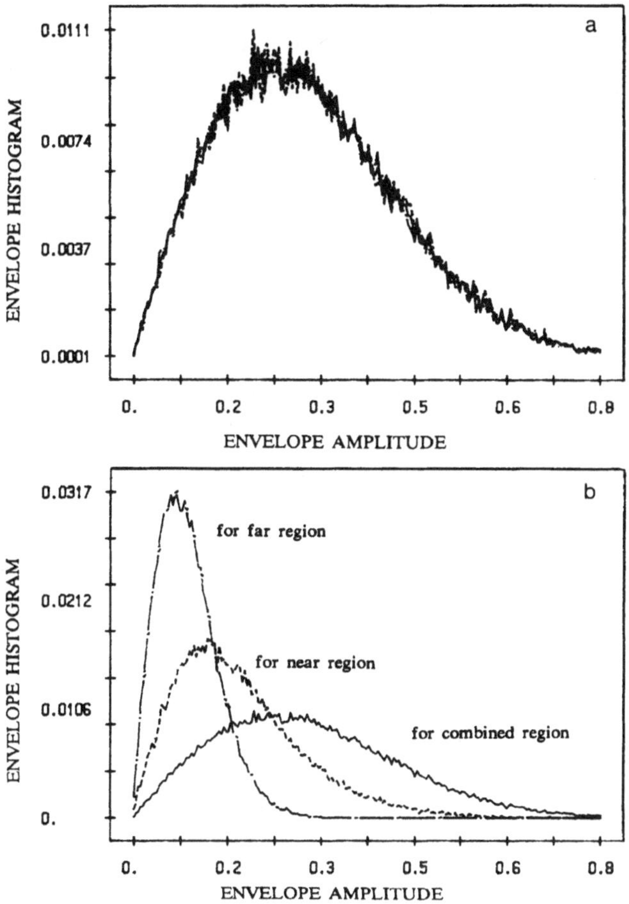

Fig. 3. The envelope histograms for two adjacent segmented regions (a) with no attenuaton, and (b) with attenuation of $\beta = 0.5\ \mathrm{dB{\cdot}cm^{-1}{\cdot}MHz^{-1}}$.

$$g(x) = xe^{2\hat{\beta}(f_0 - 4\hat{\beta}\sigma^2)\,x} \tag{21}$$

where f_0 is the center frequency, σ the signal bandwidth, and $\hat{\beta}$ the true value of attenuation coefficient of the tissue.

4. MODELLING AND SIMULATION OF THE ULTRASOUND ECHOGRAM

4.1 One-dimensional case

In practical ultrasound, the signals reflected from the body organs cannot have the ideal pulse shape. The reflecting scatterers located within the soft tissue are randomly shaped, oriented and distributed in range. Thus, we need a particular model to simulate the ultrasound signals reflected from randomly shaped interfaces for the completeness of the computer simulation. The received ultrasound echo signal from a particular region can be modelled by a filtering process corresponding to the frequency response of the ultrasound transducer for a white Gaussian random noise, which represents the random scatterers in that region. Therefore, the received echo signal e(t) can be generated as follows :

$$e(t) = F^{-1}(\ F_T(\omega)\ F_R(\omega)\ ) \text{ and } F_R(\omega) = F(\ r(t)\ ) \tag{22}$$

where F_T is the frequency response of the transducer, r(t) is a Gaussian random sequence characterized by such statistics as mean and variance. F (F^{-1}) denotes the

Fourier (inverse Fourier) transformation. Thus, the rf signal with arbitrary attenuation coefficient is generated by convolving the attenuation transfer function which is characterized as a minimum phase filter. The minimum phase filter proposed by Kuc [8] represents the physical mechanism which produces a linear loss function with frequency, and fits well with experimental data. The Rayleigh density function is generated by nonlinearly transforming the uniform distribution. The statistics of echo amplitude in simulated echoes is investigated and their histograms exhibit the typical Rayleigh probability distribution. Fig. (3a) shows that the envelope histogram between two adjacent regions has the same pdf characteristics. However, the pulse echo characteristic propagating through an attenuating medium has different stochastic properties in different regions. Histograms for the far region exhibits a relatively larger number of lower grey levels, with the histogram shape shifted toward the near region as shown in Fig. (3b). The flow chart of the 1-dimensional simulation and attenuation estimation is shown in Fig. 4. Fig. 5. shows several representative Rayleigh probability distributions which are generated with different mean values. The generated density function matches with the theoretical density function quite well. Fig. 6. shows the pdf characteristic of the echo amplitude with different attenuations, β = 0.3, 0.5, and 0.7 $dB{\cdot}MHz^{-1}{\cdot}cm^{-1}$. The grey level histograms of echo envelopes between two adjacent regions exhibit a peaked, skewed shape which is substantially different from each other.

In order to verify the validity of the above mentioned ED estimator, computer simulation was performed, in which we assumed β = 0.5 $dB{\cdot}cm^{-1}{\cdot}MHz^{-1}$ and 1024 data points were taken for one way path length of 3 cm (sampling time = 40ns, sonic velocity = 1540 m/sec, and center frequency = 2.25 MHz).

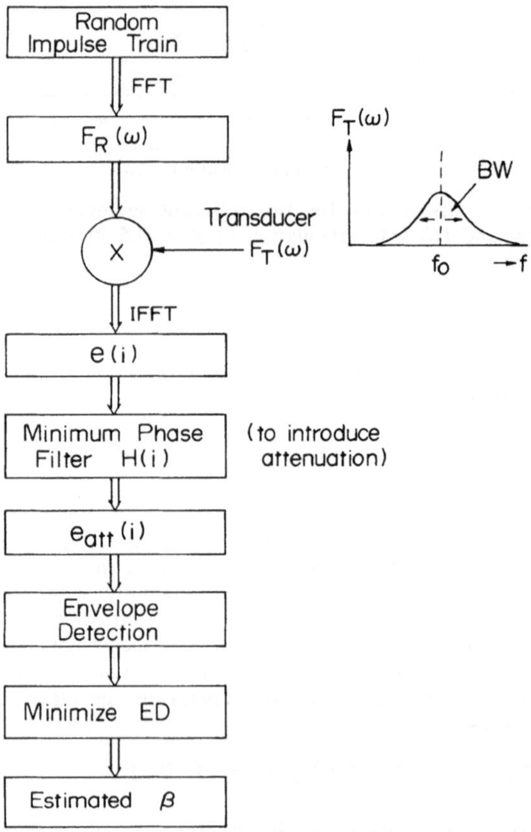

Fig. 4. Flow chart of 1-dimensional simulation

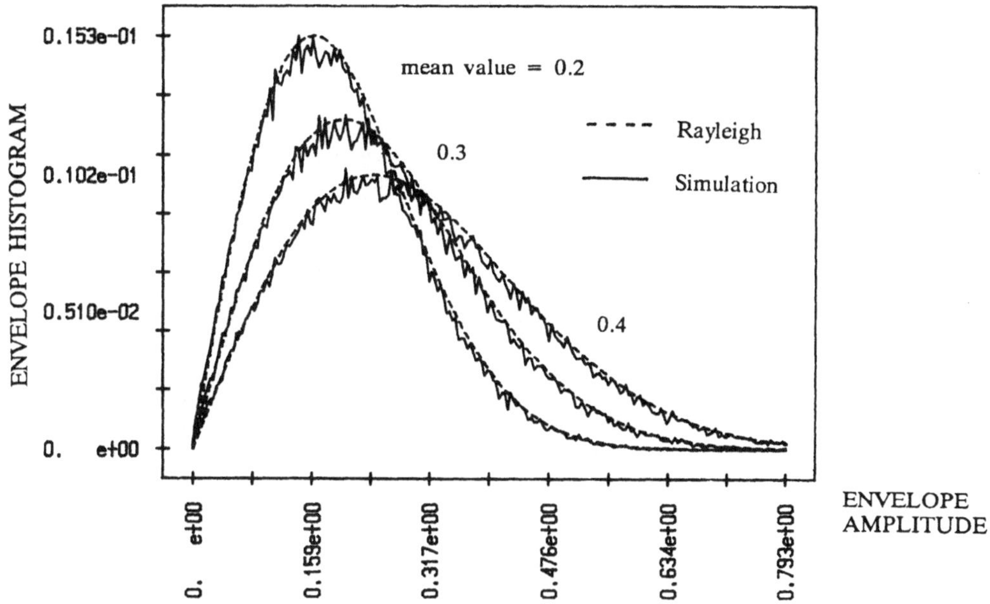

Fig. 5. Comparisons of Rayleigh probability density functions of echo amplitude for a simulated echo signal (no attenuation).

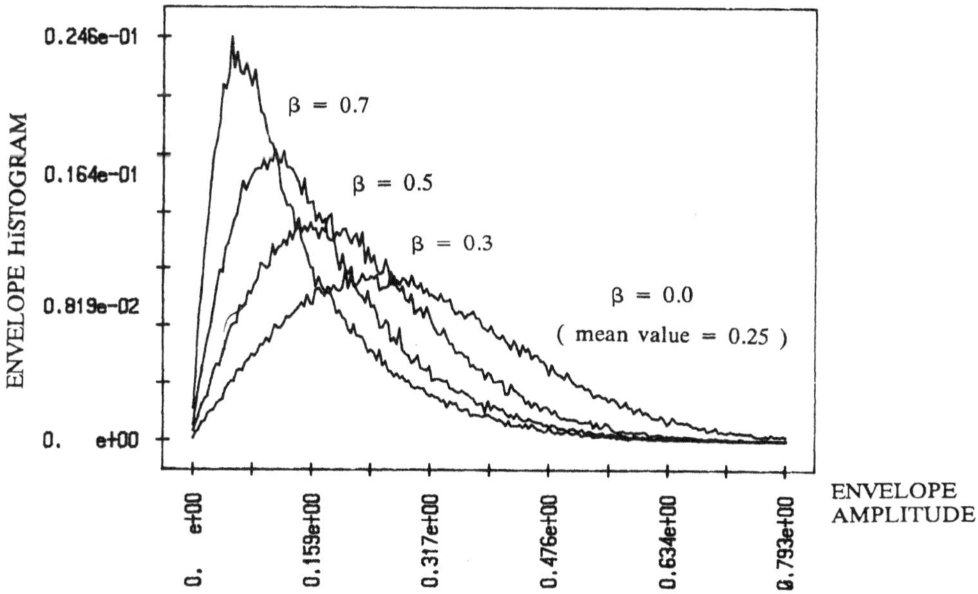

Fig. 6. Probability density functions of echo amplitudes with different attenuation $\beta = 0.3$, 0.5, and 0.7 dB $cm^{-1}MHz^{-1}$ respectively.

The algorithm for computing the attenuation coefficient, using the entropy difference method consisted of the following steps:

1) Full-wave rectification of rf data.
2) Low-pass filter the rectified rf data to get the echo envelope of one scan line.
3) Multiply the entire envelope by the previously mentioned gain function g(x)
4) Calculate the histogram of each of equally segmented two data sets.
5) Calculate the absolute ED of the two segmented histograms.
6) Increase (or decrease) β by 0.05 $dB \cdot cm^{-1} \cdot MHz^{-1}$.
7) Repeat steps 3), 4), 5), and 6) until the ED is minimized.

The outputs of the algorithm are the minimum ED and the corresponding β. Fig. 7 plots the ED versus β, which shows a sharp minimum point where we get the estimated value of β. The data length can be reduced to 7.5 mm distance resolution. Table 1 shows percent errors of the attenuation estimator with the data length of the segment successively reduced by one half and for different SNR's. The error of the estimate increases almost monotonically as the sample volume is gradually decreased by reducing the scan depth. The error is due to interaction among scatters. This effect usually necessitates an averaging process of the information from a number of scan lines.

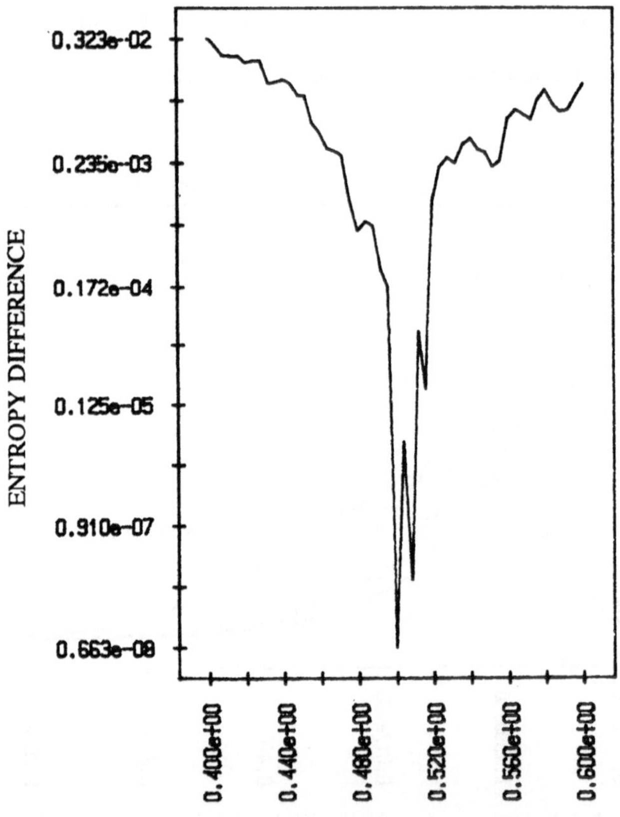

Fig. 7. Change of entropy difference with compensation of β (1-dimensional case). Total data points : 8,000 from 3 cm through 6 cm True β = 0.5$dB \cdot cm^{-1} \cdot MHz-1$.

Table 1. Simulation result of the entropy difference estimator when the data length of the segment is successively reduced by one half for different added noise levels -- one dimensional case (true $\beta = 0.5$)

Data length \ SNR		noise free	20 dB	10 dB	3 dB
3 cm	β	0.504	0.508	0.508	0.512
	error	0.8%	1.6%	1.6%	2.4%
1.5 cm	β	0.504	0.516	0.524	0.532
	error	0.8%	3.2%	4.8%	6.4%
0.75 cm	β	0.508	0.540	0.548	0.556
	error	1.6%	8.0%	9.6%	11.2%

4.2 Two-dimensional case

A two-dimensional simulation analysis of the ED method was also performed to study the accuracy of the estimation of the attenuation coefficient in tissues. First, we briefly describe the underlying theory and assumption in our analysis. An ultrasonic wave which propagates through an inhomogeneous medium is sensitive to local changes in its density and compressibility, which condition the proportions of the signal reflected, absorbed and transmitted. Since such a medium is relatively complex, a simpler tissue model is generally used, consisting of point scatterers embedded in a homogeneous attenuating medium. Also the Born's approximation is assumed, namely, the intensity of the wave reflected by a scatterer is negligible in comparison with the intensity of the incident wave (and hence the transmitted wave, behind the scatterer, is equal to the incident wave). Fig. 8 shows the flow chart of a B-mode imaging system for the simulation of the proposed attenuation estimator. This system, a typical linear array system, consists of 16 transmitter/receiver channels and an array transducer of 128 elements and acquires one frame image data as a group of 16 elements slides laterally. Also, we assumed a constant velocity c_0 throughout the medium. The transmitter produces a short pulse which excites each transducer element. The transducer converts electric energy from transmitter into and ultrasonic wave launched into the interrogated medium, and converts the returned echoes back into an electric signal. The echoes detected by the transducer have about 115 dB dynamic range [9]. The overall received signal $f_{RX,J}(\mathbf{r}, t)$ is given by [10]

$$f_{RX,J}(\mathbf{r}, t) = R(\mathbf{r})\, f_{TX}(\mathbf{r}, t) * [S_{RX,J}(\mathbf{r})\, \delta(t - \frac{R_J}{c_0}) * p_{RX,J}(t)] \tag{23}$$

$$= R(\mathbf{r}) \sum_{I=1}^{N} S_{TX,I}(\mathbf{r})\, S_{RX,J}(\mathbf{r})\, \delta(t - \frac{R_I}{c_0} - \Delta t_{TX,I}) * \delta(t - \frac{R_J}{c_0})$$

$$* [p_{TX,I}(t) * p_{RX,J}(t) * e(t)] \text{ ,where}$$

$f_{TX}(\mathbf{r}, t)$: total field of arbitrary point r

$S_{TX,J}(\mathbf{r})$: the transmisssion radiation pattern

$S_{RX,J}(\mathbf{r})$: the reception radiation pattern

$p_{TX,I}(r, t)$: impulse response of transducer on TX mode

$p_{RX,J}(r, t)$: impulse response of transducer on RX mode

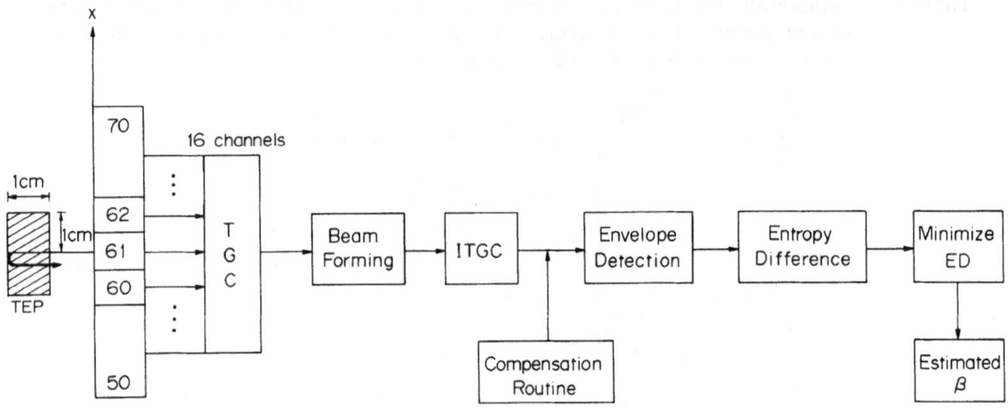

Fig. 8. Flow chart of 2-dimensional simulation

Tissue attenuation limits the maximum possible range from which echoes can be detected. Variations in backscatterers define a dynamic range of expected echoes from any particular depth range. Over the range of diagnostically used frequencies, soft tissues and muscle exhibit attenuation coefficients varying nearly linearly with frequency. The B-mode echograms were simulated by employing the impulse response method in transmission and reception using a discrete scatterer tissue model[11]with attenuation, neglecting the multiple reflections. Simulation with the 2-dimensional array is carried out under the following conditions :

f_0 = 3.5 MHz f_s = 28 MHz σ = 0.44 MHz

Transmitter focal depth = 100 mm

number of the scan line (N) = 20 scatterer density = 1000 / cm^2

β = $0.5 dB \cdot MHz^{-1} \cdot cm^{-1}$ slilding scattering volume = 21 scans x 728 points

A scan covered an area of 20 mm lateral x 10 mm axial. Again, data was obtained from only one scanning at one scan position. Table 2 shows the simulation result. We see that the error decreases in all of the methods as the number of scan line increases, but that the ED method gives much lower error. It should be mentioned that the error must decrease further as the scatterer density is increased in the simulation. (Due to the computation time, we could not increase the scatterer density, although at least 10^6 scatterers/ cm^2 are needed for adequate tissue simulation [12]).

Table 2. Simulated percentage error in β estimation by different methods --two dimensional case (true β = 0.5 and noise free situation). Axial length = 10 mm with 728 data points, and lateral scan line interval = 1 mm.

# of scans	Entropy difference	Envelope peak	Center frequency shift
1	31.4 %	89.0 %	≃ 100.0 %
5	5.0 %	42.8 %	87.1 %
10	2.4 %	28.0 %	39.6 %
21	0.0 %	10.4 %	21.7 %

5. EXPERIMENTAL RESULTS AND DISCUSSION

In order to further show the usefulness of the new attenuation estimator, experiments were performed with a TE phantom. The phantom was designed and fabricated so that uniform ultrasonic properties may be obtained. The tissue homogeneity was realized by controlling the cooling process properly. Homogeneity means constant local attenuation throughout the tissue region under consideration, which can be verified experimentally by measuring attenuation of the phantom in many different directions. The soft tissue equivalent phantom consisted of powdered graphite suspensions in a solid matrix of agar, distilled water, and n-propanol. By changing the concentration of n-propanol and powdered graphite, phantoms with different attenuation coefficients can be obtained. Fig. 9 shows the block diagram of the measurement system. For the transmission mode experiments, a 2.25MHz unfocused disk type transducer pair was used. The water path signal and the signal through the TEP were digitized and compared after FFT to calculate the attenuation.

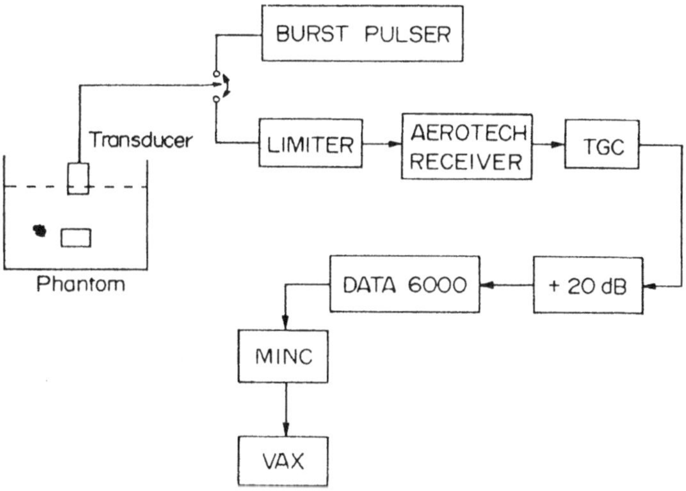

Fig. 9. Block diagram of the measurement system.

In the reflection mode, the data 6000 waveform recorder was used to obtain digitized RF data. The burst pulser was composed of 14 sinusoidal pulses which had a center frequency of 2.25 MHz. The transducer used in this experiment had a diameter of 12 mm, a focal length of 50 mm, and the backscattered RF signal was digitized and recorded by the Data 6000 waveform recorder of Data Precision Corp. The sampling rate and the sampling window were controlled by the Minc-11/23. The maximum length of the sampling window is limited by the buffer size of the Data 6000 and the sampling frequency. In this experiment, the backscattered RF-A scan signal contained 1024 samples representing a 3 cm one way path length (sampling time = 40 ns, sonic velocity = 1540 m/sec). In the above description of the β estimation approach, we tacitly assumed that the change in the propagating pulse spectrum was solely due to the frequency dependent attenuation of the tissue. However, typical diagnostic transducers do not produce ideal plane waves. Their spherically curved surfaces and finite-sized apertures produce waves that are nonplanar, which in turn leads to diffraction effect. This effect causes the propagating pulse spectrum to vary with range. Hence, in order to take the diffraction effect into account, we used the reference signal spectrum which was obtained with the beam normally incident to a glass reflector placed in the focal zone. The reflected A-scan signal through the fabricated tissue equivalent phantom (TEP) was obtained with the time gain compensation (TGC) properly adjusted. All of the echoes are of similar amplitudes as shown in Fig. 10(a). The RF waveform when the TGC effect is subtracted, gradually diminishes in intensity (Fig. 10(b)). The dotted line is the

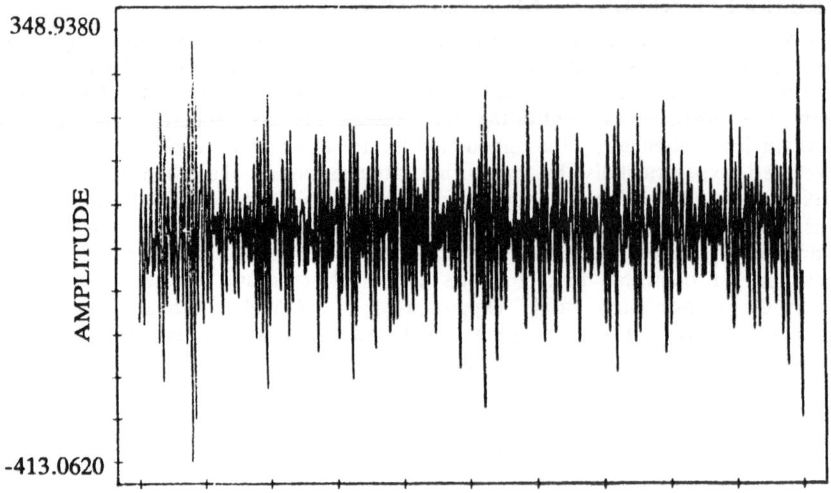

(a)

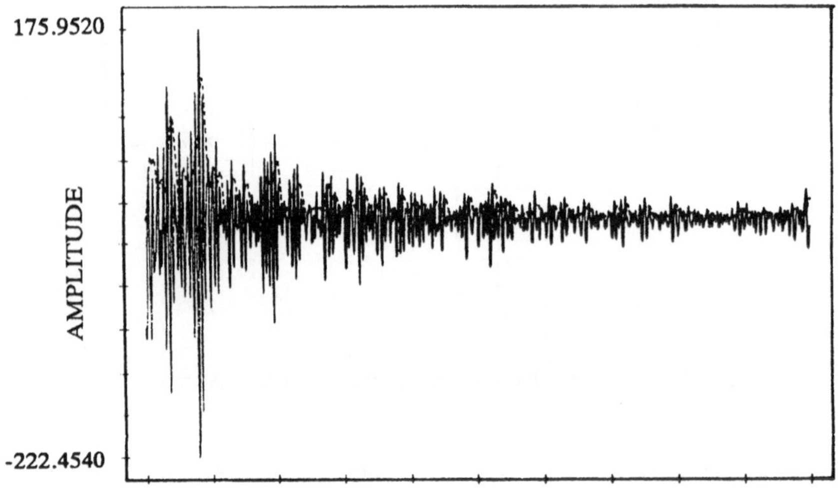

(b)

Fig. 10. (a) The reflected A-scan through the fabricated tissue equivalent phantom (TEP) with the time gain compensation (TGC) properly adjusted. (b) The RF waveform when the TGC is subtracted (the dotted line is the envelope).

envelope of an A-scan line. The transmission experiment shows that comparisons of β has a 0.585 $dB \cdot cm^{-1} \cdot MHz^{-1}$. The entropy difference between the two adjacent regions was computed. The proposed entropy difference method is a time domain, amplitude method. The ED method takes the frequency dependency of attenuation into account, therefore it is also suitable for non-narrowband signals. Table 3 shows comparisons of the percentage errors of several estimators when the data length of the segment is successively reduced by one half in the experimental results. The length of each A-line was 3 cm while the number of A-line used in the attenuation estimation was changed from 1 to 20. Note that the third row of the table corresponds to 1 cm x 1 cm resolution error. This shows that the entropy difference estimator may find a promising application in clinical diagnosis .

Table 3. Comparisons of the percentage errors of several estimators when the data length of the segment is successively reduced. (2.25 MHz disk type transducer, true $\beta = 0.585$, and depth = 3 cm, the number of A-line = 20 lines, the lateral line interval = 1 mm)

Depth	Entropy difference	Envelope peak	Center frequency shift
3 cm	2.5 %	9.0 %	22.3 %
2 cm	7.2 %	26.5 %	52.6 %
1 cm	10.4 %	50.4 %	89.0 %

6. CONCLUSIONS

The proposed method of estimating the attenuation coefficient is based on the monotonic relationship between the entropy of the envelope histogram and the statistical behaviors of the echo distribution. The computation is performed only with the echo envelope histogram, and the norm used for minimization is different from the conventional methods. The simulation results and experimental results using the proposed ED estimator shows that the data length and the number of independent data segments can be reduced significantly with relatively good noise rejection.

REFERENCES

[1] O'Donell, M., Mimbs, J. W., et al.: 'Ultrasonic attenuation in normal and ischemic myocardium', Ultrasonic Tissue Characterization, M. Linzer ed., NBS spec. publ. 525, 1979, pp. 63-71.

[2] Ophir, J., Shawker, T. H., and Maklad, N. F., et al.: 'Attenuation estimation in reflection : progress and prospects', Ultrason. Imag. 6, 1984, pp. 349-395.

[3] Flax, S. W., Glover, G. H., and Pelc, N. J.: 'Textual variations in B-mode ultrasonography - A stochastic model', Ultrason. Imag., 1981, 3, pp. 235-257.

[4] He, P., Greenleaf, J. F.: 'Application of stochastic analysis to ultrasonic echos - Estimation of attenuation and tissue heterogeneity from peaks of echo envelope', J. Acoust. Soc. Am., 1986, 79(2), pp. 526-534.

[5] Kim, S. I., Reid, J. M.: 'Estimation of attenuation coefficient by using the maximum entropy method', Int'l Confer. on Acoust. Imag., 1987.

[6] Burckhardt, C. B.: 'Speckle in ultrasound B-mode scans', IEEE Trans. Sonic. Ultrason., SU-25, 1978, pp. 1-6.

[7] Eugene, W., et al.: 'Quantiatative tissue characterization based on pulsed echo ultrasound scans', IEEE Trans. on Biomed. Eng. BME-33. NO. 7, 1986, pp. 637-643.

[8] Kuc, R.: 'Digital filter model for media having linear with frequency loss characteristics', J. Acoust. Soc. Am., 1981, 69(1), pp. 35-40.

[9] Maxwell, G. M., et al.: 'Methods and terminology for diagnostic ultrasound imaging system', Proc. IEEE, vol. 67, NO. 4, 1979, pp. 641-653.

[10] Mitchell, M. G., et al.: 'A three dimensional model for generating the texture in B-scan images', Ultrason. Imag. 5, 1983, pp. 253-279.

[11] Wagner, R. F., et al.: 'Statistics of speckle in ultrasound B-scans', IEEE Trans. on Sonic. and ultrason. vol. SU-30, 1983, pp. 156-163.

[12] Oosterveld, B. J., et al.: 'Textute of B-mode echograms : 3-D simulations and experiments of the effect of diffraction and scatterer density', Ultrason. Imag. 7, 1985, pp. 142-160.

A NEW METHOD FOR QUANTITATIVE REFLECTION IMAGING

Jian-Yu Lu and Yu Wei

Department of Biomedical Engineering
Nanjing Institute of Technology
Nanjing, China

INTRODUCTION

B-scanner is one of the most successful medical ultrasonic diagnostic equipment developed in 1970s. It can provide the outline images of the internal structures of the cross-sections of the biological soft-tissues and is useful in diagnosing the diseases of the shape abnormalities of the tissues, such as, tumors in livers, kidneys, etc.. Although great efforts have been done to further improve the quality of B-scan images[1-9], the ordinary B-scanner can not provide the quantitative images of the distributions of the acoustical parameters of the biological soft-tissues. In this paper, a new quantitative reflection imaging method (short for QRI method) is developed, which is directly based on the acoustical transmitting/receiving geometry of the ordinary B-scanner and can provide a quantitative image of the sound speed distribution of the biological soft-tissues and will be useful in tissue characterization.

The imaging procedure of the QRI method is as follows. First, the high-frequency component of one-dimensional object function (i.e., the object function on the line through which a focused pulse acoustical wave is propagated) is obtained from the rf (radio frequency) echo signals returned from the object. Then, the low-frequency component of the one-dimensional object function is determined from the high-frequency component using the GP (Gerchberg-Papoulis) frequency extrapolation technique[10] and using the a priori knowledges that the outlines of the internal structures of the object and the phases of the rf echo signals returned from these outlines are known. From the complete frequency spectrum, the one-dimensional object function can be reconstructed. By scanning the focused pulse acoustical wave in the cross-section of the object to be imaged, the two-dimensional object function in that cross-section can be reconstructed.

In addition, an experimental system which connects the B-scanner to computer is developed for the datum acquisition of the rf echo signals, and images of practical objects are reconstructed using the data obtained by this experimental system. The results show that the images reconstructed by the QRI method are more helpful in understanding the internal structures of the testing objects than the ordinary B-scan images, and to a certain extent, the images are quantitative.

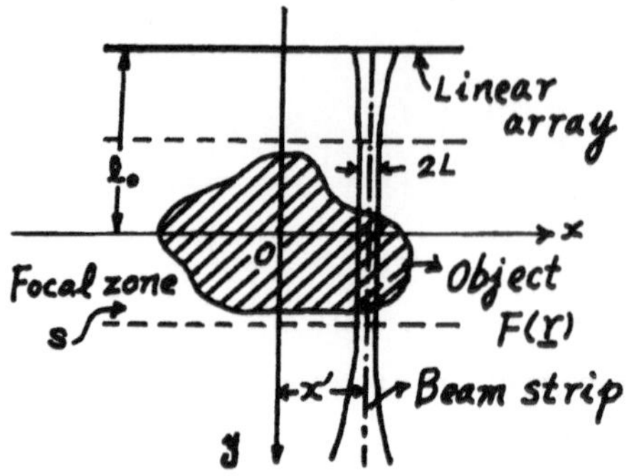

Fig.1 Datum acquisition geometry of the QRI method

BASIC PRINCIPLES

Theoretical Preliminaries

Fig.1 is the datum acquisition geometry of the QRI method. The x-y coordinates are fixed in the space; $F(\underline{r})$ represents the two-dimensional object function (objects are located in an echoless water-tanker and in the focal zone of the linear array probe); L is the half-width of the focused ultrasonic beam; x' indicates the distance of the center of the beam to the y-axis and l_0 is the distance between the surface of the linear array probe and the x-axis. The linear array probe used here is consist of many small transducer elements and can be scanned electrically, it is used both as transmitter and receiver. The focal zone of the linear array probe is defined as the region where the ultrasonic beams produced are focused so narrow and even that they can be approximately looked as a plane wave within the width of each beam strip.

In this paper, the incident wave used is a pulse acoustical wave. For the pulse acoustical wave can be decomposed into monochromatic wave with different frequencies, in the following, we will first consider the case of the monochromatic wave. Suppose that the acoustical waves interacting with the biological soft-tissues are governed by the following Helmholtz equation[11]

$$\nabla^2 U(\underline{r})+k_0^2 U(\underline{r}) = -F(\underline{r})U(\underline{r}) \tag{1}$$

where $U(\underline{r})$ represents the total sound pressure field; $k_0 = \omega/c_0$ is the wavenumber of the homogeneous medium which surrounds the object (ω is the angular frequency of the incident wave and c_0 is the sound speed of the surrounding homogeneous medium). The object function $F(\underline{r})$ is given by

$$F(\underline{r}) = \begin{cases} k_0^2[n^2(\underline{r})-1]; & \text{if } \underline{r} \text{ is in the object} \\ 0 & ; \text{otherwise} \end{cases} \tag{2}$$

where $n(\underline{r})$ represents the distribution of the refractive index of the object and is related to the sound speed distribution by $n(\underline{r}) = c_0/c(\underline{r})$. Considering the first-order Born approximation (i.e., weak scattering assumption is assumed), one obtains the following integral solution of Eq.(1)[12]

$$U_s(\underline{r}) = \int_s F(\underline{r_0})U_i(\underline{r_0})g(\underline{r}|\underline{r_0})d\underline{r_0} \tag{3}$$

where $U_s(\underline{r})$ and $U_i(\underline{r})$ represent scattered field and incident field respectively, (the total field $U(\underline{r})$ is the sum of $U_i(\underline{r})$ and $U_s(\underline{r})$); the integral area s is the focal zone of the linear array probe, as shown in Fig.1; $g(\underline{r}|\underline{r}_0)$ is assumed to be the two-dimensional free-space Green's function which can be represented by[13]

$$g(\underline{r}|\underline{r}_0) = (j/4)H_0(k_0|\underline{r}-\underline{r}_0|) \tag{4}$$

where $H_0(k_0|\underline{r}-\underline{r}_0|)$ is the zeroth-order Hankel function with first kind. If the argument of $H_0(k_0|\underline{r}-\underline{r}_0|)$ is much greater than unity, i.e., $k_0|\underline{r}-\underline{r}_0|>>1$, Eq.(4) can be simplified to

$$g(\underline{r}|\underline{r}_0) = \frac{j}{4}\sqrt{\frac{2}{\pi k_0|\underline{r}-\underline{r}_0|}}\exp[j(k_0|\underline{r}-\underline{r}_0|-\pi/4)] \tag{5}$$

As mentioned above, the incident wave is assumed to be a plane wave in the focal zone s and within the width of the beam strip (see Fig.1)

$$U_i(\underline{r}) = \begin{cases} A\exp(jk_0\underline{s}_0\cdot\underline{r}); & |x-x'|\leqslant L \\ 0 & ; \text{otherwise} \end{cases} \tag{6}$$

where A is a constant and $\underline{s}_0$ is a unit vector in the direction of the plane wave insonification.

Substituting Eq.(5) and Eq.(6) into Eq.(3), we obtain the expression of the scattered field measured on the point $(x',-l_0)$ when the center of transmitting group of the transducer elements is on the same point

$$U_s(x',-l_0;k_0) = \iint_{-\infty}^{+\infty} F(x_0,y_0)A\exp(jk_0y_0)\frac{j}{4}\sqrt{\frac{2}{\pi k_0\sqrt{(x'-x_0)^2+(-l_0-y_0)^2}}}$$
$$\cdot\exp[j(k_0\sqrt{(x'-x_0)^2+(-l_0-y_0)^2}-\pi/4)]dx_0dy_0 \tag{7}$$

where the notation k_0 in $U_s(x',-l_0;k_0)$ emphasizes the frequency dependence nature of the scattered field. Because the half-width of the incident beam L in Eq.(6) is very small, the integration in terms of the variable x_0 in Eq.(7) can be approximately looked as a constant. Therefore, from Eq.(7), one obtains

$$U_s(x',-l_0;k_0) = \int_{-\infty}^{+\infty} F(y_0)A'\exp(jk_0y_0)\frac{j}{4}\sqrt{\frac{2}{\pi k_0|-l_0-y_0|}}$$
$$\cdot\exp[j(k_0|-l_0-y_0|-\pi/4)]dy_0 \tag{8}$$

where $A' = 2LA$ is a new constant and $F(y)$ is the one-dimensional object function which is obtained by evaluating the two-dimensional object function $F(\underline{r})$ on the line $x = x'$ and is given below

$$F(y) = \begin{cases} k_0^2[n^2(x',y)-1]; & \text{if point } (x',y) \text{ is in the object} \\ 0 & ; \text{otherwise} \end{cases} \tag{9}$$

where $n(x',y) = c_0/c(x',y)$ is the distribution of the refractive index of the object on the line $x = x'$. For the convenience of the discussions in the following, two new functions which are related to the distribution of the refractive index of the object are defined

$$n_1(y) = F(y)/k_0^2 \tag{10}$$

$$n_e(y) = n_1(y)/\sqrt{l_0+y} \tag{11}$$

where $n_1(y)$ is another form of the object function and is assumed to be independent of the angular frequency ω; $n_e(y)$ is called equivalent object function. From Fig.1, one can see that l_0+y_0 is always greater than zero when the integral variable y_0 in Eq.(8) is confined to the focal zone s of the linear array probe. Therefore, from Eq.(8), we obtain

$$U_s(x',-l_0;k_0) = \frac{jA'k_0^2\exp[j(k_0 l_0-\pi/4)]}{2\sqrt{2\pi k_0}}\tilde{n}_e(-2k_0) \tag{12}$$

where $\tilde{n}_e(k)$ represents the Fourier transform of the equivalent object function $n_e(y)$.

Taking the mechanical—electrical characteristics of the transducer into account, we have

$$P_s(x',-l_0;k_0) = P(k_0)\frac{jA'k_0^2\exp[j(k_0 l_0-\pi/4)]}{2\sqrt{2\pi k_0}}\tilde{n}_e(-2k_0) \tag{13}$$

where $P_s(x',-l_0;k_0)$ represents the electrical signals produced by the returned rf echoes and $P(k)$ is the Fourier transform of the electrical—mechanical characteristic function. Eq.(13) can be written in another form

$$\tilde{n}_e(-2k_0) = \frac{2\sqrt{2\pi}\exp[-j(k_0 l_0+\pi/4)]}{k_0\sqrt{k_0}}\left[\frac{P_s(x',-l_0;k_0)}{P'(k_0)}\right] \tag{14}$$

where $P'(k_0) = A'P(k_0)$ and is calculated by

$$P'(k_0) = \frac{2\sqrt{2\pi l_0}\exp[-j(k_0 l_0+\pi/4)]}{k_0\sqrt{k_0}}P_{s\delta}(x',-l_0;k_0) \tag{15}$$

where $P_{s\delta}(x',-l_0;k_0)$ is the electrical signal produced by the echo returned from a point scatterer (for a point scatter, the object function $n_1(y)$ is of the form of Dirac—Delta function $\delta(y)$).

Frequency Extrapolation and Image Reconstruction

Because the incident wave and, therefore, the returned rf echoes are high center frequency band-pass acoustical signals, from Eq.(14), only the high-frequency component of the equivalent object function $n_e(y)$ can be obtained. To recover the low-frequency component of the equivalent object function from its high-frequency component, the Gerchberg-Papoulis (GP) frequency extrapolation technique is adopted and some available a priori knowledges are used in the frequency extrapolation.

The ordinary B-scanner can provide the outline information of the internal structures of the object, therefore, it is possible to determine the phases of the rf echo signals returned from these outlines. Fig.2 (a) and (b) are the waveforms of the rf echo signals returned from the water--agar and agar—water interfaces respectively. It is seen that their phases are different.

In order to use the phase information of the rf echo signals in the GP frequency extrapolation, it is required to know the relationship between the derivative of the equivalent object function and the measured the electrical signals produced by the returned rf echoes. Take the notations

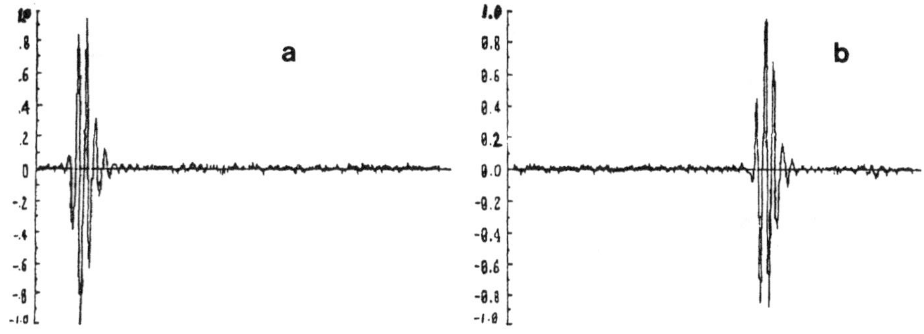

Fig.2 (a) rf echo signals returned from the water—agar interface (b) from the agar—water interface

$n_e'(y)$ and $\tilde{n}_e'(k)$ as the derivative of the equivalent object function and its Fourier transform, from Eq.(14), we obtain

$$\tilde{n}_e'(-2k_0) = \frac{4\sqrt{2\pi}\exp[-j(k_0 l_0+3\pi/4)]}{\sqrt{k_0}}\left[\frac{P_s(x',-l_0;k_0)}{P'(k_0)}\right] \tag{16}$$

The procedures of the GP frequency extrapolation is described simply as follows:

(a) Calculate $\tilde{n}_e'^{(1)}(k)$, the high-frequency component of $n_e'(y)$, using Eq.(16) ($\tilde{n}_e'^{(1)}(k)$ is used to replace the high-frequency component of each order partial reconstruction of the equivalent object function in following iteration procedures).

(b) Find $n_e'^{(1)}(y)$, the first order partial reconstruction of the equivalent object function, by taking the IFFT of $\tilde{n}_e'^{(1)}(k)$.

(c) Determine the signs of $n_e'^{(1)}(y)$ on the outlines of the internal structures of the object in accordance with the phases of the rf echo signals returned from these outlines.

(d) Find $\tilde{n}_e'^{(2)}(k)$, the Fourier transform of the spatial modified $n_e'^{(1)}(y)$, by using the FFT.

(e) Substituting $\tilde{n}_e'^{(1)}(k)$ into the high-frequency portion of $\tilde{n}_e'^{(2)}(k)$ and then taking IFFT, we obtain $n_e'^{(2)}(y)$, the second order partial reconstruction of the equivalent object function.

(f) Calculate the average of $|n_e'^{(2)}(y)-n_e'^{(1)}(y)|$ and see if it is smaller than a preset small positive value. If the condition is true, go to step (g), otherwise, go back to step (c).

(g) Obtain the equivalent object function $n_e(y)$ by integrating $n_e'^{(N)}(y)$, the Nth order partial reconstruction of the equivalent object function (the integration constant is determined using the condition that $n_e(y) = 0$ if the point (x',y) is not in the object). By scanning the acoustical beam in a cross-section of the object, the equivalent object function $n_e(y)$ on different line $x = x'$ can be determined, and thus, tomographic image of the object can be reconstructed.

DESCRIPTION OF EXPERIMENTAL SYSTEM

Fig.3 is the block-diagram of the experimental system of the QRI

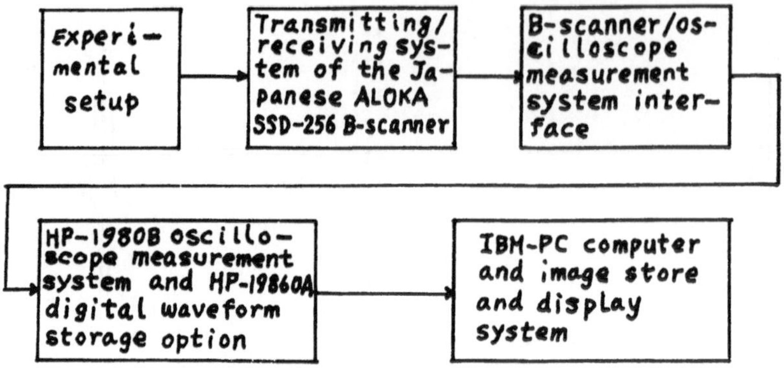

Fig.3 Experimental system for the QRI method

method. The experimental setup provides an acoustical environment for the datum acquisition. And the transmitting/receiving system of the Japanese ALOKA SSD-256 B-scanner[14] is used for the emission of the acoustical pulses and the reception of the rf echo signals returned from the testing objects. The signals received are transferred to HP-1980B oscilloscope measurement system[15] and sampled and quantized by HP-19860A digital waveform storage option[16] (to stabilize the datum acquisition, the B-scanner/oscilloscope measurement system interface which is developed by us is used). The digitized signals are then sent to IBM-PC computer through a standard IEEE-488 parallel interface for image reconstruction and the reconstructed images are transferred through a standard RS-232C serial interface to ARLUNYA TF-4000 temporal filter and image store, and, finally, displayed on the JVC high-resolution monitor.

EXPERIMENTAL RESULTS

Fig.4 shows cross-sectional figures of three agar phantoms prepared. The phantoms are even in the direction perpendicular to their transversal cross-sections. Fig.4(a) contains six small holes (one of them is of the diameter of 2 mm, and the others are of 3 mm) and is prepared from 4% agar aqueous solution. Fig.4(b) and (c) contain no hole and are prepared from 4% and 8% ager aqueous solution respectively. Fig.5 (a), (b) and (c) are the photographs of these phantoms and are corresponding to Fig.4 (a), (b) and (c) respectively.

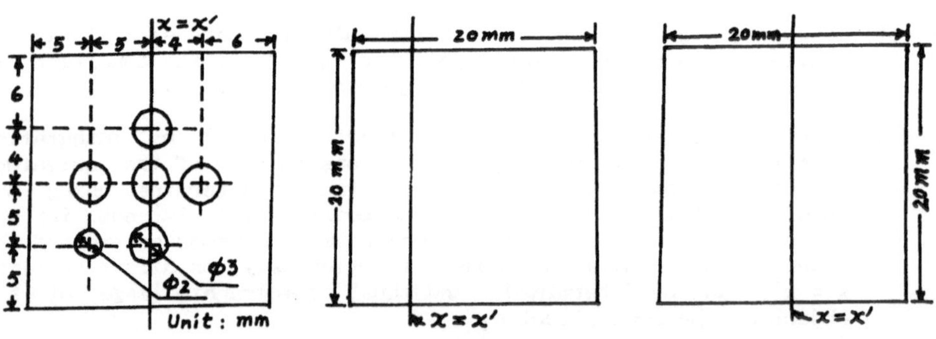

Fig.4 (a) Phantom with six holes (made of 4% agar aqueous solution), (b) and (c) Phantoms with no holes (made of 4% and 8% agar aqueous solution respectively)

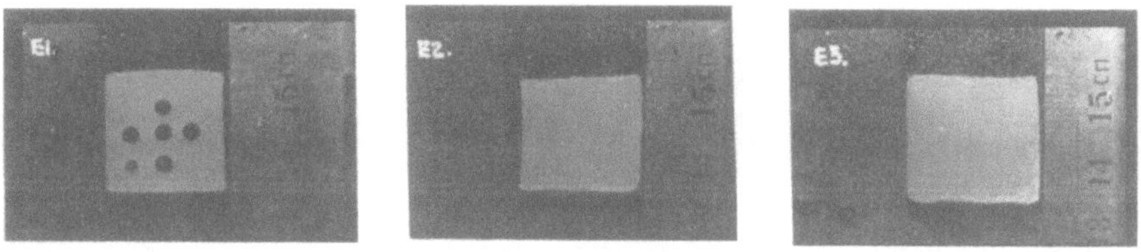

(a) (b) (c)

Fig.5 (a), (b) and (c) Photographs of the phantoms corresponding to Fig.4 (a), (b) and (c) respectively

To determine the electrical—mechanical characteristics of the transducer, a very thin silk with diameter of 0.04 λ (the center frequency of the acoustical pulses wave is about 3 MHz and the wave length λ in the biological soft-tissues is about 0.5 mm) is used as a point scatterer and is put on the focal point of the incident beam. Substituting the measured echo signals into Eq.(15), one can obtain the electrical—mechanical characteristics of the transducer.

Fig.6 (a), (b) and (c) are the images of the ordinary B-scanner and are corresponding to Fig.5 (a), (b) and (c) respectively. It is seen that these images can only provide the outline information of the internal structures of the testing objects. Fig.7 (a), (b) and (c) are the images reconstructed by the QRI method with the phantoms shown in Fig.5 (a), (b) and (c) respectively. Fig.8 (a), (b) and (c) are the comparisons of reconstructed values (real lines) of the images reconstructed by the QRI method and the real values (dashed lines) on the lines x = x' shown in Fig.4 (a), (b) and (c), respectively. Because there is an experimentally determined constant A' in Eq.(13), the reconstructed values are only of relative meaning. Here, the reconstructed values averaged on the line x = x' shown in Fig.4 (b) are taken as the reference for the comparisons. From Fig.6, Fig.7 and Fig.8, one can see that the images reconstructed by QRI method are more helpful in understanding the internal structures of the testing objects than the ordinary B-scan images, and to some extent, the images are quantitative.

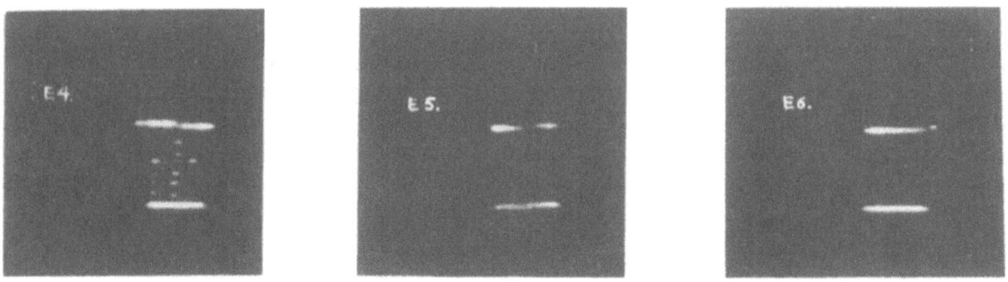

(a) (b) (c)

Fig.6 (a), (b) and (c) are the images obtained by the ordinary B-scanner and are corresponding to Fig.5 (a), (b) and (c) respectively

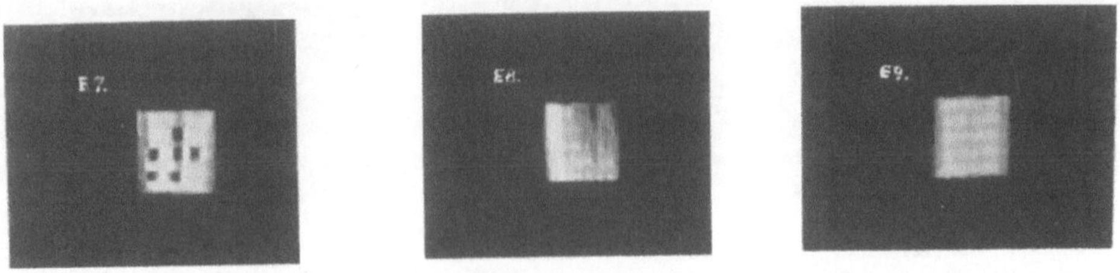

(a) (b) (c)

Fig.7 (a), (b) and (c) are the images reconstructed by the QRI method and are corresponding to Fig.5 (a), (b) and (c) respectively

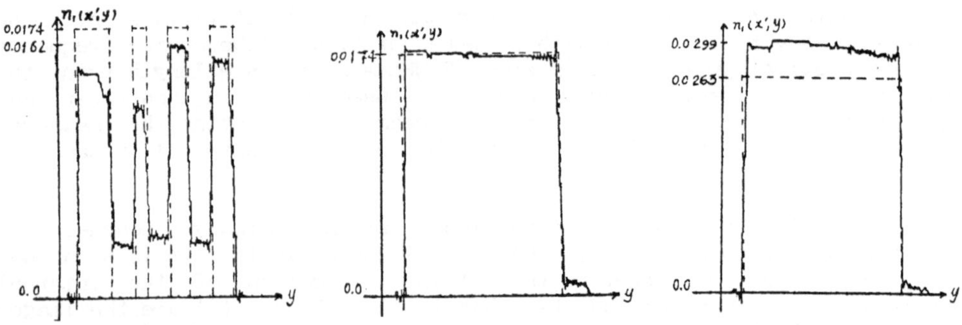

(a) (b) (c)

Fig.8 (a), (b) and (c) are the figures of the comparisons of the reconstructed values (real lines) of the images in Fig.7 and the real values (dashed lines) on the lines x = x' shown in Fig.4 (a), (b) and (c) respectively

SUMMARY

In this paper, a new quantitative reflection imaging method which employs the acoustical transmitting/receiving geometry of the ordinary B-scanner is developed. In addition to the theoretical analysis of this new imaging method, an experimental system is set up and images of the practical testing objects are reconstructed using the data obtained from this experimental system. The results show that the images reconstructed by the QRI method are more helpful in understanding the internal structures of the testing objects than the ordinary B-scan images and to some extent, the images are quantitative. Therefore, the QRI method will be useful in tissue characterization and will strengthen the abilities of the ordinary B-scanner in diagnosing the diseases.

Although the preliminary theoretical and experimental results obtained above, further researches (such as, considering the attenuation of the acoustical energy in the biological soft-tissues, taking into account the sound scattering caused by the inhomogeneities of the tissue density, improving the accuracy of the phase determination of the rf echo signals, using real time datum acquisition system and using practical biomedical soft-tissues as testing objects, etc.), must be continued if the QRI method is to be used in practical medical imaging.

REFERENCES

1. Jean-Pierre Ardouin and A. N. Venetsanopoulos, "Modelling and Restoration of Ultrasonic Phased-Array B-Scan Images", Ultrasonic Imaging, Vol.7, 1985, pp.321-344

2. M. Fatemi and A.C. Kak, "Ultrasonic B-Scan Imaging: Theory of Image Formation and a Technique for Restoration", Ultrasonic Imaging, Vol.2, 1980, pp.1-47

3. M.A. Fink and J.F. Cardoso, "Diffraction Effects in Pulse-Echo Measurement", IEEE Trans. Sonics & Ultrasonics, Vol.31, No.4, Jul.1984, pp.313-329

4. P.M. Gammell, "Improved Ultrasonic Detection Using the Analytic Signal Magnitude", Ultrasonics, Mar.1981, pp.73-76

5. G. Kossoff, "Progress in Pulse-Echo Techniques", Ultrasonics in Med., Exerpta Medica, No.309, 1974, pp.37-42

6. D.T. Kuan, A.A. Sawchuk, T.C. Strand and P. Chavel, "Adaptive Restoration of Images with Speckle", IEEE Trans. Acoustics, Speech and Signal Processing, Vol.35, No.3, Mar.1987, pp.373-383

7. M. O'Donnell, "Phase-Insensitive Pulse-Echo Imaging", Ultrasonic Imaging, Vol.4, 1982, pp.321-335

8. M. Ueda, "Computer Simulation of Artifacts in B-Mode Images", Proc. Ultrasonics Symposium, 1983, pp.718-721

9. R. Vaknine and W.J. Lorenz, "Lateral Filtering of Medical Ultrasonic B-Scans before Image Generation", Ultrasonic Imaging, Vol.6, 1984, pp.152-158

10. B. A. Roberts and A. C. Kak, "Reflection Mode Diffraction Tomography", Ultrasonic Imaging, Vol.7, 1985, pp.300-320

11. R.K. Mueller, M.Kaveh, and G. Wade, "Reconstructive Tomography and Applications to Ultrasonics" Proc. IEEE, Vol.67, No.4, 1979, pp.567-587

12. D. Nahamoo, S.X. Pan, and A.C. Kak, "Synthetic Aperture Diffraction Tomography and Its Interpolation-Free Computer Implementation", IEEE Trans. on Sonics and Ultrason., Vol.SU-31, No.4, Jul.1984, pp.218-229

13. P.M. Morse, H. Feshbach, "Methods of Theoretical Physics", McGraw-Hill, 1953

14. ALOKA Co. Ltd., Japan, "SSD-256 Training Manual".

15. HP Co. Ltd., America, "HP-1980A/B Oscilloscope Measurement System -- Operating Manual".

16. HP Co. Ltd., America, "HP-19860A Digital Waveform Storage Option -- Operating Manual".

ANALYSIS OF THE BEHAVIOR OF INTRAVENTRICULAR BLOOD FLOW IN MYOCARDIAL INFARCTION BY TWO-DIMENSIONAL VELOCITY VECTOR DISTRIBUTION

N. Endoh, M. Tanaka, A. Yamamoto, K. Takahashi,
S. Ohtsuki*, and M. Okujima*

Dept. of Med., Eng., and Cardiol., The Research Inst. for Chest Diseases and Cancer, Tohoku Univ., Sendai
*Tokyo Inst. of Technology, Yokohama, Japan

INTRODUCTION

The performance of the pump function of the heart depends on the magnitude of the force that is generated by contraction of the myocardium. The force generated in the myocardium is transmitted to the intracardiac blood through the changes of the ventricular shape and the displacement of the ventricular wall. The flow dynamic parameters are under direct influence of the myocardial function.

The two-dimensional color flow mapping technique is inadequate for the quantitation of flow dynamic parameters. For the purpose of non-invasive evaluation of local and overall myocardial functions, a new method for extracting the flow dynamic parameters of the intracardiac blood flow from the Doppler velocity data has been developed in our laboratory[1,2].

In this study, characteristics of intracardiac blood flow and local myocardial function in a case of myocardial infarction (MI) with apical aneurysm were investigated by using the new method.

METHOD

An ultrasonic pulse of 3 MHz in frequency and 4.4 kHz in repetition rate was used in this investigation. Flow velocity data was obtained from about 150 sampling points on the apical long-axis section plan, and was analyzed with FFT method in real time.

The envelope velocity data of FFT output signals in one cardiac cycle at each sampling point were put into the computer (Fig.1).

On the apical long axis section plane (x-y plane), the orientation of the x-axis agrees with that of ultrasonic beam (Fig.2).

The coordinate is denoted by (x,y,z) as following:

$$V = (u, v, w)$$

Using the components u, v and w of the velocity V on the axes x, y and z respectively, the x-component of the accelaration (Du/Dt) is expressed as following:

$$Du/Dt = \underbrace{\partial u/\partial t}_{\text{local acc.}} + \underbrace{u * \partial u/\partial x + v * \partial u/\partial y + w * \partial u/\partial z}_{\text{convective acc. = Acr}}$$

The velocity component w which is vertical to the plane is theoretically 0, because the flow is symmetric to the cross section plane (x-y plane). Accordingly, the convective acceleration is given by the under formula.

$$Acr = u * \partial u/\partial x + v * \partial u/\partial y$$

The unknown variable is v. The other variables such as u, $\partial u/\partial x$ and $\partial u/\partial y$ can be evaluated with measured Doppler velocity distribution on an observing plane.

After processing the data by applying the law of hydrodynamics (Fig.3), velocity vector (u,v) distribution is obtained.

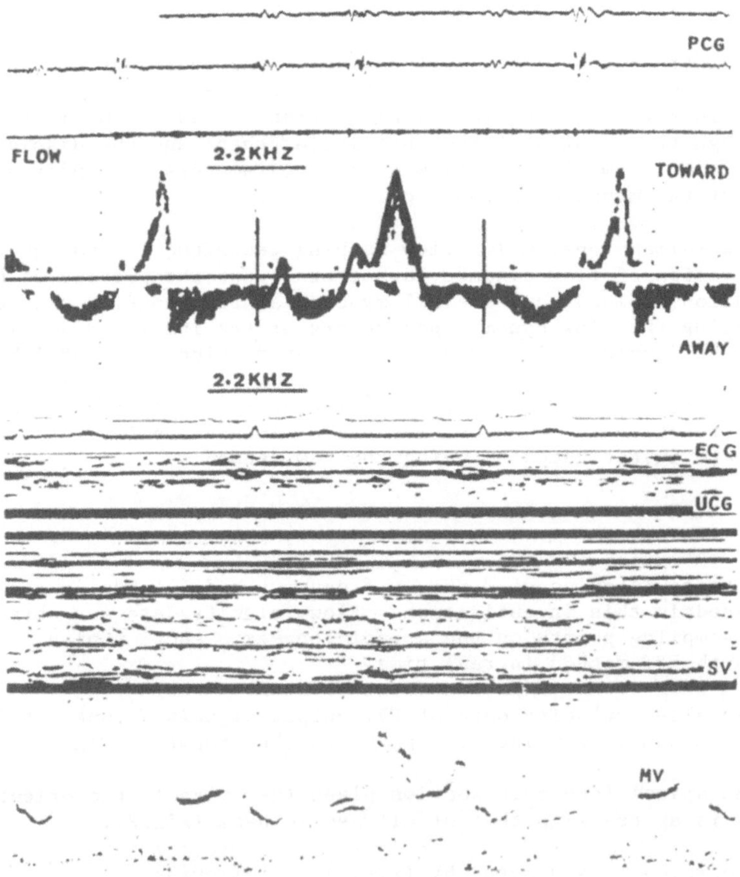

Fig.1 The envelope velocity data analyzed with FFT method in one cardiac cycle.

On the apical long axis section plane (x-y plane), the orientation of the x-axis agrees with that of ultrasonic beam.

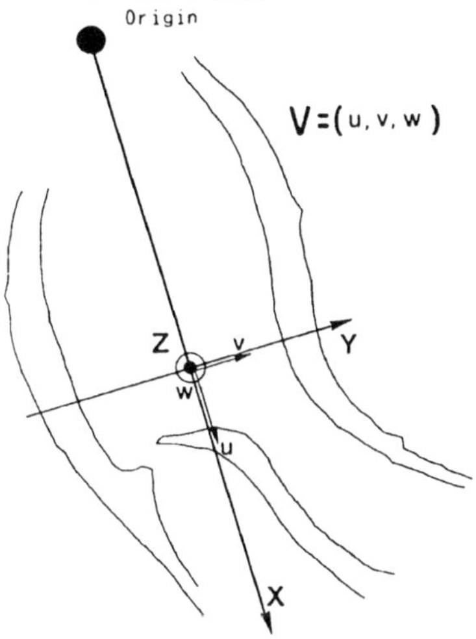

The x-component of the acceleration (Du/Dt) is expressed as following :

$$\frac{Du}{Dt} = \underbrace{\frac{\partial u}{\partial t}}_{\text{local acc.}} + \underbrace{u\frac{\partial u}{\partial x} + v\frac{\partial u}{\partial y} + w\frac{\partial u}{\partial z}}_{\text{convective acc.} = \text{Acr}}$$

If $w = 0$

$$\text{Acr} = u\frac{\partial u}{\partial x} + v\frac{\partial u}{\partial y}$$

Fig.2 A principle of the new method

$$Acr = u \frac{\partial u}{\partial x} + v \frac{\partial u}{\partial y}$$

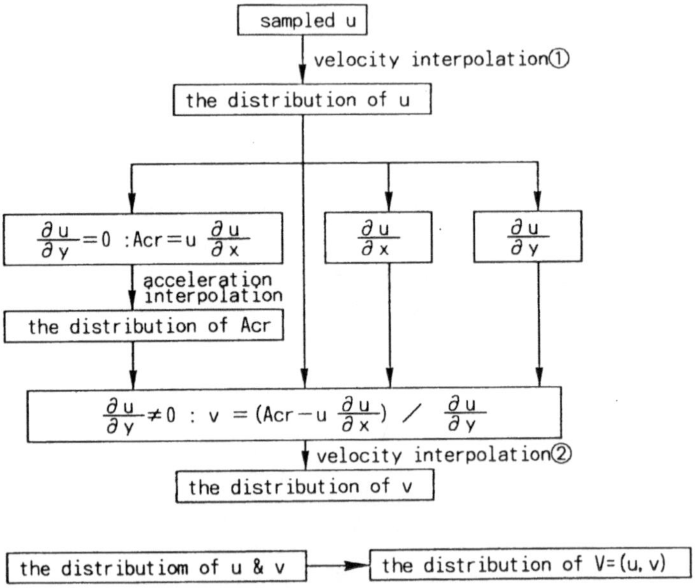

Fig.3 Processing the data by applying the law of hydrodynamics: In this way, the velocity vector (V) and its distribution by using velocity data u and v can be calculated easily.

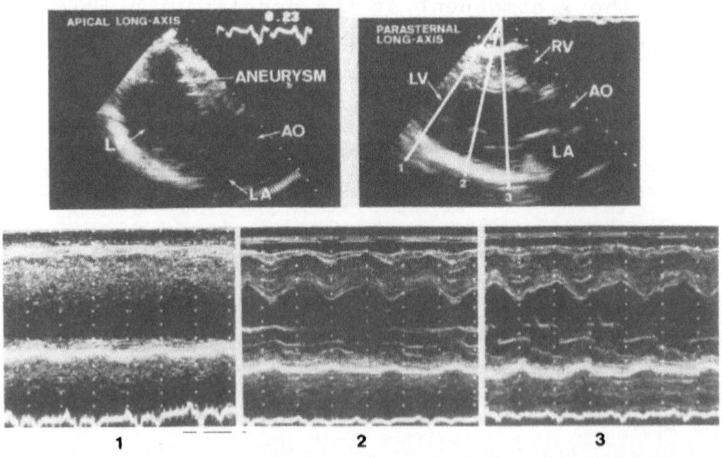

Fig.4 Wall motion: The ventricular septum shows hyperkinesis and wall movement at the basal portion of the intact free wall remains in almost normal. The movement from mid to apical portion of the free wall shows gradual decrease. At the aneurysmal portion near the apex, the movement shows dyskinesis.

RESULTS AND DISCUSSION

These panels in Fig.5,6 show the velocity vector distribution on the long-axis cross-section of the left ventricle. The length of vector indicates the absolute velocity.

On the right panels, black arrows indicate the schematic representation of the flow direction. The broken line shows the location of the ventricular aneurysm.

1) The distribution pattern of the flow vector during systole

There are two large rotating flows in the ventricle (Fig.5-2).

These separate flow are merged and become single rotating flow which gradually reduces (Fig.5-3, 5-4).

The rotating flow separated from the ejection flow is observed in the area of inflow tract (Fig.5-4, 5-5).

The facts suggest that the force produced from the intact myocardium in the basal portion of the free wall and the septum is separated into two directions. One is upward flow directed to the aorta for ejection, the other for the formation of the downward flow directed to the aneurysm.

Monotonous flow pattern observed in the aneurysm indicates that the wall of the aneurysm is not functioned for the ejection.

2) The distribution pattern of the flow vector during diastole

The intraventricular blood flow is stagnating and rotating in the aneurysm. The onset of filling of the ventricle is markedly delayed (Fig.6-2).

The inflow velocity is slow and the inflow is not observed at the apical area during the filling phase (Fig.6-3, 6-4, 6-5). The fastest inflow velocity is appeared in the atrial contraction phase. The inflow velocity in this phase is approximately 55 cm/sec (Fig.6-6).

In the apical area, the blood flow patterns show almost no change during diastole and no effects of the filling flow are observed.

The decrement of the inflow velocity at the central area suggests that the ischemic lesion decreases in extensibility and the collision of the inflow blood occurs against the stagnated blood in the aneurysm.

During the filling phase, the flow pattern is displayed monotonous pattern around the apical area, which indicates the extensibility of the damaged myocardium.

The atrial contraction compensates for the filling loss in the early stage of diastolic phase.

CONCLUSION

We demonstrate the two-dimensional distribution of flow velocity vectors in the ventricle in a case of old myocardial infarction.

The flow in aneurysm is monotonous during cardiac cycle and is separated from that in the ventricle. Some eddies and rotating flow

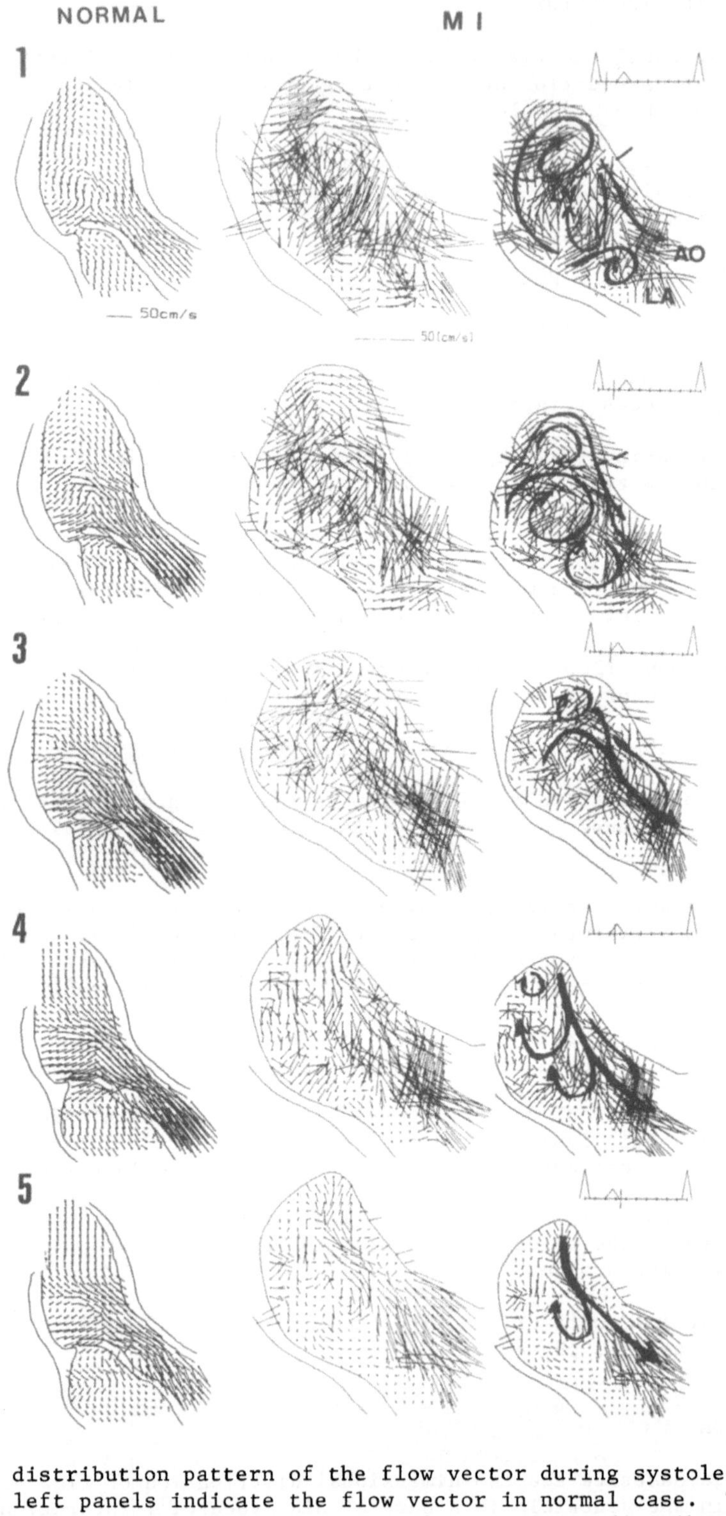

Fig.5 The distribution pattern of the flow vector during systole: The left panels indicate the flow vector in normal case. The middle and right panels shwo the velocity vector distribution in a case of myocardial infarction (MI). LV = left ventricle, LA = left atrium, AO = aorta

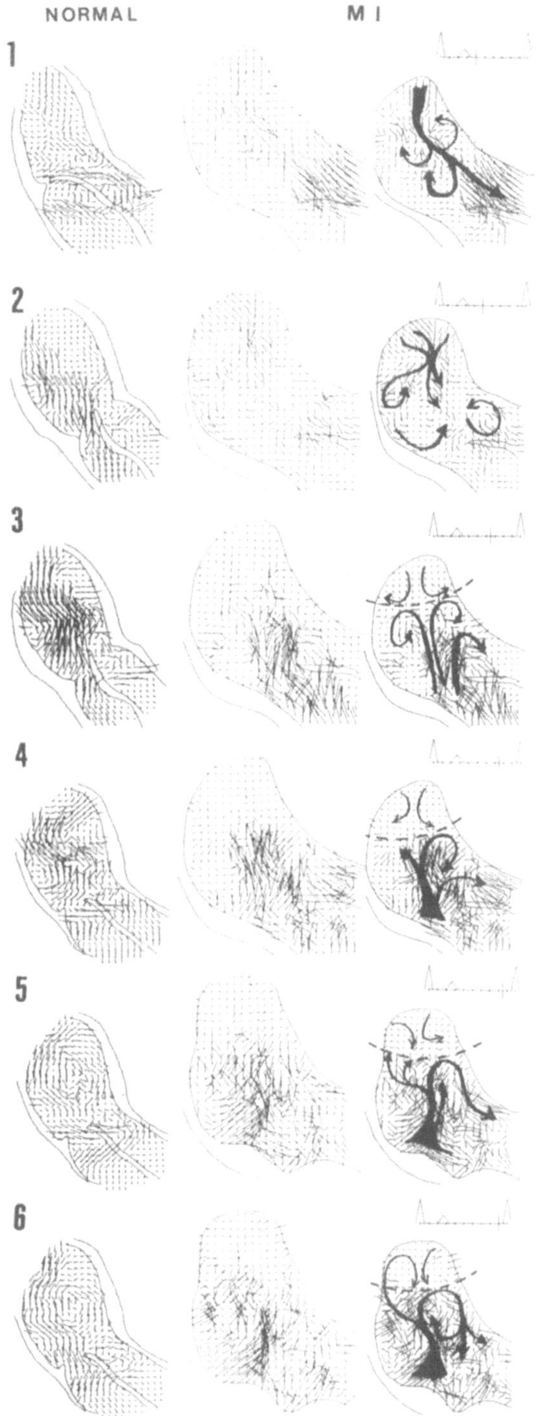

Fig.6 The distribution pattern of the flow vector during diastole

appear around the posterior wall continuously, while the flow around the ventricular septum is already normal.

Flowdynamic study of the left ventricle assessed by two-dimensional distribution of velocity vectors is very useful for comprehension of flowdynamic phenomena reflected by abnormal myocardial funcition.

REFERENCES

1) M. Tanaka et al.: Flow vector distribution patterns constructed from the informations obtained from the unidirectional tracing. Med. Ultrason. 13 (suppl. 1): 230, 1986.
2) S. Ohtsuki et al.: A method of flow vector mapping deduced from Doppler information. J.A.S.J. 43: 764, 1987.

THE FOCUSING PERFORMANCE OF ZONE PLATE TRANSDUCERS

Jing-ao Sun and Glen Wade

Department of Electrical and Computer Engineering
University of California
Santa Barbara, CA 93106, U.S.A.

ABSTRACT

Theoretical study of the focusing performance of different types of Fresnel-zone plate (FZP) transducers shows that the positive and negative patterns exhibit significantly different focusing characteristics when the number of zones is less than ten. The study also shows that for B-scan imaging, high resolution and effective sidelobe reduction can be obtained by using a single-ring annular transducer for transmitting in combination with a two-phase FZP transducer for receiving.

INTRODUCTION

In medical ultrasonic imaging, a B-scan system employing a single transducer has limited ability to gather information. Although a phased-array technique involving such transducers is capable of accomplishing some of the desired functions, such as electronic beam scanning for real-time imaging and dynamic focusing over a large depth range [1], the phased-array technique possesses basic limitations [2][3]. In order to find a better technique, we have explored the capabilities of systems using zone-plate transducers [4]-[6], or transducers with single-ring annular apertures [7]-[9]. A single annulus can be regarded as a special kind of zone plate.

The Fresnel-zone plate (FZP) has been used in such varied fields as gamma ray imaging [10] and X-ray microscopy [11]. The FZPs used in these applications have a large number of zones (15000, for example [11]). In these cases, the focusing properties of FZPs are analogous to that of refractive lenses. This is not the case in ultrasound imaging since practical limitations require that the number of zones of acoustic FZP transducers commonly used be less than 10, or even as little as 2 [5][6]. The focusing performance of FZP transducers having such a small number of zones is significantly different from that of lenses, but thorough quantitative formulations are not found in the literature. A theoretical evaluation of FZP focusing properties is given in an early paper [12], but unfortunately the results reported in that paper seem to be invalid for the conditions involved in our study here [13][14].

It is well known that a thin annular transducer is capable of providing super resolution over a large depth. The main drawback preventing an annular transducer from being widely employed in imaging is the existence of large sidelobes in its radiation pattern. A technique called J^2-synthesis [7][8] and some other approaches [17] have been proposed for sidelobe reduction, however these methods are unduly complicated.

In this paper, we present a theoretical analysis and various computations to predict the focusing performance of FZP transducers. We show that positive and negative FZP transducers have significantly different focusing characteristics when the number of zones is small. We also show that since B-scan systems operate in a transmitter/receiver mode, high resolution and low sidelobes may be obtained by using a single-ring annular transducer for transmitting in combination with a two-phase FZP transducer for receiving.

RESOLUTION OF FZP TRANSDUCERS

Three zone-plate patterns for FZP transducers are shown in Fig. 1. The transducer (a) with an active central zone is called a positive FZP (P-FZP) transducer, and its opposite, (b), a negative FZP transducer (N-FZP). In our later discussion, the number of zones, N, of an FZP is counted in a way to include both positive and negative zones. Therefore, a negative FZP having only 2 zones is actually an annulus. By referring to Fig. 1(a), we can see that the number of zones of a P-FZP transducer can only be odd because it is meaningless to count an additional zone outside the outermost active zone. For the same reason, the number of zones of an N-FZP transducers is always even. In addition, an acoustic transducer, as opposed to the optical FZP, can be operated so as to have both the positive and the negative zones active. One way to do this, shown in Fig. 1(c), is to use two signals having the same frequency but with a 180° phase difference so as to separately excite the positive and negative zones. This transducer is referred as a two-phase FZP (T-FZP) transducer.

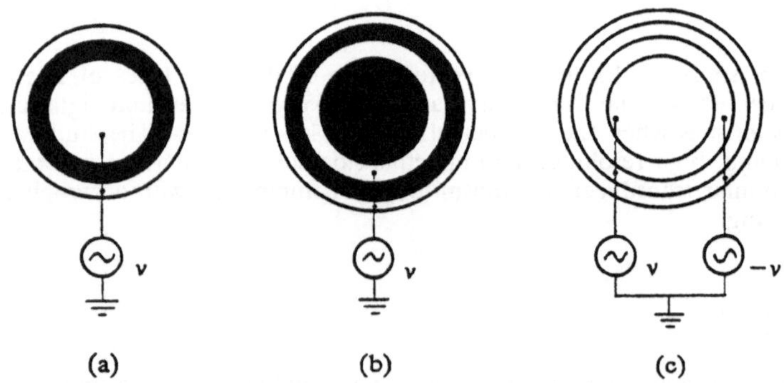

Fig.1 (a) P-FZP, (b) N-FZP and (c) T-FZP transducers. The active zones with *ac* signal excitation are shown in white.

Assuming that the wave emitted from an FZP transducer is a geometric extension of the transducer face, the spatial amplitude distribution of the radiated sound can be found by applying the Fresnel diffraction formula [16]

$$u(x_o, y_o, z) = \frac{exp(jkz)}{j\lambda z} \int\int_{-\infty}^{\infty} u(x_1, y_1) exp\left\{j\frac{k}{2z}[(x_o-x_1)^2+(y_o-y_1)^2]\right\} dx_1 dy_1 \qquad (1)$$

where

$$u(x_1, y_1) = \begin{cases} u_o & \text{over active zones} \\ 0 & \text{otherwise} \end{cases} \qquad (2\text{-a})$$

or

$$u(x_1, y_1) = \begin{cases} u_o & \text{over positive zones} \\ -u_o & \text{over negative zones} \end{cases} \qquad (2\text{-b})$$

are transmitted waves by P- (or N-)FZP transducers or T-FZP transducer respectively, $k=2\pi/\lambda$ is the wave number and λ is the wavelength of the sound in the medium. The time dependence of $u(x_1, y_1)$ has been omitted by conventional understanding. Fig. 2 depicts the coordinate system involved.

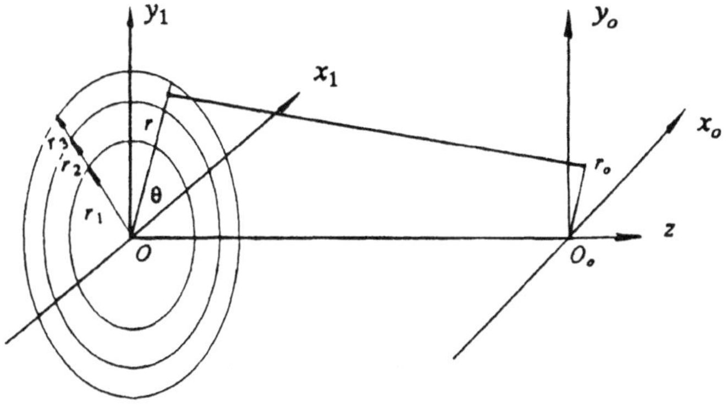

Fig.2 Coordinate system involved in Eq. (1).

Since an FZP transducer is circularly symmetrical, its radiation pattern must also be circularly symmetrical. We therefore lose nothing if we confine our observation to, for example, the line $x_o=0$. Also because of circular symmetry, the above integration can be conveniently performed using polar coordinates. Let $x_1=r\cos\theta$, $y_1=r\sin\theta$, $x_o=0$ and $y_o=r_o$. Eq. (1) then can be simplified to

$$u(r_o, z) = \frac{2\pi exp(jkz)exp(j\frac{\pi}{\lambda z}r_o^2)}{j\lambda z}\int_0^\infty u(r)rJ_o(\frac{2\pi r_o}{\lambda z}r)exp(j\frac{\pi}{\lambda z}r^2)dr \tag{3}$$

where

$$J_o(\frac{2\pi r_o}{\lambda z}r) = \frac{1}{\pi}\int_0^\pi \cos(\frac{2\pi r_o r}{\lambda z}\sin\theta)d\theta$$

is a zero-order Bessel function of the first kind. For simplicity, it is presumed that the width of separation between two adjacent zones of the transducer is negligible. Thus, when substituting Eq. (2) into Eq. (3), the integral limits are related to the FZP parameter by [17]

$$r_n = (n\lambda f)^{1/2} \tag{4}$$

where r_n and f are the radius of the nth zone and the focal length of the FZP respectively. Then, the spatial intensity distribution can be expressed as

$$I(r_o, z) = u(r_o, z)u^*(r_o, z) \tag{5}$$

where the asterisk designates a complex conjugate.

Let us define the focal-plane response of a transducer, $H(r_o, f)$, as the intensity distribution normalized by its value at the center of the focal plane, i.e.

$$H(r_o, f) = \frac{I(r_o, f)}{I(0, f)} \tag{6}$$

The intensity values at $r_o=0$ for different FZP transducers are derived in the appendix as

$$I(0, f) = \begin{cases} (N+1)^2u_o^2 & \text{for } \textit{FZP transducers } (N=odd) \\ N^2u_o^2 & \text{for } \textit{FZP transducers } (N=even) \\ 4N^2u_o^2 & \text{for } \textit{two-phase FZP transducers} \end{cases}$$

The lateral resolution, δ, of an FZP transducer can be defined as the distance between the center and the first minimum of $H(r_o, f)$.

By using Eqs. (3) to (6), the resolution of P-FZP, N-FZP and T-FZP transducers can be computed and the results are plotted in Fig. 3 as functions of the zone number N. For easy comparison, the ordinate of this figure is designated as the resolution value, δ, normalized by $\lambda f/D$, i.e. $D\,\delta/(\lambda f)$, to cancel the physical parameter dependence. D is the diameter of the FZP transducer. For example, the resolution of a lens is commonly known as $\delta = 1.22\ \lambda f/D$, so the normalized resolution is 1.22 (shown by the dashed line in Fig. 3 for comparison). The relationship of the first sidelobe levels and zone number N, for the three types of transducers is illustrated in Fig. 4. Figs. 3 and 4 show that the P-FZP transducer has diminished sidelobes but coarser resolution. The N-FZP transducer has sharper resolution but larger sidelobes. The resolution and sidelobe levels of T-FZP transducers are about the same as for a lens. Clearly, the T-FZP transducer provides a more acceptable compromise between resolution and sidelobe level when the number of zones is small (4 to 6, for example).

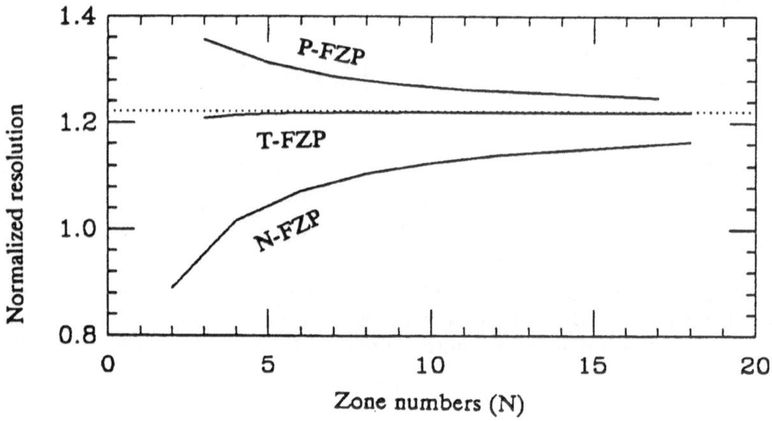

Fig.3 Resolution of P-FZP, N-FZP and T-FZP transducers *vs* number of zones (N). The dashed line shows the resolution of the lens.

SYSTEM PERFORMANCE WITH DIFFERENT TRANSDUCER COMBINATIONS

In a B-scan system the object plane is defined by the transducer axis and the scanning direction, rather than being orthogonal to the axis. The resolution in the scanning direction is poorer for structures both nearer to and further from the transducer than the focal length of the transducer. In order to ensure adequate lateral resolution over the desired depth range, it is important to design the system to have as great a depth of focus as possible. For this, a transducer having a thin annular aperture is particularly attractive because it is capable of providing high resolution over a large depth of focus. The major problem encountered in adapting this device to acoustic imaging is due to the big sidelobes.

Since an ultrasound B-scan system operates in a transmit/receive mode, the spatial amplitude response of the system for a point reflector is the product of the amplitude response of

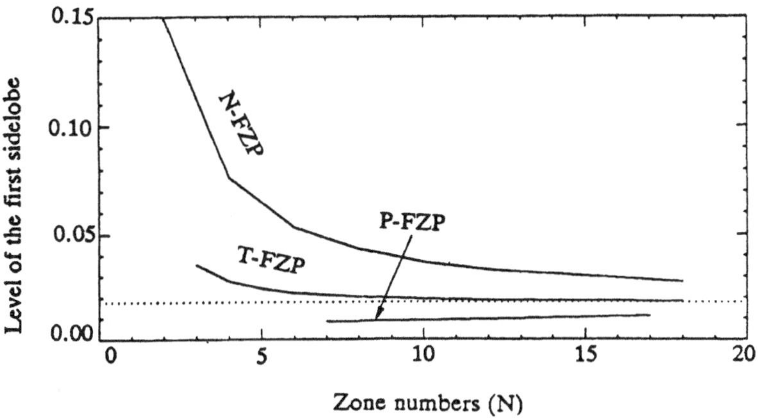

Fig.4 The first sidelobe level for P-FZP, N-FZP and T-FZP transducers *vs* number of zones (N). The dashed line shows the first sidelobe level of the lens.

the transmitter and that of the receiver. This suggests an opportunity to achieve better performance by combining an annulus with another type of transducer to form a transmitter/receiver pair [17]. The problem here is to find a good partner for the annulus.

As we have seen from Fig. 3, a P-FZP transducer has the lowest sidelobe level. This immediately suggests combining a P-FZP with an annular transducer. In practice, this can be done by using the outermost zone of a P-FZP transducer as an annular transducer only for receiving and the entire transducer for transmitting, or *vice versa*.

Let $u_a(r_o, z)$ and $u_p(r_o, z)$ be the sound amplitude distribution radiated by the outermost ring (annulus) and the P-FZP transducer respectively. The round-trip amplitude response of the imaging system will be $u_a(r_o, z)u_p(r_o, z)$. Thus, the spatial intensity response of the system can be written as

$$H_{a/p}(r_o, z) = \frac{|u_a(r_o, z)u_p(r_o, z)|^2}{|u_a(0, f)u_p(0, f)|^2} = H_a(r_o, z)H_p(r_o, z) \tag{7}$$

where $H_a(r_o, z)$ and $H_p(r_o, z)$ are the intensity distributions for the annulus and the P-FZP transducers respectively. If this P-FZP were time-multiplexed between transmission and reception, the system response would be $H_p^2(r_o, z)$, rather than that shown by Eq. (7).

Other possible combinations are annulus/N-FZP and annulus/T-FZP. The system responses for these two combinations are $H_{a/n}(r_o, z)$ and $H_{a/t}(r_o, z)$ respectively. Fig. 5 shows the system response at the focal plane for the three combinations, i.e. $H_{a/p}(r_o, f)$, $H_{a/n}(r_o, f)$ and $H_{a/t}(r_o, f)$. For comparison, the response of a system with an annulus/annulus pair, $H_a^2(r_o, f)$, is plotted in Fig. 5(a). In all cases, the number of zones, N, are 9, and the outermost ring is used as an annular transducer. Clearly, high resolution and effective sidelobe reduction are achieved with annulus/P-FZP and annulus/T-FZP combinations and, in this particular example, the latter is more effective than the former.

Now let us examine the depth of focus, defined as being the distance between the two points on a plot of system spatial response along the transducer axis where the intensity has dropped by 6 *db* from its peak value. $H_{a/p}(0, z)$, $H_{a/n}(o, z)$ and $H_{a/t}(o, z)$ are depicted in Fig. 6. The -6 *db* level is indicated by the dashed lines. Figs. 5 and 6 show that the annulus/T-FZP combination provides good sidelobe reduction and adequate depth of focus.

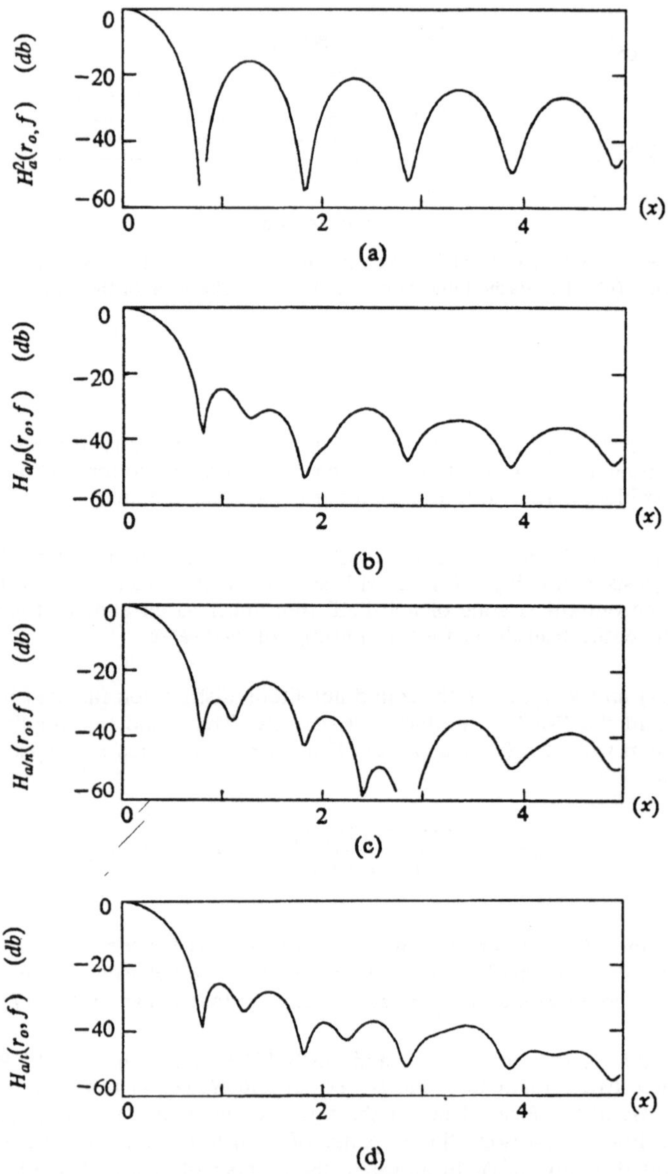

Fig.5 System response in the focal plane for (a) the single annulus transducer, (b) the single annulus/P-FZP combination, (c) the single annulus/N-FZP combination and (d) the single annulus/T-FZP combination. The abscissa in each case is $x=Dr_o/(\lambda f)$.

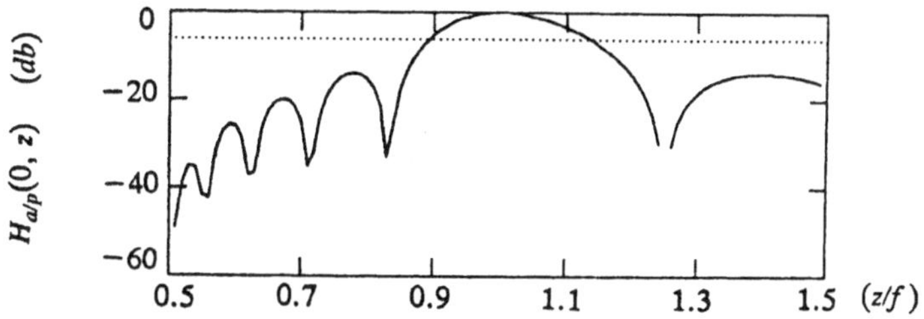

(a)

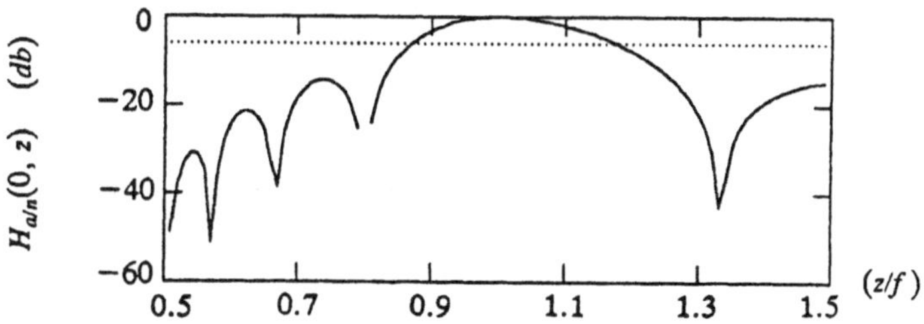

(b)

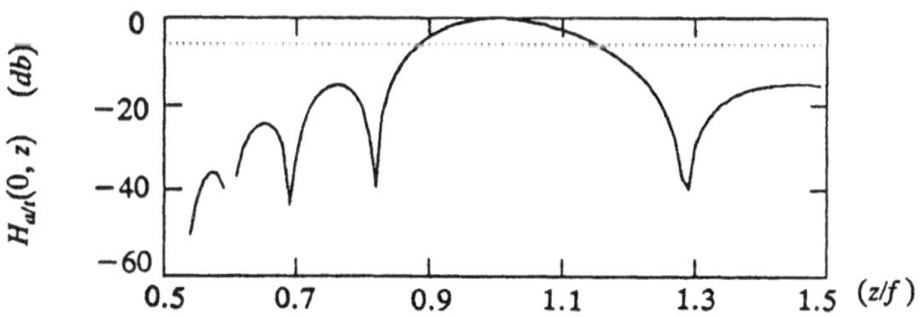

(c)

Fig.6 System response in the axial direction for transducer combinations (a) annulus/P-FZP, (b) annulus/N-FZP and (c) annulus/T-FZP. The abscissas are the distances to the transducer plane normalized by focal length *f*. The dashed line indicates the $-6db$ level.

Since the positions of the maxima in Fig. 5(a) are functions of the annulus diameter D, it is possible to further reduce the sidelobes if the suitable ring is chosen from the FZP transducer (it is not necessarily the outermost one) so that the the sidelobe maxima more closely coincide with the minima in its partner's focal plane response.

CONCLUSION

We have shown that the focusing characteristics of P- and N-FZP transducers are significantly different from each other when the number of zones is small. High resolution, adequate depth of focus and effective sidelobe reduction can be achieved by using an annular transducer combined with some kind of FZP transducer. Sidelobe reduction may be further enhanced by the correctly selecting the most appropriate ring of the FZP transducer to be used as the annulus for reception.

REFERENCES

[1] H. Lee and G. Wade, Editors, *Modern Acoustic Imaging*, IEEE Press (1986).

[2] J.F. Haflice and J.C. Tasnzer, "Medical Ultrasonic Imaging: An Overview of Principles and Instrumentation," *Proc. of the IEEE*, 67:7-28 (1979).

[3] P.N.T. Wells, and M.C. Ziskin, *New techniques and Instrumentation in Ultrasonography*, Churchill Livingstone, New York (1980).

[4] K. Wang and G. Wade, "A Scanning Focused-Beam System for Real-Time Diagnostic Imaging," *Acoustic Holography*, Ed. O.N. Booth, Plenum Press, New York, 6:213-228 (1975).

[5] D.R. Diets, S.I. Porks and M. Linzer, "Expanding-Aperture Annular Array," *Ultrasonic Imaging*, 1:56-75 (1979).

[6] M. Arditi, W.B. Taylor, F.S. Foster and J.W. Hunt, "An Annular Array System for High Resolution Breast Echography," *Ultrasonic Imaging*, 4:1-31 (1982).

[7] D. Vilkomoerson, "Acoustic Imaging with Thin Annular Apertures," *Acoustic Holography*, Ed. P.S. Green, Plenum Press, New York, 5:283-317 (1974).

[8] C.B. Burckhardt, P. Grandchamp and H. Hoffmann, "Focusing Ultrasound over a Large Depth with an Annular Transducer _ An Alternative," *IEEE Trans. on Sonics and Ultras.*, 22:11-15 (1975).

[9] M.S. Patterson and F.S. Foster, "The Improvement and Quantitative Assessment of B-Mode Images Produced by An Annular Array/Cone Hybrid," *Ultrasonic Imaging*, 5:195-213 (1983).

[10] H.H. Barrett and F.A. Horigan, "Fresnel Zone Plate Imaging of Gamma Rays; Theory," *Applied Optics*, 12:2686-2702 (1973).

[11] B. Niemann, G. Schmahl and D. Rudolph, "X-Ray Microscopy: Recent Developments and Practical Applications," *SPIE*, 368:2-8 (1982).

[12] D.J. Stigliani, R. Mittra, R.G. Semonin, "Resolving Power of a Zone Plate," *J.O.S.A.*, 57:610-613 (1967).

[13] J. Sun and G. Wade, "A Study of the Focusing Properties of Fresnel Zone Plate," to be submitted to J.O.S.A..

[14] G.S. Waldman, "Variations on the Fresnel Zone Plate," *J.O.S.A.*, 56:215-218 (1966).

[15] A. Macovski and S.J. Norton, "High-Resolution B-Scan System Using A Circular Array," *Acoustic Holography*, Ed. N. Booth, Plenum Press, New York, 6:121-143 (1975).

[16] J. Goodman, *Introduction to Fourier Optics*, Mcgraw-Hill, New York (1968).

[17] Sommerfeld, *Optics*, Academic Press, New York, Chapter 5 (1954).

APPENDIX: Intensities at the Foci of FZP Transducers

Substituting $r_o=0$ and $z=f$ into Eq. (3), the wave amplitude at the focus of an FZP transducer is calculated by

$$u(0,f) = \frac{2\pi exp(jkf)}{j\lambda f}\int_0^\infty ru(r)exp(j\frac{\pi}{\lambda f}r^2)dr \tag{A-1}$$

Let $v=r^2$ and $dv=2rdr$, then Eq. (A-1) can be changed to the form of

$$u(0,f) = \frac{\pi exp(jkf)}{j\lambda f}\int_0^\infty u(v)exp(j\frac{\pi}{\lambda f}v)dv \tag{A-2}$$

Correspondingly, by setting $v_n=r_n^2$, Eq. (4) is changed to

$$v_n=n\lambda f$$

Thus, for the P-FZP transducer, the integration of Eq. (A-2) is carried out from 0 to λf, from $2\lambda f$ to $3\lambda f$, ..., and from $(N-1)\lambda f$ to $N\lambda f$:

$$\begin{aligned} u_p(0,f) &= \frac{\pi exp(jkf)u_o}{j\lambda f}\frac{1}{j\pi/(\lambda f)}[(e^{j\pi}-1)+(e^{j3\pi}-e^{j2\pi})+\cdots+(e^{jN\pi}-e^{j(N-1)\pi})] \\ &= (N+1)u_o e^{jkf} \qquad (N=odd) \end{aligned} \tag{A-3}$$

Similarly, carring integration from λf to $2\lambda f$, $3\lambda f$ to $4\lambda f$, ... and $(N-1)\lambda f$ to $N\lambda f$ for an N-FZP and, for a T-FZP transducer, over both positive and negative zones and with the excitation signals shown by Eq. (2-b), the wave amplitudes at foci are respectively obtained as

$$u_n(0,f) = -Nu_o e^{jkf} \qquad (N=even) \tag{A-4}$$

and

$$u_t(0,f) = 2Ne^{jkf} \qquad (N=odd\ or\ even) \tag{A-5}$$

In (A-4), as expected, the phase of the sound wave is reversed compared with that in (A-3). The wave intensities at the foci can be easily written as

$$I(0,f) = \begin{cases} (N+1)^2u_o^2 & \text{for FZP transducers } (N=odd) \\ N^2u_o^2 & \text{for FZP transducers } (N=even) \\ 4N^2u_o^2 & \text{for two-phase FZP transducers} \end{cases} \tag{A-6}$$

HIGH RESOLUTION ULTRASONIC IMAGING USING A LARGE APERTURE ANNULAR ARRAY TRANSDUCER OF P(VDF-TrFE) COPOLYMER

N. Hashimoto, T. MIya, K. Yoneya, A. Ando, and H. Ohigashi*

Engineering Research Laboratories, Toray Industries, Inc.
Sonoyama, Otsu 520, Japan
* Department of Polymer Materials Engineering, Faculty of Engineering, Yamagata University, Jonan, Yonezawa 992, Japan

ABSTRACT

A 7.5 MHz concave annular array ultrasonic transducer comprising a piezoelectric film of vinylidene fluoride and trifluoroethylene copolymer, and its driving system has been developed for high resolution medical imaging. The transducer is composed of eight annular elements of equal areas and has a large aperture (70 mm) and a long radius of curvature (180 mm). With phase controlled electric driving pulses, the focal length can be varied dynamically from 150 to 200 mm. The transducer is scanned mechanically in a sector mode in a water bag container to which the objects to be examined are contacted. The acoustic field distribution, pulse waveform, and frequency response characteristics are analysed experimentally and theoretically. The consistency is quite satisfactory. The lateral resolution obtained is 0.5-1.0 mm over the focal zone, and the depth resolution is 0.3 mm. Clear and fine echotomographic images of the breasts and thyroid glands can be obtained with this imaging system.

INTRODUCTION

In ultrasonic imaging systems for medical diagnosis and non-destructive testing, the imaging quality relies largly on performance of transducers, that is, on sensitivity, impulse response characteristics, and the capability of focusing the acoustic beam over the depth of exploration. Although an imaging system equipped with a fixed focus, single element transducer is low cost and provides high quality images in a small region about the focal point, the image quality becomes poor outside the focal region. This tendency becomes more serious when these transducers have short focuses and high working frequencies, despite such transducers are necessary for high resolution imaging. Performance limitations for a fixed focus transducer can be overcome by an electrically focused annular array transducer whose focal points can be shifted along its axis by controlling the timing of electric pulses to drive the annular elements (dynamic foucussing). However, complexity in structure of an annular array transducer of piezoelectric ceramics have prevented hitherto extensive availability in medical imaging.

Recent studies on piezoelectric polymers have revealed that they are usable as effective ultrasonic transducing material, and are even the most suitable for transducers used in medical ultrasonics [1,2]. The reason is as follows: (1) their acoustic impedance is close to that of water or body tissues, which enables us to fabricate transducers operating in a broader bandwidth, or transmitting shorter pulses as compared with ceramic transducers; (2) they are flexible enough to be fabricated into transducers having complex shapes; (3) when they are used in array transducers, they are not necessary to be separated mechanically into array elements, because acoustic and electrical interference between adjacent elements are weak even in a continuous film; and (4) the area of piezoelectric film is practically unlimited, and we may construct transducers with large active areas.

We have shown that P(VDF-TrFE), copolymer of vinylidene fluoride and trifluoroethylene, has an elecromechanical coupling coefficient k_t as high as 0.3, the value being largest among piezoelectric polymers ever reported [3]. Using P(VDF-TrFE) piezoelectric film, we developed various ultrasonic transducers; single element concave transducers and linear array transducers for medical imaging [4,5,6], and very high-frequency transducers for scanning acoustic microscopy [7,8].

In this paper, we describe a medical ultrasonic imaging system using an concave annular array transducer of P(VDF-TrFE) film which has a large aperture and a large radius of curvature. With this transducer, we have succeeded in obtaining clear, high resolution B-mode images, which we believe to have the best quality ever attained.

DESCRIPTION OF THE ANNULAR ARRAY TRANSDUCER

A concave polymer transducer comprising five-element annular array has recently reported by Miyake et al. [9]. However, since their transducer has a small aperture (25 mm) and short focus (75 mm), the focal zone is restricted within a rather narrow region. To achiver higher resolution power over a wider focal zone, we have to employ an annular array transducer having larger aperture and a longer focal length. Piezoelectric polymer of P(VDF-TrFE) is presently the best material to achieve this purpose. Figure 1 shows the structure of the annular array transducer we have designed. The transducer is composed of eight annular backing electrodes with equal areas, a single sheet of P(VDF-TrFE) film 60 μm thick, and a front (ground) electrode, which is covered with a water resistive polymer film. The aperture is 70 mm, and the radius of curvature is 180 mm. The backing electrodes glued on a concave plastic substrate were processed from a thin (100 μm) copper plate using a chemical etching technique. The gap between adjacent electrodes is less than 0.2 mm. The transducer operates at a center frequency of 7.5 MHz in water. An inductor of 1 μH is inserted in series between each annular element and the source for closer electrical matching between the driving sources and the transducer elements. When eight annular elements are pulse-drived at the same time the acoustic waves from these transducer elements are of course focused on the geometrical focal point (the center of curvature), but the focal point can be varied on the transducer axis by driving each annular element with an electric pulse with appropriate delay time through an electromagnetic delay line which is selectable from a group of delay lines by electronic switches.

Figure 2 is a photograph showing the annular array transducer we have developed and, for comparison, a P(VDF-TrFE) single element concave transducer with a fixed focal length of 75 mm and an aperture of 25 mm. The latter is now used practically for ultrasonic diagnosis in many medical institutes.

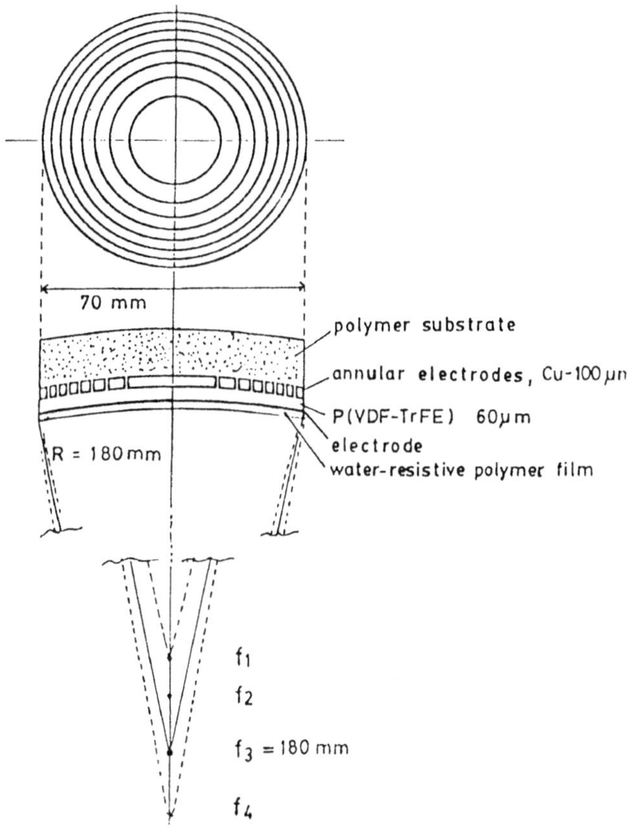

Fig. 1 Structure of P(VDF-TrFE) concave annular array transducer designed for high resolution ultrasonic imaging.

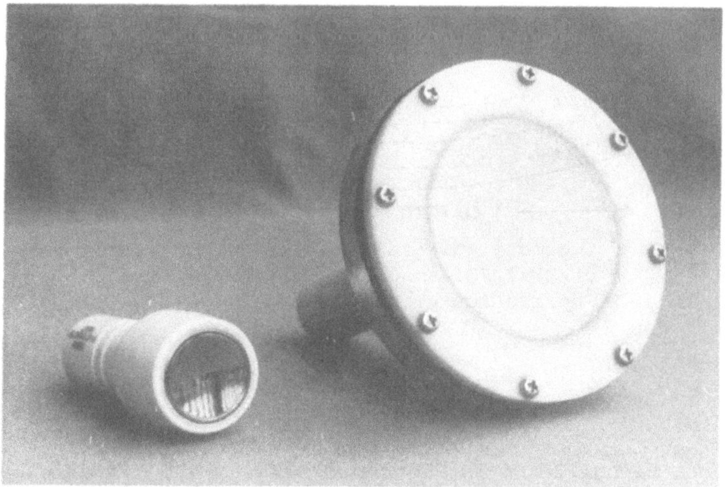

Fig. 2 The P(VDF-TrFE) concave annular array transducer (right) and a P(VDF-TrFE) fixed focus transducer (25 mmϕ)(left).

SCANNINNIG MECHANISM AND IMAGING SYSTEM

The annular array transducer is scanned mechanically in a sector plane by varing the angle θ as shown in Fig. 3. The transducer is immersed in a water bag to which the object to be examined is contacted. The center of the sector motion is at 90 mm behind the front surface of the transducer. In the present imaging system, we set four dynamic focal points. Thus, for every scanning line (at every 0.055° in θ), the transducer transmits four acustic pulses in series which are focused on these four points. Transducer output signals induced by acoustic waves scattered backwards from these four focal zones are stored in IC memories, and a resultant sector B-mode image

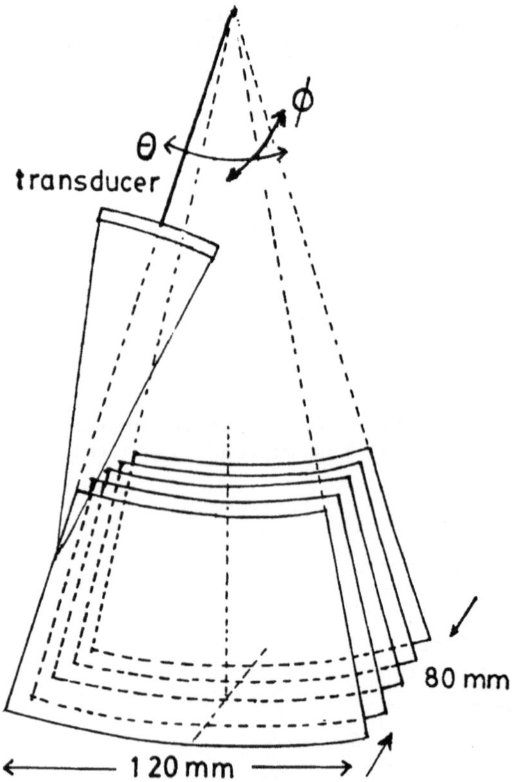

Fig. 3 Schematic diagram showing sector planes of P(VDF-TrFE) annular array transducer. The transducer is scanned in a water bag.

(exploring area being 120 mm in width x 80 mm in depth) is displayed on a CRT. The scanning time required for one B-mode image is 4s. (1s is requied for a scanning mode to survey rapidly a region to be interested, in which mode the focus is fixed on one of the four focal pionts.) A B-mode image of a desired sector plane is accessible by deflecting the transducer in the direction ϕ using a stepping motor.

Table I Material Constants of P(VDF-TrFE) [3]

Density ρ, $10^3 kg/m^3$	1.90
Sound velocity v, km/s	2.37
Acoustic impedance Z, 10^6 $kg/m^2 \cdot s$	4.49
Electromechanical coupling factor k_t	0.285
Stiffness constant c, GPa	10.6
Dielectric constant $\varepsilon/\varepsilon_o$	4.0
Mechanical loss tangent $\tan \delta_m$	0.034
Dielectric loss tangent $\tan \delta_e$	0.13

PERFORMANCE OF THE ANNULAR ARRAY TRANSDUCER

Frequency and Pulse Response

To evaluate correctly the frequency dependence of efficiency of a transducer comprising lossy piezoelectric material, we have to adopt three different losses in the transducer. These are the matching loss ML, conversion loss CL, and transducer loss TL, which are defined by ML = $-10\log(P_t/P_o)$, CL = $-10\log(P_a/P_t)$, and TL = $-10\log(P_a/P_o)$ = ML + CL. In these expressions, P_o is the maximum electric power available from the electric soruce, P_t the electric power transferred to the transducer, and P_a the acoustic power transmitted into water. ML is evaluated from the electric impedance of the transducer in water, and CL can be determined by comparing the driving voltage and the output voltage induced by reflected waves from a perfect reflector in water.

Figure 4 shows the measured frequency dependence of CL and ML and those calculated theoretically for the transducer element located at the center of the annular array. The calculated CL and ML were obtained using a Mason's equivalent circuit [10] and the material constants of P(VDF-TrFE) shown in Table I [3]. In this calculation, we introduced complex quantities, $\varepsilon^* = \varepsilon(1-j\tan\delta_e)$ and $c^* = c(1+j\tan\delta_m)$, into Mason's equivalent circuit as the dielectric constant and the stiffness constant, respectively, to fit for a transducer composed of piezoelectric polymer having large dielectric and mechanical loss factors. The measured CM and ML are in good agreement with the calculated ones. Owing to the matching inductor, the bandwidth of the transducer becomes narrower, but TL decreases considerably as compared the transducer element without the matching inductor. The bandwidth may be widened if we use a quarter wavelength acoustic matching layer of polymer film attached to the piezoelectric polymer film [11].

The resultant pulse response waveform when all of the annular elements are excited is shown in Fig. 5(a). Figure 5(b) shows the pulse response waveform calculated for the center element of the array transducer by Fourier transformation of its frequency response of TL. It can be seen that the measured and calculated pulse response waveforms are also in good agreement with each other. The pulse width defined as the duration between the times at which the signal amplitude rises and decays to 1/10 of the peak voltage, is about 0.35 us. This value corresponds to a depth resolution of 0.26mm, which is sufficiently fine for high resolution imaging.

Spacial Distribution of Acoustic Field

The spacial intensity distribution I(u,v,w) of the pulse acoustic field from the annular array transducer was measured on several dynamic focal

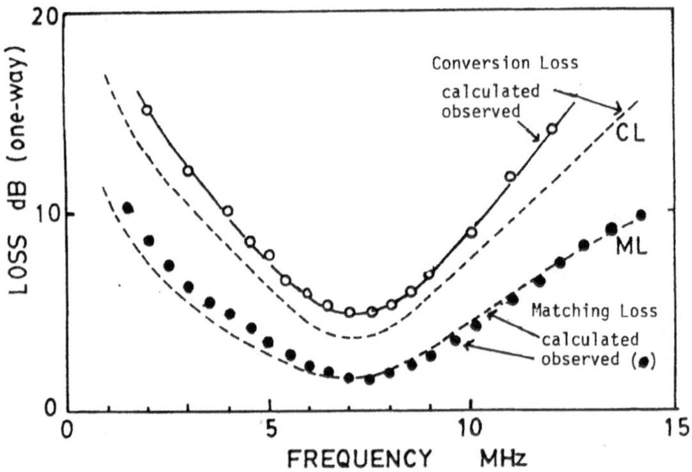

Fig. 4 Observed and calculated conversion loss (CL) and matching loss (ML) for one of the annular elements of P(VDF-TrFE).

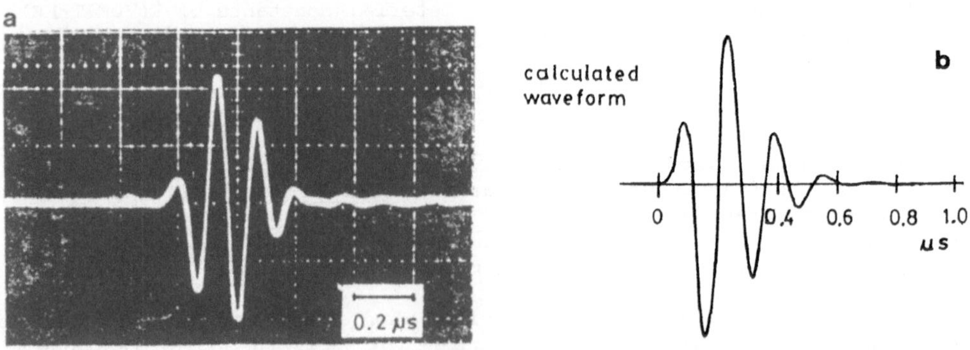

Fig, 5 (a) Observed pulse response waveform of the P(VDF-TrFE) annular array transducer.
(b) Calculated pulse response waveform of one of the P(VDF-TrFE) annular array.

planes. The field intensity was detected with a small hydrophone of P(VDF-TrFE) film coated on a semispherical metal electrodes. The result is shown in Fig. 7. The results was compared with the intensity distribution of continuous field calculated with eq.(1),

$$I(u,v,w) = |A(u,v,w)|^2 = \left| A_o \sum_{i=1}^{8} \int_{S_i} \frac{\exp[j\omega(\tau_i + r_i/c)]}{r_i} dS_i \right|^2 . \qquad (1)$$

In eq. (1), A is the amplitude of the acoustic field, ω is the angular frequency, τ_i is the delay time of driving voltage for the i-th annular element, r_i is the distance from a point (x,y,z) on the i-th element S_i to the observation point (u,v,w), and c is the sound velocity in water. The measured field distribution is fairy in good agreement with the calculated one for all electronically focal planes as shown in Fig. 6. Small difference in calculated and measured field distributions may arise from (1) geometrical deviation from spherical surface in the concave transducer, (2) the differnce in waveforms used (the one is pulse waves and the other is continuous waves), and (3) finite dimension of the aperture of the hydrophone. Neverthless, high lateral resolution can be achieved: -6db beamwidth at focal points is about 0.5-0.6 mm.

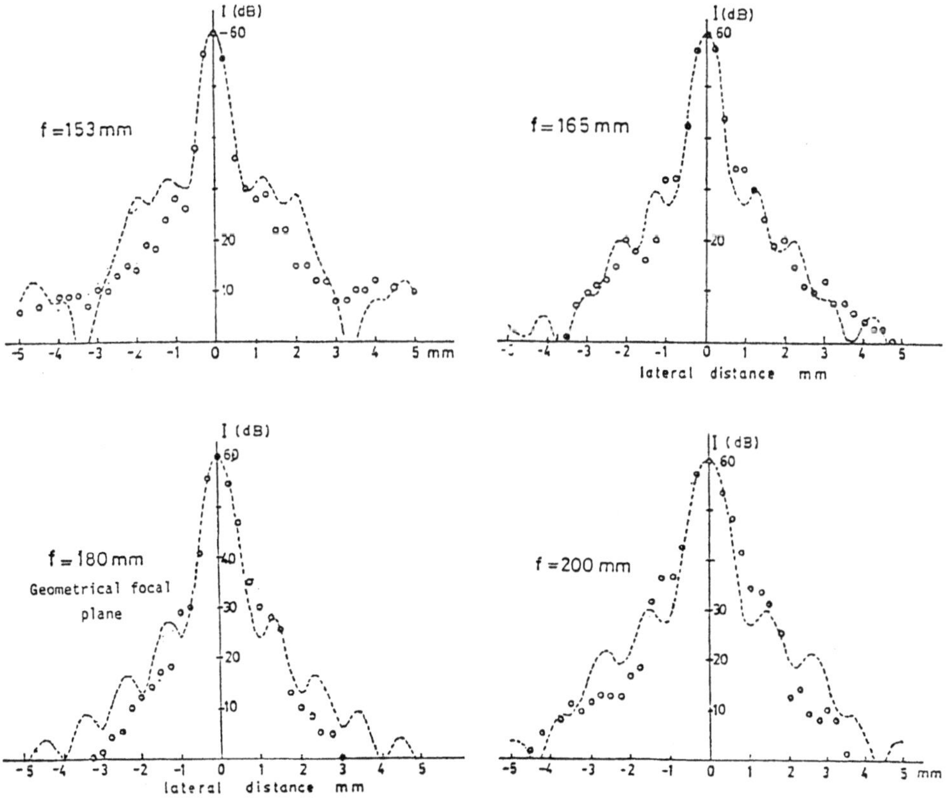

Fig. 6 Observed (open circles) and calculated (broken lines) intensity distribuion (normalized) of acousic field of the P(VDF-TrFE) concave annular array transducer on various focal planes.

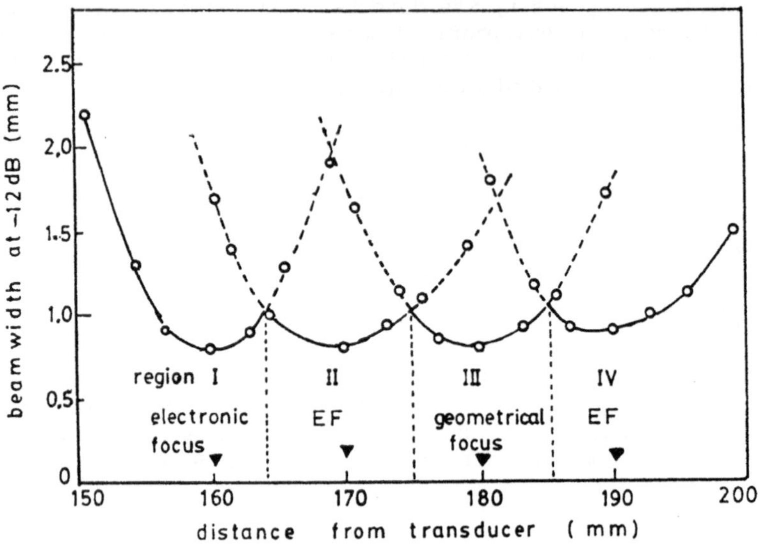

Fig. 7 -12 dB beamwidth of acoustic waves from the P(VDF-TrFE) annular transducer. The dynamic (electronic) focal points are set at 160, 170, and 190 mm from the transducer.

Beamwidth of acoustic pulses fousing on the various focal points was also measured as a function of distance from the transducer along the transducer axis. This is shown in Fig. 7: Curve a, b, c, and d correspond to the beams which are focused, in this case, on the points at 160, 170, 180, and 190 mm, from the transducer, respectively. These curves show that, when we display on a CRT the reflected signals of beam (a) from region I, the signals of beam (b) from region II, and so on, we may obtain a resultant B-mode image with lateral resolution better than 1 mm over a region of 40 mm in depth.

The results shown in Figs. 5 and 7 indicate clearly advantage for polymer piezoelectric polymer transducer as compared to ceramic transducers. With ceramic materials, it would be very difficult to fabricate such a large aperture, concave annular array transducer. Our computor simulation showed that, if we use a flat annular array transducer comprising 8-element annular ring elements with an outermost diameter of 70 mm, we could not obtain a convergent beam at all no matter how their driving phases are selected. Furthermore, the acoustic field from ceramic annular arrays would be inferior as compared to that from polymer transducers, because vibration of ceramic annular rings would be accompanied with undesired spurious vibrational modes, which deteriorates the field distribution. On the contrary, a piezoelectric polymer film vibrates in pure thickness mode and has no spurious mode, because its lateral vibrations are clamped completely by the backing electrodes.

ACOUSTICAL IMAGES

Figure 8 shows a B-mode image of the normal thyroid glands taken with the present annular array transducer in the sector scanning mode. The image quality is quite satisfactory; high resoution is realized both in lateral and depth directions. Plural structure of the surface skin, boundary portion of the thyroid glands, the carotid arteries and veins are all imaged very clearily. We believe that the acoustical images taken with this transducer are of highest quality ever obtained.

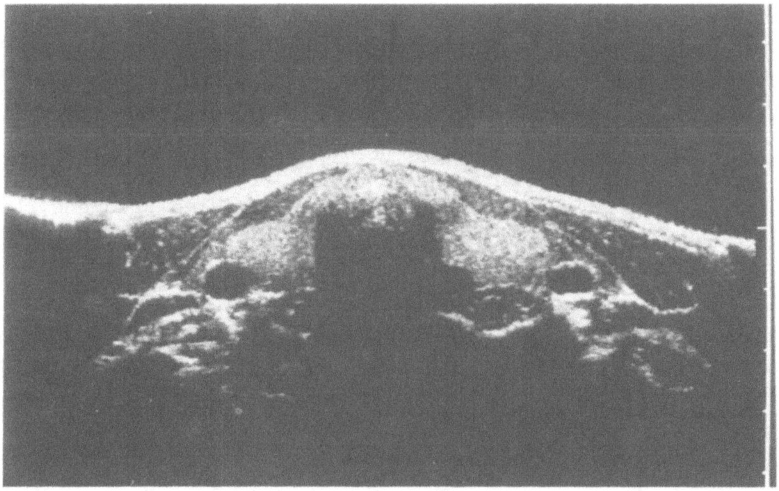

Fig. 8 A B-mode ultrasonic image of the normal thyroid glands taken with the P(VDF-TrFE) concave annular array transducer. (One division of the scale bar on the right side is 1 cm.)

The annular array transducer developed in this study is designed for diagnosis of surperfacial organs such as breasts and thyroid glands to operate at a center frequency of 7.5 MHz . However, it is quite easy to extend the frequency to lower or higher regions simply by changing the thickness of P(VDF-TrFE) film. Therefore, this imaging system may be used in examinations of abdominal organs, and in such fields as obsterics and gynacology and darmatology. Further, this transducer system is also applicable to non-destructive testing of industrial materials.

ACKNOWLEDGEMENTS

The authors are very grateful to Professor R. Omoto of Saitama Medical School and to Dr. A. Kawauchi of Showa Medical School for their valuable comments on medical evaluation of acoustic images. P(VDF-TrFE) material was kindly supplied from Daikin Kogyo Co., Ltd.

REFERENCES

1. N. Chubachi and T. Sannomiya, Composite Resonators Using PVDF Film and its Application to Concave Transducers for Focusing Radiation of VHF Ultrasonic Waves, 1977 IEEE Ultrasonic Symposium, p. 119.
2. H. Ohigashi, Ultrasonic Transducer in the MHz Range, in "Applications of Ferroelectric Polymers", T. T. Wang, J. M. Herbert, and A. M. Glass, eds., Blackies, Glasgow (1987).
3. K. Koga and H. Ohigashi, Piezoelectricity and Related Properties of Vinylidene Fluoride and Trifluoroethylene Copolymers, J. Appl. Phys. 59:2142 (1986).
4. R. Omoto, M. Maruyama, M. Kobayashi, H. Ohigashi, T. Nakanishi, and M. Suzuki, Piezoelectric Polymer Trasnsducers for High Resolution Ultrasound Imaging, Proceedings of the 26th Annual Meeting of the American Institute of Ultrasound in Medicine (1981, San Francisco) p.135.

5. H. Ohigashi, K. Koga, M. Suzuki, T. Nakanishi, K. Kimura, N. Hashimoto, Piezoelectric and Ferroelectric Properties of P(VDF-TrFE) Copolymers and Their Applications to Ultrasonic Transucers, Ferroelectrics, 60:263 (1984).
6. K. Kimura, N. Hashimoto, and H. Ohigashi, Performance of of a Linear Array Transducer of Vinylidene Fluoride and Trifluoroethylene Copolymer, IEEE Trans. Sonics & Ultrason. SU-32:566 (1985).
7. K. Kimura and H. Ohigashi, Generation of Very High Frequency Ultrasonic Waves Using Thin Film of Vinylidene Fluoride and Trifluoroethylene Copolymer, J. Appl. Phys. 61:4749 (1987).
8. H. Ohigashi and K. Koyama, S. Takahashi, Y. Wada, Y. Maida, R. Suganuma, and T. Jindo, Ferroelectic Polymer Transducer for High Resolution Scanning Acoustic Microscopy, Acoustical Imaging, 16 (1977), in press.
9. Y. Miyake, K. Fujieda, M. Hirose, and H. Hirose, High Resolutional Ultrasonic Diagnastic Equipment Using Annular Array Transducer, Jpn. Soc. Ultrasound in Medicine, Proceedings, 43:449 (1983).
10. W. P. Mason, "Electromechanical Transducers and Wavefilters", D. Van Nostrand, New Jersey (1948).
11. K. Kimura, K. Yoneya, and H. Ohigashi, A Wide-Band Polymer Ultrasonic Transducer Using Acoustic Impedance Matching Technique, Jpn. J. Appl. Phys. 27:547 (1988).

ARRAY CONFIGURATION TO ELIMINATE GRATING LOBES

Shin-ichiro Umemura and Kageyoshi Katakura

Central Research Laboratory, Hitachi Ltd.
Kokubunji, Tokyo 185, Japan

ABSTRACT

A new configuration of ultrasound array transducers is proposed whose element boundaries interlock in a fine pattern to eliminate grating lobes from the acoustic field without using a very large number of array elements. Two dimensional acoustic fields produced by interlocking element 3.5 MHz array transducers are computer-simulated. These acoustic fields are compared with those from the corresponding conventional array transducer with straight-line element boundaries. The new configuration reduces grating lobe intensity by 20 dB in a round-trip acoustic field and virtually eliminates grating lobes.

INTRODUCTION

Ultrasound array transducers are widely used in imaging systems for medical diagnosis and non-destructive testing. However, a problem in beam forming using array transducers is their unwanted grating lobe responses. Grating lobes may arise either from beam steering or from focusing. A situation in which grating lobes are formed due to focusing is schematically shown in Fig.1. Figure 1(b) is a simplified diagram of a linear array transducer and a focusing circuit. The lateral profile of the produced acoustic field is shown by a bold line in Fig.1(a). Grating lobes are seen at the angular locations between 40 and 60 degrees, and can cause problems such as artifacts in ultrasound images.

Intensity of grating lobes made by focusing in one-way response is approximately proportional to the array element pitch divided by the ultrasound wavelength and is also inversely proportional to the F number of the aperture.[1] In theory, the easiest way to eliminate grating lobes is to use a smaller pitch and a larger number of array elements. Unfortunately, more elements require a heavier transducer probe, a thicker bundle of cables, larger circuit boards, and a bigger price tag, all of which are impractical. We have tried to eliminate grating lobes without increasing the number of array elements.

To understand how we have approached this problem, examine Fig.1 for an explanation of grating lobe formation. The signals received by each transducer element are delayed for specific lengths of time and then summed so that the focusing circuit may perform as a kind of an acoustic lens to form the main beam. However, the staircase-shaped acoustic lens also forms unwanted grating lobes. The broader profile in Fig.1(a) is the directivity of a transducer element, which is obtained from the Fourier transform of the rectangular-shaped element sensitivity distribution in Fig.1(c). Grating lobes arising from focusing are located around the angles of nulls of the element directivity. This divides the original grating lobe into two lobes, which suppresses its intensity.

These grating lobes should be reduced or eliminated by making the acoustic lens shape smoother using a smoother element sensitivity distribution as shown in Fig.1(d), in which element sensitivity distribution overlaps with neighboring elements. In the Fourier domain, the grating lobes should be reduced if somewhat narrower element directivity as shown by a dotted line in Fig.1(a) is used.

As shown in Fig.2(a), an array configuration with oblique element boundaries provide good element sensitivity distribution in the array direction as shown in Fig.2(b). The grating lobes in this direction would be reduced, but grating lobes which are almost as large as those using rectangular element boundaries still exist in the direction perpendicular to the oblique element boundaries. Therefore, this configuration does not really solve the problem.

Okujima proposed the array configuration in Fig.2(c), which substantially decreases grating lobe intensity.[2] It may be possible to further reduce grating lobes by using piezoelectric materials such as ceramic-polymer composites[3] or piezoelectric polymers[4,5], which have recently become available. These provide more freedom of array transducer design.

An example of a new array configuration using interlocking element patterns on a 1-3 ceramic polymer composite is shown in Fig.3. This transducer consists of millions of piezoelectric ceramic micro rods separated by a polymer. Putting electrodes like those in Fig.3(a) on one side of the transducer and a common electrode on the other side, one can make of transducer elements in various shapes without introducing too much electrical or mechanical coupling between adjacent elements. In this case, the element sensitivity distribution shown in Fig.3(b) is available without producing any additional grating lobes. This is due to the fine interlocking pattern of the array configuration.

The following describes computer-simulated two-dimensional acoustic fields produced by these interlocking element arrays. These are and compared with the acoustic fields from conventional and oblique element arrays in the same acoustic conditions.

CONVENTIONAL ARRAY

A two-dimensional acoustic field produced by a conventional linear array is shown in Fig.4(a). An array transducer of 3.5 MHz with straight-line element boundaries, 0.6 mm element width, 50 elements in an aperture, 10 mm height, and 60 mm focal distance was simulated using the HITAC S-810 supercomputer. In this simulation, each transducer element is numerically subdivided into discrete elements with an assumed point source-receiver. Separation between neighboring point source-receivers is always less than an eighth of the ultrasound wavelength. A continuous wave acoustic field in the 60 mm radius hemispherical surface shown in Fig.4(c) was computed and axonometrically plotted on the projected disk shown in

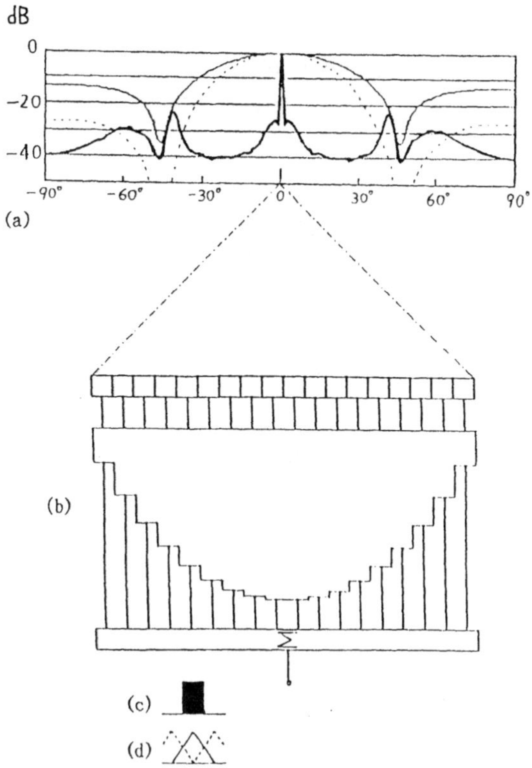

Fig.1 Grating lobe formation by array transducer.
(a) Acoustic field profile;
bold line: acoustic field from wole aperture, solid line: directivity of array element, dotted line: narrower directivity of array element.
(b) Simplified picture of array transducer and focusing circuit.
(c) Element sensitivity distribution.
(d) Smoothed element sensitivity distribution.

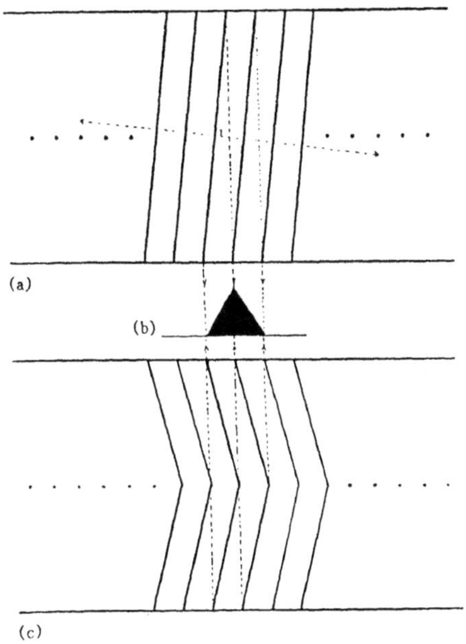

Fig.2 Oblique element array.
(a) Basic configuration.
(b) Element sensitivity distribution in array direction.
(c) Bent element configuration.

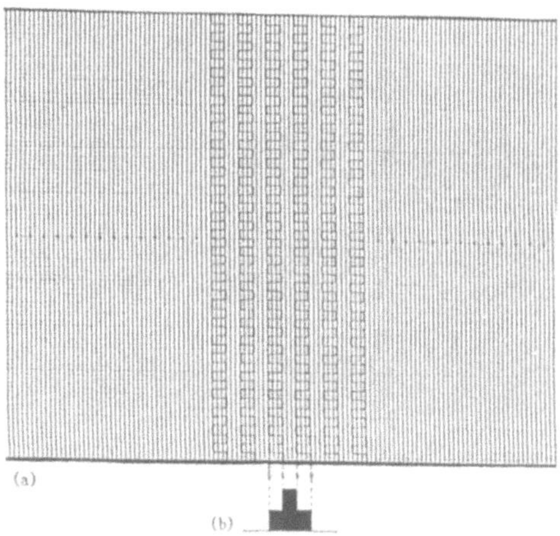

Fig.3 Interlocking element configuration using 1-3 ceramic-polymer composite.
(a) Array configuration.
(b) Element sensitivity distribution in array direction.

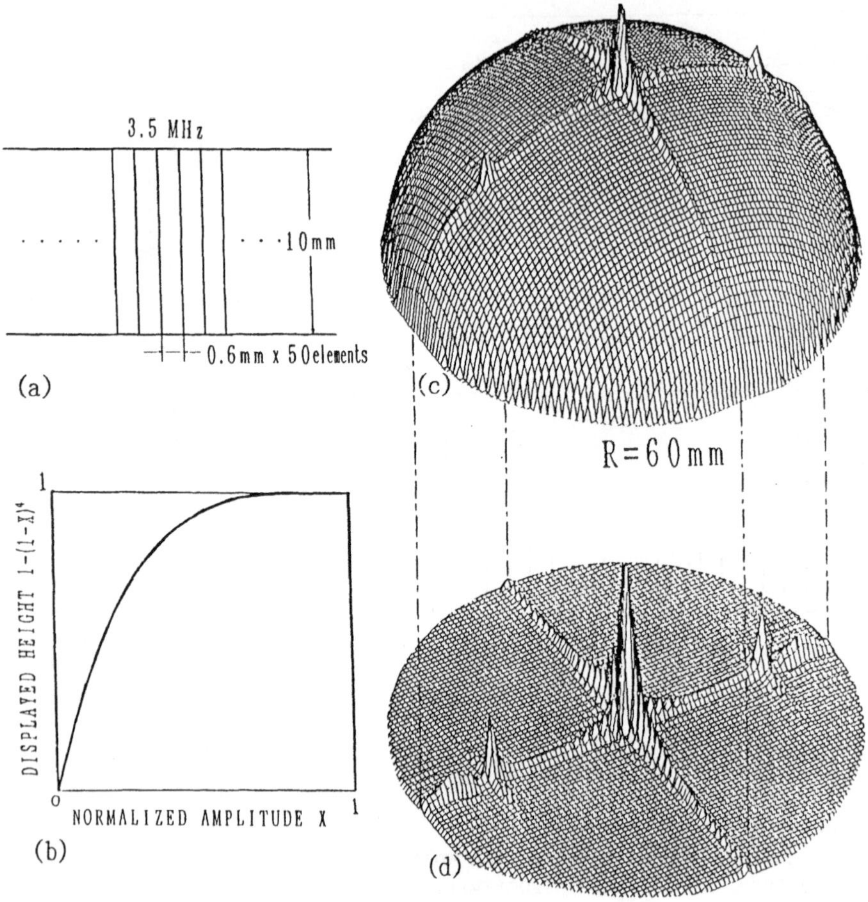

Fig. 4 Two-dimensional acoustic field from conventional array.
(a) Array configuration.
(b) Amplitude compression curve.
(c) Acoustic field on hemispherical surface.
(d) Acoustic field on projected disk.

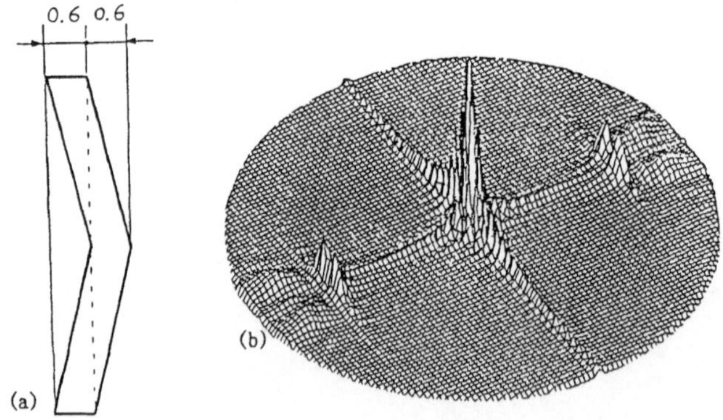

Fig. 5 Two-dimensional acoustic field from oblique element array.
(a) Element shape.
(b) Acoustic field on projected disk.

Fig.4(d). To emphasize the grating lobes, one-way acoustic field amplitudes are compressed using the curve in Fig.4(b) after normalized by the peak amplitude at the main lobe. This compression of the displayed amplitude makes the main lobe look much broader in the figures than it really is.

The grating lobe intensity is 19 dB smaller than the main lobe for this conventional array configuration.

OBLIQUE ELEMENT ARRAY

The acoustic field from the bent oblique element array shown in Fig.5(a) is axonometrically plotted in Fig.5(b). A grating lobe from the conventional array is split into two lobes along the transducer height, and intensity is reduced by 4 dB in one-way response.

INTERLOCKING ELEMENT ARRAY

Acoustic fields from interlocking element arrays with the oblique element boundaries shown in Fig.6(a) and (c) are axonometrically plotted in Fig.6(b) and (d). The repetition pitch along the height direction of the interlocking pattern of 0.8 mm does not seem to be small enough, as some additional grating lobes can be seen in Fig.6(b). With a repetition pitch of 0.4 mm, these additional grating lobes have disappeared as seen in Fig.6(d), and the grating lobe intensity is 10 dB less than that of the conventional array in one-way response.

Acoustic fields from interlocking element arrays with the rectangular element boundaries shown in Fig. 7(a), (c), and (e) are axonometrically plotted in Fig.7(b), (d), and (f). Again, the repetition pitch of 0.8 mm does not seem to be small enough, producing the additional grating lobes in Fig.7(b). With the repetition pitch of 0.4 mm, these lobes have disappeared as seen in Fig.7(c), and grating lobe intensity is 9 dB less than that of the conventional array in one-way response. Figure 7(e) and (f) show the optimized interlocking lengths for these arrays. After optimizing interlocking lengths, grating lobe intensity is 4 dB less than that when the interlocking length is 50 % of the element pitch in Fig.7(c) and (d). This grating lobe intensity is 13 dB less than that of the conventional array in one-way response, and grating lobes have been virtually eliminated.

SUMMARY AND DISCUSSION

A comparison of the different array element configurations in round-trip grating lobe intensity is summarized in Fig.8. The grating lobe intensity of 38 dB from the conventional array may not be small enough for most of the applications of medical diagnostic ultrasound. Grating lobe intensity is substantially reduced by interlocking element configurations, and the grating lobes have almost disappeared for the configuration with optimized interlocking lengths.

When comparing similar element shapes with different repetition pitches in the height direction, a smaller repetition pitch always results in smaller grating lobe intensity. When comparing oblique element boundaries and rectangular element boundaries with interlocking lengths of 50 % of the element pitch, oblique element boundaries always give slightly smaller grating lobe intensity. Optimizing the interlocking length reduces grating lobe intensity to 9 dB less than that of an interlocking length of 50% of the element pitch, which is 26 dB overall reduction from the conventional array.

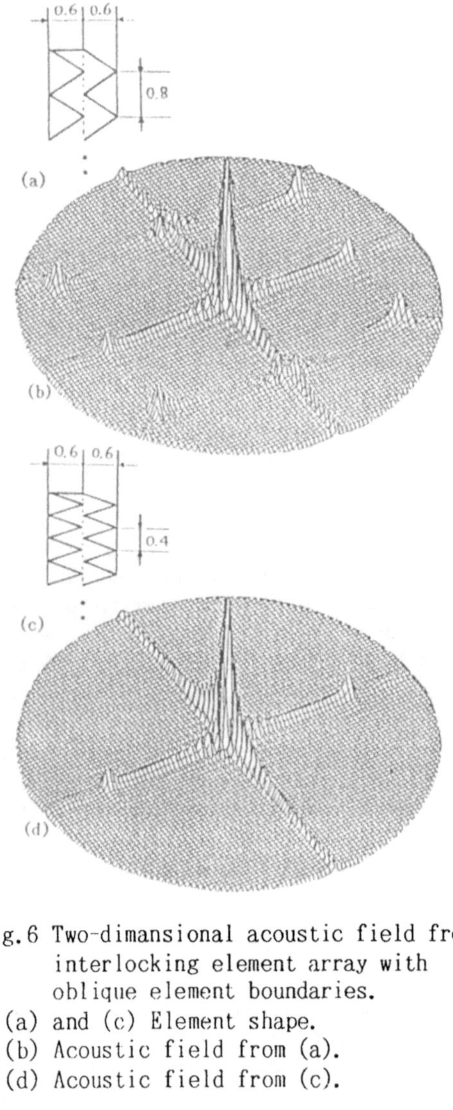

Fig. 6 Two-dimansional acoustic field from interlocking element array with oblique element boundaries.
(a) and (c) Element shape.
(b) Acoustic field from (a).
(d) Acoustic field from (c).

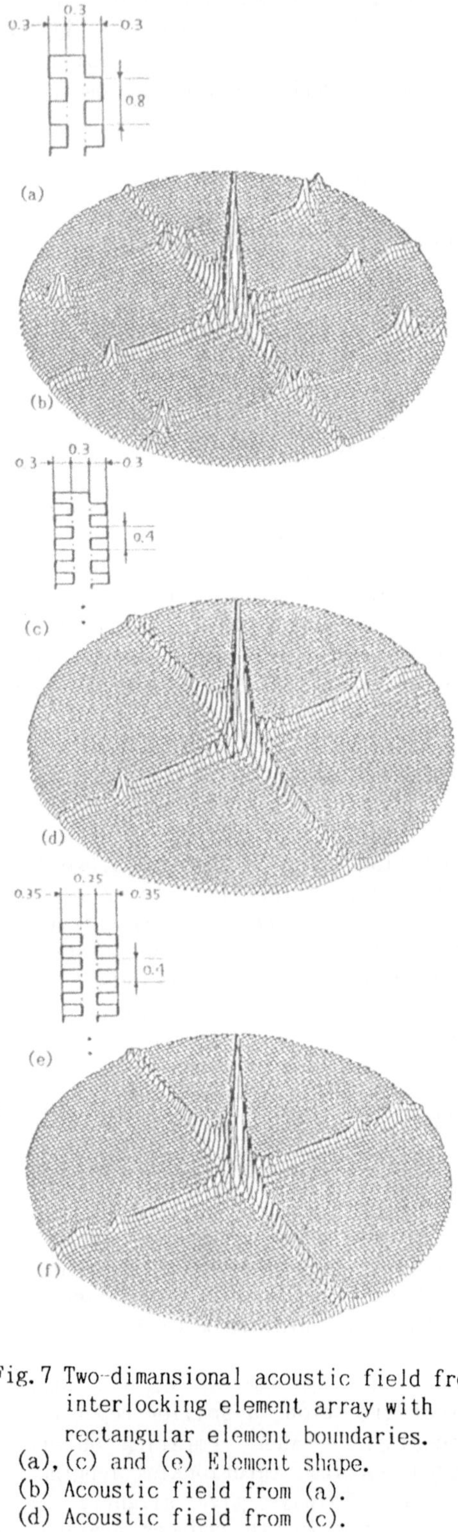

Fig. 7 Two-dimansional acoustic field from interlocking element array with rectangular element boundaries.
(a), (c) and (e) Element shape.
(b) Acoustic field from (a).
(d) Acoustic field from (c).
(f) Acoustic field from (e).

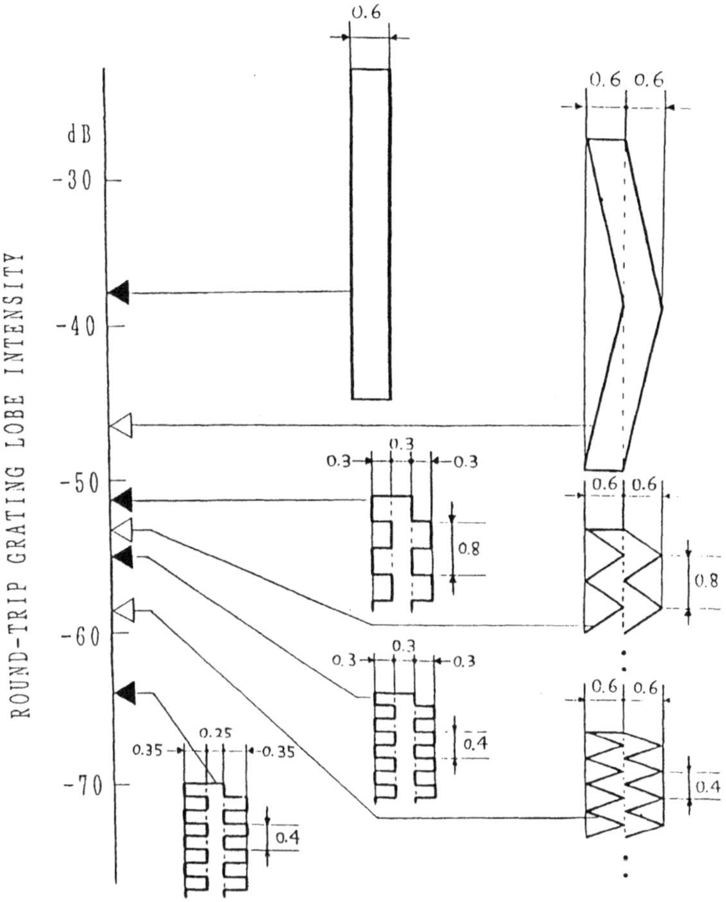

Fig.8 Suppression of grating lobe intensity by interlocking element configuration.

These interlocking element array transducers can be implemented using the ceramic-polymer composites in Fig.3. They can also be implemented using piezoelectric polymers. Separating transducer elements by cutting the piezoelectric material may not be necessary for piezoelectric polymers[4,5] because of their low mechanical Q. Using certain etching techniques to separate the elements, even piezoelectric materials having high mechanical Q such and ceramics or crystals could be used.

This interlocking element array configuration is quite effective to eliminate grating lobes arising from focusing where broad directivity of array elements is not required to form the main lobe. But it may not be so useful to reduce those arising from beam steering especially with a large steering angle, since large-angle beam steering requires relatively broad element directivity to form the main lobe itself while the interlocking configuration somewhat sacrifices broadness of the element directivity.

CONCLUSION

The interlocking element array transducer configuration is proposed to eliminate grating lobes. Two-dimensional acoustic fields produced by this array configurations are computer-simulated and compared with those from a conventional array and from a oblique element array configuration. Round-trip grating lobe intensity is more than 20 dB less than that of conventional array configurations and more than 10 dB less than that of oblique element array configurations. Grating lobes can be virtually eliminated by this array configuration.

REFERENCES

1. S. Umemura and K. Katakura, "Theoretical analysis of grating lobe intensity", 1986 IEEE Ultrasonic Symposium Proceedings, (1986), pp.659-662.
2. M. Okujima, "A method of grating lobe reduction for sector-scan linear array transducer", Proceedings of Acoustical Society of Japan, (March 1984), pp.639-640 (in Japanese).
3. C. Nakaya, H. Takeuchi, K. Katakura, and A. Sakamoto, "Ultrasonic probe using composite piezoelectric materials", 1985 IEEE Ultrasonic Symposium Proceedings, (1985), pp.634-636.
4. H. G. Nguyen, P. Hartemann, and D. Broussoux, "Single element and array PVDF transducers for acoustic imaging", 1982 IEEE Ultrasonic Symposium Proceedings, (1982), pp.832-836.
5. K. Kimura, T. Miya, and H. Ohigashi, "Performance of a linear array transducer of vinylidene fluoride trifluoroethylene coploymer", IEEE Transaction on Sonics and Ultrasonics, (1985), SU-32, 4, pp.566-573.

THE SENSITIVITY OF MINIATURE HYDROPHONES MADE OF P(VDF-TrFE) THIN FILM

K. Koyama, T. Eda, H. Suzuki, Y. Wada, and O. Ishizuka

Department of Polymer Materials Engineering
Faculty of Engineering, Yamagata University
Yonezawa 992, Japan

ABSTRACT

The miniature needle type hydrophone, the spherical hydrophone, and the line type hydrophone were constructed of vinylidene fluoride (VDF) - trifluoroethylene (TrFE) copolymer thin film. The sensitivity of the hydrophones was strongly affect by the electrical impedance mismatching. Improvement of sensitivity was examined by increasing the characteristic impedance. The hydrophone with the higher characteristic impedance has a sensitivity higher than that with 50Ω by ca. 12 dB. The spherical hydrophone was made by coating the copolymer film on the glass sphere. The hydrophone can act as a three-dimension-omnidirectional sensor. Uniform sensitivity within 10 dB was obtained over -60° to 60° in the wide frequency range. The directivity is much wider than that of needle like miniature hydrophones ($\pm 15^{\circ}$ within 10 dB at 10 MHz).

INTRODUCTION

Miniature hydrophones are used for measuring temporal and spatial variation of acoustic pressure in a liquid medium as well as frequency response, linearity and dynamic range of transmitting transducers. Needle type miniature hydrophones were made of piezoelectric of films of poly(vinylidene fluoride) (PVDF).[1-3] The PVDF hydrophones have good frequency characteristics. Although much work on PVDF transducers have been carried out during recent one decade and many superior characteristics of PVDF transducers have been revealed, the PVDF transducers are still minor in commercial ultrasonic transducers, because drawing process is required to obtain a high piezoelectricity in PVDF films and even in the drawn films the electromechanical coupling constant (k_t) is still much lower than that of the ceramic transducers. The largest value of k_t is amounts to 0.2.

Vinylidene fluoride-trifluoroethylene copolymer (P(VDF-TrFE)) shows a larger value of k_t (=0.3) than PVDF. Furthermore, the copolymer does not need drawing to obtain large k_t. The piezoelectric copolymer films without drawing are thermally stable in dimension because of no residual strain and then it is good for long term usage. The value of k_t of the copolymer depends on the VDF content and shows the maximum value at ca. 80% of VDF. In the previous paper, P(VDF-TrFE) copolymer with a comonomer ratio 78/22 was applied to a wide-band miniature hydrophones using Semi-Rigid Coaxial Cable (SRCC).[4] Although the sensitivity of the hydrophone is higher than that of conventional miniature hydrophones, the sensitivity is desired to be improved by avoiding the electrical impedance mismatching. In the present study, P(VDF-TrFE) copolymer is also used as a transducer material for highly sensitive miniature hydrophones. The improvement of sensitivity is examined by increasing the impedance of coaxial cable. The copolymer is also applied to three dimensional omnidirectional wide-band hydrophones.

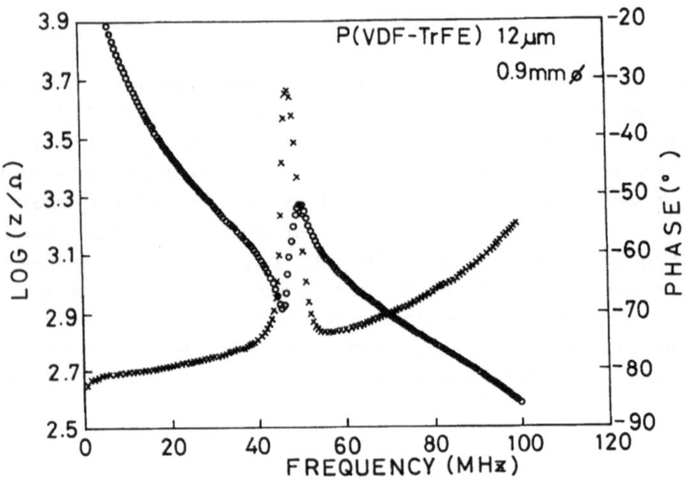

Fig.1. Absolute value and phase of electrical impedance of copolymer film.

TRANSDUCER MATERIAL FOR MINIATURE HYDROPHONES

The 78/22 P(VDF-TrFE) copolymer was used as the starting material because it was suitable to obtaining transducers with a high coupling constant. The copolymer was supplied in the forms of powder or extruded film from Daikin Kogyo Co. Thin films were prepared by solvent (dimethylformamide) casting. These films were used after heat-treatment at 143°C for 30 min. After gold electrodes had been deposited on the surfaces, the films were polarized at room temperature by applying a DC electric field higher than 100 MV/m. The coupling constants k_t of the films were estimated to be from 0.25 to 0.3.

The piezoelectric films with 12 µm were cut into a circular element with diameter of 0.9 mm. The frequency dependence of electrical impedance of the element is shown in Figure 1. The frequency of fundamental resonance of thickness mode is located at 48 MHz. The electrical impedance at 20 MHz is ca. 3,000 Ω. The circular element bonded to the end of SRCC (1.2 mmϕ, UT-47) which supports the element. The cable has characteristic impedance of 50 Ω and capacitance 96 pF/m. The electrical impedance mismatching between transducer element and cable is extremely large.

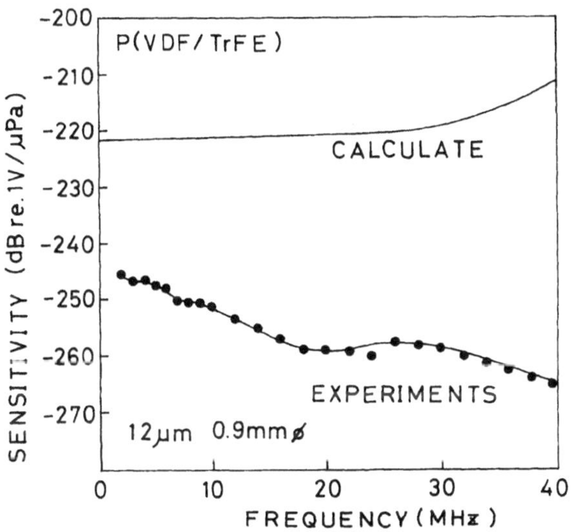

Fig.2. Absolute sensitivity of miniature hydrophone.

SENSITIVITY OF MINIATURE HYDROPHONES

Absolute sensitivity measurement of the hydrophones were carried out by a three-transducer reciprocity calibration. The calibration was carried out using the burst wave from a pulse generator. The measurements were made between a pair of hydrophone in a water tank, at a distance of 20 mm from each one. The sensitivity obtained by 50 Ω measuring system are shown

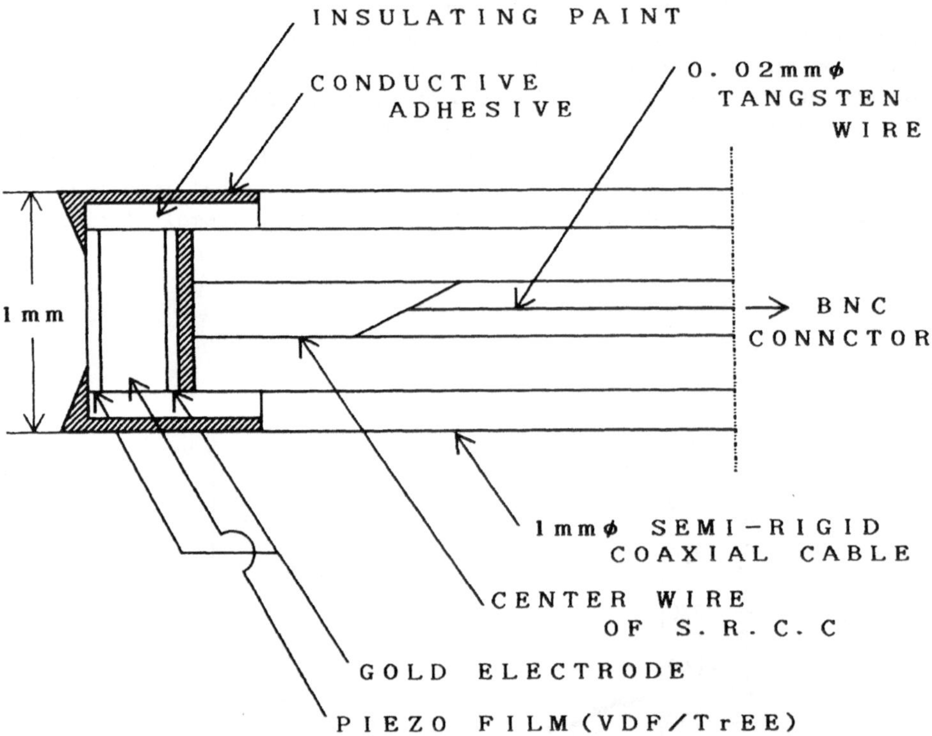

Fig.3. The construction of needle-type miniature hydrophone.

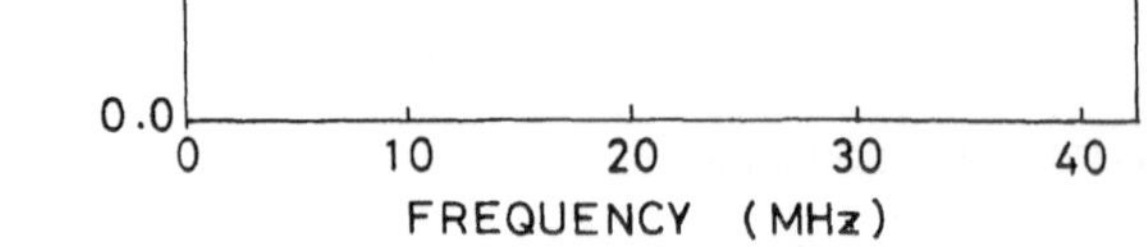

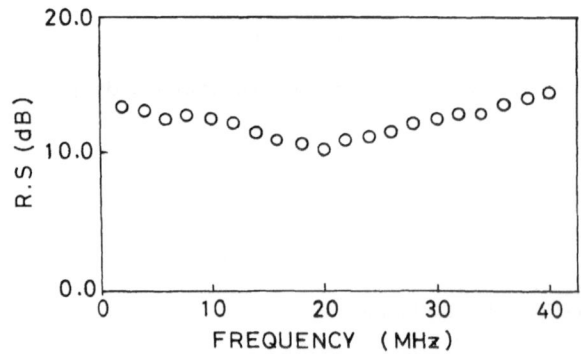

Fig.4. Relative sensitivity of high impedance cable hydrophone compared with 50 Ω cable hydrophone.

in Figure 2. The cable length was 1 m in this measurement. The sensitivity is -250 to -260 dB re.1V/uPa and is higher than that of the PVDF hydrophones. The sensitivity decreases with increasing frequency, mainly as a result of the decrease in dielectric constant and the increase in electric impedance of the copolymer elements with increasing frequency.

Figure 2 also shows the theoretical results. The sensitivity equation is derived from the Mason's equivalent circuit under the condition that the output current is zero. The sensitivity is given by the ratio of the output voltage V_o to the input acoustic stress T_{in} for the hydrophone in which the piezoelectric film is backed with a material having acoustic impedance z_{mb} and infinite thickness.[5]

$$\frac{V_0}{T_{in}} = \frac{2h}{j\omega}\left[Z_w + Z_0 \coth\gamma t + Z_0 \operatorname{cosech}\gamma t * \frac{Z_w + Z_0 \tanh(\gamma t/2)}{Z_{mb} + Z_0 \tanh(\gamma t/2)} \right]^{-1}$$

$$\gamma = j\left(\frac{\omega}{v}\right)\left(1 - j\frac{\tan\delta_m}{2}\right) \qquad (1)$$

where z_w and z_0 are the specific acoustic impedances of water and the polymer film, respectively. The difference between theoretical calculation and experimental results is larger than 30 dB. The difference results from the electrical impedance mismatching and dielectric losses of the cable.

The dielectric losses of the cable were discussed by Chivers and Lewin using an equivalent electrical circuit of hydrophone system.[6] The equivalent electrical circuit of hydrophone are composed of a element capacitance C_t, a cable capacitance C_c and stray capacitances C_s. The low sensitivity of hydrophones was explained by severe cable loading ($C_t \ll C_c$). On the other hand, the characteristic impedance of coaxial cable is given by

$$Z_0 = \frac{60}{\sqrt{\varepsilon_r}} \text{LOG} \frac{R_2}{R_1} \qquad (2)$$

where ε_r a dielectric constant of insulating material in the cable, R_1 and R_2 are the radius of conductive tube in the cable and that of center lead wire in the cable.

A high characteristic cable impedance and small cable capacitance is attained by the exchange of the center lead wire in SRCC by a tungsten wire with a thinner diameter (0.02mm). As a results, the characteristic impedance of the new cable is twice larger than the original SRCC. The construction of high cable impedance hydrophone is shown in Figure 3. The relative sensitivity of these two hydrophones with 50 Ω cable and high electrical impedance cable is shown in Figure 4. The hydrophone with high impedance cable gains 12 dB in sensitivity in comparison with that with 50 Ω cable. Although impedance converter using small size FET (2SK67A) was composed to the hydrophone, little effect of the sensitivity was obtained in the present case in contrast with the expectation.

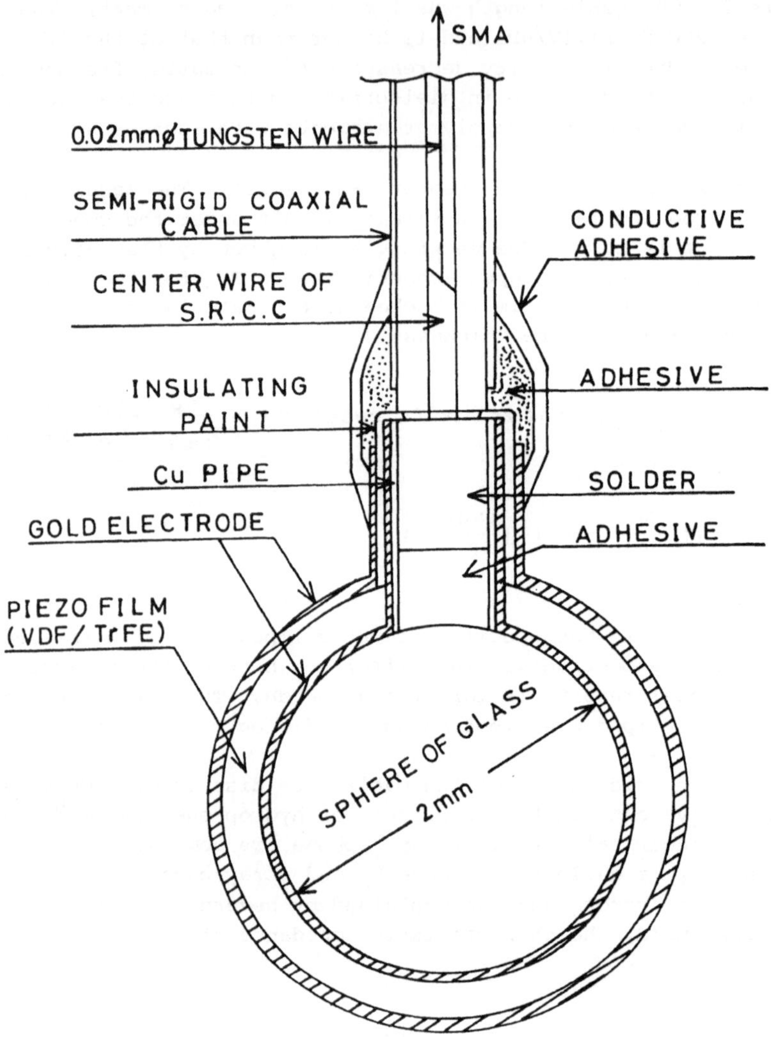

Fig.5. The construction of spherical miniature hydrophone.

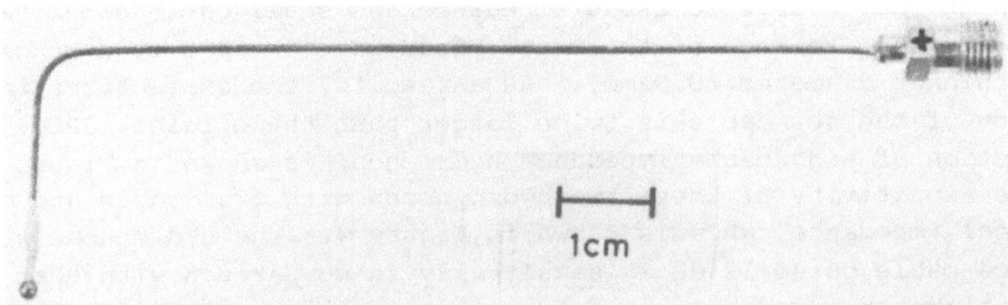

Fig.6. The spherical miniature hydrophone.

A spherical hydrophone was made of P(VDF-TrFE) copolymer and a glass sphere. The hydrophone can act as a three-dimension-omnidirectional sensor. the construction is shown in Figure 5. The glass sphere with 2 mm in diameter was glued to the end of SRCC (UT-34). Gold electrode was deposited on the surface of glass sphere. A 2-butanone solution (10 %) of P(VDF-TrFE) copolymer was coated on the surface of sphere electrode. Then, after drying, the other gold electrode was evaporated on the copolymer

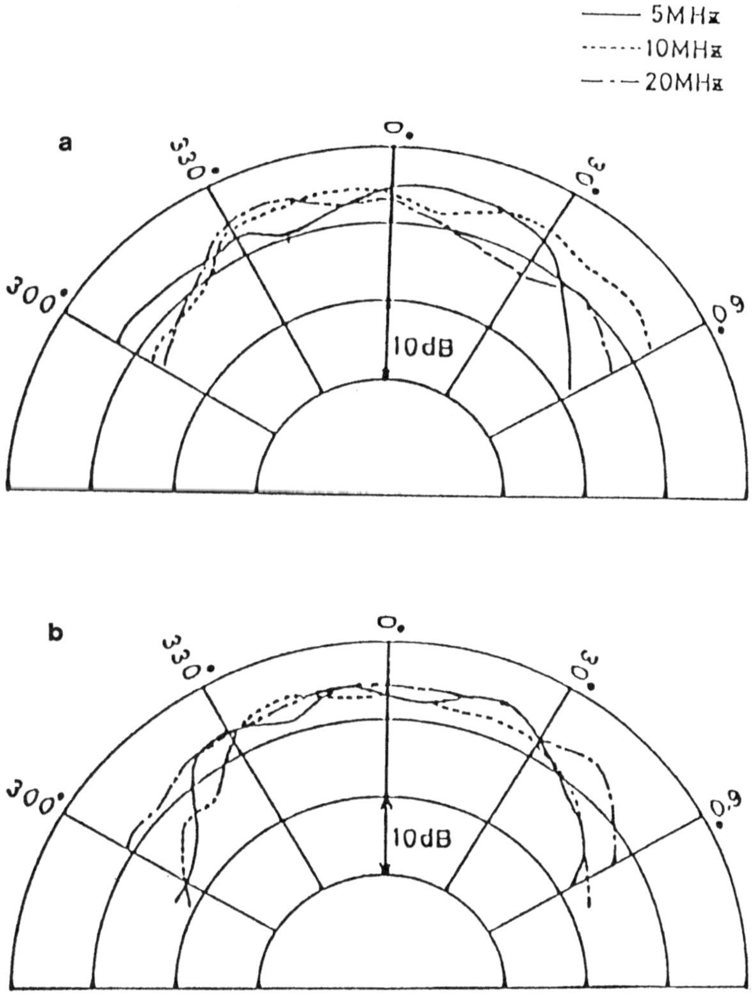

Fig.7. The directivity patterns of the spherical hydrophone at 5, 10, 20 MHz. (a) in the plane normal to the SRCC axis. (b): in the plane including the SRCC axis.

film. After heat treatment at 143°C for 30 min, the sphere shape thin film was polarized by the corona point discharge method. The distance between film and needle electrode was 1.5 cm, applied voltage was 15 kV and poling time was 5 min. Figure 6 shows an example of the spherical hydrophone. This hydrophone is easily accessible to the acoustic field to be measured, because the flexibility of SRCC. The sensitivity measured by the reciprocity method is -220 dB re. 1V/Pa. for the spherical hydrophone.

Figure 7a (in the plane normal to the SRCC axis) and 7b (in the plane including the SRCC axis) show respectively measured results of the directivity pattern of the spherical hydrophone at 5, 10, and 20 MHz. Uniform sensitivity within 10 dB was obtained over -60° to 60° in the wide frequency range. The directivity is much wider than that of needle like miniature hydrophones (±15° within 10 dB at 10 MHz).[4] The small sensitive irregularity may results from non-uniformity of film thickness.

CONCLUSION

Highly sensitive miniature hydrophones were attained by using P(VDF-TrFE) and high impedance coaxial cable. The spherical hydrophone was made by coating the copolymer film on the glass sphere. Omnidirectivity was obtained in a wide frequency range by the spherical hydrophone.

Acknowledgments

This work was partly supported by a Grant-in Aid from the Ministry of Education, Science and Culture of Japan. The authors are greatly indebted to Daikin Kogyo Co. for supplying them with samples of P(VDF-TrFE).

REFERENCES

1. M.Ide and E.Ohdaira, "Wide Frequency Range Miniature Hydrophone for the Measurement of Pulse Ultrasonic Field", Jpn.J.Appl.Phys.20:S20-3,205-208(1981).
2. P.A.Lewin, "Miniature Piezoelectric Polymer Ultrasonic Hydrophone Probes", Ultrasonics,19:213-216(1981).
3. M.Platte, "A Polyvinylidene Fluoride Needle Hydrophone for Ultrasonic Applications", Ultrasonics,23:113-118(1985).
4. S.Tsuchiya, T.Sato, K.Koyama, S.Ikeda and Y.Wada, "Application of Piezoelectric Film of Vinylidene Fluoride-Trifluoroethylene Copolymer to a Highly Sensitive Miniature Hydrophone", Jpn.J.Appl.Phys.26:S26-1,103-105(1987).
5. H.Ohigashi, Ultrasonic Transducers in the Megaherz Range, in "The Applications of Ferroelectric Polymers",T.T.Wang, J.M.Herbert, A.M.Glass, ed.,Blackie, Glasgow(1988).
6. R.C.Chivers and P.A.Lewin, "The Voltage Sensitivity of Miniature Piezoelectric Plastic Ultrasonic Probes", Ultrasonics,20:279-281(1982)

A MECHANICAL-SCANNING ULTRASONIC TRANSDUCER FOR QNDT IMAGING

Lin Wang and Yu Wei

Department of Biomedical Engineering
Nanjing Institute of Technology
Nanjing, Jiangsu, P.R.China

I.INTRODUCTION

In ultrasonic QNDT imaging, broadband or short pulse ultrasonic transducers are required to be used to improve axial resolution of detection. Scattering resulting from inhomogeneity, anisotropy and despersion in materials makes it less efficient to improve the lateral resolution by means of direct beam-focusing. So various new approaches to detection, processing and imaging have emerged to improve it indirectly. Ultrasonic transducers of broadbeam or beam-scanning in a sufficient broad range are required to realize these methods.

Presented in this paper are theoretical studies and research work about an ultrasonic transducer which has short pulse response and can scan over a wide range automatically as well. A design approach to the short pulse transducer --- the approach of matching strong damping backing --- is put forword, which can well solve the contradiction between efficiency and short pulse response or frequency bandwidth that exists in conventional ultrasonic transducers. The properties of backing and bonding layers are studied in detail, and choice of backing and bonding layers is considered and decided. A structure of a rotating mechanical fan-scanning ultrasonic tranducer is given. A software package for radiation field calculations and designs of the transducers is developed.

Based on theoretical studies, with the help of the approach, a short pulse mechanical scanning ultrasonic transducer is designed and manufactured. Measurement results about the characteristics of the transducer are given and discussed.

II.THEORIES AND DESIGNS

A. The Properties of Backings Made of Tungsten and Epoxy

The inner structure of an ultrasonic transducer is assumed to be as shown in Fig.1. A tungsten/epoxy backing is assumed to be bouned, by a thin layer, to a piezoelectric disk, which is formed from Lead-Zirconate titanate(PZT).

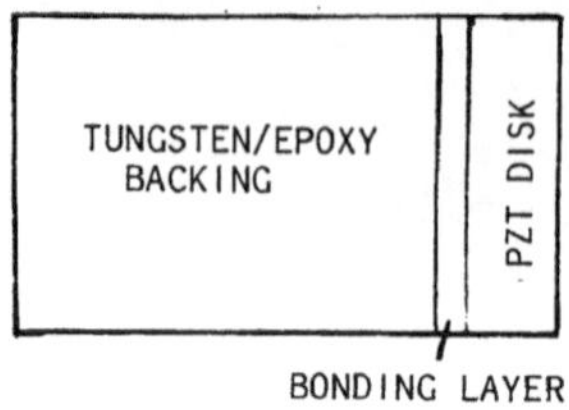

Fig.1 Stylish Diagram of a Typical Ultrasonic Transducer

There exist the following two practical problems in conventional ultrasonic transducers. The firat problem is impedance mismatch between backing and PZT disk. The second one is that the bonding layer between backing and PZT disk is not thin enough. They result in reflection at the boundary between backing and PZT disk, which influences the pulse response of the transducer. Based on analyses on backing material, a method is put forward using a special kind of high density tungsten powder to make a damping block and the mixture of tungsten/epoxy as a bonding layer in order to solve the above two problems.

The longitudinal wave velocity of epoxy is estimated to range from 2300 to 2700 m/s, the density of epoxy is thought of as 1.08 g/cc, the density of tungsten is about 19.3 g/cc. Taking reasonal mean value of their densities, velocities, and treating the tungsten powder as particulate material, the variation curve of the longitudinal wave velocity c of tungsten/epoxy mixture, as the function of the volume concentration of tungsten in the mixture, is given in Fig.2. As shown in Fig.3, the density of the mixture is linearly proportional to the tungsten concentration in the specimen. A curve of great interest in probe construction and design is shown in Fig.4, which is developed from Fig.2 and Fig.3. It shows that the acoustic impedance of the backing can match that of PZT only when the volume fraction of tungsten arrives at 85 percent. However, it is too difficult to make such a mixture, as there is a big gap between the densities of tungsten and epoxy. Commonly in conventional transducers, the weight ratio of tungsten to epoxy is 16:1 or 20:1, which corresponeds to a volume ratio (tungsten/mixture) of only 43 or 53 percent and a backing acoustic impedance of $17 \cdot 10^{6}$ Rayl. Obviously, this value is far less than that of PZT, so, there exist serious acoustic impedance mismatches between backing and PZT disk in conventional transducers.

Assuming the densities of tungsten, epoxy and their mixture in backing as ρ_1, ρ_2, ρ, the volumes of tungsten and epoxy in the backing as v1 and v2, their ratio v1/v2 as δ, we have:

$$\rho = \rho 1^* \quad + \rho 2^*(1-\delta) \qquad (1)$$

Based on the above equation, we can further get the curve of the impedance of a mixture, varying with its density ρ, which is shown in Fig.5. It can be seen from it that the impedance of the mixture can match that of PZT, only when the density of the mixture arrives at 13.6 g/cc. Actually, the density of the holeless mixture in which tungsten/mixture volume ratio is 50 percent can only get to 10.2 g/cc, which however can not meet the needs of PZT impedance satisfatorily.

To increase the density of mixture, a special kind of high density of tungsten powder is chosen to make the backing, so that the impedance of backing can match that of PZT, the pulse performance of ultrasonic transducer can be improved.

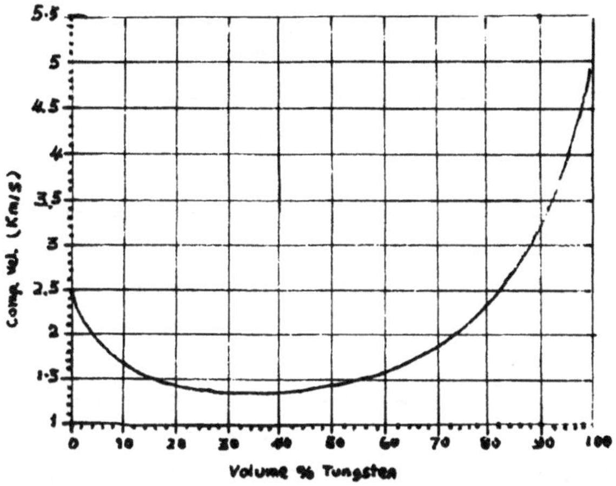

Fig.2 Wave Velocity of Tungsten/Epoxy Mixture Varying the Volume Concentration of Tungsten

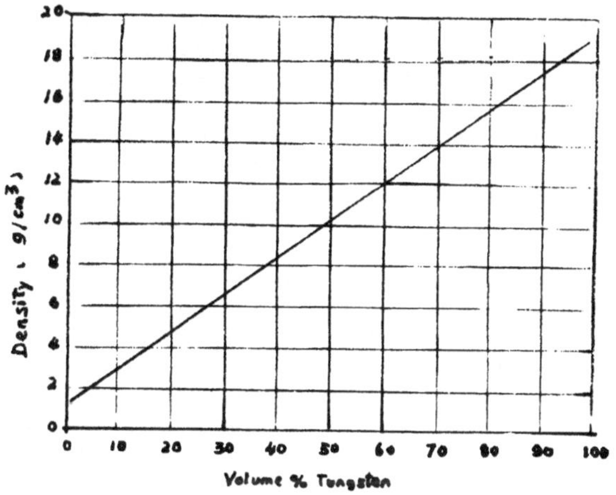

Fig.3 Density of Tungsten/Epoxy Mixture Varying the Volume Concentration of Tungsten

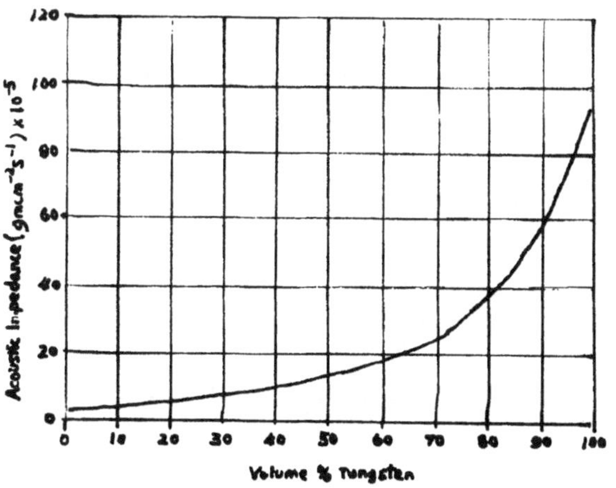

Fig.4 Acoustic Impedance of Tungsten/Epoxy Mixture Varying the Volume Concentration of Tungsten

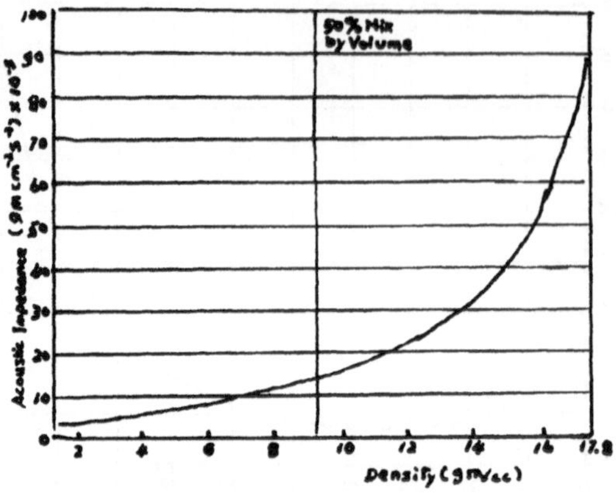

Fig.5 The curve of Relation between Impedance and Density of Tungsten/Epoxy mixture

B. Acoustic Reflection at Bonding Layer between Backing and PZT Disk

In most transducer designs, the active element such as a PZT disk is bonded to a backing. In fact, a suitable bonding layer is of great importance to the pulse performance of a transducer.

It is basically assumed that when an ultrasonic wave is incident normally into a tungsten/epoxy backing, and the backing is a perfect match of PZT, ultrasonic reflection at the boundary between backing and PZT disk will rapidly fall into zero. Then the variation of the reflectivity of ultrasonic pressure and energy with the thickniss of the bonding layer, when bonding material is water and epoxy respectively, can be shown in Fig.6(a)&(b), respectively. It can be seen from them that more than 10 percent of energy will be reflected at the boundary layer when the thickness of the layer is $\lambda/200$ (λ is the wavelengh). Correspondingly, the thickness of the bonding layer must be less than 3 um when the layer is water, and less than 5 um, when epoxy, so as to ensure less than 10 percent reflection of energy. In fact, the flatness criterion for PZT disk is no better than 15 um, and different from one to one. Besides, the diameter of tungsten powder ranges from 8 to 40 um. Therefore, it seems certain that with this type of transducer construction, reflection of energy at the back face of crystal is a major factor in preventing ultrasonic transducer damping.

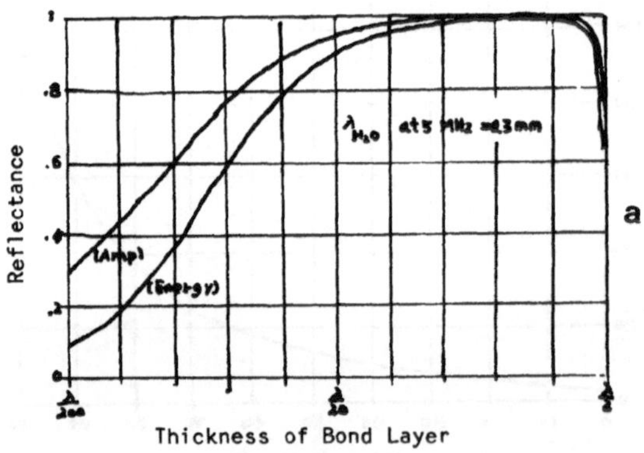

Fig.6 Curves of Relationship between Ultrasonic Reflectivities and the Thickness of Bonding Layers, Which Are Resprctively (a) Water (b) Epoxy and (c) Tungsten/Epoxy Mixture

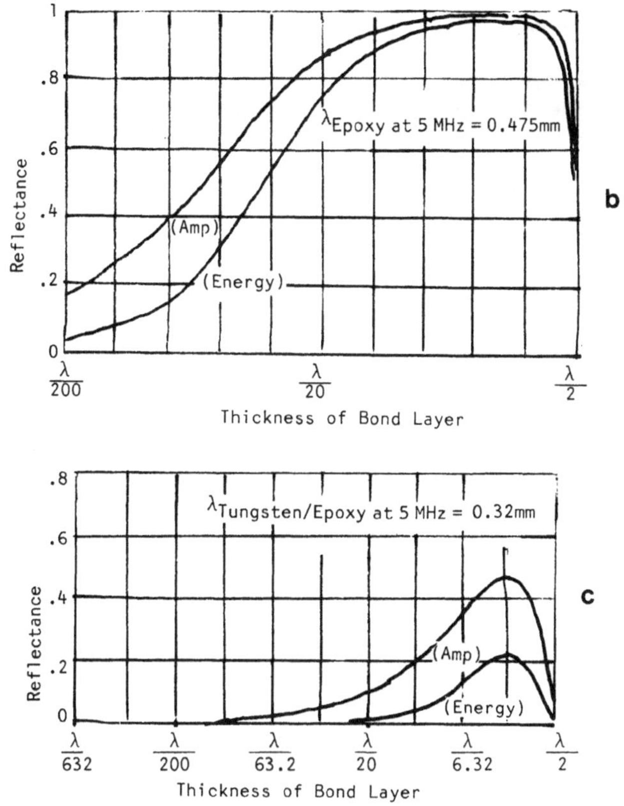

Fig.6 Curves of Relationship between Ultrasonic Reflectivities and the Thickness of Bonding Layers, Which Are Resprctively (a) Water (b) Epoxy and (c) Tungsten/Epoxy Mixture

In the design presented in this paper, the bonding layer between backing and PZT disk is also made of a tungsten/epoxy mixture. The variation of reflectivity of ultrasound with the thickness of bonding layer is shown in Fig.6(c), which indicates that there is still hardly reflection of energy when the thickness of bonding layer is $\lambda/20$. It is obvious that in this case, the requirement for the thickness of bonding layer is reduced.

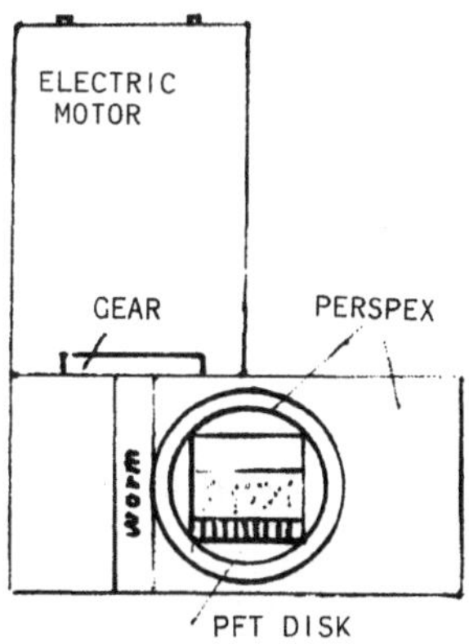

Fig.7 The Diagram of Mechanical-Scanning Ultrasonic Transducer's Structure

C. The Structure of Mechanical-Scanning Ultrasonic Transducer

The strutcure of a mechanical-scanning ultrasonic transducer is shown in Fig.7. The active part of the transducer is put in a hollow of a perspex cylinder within a perspex block. Coupled by a worm and a gear which is driven by an electric moter, the cylinder can rotate round its axial automatically. When doing detection, the rear surface of the perspex block is attached to the surface of specimen.

Based on theoretical analyses, when a detected speciman is metallic material, a little longitudinal wave can be for-backward deflected, so it is necessary for the detection to use transverse wave. With calculations, the for-backward deflected ultrasonic pressure Etl at perspex/steel boundary, when coupled by the liquid, varying with the angles is shown in Fig.8 (Etl here means the acoustic pressure amplitude ratio of the back deflected longitudinal wave in the original way of the forward deflected transverse wave into thr speciman to the incidant wave). It can been seen that critcal angles are 30 and 50 degrees for longitudinal and transverse wave respectively. The maxium for-back deflectivity is about 30 percent.

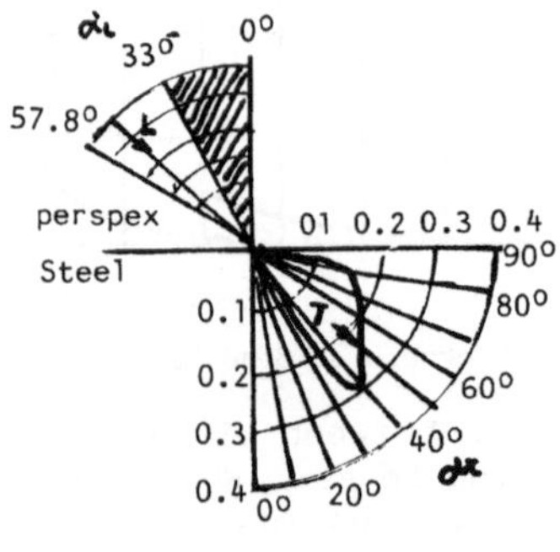

Fig.8 The Curve of For-Back-Deflectivity in the Perspex/Steel Medium

Taking into acount the factors about manufacture and the whole volume of the transducer, -45 to 45 degree scanning range is chosen, if defined the normal incidance case is 0 degree.

Driven by the electric motor, the ultraonic sensor rotates continuously. An output of an optic-electric switch is used as a trigger signal of data acquisition to control the real scanning range of the transducer. The beam-scanning angle of the transducer at any moment can be decided by the rotating speed of the moter. There are not such errors as exist in the usual fan-scanning transducers resulting from the starting and ending of mechanical motion, and the transformation from clockwise rotation to reverse one.

III. IMAGING SYSTEM AND MEASURED RESULTS

A. Imaging System

Fig.9 Shows a system of imaging and detection,in which the transducer is excited by a 0.2 us long negative pulse whose amplitude is 158 v. The received signal is sampled in a high speed sampling oscilloscope whose highest sampling rate is 100 MHz, amplified in a receiving amplifier, and transformed into a digital signal which is then transferred into a computer. Results are outputed by a printer and a ploter.

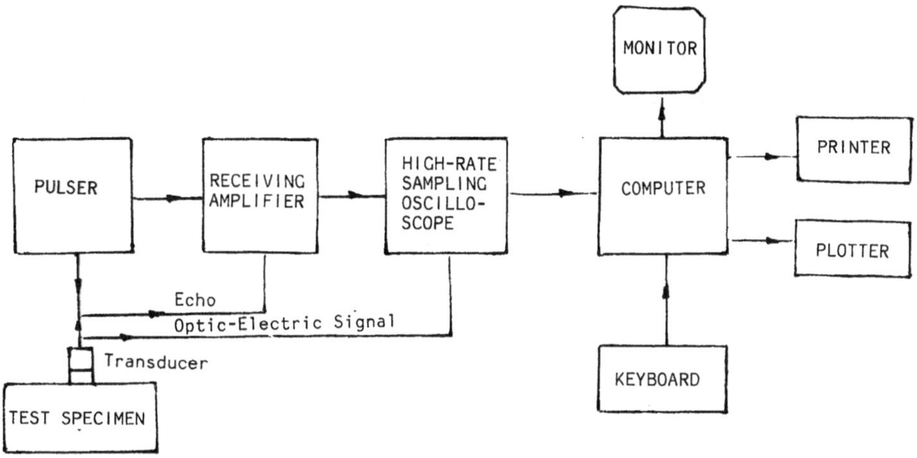

Fig.9 The Sketch of the Imaging System

B. Measured Results and Their Analyses

(1) The Relationship between Scanning Speed of the Transducer and Electric Sourse Voltage of the Motor

As shown in Fig.10, the relationship between scanning speed of the transducer and electric sourse voltage driving the motor is almost linear, and the transducer will rotate when the sourse voltage is more than 5 v. The linearity is mainly depandant upon the rotating homogeneity of the motor, the coupling between the perspex cylinder and block, and the voltage stability of electric sourse. If necessary, a speed-stablized circuit can be added.

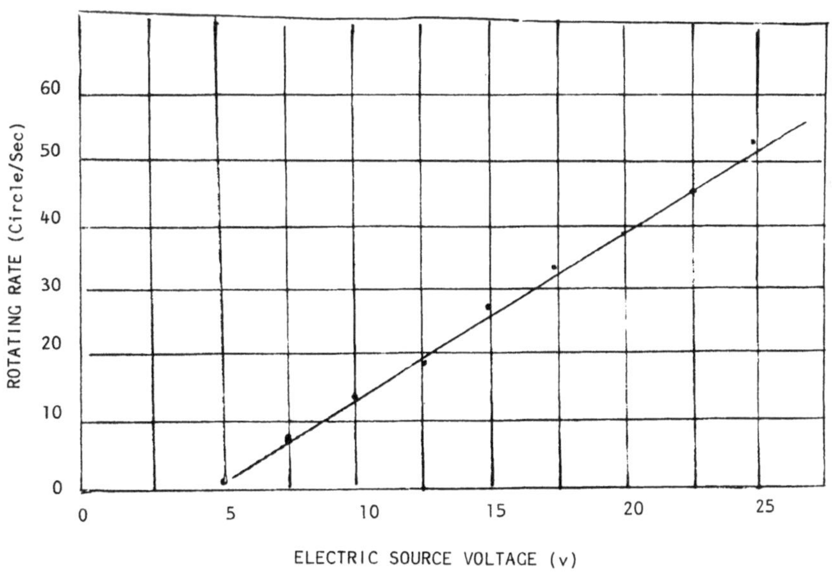

Fig.10 Curve of Variation of Rotation Speed with the Electric Sourse Voltage of the Motor

(2) Beam-Locating Signal

Fig.11 is the curve of beam-locating signals when the beam is scanning, with different source voltages. The negative edges start sampling and the positive ones end the sampling. It can be seen that the signals are steep enough and have enough amplitude.

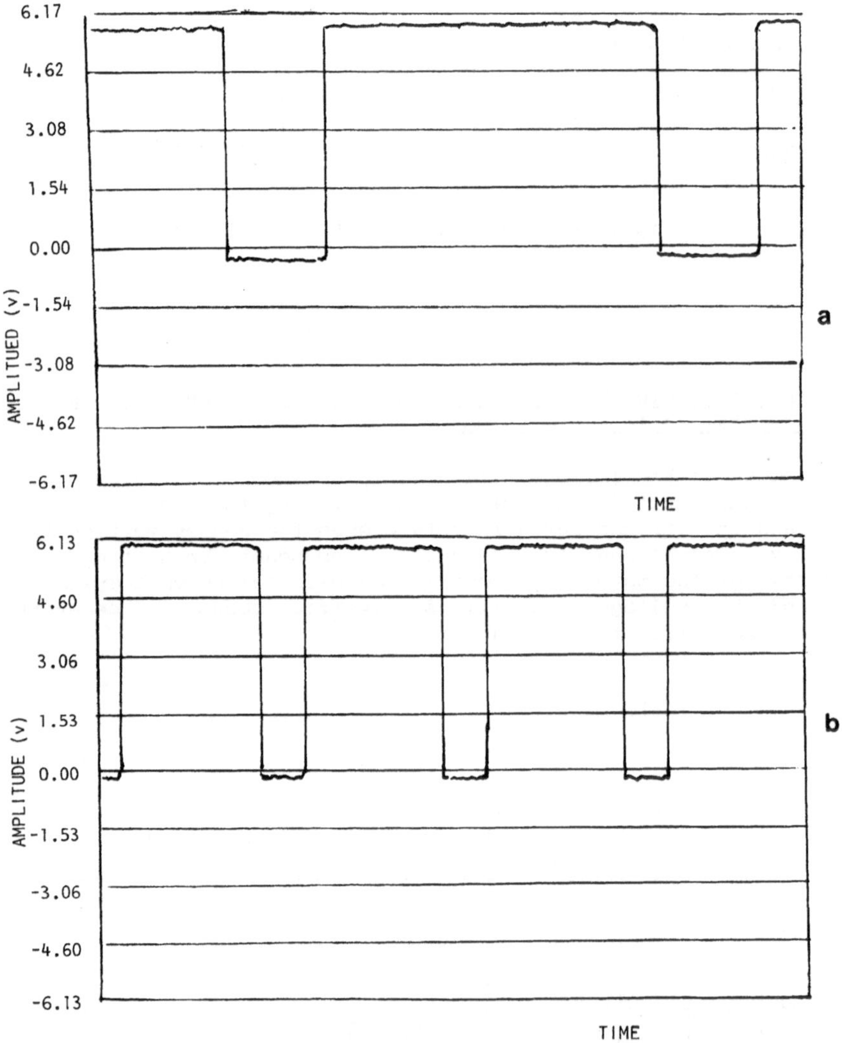

Fig.11 The Curves of Beam-Locating Signals, the Electric Sourse Voltages Are 10 v and 20 v in (a) and (b) Respectively

(3) The Pulse Response of the Transducer

As an acoustic load of the transducer, a test block is shown in Fig.12. When incident angles are 30 and 40 degrees, the transducer is at different positions on the surface of the block; received echoes against a circular hole within the block are shown in Fig.13.

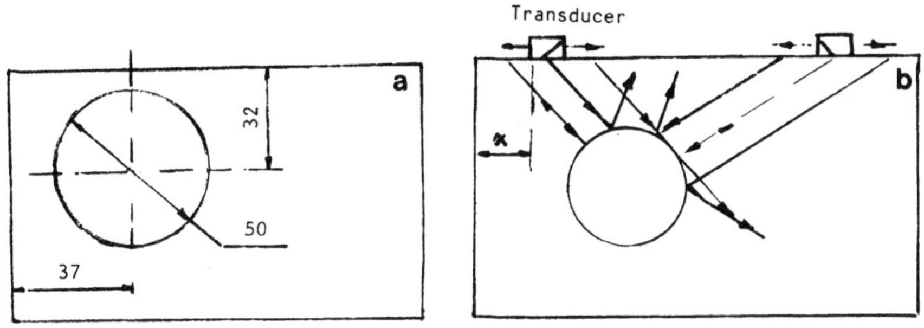

Fig.12 The Sketch of a Test Block (a) Test Block; (b) The Reflections from the Hole

In reference to Fig.12(b), the maximum reflection and shortest duration can be gotten only at the point on the rim of the hole which corresponds to the normal incidance. That is, reflected echo can arrive at the maximum value and shortest duration, only when the transducer is in a certain position, if the incident angle is fixed, which is just shown in Fig.13, and in other positions, most of energy of the reflected wave does not return in the orignal way; the received wave is scattered signal at the point on the rim. So, its amplitude is less, the waveform is more irregular and longer. The bigger the difference between incident direction and normal direction of a reflecting point, the more serious the scattering, the more irregular the waveform, and the longer the duration. Hardly any signal can be received when the incident direction is along that of the tangent about the reflecting point. In the case of fixed angle incidence, only a fraction of reflected information about flaws in material can be obtained. But by means of varying the incident angles, much more of the information can be gotten. Therefore the mechanical-scanning transducer is more advantageous to practical detection.

By the way, it should be indicated that a B-scanning image can be developed from these received echoes such as shown in Fig.13 by using certain imaging methods.

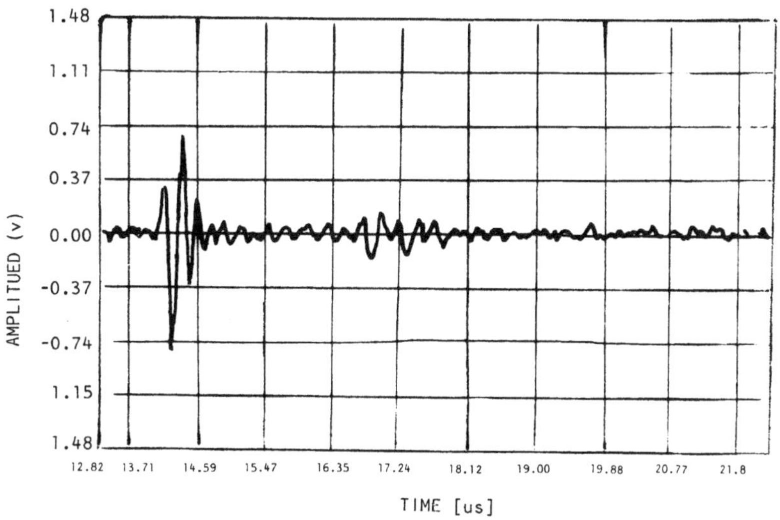

Fig.13 The Curves of Echoes against the Hole When Incident Angel is 30°

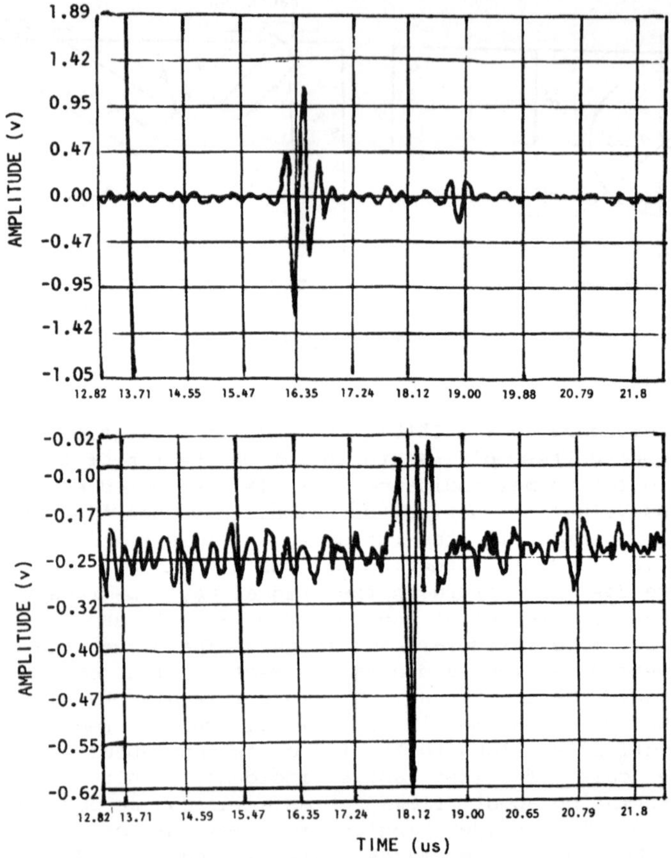

Fig.13 The Curves of Echoes against the Hole When Incident Angel is 30°

IV. CONCLUSION

The theories and designs of a short pulse mechanical-scanning transducer have been presented in this paper. The transducer has been used in an ALOK imaging system, which shows it is effective to use this transducer to improve the resolution of imaging and detection, and it can also be used for real-time imaging.

Besides, compared to phased array, the mechanical-scanning ultrasonic transducer is easier to construct and does not need complicated control circuits to realize the beam-scanning.

V. ACKNOWLEGEMENT

The authors wish to express their gratitude to Engineer Chang-Fu Wang for his some work in the manufacture of the transducer and Mr. Xin-De Li for some help about the software.

VI. REFERENCE

(1) K.F. Bainton and M.G. Silk, "Some Factors which Affect the Performance of Ultrasonic Transducers," Br.J.NDT 22,15-20

TONPILZ PIEZOELECTRIC TRANSDUCERS WITH ACOUSTIC MATCHING PLATES FOR UNDERWATER COLOR IMAGE TRANSMISSION

T. Inoue, T. Nada*, T. Tsuchiya**, T. Nakanishi**,
T. Miyama*, S. Takahashi, and M. Konno***

Material Development Center, NEC Corporation, Kawasaki 213, Japan
* Radio Application Division, NEC Corporation, Fuchu, Tokyo 183, Japan
** Japan Marine Science and Technology Center, Yokosuka 237, Japan
*** Faculty of Engineering, Yamagata University, Yonezawa 992, Japan

ABSTRACT

Tonpilz piezoelectric transducers with multiple acoustic matching plates are suitable for color image acoustic transmission, regarding the possibility of wideband low-ripple characteristics as well as high-efficiency high-power transmitting capability. The design method for the transducers was investigated on the basis of multiple-mode filter synthesis theory. For respective transducers with single, double and triple matching plates, optimum specific acoustic impedances and lengths were calculated. Moreover, based on this design method, a 24 kHz array comprising nine identical transducers with single matching plates was built and evaluated. As a result, this array showed high-efficiency, low-ripple and wide-band characteristics. Excellent agreement between theoretical values and experimental results was obtained. Then, a field test was carried out on color image transmission from 3500 m sea depth, using the fabricated array. As a result, good color images were received.

1. INTRODUCTION

Ultrasonic transducer, which can directly and distinctly transmit color images at the sea bed to a vessel on the surface, are desired for ocean subsurface investigation. Wideband low-ripple characteristics as well as high-efficiency high-power transmitting capability are required for the transducers, to transmit the color video pictures with high-quality and high-speed over a long distance, because the transducers have good pulse transfer capability and can transmit and receive a lot of information. Many attempts to broaden the transducer bandwidth have been made involving equipping piezoelectric ceramic resonators with acoustic matching plates[1-9].

Tonpilz piezoelectric transducers are widely used for high-power underwater transducers[10-11]. The Tonpilz transducers have fairly smaller mechanical impedance densities, looking from their acoustic radiation end, than those of conventional piezoelectric ceramic transducers operating in a longitudinal mode. Therefore, by forming single or multiple acoustic matching plates on the Tonpilz transducer radiation surface, a wider band transducer may be obtained, as compared with a piezoelectric ceramic transducer with single or multiple matching plates[12].

This paper presents a precise electro-mechano-acoustical equivalent circuit, including a stress-bolt assembly, for the Tonpilz transducer with multiple acoustic matching plates and its transmission matrix representation. Also, design methods for a Tonpilz transducer with multiple acoustic matching plates were investigated on the basis of multiple-mode filter synthesis theory[13-14]. For the respective transducers with single, double and triple matching plates, the optimum specific acoustic impedances and lengths of the matching plates were calculated using an iterative calculation method. The transmitting and receiving voltage sensitivities for the transducers are expressed as absolute values by MKS units. Moreover, based on this design method, a 24 kHz array comprising nine identical transducers with a single matching plate was built and evaluated. The matching plate is made of alumina-epoxy composite, whose specific acoustic impedance is 3.2×10^6 kg/m^2·s. As a result, this transducer showed high-efficiency, low-ripple and wide-band characteristics. Excellent agreement between theoretical values and experimental results was obtained.

Then, a field trial was carried out on the acoustic transmission of freeze color video pictures from 3500 m sea depth, using the fabricated array. FDM carrier systems were introduced to data transmissions and controls between video picture transmission systems on a ship and in water. As a result, clear color video pictures were obtained.

2. TRANSDUCER DESIGN

The construction of a Tonpilz piezoelectric transducer with multiple acoustic matching plates, which is considered here, is shown in Figs. 1(a) and (b). Figures 1(a) and (b) show a front view and a cross-sectional view, respectively. For the transducer structural materials, Al alloy is used for the head mass, stainless steel is used for the disk and tail mass, and high-tensile steel is used for the bolt and bolt head. The ceramic stack is composed of four piezoelectric ceramic rings polarized longitudinally and thin metal plates inserted between them, bonded together with epoxy adhesive. The adjacent ceramic rings are arranged in

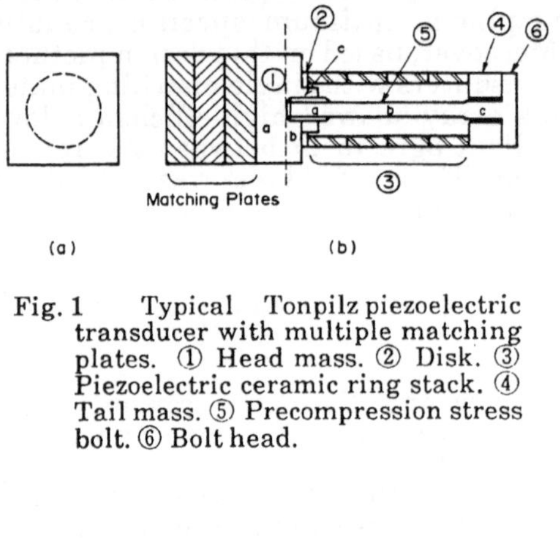

Fig. 1 Typical Tonpilz piezoelectric transducer with multiple matching plates. ① Head mass. ② Disk. ③ Piezoelectric ceramic ring stack. ④ Tail mass. ⑤ Precompression stress bolt. ⑥ Bolt head.

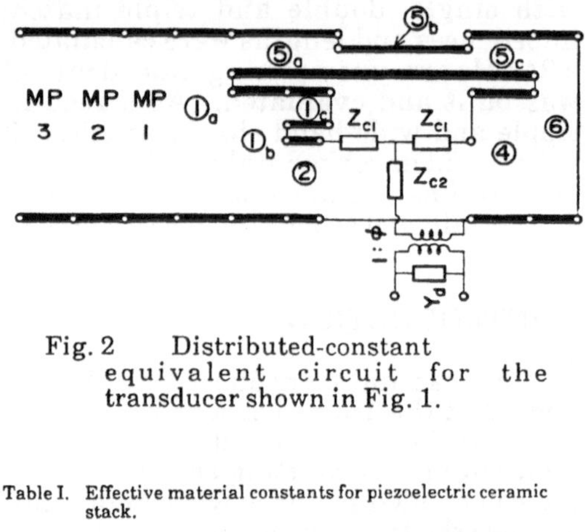

Fig. 2 Distributed-constant equivalent circuit for the transducer shown in Fig. 1.

Table I. Effective material constants for piezoelectric ceramic stack.

Parameter	Quantity
Density $\dot{\rho}_3$ (kg/m^3)	7.73
Sound velocity $\dot{c}_3$ (10^3m/sec)	2.90
Elastic compliance constant $\dot{s}_{33}^E$ (10^{-12}m^2/N)	15.4
Specific dielectric constant $\dot{\varepsilon}_{33}^T/\varepsilon_0$	1450
Electromechanical coupling factor $\dot{k}_{33}$	0.62
Dielectric loss tangent $\tan\dot{\delta}$	0.30×10^{-2}
Mechanical quality factor $\dot{Q}_m$	200

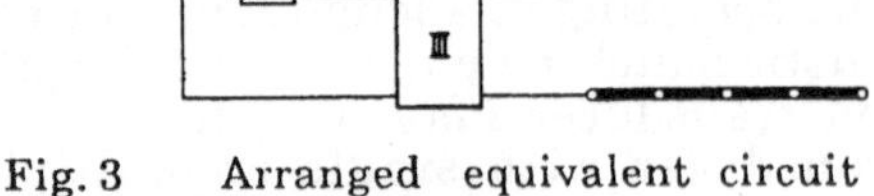

Fig. 3 Arranged equivalent circuit for the transducer.

opposite polarized directions from each other. The ceramic rings are electrically connected in parallel.

Deriving the transducer equivalent circuit, the following approximative treatment is applied, so as to simplify the analysis. For the bolt-head mass part, it is assumed that the bolt is completely incorporated with the head mass ①$_a$ part at plane 1-1' and is not mechanically contacted with the head mass ①$_b$ part[15], since the vibrational velocity distribution in the head mass is almost uniform. To increase calculating precision, it is considered that the ceramic stack is a strongly united structure of effective ceramic rings[16]. Thus, the effects of the electrode plates, the adhesive layers and the compressive bias stress caused by bolting are experimentally included into the equivalent material constants of the effective ceramic rings. The effective material constants for the piezoelectric ceramic stack are listed in Table I.

By fully examining the vibrational velocities and the forces among the structural elements in this transducer, a distributed-constant equivalent circuit is obtained from electronic terminals to an acoustic radiation surface, as shown in Fig. 2. A Martin equivalent circuit[17] is used for the piezoelectric ceramic ring stack, because each individual ring length is fairly smaller than elastic wavelengths in operated frequency range.

Then, by rearranging the equivalent circuit shown in Fig. 2, the equivalent circuit shown in Fig. 3 is obtained. In Fig. 3, four terminal circuits, I, II and III, are denoted as follows:

I Four-terminal circuit cascade-connected by structural elements ④ and ⑥;

II Four-terminal circuit cascade-connected by structural elements ⑤$_c$, ⑤$_b$ and ⑤$_a$;

III Four-terminal circuit cascade-connected by structural elements ②, ①$_c$ and ①$_b$.

To serve the facility for transducer design, it is considered that damped admittance Y_d and mechanical loss for the transducer are neglected. The F-matrix components for the bare Tonpilz transducer with acoustic matching plates removed, that is, the mechanical system from the right adjoining part of an electromechanical transformer to the head mass end, as shown in Fig. 3, are expressed as A_{mo}, B_{mo}, C_{mo} and D_{mo}.

Composite F-matrix components A_m, B_m, C_m and D_m for the transducer mechanical system with matching plates are given by

$$\begin{bmatrix} A_m & B_m \\ C_m & D_m \end{bmatrix} = \begin{bmatrix} A_{mo} & B_{mo} \\ C_{mo} & D_{mo} \end{bmatrix} \prod_{i=1}^{n} \begin{bmatrix} A_{mpi} & B_{mpi} \\ C_{mpi} & D_{mpi} \end{bmatrix} , \quad (1)$$

where the i-th matching plate from the head mass end is indicated by appending a subscript mpi ($i=1, 2, \dots, n$).

For the transducer, as shown in Fig. 1, multiple resonances can be detected at the electronic terminals, analogous to the piezoelectric ceramaic disk transducers with multiple matching layers[13]. Therefore, a design technique based on multiple mode filter synthesis theory can be applied to this transducer. To be exact, design parameters are the specific acoustic impedances z_{ompi} ($i=1, 2, \dots n$) and the lengths l_{mpi} ($i=1, 2, \dots, n$) for the respective matching plates. In this transducer design, to reduce the design parameters, length l_{mpi} is expressed, using a common nondimensional coefficient a_λ[13], by

$$l_{mpi} = a_\lambda \lambda_{mpi} / 4 , \quad (2)$$

where λ_{mpi} is the i-th matching plate wavelength prescribed by the resonant frequency f_o of the bare transducer with no matching plate. Owing to this simplification, the design parameters become z_{ompi} ($i=1, 2, \dots, n$) and a_λ.

If it is intended to obtain a low-ripple, wide-band transducer, it is necessary that image impedance Z_{02}, looking toward the transducer from the acoustic load, should equal load impedance Z_L at the center frequency. Also, the design parameters should be determined in order that Z_{02} may become the real number extending over the widest possible frequency range.

Table II. Design values for Tonpilz transducer with matching plates.

Matching Plates	Specific Acoustic Impedance (10^6 kg/m^2·s)			Length Coefficient
	z_1	z_2	z_3	a_λ
Single	3.2			1.05
Double	4.5	1.8		1.10
Triple	6.0	2.6	1.7	1.10

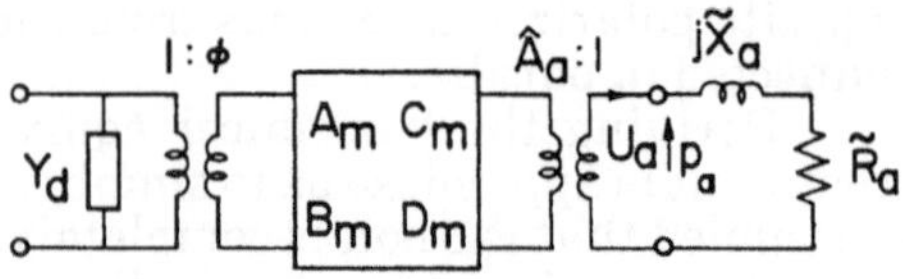

Fig. 4 Electro-mechano-acoustical equivalent circuit of water-loaded transducer.

On the basis of the design method described above, Tonpilz transducers having single, double and triple matching plates were designed, using iterative calculations. The optimum specific acoustic impedances z_{ompi} ($i=1, 2, 3$) and length coefficient a_λ values for the transducers are listed in Table II.

3. PERFORMANCE ESTIMATION

In general, Tonpilz transducers are used as matrix arrays, in order to satisfy required directivities and source levels and to reduce unavailable acoustic radiations. The array dealt with in this section is a regular square array where identical transducer elements are closely arranged into an $N\times N$ matrix.

Dielectric and mechanical losses for the transducer, except the matching plates, have been described in Sect. 2. For the matching plates, hereupon, the complex elastic modulus method is not employed, as they are usually made of high-lossy materials. Instead, the mechanical loss for the i-th matching plate is represented by loss factor η_i, in the same way as reported in Ref. (14).

An electro-mechano-acoustical equivalent circuit, including acoustic radiation impedance for a single transducer element in an array, is shown in Fig. 4. p_a and U_a are sound pressure at acoustic radiation surface and volume velocity, respectively. Also, $\tilde{R}_a$ and $\tilde{X}_a$ indicate the resistance and the reactance, respectively, in acoustic radiation impedance $\tilde{Z}_a$. F-matrix components A, B, C and D from the electronic terminals to the acoustic terminals are given by

$$\begin{bmatrix} A & B \\ C & D \end{bmatrix} = \begin{bmatrix} 1 & 0 \\ Y_d & 1 \end{bmatrix} \begin{bmatrix} \phi^{-1} & 0 \\ 0 & \phi \end{bmatrix} \begin{bmatrix} A_m & B_m \\ C_m & D_m \end{bmatrix} \begin{bmatrix} \hat{A}_a & 0 \\ 0 & \hat{A}_a^{-1} \end{bmatrix} \quad (3)$$

where $\hat{A}_a$ and ϕ denote the acoustic radiation area and the electromechanical transformer ratio, respectively.

3.1 Sound Pressure Level SL_d and Transmitting Voltage Sensitivity S_s for Square Array

Acoustic radiation power P_a for a square array comprising $N\times N$ identical transducers is N^2-times larger than that for a single transducer element in the same array. Therefore,

$$P_a = N^2\tilde{R}_a|U_a|^2 = N^2\tilde{R}_a \frac{V_{in}^2}{|A\tilde{Z}_a+B|^2}, \quad (4)$$

where V_{in} is input voltage applied to electronic terminals. Using the relation between the acoustic radiation power P_a shown in Eq. (4) and sound intensity[18], sound pressure level SL_d, generated at a point d meters from the radiation surface, are given by the following formula. Hence,

$$SL_a(\text{dB } re\ 1\,\mu\text{Pa}) = 20\log_{10}N + 10\log_{10}\frac{\tilde{R}_a}{|A\tilde{Z}_a+B|^2} + 20\log_{10}V_{in} + 10\log_{10}g_0 - 10\log_{10}(4\pi/\rho c) + 120 - 20\log_{10}d, \quad (5)$$

where ρ and c are water density and sound velocity in water, respectively, and g_0 is directivity index for the array.

Transmitting voltage sensitivity S_s is defined as the sound pressure, generated at intervals just 1 meter from the radiation surface, when 1 V voltage is applied. By substituting the conditions $d=1$ m, $V_{in}=1$ V, $\rho=1000$ kg/m^3 and $c=1500$ m/s into Eq. (5), S_s is written as

$$S_s(\text{dB } re\ 1\,\mu\text{Pa}/V)=20\log_{10}N+10\log_{10}\frac{\tilde{R}_a}{|A\tilde{Z}_a+B|^2}$$

$$+10\log_{10}g_0+170.77\ . \tag{6}$$

Acoustic radiation impedance density Z_a for the array is approximated by that for an equivalent piston disk with radiation area N^2A_a mounted in an infinite baffle[19-21]. The radius a_n of the equivalent piston disk becomes $a_n=N\sqrt{A_a/\pi}$. Then, acoustic radiation impedance Z_a for the single transducer element is given by,

$$\tilde{Z}_a=\tilde{R}_a+j\tilde{X}_a=\frac{\rho c}{\hat{A}_a}\left\{\left(1-\frac{2J_1(x)}{x}\right)+j\frac{2S_1(x)}{x}\right\}, \tag{7}$$

where $x=2a_n\omega/c$
$J_1(x)$; first kind first order Bessel function
$S_1(x)$; Struve function.

Also, directivity index g_0 values for the array are given, using the directivity index for the equivalent piston disk, by

$$g_0=\frac{x^2}{4\left\{1-\dfrac{2J_1(x)}{x}\right\}}\ . \tag{8}$$

3.2 Receiving Voltage Sensitivity M_o for Square Array

Receiving voltage sensitivity M_o is defined as the open-circuit generated voltage for a receiver system, in a free sound field for a traveling wave with a unit sound pressure, reconciling the directivity axis for the receiver system with the traveling-wave progressive direction[22]. Based on the Sitting and Meitzler theory[23-24], M_o for the array is given by

$$M_o(\text{dB } re\ 1\,V/\mu\text{Pa})=-120+20\log_{10}\left|\frac{1}{D+\tilde{Z}_aC}\right|\ . \tag{9}$$

3.3 Calculation Results

For the respective 3×3 matrix arrays comprising Tonpilz transducers with single, double and triple matching plates, designed as reported in Sect 2, the S_s and M_o frequency characteristics were calculated. The acoustic radiation area of each transducer is 2.5 cm×2.5 cm. Theoretical S_s and M_o responses are shown in Figs. 5 and 6, respectively, as contrasted with those for the 3×3 matrix array comprising Tonpilz transducers with no matching plate. The peaks and dips, in the from 46.0 kHz to 47.2 kHz frequency range, seen in all these arrays are caused by higher-order longitudinal mode resonances and antiresonances. Also, the fractional 6-dB bandwidths for the theoretical S_s and M_o responses are listed in Table III.

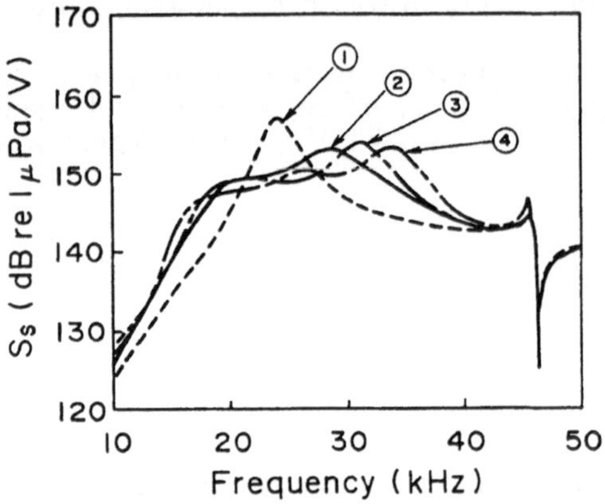

Fig. 5 Theoretical transmitting voltage sensitivity characteristics for several transducer arrays. ① No matching plate. ② Single matching plate. ③ Double matching plates. ④ Triple matching plates.

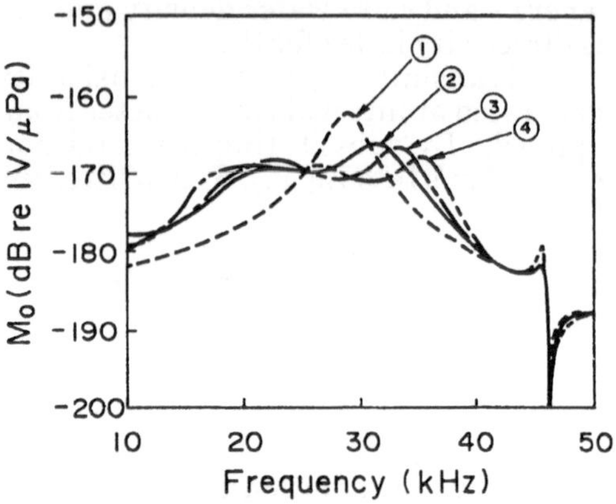

Fig. 6 Theoretical receiving voltage sensitivity characteristics for several transducer arrays. ① No matching plate. ② Single matching plate. ③ Double matching plates. ④ Triple matching plates.

Table III. Theoretical S_s and M_o fractional 6-dB bandwidths for designed transducer arrays (3×3).

Transducers	Fractional 6-dB Bandwidth (%)	
	S_s	M_o
No matching plate	19.2	18.1
Single matching plate	60.2	62.7
Double matching plates	64.4	70.9
Triple matching plates	65.2	86.3

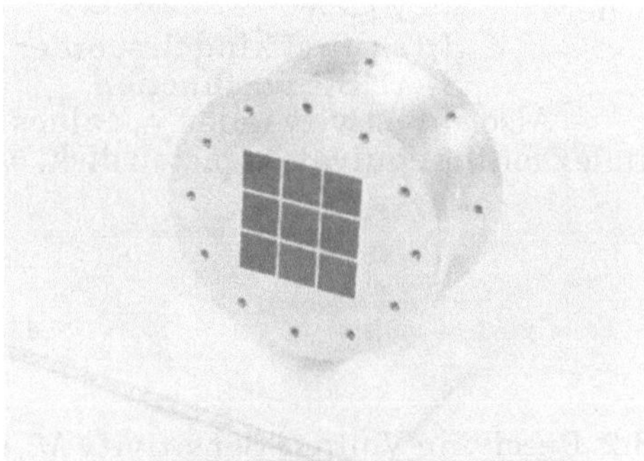

Fig. 7 Fabricated 24 kHz oil-filled matrix array comprising nine transducers.

4. EXPERIMENTAL RESULTS FOR TRANSDUCER ARRAY

An array, comprising nine identical Tonpilz transducers with single acoustic matching plates, was built on the basis of the above-described design method. The resonant frequency for the bare transducer with no matching plate is 24.2 kHz. All these transducers in the array were supported at their ceramic stack central parts, using an epoxy-FRP plate with three mounting circular apertures. The acoustic radiation surfaces of these transducer elements were arranged with 2 mm gaps. The acoustic radiation area of each transducer is equal to the value used in the calculation in Subsect. 3.3. Each individual transducer is 81 mm long overall and its matching plate is 23 mm long. The matching plate is made of Al_2O_3-epoxy composite, to realize the desired specific acoustic impedance of 3.2×10^6 kg/m^2·s. The electroacoustic conversion efficiency for the transducer, actually obtained from midair and underwater admittance loci, was over 0.85, when only the matching plate part was dipped into water.

The array was mounted in a sturdy cylindrical housing case made of Al alloy and was immersed in castor oil, as shown in Fig. 7. Frequency responses for S_s and M_o were measured as a function of frequency. Measured results for S_s and M_o are shown in Figs. 8 and 9, as contrasted with theoretical values. For these calculation, dielectric loss tangent (tan δ), effective mechanical quality factor Q_m

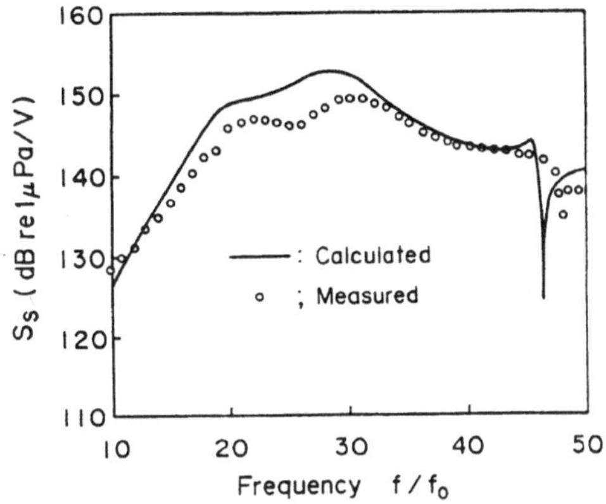

Fig.8 Transmitting voltage sensitivity S_s response for the array with a single matching plate.

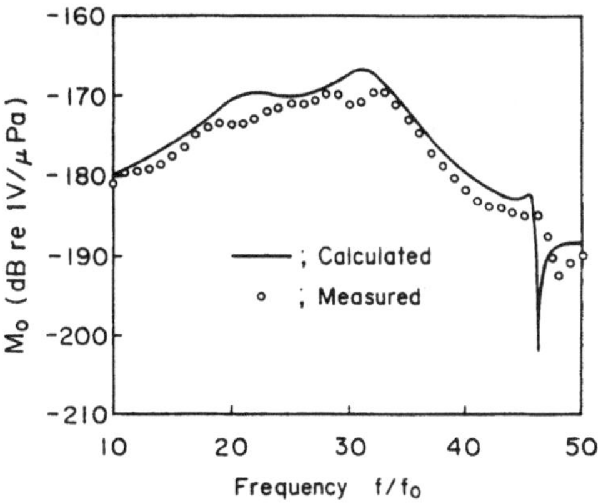

Fig. 9 Receiving voltage sensitivity M_0 response for the array with a single matching plate.

including supporting effect and loss factor η_l are determined as 0.003, 200 and 0.032, respectively. The maximum S_s and M_0 values, obtained in the test, are 149 dB and −170 dB, respectively. The actually obtained S_s value is lowered to a small extent, compared with theoretical S_s value, owing to unavailing acoustic radiation to the oil. Both measured S_s and M_0 responses exhibit low-ripple wide-band characteristics, which are similar to the theoretical characteristics. Experimentally obtained fractional 6-dB bandwidths for S_s and M_0 are 56 % and 57%, respectively. These bandwidth are nearly twice as large as those for a longitudinal mode piezoelectric ceramic transducer array with single matching plate. As can be seen in Figs. 8 and 9, the deteriorations in S_s and M_0 responses in the 47 kHz to 48 kHz range are fairly suppressed, because the ceramic stack center part for the transducer, which corresponds to the side portion of spurious vibration mode, is supported by an epoxy-FRP plate. Measurements for S_s and M_0 were carried out, using a comparative calibration measurement method[22], with a standard hydrophone and a standard transmitter, respectively.

The measured typical horizontal directivity pattern for the array is shown in Fig. 10. Actual front beam patterns are in agreement with theoretical beam patterns calculated from an array comprising equivalent square piston vibration plates[20]. Measured beam width at 24 kHz is 46° at the −6 dB points, which is narrower by 4° than theoretical beam width. However, measured rear beam patterns differ from theoretical beam patterns, because of the unfavorable acoustic radiation from the housing case through the castor oil medium.

5. COLOR VIDEO PICTURE TRANSMISSION

FDM (Frequency Division Multiplex) carrier systems were introduced to data transmissions and controls between video picture transmission systems on a ship and in water. The 20 to 30 kHz acoustic frequency band was used in this color video picture transmission. This frequency band is extremely narrower than that for ordinary TV broadcasting. Therefore, it is essential that the data quantity per unit time be reduced, while still enabling acoustically transmitting the video data. The highest priority impressed upon the authors was to prevent damage to the video picture quality as much as possible. Accordingly, the freeze picture transmission method, converted from the video data, was examined.

5.1 Transmission System

A blockdiagram of the color video picture transmission system is shown in Fig. 11. The greater part of this system is controlled by 8 bit CPUs. When a frame was composed of 256×256 pixels, it was previously found that practically satisfactory frames were obtained. This pixels number was adopted in the system.

The picture taken by a video camera forms color composite signals by the

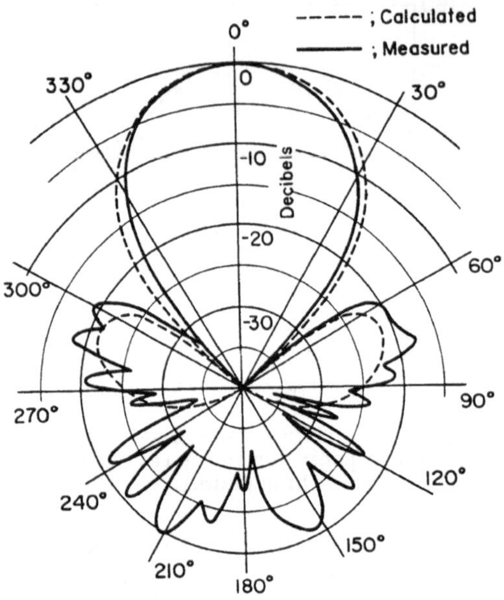

Fig. 10 Typical horizontal beam pattern (24 kHz) for the array with a single matching plate.

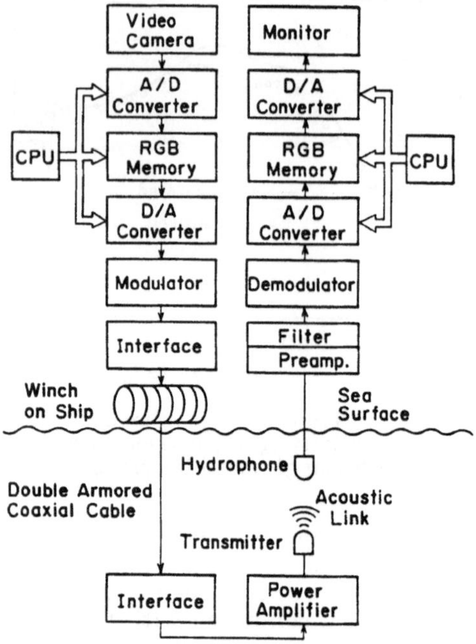

Fig. 11 Blockdiagram of color video picture transmission system.

NTSC (National Television System Committee) method. This picture is arranged into chrominance signals of R (Red), G (Green) and B (Blue) and luminance signals. Since these arranged signals are analog signals, these signals are converted to digital signals by an A/D converter, and then are recorded in the RGB memory by CPU control.

One line consist of 256 pixels transmitted in sequence, in which one pixel contains both R, G and B chrominance signals and luminance signals with 256 gradations per individual R, G and B signal. Also, a line sequential-color TV system is adopted in this system, to extremely lessen color aberration. One frame is composed by transmitting the 256 lines.

The SSB-SC (Single Sideband Suppressed Carrier Modulation) system[25] was used for the modulation system in this transmission system, so as to suppress spectrum spreading as much as possible and to maximally utilize the limited frequency band. Owing to the SSB-SC system, signals equivalent to FM signals can be obtained by suppressing the carrier wave and extracting only the upper sideband (or the lower sideband). In an actual modulation circuit, where the upper sideband was extracted, the carrier wave could be suppressed to less than −40 dB, and no mutual interference between two sidebands was observed, using a balanced modulator and a filter.

5.2 Field Test

Field test was carried out over a maximum 3500 m transmission distance, along with depth direction in a 4000 m depth sea area. The color video signals on the ship were transmitted to the transducer array in water through double armored coaxial cable. Then, the video signals were acoustically transmitted from the array.

As the result of a preparatory experiment carried out in an acoustic test pond, it was found that a less than −20 dB signal-to-noise ratio (S/N) would be necessary, to enable receiving clear color video pictures. To guarantee this S/N value, source level was set up at over 200 dB re 1 μPa, considering acoustic attenuation[26]. The hydrophone was suspended at 20 m depth below the sea surface.

Results of this field test are as follows.

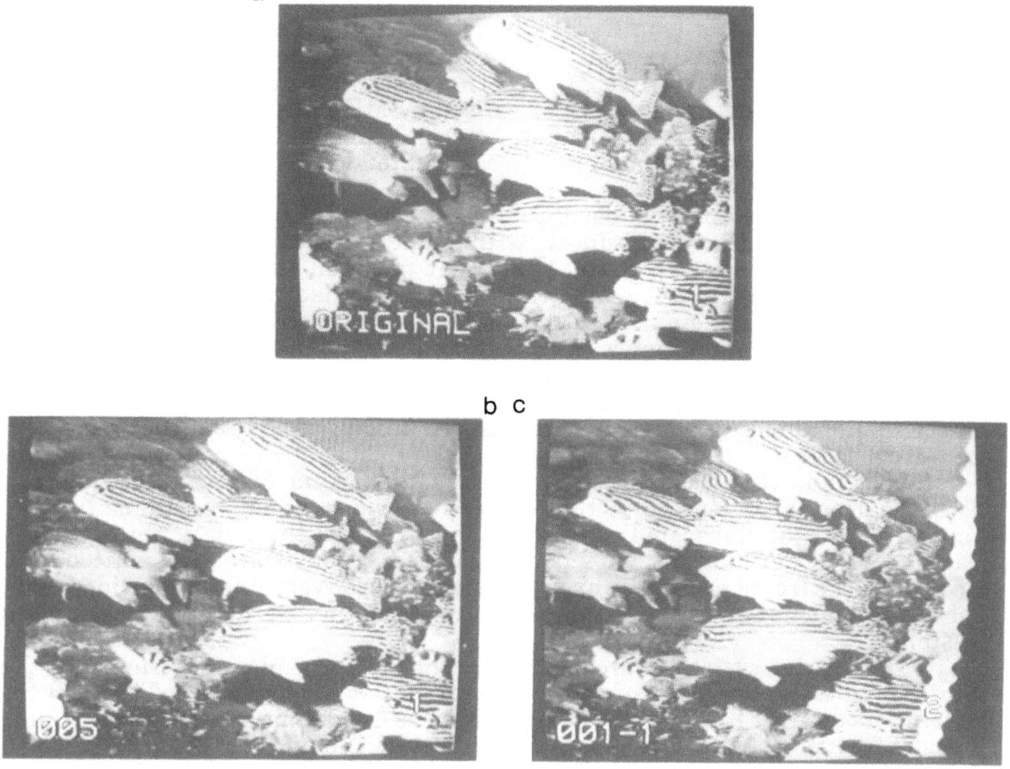

Fig. 12 Color video pictures obtained by field trial at 3500 m transmission distance along with depth direction.
(a) Original picture. (b) Received picture by transmission synchronizing method. (c) Received picture by independent synchronizing method.

The original video pictures taken by the video camera and the video pictures received using this transmission system are shown in Figs. 12(a) to (c).

(1) Clear color video pictures, as shown in Fig. 12(b), could be received throughout the selected acoustic frequency band (20-30 kHz), using the transmission synchronizing method, wherein both vertical and horizontal synchronizing signals were transmitted.

(2) For the independent synchronizing method, wherein only vertical synchronizing signals, indicating the starting points of the video scanning lines, were transmitted, the distortion of the video pictures markedly appeared, as shown in Fig. 12(c). This phenomenon is attributable to the aberration in the scanning line terminals, because the distance change between the transmitter and the hydrophone, caused by the roll and pitch of the ship, affects scanning time.

(3) The transmission velocity of the color video picture was confined to 45 s/frame, using this transmission system. However, improvement in transmission velocity is sufficiently practicable, owing to the increase in data processing velocity, using the digital transmission system (PSK; Phase Shift Keying).

6. CONCLUSION

A precise electro-mechano-acoustical equivalent circuit for a Tonpilz transducer with multiple acoustic matching plates and its transmission matrix representation have been presented. Design method for the transducer was investigated on the basis of multiple-mode filter synthesis theory. For respective transducers with single, double and triple matching plates, the optimum specific acoustic impedances and lengths for the matching plates were calculated, using an iterative calculation method. Transmitting voltage sensitivities S_s and receiving voltage sensitivities M_0 for the 3×3 matrix transducer arrays, comprising these transducers, were expressed by MKS units. It was ascertained that these arrays have high-sensitivity, low-ripple and wide-band characteristics and that these bandwidths cannot be greatly extended, although the number of matching plates has increased.

Moreover, based on this design method, a 24 kHz oil-filled square array, comprising nine identical transducers with single matching plates, was built and evaluated. As a result, excellent agreement between theoretical values and experimental results were obtained. The transducer exhibited high electroacoustical conversion efficiency of over 85 %. For the actual array, low-ripple wide-band characteristics were obtained. Fractional 6-dB bandwidths for S_s and M_0 were 56 % and 57 %, respectively. These bandwidths are nearly twice larger than those for longitudinal mode piezoelectric ceramic transducer array with a single matching plate. Also, measured beam width at 24 kHz was 46° at the −6 dB point.

Then, field test was carried out over a maximum 3500 m transmission distance along with depth direction. The color video signals on the ship were transmitted to the fabricated oil-filled array in water through double armored coaxial cable. Contiguously, the video signals were acoustically transmitted from the array. The freeze picture transmission method, converted from the video data, was investigated, to prevent damage to the video picture quality as much as possible.

FDM carrier systems were introduced to the transmission systems on the ship and in water, because of their uncomplicated construction. The SSB-SC system was used for the modulation system, so as to maximally utilize the limited acoustic frequency band. As a result, clear color video pictures could be received throughout the selected acoustic frequency band (20-30 kHz). Accordingly, there is a fair possibility for the acoustic transmission of practical color video pictures from the abyssal zone. For that purpose, it is extremely important that digital transmission systems, including data compression and image processing techniques, are introduced, so as to increase data processing velocity and to minimize transmission errors.

ACKNOWLEDGEMENT

The authors are very grateful to general manager T. Ohno, and managers K. Sugiuchi and M. Suga at NEC Corporation for their guidance and support in this work.

REFERENCES

1) T. Matsuki: J. IECE Japan, 35, pp. 560-563 (1952).
2) G. Kossoff: IEEE Trans., Sonics and Ultrasonics, SU-13, pp. 20-30 (1960).
3) J.H. Goll and B.A. Auld: IEEE Trans. Sonics and Ultrasonics, SU-22, pp. 52-53 (1975).
4) C.S. Desilets, J.D. Fraser, and G.S. Kino: IEEE Trans. Sonics and Ultrasonics, SU-25, pp. 115-125 (1978).
5) J. Souquet, P. Defranould, and J. Desbois: IEEE Trans., Sonics and Ultrasonics, SU-26, pp. 75-81 (1979).
6) J.H. Goll: IEEE Trans. Sonics and Ultrasonics, SU-26, pp. 385-393 (1979).
7) K. Shibayama, T. Matsunaka, and H. Sato: J. Acoust. Soc. Japan, E-1, pp. 113-119 (1980).
8) A.R. Selfridge, B. Baer, B.T. Khuri-Yakub, and G.S. Kino: Proc. 1981 IEEE Ultrasonic Symp., 1981, pp. 644-648 (IEEE Cat. 81CH1689-9).
9) B.V. Smith and B.K. Gazey: IEE Proceedings, 131, pt.F, pp. 285-293 (1984).
10) D.F. McCammon and W. Thompson, Jr.: J. Acoust. Soc. Am., 68, pp. 754-757 (1980).

11) J.N. Decarpigny, J.C. Debus, B. Tocquet and D. Boucher: J. Acoust. Soc. Am., 78, pp. 1499-1507 (1985).
12) M.V. Crombrugge and W. Thompson, Jr.: J. Acoust. Soc. Am., 77, pp. 747-752 (1985).
13) T. Inoue, M. Ohta, and S. Takahashi: IEEE Trans. Ultrasonics, Ferroelectrics, and Frequency Control, UFFC-34, pp. 8-15 (1987).
14) T. Inoue, T. Nada, A. Kameyama, K. Sugiuchi, S. Takahashi, and M. Konno: Trans IEICE Japan, E70, pp. 723-733 (1987).
15) S. Ueha and E. Mori: J. Acoust. Soc. Japan, 34, pp. 635-640 (1978).
16) T. Inoue and S. Takahashi: Trans. IECE Japan, E69, pp. 1180-1188 (1986).
17) G.E. Martin: J. Acoust. Soc. Am., 36, pp. 1496-1506 (1964).
18) R.J. Urick: "Principles of Underwater Sound", McGraw-Hill, New York (1975).
19) L.L. Beraneck: "Acoustics", McGraw-Hill, New York (1954).
20) Y. Kikuchi: "Magnetostrictive Vibrations and Ultrasonic Waves," Corona, Tokyo (1974).
21) T. Hayasaka and S. Yoshikawa: "Electroacoustic Vibration Theory", Maruzen, Tokyo (1974).
22) J. Saneyoshi, Y. Kikuchi, and O. Nomoto: "Handbook of Ultrasonic Technology", Daily Industrial Press, Tokyo (1960).
23) A.H. Meitzler and E.K. Sittig: J. Appl. Phys., 40, pp. 4341-4352 (1969).
24) E.K. Sittig: IEEE Trans. Sonics and Ultrasonics, SU-16, pp. 2-10 (1969).
25) Members of the Technical Staff, Bell Telephone Labs.: "Transmission Systems for Communications —5th Edition", Bell Telephone Labs., North Carolina (1982).
26) M. Schulkin and H. W. Marsh: J. Acoust. Soc. Am., 34, pp. 864-865 (1962).

PVDF-DMOS SENSORS AND ARRAY FOR UNDERWATER ACOUSTIC IMAGING

Yuan Yi-quan
Department of Radio
Nanjing Institute of
Technology, China

Shen Shoupeng and Shi Bing-wen
Shanghai Institute of Organic
Chemistry, Acadamia Sinica

ABSTRACT

In this paper, the new application of high molecule piezoelectric film in underwater acoustics is analysed by showing the example of developing the PVDF-DMOS IC and sensor plane array (8 x 8, the element area 12-18 mm^2) for underwater acoustic imaging. The vibration model and mechanical and electrical equivalent diagrams for individual sensors has been established. Various factors that could have an influence on the sensitivity are analysed and the measurement results from the work are presented in this paper. Experimentally measured transducer impulse response decays 20 dB in two cycles. The paper can be used as a reference for the design of 200-500 Hz PVDF underwater acoustic imaging arrays.

INTRODUCTION

In recent years, acoustic holographic devices[1] that can bring into view objects in muddy water are under intensified development both at home and aboroad out of the need in marine exploitation. Generally they use arrays composed of PZT piezoelectric ceramic elements. With high effective electromechanical coupling coefficient, low impedance and attractive price, PZT elements are still frequently used in underwater acoustic imaging systems. However, the acoustic impedance of PZT elements are over 20 times greater than that of water medium, which results in severe mis-matching of acoustic impedance, leading to decreased sensitivity and narrowed bandwidth. Although this can be improved by adopting front and rear matching layers[2], this results in a series of new troubles with the process technique.

The use of polyvinylidene fluoride (PVDF*) as a sensor material can obtain good sensitivity and bandwidth without the use of front and rear matching layers, as it is low in acoustic impedance (see Table 1[3]), ductile, not easy to be broken and with high piezoelectric constant g (see Table 2[4]). It can realize electromechanical transducing for quasi-state, sonic and ultrasonic spectrum up to 100 MHz, with good resistance to humidity and good time stability[5]. The reception sensitivity spec-

* PVDF was provided by Shanghai Institute of Organic Chemistrys, Acadamia Sinica.

trum response variation will not exceed 6 dB from low frequency range up to 17 MHz (for film 30 μm thick).[6] This paper discusses the research work done for underwater acoustic imaging sensors and arrays in the low frequency spectrum of 200-500 kHz.

Table 1 Acoustic Characteristics of Different Materials [8]

Material	Density $10^3 kg/m^3$	Acoustic Speed $10^3 m/s$	Acoustic Impedance $10^6 Pa/s \cdot m$
Water	1.00	1.50	1.5
PVDF	1.80	2.10	3.7
PZT-5A	7.75	4.35	33.7
Bronze	8.5	4.40	37.0
Epoxy Resin	1.20	2.50	3.0
Ceramic	2.41	4.88	11.7
Teflon	2.20	1.38	3.05

Table 2. Physical Constants of PVDF, Crystal and Ceramic

Material	Tangi	Density $10^3 kg/m^3$	Elastic Modulus $10^9 Pa$	Dielectric Constant $\varepsilon/\varepsilon_0$	Piezoelectric constant $d(10^{-12} C/n)$	$g(10^{-3} V \cdot m/N)$	Coupling Coefficiant K%
Quatz	χ	2.65	77.2	4.5	2	50	10
	$45^\circ \chi$	1.77	17.7	350	275	90	73
$BaTio_3$	Z	5.7	110	1700	78	5.2	21
PZT	Z	7.5	83.3	1200	110	10	30
PVF_2/PZT	Z	3.5	4.0	55	69	47.1	6.6
PVDF	Z	1.76	3.0	13	20	174	19

THE CONSTRUCTION OF PVDF SENSOR ARRAY AS AN ACOUSTIC IMAGING

The PVDF-DMOS trunsducer discussed in this paper consists of acoustic and DMOS IC electric structures. As it has large array dimension due to the low frequency, it is assembled from two entirely independent parts, which has the advantage of simplified process, great flexibility and high receiving sensitivity.

Fig. 1 shows the acoustic imaging array structural principle of a PVDF-DMOS sensor. The front of the PVDF is covered with a layer of aluminium as an electrode, which is electrically grounded, and its back is glued to the ceramic plate, the thickness of which can roughly satisfy the physical boundary conditions for rigid barrier plates. In the ceramic plate has been laid an extremely thin metal bottom electrode, which is a disc of 3.5-5mm in diameter for the four shape with a design, based on the acoustic directional characteristic requirements for acoustic imaging. The bottom electrode is connected to the grid of the DMOS tube. When the sound wave is transferred to the piezoelectric film, the film is modulated by the alternately changing vertical wave acoustic signal, and due to the piezoelectric effect, a corresponding, linear alternating electric signal is induced on the bottom electrode of the piezoelectric film and is fed directly to the grid of the DMOS tube.

The DMOS tube (see Fig. 2) has a high input impedance of 10^9-$10^{10}\,\Omega$, and is integrated with a linear voltage amplification circuit of over 20 multiplies. This DMOS integrated electronic system is characterized by high input impedance impact volume, low self noise and low output impedance.

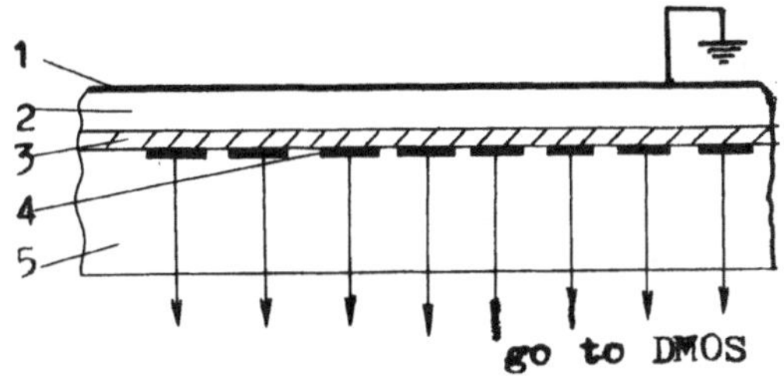

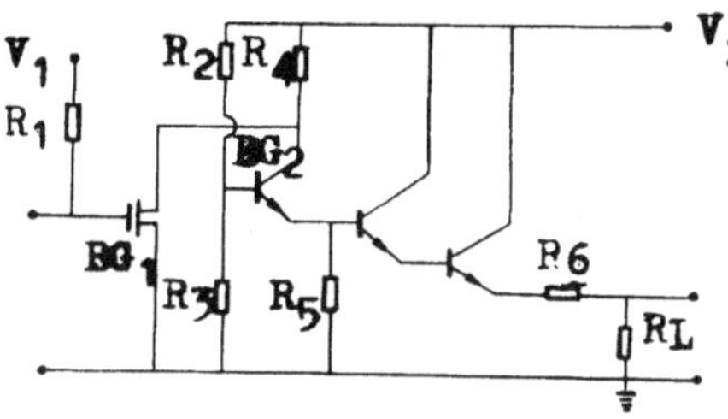

Fig.1 8x8 Plane array of PVDF-DMOS Fig.2 The DMOS integrated circuilty
1.Electrode 2.PVDF 3.Glue layer
4.Bottom electrode 5.Ceramic base

ANALYSIS AND MODELING

As PVDF is flexible and ductile and its elastic modulus is 1/28 that of PZT, as shown in Table 2. It generates very little transverse strain, with very pure vibration mode of activiation and sensing. When the whole PVDF piezoelectric film with electrode deposited on one side is overlapped with an array of several hundred small electrodes, it is activated because of the presence of bottom electrodes, and the transerse coupling between arrays is so small that it can be neglected. Only the disc vibrator arrays, formed by bottom and top electrodes have real piezoelectric response.

Furthermore, the activiated disc vibrators have a very small ratio of thickness (54 μm) to diameter (3.5-5 mm) and therefore can be simplified as a thickness oriented one-dimensional vibrant model for which the "thickness" model of standard thinplate, i.e, Mason Model, can be adopted. The film is bonded to the high impedance ceramic plate, with the acoustic sound receiving side being the low impedance water medium (the influence of glue and protection layers is neglected for a time being), which corresponds to the boundary conditions of one fixed end and one free end. It will show the λ/4 vibration type when analyzed from the principle of acoustic transmission lines. It's equivalent electromechanical circuit is shown in Fig.3.

In Fig.3 Za is the acoustic impedance of water medium. A the surface area of bottom electrode, T_{in} the acoustic input stress, k the number of waves in the PVDF, L the piezoelectric film thcikness, Zo the acoustic impedance of PVDF, C_0 the static capacity of film disc, V_{out} the equivalent output voltage, h the electric conversion coefficient. From Fig. 3, the expressions for receiving sensitivity can be obtained as follows:

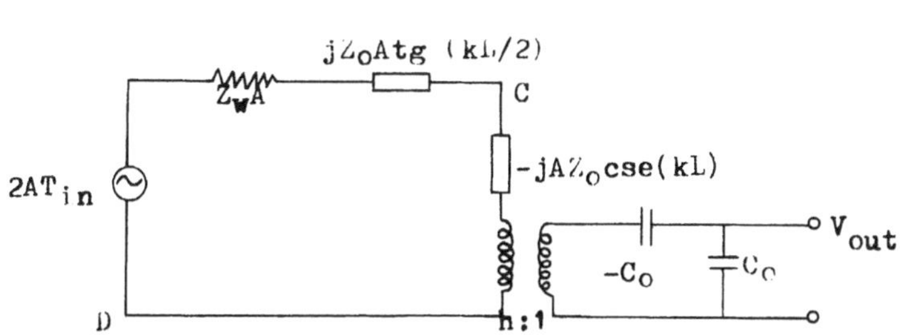

Fig.3 Equivalent electromechanical circuit of the sensor

$$M_1 = \frac{V_{out}}{T_{in}} = \left(\frac{2h}{j\omega C_0}\right)\left\{\frac{1}{Z_M(1-j(Z_0/Z_M)\cot(kl))}\right\} \qquad (1)$$

$$h = \frac{A}{l} e_{33} \qquad (2)$$

where, e_{33} = piezoelectric constant (here e_{33} = 0.15 c/m^2)
$\omega = 2\pi f$ (f being the working frequency 200 kHz)
$k = 2\pi/\lambda = \omega/c$ (C being the sound speed in PVDF)
$C_0 = \varepsilon_r(A\varepsilon_0/l)$(ε_r being the dielectric constant of PVDF)
$A = \pi d^2/4$ (d = 3.5mm), l = 54μm.

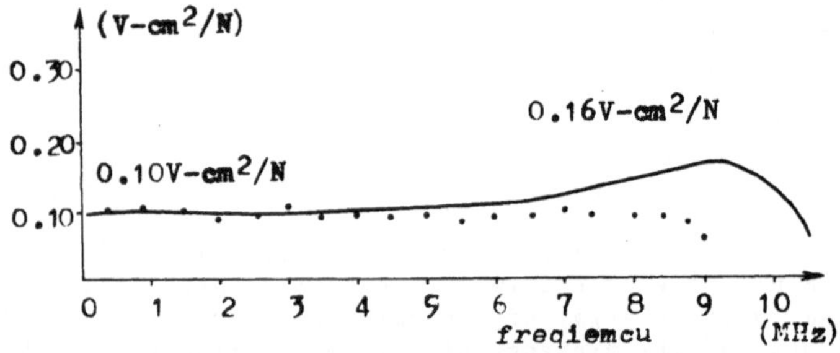

Fig.4 Receiving Sensitivty Response of the Sensor
(———— ideal; · · · · · · experiment)

Substitute the above parameters into equation (1) and normalize it at the $\lambda/4$ resonance peak value (about 9 MHz) of the film thickness direction, the ideal response curve of sensitivity vs. frequency as calculated by the computer is shown in Fig.4. The calculated value at 200 kHz being 0.10 V-cm^2/N.

PRACTICAL RECEIVING SENSITIVITY

It has been proved in practice that as the PVDF film is bonded to the ceramic plate by glue, the thickness of the glue layer should be very small as the film itself is very thin, or there would be great impact on the sensivitity. The glue layer thickness reducing process plays a very essential role in the performance of the element due to the following main reasons.

Acoustic Reason. The acoustic impedance of the glue layer is different from that of the ceramic plate and the PVDF. From the point of view of the acoustic transmission principle, it corresponds to a vibration rate "shunt" acoustic impedance element between points C and D in Fig.3, with the value.

$$Z_A = (AZ_a/j\,tgk_1 t) + jAZ_a\,tg(kl/2) \qquad (3)$$

Where Z_a is the acoustic impedance of the glue layer, t the thickness of the glue layer and k_1 the number of waves in the glue layer.

With smaller t and higher Z_a, the shunt will become smaller, hence will lose less in sensitivity. When $t=\lambda/4$ (λ being the wavelength in the glue layer), with an epoxy glue layer, where t=2.6 mm, the first term in the formula (3) becomes zero, the vibrator will show half wavelength vibration and the decrease in sensitivity will reach the maximum.

Electric Reason. The series coupled capacitor C_s in the glue layer will produce a shunt in voltage, hence decreasing the sensitivity.

$$C_s = \varepsilon_s \cdot \frac{\varepsilon_0 \cdot A}{t} \quad (4)$$

where ε_s is the dielectric constant of the glue layer. With smaller t and greater dielectric constant ε_s, the C_s would be greater, the shunting action would decrease and so the loss in sensitivity. Practice has shown that C_s would give the most serious loss when t 54μm.

Mechanical Reason. The glue layer has a different elastic modulus than the PVDF, and the equivalent vibrator non-activated portion will increase as the glue layer gets thicker, and the effective electromechanical coupling coefficient will be decreased, which requires correction of $K_{eff}^{'2}$.

$$K_{eff}^{'2} = K_{eff}^{2}/[1+(1-K_{33}^{2})S_b/S_s] \quad (5)$$

where $K_{eff}^{'}$ is the effective electromechanical coupling coefficient after correction, K_{eff} is the effective electromechanical coupling coefficient before correction. $S_b = t/AE_{33}$, E_{33} the elastic modulus of PVDF, $S_s = t/AE_s$, E_s being the elastic modulus of the glue layer.

$$K_{33}^{2} = d_{33}^{2}C_{33}^{E}/\varepsilon_{33}^{T} \quad (6)$$

where K_{33}, d_{33}, C_{33}^{E}, ε_{33}^{T} are respectively the electromechanical coupling coefficient, the piezoelectric coefficient, the material stiffness and the free dielectric constant.

A rough estimation shows that for the epoxy glue layer, the effective electromechanical coupling coefficient will decrease by about 10%. if its thickness reaches 1/3 that of PVDF. Secondly, the capacitor C_{gs} between the grid (DMOS tube) and the source will influence the receiving sensitivity, with better results in the low frequency range than in the high frequency range. As this study separates the electric structure from the acoustic structure, the disadvantage of "POSFET" structure has been overcome to the great extend and the sensitivity increased by sixfolds (that C_{gs} of the DMOS CI element used for this study has only 10-20 PF).

Taking into consideration of all the above-mentioned electric influences, the model as shown in Fig.5 can be used, with the following correction,

$$V_c/V_{out} = C_0 C_s/[C_0 C_s + C_0 C_{gs} + C_s C_{gs}] \quad (7)$$

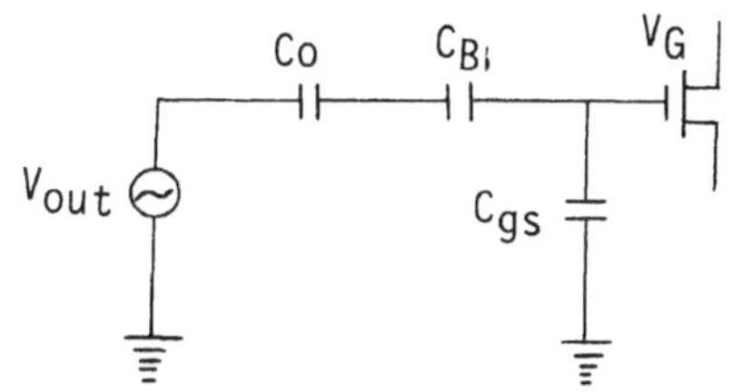

Fig.5 Equivalent Structure Diagram of PVDF-DMOS

Formula (7) is effective in the range from DC up to the frequency range of the resonance point. It can be concluded from above that decreasing C_{in}, increasing C_g, Z_m as well as k''_{eff}, is the key to upgrading the receiving sensitivity of this unit system, and the main determining factor is to decrease the thickness of the glue layer. Finally, the complete expression for the sensitivity of the whole system can be approximately written as;

$$M = M_1 \frac{K''_{eff}}{K_{eff}} \cdot \left[\frac{C_0 C_g}{C_0 C_g + C_0 C_{in} + C_g C_{in}}\right] \cdot K \cdot K_d \cdot K_p \qquad (8)$$

where K is the voltage gain of the integrated amplifier (K is about 20), K_d the correction factor brought about by the vibration rate "shunting" (usually K_d=1-0.5), K''_{eff}/K_{eff} is about 1-0.8, K_p the correction factor for different protection layer thickness of PVDF film (K_p is about 1-0.5). This makes the total correction as;

$$G = (K''_{eff}/K_{eff}) \cdot \left(\frac{C_0 C_g}{C_0 C_g + C_0 C_{in} + C_g C_{in}}\right) \cdot K_d \cdot K_p \qquad (9)$$

and formula (8) can be put into

$$M = M_1 \cdot K \cdot G \qquad (10)$$

Table 3. Performance for Different Glue Thickness

Number	Glue thickness (μm)	Amplification (K)	Receiving value before ampli. (mv)	Receiving value after ampli. (mv)	Sensitivity (dB)
1	22	20	0.038	21.6	-224.0
2	15	22	0.030	34.4	-220.0
3	8	22.5	0.083	82.2	-212.3
4	15	21.5	0.029	31.2	-220.7

The measurement showed that the PVDF-DMOS sensor as mentioned in this work can reach the miximum G value of 0.33, and the actual measured values are listed in Table 4 of this paper.

Table 3 shows the characteris.ics of each PVDF-DMOS sensor in the plane equal-spaced four-element array as measured with comparative method. This plane four-element array has a center-to-center distance of 9 mm, element diameter of 3.5 mm and is measured at a frequency of 200 kHz. As the glue thickness of these four elements is not just the corresponding sensors, the standard hydrophone is (0dB=1VμPa) at 200 kHz measurement and comparisons are made in the vertical direction and at the same distance. All conditions are the same for four transducers.

The measured values in table 4 are the characteristics for PVDF-DDMOS sensors with an additional 1 mm thick teflon protection and with the glue layer coupled with watch oil. The other aspects of the structure are the same as before.

Fig. 6 is the actually measured directions of the sensor. It is measured by fixing the transmitting transducer and rotating the PVDF-DMOS transducer that is being measured. The theoretical values shown in Fig. 6 are detailed in Reference (9).

Table 4. Measured Parameters for PVDF-DMOS Sensors

Dia. (mm)	Glue thickness (μm)	Frequency (kHz)	Sensitivity (dB)	Fluctuation within 38° (dB)	Amplifier bandwidth (MHz)	Electric leakage (dB)	Self noise (μV)
3.5	4	200	-205	<±1.2	Flat below 1.5	<-50	18

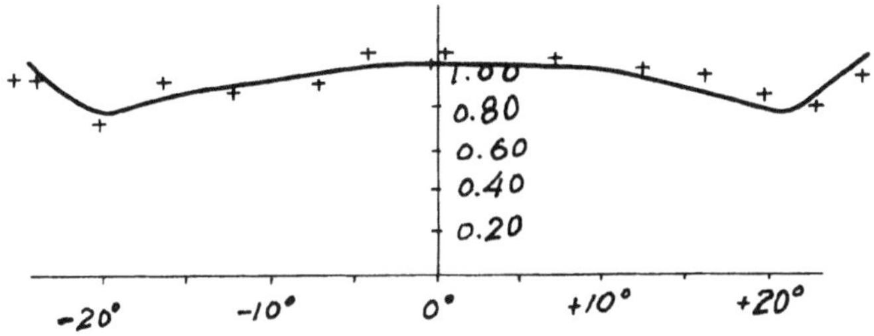

Fig.6 Directivity Pattern
—— theory ++++ measured value

Fig. 7 The photograph of the 8x8 plane array

THE ARRAY STRUCTURE OF PVDF-DMOS PIEZOELECTRIC FILMS

A blank 64-element receiving sensor array has been built and evaluated. Array dimensions are 100 x 100 mm.

The structural principle of the 8 x 8 plane array of the PVDF-DMOS piezoelectric film is shown in Fig.1 and Fig.7 the patterns of the array elements (i.e. the shape of the receiving plane of the sensor in Table. 5.6)

The maximum aperture is the same for the above four receiving planes, about dimension 5 mm.

Table 5 Sensitivity Uniformity Measured on June 25,1984

Pattern number		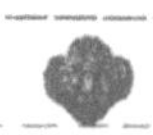	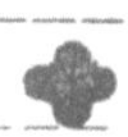	
Average sensitivity	-207.8 dB 0.76 mv	-205 dB 0.92 mv	-207 dB 0.82 mv	-208 dB 0.74 mv
Maximum deviation	±17%	±10%	±27%	±23%
Position	Upper row	Upper middle row	Lower middle row	Lower row

Table 6. Sensitivity Uniformity Measured on September 7, 1984

Pattern number				
Average Sensitivity	-213.2 dB 21.4 mv	-207.0 dB 43.2 mv	-206.5 dB 46.0 mv	-205.0 dB 49.7 mv
Maximum deviation	±50%	±29%	±40%	±37%
Position	Upper row	Upper middle row	Lower middle row	Lower row

* (0dB=1V/μpa)

The impulse response was obtained by driving a discrete PVDF transmitting transducer with a current impulse (1μs time). A water medium was used between the transtter and array. This impulse and the array output signal are shown in Fig.8. It is seen that the response has decayed 20 dB below the peak after 3μs. This translates to an axial resolution in water of 3 mm.

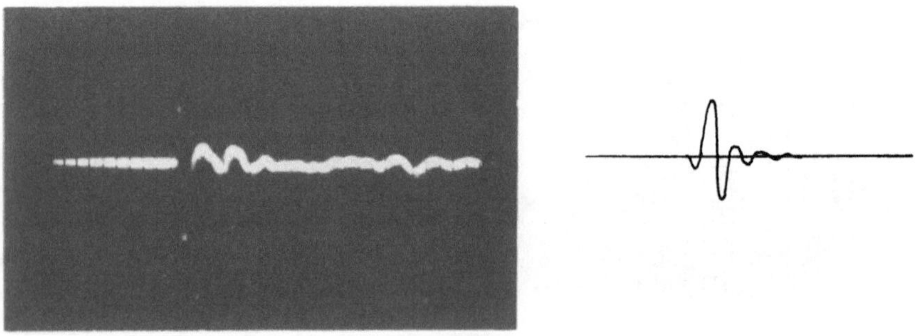

Fig. 8 Impulse response of PVDF underwater imaging array element

CONCLUSION OF EXPERIMENT FOR PVDF-DMOS SENSERS ARRAY

1. The direction sight angle of PVDF-DMOS piezoelectric film array is 38°-40°(±3 dB) at 200 kHz, with some difference according to the patterns of the receiving elements, so it can basically satisfy the requirements for underwater acoustic imaging. It can also satisfy the requirements at 500 kHz.

2. The average receiving sensitivity of PVDF-DMOS piezoelectric film array is-207-208 dB (0dB=1V/μpa), which can basically satisfy the requirements for underwater acoustic imaging.

3. The element receiving frequency response of the PVDF-DMOS piezoelectric film array is basically flat from 100 kHz to 500 kHz (with a fluctuation smaller than ±3 dB), i.e. it can be used in the working frequency range of 100 kHz to 500 kHz.

4. The installation tolerance of PVDF-DMOS piezoelectric film array can be controlled to within ±5μm.

5. The PVDF-DMOS piezoelectric film array has displayed good phase uniformity.

REFERENCES

[1] J.L.Surron, Proceedidngs of the IEEE, 67-4 (1979), 554.
[2] C.Desilers, J.Fraser, IEEE Trans. Sonics and Ultrsonics, SU-25-3 (1978), 115.
[3] H.R.Gallantree, Marconi Rev., 45 (224) (1982), 49.
[4] N.Muragama, K.Nakamura, Ultrsonics, 14-1 (1976), 15.
[5] Chen Bingqi, Organic Chemistry, 3 (1983), 191.
[6] R.G.Swartz and J.D.Plummer, IEEE Trans. Electron Devices, ED-26-12 (1979), 1921.
[7] D.Berlincourt, "Piezoelectric Crystals and Ceramics" in Ultrsonic Transducer Materials, O.E.Mattiat, Ed. New York; Plenum, 1971,74.
[8] R.S.Wollett, J.Acoust. Soc. Am, 40 (1966), 1112.
[9] Yuan Yi-quan, Zeng Wenhui, Journal of Nanjing Institute of Technology, 1(1985), 81.

STUDY ON QUARTZ THICKNESS-SHEAR RESONATOR IMMERSED IN LIQUID

AND ITS BIOSENSOR APPLICATION

Takamichi Nakamoto, Kunihiro Inadama* and Toyosaka Moriizumi

Cooperation Center for Science and Technology
Tokyo Institute of Technology
*)Toppan Company, Tokyo 101, Japan

Introduction

A quartz oscillator is well known because of its stable oscillation frequency, and widely used in the communication and electronics fields. It is also used for a sensor. Its resonant frequency is very sensitive to the mass of the deposited material, and is shifted due to the mass variation following the equation given by Sauerbrey[1].

$$\Delta f = -2f^2/(V \cdot S \cdot \rho_q) \times \Delta M \quad (1)$$

where f is the fundamental frequency, v the bulk wave velocity, S the electrode area, ρq the quartz density and ΔM the mass variation. That phenomenon is called mass loading effect and its typical application is a microbalance whose minimum detectability is approximately 10^{-9}g. This microbalance is called QCM (Quartz Crystal Microbalance) and is used for monitoring the thickness of thin films during vacuum evaporation.

In recent years, the behavior of the quartz resonator in liquid was investigated by a few workers. Although it was believed that the oscillation stopped when the quartz resonator was immersed in liquid, Nomura found that an AT-cut quartz resonator could cause oscillation even in liquid.[2] He aimed to quantify the various kinds of ion concentration using the quartz resonator and derived the empirical relationship between the frequency shift and the ion concentration [3]-[4]. Besides Nomura's work, Konah applied the quartz resonator to the detector for liquid chromatography.[5]

Another application of the quartz resonator immersed in liquid is a biosensor. A typical one is an immunosensor for the measurements of antigen or antibody in our body fluid. Muramatsu et al. reported they fabricated the biosensor by modifying the resonator surface with Protein A[6]. Immunoglobulin concentration between 10^{-6} and 10^{-2} mg/ml could be determined using that sensor. In spite of its high sensitivity, the frequency shift was greater than the value estimated by the conventional mass-loading-effect equation, eqn.(1). Thompson et al. described the time-dependent frequency response of quartz resonators when hydrophilic and hydrophoblic surface treatments were used[7]. They observed that the oscillation frequency increased extraodinarily several hundreds of minutes after the immersion in liquid in case of hydrophoblic surface treatment. It should be noted, therefore, that what happened to the quartz resonators is not clear when they are immersed in liquid and no reliable results have not been obtained concerning biosensor application of quartz resonators.

A few workers challenged to analyze the behavior of the quartz resonator in contact with liquid. Kanazawa et al. [8] derived the simple relationship between the frequency shift and the liquid viscosity shown in eqn.(2).

$$\Delta F = F^{1.5} \times \sqrt{\frac{\rho_l \eta_l}{\mu_q \rho_q \pi}} \qquad (2)$$

where ρ_l is liquid density, η_l liquid viscosity μ_q the shear elastic modulus.

Yao et al. reported that the frequency shift was dependent on the capacitance value of the oscillator circuit [9]. Nomura reported that the shift was dependent on the types of the oscillator devices[4]. Their results mean eqn.(2) is not always valid.

Mass loading effect in liquid was investigated by Bruckenstein et al. using electroplating.[10] They reported that eqn.(1) was valid even if the quartz resonator was immersed in liquid. As described above, however, it was reported that eqn.(1) was not valid in case of immunoglobulin adsorption.[6]

The aim of the present study is to make the ambiguous points clear, which are concerned with a quartz resonator behavior in liquid, especially when it is applied to a biosensor. For that purpose, the impedance characteristics of the quartz resonator immersed in liquid were studied. First, we describe the theoretical analysis using the electrical equivalent circuit, and then compare the calculated characteristics with experimental ones. Finally, the biosensor application will be discussed.

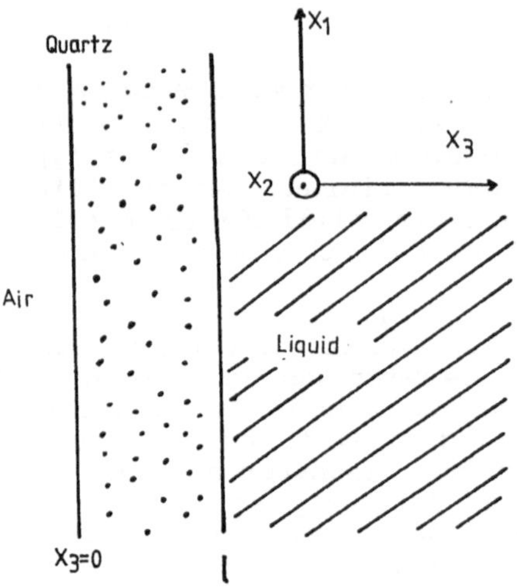

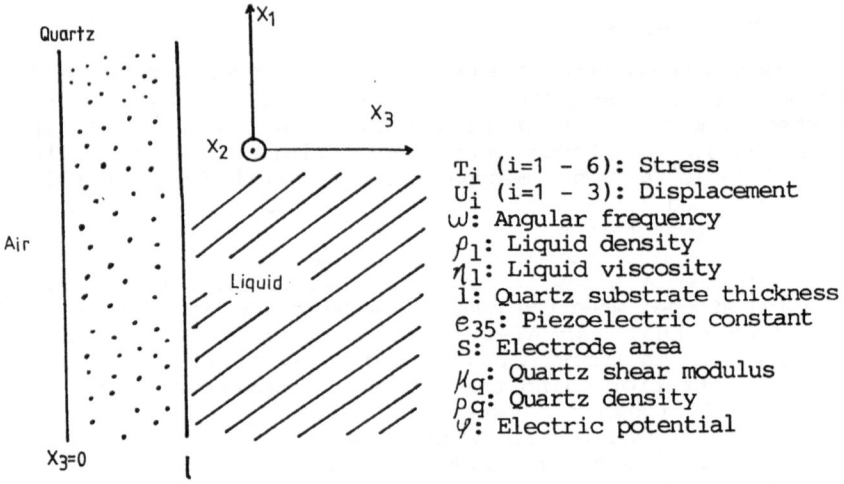

T_i (i=1 - 6): Stress
U_i (i=1 - 3): Displacement
ω: Angular frequency
ρ_l: Liquid density
η_l: Liquid viscosity
l: Quartz substrate thickness
e_{35}: Piezoelectric constant
S: Electrode area
μ_q: Quartz shear modulus
ρ_q: Quartz density
φ: Electric potential

Fig.1 Coordinate system and notation.

Equivalent circuit analysis

The notation and the coordinate system are shown in Fig.1. If the both sides of the quartz resonator are in contact with liquid, the liquid permittivity and conductivity influences the quartz resonator impedance. That situation is so confusing that the only one side of it is in contact with liquid.

An AT-cut quartz resonator is used in the present study. As its vibration mode is thickness shear and there is no displacement perpendicular to its surface, it can cause oscillation even in liquid. There are three types of the bulk waves propagating along the X_3 direction. One of them is a longitudinal wave and the other two are shear ones. Using the Newton's equation and Hooke's law, the following equation, which describes the behavior of the particle displacements, is derived.

$$\begin{bmatrix} A_{11} & 0 & 0 & A_{14} \\ 0 & A_{22} & A_{23} & 0 \\ 0 & A_{32} & A_{33} & 0 \\ A_{41} & 0 & 0 & A_{44} \end{bmatrix} \begin{bmatrix} U_1 \\ U_2 \\ U_3 \\ \varphi \end{bmatrix} = \begin{bmatrix} 0 \\ 0 \\ 0 \\ 0 \end{bmatrix} \qquad (3)$$

where A_{ij} is the component of the Christoffel equation matrix. As only the slow shear wave couples with the electric field and there is no coupling between that wave and other waves, only that one should be considered.

For liquid, in contrast to solid, there is no concept of strain, but strain velocity is used in wave equations. The equation of motion in liquid can be obtained by replacing strain with strain velocity in Hooke's law for solid[11]. Therefore, eqn.(4) is derived.

$$\rho_l \frac{\partial^2 u_1}{\partial t^2} = j\omega\eta_l \frac{\partial^2 u_1}{\partial x_3^2} \qquad (4)$$

Eqn.(4) can be also derived from Navier-stokes equation. In liquid near the resonator surface, the transverse damped wave given by eq.(5) exists.

$$u_1 = C' \cdot e^{-j\frac{\omega}{v_l}x_3} \qquad x_3 \geq l \qquad (5)$$

where

$$v_l = \sqrt{\frac{j\omega\eta_l}{\rho_l}} \quad \text{and} \quad C' \text{ is constant} \qquad (6)$$

It is convenient to define the acoustic impedance of shear wave in liquid, z_{as}, like the shear and longitudinal acoustic impedances in solid. z_{as} is given by

$$z_{AS} = -\frac{T_5}{j\omega U_1} = \sqrt{j\omega\eta_l\rho_l} \qquad (7)$$

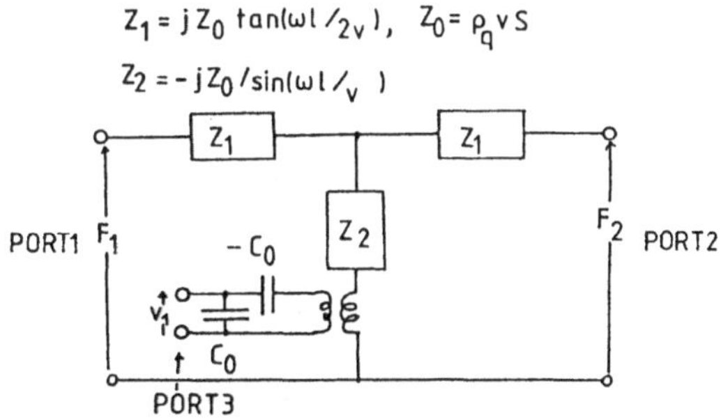

Fig.2 Mason equivalent circuit.

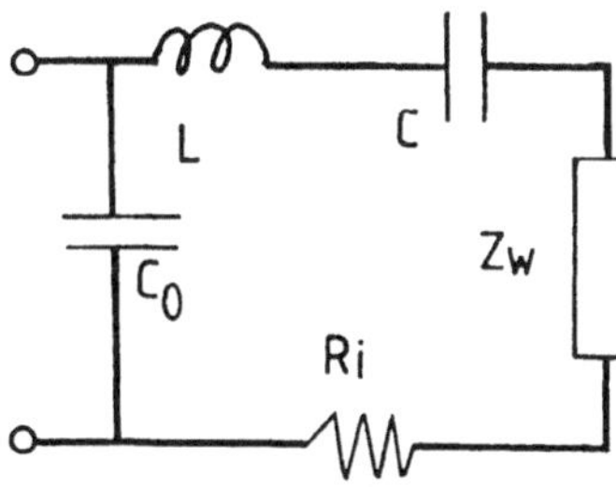

Fig.3 Simplified equivalent circuit of the quartz resonator whose one side is immersed in liquid.

In order to determine the equivalent circuit of the quartz resonator immersed in liquid, Mason equivalent circuit shown in Fig.2

is used.[12] Ports 1 and 2 correspond to its surfaces in contact with liquid and air, respectively. Z_{as} shown below is connected to port 1,

$$Z_{AS} = S \times z_{AS} \qquad (8)$$

and port 2 is made short-circuited. The equivalent circuit in Fig.2 is simplified to the two-port circuit in Fig.3 near the resonant frequency. L is an inductor, C a capacitor, R_i the internal resistance with no mechanical loading, and C_0 the capacitance between two electrodes. Z_w in Fig.3 is given by eqn.(9).

$$Z_W = \frac{l^2}{4e_{35}^2 S}\left[\sqrt{\frac{\omega\eta_1\rho_1}{2}} + j\sqrt{\frac{\omega\eta_1\rho_1}{2}}\right] \qquad (9)$$

The equivalent circuit in Fig.3 reduces to the well known equivalent circuit of the quartz resonator with no mechanical loading when Z_w is zero. The real part of Z_w expresses the loss due to the liquid viscosity. Therefore, the viscosity resistance R_v can be defined as follows.

$$R_V = \frac{l^2}{4e_{35}^2 S}\sqrt{\frac{\omega\eta_1\rho_1}{2}} \qquad (10)$$

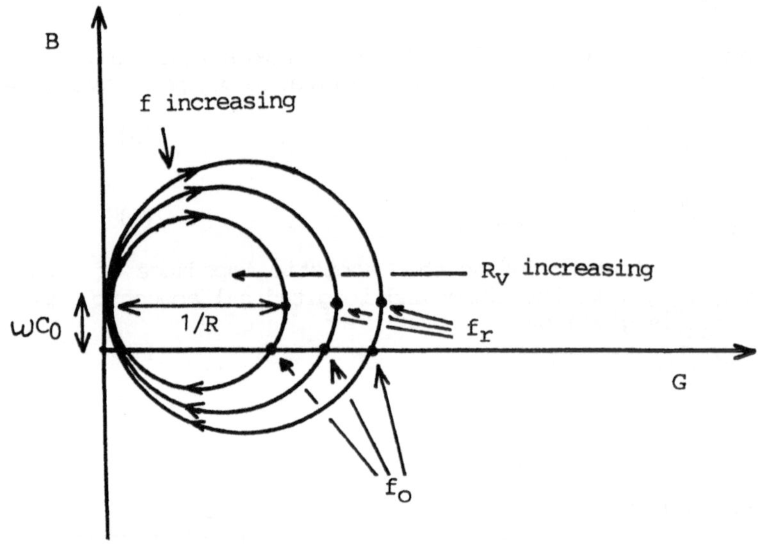

Fig.4 Motional admittance locus of quartz resonator.

The imaginary part of Z_w gives rise to the resonant frequency shift of the quartz resonator. When the parasitic capacitance C_0 is ignored, the same equation as Kanazawa's one, eqn.(2), can be derived. It means that eqn.(2) is only valid when C_0 is sufficiently small. Later the experimental result concerned with that equation will be described.

Motional admittance locus and experimental setup

The overall properties of a quartz resonator immersed in liquid can be discussed using motional admittance. Its admittance with liquid loading is given by

$$Y = j\omega C_0 + \frac{1}{j(\omega L - 1/\omega C) + R_i + Z_w} \qquad (11).$$

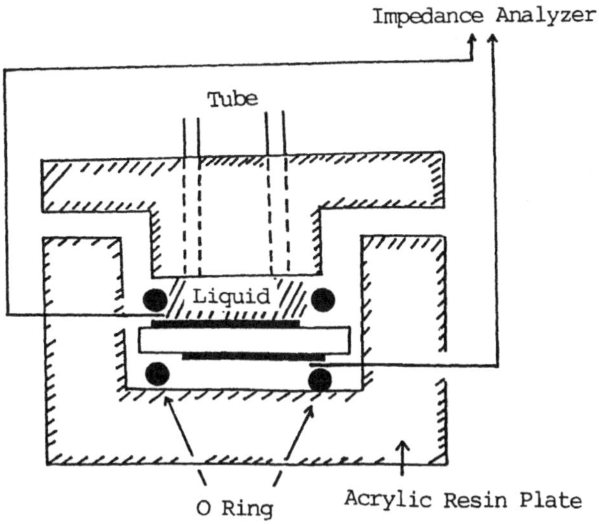

Fig.5 The measurement system.

The motional admittance circle is drawn as shown in Fig. 4 The circle diameter is a reciprocal number of the resistance component which exists in series with LC resonant circuit. In case of liquid loading, the resistance component is R_i+R_v, while it is R_i in case of no mechanical loading. Therefore, R_v can be obtained by subtracting the resistance component without loading from the one with liquid loading.

Z_w imaginary part causes the resonant frequency shift, and the two methods are adopted here for measuring the frequency shift. In the first method, frequency f_0 in Fig.4, where the admittance phase is equal to zero, is measured. f_0 is equal to the resonant frequency, assuming that the parasitic capacitance C_0 is ignored. The shift of that frequency caused by liquid loading is designated as df_0. In the second method, frequency f_r in Fig.4, where the conductance becomes maximum, is measured. f_r gives the accurate resonant frequency without being influenced by C_0 and R_i+R_v. The shift of f_r caused by liquid loading is designated as df_r.

As the resistance value becomes larger, the diameter of the admittance circle becomes smaller as shown in Fig.4. Therefore the difference between f_r and f_0 is expanded when the resistance increases.

R_v, df_0 and df_r were measured using the impedance analyzer which was controlled by a personal computer. The flow cell made of acrylic resin shown in Fig.5 was used for the measurement. The volume of the measurement cell was 80 µl. The AT-cut quartz resonator with the fundamental resonant frequency of 7.9 MHz was used. Chromium/gold electrodes were evaporated on both sides of the quartz plate. The quartz plate diameter is 10mm and the electrode one is 6mm. The aqueous glycerol solutions with various kinds of concentrations, and with different viscosities and densities are prepared as liquid samples.

Experimental result

Figs.6(a) and (b) show the frequency characteristics of the quartz resonator impedance. As the glycerol concentration increased, the dip near the resonant frequency in the amplitude-frequency plot became shallow because of the viscosity resistance increase.

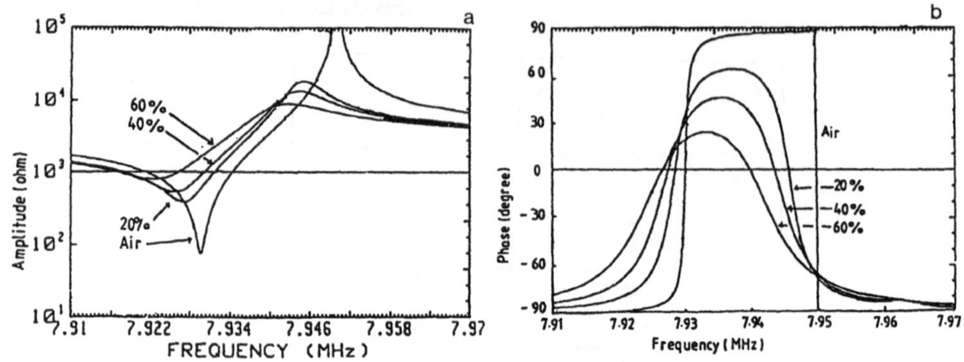

Fig.6 Impedance characteristics of the quartz resonator immersed in the glycerol solution.
(a) The impedance amplitude (b) The impedance phase

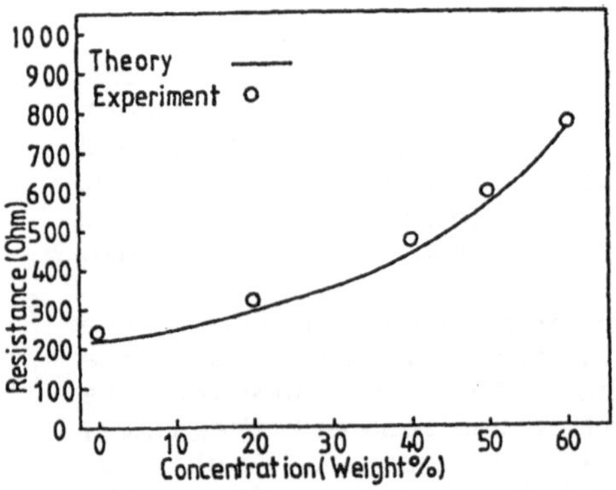

Fig.7 Relationship between glycerol concentration and viscosity resistance.

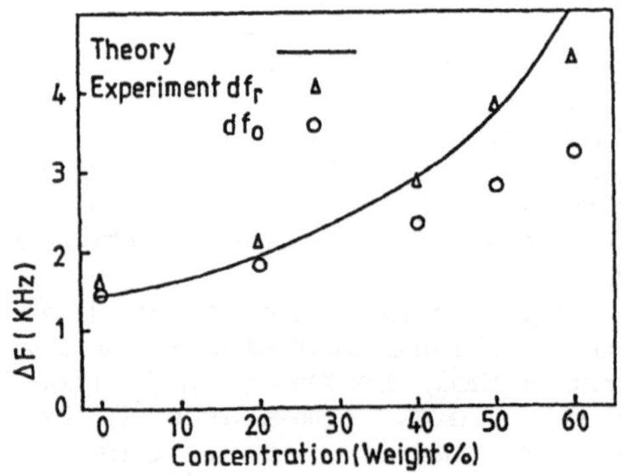

Fig.8 Relationship between glycerol concentration and and frequency shift.

As the glycerol concentration increased, the phase variation became dull. Figs.6(a) and (b) suggest that the quartz resonator Q decreased when the glycerol concentration increased.

Fig.7 is the relationship between the glycerol concentration and the viscosity resistance. Solid line is the theoretical curve given by eqn.(10). The values of the liquid viscosity and density in eqn.(10). were cited from the reference.[13] The measurement value agreed well with the theoretical one.

Fig.8 is the plot of the glycerol concentration vs. the frequency shifts, df_0 and df_r. The theoretical curve was plotted from eqn(2). df_r agreed well with the theoretical curve, while, df_0 deviated from the theoretical curve as the glycerol concentration increased. df0 deviation in high glycerol concentration was due to the influence of the parasitic capacitance C_0.

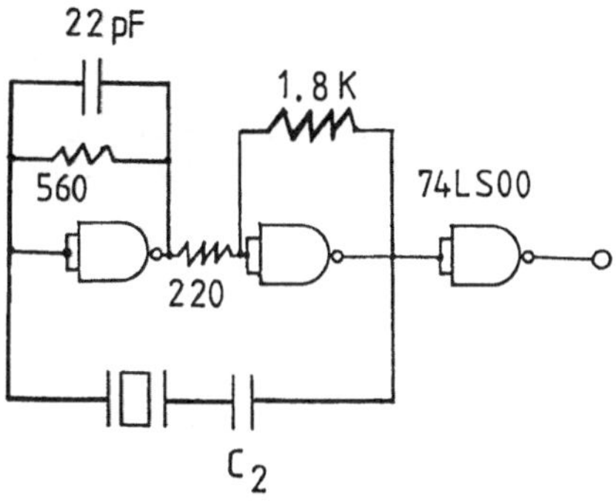

Fig.9 TTL oscillator circuit used in this experiment.

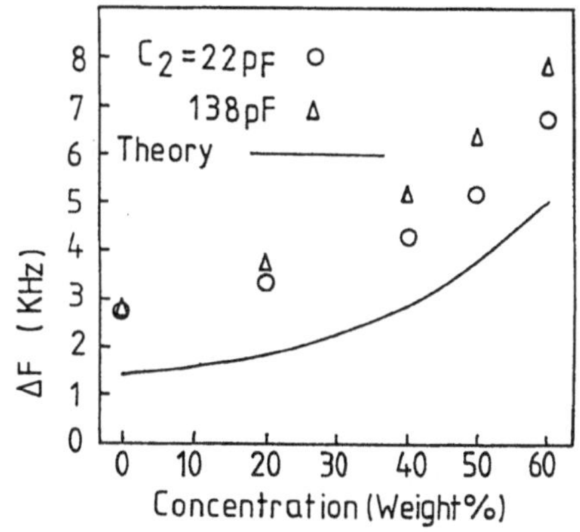

Fig.10 Relationship between glycerol concentration and the frequency shift.

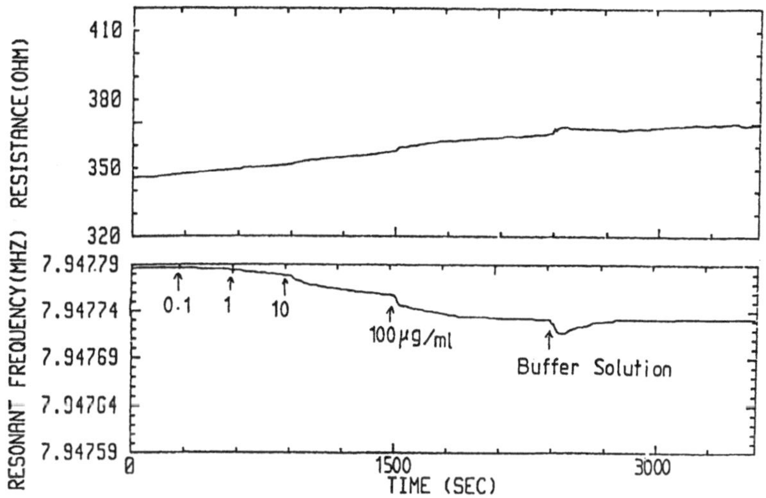

Fig.11 The frequency shifts caused by the protein injection.

The quartz resonator behavior in liquid was well understood when its impedance was examined. However, almost all the researchers measured the frequency shifts of oscillator circuits. In order to compare the frequency shift of a oscillator circuit with the theoretical one, we used the TTL oscillator circuit shown in Fig.9, which is the typical one used in the various kinds of digital circuits. Using the circuit and measuring the frequency shift, the output demonstrated the frequency shift in Fig.10. When the value of C_2 in Fig.9 increased, the frequency shift became larger. The solid line was calculated using eqn.(2). The measurement values were larger than the theoretical ones when the TTL oscillator circuit was used. The frequency shift of the oscillator circuit did not agree with its theoretical value. Although there have been several reports about the frequency shift of the oscillator, it is not clear by what origins the frequency shift is caused. Using our method, however, more exact frequency shift was found to be obtained from f_r, and the measured value agreed well with its theoretical prediction.

Finally, the experiments about protein adsorption was performed. The phosphate buffer solution with PH 7.2 and 1mM flew through the measurement cell, and the protein was injected into the solution. The protein used in this experiment was BSA(Bovine Serum Albumin). The resonant frequency decreased when BSA was injected. The reason for the frequency decrease was BSA adsorption on the electrode surface. Maximum frequency shift was about 50 Hz and about 0.1μg of BSA was estimated to be adsorbed on its surface using eqn.(1). That mass of BSA was equal to the one if BSA monolayer were closely packed on the surface and it seemed to be too large. The same result was reported by Muramatsu et al[6]. as described above. Bruckenstein reported that eqn.(1) was valid even in case of liquid loading[10]. It is, however, not clear if eqn.(1) is valid when a nonuniform deposition occurred in liquid. In case of the biosensor application, nonuniform adsorption often occurs. Therefore we are now investigating whether eqn.(1) is valid when a nonuniform adsorption occurs.

Conclusion

Several ambiguous points of the quartz resonator behavior in liquid were discussed by deriving the equivalent circuit and examining the impedance characteristics, and the experimental results could be explained using that circuit and its admittance locus. It was found that the best method for determining the resonant frequency shift is to measure the shift of maximum-conductance frequency. The obtained shift agreed well with its theoretical value. On the other hand, however, the frequency shift of TTL oscillator deviated from the theoretical one. It is concluded that a quartz-oscillator sensor often gives rise to wrong results when immersed in liquid, and should be used paying such attentions as mentioned in the present study.

References

[1] Gunter Sauerbrey: Zeitschrift fur Physik 155,206-222(1959).
[2] T.Nomura and O. Hattori: Anal. Chim. Acta, 115, 323 (1980)
[3] T.Nomura and A.Minemura: Nippon Kagaku Kaishi 1980, 1980,1261.
[4] T.Nomura and M.Watanabe: Anal. Chim. Acta,175(1985) 107-116.
[5] P.L.Konah and G.J. Bastiaans: Anal. Chem. 1980,52,1929-1931.
[6] H.Muramatsu, J.M.Dicks, E.Tamiya and I.Karube: Anal. Chem. 1987,59,2760-2763.
[7] M.Thompson, G.K.Dhaliwal, C.L.Arthur and G.S.Calabrese: IEEE Trans. on Ultrason., vol. UFFC-34, No.2, Mar., 1987.
[8] K.Kanazawa and J.G.Gordon: Anal. Chim. Acta, 175(1985) 99-105
[9] Yao Shou-Zhuo and MO Zhi-hong: Anal., Chim., Acta, 193(1987) 97-105.
[10] S.Bruckenstein and M.Shay: Eletrochimca Acta, vol.30, No.10, pp.1295-1300, 1985.
[11] T.Kikuchi and T.Moriizumi: Japanese Journal of Applied Physics, vol.25(1986) Supplement 25-1,pp.43-45.
[12] B.A.Auld: Aouctic fields and waves in solids, vol.1, p.340.
[13] R.C.West(Ed.): Handbook of Chemistry and Physics,67th Edition, CRC Press.

ALTERNATIVE ERROR MEASURES FOR NON-DIFFRACTION

AND DIFFRACTION TOMOGRAPHY

A. A. Vassiliou, J. H. Justice, and N. J. Guinzy

Mobil Res. and Dev. Corp.; Dallas Research Laboratory
13777 Midway Road
Dallas, Texas, U.S.A. 75244

ABSTRACT

Most algebraic reconstruction techniques for non-diffraction and diffraction tomography are based on minimization of the l_2 norm residual error. We investigate the properties and characteristics of tomographic reconstructions based on the l_1 norm using both Simplex and Karmarkar algorithms as well as the l_p norms for $1<p<2$ using reweighted least squares, and results are compared to l_2-norm reconstructions. We also consider a total least squares error criterion.

I. INTRODUCTION

Tomographic image reconstruction has found important applications in several areas of inquiry, including medical diagnosis and non-destructive testing. Recently, attention has been focused on the possible applications of tomographic image reconstruction in geophysics.

Non-diffraction algorithms of the type used in medical image reconstruction, for example, can be considered if variations in refractive index are very small. In many geophysical imaging problems of interest, variations in refractive index are large, and it is necessary to consider diffraction algorithms. We shall confine our attention here exclusively to algebraic techniques for both non-diffraction as well as diffraction tomography.

Most algebraic algorithms for tomographic reconstruction are based on minimization of the l_2 norm. Our initial interest was in the comparison between the l_2 norm and the l_1 norm, and it was this interest which motivated this work. We were also interested in the possible advantages of the use of a total least squares error criterion in place of the usual least squares error criterion. The resulting investigations led us to consider the general l_p norms as well, with emphasis on $1<p<2$. This paper is intended to represent a very brief survey of our results which will be published in more complete form elsewhere.

II. ALGEBRAIC RECONSTRUCTION FOR NON-DIFFRACTING AND DIFFRACTING MEDIA

In this section we shall briefly outline the approach to tomographic image reconstruction in non-diffracting and diffracting media which will be used in the examples and discussion to be presented later. Let us begin by considering the algebraic formulation of the tomographic reconstruction problem in non-diffracting media. Let the zone of interest, which is to be imaged, be partitioned into regular grid cells or pixels which are indexed. In the solution, a value of the unknown refractive index will be assigned to each cell. We index the raypaths connecting all pairs of source and receiver locations. Define a matrix S by the definition of its elements

$$S_{jk} = \text{arclength of raypath } j \text{ in cell } k \tag{1}$$

For convenience in the remainder of the paper, we shall assume that the unknown parameter to be imaged is slowness (reciprocal of velocity) and that the measured data is traveltime. It is then apparent that the data vector, T, is related to the (unknown) slowness vector, P, by the equation

$$T = SP \tag{2}$$

In general, this system of equations is overdetermined and inconsistent due to data errors. It is common to consider an l_2 norm solution given by

$$P \simeq (S^TS + \lambda I)^{-1} S^TT \qquad (3)$$

This solution may be conveniently approximated using conjugate gradient search, and if the initial estimate of the solution is taken to be zero, the conjugate gradient algorithm converges to the minimum norm solution (Hestenes, 1975) which is often a good choice. Further, this is the solution obtained by singular value decomposition (SVD).

In diffracting media, the matrix, S, becomes itself a function of the unknown solution, P, so that equation (2) takes the form

$$T = S(P)P \qquad (4)$$

Given an initial estimate, P_0, of the slowness vector, we may compute a first order Taylor expansion of T about P_0

$$T = T_0 + \left.\frac{\partial T}{\partial p}\right|_{P_o} (P - P_o) \qquad (5)$$

Denoting the Jacobian computed from the slowness field, P, by J(P), we may use the relationship (5) as the basis for an iterative algorithm for approximating the required solution. This algorithm takes the form

$$\Delta T_k \cong J(P_k)\Delta P_{k+1} \qquad (6)$$

where ΔT_k and $J(P_k)$ are computed by tracing rays through the previous estimate of the slowness field, P_k.

The updated estimate of the solution, P_{k+1} is then computed by an algorithm which minimizes the norm (if a norm error criterion is being used)

$$\left\| \Delta T_k - J(P_k)\Delta P_{k+1} \right\| \qquad (7)$$

and iterations cease when satisfactory convergence has been obtained.

We use a continuous velocity field in each iteration. It is very important that the method of ray tracing be sufficiently sophisticated that the observed arrivals which are required in the inversion are included in the forward modeling and are accurately detected by the picking algorithm.

III. THE KARMARKAR ALGORITHM

One of the most powerful algorithms to appear for linear programming applications is the Karmarkar algorithm developed at Bell Laboratories. For convenience in pursuing this idea, we decided to begin with the simpler non-diffraction reconstruction problem, and if it seemed warranted, to then proceed on to the diffraction problem. For the algebraic non-diffraction tomography problem as formulated in Section II, the l_1 - norm tomographic reconstruction problem can be posed in the linear programming form (Luenberger, 1984)

Minimize C^TY

Subject to the constraints

$BY = T$

$Y \geq 0$

Plus any additional constraints desired on the solution vector, Y.

The vector Y contains the unknown slowness vector, P, as well as the residual error whose l_1 norm is to be minimized. The matrix, B, contains the raypath matrix, S, and T is the vector of observed traveltimes.

Two different velocity models were used to test the algorithm. The first is a 900 grid point model (30 X 30) shown in Figure 1. The velocity range was between 2000 and 2600 m/sec. The data set consisted of 2209 traveltimes. Three different tests were done on this data set. First the algorithm was tested in an unconstrained mode; second, bounds were placed on the velocities; and third, outliers were added to the traveltimes with the velocities being bounded again. The result with bounds and no outliers is shown in Figure 2. For comparison, the l_2-norm solution without any outliers is shown in Figure 3. The summary of these results in terms of accuracy and CPU time is presented in Table 1.

TABLE 1

Comparison of the Karmarkar and the l_2-norm algorithm for Accuracy and CPU time on the small data set

	Karmarkar Unconstrained	Karmarkar Constrained (no outliers)	Karmarkar Constrained (outliers)	l_2-norm
Mean Absolute Traveltime Error (msec)	0.62	1.16	1.16	0.97×10^{-1}
Mean Squared Traveltime Error (msec)	0.91	1.82	1.82	0.12×10^{-1}
Mean Squared Velocity Error (m/sec)	0.145×10^{9}	0.110×10^{3}	0.110×10^{3}	0.575×10^{2}
CPU Time (secs)	290 (1)	240 (1)	240 (1)	12.5 (2) 23.5 (3) 85.0 (4)

(1) : CPU time given on Alliant FS/8 supercomputer
(2) : CPU time given on Cray X-MP/14 supercomputer
(3) : CPU time given on Convex C-2 supercomputer
(4) : CPU time given on VAX 8600 scalar computer

The second velocity model used to test the Karmarkar algorithm has 10000 pixels. The data set for this model consisted of 10609 traveltimes, corresponding to only crossborehole rays. The results from the bounded Karmarkar algorithm reconstruction, as well as from the l_2-norm reconstruction, are given in Table 2

IV. THE SIMPLEX ALGORITHM

For the sake of comparison, we implemented the reconstruction problem with the Simplex algorithm running on a Cray supercomputer, using the formulation sketched in Section III, exactly as we had done for the Karmarkar algorithm. As before, we began with the simpler non-diffraction algebraic reconstruction problem.

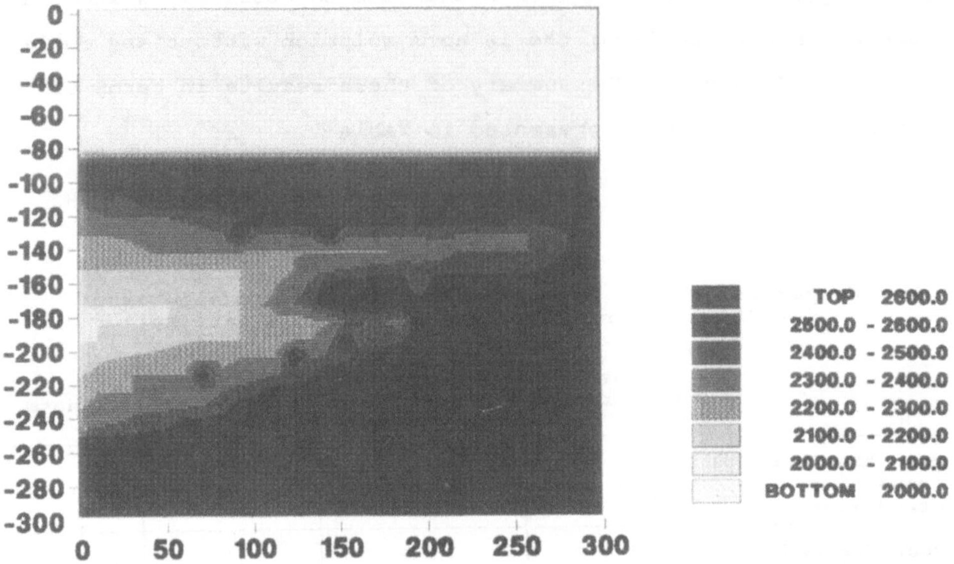

Figure 1. Velocity model (900 pixels)

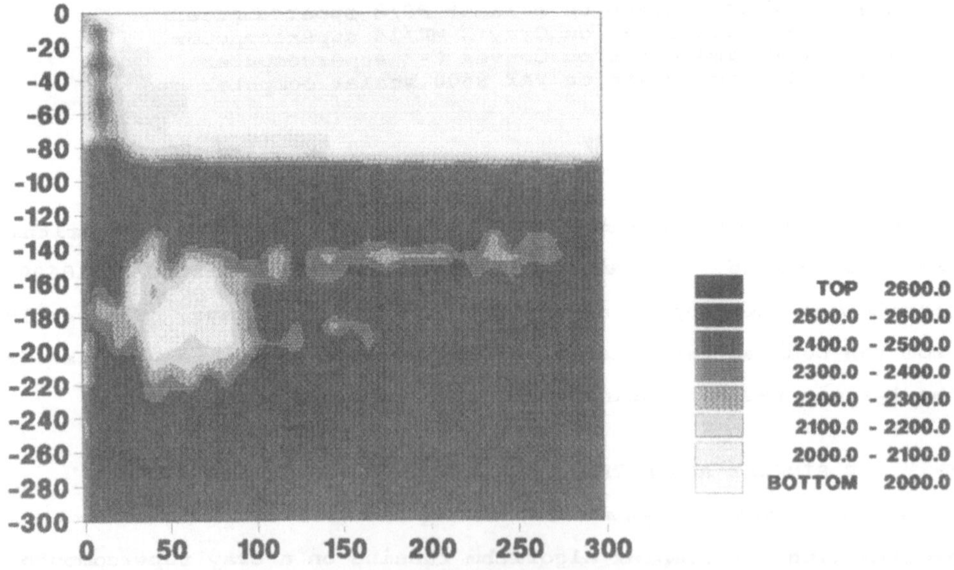

Figure 2. Constrained Karmarkar algorithm reconstruction. No outliers in the data.

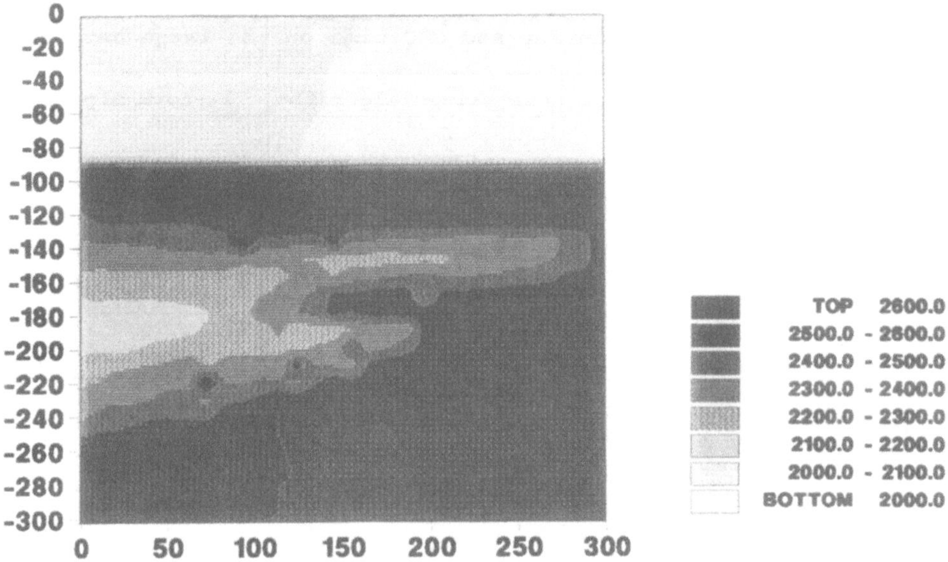

Figure 3. l_2-norm reconstruction of the velocity model in Figure 1. No outliers in the data.

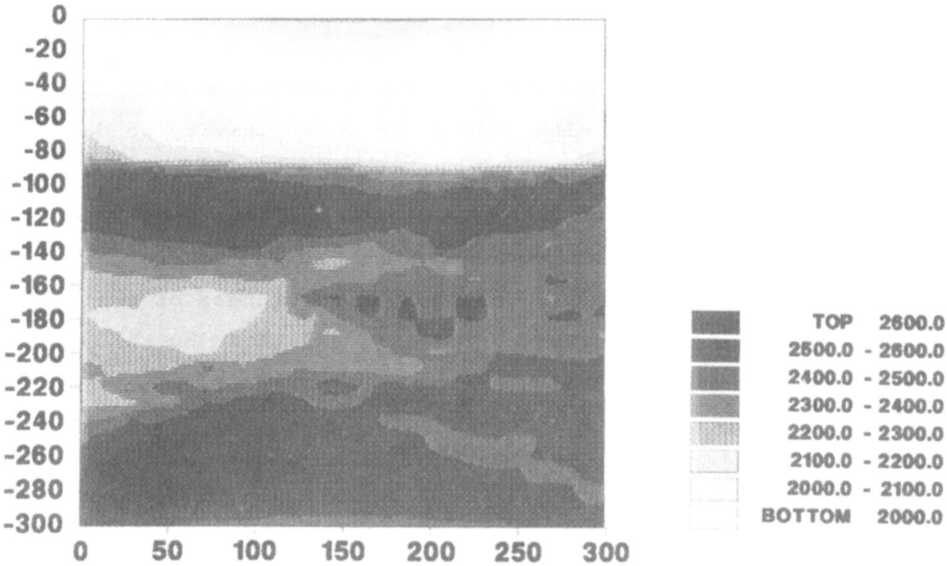

Figure 4. l_2-norm reconstruction of the velocity model in Figure 1. Outliers in the data.

TABLE 2

Comparison of the Karmarkar algorithm and of the l_2-norm algorithm in terms of accuracy and CPU time on the large data set

	Karmarkar constrained algorithm	l_2-norm algorithm
Mean Absolute Traveltime Error	0.758×10^{-2}	0.785×10^{-2}
Mean Squared Traveltime Error	0.282×10^{-1}	0.133×10^{-1}
Mean Squared Velocity Error	0.106×10^{3}	0.75×10^{2}
CPU Time	3050 (1)	130 (2) 416 (3)

(1) : on Alliant FS/8 supercomputer after 17 iterations
(2) : on Cray X/MP-14 supercomputer
(3) : on Convex C-210 supercomputer

The small data set with 2209 measurements and 900 unknowns was used to test the simplex algorithm. Exact upper and lower velocity bounds were used to constrain the solution. The simplex algorithm produced a reconstruction with many values at the upper bound, fewer values at the lower bound, and very few in between.

V. THE l_p NORMS, $1<p<2$

The l_1 norm solutions obtained with the Karmarkar algorithm did show good characteristics with respect to insensitivity to data errors. On the other hand, our approach to l_2 solutions exhibited greater sensitivity to data errors and outliers, but required no concern for constraints. It appeared that perhaps some middle ground might be found, offering the advantages of robustness of the l_2 algorithm with the insensitivity to errors exhibited by the l_1 algorithms based on linear programming. In particular, it seemed clear that algorithms based on an l_p minimization criterion would be worth considering.

The validity and usefulness of the l_p norm was tested on the small data set contaminated with outliers. Results of the reconstruction using the l_2 and the l_p ($p=1.02$) norms shown in Figures 4 and 5, respectively, demonstrate the superiority of the l_p norm when outliers exist.

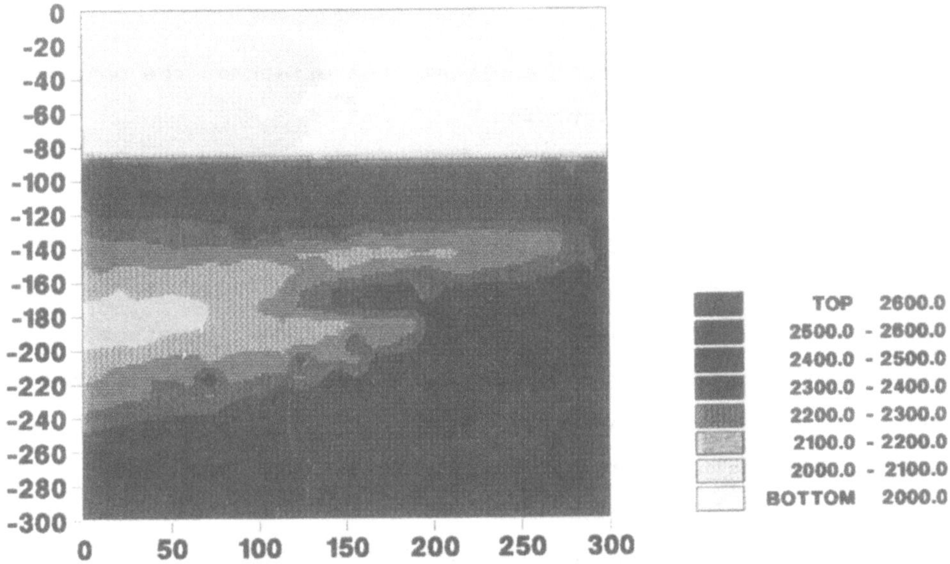

Figure 5. l_p(p=1.02) reconstruction of the velocity model in Figure 1. Outliers in the data.

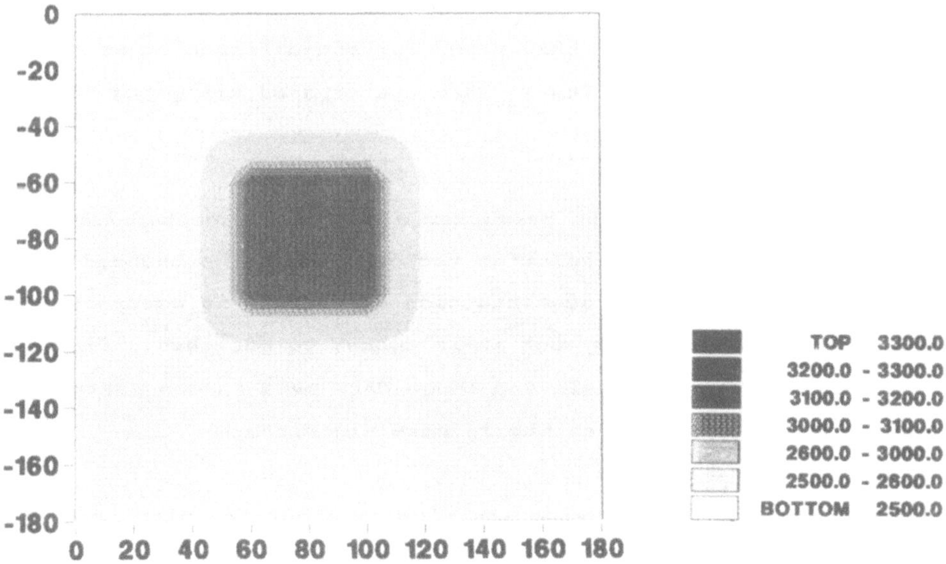

Figure 6. Velocity model Diffraction Tomography

The l_p norm was applied to diffraction tomography using a real geophysical data set. The data set used consisted of 306 observations and the number of unknown velocities was 182. Two outliers were added to the traveltimes, and the l_p norm (p=1.1) was again able to remove almost completely the effect of the outliers from the velocity reconstruction.

VI. TOTAL LEAST SQUARES

We tested the total least squares algorithm on a simulated diffraction tomographic reconstruction problem, where the velocity model was known. The model had 196 pixels, and it consisted of a constant velocity background in which a reef was imbedded. The model, the l_2-norm reconstruction, and the TLS reconstruction are shown in Figures 6, 7, and 8, respectively. The comparison of these figures suggests that TLS has the ability to correct for the errors in the Jacobian which is only accurate to first order.

VII. CONCLUSIONS

The possible use of the l_1-norm, as well as the use of the l_p norms ($1<p<2$), and finally the use of the total least squares criterion were studied in this paper for non-diffraction and diffraction tomography. The l_1 norm was studied using the Karmarkar and the Simplex algorithm.

Our experience with the Karmarkar algorithm suggested several conclusions. First, the solution vector needs to be bounded very accurately to produce a reconstruction comparable in accuracy and resolution to the l_2 norm, but it is robust to outliers. Finally, the algorithm for both small and large data sets is considerably slower computationally than the l_2 norm algorithm.

The l_1 norm solution using the Simplex algorithm also needs to be bounded. The Simplex algorithm yields slightly less accurate reconstructions than the Karmarkar algorithm.

The attempt to find a norm which combines the advantages of l_1 and l_2 norm led us to the study of the l_p norms ($1<p<2$). The tests

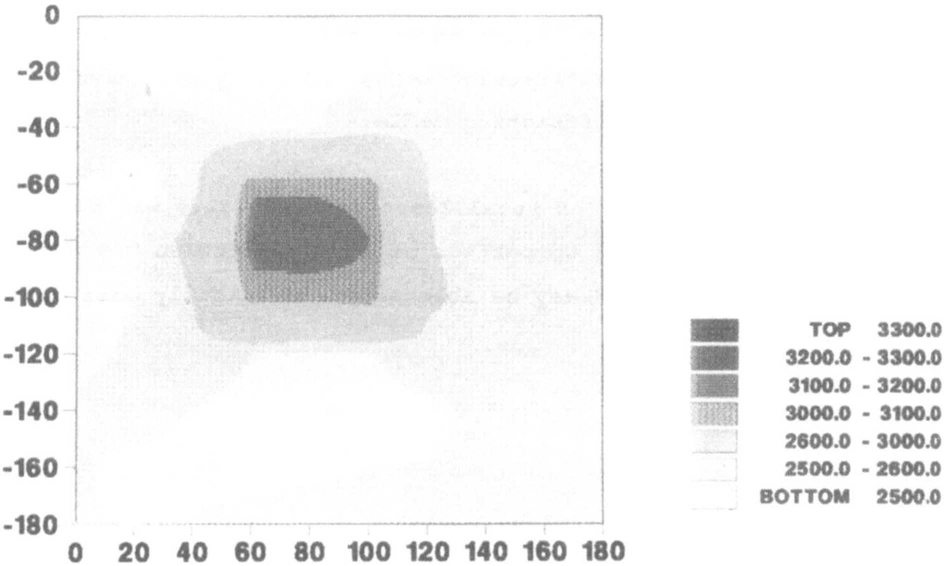

Figure 7. l_2-norm reconstruction of velocity model in Figure 6.

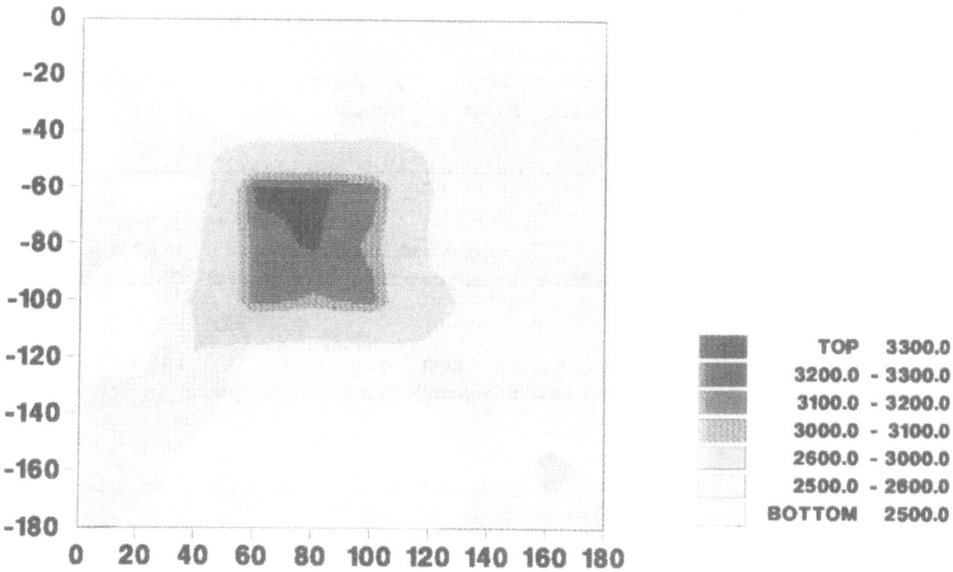

Figure 8. Total Least Squares reconstruction of velocity model in Figure 6.

using l_p norms pointed to two main conclusions. First, the l_p norms yield the same reconstructions as the l_2 norm in the absence of outliers. Second, when outliers occur, the l_p norms ($1<p<2$) are robust to them for non-diffracting media, and they are somewhwat sensitive to them for diffracting media.

Finally, the concept of total least squares (TLS) was tested on diffraction tomography. Comparison of results between TLS and least squares suggest that TLS may be able to yield slightly more accurate reconstructions.

REFERENCES

Bois, T. N., La Porte, M., Lavergne, M., and Thomas, G., 1972, Well-to-well seismic measurements, Geophysics, 37:471.

Byrd, R. H., and Payne, D. A., 1979, "Convergence of the iteratively reweighted least squares algorithm for robust regression," The Johns Hopkins Univ. Tech. Rep. 313, Baltimore, Md.

Golub, G. H., and Van Loan, C. F., 1980, An analysis of the total least squares problem, SIAM J. Numer. Anal., 17:883.

Hestenes, M. R., 1975, Pseudoinverses and conjugate gradients, Comm. ACM, 18:40.

Justice, J. H., Nguyen, D. T., and Vassiliou, A. A., 1987, A study of crossborehole tomographic inversion, CSEG meeting, Calgary.

Justice, J. H., Vassiliou, A. A., and Nguyen, D. T., 1988, Geophysical transmission tomography, to appear in Signal Proc.

Karmarkar, N., 1984, A new polynomial-time algorithm for linear programming, Combinatorica, 4:373.

Luenberger, D. G., 1984, "Linear and nonlinear programming", Addison-Wesley Publishing Company, Reading, Massachusetts.

ALGORITHM COMPARISON AND PARAMETER CONSIDERATION IN TOF TOMOGRAPHY

H. W. Jones and J. S. Meng

Acoustics Institute, Technical University of Nova Scotia
P.O.Box 1000, Halifax
Nova Scotia, Canada B3J 2X4

ABSTRACT

This paper reports the final phase of a study of the accuracy which can be obtained in the reconstruction of ultrasonic tomograms of biological specimen. The study has shown that it is difficult to simulate phantoms to the required degree of accuracy by numerical methods and that it is necessary to use a real phantom and obtain experimental data. The errors arising from experimental uncertainties are well defined and less than those obtaining in numerical studies. It is noted in passing, however, that numerical simulation has the advantage of allowing the reconstruction errors to be associated with particular parameters. The difficulty with real phantom is they involve spatial frequencies which are higher than those which can be obtained in reconstructions because of sampling theorem limitations. Similarly it is difficult to assign a sufficiently accurate value to the acoustic velocity of a pixel containing the boundary of two materials with different velocities. Details of the errors arising from these causes and the effectiveness of these methods of reconstructions, i.e. filtered backprojection, backpropagation and ray tracing iterative algorithms, are discussed. Conclusions are presented on the accuracy limitations. Some discussion is given on the computing times required to obtain a reconstruction.

INTRODUCTION

In earlier papers [1-4] we discussed the accuracy of reconstruction, which might be obtained in certain ultrasonic tomograms. In those papers we explained the problems which arose from purely digital studies, i.e. the difficulties and limitations of numerical phantoms; difficulties which are associated with the inability to simulate experimental situation to the required degree of accuracy without very extensive computational facilities. Essentially, the need in such work is for the full and detailed solution of the diffraction integrals over the sampling range so that the correct received ultrasound waveform can be properly modelled. It became obvious to us[4] that the numerical modelling was beyond our resources except for trivial symmetrical phantoms.

We had conducted experimental work on a real phantom[4] but we found two difficulties associated with the experimental work. First, over the time taken to obtain the readings, which was several days, the edges of the gelatine absorbed water and the velocity of sound in these areas changed as a result. Second, we were unable to measure the sound velocities in these areas to the degree of accuracy required because of our inability to determine the linear measurements of the samples to the required degree of accuracy. This situation led us to repeat our experiments with a new phantom. The new gelatine phantom was of an accurately known size i.e. its edge contours were measured by travelling microscopes to ± 1/80 mm. The gelatine mix provided for a sample which was poured into an accurately machined rectangular mold. The velocity of sound was determined from this gelatine sample over a range of temperatures. Measurements of the time of flight were performed in the same way as was described on earlier occasions except that the total time the gelatine was immersed in the water was about 1 hour . The results of these experiments are reported later in this paper.

The experimental results were used to test the accuracy of the tomographic reconstructions obtained by several methods. There is difficulty in obtaining satisfactory comparisons with the original phantoms for several reasons. Not least of these difficulties is that of the limitations due to the spatial frequency filtering arising from the chosen (31 x 31) pixel array. An associated difficulty is that concerned with assigning a "best value" to a pixel which may, in fact, be partly occupied by gelatine and partly by water; even minor fractional changes in the areas occupied by the water and the gelatine lead to substantial errors.

The computing time required to obtain reconstructions is of concern to us. Although we have directed our attention, first, to the problems of obtaining accurate reconstructions, we have been mindful of the need to obtain results in the time period which allows the process to have useful, practical applications. The most accurate method of obtaining reconstructions does not at the present give us satisfactory reconstruction times. Our discussion of this problem shows that it should be possible to obtain reconstructions in times which are likely to be acceptable.

THE RECONSTRUCTION METHODS

There are two methods of reconstruction, other than those used in x-ray tomography, which are applicable to the problem of ultrasonic tomograms in biological tissue. If we accept that the x-ray tomographic methods (i.e. backprojection) are not sufficiently accurate, then we need to look at the other methods. It should be mentioned that we attempted second order correction of the reconstruction. The reconstruction obtained from backprojection allows us to obtain error data by comparing the data from the reconstruction to that which was obtained experimentally. This error data allows us to obtain a second order correction of the reconstruction.

The difficulty with the approach described in the previous paragraph can be supposed to arise from neglect of a treatment of the problems in terms of the scattering and diffraction of the wave fronts. Mathematically, the effect is that in the convolution treatment we obtain in Fourier space a series of data values lying on straight lines which are inclined at angles ϕ corresponding to the data sampling angles. Provided we have sufficient data in Fourier space we can, in principle, by inverse transformation obtain a reconstruction within any constraints implied by the sampling theorem. This process inevitably involves interpolation in Fourier space[5,6]. We may use whatever process we wish to deal with the interpolation and use second order procedures to correct the initial interpolation. It appears from our experience, however, the method has its limitations and as the remaining errors are unacceptable we are forced to reconsider the physical problem.

If we now suppose the neglect of the refractive index changes is associated with the unwanted errors we can proceed either to direct solution of the wave equation or to a ray tracing routine[7]. Considering the wave equation solutions; we are naturally inclined by the experience of others [5,6,9] to choose the Rytov approximation and use related solutions. We appreciate the objections to this process as we have used it. They are:

i) we have defined the problem in terms of real velocity only i.e. amplitude effects are ignored.

ii) we accept a simple aperture function which will only approximate to that which really exists in order that we may have a suitable analytic form.

iii) we have supposed a 3 dimensional problem can be reduced to one in 2 dimensions.
This means that *a priori* we suppose the errors introduced by (i)-(iii) will be smaller than those introduced by the first reconstruction method. The immediate effect of this assumption is that we obtain data in the Fourier space which may be somewhat different in magnitude to that obtained hitherto and which lie on circular arcs.[8]

Finally, we can suppose that we can use a ray tracing routine to correct the reconstructions. The beam pattern for the transmitted and received ultrasound is as shown in fig. 1; this is a well-defined pencil and gives credibility to the ray tracing treatment. Clearly the transmitted beam pattern will become more diffuse during transmission due to the diffractive and refractive effects. However the receiver aperture is that associated with a narrow beam and this lends credibility to the ray tracing approach. Another difficulty lies with the nature of the calculation used in the ray tracing routine. During the correction process, we are obliged to have a reconstruction before a ray can be traced. If we used, for example, the backprojection method to obtain an initial reconstruction then we have two errors. First, is the error associated with the incorrect reconstruction and second, that associated with ill defined edges due to the spatial filtering which is implicit in the reconstruction. The method would be unsuccessful if the equivalent refractive index is large. As we pointed out in reference 4, the refractive index is small (≈ 1.07) in biological tissue. This small value gives some possibility that reasonably correct solutions can be obtained, in that the bending of a ray in the "ill-defined edges" will allow us to approximate closely to the actual state of affairs. Given the comments at the start of this paper, the viability of this approach can only be tested by experiment; certainly the inadequacies of other methods provide us with an incentive for a test of the ray tracing routine.

EXPERIMENTAL RESULTS

Fig. 2 shows the phantom which was used to obtain the experimental measurements. These measurements were obtained using the time of flight technique mentioned in reference 2 with 69 samples along each section, the sampling interval being 2.81 mm. Measurements were made at 96 angles around the phantom. The velocity of sound in the water was known to an accuracy of $\pm$ 0.5 m/sec and in the gelatine to $\pm$ 2.0 m/sec[4]. The results were used to reconstruct the phantom by a backprojection method. The known phantom (shape and velocity) data was stored in the computer and it was used to provide a reference for later error analysis. As it was possible that this data could relate to a reconstruction which was not exactly aligned with the reconstruction coordinates, the data was modified so that the known

Fig 1 Beam pattern

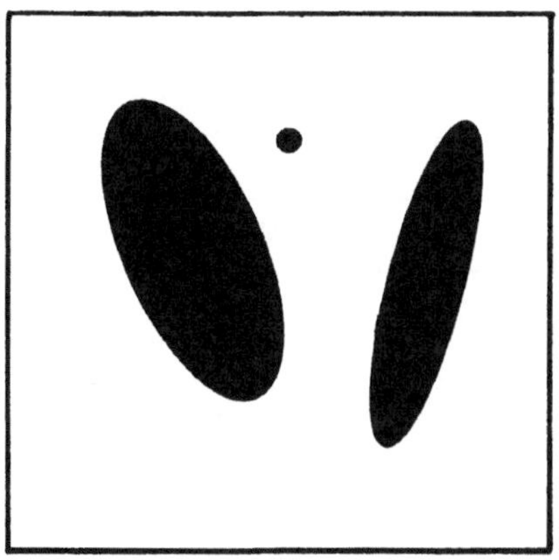

Fig 2 Gelatine phantom

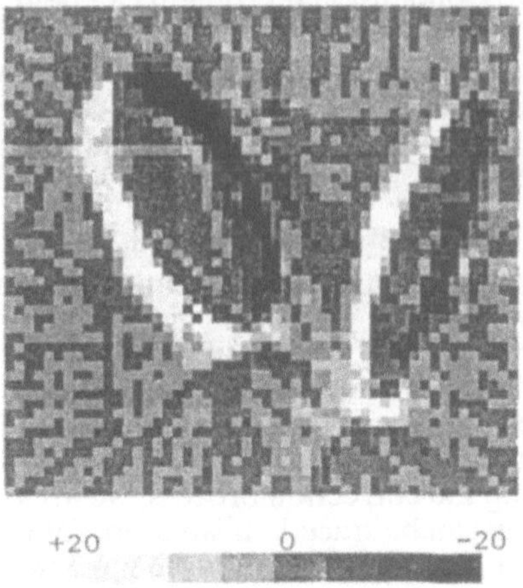

Fig 3 Reconstruction error of a "misaligned" tomogram

phantom was oriented by an iterative process until it coincided most closely with the reconstruction. The coincidence of the two phantoms was easy to see from the error reconstructions i.e. a tomogram showing the differences between the two reconstructions; fig. 3 shows a "misaligned" tomogram and proves that this an acceptable process. Clearly, no matter how carefully this iterative process is carried out there will be some remaining errors. When accurate registration was obtained, we are in a position to obtain detailed error analysis and to compare the various reconstruction techniques.

First we give the results obtained from wave theory using the Rytov approximation as shown in tables 1 and 2.

Next we obtained data for the ray theory reconstruction. In this reconstruction method, we took the original reconstruction and traced the rays through it to obtain calculated times of flight. We subtracted data so obtained from the experimentally obtained data and used the difference to correct the first reconstruction. Successive corrections by this technique are possible. The successive corrections do not produce a significant improvement in the observed errors after two iterations.

Table 1. The errors in the Rytov reconstruction by comparison with the "original" phantom for different assumed velocities of sound in gelatin (units m/sec).

$C_{gelatin}$	Total	Water	Boundary	Gelatin
1526.0	4.59	2.30	9.20	4.69
1525.5	4.52	2.30	9.03	4.28
1525.0	4.45	2.30	8.86	3.89
1524.5	4.39	2.30	8.70	3.53
1524.0	4.37	2.30	8.55	3.22

Table 2. The errors in the Rytov reconstruction by comparison with a low-pass-filtered phantom for different assumed velocities of sound in gelatin (units m/sec).

$C_{gelatin}$	Total	Water	Boundary	Gelatin
1526.0	1.86	1.06	3.08	4.71
1525.5	1.80	1.06	2.98	4.23
1525.0	1.75	1.06	2.89	3.74
1524.5	1.75	1.09	2.87	3.33
1524.0	1.79	1.11	2.98	3.33

When these processes have been completed, we have a reconstruction which can be compared with the original phantom – first, without any spatial filtering of the original and then with spatial filtering, which is consistent with the pixel array used. Tables 3 and 4 show the results which were obtained. It is to be noted that there is an uncertainty of ± 2 m/sec in the velocity of sound in gelatine. We have explored the effects of changing the assumed best value velocities in steps of 0.5 m/sec over the range of the experimental uncertainty. We observed that the least errors in the reconstruction occurred for the lower velocities. and consequently we have omitted data for the higher velocities.

Finally, we present data in tables 5 and 6 which were obtained for the backprojection reconstruction corrected for sound velocities in the gelatine over the experimental range referred to.

Fig. 4 shows the reconstructed tomograms and fig. 5 shows the errors for the different reconstructions algorithms.

RECONSTRUCTION COMPUTING

Table 7 shows the computer times for the various reconstruction techniques using the computers immediately available to us; we did not use main frame machines because of the possible interfacing problems, which might occur if the method became generally useful. We did not have the co-processor board installed in the PS2 computer

Table 3. The errors in the ray-tracing-corrected reconstruction by comparison with the "original" phantom for different assumed velocities of sound in gelatin (units m/sec).

$C_{gelatin}$	Total	Water	Boundary	Gelatin
1526.0	4.37	2.12	8.99	3.72
1525.5	4.28	2.12	8.78	3.27
1525.0	4.19	2.12	8.55	2.82
1524.5	4.10	2.12	8.29	2.54
1524.0	4.03	2.12	8.11	2.11

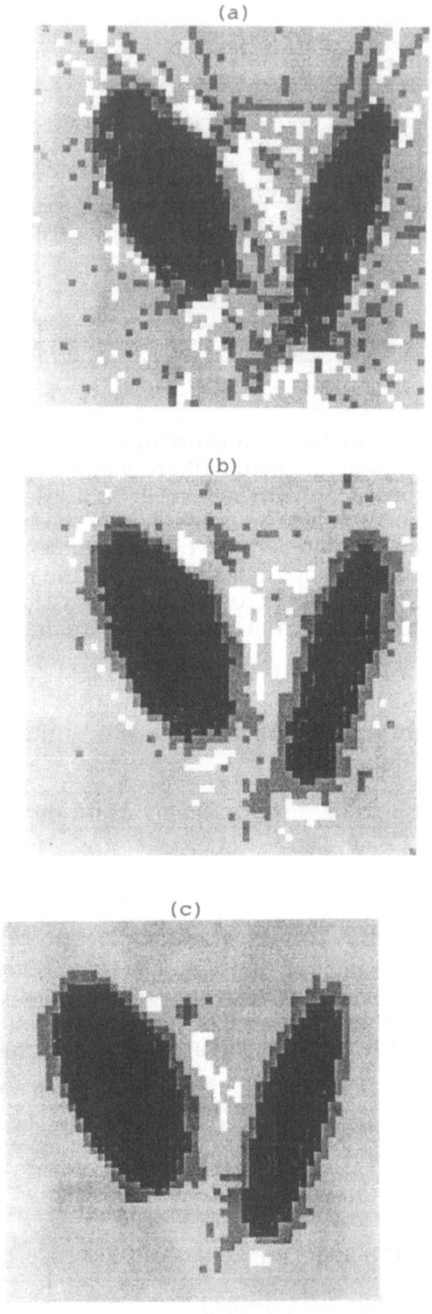

Fig 4 Reconstructed tomograms by
(a) backprojection
(b) back-propagation
(c) ray-tracing iterative correction

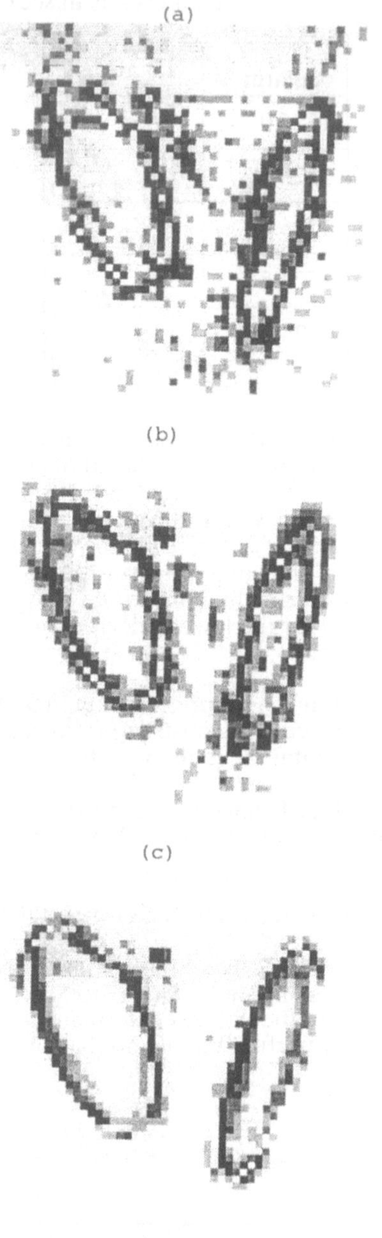

Fig 5 Reconstruction errors by comparison with the original phantom
(a) backprojection
(b) back-propagation
(c) ray-tracing iterative correction

at the time of writing. As a result, we used the most conservative figures given to us by the local computing experts in the determination of the PS2 plus co-processor computing times. Clearly the time of 8.5 min. is too long. We believe, however, that we now have a proper understanding of the reconstruction processes required for this problem and that we can find methods to obtain something approaching an order of a magnitude improvement in this time.

Table 4. The errors in the ray-tracing-corrected reconstruction by comparison with a low-pass-filtered phantom of different assumed velocities of sound in gelatin (units m/sec).

$C_{gelatin}$	Total	Water	Boundary	Gelatin
1526.0	1.70	0.77	3.28	4.11
1525.5	1.61	0.77	3.09	3.70
1525.0	1.54	0.77	2.91	3.26
1524.5	1.46	0.77	2.70	2.83
1524.0	1.40	0.77	2.56	2.42

Table 5. The errors in the back-projected reconstruction by comparison with a "original" phantom of different assumed velocities of sound in gelatin (units m/sec).

$C_{gelatin}$	Total	Water	Boundary	Gelatin
1526.0	5.81	3.41	10.85	4.18
1525.5	5.53	3.41	10.36	3.91
1525.0	5.27	3.41	9.24	3.63
1524.5	5.19	3.41	8.99	3.24
1524.0	5.15	3.41	8.90	3.19

Table 6. The errors in the back-projected reconstruction by comparison with a low-pass-filtered phantom for different assumed velocities of sound in gelatin (units m/sec).

$C_{gelatin}$	Total	Water	Boundary	Gelatin
1526.0	4.94	3.02	8.96	3.96
1525.5	4.89	3.02	8.81	3.77
1525.0	4.81	3.02	8.59	3.49
1524.5	4.72	3.02	8.32	3.20
1524	4.69	3.02	8.24	3.13

Table 7. Computational times (units min.).

Algorithm	IBM-PS2	IBM-PS2*	SANYO-895
Back-Projection	4.7	1.2	24.8
Backpropagation	6.0	2.0	29.5
Ray Tracing	25.5	8.5	61.5

* with co-processor

CONCLUSION

We have conducted a series of investigations into acoustical tomography related to a velocity range associated with soft biological tissue. We have demonstrated that it is possible to obtain reconstructions in which the average error in the reconstruction is about 2 m/sec (with the velocity in the medium approximately 1500 m/sec) i.e. errors of less than 0.15%. We conclude that for these circumstances a ray tracing routine used as a correction to a backprojection reconstruction is the most effective way of obtaining accurate results. We have shown that it is difficult to conduct an accurate error assessment either by direct experiment or by computer modelling. Clearly, there is a link between the errors and the spatial resolution in the tomographic reconstruction, we have not, at the present, been able to define this relationship.

We observed that the reconstruction times are significantly long in the routines we have used for ray tracing reconstructions and we are attempting to reduce this time to 1 minute or less.

ACKNOWLEDGEMENT

We express our appreciation to J. Berry for his work in obtaining the data from the phantom and for later assistance. This work was supported by the Natural Sciences and Engineering Research Council of Canada and Esso Resources Ltd. of Canada.

REFERENCES

1. H. W. Jones, "Proc. Third Spring School on Acoustooptics and Applications", Gdansk, Poland(1986), pp 46-63.
2. H. W. Jones et al., "Acoustical Imaging", vol. 14:617-628, Plenum, New York(1985).
3. J. A. Berry et al., "Acoustical Imaging", vol. 15:79-89, Plenum, New York(1987).
4. A. Markiewicz et al., "Acoustical Imaging", vol. 16, Plenum, New York(1987), (to be published).
5. A. J. Devaney, Ultrasonic Imaging 4:336-350(1982).
6. S. X. Pan et al., IEEE Trans. on Acoustics, Speech and Signal Processing, ASSP-31:1262-1275(1982).
7. A. H. Andersen et al., J. Acoust. Soc. Am., 75:1593-1606(1982).
8. R. K. Mueller et al., Proc. LEEE 67:567(1979).
9. C. F. Schueler etal., IEEE Trans. on Sonics and Ultrasonics, SU-31:195-217(1984).

FLOW VELOCITY FIELD TOMOGRAPHY USING MULTIPLE ULTRASONIC BEAM DETECTORS AND HIGH ORDER CORRELATION ANALYSIS

Kyung-Young Jhang and Takuso Sato

The Graduate School at Nagatsuta
Tokyo Institute of Technology
4259 Nagatsuta, Midori-ku
Yokohama-shi, 227 Japan

ABSTRACT

In this paper, a new method for tomographic reconstruction of velocity field which consists of a set of streams with different velocities, directions, positions and radiuses of curvatures with statistically independent fluctuations is proposed .

In this method, five ultrasonic beams which are arranged in proper positions and directions are used as the line detectors and the fifth-order correlation analysis is applied to the detected data. Then information about each stream is transformed as a corresponding peak in the fifth-order correlation space, hence the velocity field can be reconstructed by finding flow parameters from the measured correlation data in connection with the geometries of the measuring system.

The usefulness of this method is confirmed by computer simulations as well as experiments which are carried out by using 4.5 MHz ultrasonic wave system.

1. INTRODUCTION

2-D velocity field measuring techniques are required in many fields such as plant industries or combustion systems.

The conventional three basic means to measure the fluid flows are, i) Doppler velocimetry, which has been extensively used although it relies on the presence of some kinds of seeds in the flow [1] , ii) the time-of-flight velocimetry which measures the average velocity along the path [2,3] and, iii) the cross-correlation velocimetry which measures also the average velocity by using the random disturbances in the flow [4] .

In this paper, we propose the utilization of multiple line detectors and corresponding high order correlation analysis to the last method, so that the reconstruction of velocity fields which consist of a set of streams with different velocities, radiuses of curvatures, locations and directions is possible.

Details of the principle, numerical results and experiments are shown in the following.

2. PRINCIPLE

First, let us start from a simple case[5] where two streams exist in the same direction as is shown in Fig.1. In this case, the parameters required to identify the velocity field are the locations of the streams and the corresponding velocities. Three line detectors(LD_1, LD_2, LD_3) are arranged along x=0, x=d and y=Ax (A = $\tan\beta$ > 0), respectively. They detect the fluctuations accumulated along each lines induced by the flows as follows,

$$a_i(t) = \int_0^{L_i} \zeta_i\,(l_i,t)\,dl_i \qquad (i=1,2,3) \tag{1}$$

where, $\zeta_i(l_i,t)(i = 1,2,3)$ show the fluctuations on each detector.

Now, let us derive the third order moment function $m_3(\tau_1,\tau_2)$ of the detected signals as follows,

$$m_3(\tau_1,\tau_2) = \frac{1}{T}\int_0^T a_1(t)\,a_2(t+\tau_1)\,a_3(t+\tau_2)\,dt \tag{2}$$

Then if the disturbances distributed along each stream are statistically independent random processes with δ-function-like moment functions up to third-order ones, the moment function $m_3(\tau_1, \tau_2)$ will have peaks P_A and P_B on the lines $\tau_2/\tau_1 = d_A/d$ and $\tau_2/\tau_1 = d_B/d$ as is shown in Fig.2. That is, the locations of two streams can be distinguised in the third-order correlation space and the relations between the parameters of streams and the peaks in the correlation space are given as follows,

$$y_A = d_A(\tan\beta) = d\,\frac{\tau_{2A}}{\tau_{1A}}\tan\beta\,,\quad y_B = d_B(\tan\beta) = d\,\frac{\tau_{2B}}{\tau_{1B}}\tan\beta \tag{3}$$

and

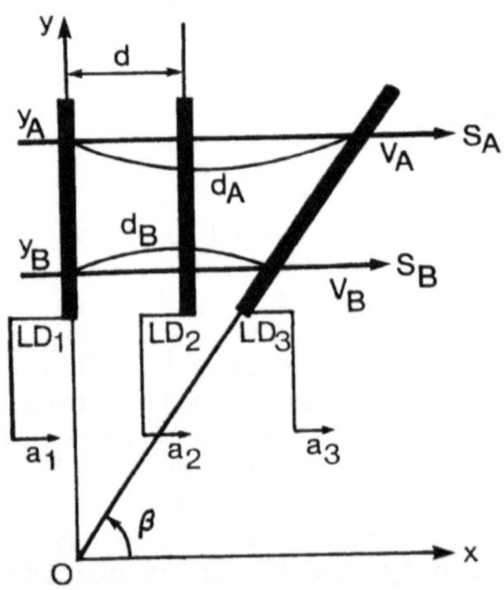

Fig. 1. Velocity field which consists of two streams of the same direction and the arrangement of three line detectors (LD_1, LD_2, LD_3).

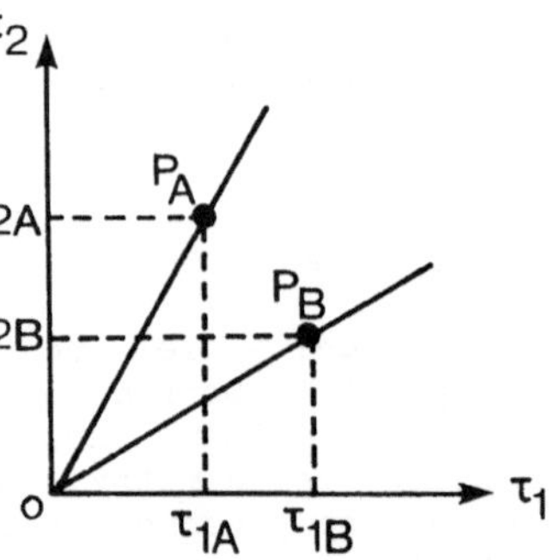

$$\tan\theta_A = \frac{\tau_{2A}}{\tau_{1A}} = \frac{d_A}{d}\,,\quad \tan\theta_B = \frac{\tau_{2B}}{\tau_{1B}} = \frac{d_B}{d}$$

$$v_A = \frac{d}{\tau_{1A}} \qquad v_B = \frac{d}{\tau_{1B}}$$

Fig. 2. Peaks of third-order moment function for the stream S_A and S_B in Fig. 1.

$$V_A = \frac{d}{\tau_{1A}} = \frac{d_A}{\tau_{2A}} \quad , \quad V_B = \frac{d}{\tau_{1B}} = \frac{d_B}{\tau_{2B}} \tag{4}$$

In this process the information of the velocity field is detected as an accumulated ones by the line detectors just like the projections in the conventional tomography[6] , and the distribution is reconstructed by applying the third-order cross-correlation analysis for the detected signals in connection with the geometries of the detectors. Hence, it may be said that it is a kind of tomographic process. Fig.3 shows the conceptional flow diagram of our tomographic method.

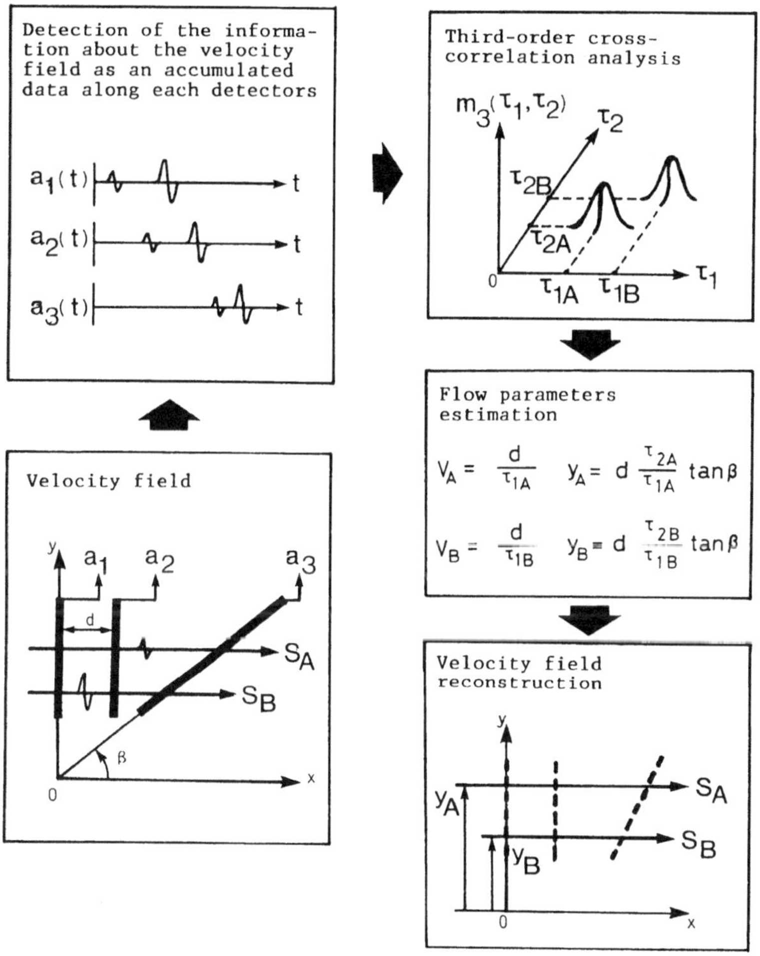

Fig. 3. The conceptional flow diagram of the proposed method.

By the way, if we use only two line detectors and apply the second-order correlation analysis as in the conventional cross-correlation flow meter, then we can know the existance of the two streams but can not distinguish the locations of the streams.

By extending the concept, if we use five line detectors we can detect the velocity fields which consist of a set of streams with different velocities, radiuses of curvatures, locations and directions as is shown in Fig.4.

In this case, fifth-order moment function $m_5(\tau_1, \tau_2, \tau_3, \tau_4)$ is derived as follows,

$$m_5(\tau_1,\tau_2,\tau_3,\tau_4)=\frac{1}{T}\int_0^T a_1(t)\cdot a_2(t+\tau_1)\cdot a_3(t+\tau_2)\cdot a_4(t+\tau_3)\cdot a_5(t+\tau_4)\,dt \quad (5)$$

Then, peaks corresponding to each streams will be observed in the fifth-order correlation space, where the following relations are satisfied, provided the fluctuations along each streams are statistically independent and have δ-function like moment functions up to fifth-order ,

$$\frac{\rho\,\alpha_1}{V} = \tau_1 \ , \quad \frac{\rho\,\alpha_k}{V} = \tau_k - \tau_{k-1} \quad (k= 2,3,4) \qquad (6)$$

From one peak in the correlation space, four parameters (radius of curvature ρ, velocity V, position Y_3, direction Θ_3) of the corresponding one stream are derived.

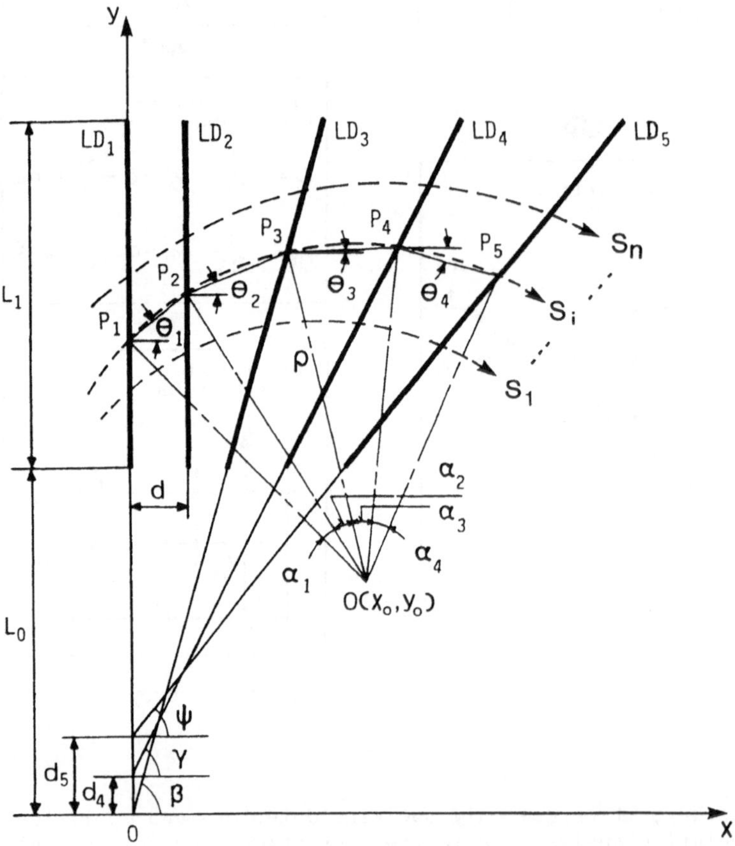

Fig. 4. Velocity field which consists of curved streams and the arrangement of the five line detectors.

First, we get the following five relations,

$$\alpha_1 = 2\,\Omega\left(\frac{\tau_1}{\tau_2}\right) \tag{7}$$

$$\rho = \frac{d}{2\cos\Theta_1\,\sin(\alpha_1/2)} \tag{8}$$

$$V = \frac{\rho\,\alpha_1}{\tau_1} \tag{9}$$

$$y_3 = \frac{d + 2\cos(\Omega - \Theta_1)\,\sin(\Omega - \alpha_1/2)}{\tan\beta} \tag{10}$$

$$\Theta_3 = \Theta_1 - \Omega\left(\frac{\tau_3 + \tau_2 - \tau_1}{\tau_2}\right) \tag{11}$$

where $\Omega = \Theta_1 - \Theta_2$.

These relations can be reduced to the following two ones which contain two variables Θ_1 and Ω ,

$$\sin(\gamma + k_1\Omega) - Z_1 = \tan\Theta_1\,[\cos(\gamma + k_1\Omega) + Z_2] \tag{12}$$

$$\sin(\psi + k_2\Omega) - Z_3 = \tan\Theta_1\,[\cos(\psi + k_2\Omega) + Z_4] \tag{13}$$

where,

$$k_1 = (\tau_3 + \tau_2 - \tau_1)/\tau_2, \qquad k_2 = (\tau_4 + \tau_3 - \tau_1)/\tau_2 \tag{14}$$

Finally, by eliminating Θ_1, we get

$$\begin{aligned} &Z_1\cos(\psi + k_2\Omega) + Z_2\sin(\psi + k_2\Omega) - Z_3\cos(\gamma + k_1\Omega) \\ &- Z_4\sin(\gamma + k_1\Omega) - \sin(\gamma - \psi + k_1\Omega - k_2\Omega) + Z_1Z_4 - Z_2Z_3 = 0 \end{aligned} \tag{15}$$

where,

$$\begin{aligned} Z_1 = \frac{\sin\gamma\,\sin[(\tau_1/\tau_2)\Omega]}{d\,\sin[((\tau_3 - \tau_2)/\tau_2)\Omega]}\Bigg[\left(\frac{\tan\beta}{\tan\gamma} - 1\right) d\,\Bigg\{ 1 \\ + \cos\Omega\,.\,\frac{\sin[((\tau_2 - \tau_1)/\tau_2)\Omega]}{\sin[(\tau_1/\tau_2)\Omega]}\Bigg\} - \frac{d_4}{\tan\gamma}\Bigg] \end{aligned} \tag{16}$$

$$\begin{aligned} Z_2 = \frac{\sin\gamma\,\sin[(\tau_1/\tau_2)\Omega]}{\sin[((\tau_3 - \tau_2)/\tau_2)\Omega]}\Bigg[\left(\frac{\tan\beta}{\tan\gamma} - 1\right) \\ \sin\Omega\,.\,\frac{\sin[((\tau_2 - \tau_1)/\tau_2)\Omega]}{\sin[(\tau_1/\tau_2)\Omega]}\Bigg] \end{aligned} \tag{17}$$

$$Z_3 = \frac{\sin\psi\,\sin[(\tau_1/\tau_2)\Omega]}{d\,\sin[((\tau_4 - \tau_3)/\tau_2)\Omega]}\left[\left(\frac{\tan\gamma}{\tan\psi} - 1\right) d\,\left\{ 1\right.\right.$$

$$+ \cos(k_2 \Omega) \cdot \frac{\sin[((\tau_3 - \tau_2)/\tau_2)\Omega]}{\sin[(\tau_1/\tau_2)\Omega]}$$

$$+ \cos\Omega \cdot \frac{\sin[((\tau_2 - \tau_1)/\tau_2)\Omega]}{\sin[(\tau_1/\tau_2)\Omega]} \Big\} - \frac{d_5 - d_4}{\tan\psi} \Big] \tag{18}$$

$$Z_4 = \frac{\sin\psi \sin[(\tau_1/\tau_2)\Omega]}{\sin[((\tau_4 - \tau_3)/\tau_2)\Omega]} \Big[\Big(\frac{\tan\gamma}{\tan\psi} - 1 \Big) \Big\{$$

$$\sin(k_2 \Omega) \cdot \frac{\sin[((\tau_3 - \tau_2)/\tau_2)\Omega]}{\sin[(\tau_1/\tau_2)\Omega]}$$

$$+ \sin\Omega \cdot \frac{\sin[((\tau_2 - \tau_1)/\tau_2)\Omega]}{\sin[(\tau_1/\tau_2)\Omega]} \Big\} \Big] \tag{19}$$

where, β, γ, ψ are as shown in Fig.4. Therefore, by solving the nonlinear eq.(15) by proper numerical method, four parameters ρ, V, Y_3 and Θ_3 can be determined .

When the flow field consists of many independent streams, each stream can be reconstructed one by one from the corresponding peaks in the fifth-order correlation space.

3. COMPUTER SIMULATIONS

Numerical analysis is carried out to verify the proposed method. Sequences of independent Bernoulli trials are used as the random disturbances. Some examples are shown in Fig.5. The parameters of the measuring field of Fig.4 are as follows: L_0 = 180, L_1 = 180, D = 30, D_4 = 20, D_5 = 40 (mm), β=1.3, γ= 1.107, ψ= 0.905 (rad) and the conditions of three streams S_1, S_2, S_3 are given in Table.1.

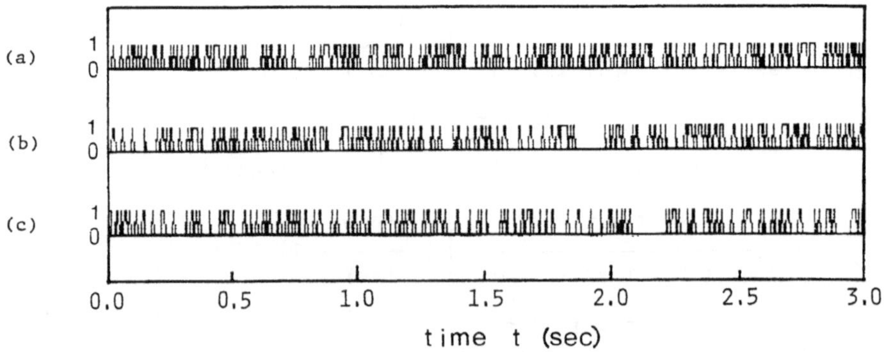

Fig. 5. Examples of Bernoulli random processes used as the random perturbations, $P(t_i=1) = p$ and $P(t_i=0) = 1-p$. (a) for the case of p=0.45, (b) for the case of p=0.40, and (c) for the case of p=0.30.

Table 1. Specific parameters given to the streams and the values in the brackets show the corresponding results of the computer simulations.

STREAM	S_1	S_2	S_3
ρ (mm)	-278.0 (-278.1)	-294.0 (-294.1)	-299.0 (-298.9)
V (mm/s)	149.0 (148.9)	180.0 (179.4)	205.0 (204.9)
y_3 (mm)	290.0 (289.7)	282.3 (281.8)	246.0 (245.6)
θ_3 (rad)	-0.1912 (-0.1912)	-0.3432 (-0.3432)	-0.4804 (-0.4799)

Table 2. Observed peaks in the fifth-order moment function of the signals obtained from the given random processes.

PEAK	τ_1	τ_2	τ_3	τ_4 (sec)
1	0.205	0.435	0.630	0.805
2	0.215	0.525	0.790	1.060
3	0.235	0.600	0.940	1.335

Table.2 shows peaks observed in the fifth-order correlation space. The parameters estimated by the proposed algorithm from the peaks are shown in the brackets of Table.1. They show fairly good agreement with the given ones.

4. EXPERIMENTS

Preliminary experiments were carried out by using 4.5 MHz ultrasonic wave system in a water tank. The arrangement and the signal processings used are shown in Fig.6. Five ultrasonic beam line detectors are used. In the following, the results of observation for a simple case are shown.

Fig.7(a) shows an example of acquired data by one of the ultrasonic beam line detectors as the phase fluctuation when two streams exist. Fig.7(b) is the second-order auto-correlation function of the data. The fluctuation seems random and its auto-correlation function has a sufficiently δ-function-like form. The signal has a lot of impulse-like variations in either positive or negative sides, hence we picked up only peaks of the positive side and the correlation analysis is applied, thus we can get δ-function form moment functions up to fifth-order ones.

Fig.8 shows the results at each stages of the velocity field reconstruction, (a) indicates the conditions of the given two streams, and (b) shows the second-order moment functions among the detected signals $a_1(t)$, $a_2(t)$ and $a_3(t)$. Two peaks exist in either $m_2(\tau_1)$ and $m_2(\tau_2)$ corresponding to the two streams, but we can not distinguish their locations from the second-order correlation analysis. On the other hand,

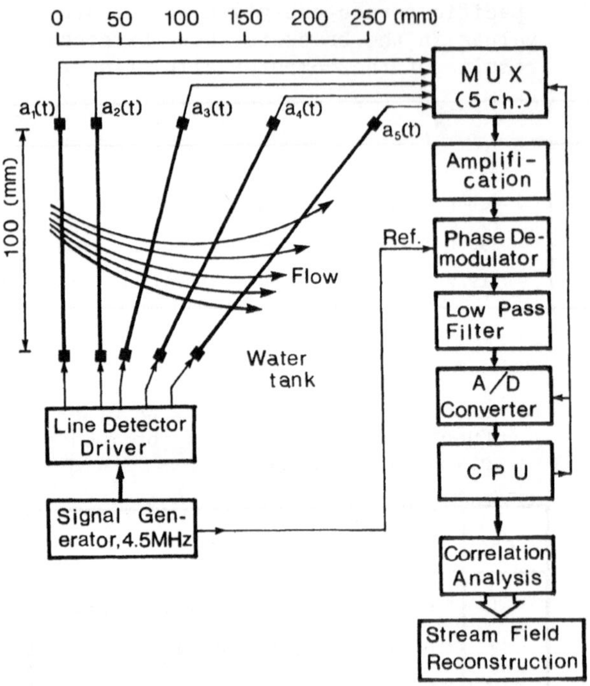

Fig. 6. Schematics of the experimental set-up which uses 4.5 MHz ultrasonic beam line detectors and the flow daigram of the signal processings.

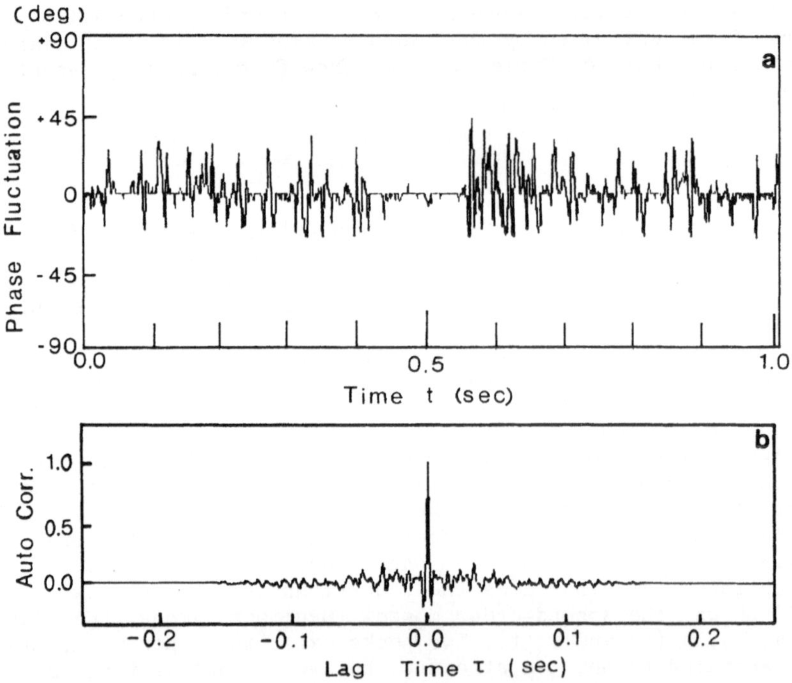

Fig. 7. (a) An example of the acquired time data of the phase fluctuation of one of the ultrasonic beam detectors when independent two streams exist , (b) second-order auto-correlation function of (a).

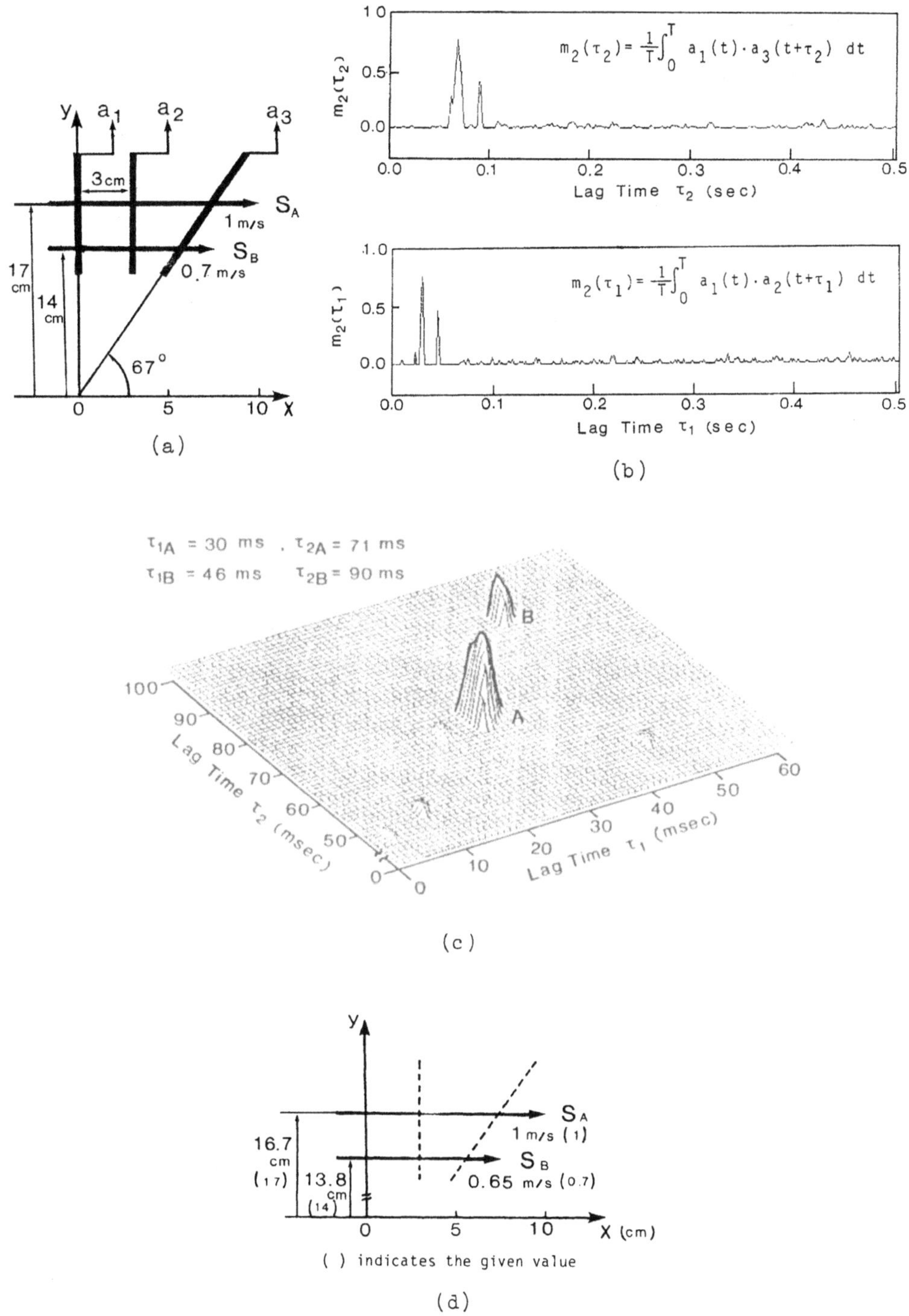

Fig. 8. Experimental results for the flow which consists of independent two streams with different velocities and positions.
(a) Given velocity field.
(b) Second-order moment functions among detected signals.
(c) Third-order moment function of the detected signals.
(d) Reconstructed velocity field.

the third-order moment function gives two distinguishable peaks one corresponding to each stream as is shown in (c). Therefore, we can estimate the required parameters of the corresponding streams from those peaks according to the proposed processes and finally the velocity field was reconstructed as is shown in (d). The reconstructed velocity field showed very good agreement with the given ones.

The experiments are also carried out for the cases where curved streams exist using five ultrasonic beam detectors.

5. CONCLUSIONS

A new tomographic velocity field reconstruction method which uses multiple ultrasonic beam detectors and high order correlation analysis was proposed. Concrete reconstruction algorithm of the velocity field which consists of curved streams is shown, and its validity was confirmed by computer simulations. The conceptional analogy between the conventional tomographic process and this method was also mentioned. The usefulness of this method was also verified by preliminary experiments.

REFERENCES

1. H. D. Thompson and W. H. Stevenson, Developments in Instrumentation, in: "Laser Velocimetry and Particle Sizing", Hemisphere Pub. Corp. , New York (1978)
2. W. H. Munk and C. Wunsch, UPIDOWN Resolution in Ocean Acoustic Tomography, Deep-Sea Res. 29: 1415 (1982)
3. Takuso Sato and Makoto Shiraki, Tomographic Observation of Flow in a Water Tank, J. Acout. Soc. Am., 76: 1427 (1984)
4. M. S. Beck and A. Plaskowski, Analytical and Physical Models for Cross Correlation Flowmeter Design, in: "Cross Correlation Flowmeters" , Adam Hilger, Bristol (1987)
5. Osami Sasaki and Takuso Sato, Two-Dimensional Velocity Distribution Measuring System Using Multiple Laser Beam Line Detectors and High Order Correlation Analysis, Applied Optics, 18: 3522(1979)
6. S. A. Johnson, J. F. Greenleaf, W. A. Samayoa, F. A. Duck and J. Sjostrand, Reconstruction of Three-Dimensional Velocity Fields and Other Parameters by Acoustic Ray Tracing, IEEE Utrasonic Symposiums Proceedings: 46 (1975)

LIMITED-ANGLE PLANAR DIFFRACTION TOMOGRAPHY USING OPTIMAL INCIDENT ANGLES

C. H. Chen*, A. Meyyappa** and G. Wade**

* Department of Electrical Engineering, National Cheng Kung University, Tainan, Taiwan

** Department of Electrical and Computer Engineering, University of California Santa Barbara, CA 93106, U.S.A.

ABSTRACT

We have shown in an earlier paper that the expected mean-squares error for a planar imaging system using optimal incident angles is lower than that for the uniformly-spaced incident angles when a full 180° range of incident angles is available.

In this paper, we present an analysis of the expected mean-squares error for the case when the range of the incident angles is small than 180°. The range of incident angles over which useful projection data are not availabe is employed as a parameter. A first-order solution and computer simulation are presented. They both show that the expected mean-squares error increases as the range of incident angles decreases.

INTRODUCTION

It is frequently necessary to reconstruct tomograms from limited projection data. In many tomographic systems the angular range of the projections is restricted by obstacles or by weak response from the detection circuits. This gives rise to what is known as the "limited-angle problem". Techniques for reconstructing images from limited-angle data have been widely reported in the literature.[1-3]

In this paper, we consider an imaging system, the scanning tomographic acoustic microscope(STAM), where significant projection data are difficult to obtain at highly oblique incident angles.[4] STAM is a system which utilizes the principles of tomography and incorporates digital signal processing to obtain high-resolution subsurface imaging.[5,6] It makes use of elements from an existing ultrasonic microscope, the scanning laser acoustic microscope (SLAM), which detects the acoustic field with a scanning laser as shown in Fig. 1. In STAM, the data are acquired on a fixed receiving plane while the propagation direction of the insonifying plane wave is varied. The system is most suitable for image objects with planar structure. Because of the relatively long wavelenght of ultrasound, diffraction takes place as the radiation propagates through the object. To reconstruct images, we must take diffraction into account.

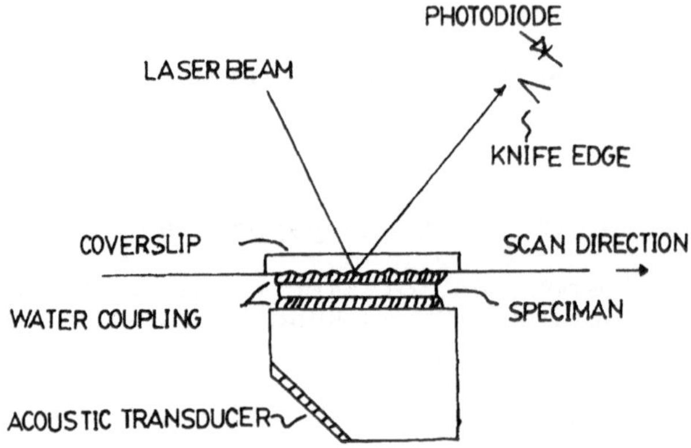

Fig. 1 The scanning laser acoustic microscope (SLAM).

An algorithm based on back-and-forth propagation has been used to reconstruct tomograms in STAM.[7,8] In back-and-forth propagation , the received wavefield and source wavefield are computationally propagated backwards and forwards to a specific plane. A least-squares estimate of the planar distribution within the object is then made to produce the tomogram of a specific layer of interest.

It has been shown that the eigenvalue distribution of the source matrix affects the expected mean-squares error in back-and-forth propagation.[9] The theoretical lower bound of the expected mean-square error is obtained when the eigenvalues of the source matrix are all the same. However, this optimal condition of "equal-valued eigenvalues" requires incident angles over a range of 180°.[9]

For the case when the range of the incident angles is smaller than 180°, the source matrix becomes nondiagonal. A first-order approximation in which the source matrix is approximated by an off-diagonal matrix has been obtained by Chen et al..[10] The correlation coefficient of the off-diagonal matrix is determined by the range of incident angles which give useful information for reconstruction. This paper presents both the analysis of the expected mean-squares error and computer simulations showing how the error increases as the range of incident angles decreases.

PROBLEM FORMULATION

Consider an object of planar structure insonified by acoustic waves coming from below as shown in Fig. 2. The object is homogeneous to the ultrasound except at plane z_2 which contains elements of different elastic composition.

For each projection, the acoustic waves travel through the object from z_1 to z_2. They are modified by the transmittance T associated with the layer at z_2 and are received at z_3. The received wavefield will contain undesired noise-like signal components, $\bar{n}$, caused by scattering due to unknown wave interactions within the object. The quantities $\bar{v}$ and $\bar{u}$ represent the computationally backward-propagated and forward-propagated wavefields respectively at z_2. They can be related to each other by [9]

$$\bar{v} = \bar{u}T + \bar{n} \tag{1}$$

where $\bar{v}, \bar{u}$ and $\bar{n}$ are 1XN row vectors and NXN transmittance matrix.

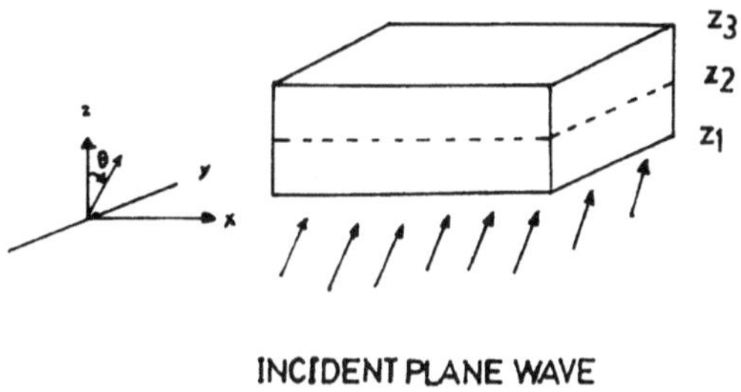

Fig. 2 An object of plannar structure insonified by ultrasound coming from below.

For M projections generated by the planar sources with M different incident angles, we obtain a set of equations

$$\bar{v}_m = \bar{u}_m T + \bar{n}_m, \qquad m=1,2,\ldots\ldots,M \tag{2}$$

The row vectors in Eq. (2) can be combined into the matrix formulation as follows

$$V = UT + N \tag{3}$$

where V, U, N are MXN matrices consisting of M row vectors. In the case where no a priori information concerning the transmittance is known and the undesired signal components are uncorrelated from projection to projection, the minimum-variance estimate of T, denoted by $\hat{T}$, is given by [9]

$$\begin{aligned} \hat{T} &= (U^{*T}U)^{-1}U^{*T}(V-N_0) \\ &= (A_N)^{-1}U^{*T}(V-N_0) \end{aligned} \tag{4}$$

where $A_N = U^{*T}U$ is the source matrix, $N_0 = E(N)$ and U^{*T} represents the adjoint matrix of U. The expected mean-squares error, E_N, is given by [9]

$$\begin{aligned} E_N &\equiv E\left[\sum_{n=1}^{N} |\hat{t}_{nn} - t_{nn}|^2\right] \\ &= \frac{1}{\lambda_1} + \frac{1}{\lambda_2} + \ldots\ldots + \frac{1}{\lambda_N} \end{aligned} \tag{5}$$

where $\hat{t}_{nn}$ is the estimate of t_{nn}, t_{nn} is the element of matrix T at position (n,n) and represents the image value at that pixel; $_1$, $_2$, ..., $_N$ are eigenvalues of A_N.[9] The source matrix A_N is positive definite and is given by

$$A_N = [a_{pq}] = \begin{bmatrix} M\,a_{12} & & a_{1N} \\ a_{21}\,M & & \cdot \\ a_{31}\,a_{32} & & \cdot \\ \vdots & \cdots\cdots & \vdots \\ a_{N1}a_{N2} & & M \end{bmatrix} \tag{6}$$

with

$$a_{pq} = \sum_{m=1}^{M} \exp[j\vec{k}_m \cdot (\vec{r}_q - \vec{r}_p)] \qquad p,q = 1,2,\ldots\ldots,N \tag{7}$$

In Eq. (7) $\bar{r}_p$ and $\bar{r}_q$ are the position vectors of the p-th and q-th pixels of the image and k_m is the wavevector of the m-th incident plane wave. For the two-dimensional case in which no variation in the y-direction is assumed, Eq. (7) can be expressed as

$$a_{pq} = a_{q-p} = \sum_{m=1}^{M} \exp[jk(q-p)\sin\theta_m \Delta x] \qquad p,q=1,2,\ldots\ldots,N \tag{8}$$

where θ_m is the angle that the wavevector k_m makes with respect to the positive z-axis and x is the interval between adjacent pixels in the x-direction. In order to satisfy the Nyquist criterion we choose x= /2. Eq. (8) then becomes

$$a_{q-p} = \sum_{m=1}^{M} \exp[j(q-p)\pi\sin\theta_m] \qquad p,q=1,2,\ldots\ldots,N \tag{9}$$

From Eq. (9), we can obtain a set of incident angles such that the source matrix is diagonal and its eigenvalues are all the same. The incident angles which possess this property are called "optimal." For the case when the range of incident angle is full 180°, the optimal incident angles are given by[5]

$$\theta_m = \sin^{-1}\frac{2m}{M} \qquad m = \pm1,\pm2,\ldots\ldots,\pm\frac{M}{2} \tag{10}$$

Now, consider the symmetrical case for which the incident angles are limited to a region, -(90°- θ) to (90°- θ). Under this condition,the optimal incident angles are given by[10]

$$\theta_m = \sin^{-1}\frac{2m\cos\Delta\theta}{M} \qquad m = \pm1,\pm2,\ldots\ldots,\pm\frac{M}{2} \tag{11}$$

For the set of angles defined by Eq. (10), the source matrix is the identity matrix. However, the set of incident angles given in Eq. (11) results in a nondiagonal source matrix given by

$$A_N = \begin{bmatrix} 1 & a_1 & a_2 & & . & a_{N-1} \\ a_1^* & 1 & a_1 & & : & : \\ a_2^* & a_1^* & 1 & \cdots\cdots & : & : \\ a_3^* & a_2^* & a_1^* & & a_1 & a_2 \\ \vdots & \vdots & \vdots & & 1 & a_1 \\ a_{N-1}^* & \vdots & \vdots & & a_1^* & 1 \end{bmatrix} \quad (12)$$

with

$$a_n = \frac{1}{M} \sum_{-\frac{M}{2}}^{\frac{M}{2}} \exp\left(\frac{j2\pi mn\cos\Delta\theta}{M}\right) \qquad n = 1,2,\ldots\ldots,N-1 \quad (13)$$

From Eq. (13) we see that $|a_1| > |a_2| > |a_3|$. When M is very large, Eq. (13) can be written in an integral form as

$$a_n = \frac{1}{2\cos\Delta\theta} \int_{-\cos\Delta\theta}^{\cos\Delta\theta} \exp[j\pi nt]dt$$
$$= \frac{\sin(n\pi\cos\Delta\theta)}{n\pi\cos\Delta\theta}, \qquad n=1,2,\ldots\ldots,N-1 \quad (14)$$

Fig. 3 shows the values of a_n given by Eq. (14) assuming $\theta = 0^o$ and $\theta = 30^o$. From this figure we see that the limited-angle restriction causes the values of a_n to be other than zero. We can conclude that in general the coefficients a_1, a_2, ..., a_{N-1} are nonzero whenever $\theta \neq 0$. Fig. 4 shows how a_1 varies with the range of incident angles.

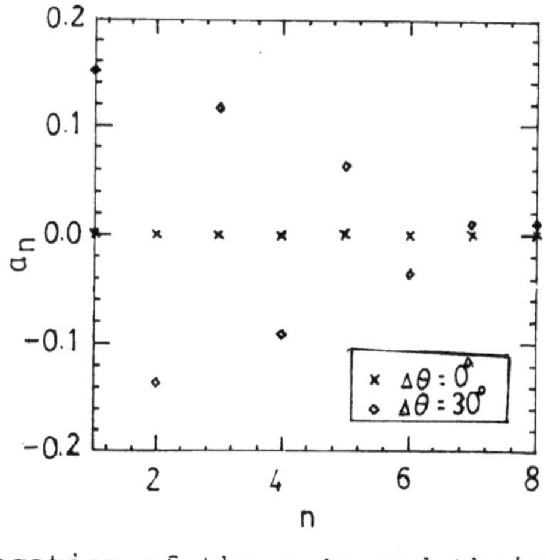

Fig. 3 Location of the a_n's and their values as given by Eq.(14).

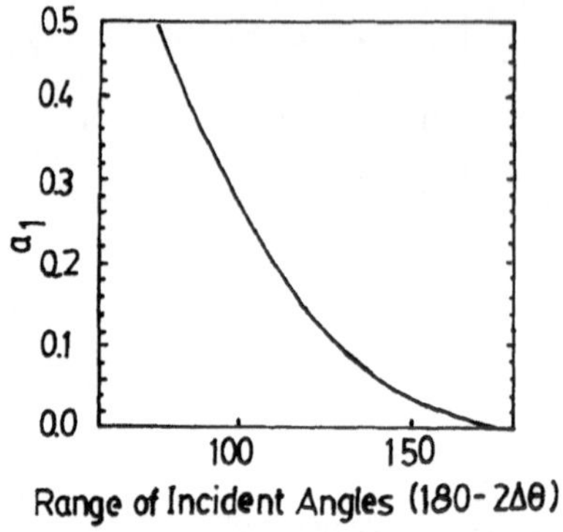

Fig. 4 Variation of first-order coefficient, a_1, with range of incident angles.

FIRST-ORDER SOLUTION

A first-order solution in which A_N is approximated by an off-diagonal matrix has been obtained by Chen et al..[10] Under this approximation, the normalized expected mean-square error e(θ) is given by[10]

$$e(\Delta\theta) \equiv \lim_{N\to\infty} \frac{E_N}{N} = \frac{1}{\sqrt{1-4|a_1|^2}} \tag{15}$$

Substituting Eq. (14) for n=1 into Eq. (15), we obtain

$$e(\Delta\theta) \equiv \lim_{N\to\infty} \frac{E_N}{N} = \frac{1}{\sqrt{1-\frac{4\sin^2(\pi\cos\Delta\theta)}{\pi^2\cos^2\Delta\theta}}} \tag{16}$$

From Eq. (16), we notice that e(θ) is a monotonically increasing function of θ for θ $\leq$ 52.89° and e(0) = 1. Fig. 5 shows how the expected mean-square error increases as the range of incident angles decreases from the ideal value of 180°.

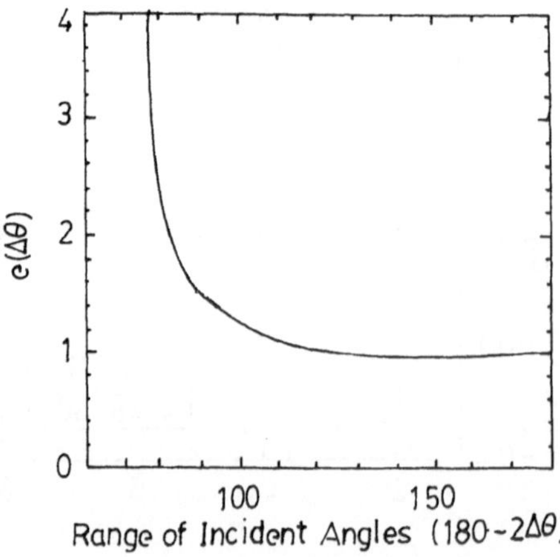

Fig. 5 Normalized expected mean-squares error versus range of incident angles.

COMPUTER SIMULATIONS

An object with planar structure is used as the simulated specimen. The object is assumed to be attenuation-free with respect to the incident wave expect for one layer ten wavelenghts away from the receiving plane. Fig. 6 shows the pattern of the layer.

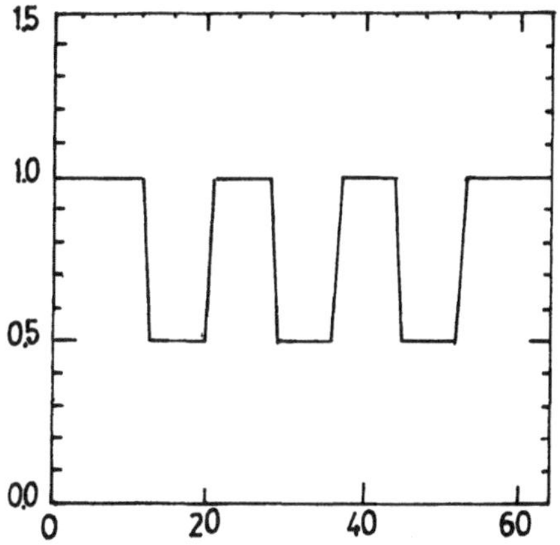

Fig. 6 Simulated pattern for the layer of interest within the object

For the first scheme, the uniform scheme, ten simulated projections of the incident wave transmitted through the object are generated sequentially. The incident angles are obtained by increasing the angle from -90° to 90° according to the uniform spacing. A 10% noise-like signal component is added to each projection. The projections are then processed to reconstruct the image using back-and-forth propagation algorithm. Fig. 7 shows the reconstructed image.

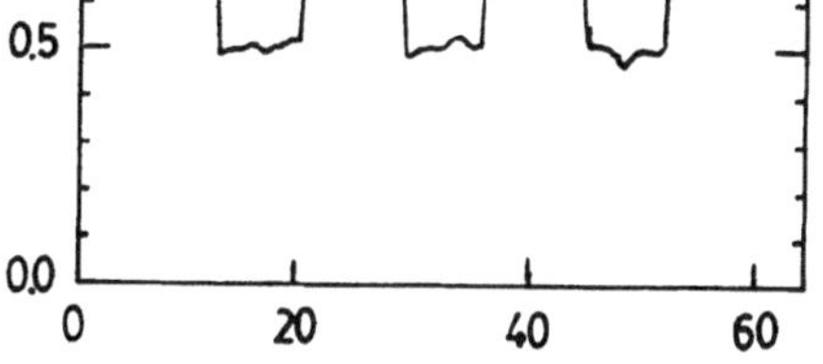

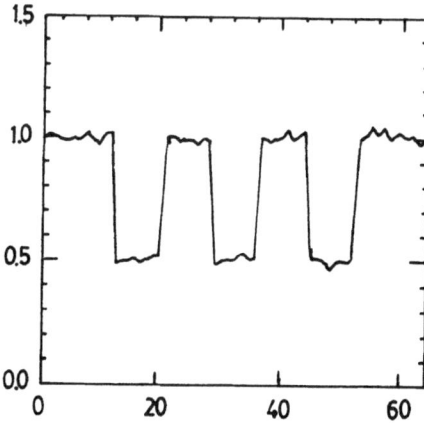

Fig. 7 Image reconstructed from ten projections using the uniform scheme over a full rang of 180°

For the second scheme, the optimal scheme, the same number of projections are generated. The incident angles this time are obtained by increasing the angle from -90° to 90° according to the optimal spacing. As in the first scheme, a 10% noise-like signal component is added to each projection. The projections are then processed to reconstruct the image by using the same algorithm. Fig. 8 shows the reconstructed image. Fig. 9 shows the one with incident angles varying from -75° to 75° according to the optimal spacing. Fig. 10 shows the one from -60° to 60°. Fig. 11 shows the one from -45° to 45°. Fig. 12 shows the one from -22.5° to 22.5°.

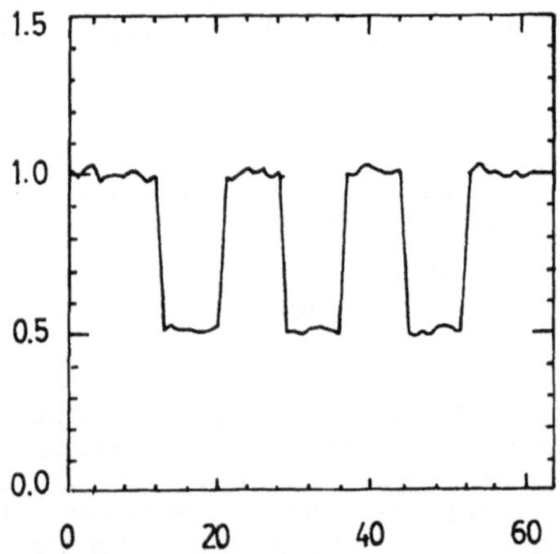

Fig. 8 Image reconstructed from ten projections using the optimal scheme over a full rang of 180°

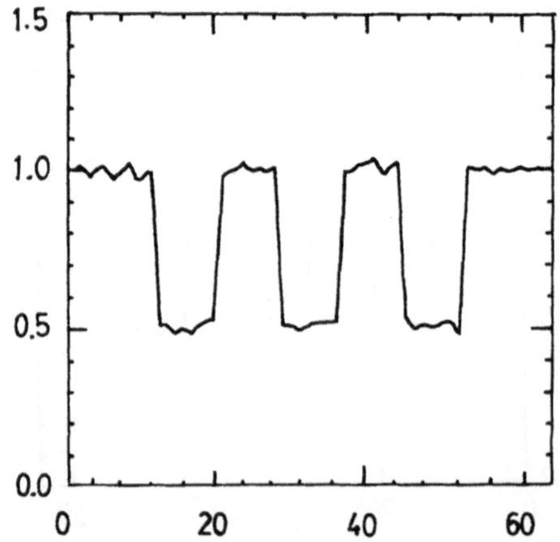

Fig. 9 Image reconstructed from ten projections using the optimal scheme with angles varying from -75° to 75°

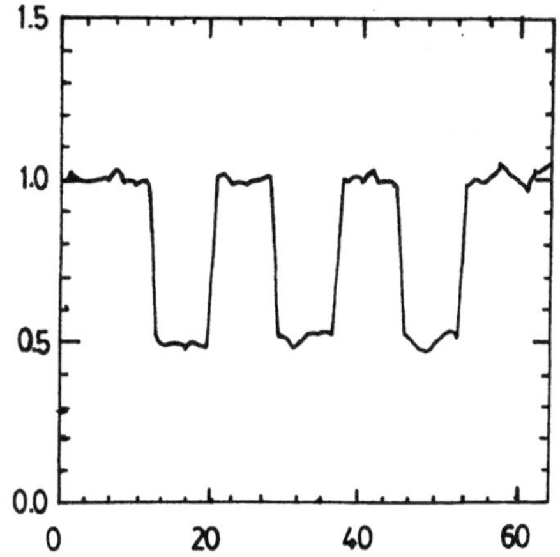

Fig. 10 Image reconstructed from ten projections using the optimal scheme with angles varying from -60° to 60°

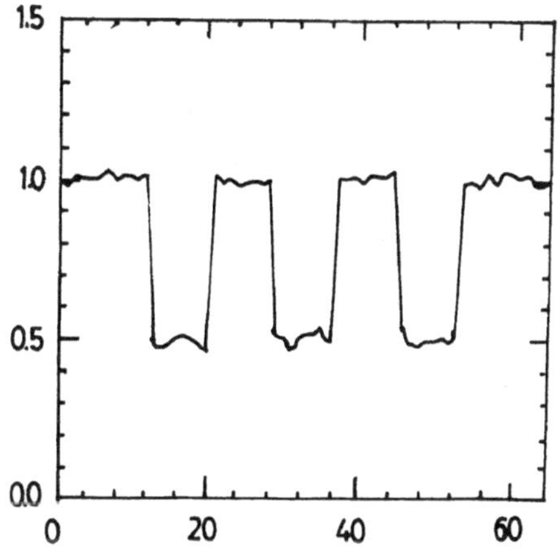

Fig. 11 Image reconstructed from ten projections using the optimal scheme with angles varying from -45° to 45°

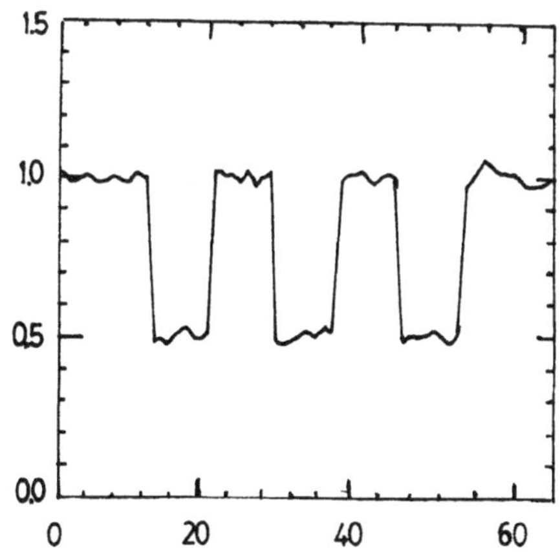

Fig. 12 Image reconstructed from ten projections using the optimal scheme with angles varying from -22.5° to 22.5°

The mean-square errors for the images in Fig. 7, 8, 9, 10 ,11 and 12 are calculated. They are 1.38, 1, 1.286, 1.548, 1.339 and 1.666 respectively. Fig. 13 shows these calculated mean-squares error as function of the range of incident angles. The dot in Fig. 13 indicates the calculated values for the uniform scheme. The dashed curve is the variation predicted by the first-order solution.

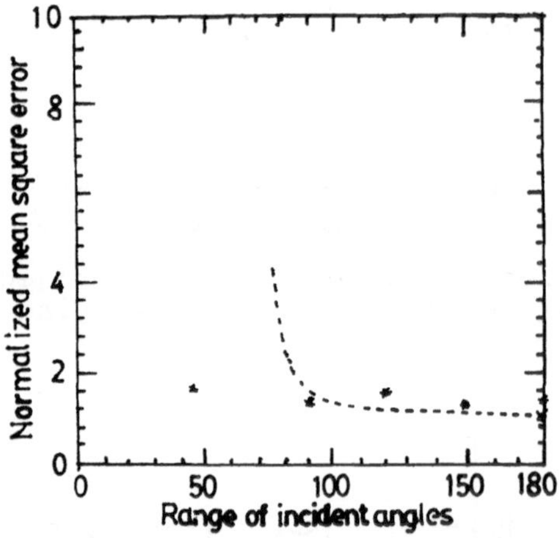

Fig. 13 The calculated mean-square error as function of the range of incident angles. The stars indicate the calculated values for the optimal scheme and the dot for the uniform scheme. The dashed curve is the variation predicted by the first-order solution.

ACKNOWLEDGMENT

We would like to express great appreciation to R. N. Chiou for his patience and skill in typing the several drafts that were necessary for producing the paper.

REFERENCES

1. Sato, T., Norton, S. J., Linzer, M., Ikeda, O. and Hirama, M. 'Tomographic Image Reconstruction from Limited Projections Using Iterative Revisions in Image and Transform Space'. Applied Optics, Vol. 20 (1981), pp. 395-399.
2. Lan, Q.C. and Wade, G. 'Limited-Angle Diffraction Tomography and its Application to Planar Scanning System'. IEEE Trans. on Sonics and Ultrason., Vol. SU-32 (1985), pp. 9-16.
3. Tominaga, S. 'Analysis of Experimental Curves Using Singular Value Decomposition'. IEEE Trans. on ASSP, Vol. ASSP-29 (1981), pp. 429-433.
4 Lin, Z. C. 'A Planar Ultrasonic Tomographic Imaging System'. Ph. D. Dissertation, University of California, Santa Barbara, (1984).
5. Lin, Z. C., Wade, G. and Schueler, C. F. 'Computer-Assisted Tomographic Acoustic Microscopy for Subsurface Imaging'. Acoustical Imaging, Plenum, New York, Vol. 13 (1983) pp. 91-105.

6. Lin, Z. C., Wade, G. and Lee, H. 'Scanning Tomographic Acoustic Microscope: A Review'. IEEE Trans. on Sonics and Ultrason., Vol. SU-32 (1985), pp. 168-180.
7. Lee H., Schueler, C.F., Flesher, G. and Wade, G. 'Ultrasonic Planar Scanned Tomography'. Acoustical Imaging, Plenum, New York, Vol. 11 (1982), pp. 309-323.
8. Lin, Z. C., Wade, G., and Lee, H. 'Back-and-Forth Propagation for Diffraction Tomography'. IEEE Trans. on Sonics and Ultrason., Vol. SU-31 (1984), pp. 626-634.
9. Chen, P. C. H., and Wade, G. '"Optimal Observation Angles for Planar Tomography in STAM System'. Acoustical Imaging, Plenum, New York, Vol.14 (1985), pp. 329-342.
10. Chen, C. H., Meyyappan, A. and Wade, G. 'Limited-Angle Diffraction Tomography in STAM: An Analysis'. Proc. of IV Mediterranean Conference on Medical and Biological Engineering'. Sevilla, Spain (1986), pp. 3-6.

A UNIFIED THEORY FOR ACOUSTICAL HOLOGRAPHY AND DIFFRACTION TOMOGRAPHY

W. S. Gan

Acoustical Services Pte Ltd
29 Telok Ayer Street
Singapore 0104, Republic of Singapore

INTRODUCTION

In optical holography, the object is assumed to be such that a considerable part of the wave penetrates undisturbed through it, a diffraction pattern called hologram which is formed by the interference of the secondary waves arising from the presence of the object scattering with the strong background wave, is recorded on a photographic plate. This means that, if the wave is expressed as the sum of the incident wave and a diffracted secondary wave, the scattering of the secondary wave and interaction within the object is neglected. This negligence represents what is usually known as Born's first approximation. In acoustical holography, where the sound wavelength is about the same order as the obstacle in the medium that causes the scattering, diffraction takes place and the scattering of the secondary wave and the interaction within the object can no longer be neglected. This is what is done in diffraction tomography (DT), but is neglected in acoustical holography (AH), sofar. So one has to combine DT and AH. Hence a unified theory for AH and DT is proposed.

FORMULATION OF THE UNIFIED THEORY

To formulate the unified theory for AH and DT, one starts form the inhomogeneous Helmholtz equation

$$(\nabla^2 + k^2)\Phi = -q \tag{1}$$

where Φ is some appropriate potential, k is the wavenumber being proportional to frequency, and q accounts either for a prescribed source or a scattering potential as far as varying material parameters are concerned.

Unlike previous theories on AH all make use of Kirchhoff approximation, in this paper, the exact diffraction integral is used. This is because Kirchhoff approximation is only appropriate only for the situations when the representative wavelengths are not too long compared with the scatterer and also is valid if interaction within object can be neglected. On top of this, Kirchhoff approximation can be used only for the imaging of plan surfaces not curved surfaces which will be the object used in this paper.

To derive the exact diffraction integral, eqn(1) is rewritten as:

$$(\nabla^2 + k^2)\, u_s(\vec{r}) = -\, u(\vec{r})O(\vec{r})\, k^2 \qquad (2)$$

where $u(\vec{r}) = u_i(\vec{r}) + u_s(\vec{r})$ (3)

$u(\vec{r})$ = total field at any position and $u_i(\vec{r})$ = incident field. The quantity $O(\vec{r})$ is the object function. The reconstruction problem consists of estimating the object function $o(\vec{r})$ given measurements of the wavefield $u_s(\vec{r})$ performed over some set of surfaces Σ for some set Ω of incident wavefields. The set of surfaces Σ and set of incident wavefields Ω depend on the specific application and experimental configuration employed. Each surface is required to lie outside or on the boundary of the support volume of the object.

The unified theory has to include both aspects of AH and DT. Although in DT, the object is illuminated from many different directions, the same equation (2) can be used for all cases. Eqn.(2) can also be written as:

$$(\nabla^2 + k^2)u(\vec{r}) = -O(\vec{r})u(\vec{r}) \qquad (4)$$

where $O(\vec{r}) = k^2\left[n^2(\vec{r}) - 1\right]$ (5)

The term $O(\vec{r})$ will be used to represent all inhomogeneities of the object. Later the object will be reconstructed in terms of the object function $o(\vec{r})$.

The incident field as indicated in Eqn.(3) is the field present without any inhomogeneities or a solution to the equation

$$(\nabla^2 + k^2)\, u_i(\vec{r}) = 0 \qquad (6)$$

That leaves the scattered field, $u_s(\vec{r})$, as that part of the field due to the object inhomogeneities, referring to Eqn.(3). The wave equation then becomes Eqn.(2). The scalar Helmholtz equation (2) cannot be solved for $u_s(\vec{r})$ directly but a solution can be written in terms of the Green's function. The Green's function, which is a solution of the different equation

$$(\nabla^2 + k^2)G(\vec{r}/\vec{r}\,') = -\,\delta\,(\vec{r} - \vec{r}\,') \qquad (7)$$

is written in three-space as

$$G(\vec{r}/\vec{r}\,') = \frac{e^{jkR}}{4\pi R} \qquad (8)$$

with $R = |\vec{r} - \vec{r}'|$ (9)

In two dimensions the solution of Eqn.(7) is written in terms of a zero-order Hankel function of the first kind, and can be expressed as

$$G(\vec{r}/\vec{r}') = \frac{j}{4} H_o^{(i)}(kR) \tag{10}$$

In both cases, the Green's function, $G(\vec{r}/\vec{r}')$, is only a function of the difference $\vec{r} - \vec{r}'$ so the function will often be represented as simply $G(\vec{r} - \vec{r}')$. Because the object function in Eqn.(7) represents a point in homogeneity, the Green's function can be considered to represent the field resulting the field resulting from a single point scatterer.

It is possible to represent the forcing function of Eqn.(4) as an array of impulses or

$$q = o(\vec{r})u(\vec{r}) = \int O(\vec{r}')u(\vec{r}') \quad \delta \quad (\vec{r}-\vec{r}')d\vec{r}' \tag{11}$$

In this equation, the forcing function q of the inhomogeneous wave equation is represented as a summation of impulses weighted by $o(\vec{r})u(\vec{r})$ and shifted by $\vec{r}$. The Green's function represents the solution of the wave equation for a single delta function; because the left hand side of the wave equation is linear, a solution can be written by summing the scattered field due to each individual point scatterer.

Using this idea, the total field due to the impulse $O(\vec{r}')u(\vec{r}')$ $\delta(\vec{r}-\vec{r}')$ is written as a summation of scaled and shifted versions of the impulse response, $G(\vec{r})$. This is a simple convolution and the total radiation from all sources on the right hand side of Eqn.(2) must be given by the following superposition:

$$u_s(\vec{r}) = \int G(\vec{r}-\vec{r}')O(\vec{r}')u(\vec{r}')d\vec{r}' \tag{12}$$

So an integral equation for the scattered field $u_s(\vec{r})$, has been written in terms of the total field, $u = u_i + u_s$. This equation needs to be solved for the scattered field. The exact diffraction integral is thus given by

$$u(\vec{r}) = u_i(\vec{r}) + \int G(\vec{r}-\vec{r}')O(\vec{r}')u(\vec{r}')d\vec{r}' \tag{13}$$

SOLUTION OF THE EXACT DIFFRACTION INTEGRAL

The object used is highly inhomogeneous and strong scattering is considered. The solution will be based on the exact diffraction integral and no Born or Rytov approximations will be used. The relationship in the spatial-frequency plane between the scattered pressures on the probing line and the induced sources within the unknown target follows from a plane wave expansion of the integration kernel. The main features of the direct and inverse solutions of our scattering problem will be depicted here. Images retrieved from synthetically generated are then given. We first discuss the spatial resolution of the imaging procedure from nono- and multi-view images of an almost punctual target. This is done briefly, for images of a complex scatterer cannot generally be reduced to the superpostion of such point-spread functions. Secondly, the influence of the images formation parameters and of the geomerical and acoustical parameters of the target and its environment, using typical examples of such complex scatterers. We emphasize the fact that targets of different shapes and /or acoustical properties can be clearly discriminated and at a lesser extent depicted using this multiview diffraction tomography approach. A brief theoretical analysis will be given below.

We consider two homogeneous fluid half-spaces S_1 and S_2 on both sides of the x = 0 plane with different sound velocities c_1 and c_2 and attenuations α_1 and α_2 at the operating frequency f. In the lower half-space the fluid scattering inhomogeneous target is emdedded with crosssection S in the r-y plane. The sound velocity c_ℓ and the attenuation α_ℓ within the target vary with position. In the upper half-space a plane P-wave is generated and propagates downwards with θ_1 angle of incidence on the interface. (See Figure 1).
The density is assumed constant. We denote $k_\ell = \omega/c_\ell + i\alpha_\ell (\ell = 1,2,\ell)$ as the respective complex-valued propagation constants. Time dependence $e^{-i\omega t}$ is assumed. The complex amplitude p of the scalar acoustic pressure satisfies the inhomogeneous Helmholtz equation

$$\nabla^2 p(x,y) + k^2(x,y)p(x,y) = -q \tag{14}$$

The known boundary conditions (continuity of the pressure and of its normal derivative) are then implied. The Green function $G_{\ell,m}(x,y,x',y')$ in the layered medium represents the field observed at (x,y) in S_ℓ when a line source is located at (x',y') in S_m, the target being absent. This function satisfies Eqn.(7) in the distribution sense:

$$\nabla^2 G_{\ell,m}(x,y,x',y') + k_\ell^2 G_{\ell,m}(x,y,x',y')$$

$$= - \delta(x-x',y-y')\ \delta_{\ell,m}, \quad \ell = 1,2, m = 1,2 \tag{15}$$

where δ is the Dirac distribution and $\delta_{\ell,m}$ the Kronecker symbol. $G_{\ell,m}$ implicity satisfies the same boundary conditions as p. We now expand $G_{\ell,m}$ into a plane waves spectrum:

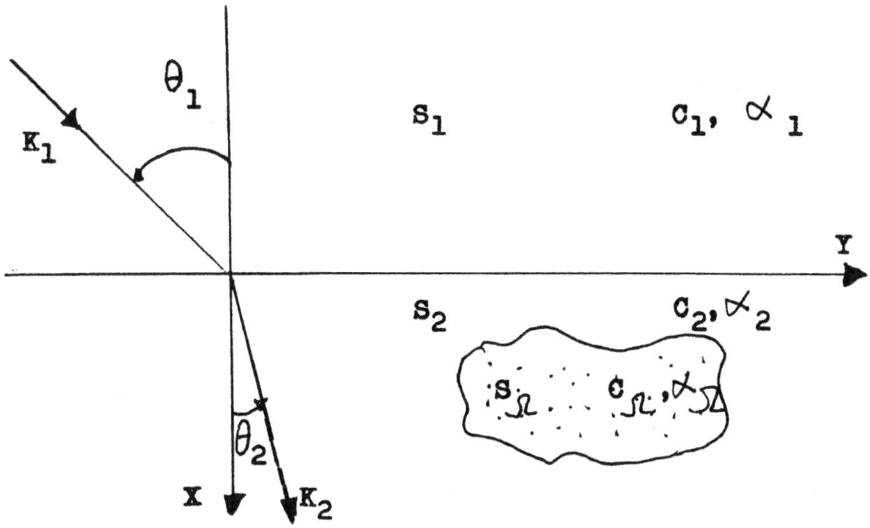

Fig.1. Geometry for the scattering problem.

$$G_{\ell,m}(x,y,x',y') = \frac{1}{2\pi}\int_{-\infty}^{+\infty} G_{\ell,m}(x,\alpha,x',y')e^{i\alpha y}d\alpha, \tag{16}$$

With the above boundary conditions, Eqn.(13) is rewritten as

$$p(x,y) = P_o(x,y) + \int_{S_\Omega} (k_\Omega^2(x',y')-k_2^2)p(x',y').G_{\ell,2}(x,y,x',y')\, dx'dy' \tag{17}$$

$\ell = 1$ when (x,y) lies within S_1 region and $\ell = 2$ when (x,y) lies within S_2 region and p_o is given by

$$p_o(x,y) = \exp\left[ik_1(x\cos\theta_1 + y\sin\theta_1)\right] + R\exp\left[-ik_1(x\cos\theta_1 - y\sin\theta_1)\right]$$

$$\text{and } p_o = T\exp\left[ik_2(x\cos\theta_2 + y\sin\theta_2)\right] \tag{18}$$

R, T, and θ_2 are, respectively, the reflection coefficient, the transmission coefficient and the transmission angle of a plane wave which falls onto the interface under incidence θ_1. We assume that Fresnel's laws hold:

$$R = \frac{k_1\cos\theta_1 - k_2\cos\theta_2}{k_1\cos\theta_1 + k_2\cos\theta_2}$$

$$T = \frac{2 k_1 \cos \theta_1}{k_1 \cos \theta_1 + k_2 \cos \theta_2} \tag{19}$$

and $k_1 \sin \theta_1 = k_2 \sin \theta_2$ (20)

P_o is the field that exists in the absence of the target. We let J be the density of the induced sources within the target, given by

$$J(x',y') = (k^2_{\Omega}(x',y') - k_2^2)p(x',y') \tag{21}$$

$(p-p_o)$ is the anomalous field related to the target pressure and we will collect it for imaging the target. To compute p, two types of approaches are used. The first approach, for small objects of high velocity or attenuation contrast with respect to their environment, the Gauss-Jordan direct algorithm is used. For large objects with lower contrasts, an interative approach is used. Here the pressure within the object is expanded into Neumann's series[2] with active Aitken's extrapolation[3] to speed up the convergence. Such an iteractive numerical solution is much faster, when convergent, than the Gauss-Jordan solution and of the same accuracy. It does not have the problem of numerical instability of the algebraic method proposed by M. Slaney and A. C. Kak[1] which limits the approach to objects with a refractive index change of less than 20 to 30%.

To analyze the imaging theory, we refer to Fig. 1. The probing line is located at $x = x_o (x_o < 0)$ above the interface. We introduce the anomalous field ψ and the normalized induced sources ϕ by

$$\psi(x_o,y) = P(x_o,y) - P_o(x_o,y)$$

$$\phi(x,y) = J(x,y)/P_o(x,y) \tag{22}$$

and their respective one- and two-dimensional Fourier transforms $\hat{\psi}$ and $\hat{\phi}$ by

$$\hat{\psi}(X_o,\alpha) = \int_{-\infty}^{+\infty} \psi(X_o,y)e^{-i\alpha y}\, dy$$

$$\hat{\phi}(\alpha,\beta) = \int_{-\infty}^{+\infty} \phi(x,y)e^{-i(\alpha x + \beta y)}\, dxdy \tag{23}$$

We now deal with fields formulation. We consider an observation point (x_o,y) on ℓ in S_1, the integration being performed on Ω. We replace the Green 's function $G_{1,2}$ by its plane-wave expansion in Eqn.(16)

$$G_{1,2}(x_o, \alpha, x', y') = i \frac{e^{i(\beta_2 x' - \beta_1 x_o)}}{1 + 2} e^{-i\alpha y'}$$

where $\beta_i = \sqrt{k_i^2 - \alpha^2}$, $i = 1,2,$ (24)

After some transformation of Eqns.(17) and (21), we obtain a relationship in the spatial-frequency plane between Fourier transforms $\hat{\psi}$ and $\hat{\phi}$:

$$\hat{\phi}(\mu_2, \nu_2) = \frac{-i}{T} (\beta_1 + \beta_2) e^{i \beta_1 x_o} \hat{\psi}(x_o, \alpha)$$

where $\mu_2 = - k_2 \cos \;_2 - \beta_2$

$$\nu_2 = - k_2 \sin \theta_2 + \alpha$$

$$/\alpha/ < k_2 < k_1 \quad \text{and} \; /\theta_2/ < \pi/2 \qquad (25)$$

This relationship does not hold or the evanescent spectrum and is exact only when S_1 and S_2 are lossless. Another one should be written k_2 larger than k_1. We note that points (μ_2, ν_2) belong to a circular arc with radius k_2 centred on $(-k_2 \cos \theta_2 - k_2 \sin \theta_2)$.

IMAGE RECONSTRUCTION

For reconstruction, the backpropagation algorithm[4] will be used. Our purpose is to image curved surfaces also and this requires axial resolution.

We start from the synthetic aperture radar (SAR) and Synthetic Aperture Focussing (SAFT) technique. One considers a single point scatterer illuminated by an incident pulse u_i, which takes a travelling time t_1 to reach the scatterer. The scattered field is then collected within some synthetic aperture defined by some linear coordinate x as function of time. Let t_s denote the travelling time of the scattered pulse to some specific data point within the aperture. Then a reconstruction of the point scatterer is defined heuristically through a time domain backpropagation algorithm:

$$u_R(x', y', z') = \int_{-\infty}^{+\infty} u_s(d, y_o, z_o, t_i(x', y', z') + t_s(x', y', z'; x, y_o, z_o))\,dx \qquad (26)$$

where x', y', z' denote reconstruction space coordinates (pixel coordinates) and y_o, z_o represent fixed cartesian coordinates of the linear aperture. Eqn.(26) can be extended to two-dimensional apertures. Varying x',y',z' through pixel space yields a sudden peak where the point scatterer resides.

It has been proved that the procedure also works for arbitrarily shaped scatterers (5) for ultrasonic NDT applications, whereas for SAR-imaging the scenery is acutally composed of point scatterers.

Nevertheless, SAR and SAFT yield axial resolution too. The only restructions apply to the situation of the aperture with respect of the main lobe scattering directions. Again, specular reflection has to be included. All these questions of resolution can be well accounted for transposing the imaging algorithm to a so-called k-space, which is the Fourier space with respect to the spatial object domain.

Sofar the above SAR and SAFT imaging procedures do not imply far field conditions, we now derive a diffraction theoretical far field counterpart to Eqn.(26). We have using Eqn.(17),

$$u_s(x,y,z_o,k) = \int_{-\infty}^{+\infty}\int_{-\infty}^{+\infty}\int_{-\infty}^{+\infty} \gamma(x',y',z') \cdot \frac{e^{2ik\sqrt{(x'-x)^2+(y'-y)^2+(z'-z_o)^2}}\,dx'dy'dz'}{(x'-x)^2+(y'-y)^2+(z'-z_o)^2} \tag{27}$$

where the characteristic function γ of the scatterer has been introduced explicitly. We assume a twodimentional planar xy- aperture at a distance z_o from the scatterer's origin and point source illumination, accounting for the factor 2 in the exponential. In Eqn.(27), only the principle is outlined and any prefactors are ignored.

Fourier transforming Eqn.(27) with respect to x and y yields

$$\tilde{u}_s(k_x,k_y,z_o,k) = \int_{-\infty}^{+\infty} \hat{\gamma}(k_x,k_y,z')e^{-i(z'-z_o)\sqrt{4k^2-k_x^2-k_y^2}}\,dz' \tag{28}$$

where the " " indicates Fourier transformed quantities. Defining a third Fourier variable k_2 through the disperation relation

$$k_2 = \sqrt{4k^2 - k_x^2 - k_y^2} \tag{29}$$

which is actually similar to the Cagniard-de Hoop technique of solving pulse problems, we obtain

$$e^{-iz_ok_z}\ u_s(k_x,k_y,z_o,k(k_x,k_y,k_z,))\ 1 \qquad (k_x,k_y,k_z) \tag{30}$$

where the second tilde above $\hat{\gamma}$ accounts for an additional Fourier transform with respect to z'. Therefore, solving Eqn.(30) for γ (x',y',z') yields

$$\gamma(x',y',z') = \int d\vec{k}\ \tilde{u}_s(k,\vec{r}_a)e^{-i\vec{r}_a\cdot\vec{k}}e^{i\vec{k}\cdot\vec{r}'} \tag{31}$$

with the aperture vector $r_a = (0,0,z_o)$. Eqn.(31) is the desired reconstruction algorithm, which, according to our farfield considerations, should be equivalent to the SAFT procedure in Eqn.(26) operating directly in the time domain, whereas here, time domain data have first to be Fourier transformed to the frequency domain.

It has to be pointed out here that u_s in Eqn.(31) will have to be given by the solution if the exact diffraction integral derived in Eqn.(17).

REFERENCES

1. M. Slaney and A.C. Kak, "Imaging with Higher Order Diffraction Tomography", paper presented at the 1985 IEEE Ultrasonics Symposium, San Franisco, USA, October 1985.
2. B. Duchene, D. Lesselier and W. Tabbara, "Acoustical Imaging of 2D Fluid Targets buried in a Half-Space: a Diffraction Tomography Approach", IEEE Trans, on Ultrasonics, Ferroelectrics and Frequency Control, UFFC-34:540 (1987).
3. G. Dahlquist and A. Bjorck, "Numerical Methods", Prentice Hall, Englewood Cliffs, NJ (1974).
4. K. J. Langenberg and V. Schmitz, "Generalized Tomography as a Unified Approach to Linear Inverse Scattering: Theory and Experiment", paper presented at the 14th International Symposium on Acoustical Imaging, The Hague, The Netherlands, April 1985.

MEASUREMENT OF INTERNAL TEMPERATURE DISTRIBUTION USING ULTRASONIC COMPUTED TOMOGRAPHY

Yoshiro Tomikawa*, MItsuharu Numata*,
Hiroaki Yamada**, and Masashi Nakamura**

*Dept. of Electrical Engineering, Yamagata University
Yonezawa, Yamagata, Japan 992

**Dept. of Electrical Engineering, Nihon University
Narashino, Chiba, Japan 275

INTRODUCTION

There is an engineering demand to measure the internal temperature distribution of objects non-destructively. If the measurement can be made from the outside by some methods without inserting a thermometer into the object, such applications as measurements of the internal temperature of the object under a certain chemical reaction can be considered. The measurement of the internal temperature of a living body under cancer hyperthermia is a typical example [1]. To such demands, we have invented the application of ultrasonic CT [2][3]. At first, for the purpose of investigating how small a difference in temperature could be discriminated using the ultrasonic CT, we assumed water as the medium of ultrasound and simulated the measurement process. When ultrasound is propagated in water, the propagation velocity differs with temperature. Therefore, the measurement of internal temperature distribution was considered to be possible if we record this ultrasonic propagation time.

Such temperature measurement was recorded in the field of hyperthermia, and measurements using both X-ray CT and NMR-CT have already been reported [4]. In this meaning, the measurement using ultrasonic CT dealt with herein is not new basically. However, because the measurement by ultrasonic CT is easy and safe to handle, it is considered to be practical, not only in the medical field but also in the industrial field.

It was found from the results of simulation that about 1 °C difference in temperature could be discriminated by applying a smoothing treatment to the CT results and a new CT method of subtracting the CT results before heating up from those after heating up.

The simulation results of ultrasonic CT are included in the first part of this paper. The second part contains the result of experimentation.

REDUCTION OF NOISES FROM CT IMAGING

We previously investigated the new ultrasonic CT method for the non-destructive inspection of rotten parts in a wooden pole, where CT imaging depended on the propagation time, as shown in Figs.1, 2 and 3 [2][3]. Therefore, we devised a new application of this method to measure the internal temperature distribution aimed in this paper and considered how to reduce the noises as much as possible. Namely, we adopted the smoothing treatment considered to be the most fundamental method in noise reduction, and tried to improve the CT images [5][6].

A CT image is made up of pixels, as shown in Fig.4. Because the CT images in this study are circular, there are pixels that are cut by a circular line partially. In classifying their states of arrangement roughly, two states, as shown in Figs.5(a) and (b) were considered. The state of pixels within the inner circle is shown in Fig.5(c). The following methods were applied to reduce these noises. That is, we applied Eq.(1) to (c), Eq.(2) to (b) and Eq.(3) to (a).

$$X' = (N+S+E+W+4X)/8 \quad (1)$$
$$X' = (2S+E+W+4X)/8 \quad (2)$$
$$X' = (2S+2E+4X)/8 \quad (3)$$

SIMULATION FOR THE CASE OF COMPARATIVELY LARGE DIFFERENCES IN TEMPERATURE

As the propagation medium of ultrasound, we assumed three-layers of concentric circular ring columns of "water", the total radius being 12 cm, as shown in Fig.6, and then we set up the temperature of each layer to be the three cases (1)-(3) in the Figure. That is, we assumed 15°C, 10°C and 5°C differences in temperature between each concentric circle, respectively. As for the propagation velocity of ultrasound as a function of temperature, we referred to reference [7], and the values shown in Fig.6 were used. Using these, we calculated 18 data for the propagation time between point A and point M, the latter being chosen with intervals of 10 degrees from point A to B. The CT images for the case of 5°C difference in temperature are shown in Fig.7 (CT images for 15°C and 10°C are abbreviated herein), where the smoothing of 5 repetition were accomplished. It was found from this result that the internal temperature distribution could be distinguished sufficiently even in the case of 5°C difference in temperature as shown in Fig.7 [5]. According to this simulation of other case, the limit for the distinction of temperature difference by this method was about 3°C [6].

SIMULATION FOR THE CASE OF 1°C DIFFERENCE IN TEMPERATURE

An accuracy better than 1°C is required in medical applications. So we investigated methods for improving the accuracy of measuring the internal temperature. The result was that a 1°C difference in temperature could be discriminated by the following CT technique [6].

We considered the case in which the difference in temperature between each layer was 1°C. To begin with, we examined CT results in case that the temperature within the three layers was uniformly 37°C. However, it was found that even in the case of no difference in temperature, periodical errors were included in the calcualted CT values and uniformly reasonable results could not be obtained. These are caused by the projection method and the errors are unavoidable. So, we devised a new CT method in which the CT values for no difference in temperature (that is, each layer was 37°C constant) was subtracted from the CT values for 1°C temperature difference of each layer, in order to remove the fundamental error mentioned above. Since our CT technique depends upon the propagation time, in case there are foreign substances which hinder ultrasonic propagation, it becomes necessary in the procedure for our new CT method to subtract the CT values caused by them, and as a result the remaining values can be considered to include only temerature information. Namely, from this viewpoint, it is necessary to use the relative propagation time. After all, we obtained the distribution of the propagation time on the circle diameter, as shown in Fig.8, together with that 10 repetitive smoothing was adopted. Using this method, we were able to obtain a result similar to CT imaging under a 5°C difference in temperature, as shown in Fig. 9, and were also able to discriminate the distribution under 1°C difference in temperature. Hence, it is considered that our CT technique can be expected to be utilized in medical applications.

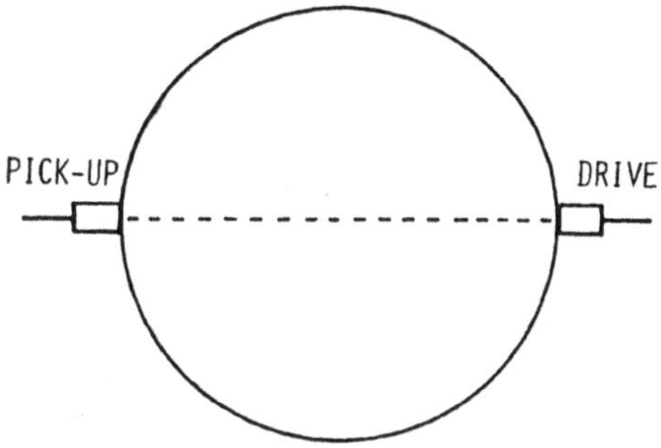

Fig.1 Propagation route

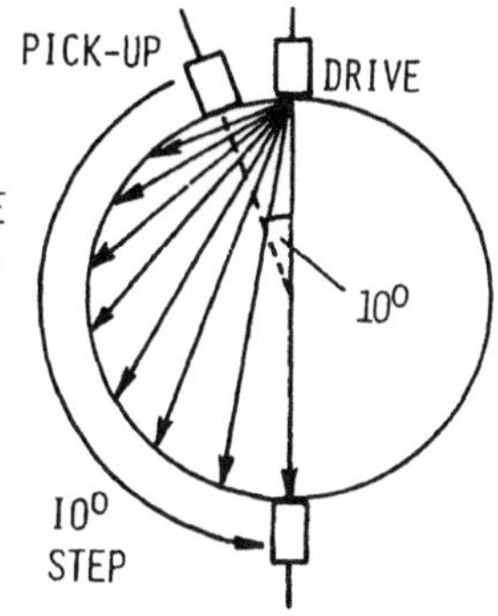

Fig.2 Measurement

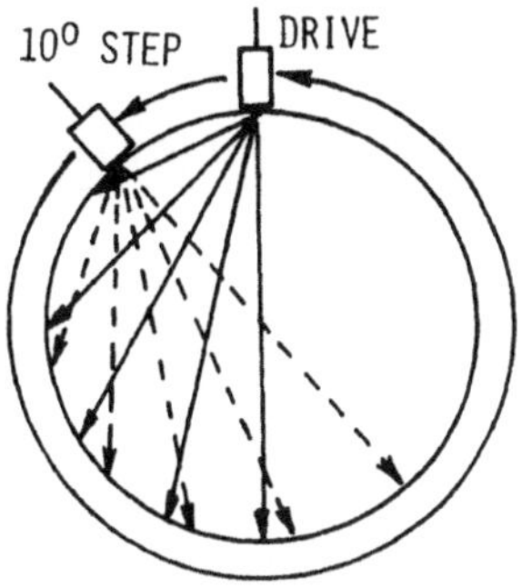

Fig.3 Measurement

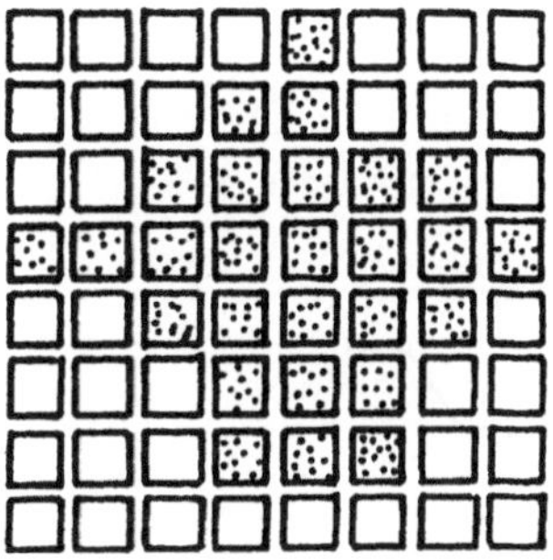

Fig.4 Model of pixels

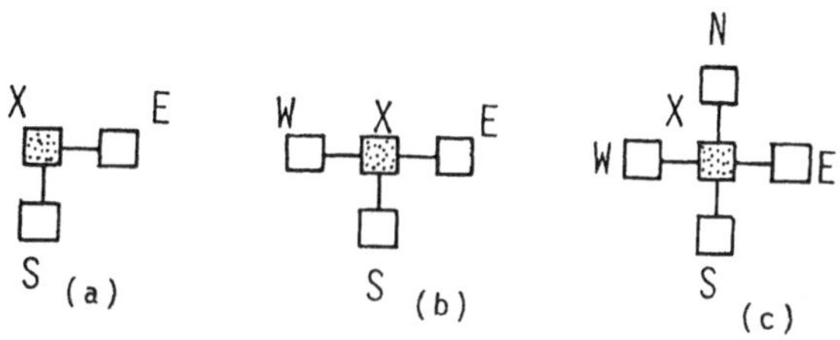

Fig.5 Arrangement of pixels

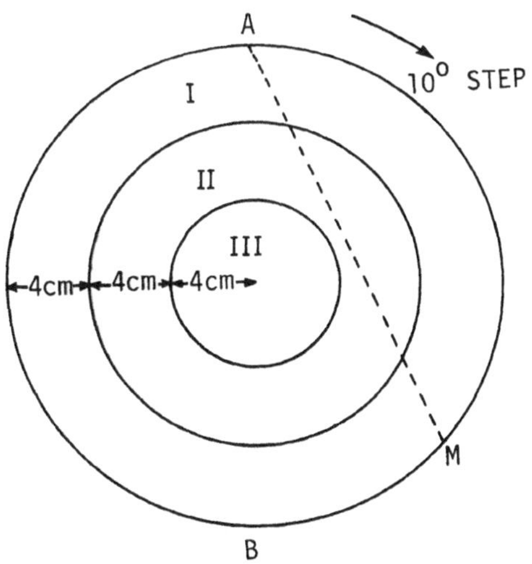

Fig.6 Model for measurement of temperature distribution

(1) Temperature difference : 15 ℃
Layer Ⅰ 37 ℃ (1523.93 m/s)
Layer Ⅱ 52 ℃ (1544.95 m/s)
Layer Ⅲ 67 ℃ (1554.43 m/s)

(2) Temperature difference : 10 ℃
Layer Ⅰ 37 ℃ (1523.93 m/s)
Layer Ⅱ 47 ℃ (1539.34 m/s)
Layer Ⅲ 57 ℃ (1549.28 m/s)

(3) Temperature difference : 5 ℃
Layer Ⅰ 37 ℃ (1523.93 m/s)
Layer Ⅱ 42 ℃ (1532.37 m/s)
Layer Ⅲ 47 ℃ (1539.34 m/s)

(4) Temperature difference : 1 ℃
Layer Ⅰ 37 ℃ (1523.93 m/s)
Layer Ⅱ 38 ℃ (1525.74 m/s)
Layer Ⅲ 39 ℃ (1527.49 m/s)

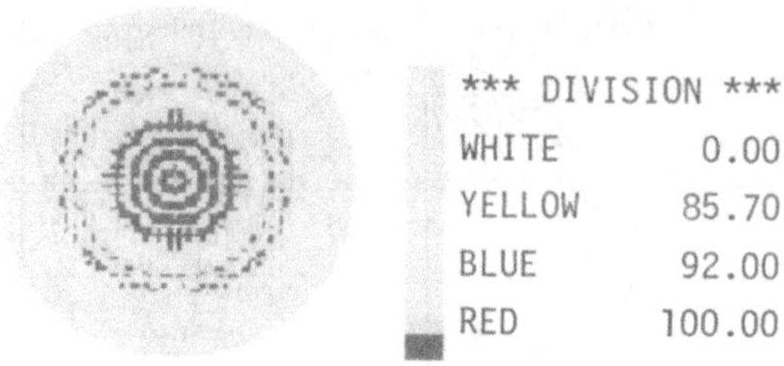

(a) Before smoothing treatment

F.7 CT results (Difference in temperature: 5°C)

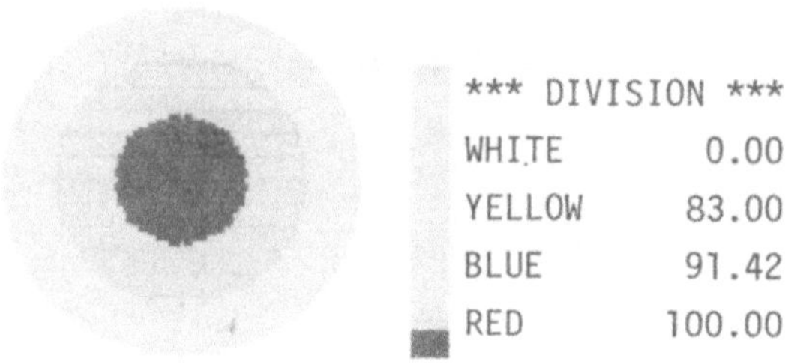

(b) after smoothing treatment

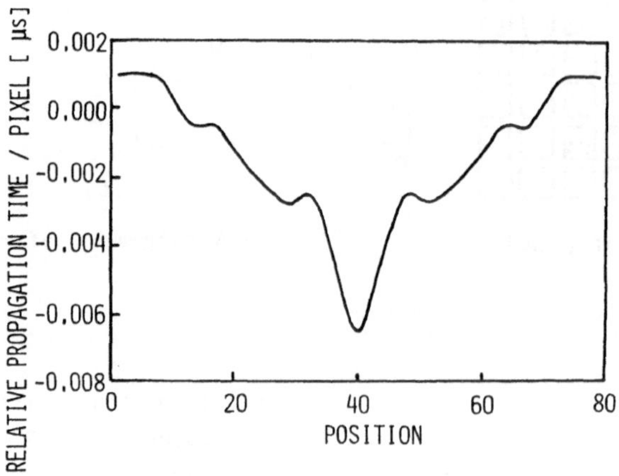

Fig.8 Relative propagation time/pixel after CT smoothing treatment

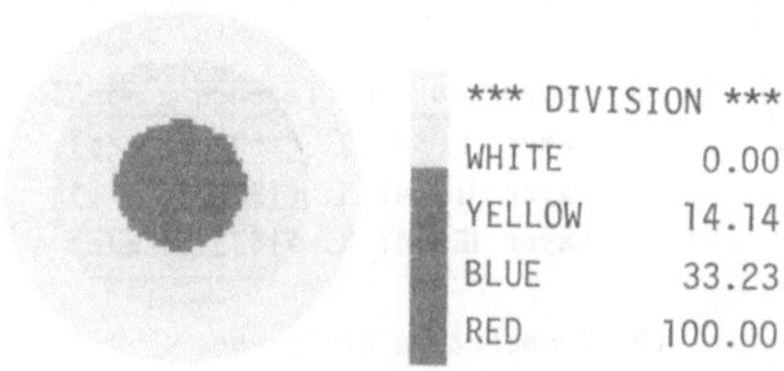

Fig.9 CT results (Difference in temperature: 1°C)

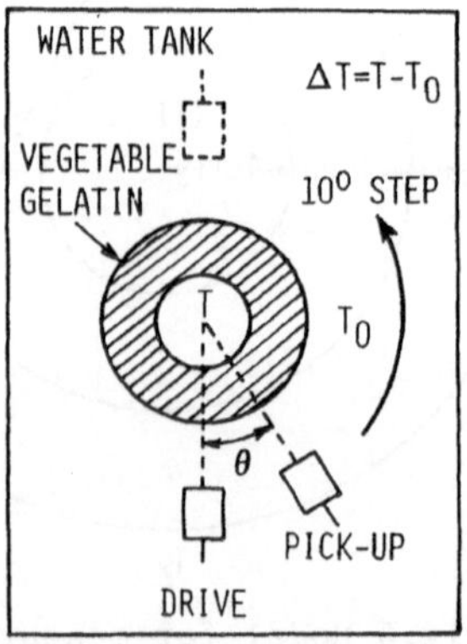

Fig.10 Experimental apparatus

EXPERIMENT 1— A CASE WHEN THE HEAT SOURCE EXISTS IN THE CENTER OF AN OBJECT

The experimental apparatus is shown in Fig.10. The phantom used in this experiment was made of vegetable gelatin and formed as a pillar of 6.5 cm radius, having a cylindrical hole of 3.5 cm in diameter at its center. We put it into a water tank, kept the water temperature around it constant and set a temerature gradient inside it by heating up the water in the hole. We experimented on four cases in which the temperature differences (ΔT) between the inside and outside of the cylindrical hole were 0°C, 5°C, 10°C and 15°C, respectively (standard temperature T_0=22.5°C). We treated the CT data as being caused only by temperature difference in applying the new CT method mentioned above. The measurement of the propagation time was done as follows: fix transmitting and receiving ultrasonic transducers as shown in Fig.10, measure the ultrasonic propagation time, and then move the receiving transducer from 10 degree to 180 degree in 10 degree increments, so each of the 18 values of the propagation time from the basic point to each receiving position could be obtained. Because the object used in this experiment was symmetrical, we treated the measurement data for the opposite symmetrical portions as being equivalent to the 18 projection data. We repeated such measurements for each case of temperature difference. The center frequency of the transducers was 5 MHz.

The CT propagation time for various pixel positions along the diameter of the phantom (θ=180°) are shown in Fig.11. The relative propagation time/pixel, which were obtained by subtracting the CT data for ΔT=0°C from that for ΔT=5°C, 10°C and 15°C, are shown in Fig.12. The minimum for each case isn't in proportion to the temperature difference. This is because the ultrasonic velocity in water and in vegetable gelatin isn't proportional to the temperature. Figure 13 shows the result of CT imaging in case of ΔT=5°C temperature difference (CT imagings in the case of ΔT=10°C and 15°C are abbreviated herein), based on the relative propagation time as shown in Fig. 12. Each color region in Fig.13 shows 3°C, 2°C and 1°C differences in temperature, respectively. Namely, we were able to verify that our CT method could be satisfactorily applied to the measurement of internal temperature distribution experimentally. However, because, as mentioned above, the ultrasonic velocity in water and vegetable gelatin isn't in proportion to temperature, it is necessary to refer to the data for a known medium in esimating an unknown temperature distribution from CT imaging.

EXPERIMENT 2— A CASE WHEN THE HEAT SOURCE DOESN'T EXIST IN THE CENTER

The phantom used in this experiment was vegetable gelatin having a cylindrical hole of 3 cm radius in a position 1.8 cm distant from the center of a pillar of 6.5 cm radius. We experimented on four cases in which the temperature differences(ΔT) between the inside and outside of the phantom were 0°C 5°C. 10°C and 15°C, respectively like the experiment 1. However, the number of data measured in this experiment was much larger than that of data in the experiment 1. Therefore, it took more time to measure, and the standard temperature difference couldn't be unified. The standard temperature was T_0=17.2 °C for ΔT=5°C, T_0=21.5°C for ΔT=10°C, and T_0=17.0°C for ΔT=15°C. In measuring the propagation time, we rotated the phantom itself at every 10 degree interval in addition to the operation explained for the experiment 1.

Figures 14,15 and 16 show the distribution of the propagation time/pixel for each temperature difference at θ=180°. Figure 17 shows the distribution of the relative propagation time/pixel. In case of the experiment 1, we experimented while keeping the standard temperature T_0=22.5°C constant. However, since it was not constant for the experiment 2, the difference in the standard temperature brought about a difference in the minus peak value (it was guessed that the temperature at this point was the highest) for the relative propagation time. That is, if the standard temperature was low, this peak value became smaller even for the same temperature difference. This is because the ultrasonic velocity in water isn't directly proportional to the tempera-

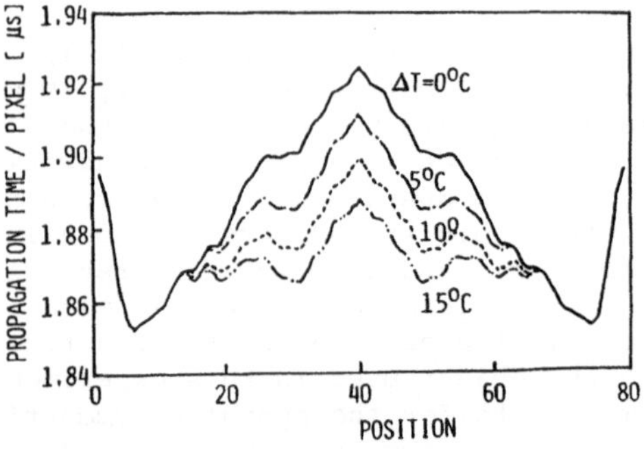

Fig.11 Propagation time/pixel characteristics

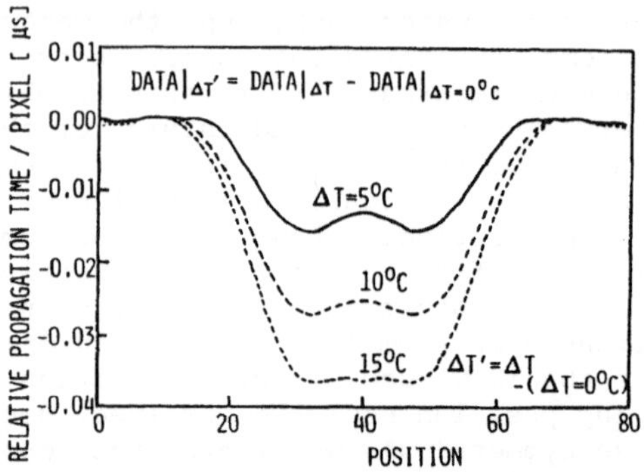

Fig.12 Relative propagation time/pixel characteristics

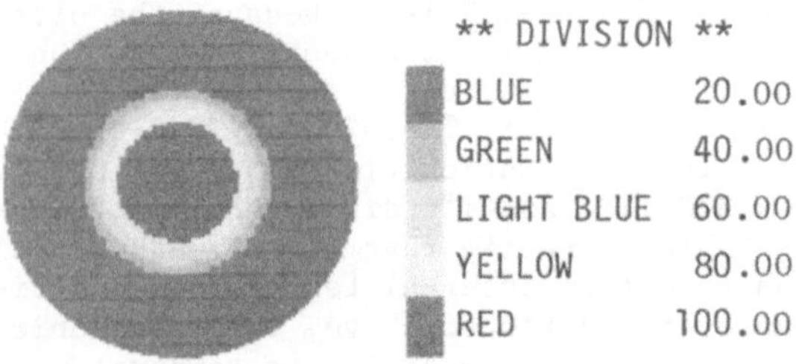

Fig.13 CT results (ΔT=5°C)

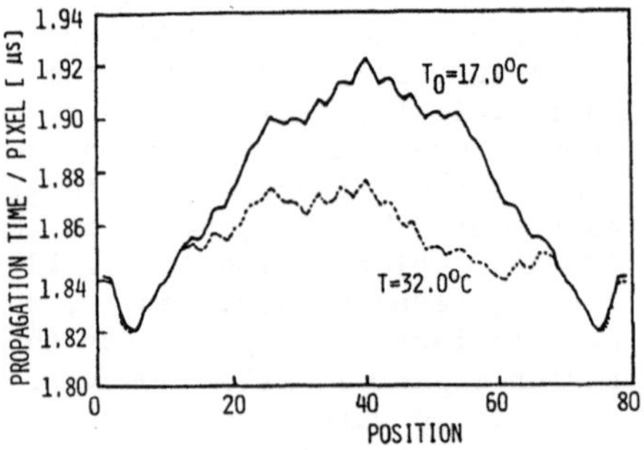

Fig.14 Propagation time/pixel characteristics (ΔT=15°C)

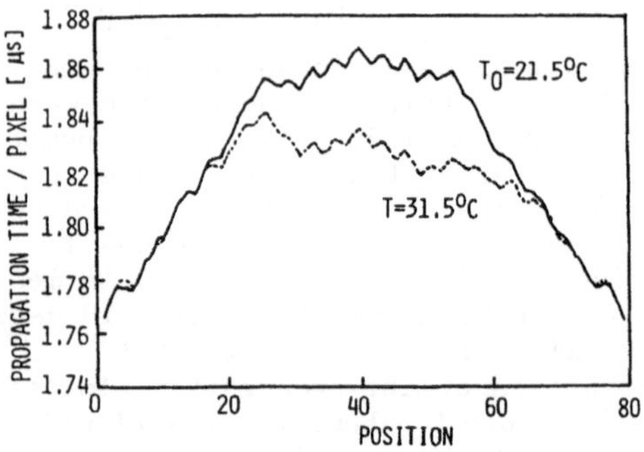

Fig.15 Propagation time/pixel characteristics (ΔT=10°C)

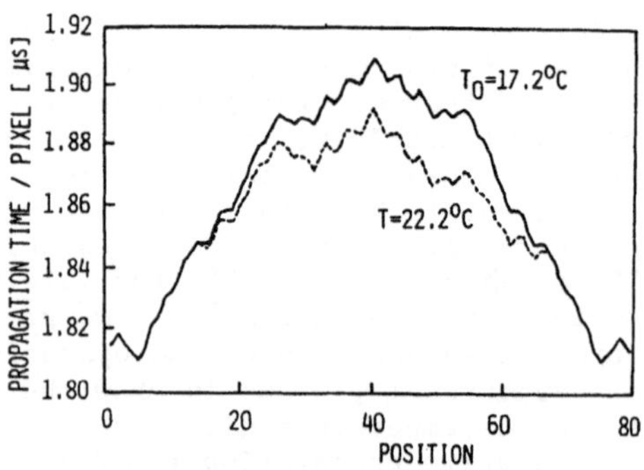

Fig.16 Propagation time/pixel characteristics (ΔT=5°C)

ture, as shown in Fig.18. Considering this point, we adjusted each standard temperature to be T_0=22.5°C, and calculated the peak value for each temperature difference. The result became as follows.

(1) in case of 5°C temperature difference
before correction $\Delta T'$=-0.0225 [μS]
after correction $\Delta T'$=-0.0196 [μS]

(2) in case of 10°C temperature difference
before correction $\Delta T'$=-0.0355 [μS]
after correction $\Delta T'$=-0.0354 [μS]

(3) in case of 15°C temperature difference
before correction $\Delta T'$=-0.0509 [μS]
after correction $\Delta T'$=-0.0426 [μS]

Figure 19 shows the minimum values for the relative propagation time, which are values corrected as mentioned above and those obtained from the experiment 1. Because the standard temperature was adjusted to be T_0=22.5°C, it should be desired that the relative propagation time conforms at each temperature difference. However, they don't conform as shown in Fig.19. The error becomes the least in case when the temperature difference is small in the experiment 1, and when it is high in the experiment 2. As for the present stage, there is about 10 percent error. However, it seems that if the problems mentioned below are solved, the accurracy will be improved. Figures 20 and 21 show the results of CT imaging in case of $\Delta T'$=10°C and 5°C temperature difference(CT imaging in case of $\Delta T'$=15°C is abbreviated herein). In case of 5°C temperature difference, the center of the heat source isn't a concentric circle. It seems that some errors were included. However, from the results mentioned above, it was verified basically that our CT method could also be applied to the case when the heat source didn't exist in the center.

As for the method of measuring internal temperature distribution by a CT image proposed in this paper, improvement of accuracy must be made hereafter. That is, the method has the following problems:

(1) Because of the narrow directivity of the used ultrasonic transducers, enough gain to measure the ultrasound propagation time could not be obtained as the included angle between the pair of ultrasonic transducers became small. Therefore, in the experiment, the surface of the transmitting transducer had to be adjusted to face one of the receiving transducer in order to obtain the gain.

(2) The diameter of the used ultrasonic transducers (1.5 cm) was much bigger than the wavelength of the ultrasound. A transducer with a small face diameter is necessary.

(3) Because relatively much time was taken for the experiment, a measuring system is necessary in which CT data can be obtained in a much shorter time while keeping the established condition unchanged by increasing the number of ultrasonic transducers.

If the above points are improved, the accuracy will be improved, and this method will be practically used.

CONCLUSION

We have reported on the results of simulations and experiments on measurements of internal temperature distribution of objects using the time of flight ultrasonic CT. The following conclusions were derived from the results.

(1) As for the simulation, even in case of a 5°C temperature difference, in other words, about 10 m/s difference in ultrasound velocity, the internal temperature distribution could be discriminated sufficiently, by only adopting simple image processing.

(2) In addition, we were able to discriminate satisfactorily even a 1°C temperature difference by subtracting the characteristics from those for a standard temperature.

(3) The temperature distribution was also able to be measured satisfactorily experimentally.

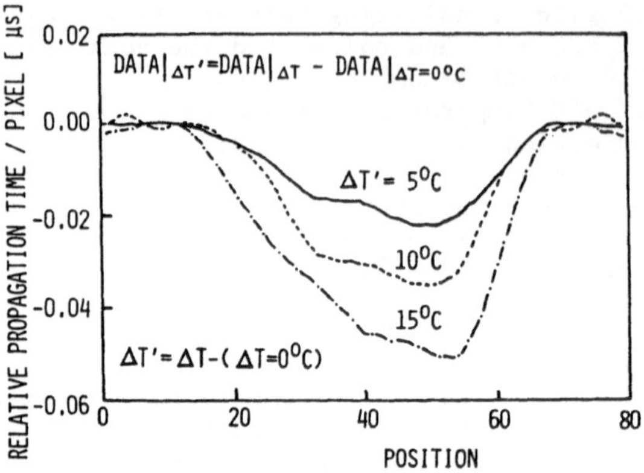

Fig.17 Relative propagation time/pixel characteristics

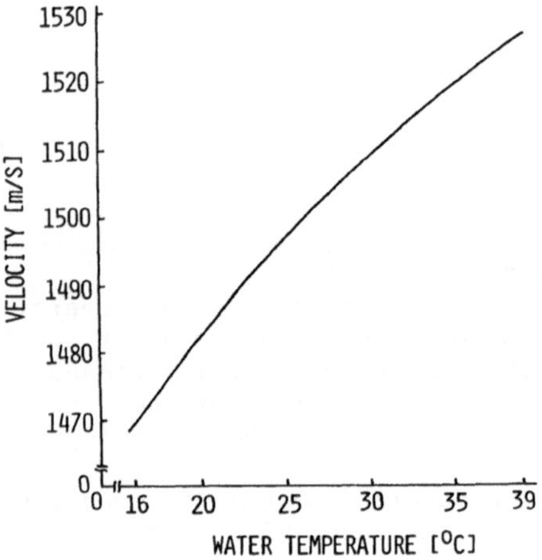

Fig.18 Ultrasound velocity in water vs temperature characteristics

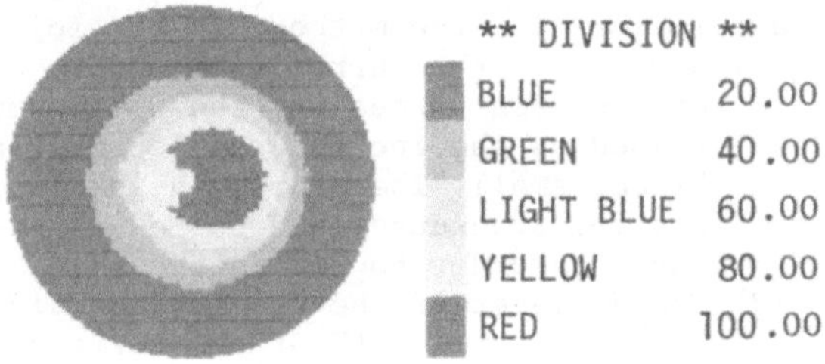

Fig.20 CT result (ΔT'=10°C)

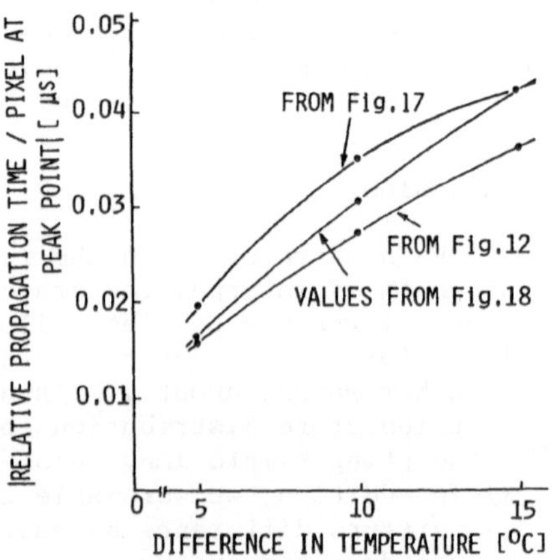

Fig.19 Relative propagation time/pixel for temperature difference

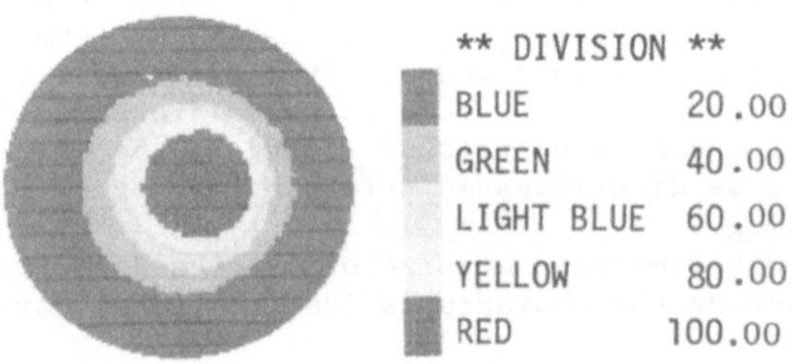

Fig.21 CT result (ΔT'=5°C)

(4) However, we must be careful of the threshold level for the color division of the CT images.

ACKNOWLEDGMENTS

The authors wish to thank Mr. K. Arita and Dr. J. Masuda of the NTT Corporation Electronics and Mechanics Engineering Laboratory and Mr. H. Honda of the Toshiba Research and Development Center for giving the authors the need for this research and technical advices.

REFFERENCES

[1] K. Hirama, T. Oshima and Y. Tomikawa,"Wireless temperature sensing system using quartz sensor for hyperthermia", IEEE US Symp. 1985, pp.917-920, Oct. 1985.

[2] Y. Tomikawa, Y. Iwase, K. Arita and H. Yamada,"Nondestructive inspection of a wooden pole using ultrasonic computed tomography", IEEE, Vol.UFFC-33, No.4, pp.354-358, July 1986.

[3] Y. Tomikawa, M. Numata, K. Arita and H. Yamada,"Nondestructive inspection of hidden knots in Japanese cedar log using ultrasonic computed tomography", Ultrasonic Technology 1987, Toyohashi International Conf. on Ultrasonic Technology, pp.90-100, May 1987.

[4] H. J. Schudder,"Introduction to computer aided tomography", Proc. IEEE, Vol. 66, No. 6, pp.628-637, June 1978.

[5] Y. Tomikawa, M. Numata, H. Yamada and M. Nakamura, "Simulation for measurement of internal temperature distribution using ultrasonic CT", Report of the 1987 Spring Meeting, The Acost. Soc. of Japan, No.2-8-11, pp.715-716, March 1987.

[6] Y. Tomikawa, M.Numata, H. Yamada and H. Nakamura, "Measurement of Internal temperature distribution using ultrasonic CT", Paper of Technical Group on US-33 IECE Japan, pp.41-48, Sept. 1987.

[7] J. Saneyoshi, et al., "Ultrasonic technical book", Nikkan Kogyo Shinbun Sha, pp.1202, Oct. 1966 [in Japanese].

[8] Y. Tomikawa, M. Numata, H. Yamada and H. Nakamura, "Measurement of internal temperature distribution using ultrasonic CT— in case of eccentric heat source existance", Report of the 1988 Spring Meeting, The Acoust. Soc. of Japan, No.2-P-20, pp.777-778, March 1988.

A UNIFIED THEORY FOR ACOUSTICAL HOLOGRAPHY AND DIFFRACTION TOMOGRAPHY

W. S. Gan

Acoustical Services Pte Ltd
29 Telok Ayer Street
Singapore 0104, Republic of Singapore

INTRODUCTION

In optical holography, the object is assumed to be such that a considerable part of the wave penetrates undisturbed through it, a diffraction pattern called hologram which is formed by the interference of the secondary waves arising from the presence of the object scattering with the strong background wave, is recorded on a photographic plate. This means that, if the wave is expressed as the sum of the incident wave and a diffracted secondary wave, the scattering of the secondary wave and interaction within the object is neglected. This negligence represents what is usually known as Born's first approximation. In acoustical holography, where the sound wavelength is about the same order as the obstacle in the medium that causes the scattering, diffraction takes place and the scattering of the secondary wave and the interaction within the object can no longer be neglected. This is what is done in diffraction tomography (DT), but is neglected in acoustical holography (AH), sofar. So one has to combine DT and AH. Hence a unified theory for AH and DT is proposed.

FORMULATION OF THE UNIFIED THEORY

To formulate the unified theory for AH and DT, one starts form the inhomogeneous Helmholtz equation

$$(\nabla^2 + k^2)\Phi = -q \qquad (1)$$

where Φ is some appropriate potential, k is the wavenumber being proportional to frequency, and q accounts either for a prescribed source or a scattering potential as far as varying material parameters are concerned.

Unlike previous theories on AH all make use of Kirchhoff approximation, in this paper, the exact diffraction integral is used. This is because Kirchhoff approximation is only appropriate only for the situations when the representative wavelengths are not too long compared with the scatterer and also is valid if interaction within object can be neglected. On top of this, Kirchhoff approximation can be used only for the imaging of plan surfaces not curved surfaces which will be the object used in this paper.

RECONSTRUCTION TECHNIQUES

Extensions to existing methods

(i) Convolution back-projection

The primary requirement in this work was to have available a range of reconstruction algorithms, suitable for use in cases of irregular geometry and/or ray bending. To this end, the convolution back-projection method has been modified to reconstruct data from an arbitrary set of source and receiver locations, with no prior restrictions on geometry.

In the conventional convolution algorithm, it is assumed that the object to be imaged has a slowness field f(x,y). Data is assumed to be taken in the form of acoustical travel times, over a series of parallel ray paths, to form a projection. A series of these projections is then obtained at various angles (θ), each projection being defined by

$$P_\theta(s) = \int_{-\infty}^{\infty} f_\theta(s,t)dt \qquad (1)$$

The complete set of parallel projections taken over all θ is related to the desired image of f(x,y) by the projection theorem. This theorem states that the 1-D Fourier transform of a projection is the center cross-section of the 2-D Fourier transform of the image, namely

$$P_\theta(s) = F\{p_\theta(s)\} = F(S \cos \theta, S \sin \theta) \qquad (2)$$

This theorem can be used to write the expression for the inverse transform of the required image f(x,y) as

$$f(x,y) = \int_{-\infty}^{\infty} \int_{-\infty}^{\infty} F(X,Y)e^{j2\pi(xX+yY)}dXdY \qquad (3)$$

and in polar coordinates this becomes

$$f(x,z) = \int_0^{\pi} d\theta \int_{-\infty}^{\infty} P_\theta(R)|R|e^{j2\pi rR\cos(\theta-\phi)}dR \qquad (4)$$

It is now evident that the inner integral is a convolution of the projection p_θ with a kernel function κ, whose Fourier transform is |R|. The outer integral is then a backprojection of all the convolved projections associated with all rays passing through the point (x=rcosϕ, y=rsinϕ).

Practical implementation of this result is limited by the discrete nature of the data. The finite sampling period of the projections Δs band-limits the convolution integral to the Nyquist bandwidth B=1/2Δs. As this can result in aliasing, it is usual to apply a filter W(R). The result of reconstruction process is an image, containing values of slowness f(x,y) at specific positions, with the resolution of this image depending upon the spacing of the parallel ray paths Δs and the angular spacing of the projections in θ.

Extension of the convolution-backprojection to an arbitrary geometry involves several steps. First, the source and receiver locations are used to define the boundary of the object in question. A regular grid of n projections, comprised of m ray paths is then superimposed onto the object. To reconstruct an image, it is necessary to use slowness values for all these projections. This is achieved by interpolation from the experimental ray paths. The nearest ray path to the desired projection is chosen and then corrected for any difference in path length across the object.

(ii) ART reconstruction

In this technique, the object plane is discretized into a grid of "pixels", and the slowness of each pixel is reconstructed from a series of path integrals in the form of travel times. The travel time for the ith ray path (t_i) may be written

$$t_i = \sum_{j=1}^{n} G_{ij} m_j \tag{5}$$

where m_j is the average slowness value in the jth pixel, and G_{ij} is the length of the ray path segment of ray path i contained within pixel j. The ART method then finds a solution to a set of simultaneous equations, of the form

$$\underline{t} = \underline{\underline{G}}\,\underline{m} \tag{6}$$

Where $\underline{m}$ is the slowness image to be reconstructed, and where $\underline{\underline{G}}$ is now known as the projection matrix. This equation is usually solved iteratively, the aim being to reduce the travel time residual (Gm-t) along each row of the matrix $\underline{\underline{G}}$, until some form of convergence is reached. The standard ART approaches have been described by Censor [5] and further details will not be given here. In the present work, we have used a variation known as SIRT (Simultaneous Iterative Reconstruction Technique), due to Gilbert [6].

The ART approach is simple to implement for an arbitrary geometry, making it attractive in the present study. In addition, we have extended the approach to account for ray bending. The solution of equation (2) above may be obtained using straight rays to give an approximate image uncorrected for ray bending. Our approach is to recalculate ray paths through this first image, using the slowness values alocated to each pixel. The image is then reconstructed again, using the new ray paths. This procedure is then repeated, until convergence to a final image occurs.

The ray path calculation was designed to be efficient, to reduce computation time. The first step involves preparing a pixel adjacency list. An estimate of the time required to travel between a given pixel and each of its neighbours is then formed from the average slowness of each pixel and their separation. For a given source location (pixel), a path is then determined which gives the minimum travel time to a selected receiver location. The calculated time of flight is then determined, for a given path, by summing the individual interpixel times. Thus, upon exit from the ray path subroutine, the $\underline{\underline{G}}$ matrix and the new calculated travel time vector $\underline{t}$ for the new ray paths have been determined, and the literature reconstruction to find $\underline{m}$ may start. The result is a new image, containing slowness variations $\underline{m}$, which includes a correction for ray-bending.

(iii) Images using synthetic data

A synthetic image was used, to test the above reconstruction methods in the presence of an acoustic velocity gradient. The synthetic data is plotted in Fig. 1(a), and represents a linear decrease in slowness (corresponding to an increase in velocity as the reciprocal of distance) across the image. This synthetic image was chosen, as the curved acoustic ray paths that would be present through such a velocity structure are easily calculated theoretically. Data was assumed to be taken at six locations, symmetrically spaced along each of the four sides of the object. The travel times occuring through such a structure were calculated, and then input to the reconstruction programs to test their performance.

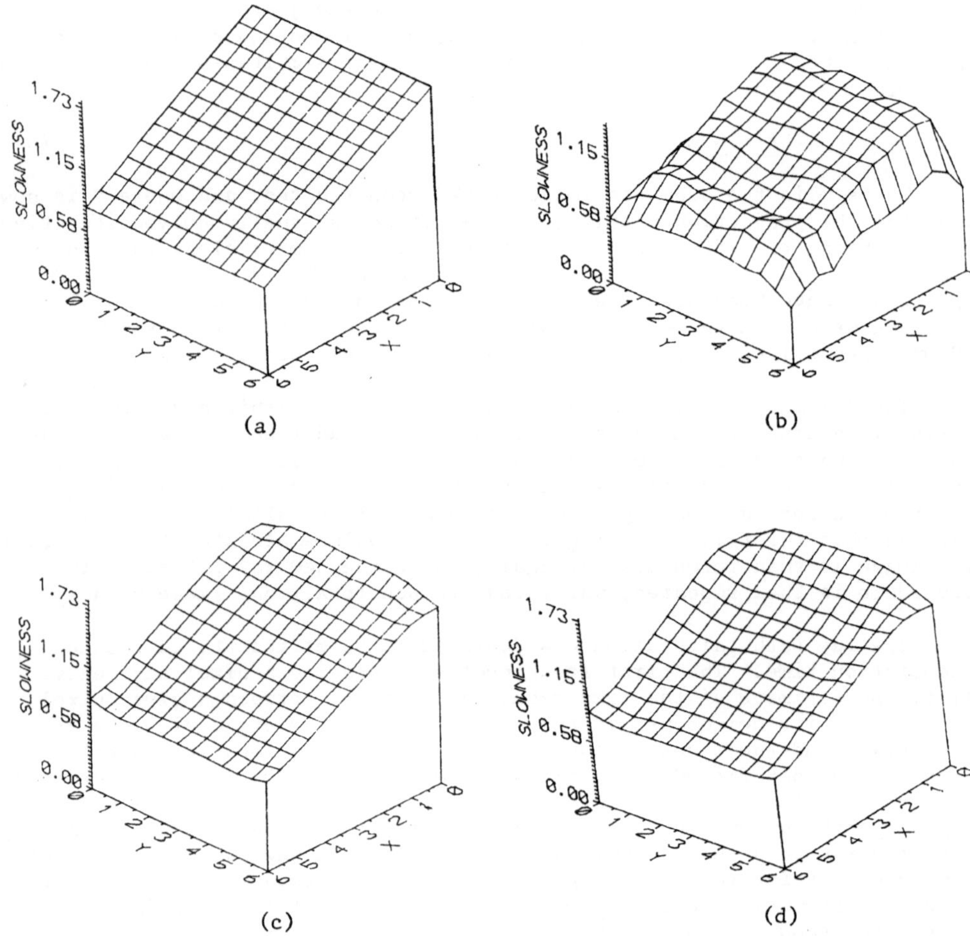

Fig. 1 Results of reconstruction using synthetic data and curved ray paths. (a) Representation of synthetic data, in the form of a constant slowness gradient. (b)-(d) reconstructions using convolution - backprojection, ART and ART with ray bending.

The results from the convolution-backprojection algorithm are presented in Fig. 1(b). Note that the correct general features are produced, but that filtering effects cause artifacts to be present at the edges of the image. These are not present to the same extent on an ART reconstruction assuming straight rays, as is evident from Fig. 1(c). Finally, Fig. 1(d) presents results from an ART reconstruction, after correcting for ray bending effects, where it is evident that little improvement over the straight - ray reconstruction resulted.

Further simulations have shown that the most robust algorithm for velocity variations of less than ≅ 10% is ART, uncorrected for ray bending. Discretization for the ray bending introduces some distortion, which produces no net advantage for small variations, but becomes useful for fields with more marked velocity structures. For irregular geometries, ART is also more readily applied than convolution - backprojection, although in general the latter is more computationally efficient.

LABORATORY EXPERIMENTS

Experiments were conducted at ultrasonic frequencies, to image the internal structure of a 60mm cube of plexiglass, containing an irregularly-shaped mineral sample (halite). Data was collected using 2.5mm diameter transducers, operating at a centre frequency of 2.5MHz, which were scanned manually around the specimen in one horizontal plane. The transducers were driven with a Panametrics pulse generator, and detected using a Nicolet 12-bit digital transient recorder. Fig. 2(a) shows a schematic diagram of the object in plain view, the longitudinal velocities being 4,500ms^{-1} in the halite and 2,700ms^{-1} in the surrounding plexiglass.

Data collected experimentally was reconstructed using the algorithms described, the results being presented in Figs. 2(b)-(d). The reconstructions are shown in the form of contours of velocity. The convolution image, Fig. 2(b), shows contours of the correct general form, but as would be expected for a sparse data set, the velocity variations do not represent a sharp edge at the halite/plexiglass interface. The image is somewhat improved using the straight ray ART method, Fig. 2(c). The use of ray-bending in the ART algorithm was found to improve the overall image and in particular to correct for the shape of the contours and the overall magnitude of the reconstructed contours. This agrees with the statement in an earlier section, namely that ray bending is of benefit in situations involving large changes in acoustic velocity.

FIELD EXPERIMENTS

The purpose of field experiments was to investigate the usefulness of acoustical tomography in the characterization of large volumes of rock, in a mining situation. A series of preliminary experiments were undertaken, the results of which are presented here, to image the internal structure of a pillar of rock, situated in a Canadian hard rock mine. Further work [7] hasshown that this technique can be used to monitor changes in rock structure, as mining near the imaged zone takes place.

The present experiments were conducted on a rock pillar, at the 762m level of the Falconbridge Strathcona mine. The pillar, which was approximately 100 m x 200m in shape, consisted primarily of metamorphic rock, with a large pocket of backfill where mining had already occurred (Fig. 3(a)). An array of 36 single component ADR 711 accelerometers was fixed to the

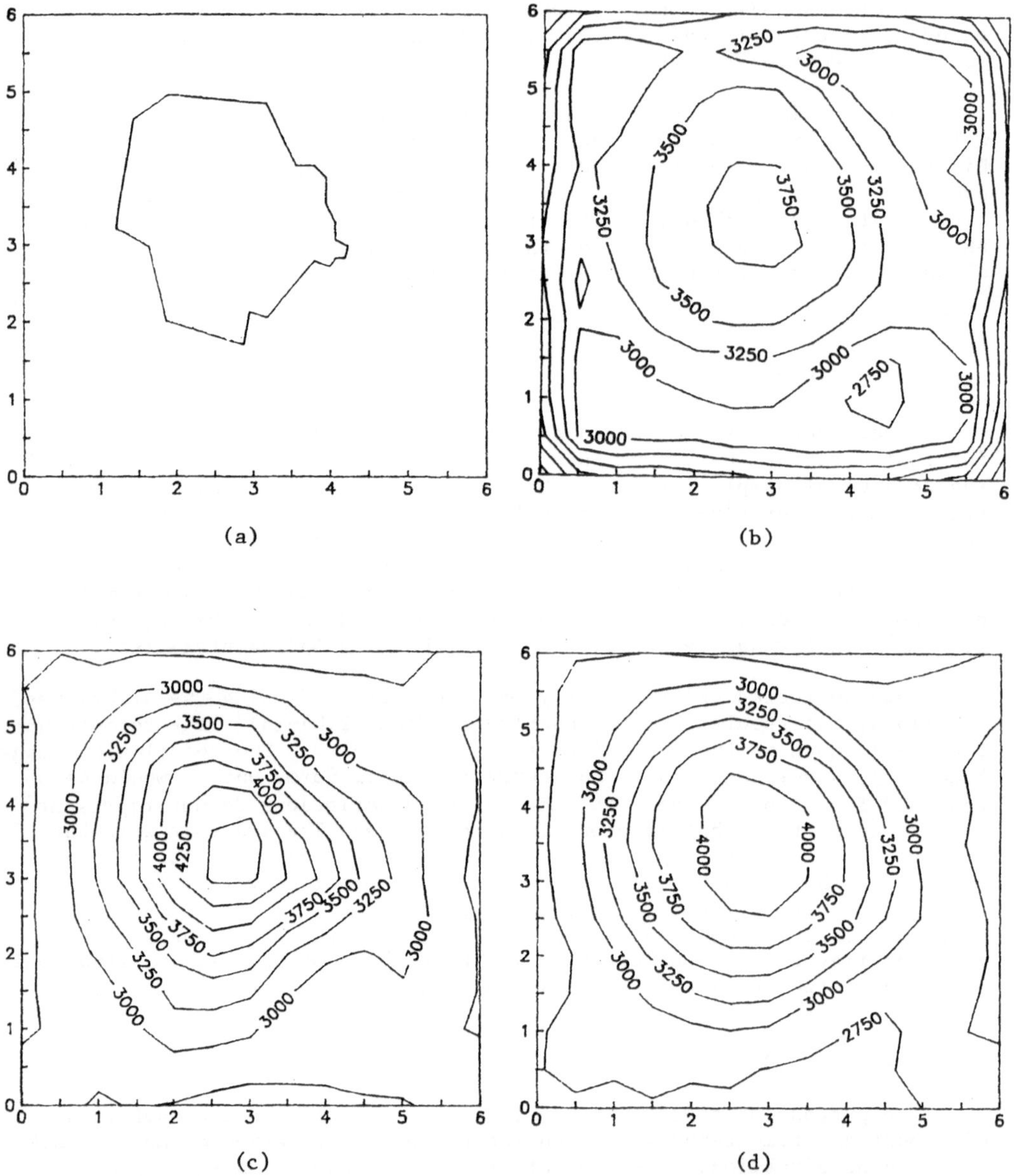

Fig. 2 Results of a laboratory experiment at 2.5MHz, on a 600mm plexiglass cube containing a halite sample. (a) Representation of object. (b) - (d), reconstructions using convolution - backprojection, ART and ART with ray bending respectively as velocity contours.

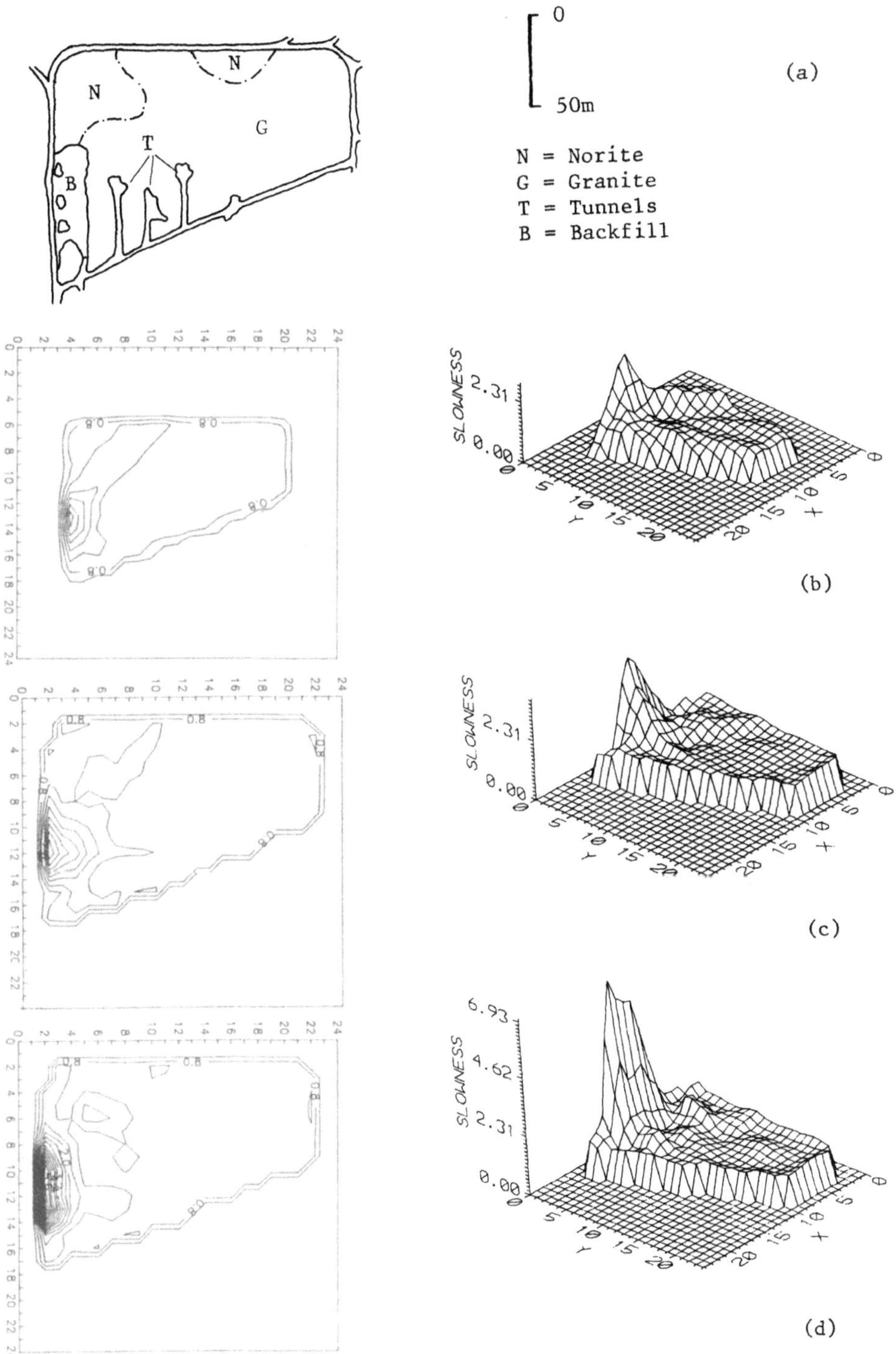

Fig. 3 Results of a field study in a Canadian hard rock mine. The rock pillar is shown in (a), whereas (b)-(d) are reconstructions using convolution-backprojection, ART and ART with ray bending respectively. Contours of slowness are to the left, topographic plots to the right.

pillar wall, such that they surrounded the pillar. Blasting caps and boosters were used to create a source, close to each sensor location, and the acoustic signal arriving at each sensor was recorded, using a 12 channel, 12 bit engineering seismograph. The data was subsequently transferred to a microcomputer for analysis.

The arrival time of the P (longitudinal) wave was picked from each record, and an image of the pillar reconstructed in terms of slowness. The results are presented in Figs. 3(b)-(d) for reconstruction by convolution -backprojection, ART and ART with ray bending respectively. As in the previous simulations and laboratory studies of Figs. 1 and 2, all three algorithms gave reasonable reconstructions, with the backfill area being indicated at the left of each image by a low velocity (i.e. high slowness) region. Note that backfill, being a loosely consolidated mixture with many air pockets, would be expected to have a slow acoustic velocity. Conversely, the central area of the pillar is characterized by a low slowness, suggesting a fairly homogenous and intact rock mass.

It is interesting to note that ray bending in the ART reconstruction has increased the image resolution somewhat, as a comparison of Figs. 3(b)-(d) will indicate. This agrees with previous statements, i.e. that a ray bending correction is useful in situations where large changes in acoustical velocity are expected.

ACKNOWLEDGEMENTS

The authors would like to thank J. McGaughey, S. Falls and M. Bojtok for their assistance. This work was funded in part by the Natural Science and Engineering Research Council of Canada.

REFERENCES

1. R.K. Mueller, M. Kaveh and G. Wade, "Reconstructive tomography and applications to ultrasonics", Proc. IEEE 67:567 (1979).
2. R.M. Lewitt, "Reconstruction algorithms: transform methods", Proc. IEEE 71: 390 (1983).
3. H.J. Scudder, "Introduction to computer-aided tomography", Proc. IEEE 66: 628 (1978).
4. R. Gordon, "A tutorial on ART", IEEE Trans. Nucl. Sci. 21:78 (1974).
5. Y. Sensor, "Finite series-expansion reconstruction methods", Proc. IEEE 71: 409 (1983).
6. P. Gilbert, "Iterative methods for the three-dimensional reconstruction of an object from projections", J. Theor. Biol. 36: 105 (1972).
7. R.P. Young, D.A. Hutchins, W.J. McGaughey and D. Jansen, "Seismic imaging ahead of mining in rockburst prone ground", in Proc. 2nd Int. Synp. on Rockbursts and Seismicity in Mines, Minneapolis, U.S.A. June 1988.

FRESNEL DIFFRACTION HOLOGRAPHY SIMULATION BY A PERSONAL COMPUTER

Heiji Okada, Yukihiro Asari and Shyouzoh Sumihiro

Department of Communication Engineering
Shibaura Institute of Technology
3-9-14 Shibaura Minato-ku Tokyo 108, Japan

INTRODUCTION

Fresnel diffraction problems are including complicated equations that are not familiar to the students. In order to teach this problems, we have developed a computer assisted Fresnel education program by the personal computer that displays visual pictures on the screen and the beginners can simulate with changing some parts of parameters and seeing what the visual presentation does.

Conventionally, Fresnel integral has been processed by a large host computer which includes a corresponding library, then the numerical calculations are handled very easy.

On the other hand, a personal computer has no library in its own system, therefore the calculation consumes a lot of CPU time, but its graphic utility is excellent in comparison with a host computer. To decrease the calculation time, a modified (offset) Simpson method has developed, because of the original method requires a number of dividing points between lower and upper limit of the integral region.

All of the programs are written in F-BASIC and as the applications of these programs, we introduce several graphic patterns with respect to the Fresnel diffraction. There are Cornu spiral, one dimensional holography and two dimensional holography.

FRESNEL DIFFRACTION AND CORNU SPIRAL

In Fresnel diffraction, source and observing point are closer to the diffraction aperture than Fraunhofer's infinite distances. Fresnel diffraction pattern by the edge of a infinite aperture is a fundamental problem. As an example, one dimensional Fresnel diffraction geometry is shown in Fig.1, where diffracted wave's complex amplitude P(x) is observed on X axis. To avoid the complexities, paraaxial approximation is adopted here, then P(x) is written as

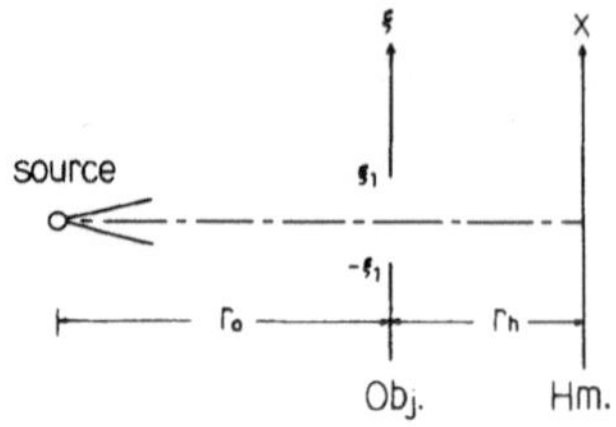

Fig.1 Geometry of one dimensional Fresnel diffraction

$$P(x) = \frac{u_0 \exp(ikD_0)}{i\lambda \bar{f}} \int_{-\infty}^{\infty} f(\xi) \times \exp\left| ik\left\{\frac{1}{2\bar{f}}(\xi - \bar{x})^2\right\}\right| d\xi \tag{1}$$

where,

$$u_0 = \frac{A}{ro + rh} \exp\{ik(ro + rh)\}$$

A:amplitude at unit distance

$$D_0 = \frac{x^2}{2(ro + rh)}$$

$$\bar{f} = \frac{ro\ rh}{ro + rh}$$

$$\bar{x} = \frac{ro}{ro + rh} x$$

In this expression, integral part is consisted with real and imaginary part as follows.

$$C + iS = \int_{-\xi_1}^{\xi_1} \exp\left\{i\frac{k}{2\bar{f}}(\xi - \bar{x})^2\right\} d\xi = \sqrt{\frac{\lambda \bar{f}}{2}} \int_{v_1}^{v_2} \exp\left(i\frac{\pi}{2}v^2\right) dv$$

The right hand side integral part is denoted as,

$$C(v) + iS(v) = \int_{v_1}^{v_2} \exp\left(i\frac{\pi}{2}v^2\right) dv \tag{2}$$

where,

$$v = \sqrt{\frac{2}{\lambda \bar{f}}}(\xi - \bar{x})$$

Equation (2) is well known as the Fresnel integral, and the Cornu spiral indicates a graphic expression of this integral. This spiral converges to the two limit points (1/2+i/2) and -(1/2+i/2) corresponding to the plus minus infinite values of v.

To calculate the integral, we have preliminary selected appropriate dividing points of v, and corresponding offset values of C(v) and S(v) are initially written in the BASIC program. Therefore, if an arbitraly value of v is given, the Simpson integral starts from the nearest offset value to that v. This method can decrease the calculation time by a computer very much.

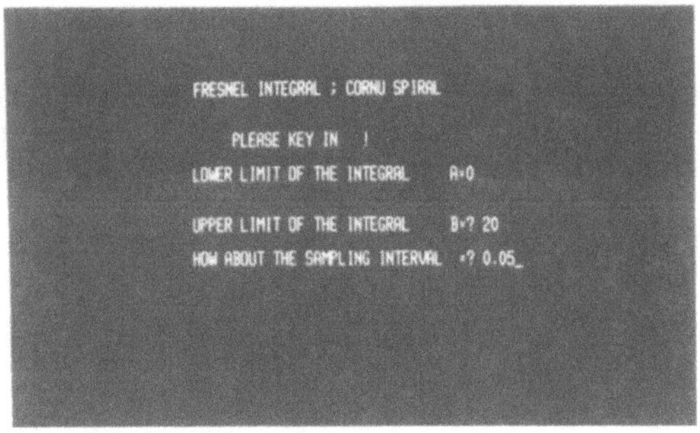

(a)

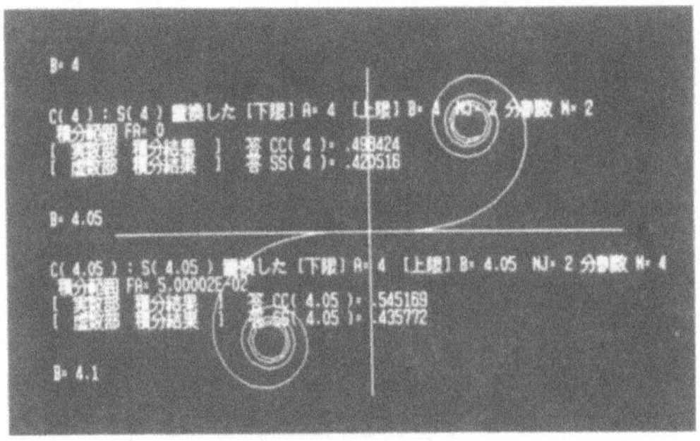

(b)

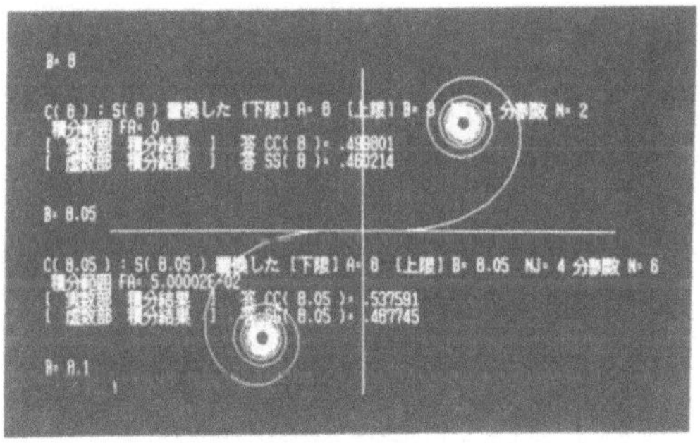

(c)

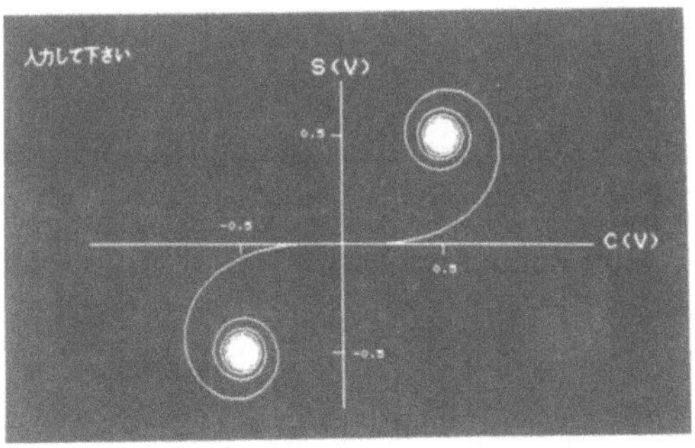

(d)

Fig.2 Cornu Spiral (a)Input data ;(b),(c) Midway data of the calculation;(d)Cornu Spiral

Fig.2 shows an example of Cornu spiral. The integral region is v=0 to 20 and step is 0.05, then the total sampling points are 400. If the sampling numbers are too small, zig-zag lines should appear on the screen instead of the Cornu spiral. In order to get a certain values of C(v) or S(v), midway computing data are scrolling on the display. When the escape key is typed, then the scroll stops and certain values of C(v) and S(v) are observed.

TWO DIMENSIONAL HOLOGRAPHY

An application of Fresnel integral, double diffraction type holograms are constructed. Assume that the object and hologram plane are paralleled each other and incident wave axis is perpendicular to the origin of the object plane, as shown in Fig.3.

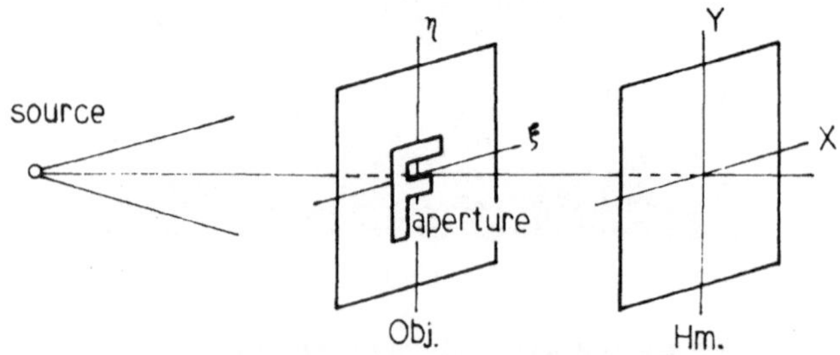

Fig.3 Geometry of two dimensional Fresnel diffraction

There are some apertures on the object plane, which diffract waves to the hologram plane (observation plane). On the hologram plane, Fresnel values on a certain sampling point requires eigh times of Fresnel calculation, because of two dimensional values are denoted as C(v1),C(v2),S(v1),S(v2) and C(w1),C(w2),S(w1),S(w2) corresponding to the X and Y.

IMAGE RECONSTRUCTION

In addition to these calculations, so many convolution processings are also required for image reconstruction of a hologram. Therefore, hologram reconstruction process is still hampered by the convolution process in the case of a 16 bit low speed personal computer. In the reconstruction program, a spherical wave is used for back ward propagation wave.

SIMULATIONS

The program is written in F-BASIC, which can handle two dimensional square, rectangular or one dimensional line type hologram, and sampling numbers on the X-Y plane are arbitrarily decided by the input type-in commands. Every distance values are also decided by the type-in commands with the wave lengthen unit.

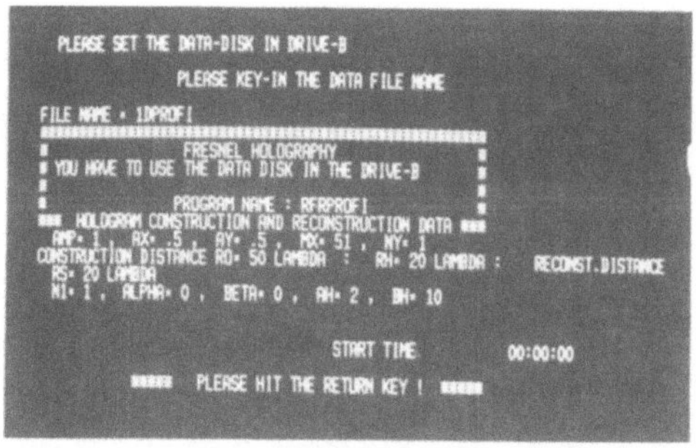

(a)

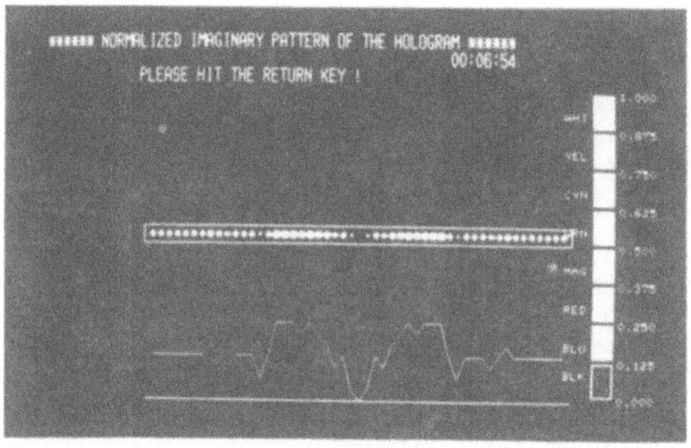

(b)

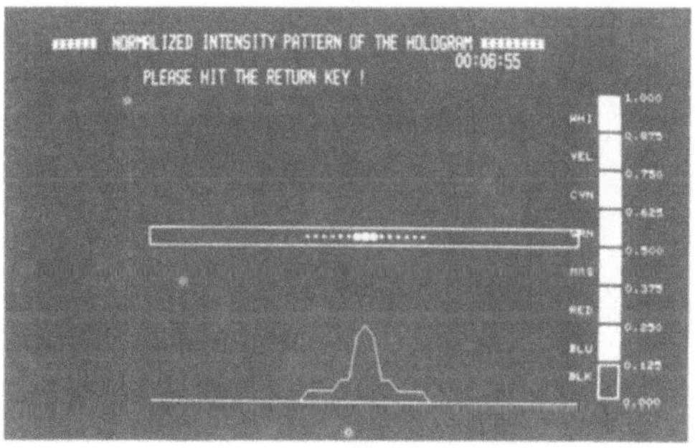

(c)

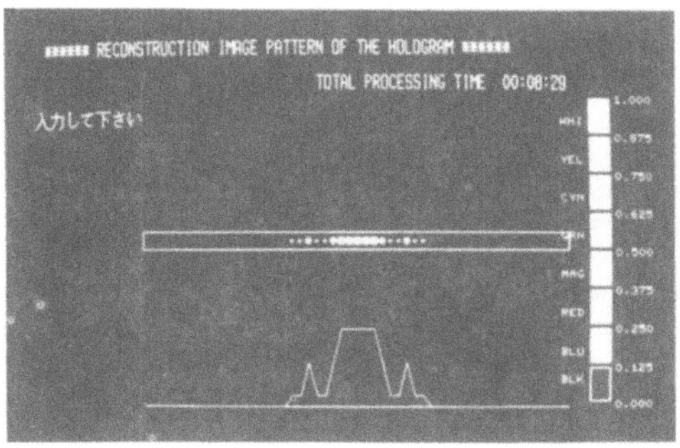

(d)

Fig.4 One dimensional holography (a)Input data; (b)Imaginary part of the hologram;(c)Intensity pattern;(d)Reconstructed pattern of " slit " type aperture

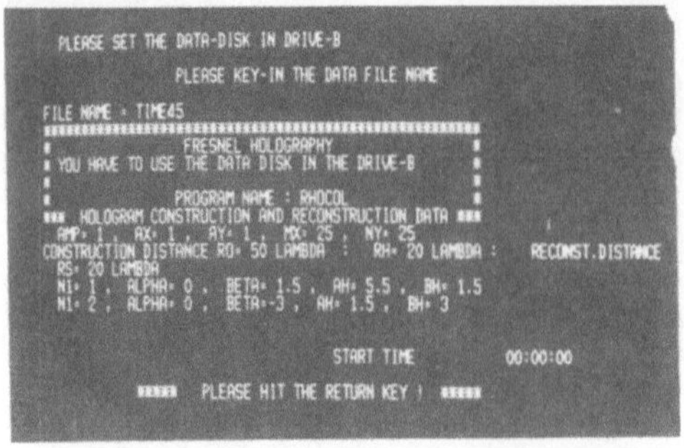

(a)

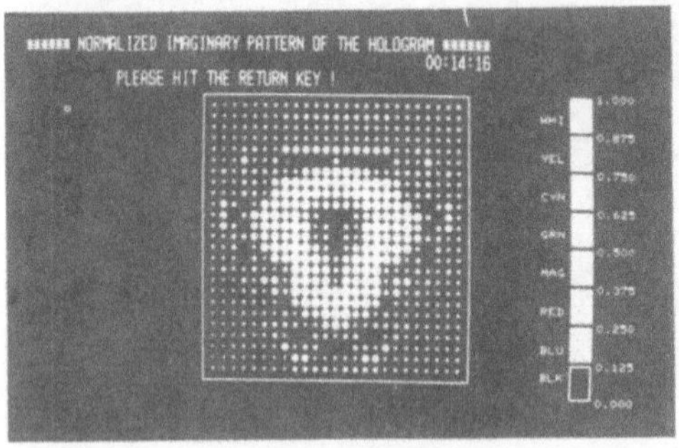

(b)

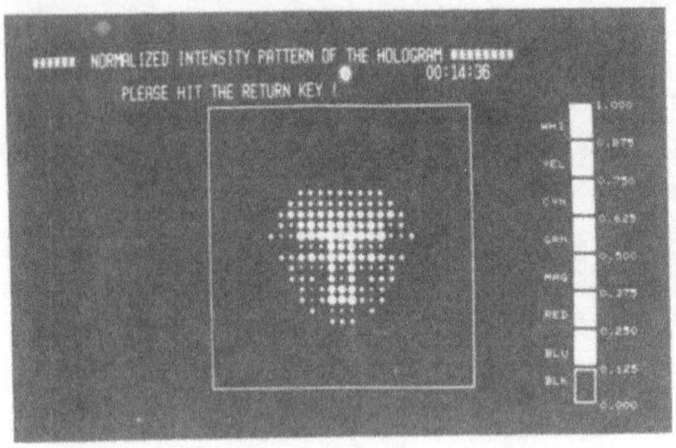

(c)

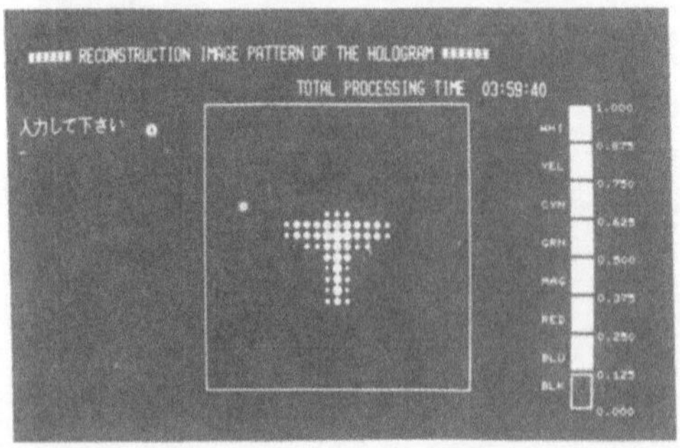

(d)

Fig.5 Two dimensional holography (a)Input data; (b)Imaginary part of the hologram;(c)Intensity pattern;(d)Reconstructed pattern of letter " T " type aperture

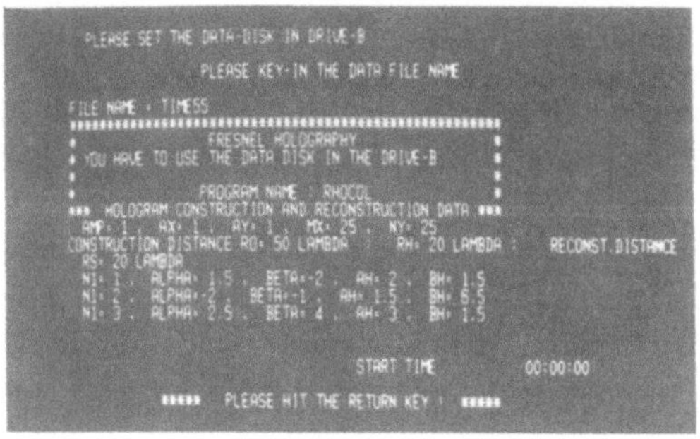

(a)

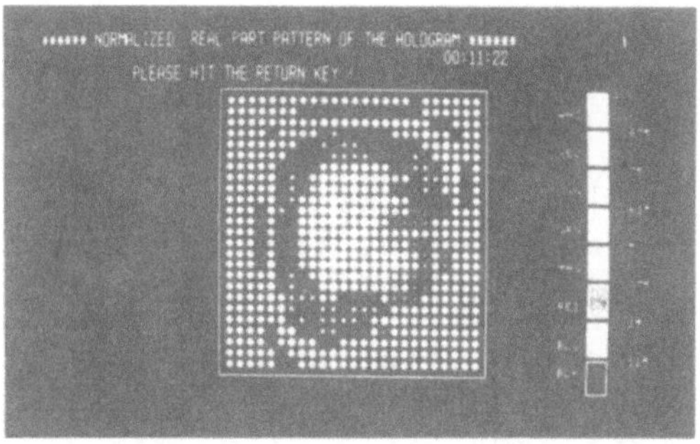

(b)

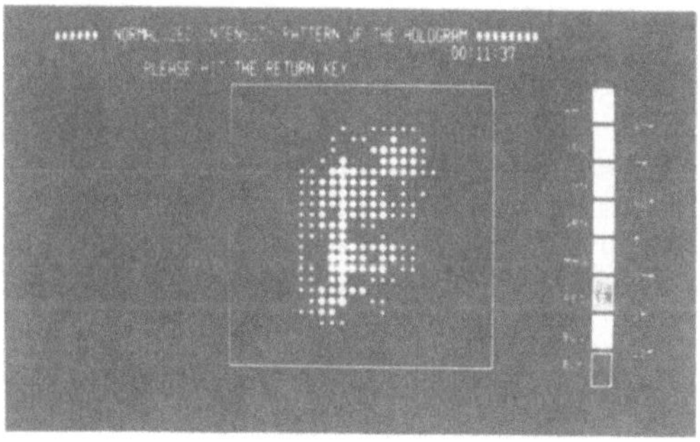

(c)

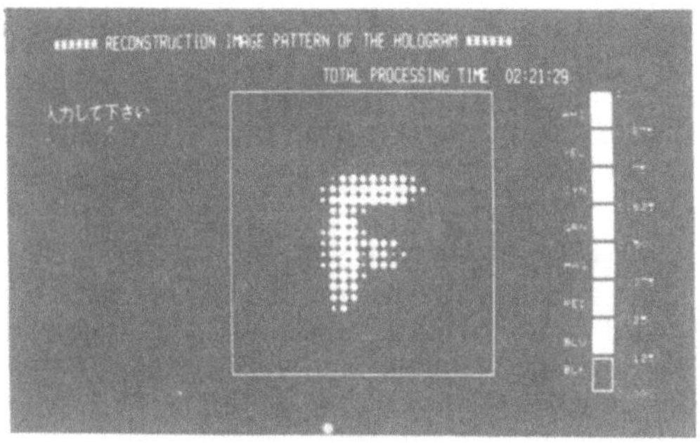

(d)

Fig.6 Two dimensional holography (a)Input data; (b)Real part of the hologram;(c)Intensity pattern;(d)Reconstructed pattern of " F " type aperture

Resultant holograms (real and imaginary part), intensity and reconstructed images are displayed on a CRT screen with circular dotted pattern as shown in Fig.4 to Fig.6. For the hard copy compatibility, each circular dot area is proportional to the amplitude or intensity of the sampling point, and color orders are chosen white, yellow, cyan, green, magenta, red, blue and black, as like as a rainbow test pattern of the conventional color TV. Then we can use any display device not only color CRT screen but also dot matrix printer or mono-chrome photograph.

CONCLUSIONS

In addition to the super-computer, many kind of 16 bit personal computers have been used for education and research in our school. But almost of students are using 8 or 16 bit personal computer for home studies.

The purposes of the programs are to teach them "What is the Cornu spiral?" and "what is the holography?" with color graphic display. One student interested in the Cornu's spiral, he tried at home with 8 bit computer. It consumes a lot of time(three days) for construction of the spiral.

Meanwhile, a host computer(6 MIPS) processed it only one second in CPU time, and the same machine processed two dimensional hologram (as like as Fig.4) less than 60 second in CPU time. Since the super computer(110 MFLOPS) 50 times more fast than 6 MIPS machine, then the same hologram would be processed one or two second in CPU time.

In spite of these circumstances, a personal computer has a lot of merit in its graphic ability and utility. These programs would be useful to analyze the diffraction effect of mask pattern and image reconstruction from the near field diffraction pattern after the data acquisition by a A/D converter.

ACKNOWLEDGEMENT

The authors would like to acknowledge the generous suggestion of Prof.K.Kido and Prof.K.Yamanouchi(Tohoku Univ.) for this work.

REFERENCES

H.Okada, A decision method of conditions for complex amplitude Fresnel hologram construction depending on one dimensional holography simulations,JASJ,Vol.38,No.6,pp339-349,(1982).

W.C.Elmore and M.A.Heald,THE PHYSICS OF WAVES, McGRAW-HILL,(1969).

DETECTION OF SHALLOW UNDERGROUND CAVITIES BY SEISMIC METHODS: PHYSICAL MODELLING APPROACH

P. Pernod, B. Piwakowski, B. Delannoy, and J.C. Tricot

Laboratoire de Physique des Vibrations et d'Acoustique
(C.N.R.S. U.A. 832 - Valenciennes)
Institut Industriel du Nord, B.P.48
59651 Villeneuve d'Ascq, Cédex, France

INTRODUCTION

In some previous papers[1,2], we have pointed out the difficulties of the detection, by seismic methods, of the shallow (8-10m) underground tunnels appearing in the North of France, which are the remains of old chalk pits now unexploited: the difficulties result from the impossibility of the resolution of the potential tunnel reflections from the background of groundroll and refraction signals. The real structure is shown Fig.1a. Fig.1b presents an example of the seismic section obtained by walkaway-type test[3] (source gather), when the source (hammer type) was placed at position x=0 and receivers (accelerometers) regularly spaced (0.66m) along the x-axis. The complexity of the seismograms is obvious: apart the direct P-arrival P_0, the refraction P_1, and the primary reflection in the first layer R_1 which are clearly visible, the rest of the section is composed essentially of a mixture of multiples of the refraction, multiples of the first layer reflections and the groundrolls G. In particular, no sign of the tunnel reflection that should appear as a hyperbole (discontinuous curve on Fig.1b) is observed.

To help search for a solution to the detection of the tunnels, we have developed a physical model (Fig.2a) inspired by the real field structure of Fig.1a. To maintain a comparison with the real seismic characteristics, special attention was paid to:
- the source[4]: we have developed an original type of impulsive transducer, a "mini-sparker", which combines all the characteristics required to model a seismic source: omni-directivity, wide band properties (20-300kHz), sufficient output power to model application and good repetitivity.
- the model structure: the materials chosen were easily available; their P-velocities, and dimensions were designed to maintain the same relation of the layers thicknesses and tunnel depth to the wavelength λ in the real and modelled structures. The scaling factor of the model (ratio of the model vertical dimension to the corresponding one of the real structure) is hence given for each layer by the ratio of the averaged wavelengths occurring respectively in the modelled and real field structures.
Finally, the model presented (Fig.2a) is composed of two layers: the upper one, made of rubber 3cm thick with acoustic P-velocity V_{p1}=1535m/s aims to model the surface clay layer 2m thick occurring in the real site, the lower is a plexiglass plate (V_{p2}=2758m/s), modelling chalk surrounding the tunnels. The tunnel is modelled by a cylindrical hole of radius R at depth d measured from the rubber-plexiglass interface. According to the scaling factor and the real field situation, this depth corresponds in the model to d=5.3cm. As the chalk in the real field structure is much deeper than the tunnel depth, and the plexiglass layer thickness should be limited in practice, its vertical dimension was fixed arbitrarily at 12 cm.

The "seismic" record obtained in the model without tunnel, in similar conditions as in Fig.1b, is presented Fig.2b. The space tracing according to the scaling factor is now 1cm. All the common features characteristic of the two layered structure are observed: first layer direct P-type arrival P_0 with velocity V_{p1}, refraction on the layer's interface P_1 corresponding to the velocity V_{p2}, P-type reflection in the first layer and its multiple R_1 and R_2, reflection from the plexiglass bottom and its multiple R_3, R_4. It should be pointed out that these two latter events, which are due to the limited thickness of plexiglass, do not exist in reality. Nevertheless, the presence of the clear and strong signal R_3 seems to be a good reference for model interpretation. In comparison with the real field situation (Fig.1b), the seismograms are simpler: the seismic phenomena are well separated in time and easily interpretable. There are two reasons of these differences: the S-type waves are strongly atte-

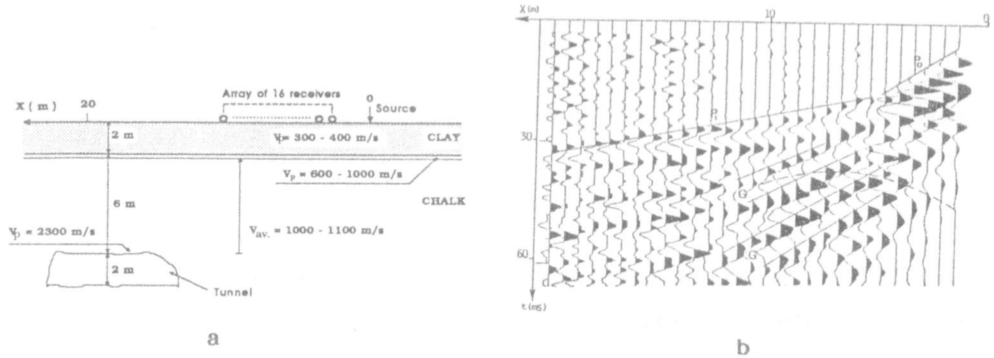

Fig.1. Real field data. (a) Geological cross section of the profile; the tunnels occur typically at a depth of 6m in the chalk; the terrain is covered by a 2m thick clay layer; V_p denotes the P-wave velocities, V_{av}. the average velocity in the chalk. (b) Walkaway noise test record for source position x=0 and space tracing 0.66m.

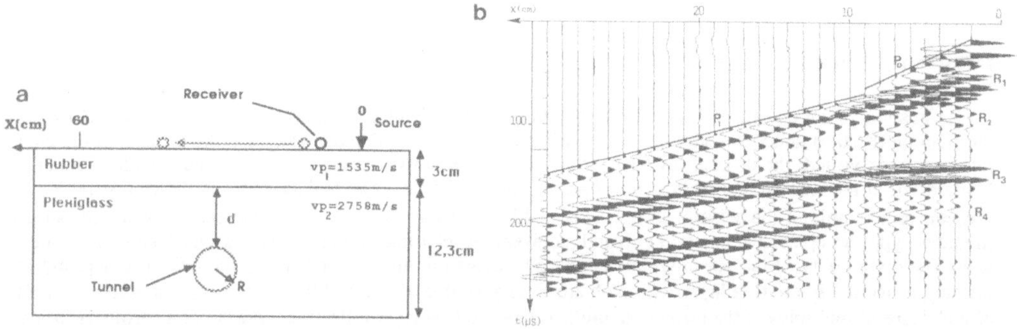

Fig.2. Physical model. (a) Model structure and parameters. (b) "Walkaway" noise test for source position x=0 and space tracing 1cm.

nuated in the rubber type materials and therefore, the groundroll signals, surface waves as well as other S-type events are not displayed; the absorption type attenuation of the real structure is much stronger than in the modelled one. This last remark explains the lower frequencies of the real field data in comparison with the modelled ones. The differences between the real and modelled phenomena described above are the disadvantages of our modelisation. Nevertheless, for a first approach, they can be considered as advantages because the seismograms will be simpler to interpret and this will facilitate the further study with presence of tunnels.

In this paper, we present theoretical considerations on the tunnel reflection level occurring in the model as a function of the tunnel's size and depth. Then, we show the experimental measurements made on the model for several types of surveys common in seismic methodology. Different sizes and orientations of the tunnel are considered. The geometry of the measurement system to obtain the best conditions of detection is discussed.

THEORETICAL EVALUATION OF THE TUNNEL REFLECTION LEVEL

The signals presented in Fig.2b represent the signals recorded without tunnel and may thus be considered as the disturbances of our problem of the tunnel's detection. Their spectral characteristics are similar to these of the suspected tunnel reflections; they may be referred to as the coherent noise. We define their amplitude as CNL(x,t) (coherent noise level) as a function of the receiver position x and the time t, and express it in decibels (dB), taking as the 0 dB reference the instrumentation setup background noise level. We can also define $EL_T(x,t_d,R)$ as the amplitude, expressed also in decibels, of the signal reflected on the cylindrical tunnel vault of radius R at depth d, where t_d denotes the suspected arrival time of this reflection. The difference between the quantities EL_T and CNL observed at the instant t_d, gives the signal to coherent noise ratio SNR for our tunnel detection problem:

$$SNR(x,t_d,R)=EL_T(x,t_d,R)-CNL(x,t=t_d) \qquad (1)$$

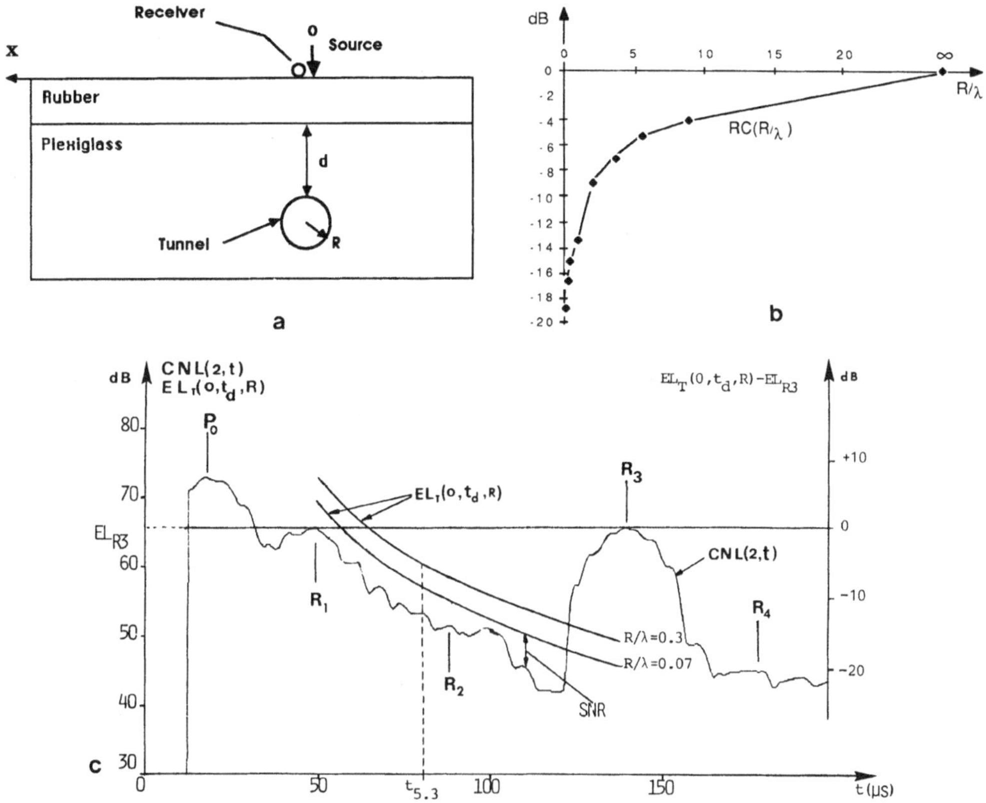

Fig.3. (a) Geometry assumed for the tunnel reflection level evaluation. (b) Reflection coefficient, for normal incidence, on a cylindrical tunnel vault, as a function of the ratio R/λ (the 0 dB reference is taken for $R=\infty$). (c) Logarithmic amplitude of the coherent noise signal CNL(2,t) recorded at position x=2cm in similar conditions as in Fig.2a, and theoretically estimated tunnel reflection level at normal incidence $EL_T(0,t_d,R/\lambda)$ as a function of the ratio R/λ and the tunnel reflection arrival time t_d.

Using the known model parameters and structure, we may find the difference between the reflection level from the bottom of the plexiglass EL_{R3} and the tunnel reflection level EL_T by means of the formula:

$$EL_T(t_d,R/\lambda)-EL_{R3}=[Ap(2d)-Ap(2*12.3)]+20\log(\frac{r'}{r})+RC(R/\lambda) \qquad (2)$$

where: - Ap(x) is the attenuation in decibels (dB/cm) of the signal propagating in the plexiglass over the distance x

- r and r' are the radii of curvature observed at the surface of the rubber of the reflected waves coming respectively from the tunnel boundary and from the plexiglass bottom
- $RC(R/\lambda)$ is the reflection coefficient on the tunnel vault expressed in dB as a function of the tunnel radius/wavelength ratio (the 0 dB reference is taken for $R=\infty$). This relation was obtained experimentally by additional experiments and is presented Fig.3b.

The calculations of the quantities EL_T-EL_{R3} were performed for the source and receiver both placed at x=0, vertically above the tunnel at depth d from the rubber-plexiglass interface (Fig.3a), for two different values of R relatively to the wavelength λ ($R=0.3\lambda$ and $R=0.07\lambda$), and for depth d varrying from 0cm (rubber-plexiglass interface) to 12cm (plexiglass bottom). The results are presented on Fig.3c as a function of the arrival time t_d of the tunnel reflection. On the same figure, we represent the CNL(x,t) curve taken for x=2cm. We suppose this case as an approximation of the vertical case x=0, because of the impossibility of placing the source and the receiver in the same location during the experiment. For the found value of EL_{R3}, it is possible to find the tunnel reflection EL_T and compare it to the coherent noise level signal. This comparison shows that the signal to noise ratio SNR (relation (1)) may be expected to be positive for all the ratio $R/\lambda>0.07$, for tunnel arrival time $t_d<125\mu s$. This means that the vertical detection of even small tunnels is theoretically possible without any additional processing. Similar types of calculation have been conducted for the real field situation[2] and have

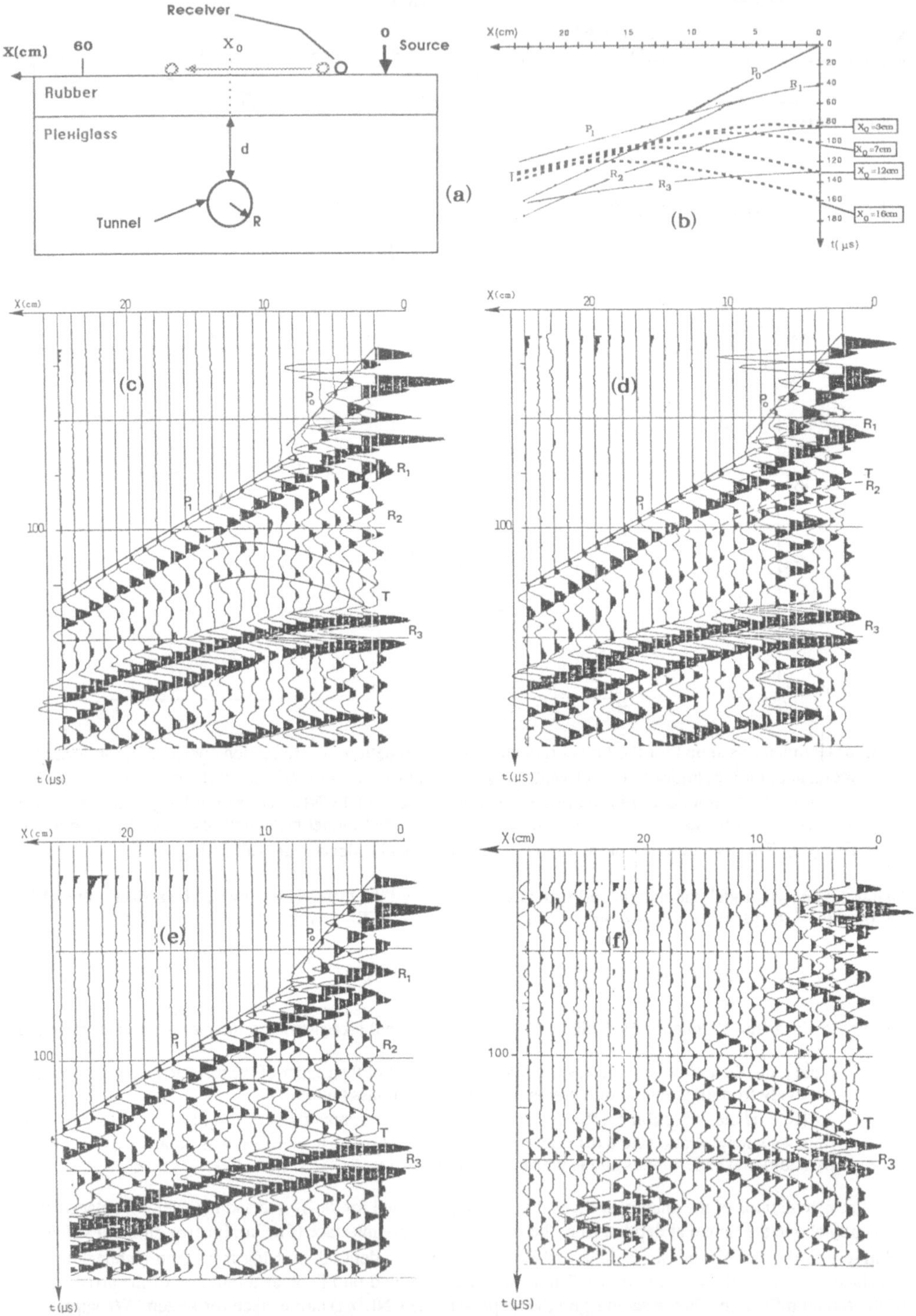

Fig.4. Multichannel acquisition type experiments. (a) Symbolic illustration of the geometry of the measurements. (b) Tunnel reflection arrival times calculated for different horizontal positions x_0 of the tunnel. (c) Results obtained for x_0=12cm, R=0.07λ. (d) Results obtained for x_0=3cm, R=0.07λ. (e) Same conditions as in 4c but for R=0.3λ. (f) Signals from section (e) processed with two-dimensional spatio-temporal filter to weakened the coherent noise.

shown that the value of the SNR is there of about -10 to -20 dB. These differences come from the simpler characteristics of our model.

EXPERIMENTAL RESULTS

This section presents the experimental results obtained in the model for the presence of tunnel. The tunnel depth is fixed at d=5.3cm; the two radii $R=0.3\lambda$ and $R=0.07\lambda$ are considered. The detection possibility is discussed for three particular types of surveys commonly applied in seismics: multichannel reception with the fixed source position, common offset and common depth point[3].

Multichannel reception - Fixed source

The geometry of the experiment is the same as in Fig.2b. The horizontal position of the tunnel is $x=x_0$ and its direction is perpendicular to the x-direction. Since we have only one receiver, the multichannel acquisition record is simulated by moving gradually the receiver along the x-axis (Fig.4a); here we benefit from the good repetitivity of the source.

The influence of the changes of the tunnel position x_0 upon the tunnel reflection arrival times is shown on the Fig.4b. Four cases are studied: x_0=3cm, x_0=7cm, x_0=12cm and x_0=16cm. The best geometry of the measurement system is realised when the hyperbole of the tunnel reflection arrival time is well-separated from the background of the other events. Taking into account the signature length of the signals being of 10μs (see Fig.2b), we observe that the best condition occure for x_0=12cm, where the resolved part of the hyperbole is the longest (0cm<x<10cm). Fig.4c,d present the results obtained for tunnel radius $R=0.07\lambda$ in the most favorable case x_0=12cm and in the unfavorable one x_0=3cm: the tunnel reflection T is more visible in the first case; in the second case, it is mixed with the reflections R_1 and R_2. Fig.4e, presents the same case as presented in Fig.4c but for $R=0.3\lambda$ - the tunnel reflection amplitude is obviously stronger. These experiments prove that application of the "optimum window technique"[5] is successful in detecting the tunnel, even of small radius, provided that it is possible to find a temporal window where the desired reflection is well separated from the coherent noise.

The previous conclusions on the best geometry of the measurement system can not be extrapolated for our real field situation without any additional considerations. Indeed, in this case, since the SNR ratio is always negative, the tunnel reflection can never be separated from the coherent noise. Furthermore, classical frequency filtering does not permit the enhancement of the SNR, for the spectra of the tunnel reflected signal and the coherent noise may be similar. The geometry of the measurement system must hence be adjusted according to another criterium, as a function of the processing applied. As we have shown in Fig.4b, for greater values of x_0, the hyperbole of the tunnel reflection displays an inversed slope, with regard to the direction of the coherent noise signals. This particularity can be exploited to extract the tunnel reflection by the application of the spatial processing technique[7]. The action of this can be used to enhance the amplitude of the signals presenting a particular preselected slope. Its efficiency is increased when this slope differs significantly from the undesired signal slopes. Greater values of x_0 are hence more favorable in this respect. But, as soon as x_0 increases, the distance covered by the wave is more important and the tunnel reflection is weakened: a compromise needs to be found. An example of the results of such a filtering is presented in the Fig.4f. The tunnel reflection T is well separated from reflections R_1, R_2 and the refraction P_1. The figures 4e,f show additionally that diffracted signals T from the tunnel is stronger for the receivers installed on the same side as the source in regard to the tunnel's position x_0.

Common offset type measurements - Tunnel's axis perpendicular to the x-direction

Diagrams and sections concerning this type of survey are presented Fig.5. During the experiment, the distance dx_0 between the source and the receiver is kept constant and the ensemble is moved along the x-axis (Fig.5f). The tunnel's position is fixed at $x=x_0$ and the tunnel's axis is prependicular to the x-direction.

The study of the optimum window for these conditions is presented Fig.5a,b for two source-receiver offsets dx_0=3cm and 9cm respectively. Since the source-receiver offset is kept constant, all the events independant of the tunnel's position appear as horizontal lines, whereas the tunnel reflection displays the curved line. Since the double reflected wave in the first layer R_2 has little amplitude and does not disturb the tunnel reflection T, we are interested to obtain the tunnel reflection as late as possible in regard with the arrival times of the group P_0, P_1, R_1. Finally, the delay marked as TW on the figures should be maximised. The analysis (not presented here) shows us that the offset dx_0=0cm is optimum. Fig.5c presents the record obtained for offset quasi-optimum dx_0=3cm (this offset was chosen for practical reasons; TW(0cm)≈TW(3cm)). For comparison, Fig.5d shows the record obtained for the case dx_0=9cm. The resolution of the tunnel reflection is obviously poorer. In the Fig.5c we observe the diminution of amplitude and some perturbations in the arrival times of the reflection

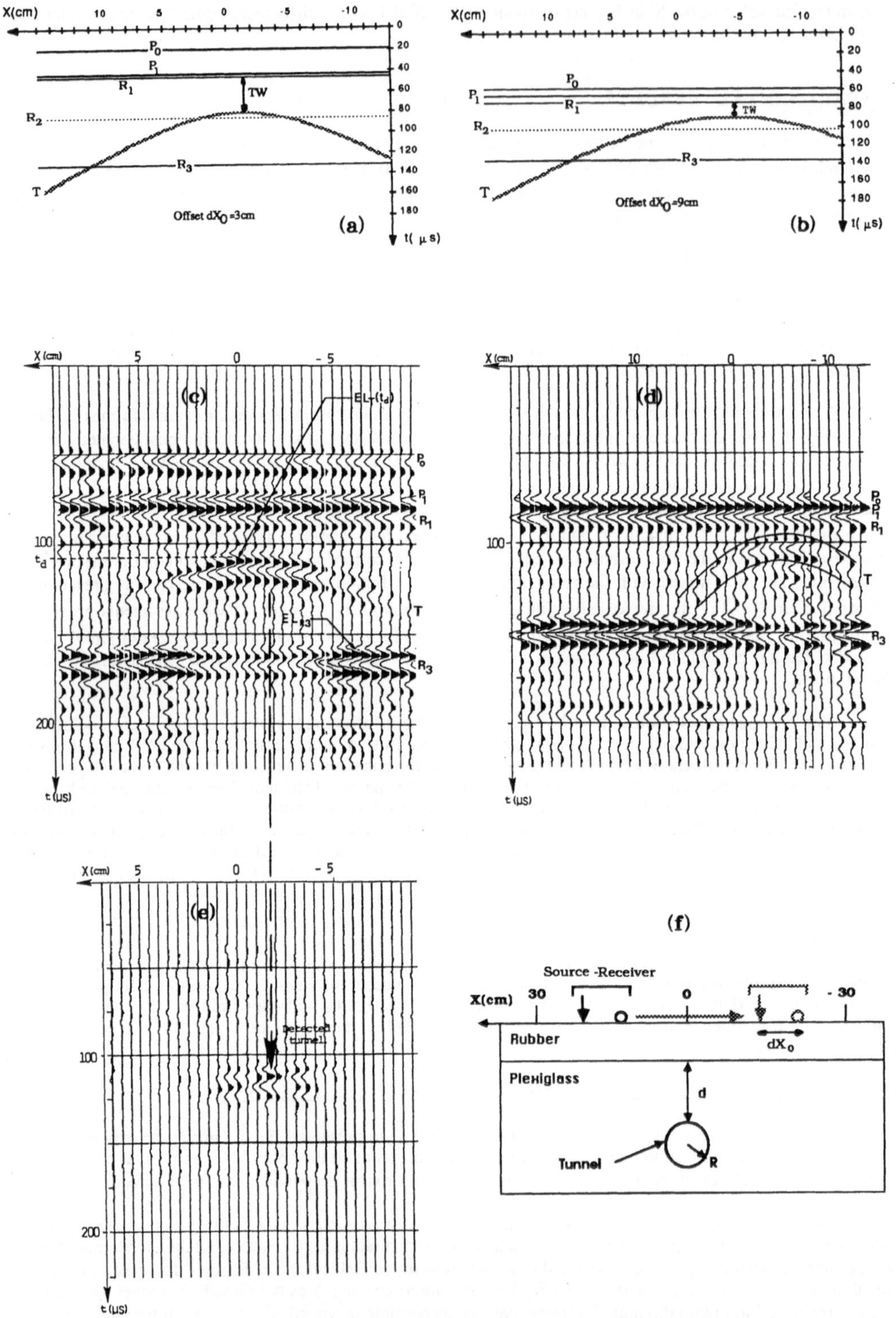

Fig.5. Common offset type experiments with tunnel axis perpendicular to the x-direction. (a) Arrival times for the source-receiver offset dx0=3cm. (b) Arrival times for the source-receiver offset dx0=9cm. (c) Experimental results for dx0=3cm. (d) Experimental results for dx0=9cm. (e) Results of the processed signals from section 5c. (f) Symbolic illustration of the geometry of the experiment.

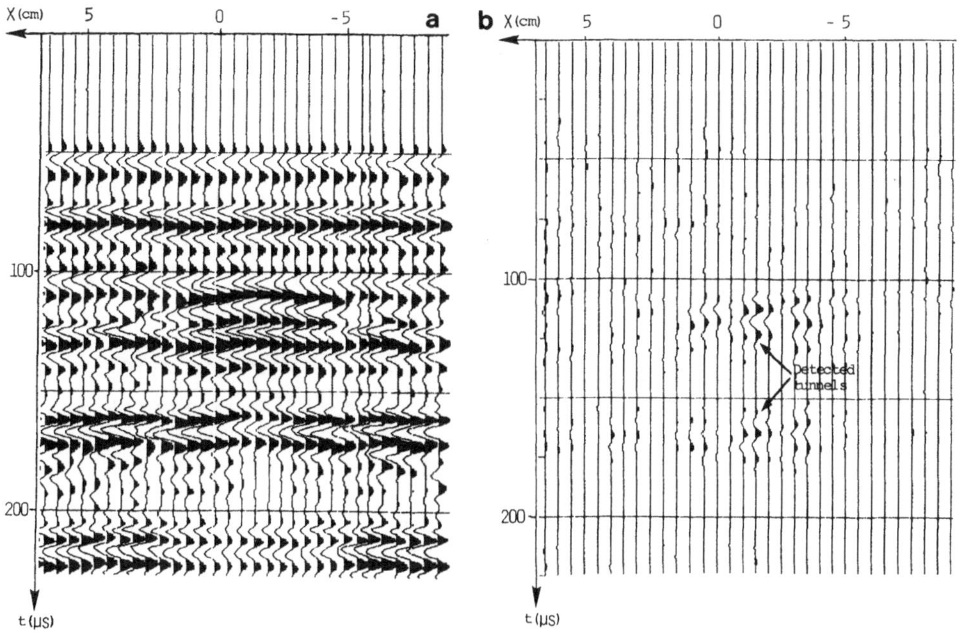

Fig.6. (a) Synthetic common offset section obtained from the section 5c to create the situation where the tunnel reflection is masked. (b) Processing version of the previous section.

R_3 correlated with the tunnel position. This confirms the tunnel detection possibility of the tunnel by the analysis of the reflections from the deeper interfaces. Such a means of detection was already proposed by Cook and al.[6]

The section of Fig.5c allow also us to compare the experimentally measured tunnel reflection level EL_T to the theoretical one found before. Indeed, we can consider the case dx_0=3cm where the source and the receiver are positionned symetrically to the tunnel's position, as an approximation of the vertical case. The experimental result is EL_T=62.1 dB whereas from Fig.3c we found EL_T=59.6 dB. The two results are of the same order, confirming the validity of the theoretical considerations.

The Fig.5e shows an example of the processing possibilities of the common offset record data. The synthetic aperture section consisting of the 21 channels was filtered with two-dimensional spatio-temporal filter[7], to remove all horizontally displayed events. Then, the dynamic focusing was used to improve the lateral resolution. Such a processing was applied to this aperture moving along the data from Fig.5c. The part of the results (for -16cm<x<16cm) displaying the detected tunnel position is presented Fig.5e; the coherent noise signals are eliminated with the efficiency greater than 20 dB and the obtained lateral resolution is about 1cm. The cross-form pattern displayed by tunnel detected signal is the spatio-temporal response of the filter applied.

As it was mentioned before, in the real conditions, the tunnel reflection is masked by the coherent noise. Such a situation was simulated: the signals from section 5c was added to its version delayed by 50µs to obtain the record which is presented in the Fig.6a, and where the tunnel reflection is evidently masked by the coherent noise. The Fig.6b shows the results of the same processing as applied before to the signals from section 5c. The tunnel is detected and located with the comparable precision. Note the second 50µs delayed tunnel location being discriminated from the signals from Fig.6a. This example shows that adequate processing can discriminate the diffracting type objects from the coherent noise (20 dB stronger for the example considered), when common offset seismic section is processed.

<u>Common offset measurements - Tunnel's axis parallele to the x-direction.</u>

In this section, the common offset measurements of the same type as before are conducted for the tunnel's direction parallel to the x-axis (Fig.7a). The extremity of the cylindrical tunnel of radius R=0.3λ is conical. The source-receiver offset applied is dx_0=9cm.

The experimental results are presented Fig.7b. The reflection of the horizontal upper boundary of the tunnel is observed for x<16cm (horizontal line), simultaneously with the shadowing effect on the reflected si-

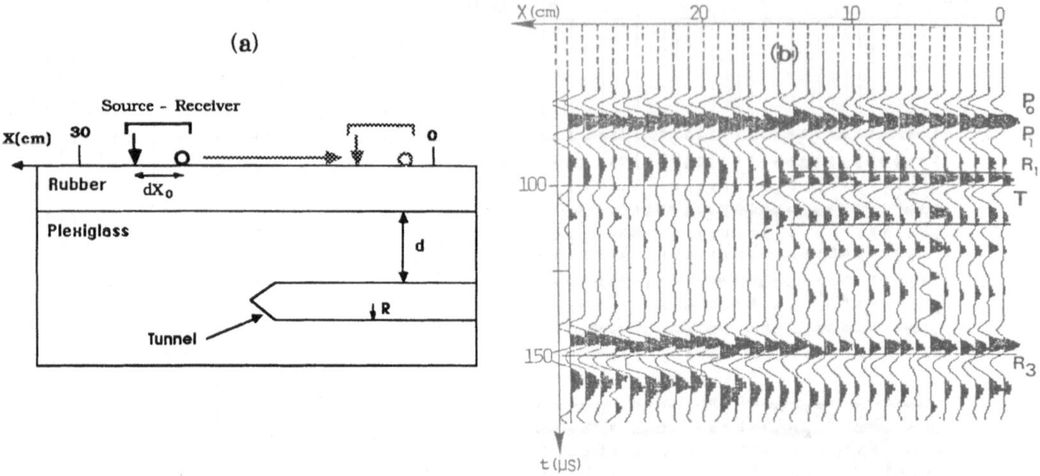

Fig.7. Common offset type experiments with tunnel axis parallel to the x-direction. (a) Symbolic illustration of the geometry of the measurement system. (b) Experimental results.

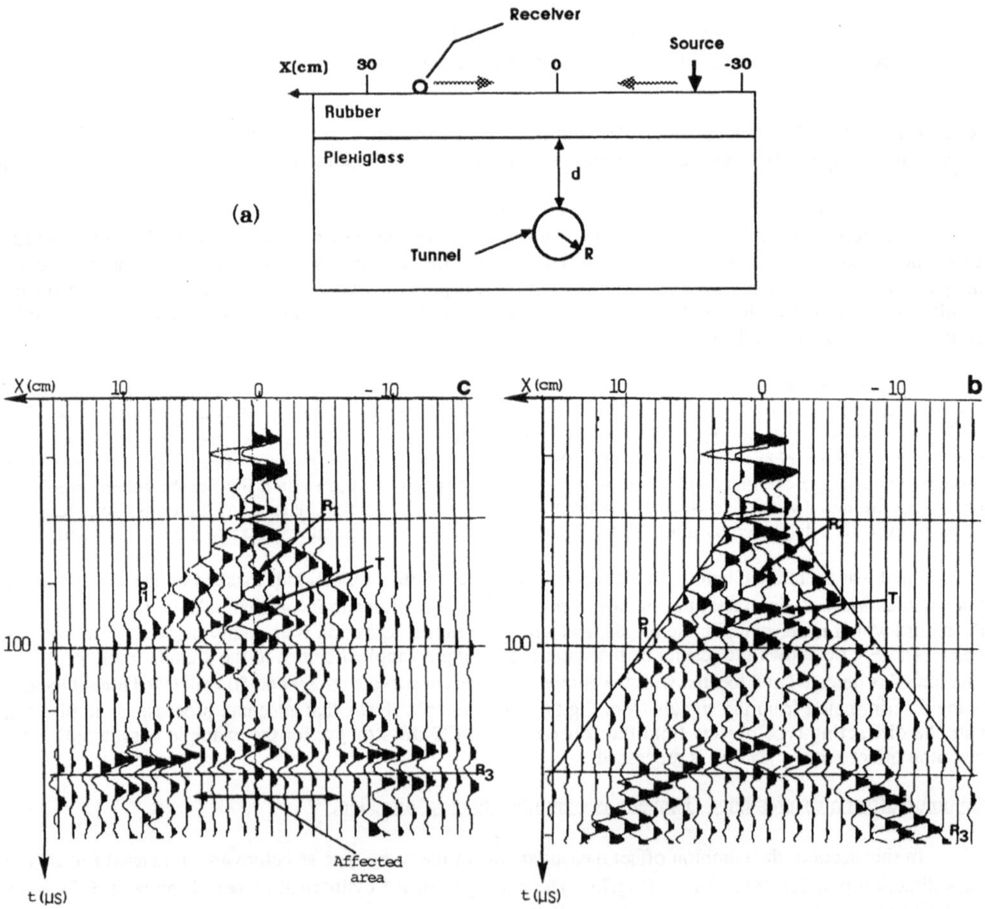

Fig.8. Common depth point gather. (a) Symbolic illustration of the geometry of the experiments. (b) Results for the tunnel's radius R=0.3λ. (c) The obtained section 8b after the NMO corrections.

gnal from the plexiglass bottom R_3. It should be noted that no diffraction from the extremity of the tunnel is observed as it was in the previous cases (Fig.5c). This lack of information about the tunnel's end is probably due to its conical form. This experiment shows the potential sensibility of the tunnel diffracted pattern on the tunnel form.

Common depth point stack

This type of seismic gather is commonly used in seismics to find plane interfaces[8]. We have tested its efficiency in application to the detection of the tunnel. The conditions of the experiments are presented Fig.8a. The starting position of the receiver and the source are respectively 15 and -15cm. The radius of the tunnel is $R=0.3\lambda$, trace spacing is 1cm. The seismic CDP gather is presented in the Fig.8b for CDP position equal to the position of the tunnel (x=0). We observe that tunnel reflection T is not marked as clear as we could presume from presented already common offset section (Fig.5c). On the contrary the signals R_1, P_1, R_3 are marked clearly. As we have shown on Fig.4e the reflections recorded on the same side as the source are privileged. Since the CDP gather searchs for the opposite situation, the recorded tunnel reflection is relatively weak. This proves that CDP gather might be not efficient in the detection of diffracting type objects. In the same time, we observe the influence of the tunnel's presence on the reflection R_3. As in the case of common offset data (Fig.5c), the masking action of the tunnel involves the anomalies in the amplitudes and the arrival times of the reflections from the point of plexiglass bottom situated below the tunnel. The NMO[8] corrected section from Fig.8b is presented Fig.8c. The anomalies observed in affected area should disturb the signal obtained from the stacking of the section Fig.8c and by these means the presence of the tunnel can be detected.

CONCLUSION

- The physical model, even simplified with regard to the real case, seems to us a very useful tool in the research in shallow seismic detection problems. In comparison with the commonly applied computer simulations, the recorded signals have the advantage of being quasi-real. The a-priori knowledge of the modelled structure characteristics enables theoretical analysis of the problem and then its verification in practique.

- We have shown that even very small tunnels compared to the wavelength can be detected by the seismic methods provided that "optimum temporal window" can be found for the given case. Since the problem of this window existence depends only upon the resolution of the seismic records, the principal way to detect the tunnel is to use the signals of sufficiently high frequency. Such a conclusion agrees with the experience reported already in this field[9].

- The common offset type measurements seems to us to be well adaptated for diffracting type object detection. The adequate signal processing could be very efficient in the signal to coherent noise ratio improvement.

- The CDP gathering seems to us as not adaptated to locate the diffracting type objects. On the contrary, the CDP method applied to detect the deeper interface can be useful for detecting the tunnel placed above.

REFERENCES

1. J. C. Tricot, B. Delannoy, B. Piwakowski, P. Pernod, Some problems and experimental results of seismic shallow propecting, 15 th Int. Symp. on Acoustical Imaging, Halifax, Canada, 1986.
2. B. Piwakowski, J. C. Tricot, B. Delannoy, P. Pernod, Shallow seismic prospection of underground tunnels-Evaluation of general detection conditions, deposed for acceptance for Geophysics in March 1987.
3. M. B. Dobrin, Introduction to geophysical prospecting, Mc Graw-Hill Inc.
4. P. Pernod, B. Piwakowski, J. C. Tricot, B. Delannoy, Mini-sparker as a source in seismic models, 16 th Int. Symp. on Acoustical Imaging, Chicago, 1987.
5. J. A. Hunter, S. E. Pullan, R. A. Burns, R. M. Gagne, R. L. Good, Shallow seismic reflection mapping of the overburden-bedrock interface with the engineering seismograph-Some simple techniques, Geophysics, 49, 8, 1984.
6. J. C. Cook, Seismic mapping of underground cavites using reflection amplitudes, Geophysics, 30, 1965.
7. B. Piwakowski, J. C. Tricot, B. Delannoy, P. Pernod, Spatial filtering in seismic shallow prospecting, 16 th Int. Symp. on Acoustical Imaging, Chicago, 1987.
8. W. A. Schneider, The common depth point stack, Proceedings of the IEEE, 72, 10, 1984.
9. R. W. Knapp, D. W. Steeples, High resolution common-depth-point seismic reflection profilling: Field acquisition parameter design, Geophysics, 51, 1986.

ULTRASONIC IMAGING FOR THE SEISMIC MODEL FAULT PLANE

Yuan Yi-quan
Department of Radio Engineering, Nanjing Institute of Technology, China

Yin Ging-rui
Shanghai Institute of Ceramic, Academia Sinica

Shi Bing-wen and Shen Shou-peng
Shanghai Institute of Organic Chemistry Academia Sinica

ABSTRACT- This paper mentions the principle and the hardware system of ultrasonic imaging. A series of experiments is also presented.

In this paper, two new types of piezoelectric material (PVDF and 3-3 connectivity pattern composite piezoelectric ceramic) are presented. By use of them, two new types of broad bandwidth narrow impulse, passive, reversible sensors have been manufactured. These sensors are successfully employed in ultrasonic imaging for the seismic model fault plane from 150 KHz to 1 MHz frequency range.

Application of these new types sensors and hardware system can distinguish 6 mm aluminium plank in water, even clearly they can distinguish multilayer model body.

INTRODUCTION

People have been making experiments on ultrasonic imaging of the seismic model fault plane for more than 30 years. It has been developed from pure velocity measurement to model fault plane imaging for multilayer complex models, and from indoor theoretical research to an important means of seismic exploitation and production as well as research work. An ultrasonic imaging experiment research center with full scale and advanced equipment has been established by the well-known Seismic Acoustic Laboratory, Houston University, USA, through investment from dozens of oil companies.

It can be known from Fig.1 that the system is composed of the water pool, ultrasonic transreceiver transducer, Model SYC-III ultrasonic instrument, microcomputer and plotter, etc. The model to be studied is put into the water pool, electric signals in narrow pulses are emitted from the ultrasonic instrument, activating the ultrasonic transmitting transducer. This sound wave is received by the ultrasonic receiving transducer after penetration and reflection. The received signal after A/D conversion, is logged on the microcomputer for processing, and plotted by the plotter. It can eventually reflect the structural fault of the seismic model, which itself can simulate the structure of the geological layer for exploitation. When the experiment logged data are in conformity with the field aquired data, they can assist the specialists to judge the geological conditions in tapping the underground

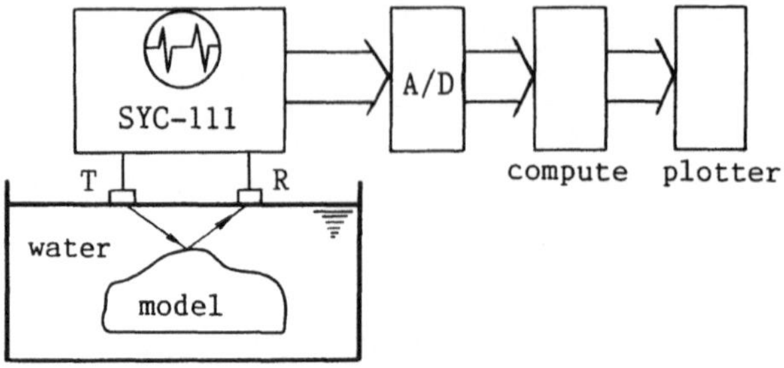

Fig.1 Principle of ultrasonic imaging for seismic model fault plane

oil and gas resources and various minerals reserves. Ultrasonic seismi model experiment is a branch of science which combines the ultrasoni study with the geological science, and is of great significance in the development of the national economy.

PROBLEMS WITH THE OLDER TRANSDUCERS IN ULTRASONIC SEISMIC MODEL IMAGING

The traditional solid PZT piezoelectric ceramic probes (or transducers) have been used in observing the seismic models for a long time and a number of development[1] have been obtained. However, the following problems do exist.

First. The solid PZT has an acoustic impedance 20 times greater than the water medium, which will cause serious mis-matching in acoustic impedance when working in the water, leading to decreased receiving sensitivity and transmitting frequency response, narrowed frequency bandwidth, distorted pulse waveform and increased residual waves, making it difficult to observe and analyze complicated seismic models. Fig. 2 shows the fault model experiment record obtained by using short residual wave probes with four vibration periods only - PZT probes with tungsten (powder strong absorber) as a back-liner matching layer. It can be seen from Fig.2 that it is not possible to exploit thin layer structures by using the PZT solid probes, essentially because of the serious mis-matching in acoustic impedance between piezoelectric solid ceramic

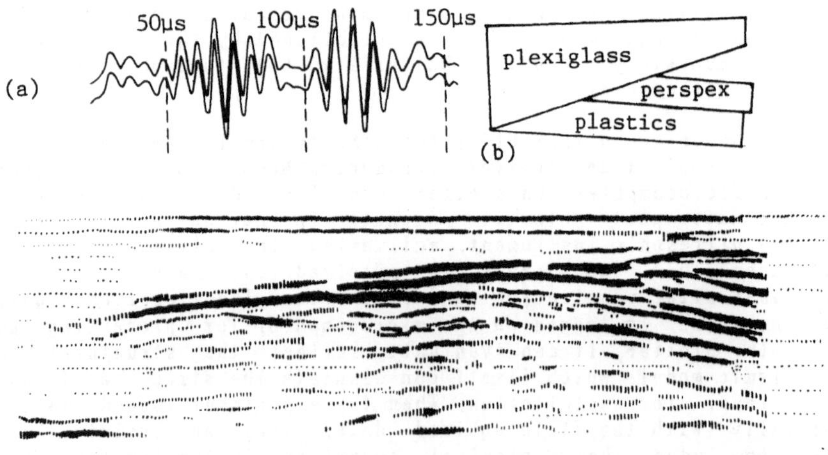

Fig.2 Seismic model record obtained by using older solid probes

($33.7 \cdot 10^6 Pa/s.m^2$) and water $1.5 \cdot 10^6 Pa/s.m^2$), producing prolonged sustaining waveforms.

Second. The older PZT transducers used in seismic model experiment are limited to working frequencies between 100 KHz - 200 KHz. Therefore the semicycle narrow pulse width of the probes is greater than 5 μs, and the shortest sustaining time of the transient waveform is about 25-30 μs. This shows that this kind of probe cannot distinguish the fine boundaries in seismic models. If we reduce the size of the existing probes and increase their working frequency, and at the same time try to satisfy the characteristics of the broad angles, then the PZT transducer will inevitably lead to two-dimensional vibrations and their mutual coupling, making it difficult to put it into practice.

Scientists involved in seismic model experiments are in urgent need of ultrasonic probes which are fit for their study purpose. But no special study has been made for a long time to solve their problems.

NEW TYPE NARROW PULSE PROBES FOR ULTRASONIC IMAGING FOR SEISMIC MODEL FAULT PLANE

The use of narrow impulse ultrasonic transducers is an important key to improving the axial resolution for the seismic model experiment. Two new types of piezoelectricity material and sensor are used.

Two type of new piezoelectric material

PVDF high molecular piezoelectric film material. Since the discovery of polyvinylidene fluoride (PVDF)[2] in 1969, numerous studies on its basic theory and application have been carried out in various countries. Through efforts made over nearly 20 years, the piezoelectric constant has been increased by nearly 20 times, and the material has found extensive applications in electric-acoustic, underwater acoustic, ultrasonic, vibration, impacting, medical diagnosis, etc. The characteristics of the China-made piezoelectric film are shown in Table. 1.

PZT ceramic composite material with 3-3 connectivity. In 1970s, Professor R. E. Newnham of Pennsylvania, USA began the study of twophase composite materials[3], and more than a dozen of different connective patterns were proposed, among which 3-3 and 3-1 type were the most successful in applicatiion. Some research have also been carried out in China, see Table 2, which is provided by the Shanghai Institute of Ceramics, Acadamia Sinica. Compared with PZT solid piezoelectric ceramic, the PZT silicon-rubber with 3-3 connectivity has a lot of advantages.

Table 1 Characteristics of China-made PVDF piezoelectric film

Thickness (t) (μm)	Piezoelectric strain constant (d_{31}) (PC/N)	Piezoelectric voltage constant (g_{31}) (Vm/N)	Elastic constant (Longitudinal) (c) (N/M^2)	Relative dielectric constant ($\varepsilon/\varepsilon_0$)	Dielectric constant (tgδ)	Density $10^3 kg/m^3$	Fracture strength (longitudinal) (N/m^2)	Resistance (Ω-m)
30-130	24-29	200-260	3×10^9	12-14.5	0.01-0.02	2.2	330-370 $\times10^6$	3×10^8

Note, These data are provided by Shanghai Institute of Organic Chemistry, Academia Sinica.

Table 2 Comparison Between 3-3 connectivity ceramic and PZT solid ceramics

Material	density ρ (10^3 kg/m^3)	$\varepsilon^T_{33}/\varepsilon_0$	tgδ %	d_{33} (PC/N)	g_{33} ($\times 10^{-3}$ Vm/N)	g_h (10^3 Vm/N)	$d_h \cdot g_h$ (10^{-15} m^2/N)	K_p	K_t	Q	Acoustic speed (vertical wave)
3-3 compound	3.2	206	1.4	209	114	88	14168	0.17	0.45	9-12	1496
PZT solid ceramics	7.8	1150	0.4	215	21	4.2	178	0.49	0.45	580	4500

Note, The data in this table are provided by Shanghai Institute of Ceramic, Acadamia Sinica.

Two Types of New Narrow Pulse Sensors

Probes Invertible PVDF wide band piezoelectric probe. The PVDF is bonded to the surface of a high impedance spherical subtrate in several layers. On the back of the subtrate, there is a high damping acoustic absorbing matching layer. As a result of this impedance difference, the transducer will be acoustically resonant at a frequency where its thickness is $\lambda/4$[4] in the normal direction of the boundary because of the piezoelectric effect. As the front end is in water medium, it can be known from the acoustic transmissison principle that the transducer resonant are in $\lambda/4$ mode. The multilayer overlapping structure is adopted in order to reduce the input impedance and increase the thickness of PVDF film. Table 3 shows the performance of there multi-layer PVDF invertible (for both ways as transmitter and receiver) wide band transducers. And Fig.3 (a) shows the outline of the probe.

Table 3 Performance of PVDF Wide Band Probes

Transmitting response at max.* (dB)	Receiving sensitivity at max.** (dB)	Frequency respone Q value	Input electric impedance Z (KΩ)	Acoustic impedance (10^6 Pa/m.s)	Resonance frequency range (kHz)	max.p-p voltage (10^3V)
120-140	210-205	1-1.5	1.6-5.0	3.7	500-1000	0.1-0.4

* odB = 1μPa/V
** odB = 1V/μPa

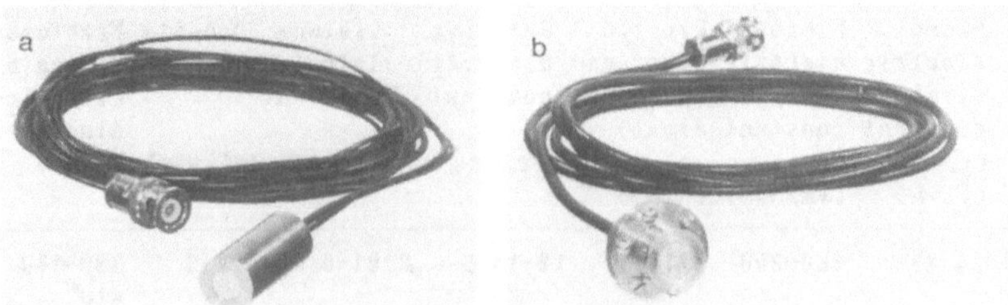

Fig.3 New types of probes, (a) PVDF probes; (b) 3-3 connectivity probes.

Invertible 3-3 connectivity piezoelectric probes. The PZT/silicon-rubber probes with 3-3 connectivity consist of composite piezoelectric-discs, protected by a very thin sound penetrating layer and lined on the back with strong damping sound absorbing matching material. It is very easy to develop them into invertible wide band piezoelectric ultrasonic probes [8] for seismic models. They have better characteristics than the solid PZT piezoelectric ceramic probes when operating in the water (for details see table 4). Fig.3 (b) illustrates the outline of the new probes.

Table 4 Characteristics of invertible PZT/ silicon-rubber with 3-3 connectivity probes (in water)

Transmitting response at Max. (dB)	Receiving sensitivity (flat zone) (dB)	Resonance frequency range (kHz)	Q value	Input resistance (kΩ)	Input capacitance (PF)	Dia. of dement effective area (mm)	Acoustic impedance (10^6 Pa/m²S)	Max. P-P voltage (10^3V)
120-140	-210	160-500	1-2	4-15	200-300	Φ 12-15	4.0-5.1	1-2

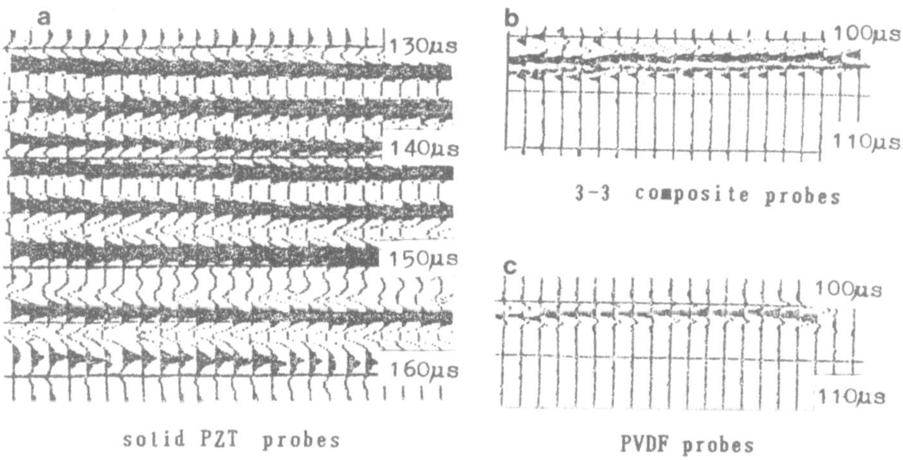

Fig.4 Recorded Waveforms on discontinuous boundaries by different probes

EXPERIMENT RESULTS FOR ULTRASONIC IMAGING FOR SEISMIC FAULT FLANE

Recorded imaging of boundaries. In seismic model tests, the reflection waves obtained from discontinuous boundaries by traditional PZT solid probes have too long sustaining vibration waveform, hence have very low resolution, as shown in Fig.4 (a); those obtained by 3-3 connectivity composite piezoelectric probes have a shorter impulse response (it decays 20 dB in 2.5 cycles), hence have higher resolution, as shown in Fig. 4 (b) and those obtained by PVDF probes are the simplest, usually with cycles, as shown in Fig. 4 (c).

Recorded imaging for thin layer physical models. In a seismic model body observed under water, the PZT solid probes cannot while both PVDF and 3-3 connectivity composite probes can distinguish the two boundaries. Fig.5 is the record obtained by 3-3 connectivity composite probes, where the pulse sound wave traveled across the 1 cm thick-plexiglass plate, so

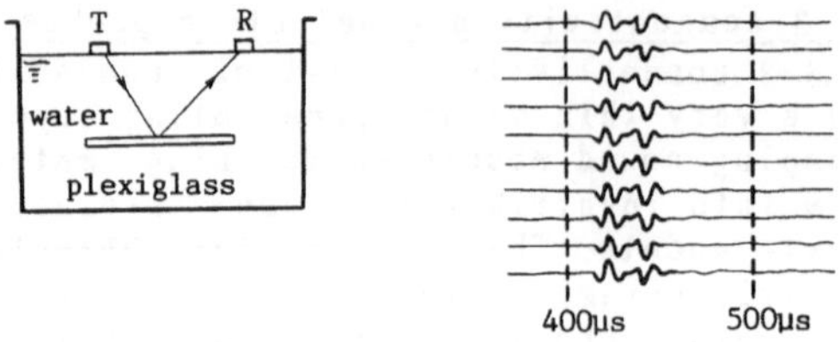

Fig.5 Recorded imaging from 3-3 connectivity composite probes observing under water a 1 cm thick-plexiglass plate

it can clearly distinguish the upper and bottom boundaries of the thin plate. Even clearly it can distinguish a 6 mm aluminium plank in water.

Recorded imaging of multi-layer complicated physical models. Fig. 6 (b) and (c) show respectively the recorded images obtained by using PVDF and 3-3 connectivity composite probes in observing seismic models made of plexiglass, rubber and perspex (as shown in Fig. 6 (a)). Both them shows good agreement with the combined record of theoretical calculation. It is more clear than the above-shown Fig.2. It can be known from Fig.6 that the resonant frequency of 3-3 connectivity composite probes is 160 kHz-500 kHz, with an impulse duration of 9-16 μs; and the resonant frequency of PVDF probes is 500 kHz-1MHz, with the impulse duration of 5-8 μs. Therefore they give the best resolution, hence the clearest recorded imaging for the seismic model fault plane.

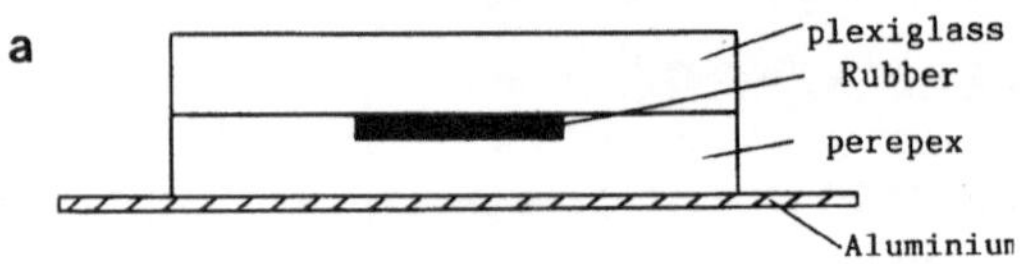

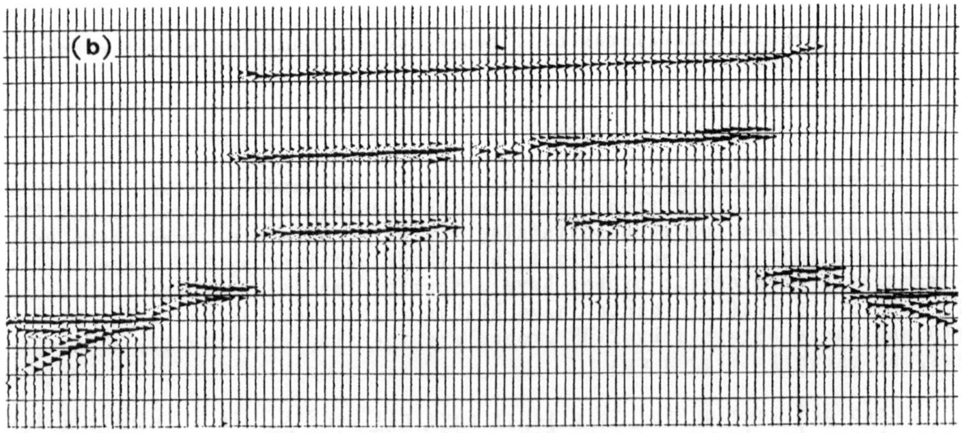

Fig.6 Recorded imaging for complicated seismic models (Models are made of plexiglass, rubber and perspex)

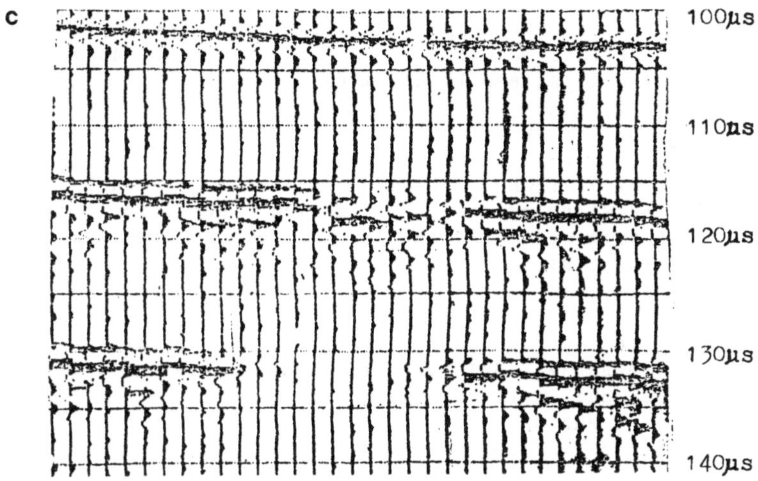

Fig.6 Recorded imaging for complicated seismic models (Models are made of plexiglass, rubber and perspex)

CONCLUSION

The basic conditions for the above-mentioned space field, frequency range and time range in the frequency band from 100 kHz to 1 MHz have been realized. The type SYC-III acoustic lithologic parameter tester and the self-developed narrow pulse adaptor are used, and under the same water pool conditions, two new types of probes, PVDF and 3-3 connectivity composite, are activiated by a 1 µs width electric signal. The imaging results obtained are summed up as follows:

1. Comparison in amplitude-frequency characteristics: The optimum frequency band Q value of PVDF probes can reach 1, and its input impedance in the 500kHz-1MHz frequency range is about 2kΩ, so it is connected to input 15 kΩ step of the adaptor and the activiation condition of constant current source can be obtained. The optimum frequency band Q value of 3-3 connectivity composite probes is as high as 1.5, and its input impedance in the 200 kHz-500kHz frequency range is about 25kΩ, so it is difficult to realize the constant current source activiation condition by an adaptor. By comparison, the former has a wide amplitude-frequency characteristics than the latter, so it can provide acoustic images with higher resolution.

2. Comparison in acoustic transmission power: Both probes under development at present show that the transmission voltage can be nearly 150 dB (with reference being 1 µ Pa/V). However, as the 3-3 connectivity composite probes have higher Curier point, their achievable potential acoustic power is also higher. On the other hand, the PVDF probes have low electric impedance, and their acoustic output power activiated by unit voltage is higher than that of the 3-3 connectivity composite probes.

3. Comparison in self-noise and anti-interference capacity: As multi-layer structure is adopted for the PVDF probes, they can work independently without depending on the MOS-IC, and can be made into its self-invertible probes. The 3-3 connectivity composite probes are also invertible probes. Therefore, both are passive probes, not

likely to be influenced by the external electric noise, featuring stable and reliable operation. For both the self-noise calculated to the input end is less than 2 μV.

4. Comparison in beam width. The transmission surface of the PVDF probes can be made into spherical shape, while for the 3-3 composite probes, the vibration amplitude distribution structure of fixed periphery and weighted center is adopted. Therefore, both types of Probes have beam width larger than that of the traditional PZT probes (uniform piston vibration type). This is very important to seismic model tests.

5. Comparison in narrow pulse duration and residual waveform of the transient response: Both the PVDF and 3-3 composite probes feature narrow pulses and short residual waveforms. However, the former work in 500 kHz - 1 MHz and the latter in 150 kHz - 500 kHz, and the former have lower frequency band Q value. So the PVDF probes provide the highest resolution, with the total duration of narrow pulse and the residual wave being 5-8 μs. The resolution of the 3-3 composite probes in next to that of PVDF probes, with the above total time of 9 - 16 μs.

Under present conditions, these two types of new probes are being used in combination and mutual complement, covering the experiment frequency range from 150 kHz to 1 MHz necessary for all the seismic physical model tests, thus creating mew conditions for the development of ultrasonic imaging for the seismic fault model plane.

REFERENCES

[1] Desilers C.S, Fraser J.D. and Kino G,S. IEEE Trans. sonics ultrason, vol, SU-25, 3(1978), 115.

[2] Sessler G.M., Acoust J.Sec.Am 70(1981), 1596.

[3] Newnham R.E., Bowen L.J., klickev K.A. and Cross L.E.: The materials research laboratory Annual report, 1980.

[4] Yuan Yi-quan, Shao Yaomei, Shi Bingwen, Applied Acoustics (QUARTERLY) Vol. 6, No. 1, January (1987), 25.

[5] Yuan Yi-quan, Acoustics and Electric Engineering No.1 (1987). 1.

INCOMPLETE PROJECTION IMAGING BETWEEN BOREHOLES FOR GEOPHYSICAL EXPLORATION

Feng Yin and Yu Wei

Department of Biomedical Engineering
Nanjing Institute of Technology
Nanjing, China

1. Introduction

It is known that Cross borehole imaging can be used to detect hazardous regions in the mine, search fault in earth and determine the location and volume of oil fields.[1,2,3] Therefore, it is one of attractive research subjects in recent years.

Because transmitters and receivers are set in two boreholes respectively, the way of getting the data is limited. It is known that cross borehole imaging problem belongs to spatially truncated and limited angle projection imaging. In order to solve imaging problems of cross borehole for geophysical exploration, the iterative reconstruction-reprojection algorithm is used to get underground imaging reconstruction between boreholes under the condition of approximation of straight-line transmission of energy. Although the ray path is usually bending, Geophysist can get the useful information between borehole from the reconstruction picture when the parameter between the background and the anomaly is small. In addition, the image derived by this algorithm can be used as the initial value for iteration when the ray bending is considered in order to improve the quality of reconstruction and develope the new reconstruction algorithm.

2. The Imaging Model

Now, the imaging model, as shown in fig.1, is considered: there are two boreholes, A1, A2, the transmitters and receivers are set as that in fig. 1. When the transmitter at the h(i) transmits the signal, then the signal going through the imaging region can be received by the receiver in borehole A2.

Let the distribution of velocity of the acoustics between boreholes be $c(x,y)$ and $h(i)$ and $r(j)$ be the position of transmitter and receiver, respectively. Then the time of fly of the signal from $h(i)$ to $r(j)$ is:

$$Ph(i)r(j)=\int_{h(i)r(j)} 1/c(x,y) * dl \qquad (2.1)$$

where $Ph(i)r(j)$ is the time of fly of signal from the $h(i)$ to $r(j)$.

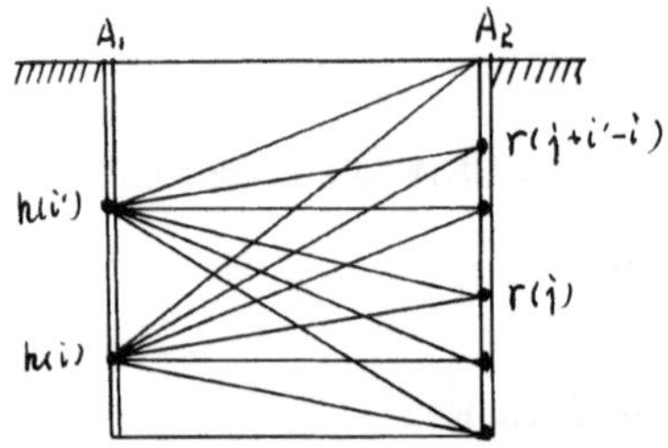

Fig.1 Cross-borehole scanning geometry for geophysical tomography

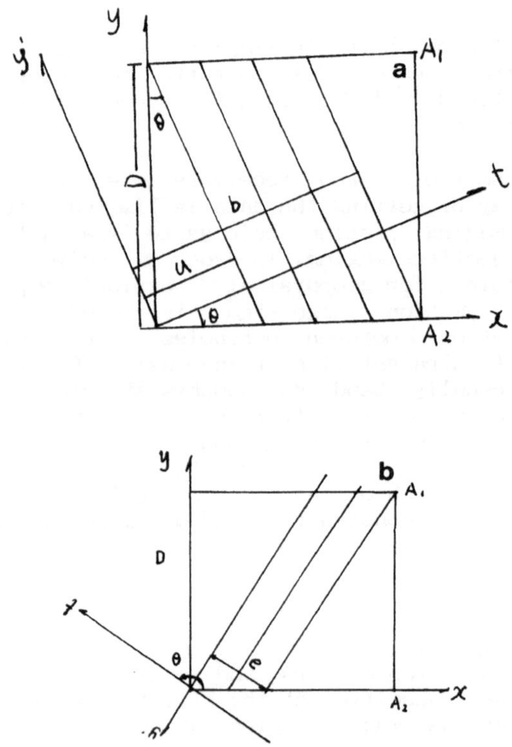

Fig.2 The measurable projection.(a) when 0 ⩽ Θ ⩽ 45 ; (b) 135 ⩽ Θ ⩽ 180.

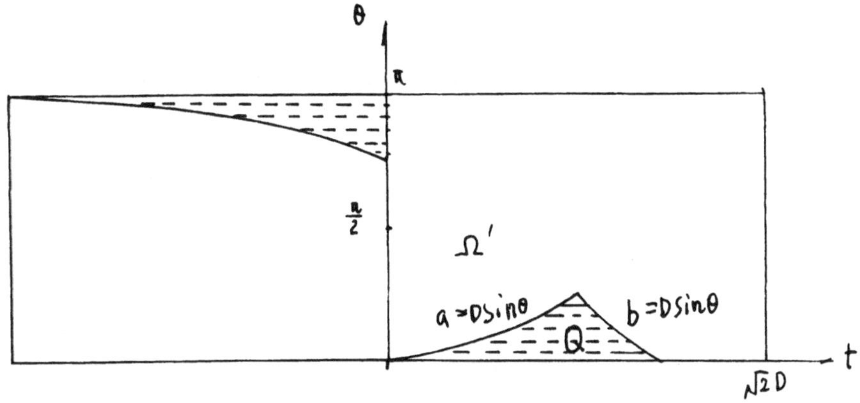

Fig.3. It is the region where the projection can be measured and that with dot lines is region where the projections can be measured in θ-t domain

The equation (2.1) is the so called Radon transform.Therefore, c(x,y) can be inversed from a series of Ph(i)r(j).It is known that the measurable projection data are incomplete,subject to constraint of imaging model. The region in which the projection can be measured on the θ-t domain can be got by the below analysis.

When the transmitters and receivers are set as that in fig.1 , the projection h(i)r(j) is parallel to h(i)r(j+i-i),the below relations can be got from fig.2

$$a=d*\sin(\theta) \tag{2.2}$$

$$b=d*\cos(\theta) \tag{2.3}$$

$$e=d(\sin(\theta)+\cos(\theta)) \tag{2.4}$$

when $0 \leqslant \theta \leqslant 45^\circ$, $a \leqslant t \leqslant b$,the projection can be measured; $45 < \theta < 135$,the projection can not be measured; $135 \leqslant \theta \leqslant 180$, $e \leqslant t \leqslant 0$, the projection can be measured.

The measurable region of projection can be got through the above analysis,as shown in fig.3

3. Iterative Reconstruction and Reprojection Algorithm

The iterative reconstruction-reprojection algorithm was used to get the reconstruction image between boreholes in this section. The main idea is to choose the ray mutually parallel among all the projections. Then convolution filtered-back projection formula is used directly. But it is known that measurable projections are incomplete from above analysis. In order to get the reconstruction from the projections, the missing projection in the incomplete domain should be estimated. In addition,prior information is taken into account to improve the quality of the image.These prior information includes that object function is nonnegative ,that is,constraint condition of object function and projection are nonnegative, that is constraint condition of projection;the image is confined within a boundary, etc.

In order to analyze conveniently, define P to be consistent full set of projection: if and only if P satisfies the constraint condition of projection; (2) the object function can be reconstructed from P and this object function satisfies the constraint condition of object function;(3) integral of this object function equals to P.

Z is called consistent estimation of missing projection,if and only if combination of it and measurable projection can form consistent complete set .

Let C^Z denote constraint operator;the minus projections are set to be zero when it operates on the projection.

Let C^I denote constraint operator of imaging function;the minus object functions are set to be zero when it operates on the object function;

Let R denote the Reconstruction operator and S denote the projection operator. The complete projection set was divided into two part;one part of it is measurable projection,w,the other is estimate of the projection $\tilde{Z}$,This will be represented as $P=[\hat{Z}_W]$.

If $\hat{z}$ is the consistent estimation of missing projection, it must satisfy the constraint condition of projection from above definition.That is

$$\hat{Z}=C\hat{Z} \tag{3.1}$$

In addition,the object function can be reconstruction from the consistent estimation of projection $\hat{Z}$ and measurable projection W. The reconstruction object function satisfies the constraint condition of object function:

$$R\begin{bmatrix}\hat{Z}\\W\end{bmatrix} = C^{I}R\begin{bmatrix}\hat{Z}\\W\end{bmatrix} \tag{3.2}$$

The Line integral of object function is equal to the complete projection which consists of measurable and estimated projection.That is

$$S\,R\begin{bmatrix}\hat{Z}\\W\end{bmatrix} = \begin{bmatrix}\hat{Z}\\W\end{bmatrix} \tag{3.3}$$

From above consistent condition (3.1)-(3.3), there exists the following relation

$$\begin{bmatrix}\hat{Z}\\W\end{bmatrix} = S\,R\begin{bmatrix}\hat{Z}\\W\end{bmatrix} = S\,C^{I}R\begin{bmatrix}\hat{Z}\\W\end{bmatrix} = S\,C^{I}R\begin{bmatrix}C^{Z}\hat{Z}\\W\end{bmatrix} \tag{3.4}$$

Let S^Z denote the operator which calculates the missing projection. Let R^Z denote the reconstruction operator which only operats on the missing projection and R^W denote the reconstruction operation which only operats on the measurable projection. The following relations should exist from the constraint condition.

$$\hat{Z} = S^{Z}C^{Z}R\begin{bmatrix}C^{Z}\hat{Z}\\W\end{bmatrix} = S^{Z}C^{I}R^{Z}C^{Z}\hat{Z} + S^{Z}C^{I}R^{W}W = T(\hat{Z}) \tag{3.5}$$

$\hat{Z}$ is called fixed point of operator T.[4]

The core of our iterative algorithm is to calculate the fixed-point of T. The general form of iteration is

$$\hat{Z}_{i+1} = \hat{Z}_i + \alpha_i\hat{V}_i \tag{3.6}$$

where $\hat{Z}_i$ is the recent estimated vector of projection,$\hat{V}_i$ is the search direction,α_i is the search step.

There are different algorithms while different the $\hat{V}i$ and αi are choosed. Let ith the residual vector be

$$ri=T(\hat{Z}i)-\hat{Z}i$$

$$=S^{Z}C^{I}[\ R^{Z}C^{Z}Zi + R^{W}W\ ] - \hat{Z}i \qquad (3.7)$$

If α_i is choosen to be 1,$\hat{V}i$ to be ri,then the following iterative form can be got

$$\hat{Z}i+1=\hat{Z}i + ri$$

$$=T(\hat{Z}i)=S^{Z}C^{I}[\ R^{Z}C^{Z}\hat{Z}i + R^{W}W\ ] \qquad (3.8)$$

That is, (i+1)th fixed point $\hat{Z}i+1$ is equal to the result of operator T operating on recent estimated vector of projection.The equation (3.8) is the so called successive substitution method.

The iterative algorithm of ours in detail is as follows :

1. choosing initial value $\hat{f}ij^{(0)}$

$$\hat{f}ij^{(0)} = \begin{cases} c & \text{when } (i,j)\in\Omega \\ 0 & \text{when } (i,j)\notin\Omega \end{cases} \qquad (3.9)$$

where Ω is imaging area,C is the average velocity of borehole A1 and borehole A2.

2. calculate the projection $P_{kl}^{(q)}$ when recent distribution of object function is equal to $\hat{f}ij^{(q)}$

$$P_{kl}^{(q)} = S\,\hat{f}ij^{(q)} \qquad (3.10)$$

3. complete projection

$$P = \begin{cases} C^{Z}S^{Z}\hat{f}ij^{(q)} & \text{when } (k,l)\in\Omega' \\ Wkl & \text{when } (k,l)\notin\Omega' \end{cases} \qquad (3.11)$$

where Wkl is measurable projection, Ω' is incomplete domain

4. calculate $\hat{P}kl^{(q)}$ from $\hat{f}ij^{(q+1)}$

$$\hat{f}ij^{(q+1)} = R\,\hat{P}kl^{(q)} \qquad (3.12)$$

5.set the minus object function to be zero

$$\hat{f}ij^{(q+1)} = C^{I}\hat{f}ij^{(q+1)} = \begin{cases} \max[0,fij\ ]: & \text{when } (i,j)\in\Omega \\ 0 & \text{when } (i,j)\notin\Omega \end{cases} \qquad (3.13)$$

6. caclulate the difference between measurable projection $W(t,\theta)$ and estimate of projection $\hat{Z}$ in the measurable domain,the difference is

$$\varepsilon^{(q+1)} = \left[\frac{\iint_Q [\hat{Z}^{(q+1)}(t,\theta) - \omega(t,\theta)]^2 dt\, d\theta}{\iint_Q [\omega(t,\theta)]^2 dt\, d\theta} \right]^{1/2} \quad (3.14)$$

when $|\varepsilon^{(q+1)} - \varepsilon^{(q)}| \ll 1$, iteration stops,otherwise,go to step 2, and next iteration continues until $|\varepsilon^{(q+1)} - \varepsilon^{(q)}| \ll 1$.

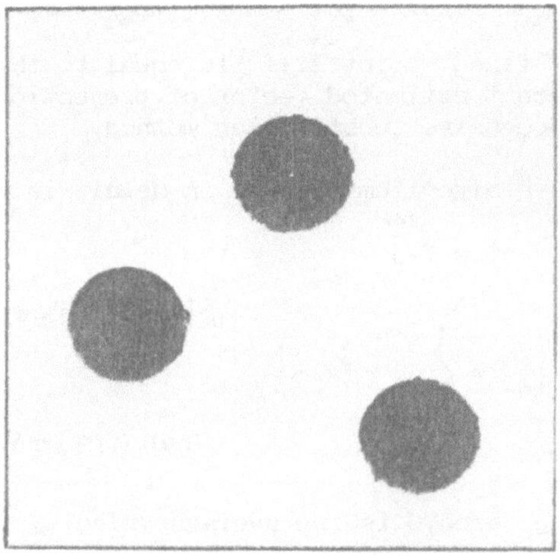

Fig.4. It is the imaging model of reconstruction.There are three circles in the background.

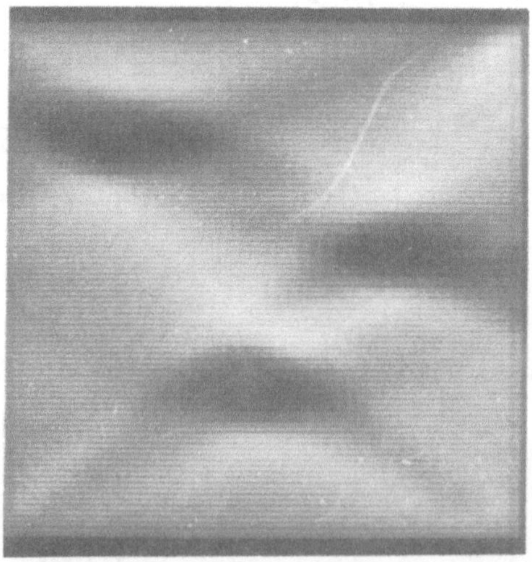

Fig.5. The image of reconstruction got by iterative reconstruction-reprojection algorithm for three iterations

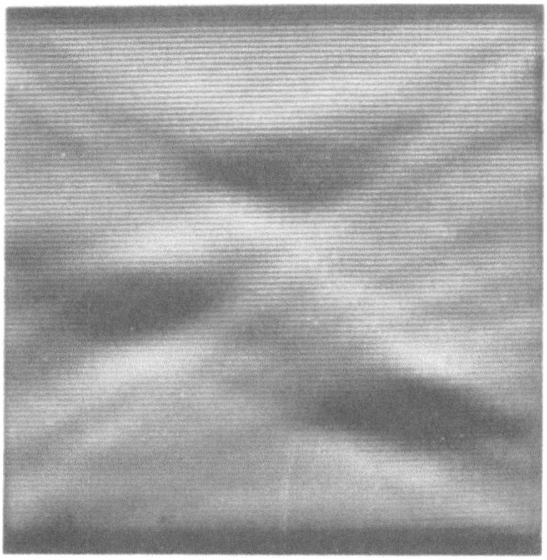

Fig.6. The image of reconstruction got by same algorithm for three iteration,but with 5 noise in data

4. The Results

The succesive substitution method was used to get reconstruction imaging between boreholes for the cross hole model. Fig.4 is the original model. The size is 30 cm X 30 cm. There are three circles in the background. The velocity in circle and in background are V1 and V2,respectively.V2/V1=1.5.

Fig. 5 is the image of reconstruction for three iterations;

Fig.6 is the image of reconstruction with 5 percent noise in data for three iterations.

5. Conclution

The image of reconstruction got by reconstruction and reprojection iteration is satisfied under assumption of straight line transmision of energy.The results show that stability and convergency of our algorithm is superior

6. References

1. J. Wong, P. Hurley, G.F. Andwest, " Crosshole Seismology and Seismic Imaging In Crystallone Rocks"Geophysics Research Letter,110,686-689,1983.
2. J.A. Forcett, and R.W. Clayton,"Tomographic Reconstruction of Velocity Monolies"Bull. Seis.Soc.Am.,74,1983
3. R.W. Clayton, and R.P. Comer, "A Tomographic Analysis of Mantle Heterogeneities From Body Wave Traval Times" Eos. Trans. Am., Geophys. Union,62,772,1983.
4. B.P.Medoff, W.R.Brody and A.Macovski, " Image Reconstruction From Limited Data" Proc. of the Internat'l Workshop on Physics Eng. In Medical Imaging,March 15-18,1982,Asilamar.CA.

5.Yu Wei, et al., " Modulation Studies of Convolution-Back Projection CT Imaging Reconstruction " Journal of Nanjing Institute of Technology,No.2,41-46,1983.
6.Koichi Ogawa, et al., " A Reconstruction Algorithm From Truncated Projections,IEEE Transactions On Medical Imaging, Vol.mi-3,No.1, 34-40, march 1984

ACOUSTICAL IMAGING OF EARTH NEAR-SURFACE INHOMOGENEITIES BY AN ECHOGRAPHIC TECHNIQUE

G.B. Cannelli and E. D'Ottavi

Istituto di Acustica "O.M. Corbino"
CNR, Rome, Italia

INTRODUCTION

The exploration of near-surface structure and inhomogeneities of the earth presents further difficulties in comparison with deeper prospectings such as those performed for oil explorations. A difficulty concerns the seismic source itself, and is due to the problem of transmitting acoustic waves into the ground according to suitably high frequencies. Another difficulty is caused by the inadequacy of common receiving devices in detecting P-waves reflected from small shallow anomalies.

When considering traditional acoustic sources for seismic prospecting, it is well known they have insufficient resolution due both to the earth's selective absorption of seismic waves in the higher frequencies band and to their intrinsic limits in generating higher frequencies waves. In fact the cut off frequency of this type of equipment is low, generally within the range 150 - 250 Hz, but most sources work at frequency values much lower than the above limit (Mac Quillin et al.,1980). Such frequencies are unusable when high resolution is required in order to detect small underground inhomogeneities.

Furthermore, receiving systems based upon the usual type of geophones don't seem suitable for imaging the shallow anomalies of the earth, although some field techniques have been proposed to enhance signal to noise ratio (Knapp,1986; Lombardi,1955). In fact surface waves, always present and amplified by geophones, mask useful signals constituted by P-waves reflected from underground inhomogeneities (Dobrin,1960; Ari Ben-Menahem et al.,1960).

The technique proposed (Cannelli et al.,1987) in the present work suggests that underground small anomalies can be resolved only by means of a properly assembled echographic system like the one here described, with high directivity and using suitably high frequencies.

This system can be considered a valid means in a wide range of applications such as archaeological exploration, location of water tables and study of characteristics and behaviour of soils in areas planned for important civil works, e.g. foundations, tunnels, dams and nuclear power plants. Furthermore, it can also be used in submarine prospecting when properly improved and insulated electrically.

ECHOGRAPHIC SYSTEM AND TECHNIQUE

The mechanical part of the echographic system consists of two similar aluminum paraboloidal transducers, with height = 52.1 cm, focal length = 3 cm and inside base diameter = 50 cm (Fig.1).

The two parts, corresponding to the transmitting and the receiving transducer respectively, are fixed on the ground or dipped into the water according to whether the prospecting is underground or under water. Their mutual distance suitably accounts for the depth of exploration.

The transmitting transducer, already described in detail in previous works (Cannelli et al., 1985, 1987), is an electroacoustic pulse source in which the seismic wave is produced by means of an electric spark-gap. For the reader's convenience, a brief description of the above transducer follows. In the transducer a high energy discharge is generated between a pair of electrodes connected to a capacitor bank, and set at the focus of the metallic paraboloid filled with a proper liquid. If the liquid is insulating (e.g. vaseline oil), the discharge is primed by a very high-voltage preliminary spark, via a third electrode between the principal ones, which ionize the liquid. Otherwise, a water - spark with an auxiliary air spark - gap can be adopted. The insulating liquid, in conjunction with a third electrode to trigger the discharge, is adopted to reduce the danger of electric shock as well as to avoid loss of electrostatic energy.

The acoustic impulse is transmitted to the soil, through a neoprene diaphragm which couples the paraboloid to the earth or directly to the water in underwater prospecting.

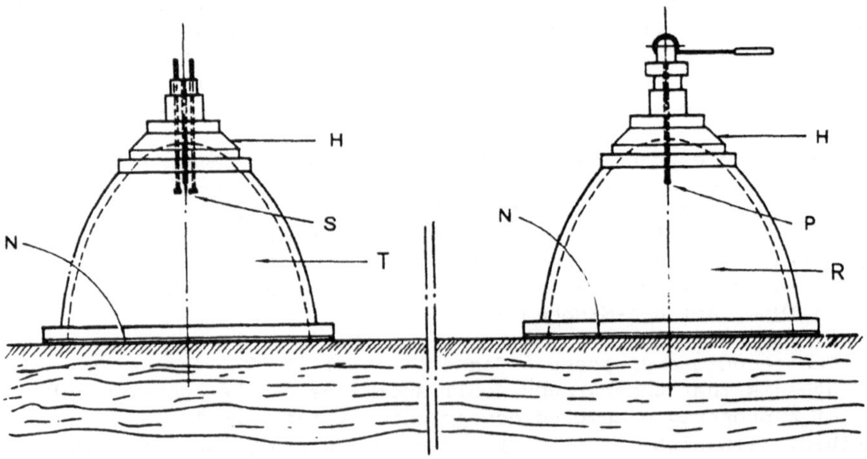

Fig.1. Schematic comprehensive view of the two paraboloidal transducers of the echographic apparatus.

T: transmitting paraboloid, S: spark - gap, R: receiving paraboloid, P: piezoelectric - ceramic, H: demountable head of the paraboloid, N: neoprene diaphragm.

A block diagram of the electronic system used to power and control the spark - gap in the transmitting paraboloid is given in Fig.2.

The discharge energy is supplied by an high - voltage generator obtained from a 12-V battery through a dc-dc converter. The generator charges a variable set of 40 μF capacitors, so that the capacitance Co can be adjusted to values within the range 40 - 360 μF, depending on experimental requirements. The maximum value of the electrostatic energy which can be stored in this prototype is about 1.1 kJ.

An important feature of the electroacoustic source is the possibility of changing the frequency spectrum of the pulse by variation of electrical capacitance. Thus, higher frequencies can be produced to resolve small near - surface inhomogeneities, or lower frequencies can be used to explore anomalies to a greater depth.

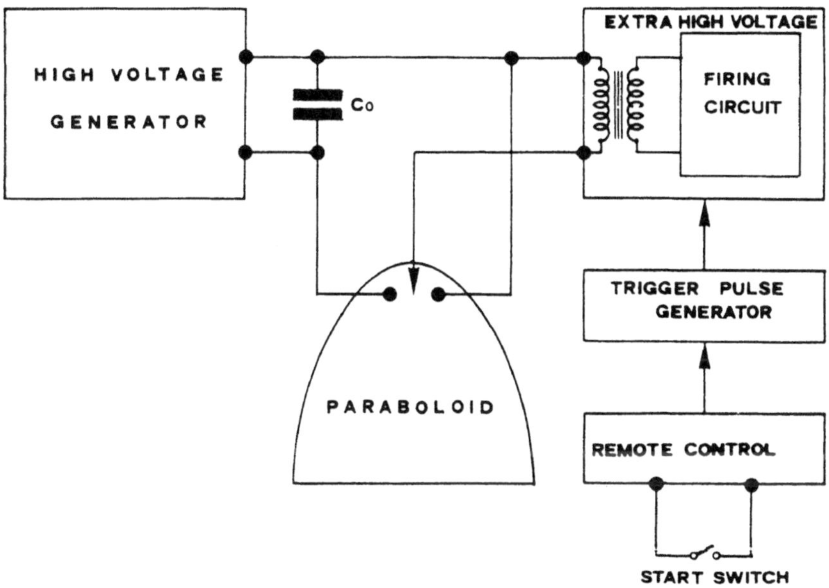

Fig.2. Block diagram of the electronic system used to power and control the spark - gap in the transmitting transducer.

A section of the head of the transmitting transducer with the mechanism for adjusting the electrodes around the focus is shown in Fig.3. The electrode bearings CC can slide across the top T of the parabolic dome by turning the knurled part of A. In this way, the position of the electrodes EE can be inched up or down with reference to the focus and a consequent change of the transducer directivity pattern is possible for suitable ratios of base diameter to wavelength.

The receiving transducer consists of a metallic paraboloid similar to the transmitting one, filled with the same liquid and placed to a suitable distance from the latter. A piezoelectric - ceramic is set at its focus to detect seismic waves.

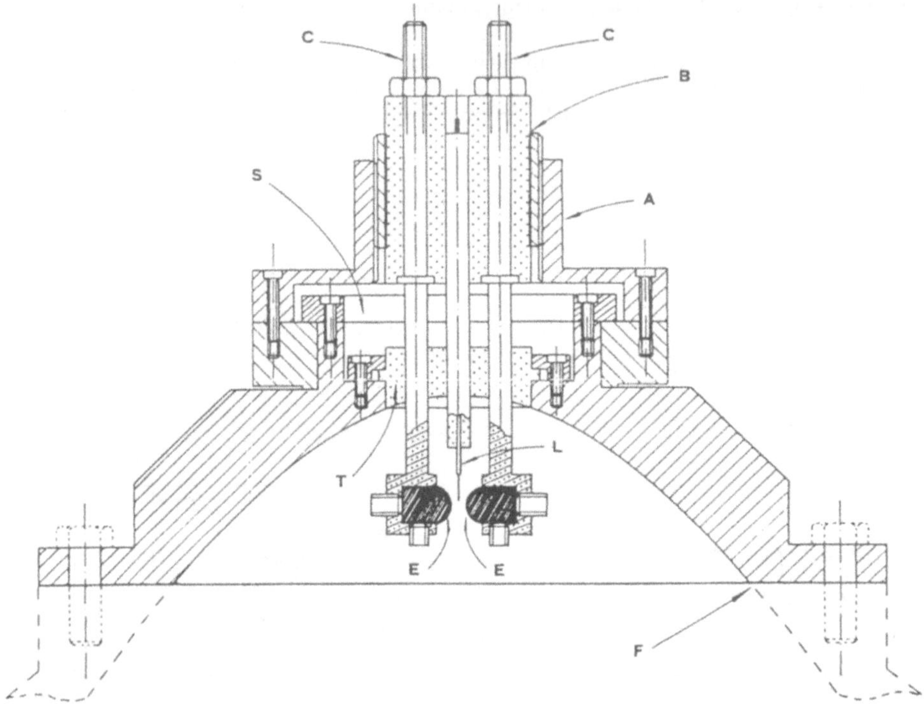

Fig.3. Section of the head of the transmitting paraboloid with the setting mechanism for the electrodes.
A: knurled ring, B: outside threaded anticorodal ring coupled to A, and inside insulating material cylinder, CC: copper bearings of electrodes, EE: principal tungsten electrodes, F: flange, L: auxiliary spark tungsten electrode, S: flowoff chamber, T: top insulating component of the paraboloid.

The head of the receiving paraboloid is shown in Fig.4. It also has a mechanism to allow displacement of the detector D around the focus in order to optimize the transducer performance. In particular, three different movements are possible. The first one is vertical along the paraboloid axis and is obtained by a device similar to that used in the transmitting paraboloid. By turning the knurled hub A around the paraboloid axis, the box coupling B can slide vertically together with the double pin joint S_1 and S_2, which fit on it and are designed to perform the other two movements. The detector bearing P, hinged on C, can move to any angle, by turning the joint S_2 around the axis bb and the joint S_1 around the paraboloid axis, as shown by the detail of the double pin joint and the view from above in the same Fig.4. The two possible rotations of P are blocked by means of the clutch - arm L and the clutch - screw V respectively.

The combination of these rotations, together with the vertical positioning, allows the piezoelectric - ceramic to sound the space around the focus along spherical dome - shaped surfaces of centre C.

The need for this mechanical adjustment is due to the fact that the position of optimal focusing of the acoustic beam, although very

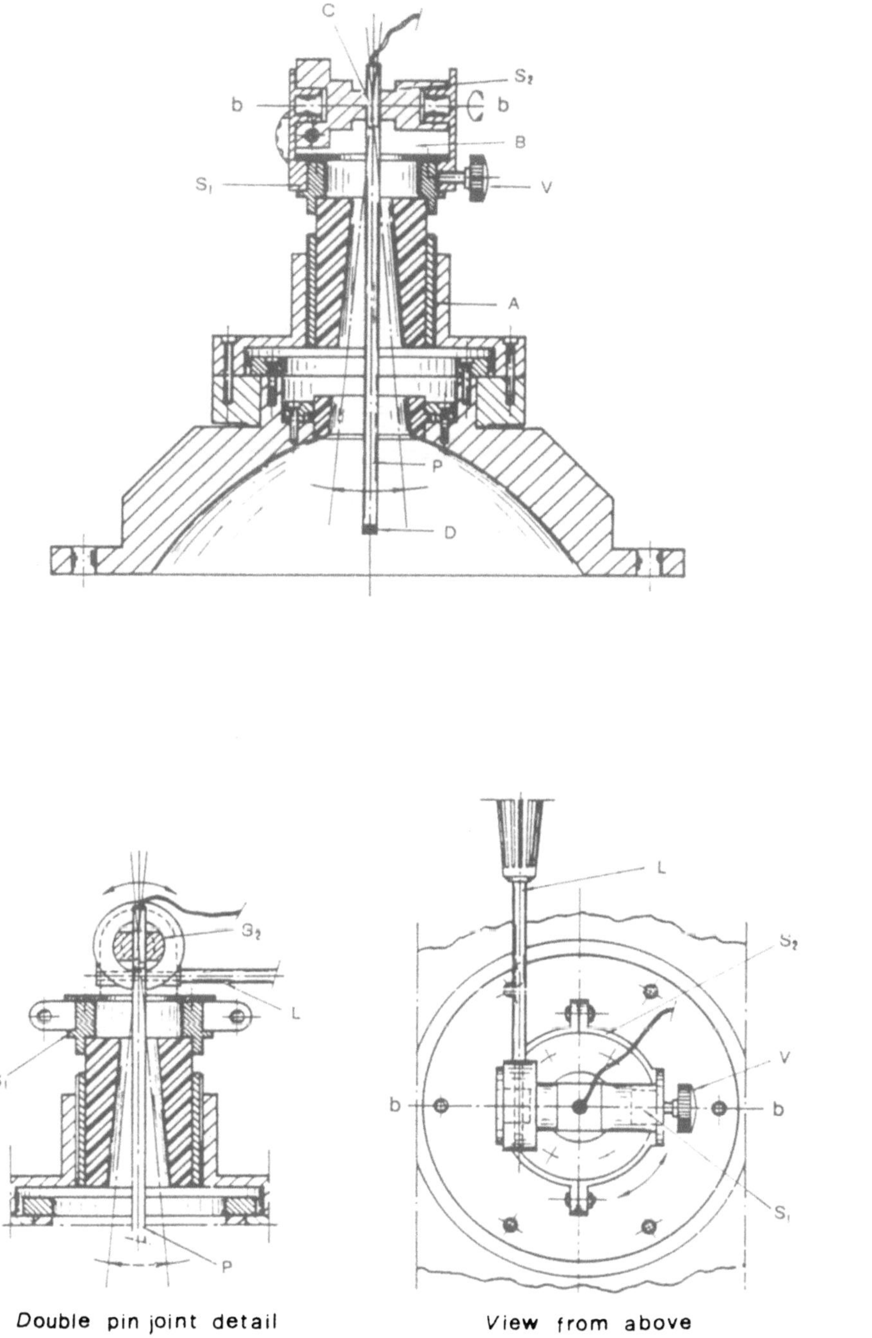

Fig.4. Section of the head of the receiving paraboloid and details of the setting mechanism for reaching the optimal focusing point.
A: knurled ring, B: box coupling, bb: rotation axis of S_2, C: rotation centre of P, D: piezoelectric- ceramic, L:clutch arm, P: detector bearing, S_1 and S_2 : pin joints, V : clutch screw.

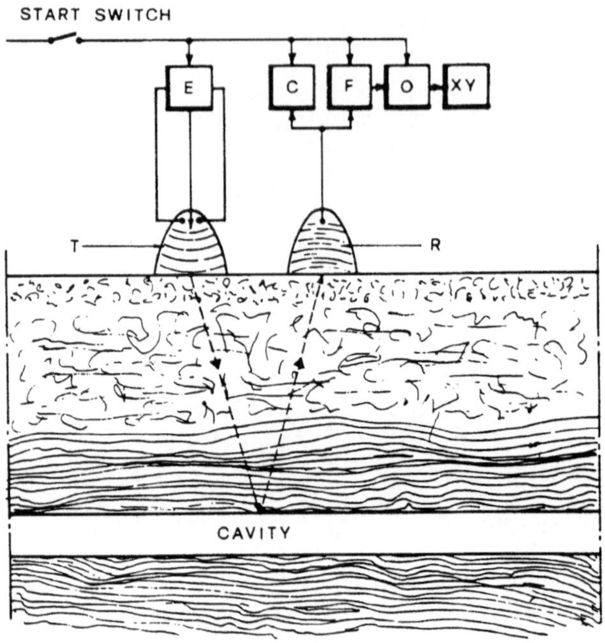

Fig.5. Schematic diagram of the complete echographic apparatus in the field.
T: transmitting paraboloid, R: receiving paraboloid,
E: electronic system to power and control the transmitting transducer, C: computer, F: filter, O: storage oscilloscope,
XY : paper graphical recorder.

close to the theoretical focus, does not coincide necessarily with it, but often depends on the particular subsoil structure being explored.

A schematic diagram of the complete echographic system in the field is shown in Fig.5. The acoustic wave, generated by the transmitting paraboloid T , travels in the subsoil as far as the earth's irregularity (e.g. a ground-cavity interface), from which it is reflected and sent back to the surface where it is detected by the piezoelectric-ceramic set at the focus of the receiving paraboloid R. The piezoelectric-ceramic can detect signals in the frequency range 0.1 - 125 kHz with accuracy 2 dB. The seismic signal is then passed through a filter F and sent to a computer C for preliminary processing. Also, the signal can be shown on a control oscilloscope O and plotted on a XY paper recorder.

EXPERIMENTAL TESTS AND RESULTS

Some prospectings were carried out in an archaeological area of Roma where various near-surface inhomogeneities are present in the form of cavities and tunnels. First a test in the field was performed to make sure of the presence of suitable high-frequencies in the seismic signal, by checking underground the signature and the spectrum of the acoustic wave generated by the transmitting transducer.

The signal was detected by means of an hidrophone placed at a depth of 3 m on the paraboloid axis. A schematic diagram of this experiment is given in Fig. 6.

The hidrophone is kept in water inside a hollow at the opened end of the pipe which conveys water from a surface tank.

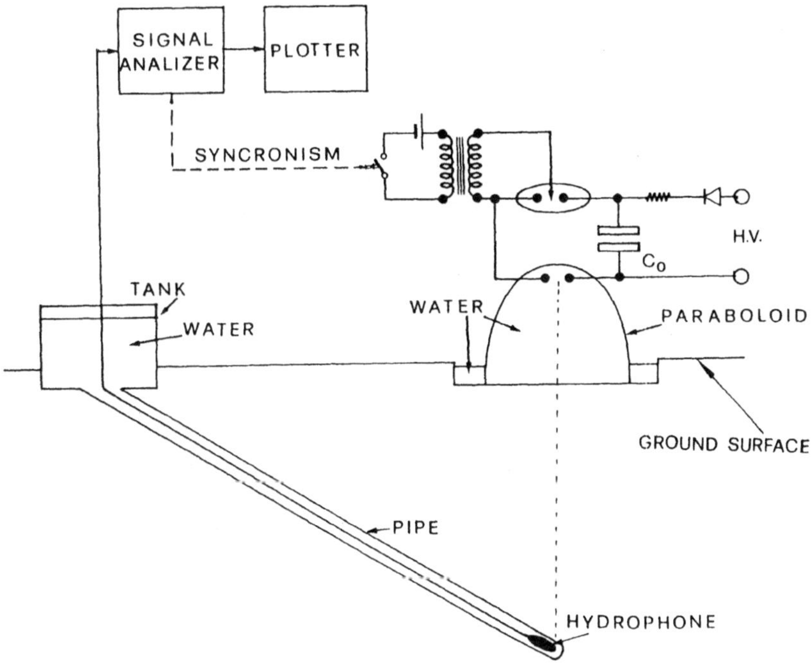

Fig.6. Schematic diagram of the experiment for picking up the the seismic signal at a depth of 3 m underground.

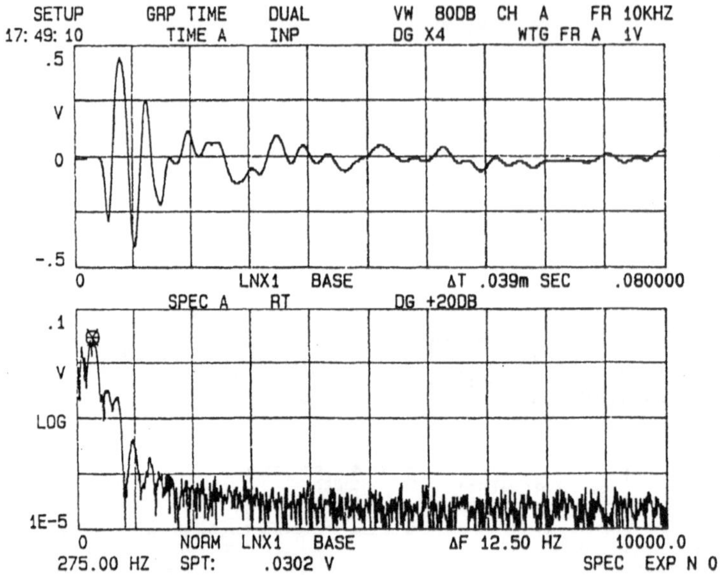

Fig.7. Acoustic pulse generated by the paraboloidal source at a depth of 3 m underground.
Upper diagram: time domain
Lower diagram: frequency domain

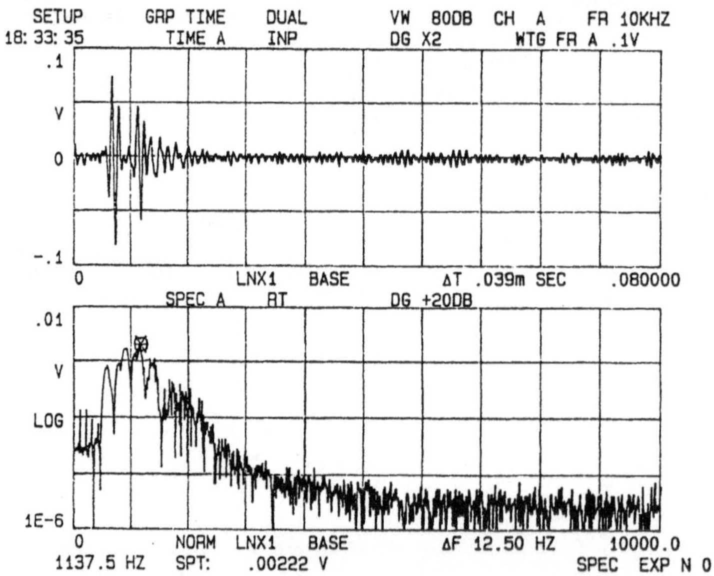

Fig.8. The same signal of Figure 7 passed through a 1000 - 2000 Hz band - pass filter.

The above signal both in the time and frequency domains is shown in Fig.7. Figure 8 shows the same signal passed through a 1000 - 2000 Hz band - pass filter.

The use of the frequency band 1000 - 2000 Hz allowed detection of a cavity as shown by the seismogram records of Fig.9. This prospecting was carried out above a tunnel between 3 and 5 meters deep.

A linear scanning was made, by insonifying the soil every 1.5 m and moving both paraboloids from the initial position above the cavity out to 6 m sideways. The optimal distance between the centre of the transducers was found to be about 1 m in this case. The scanning was normal to the axis of the cavity. A maximum of the reflected signal amplitude is clearly visible in the initial position above the cavity. Gradually decreasing amplitude are shown by the seismogram records when the paraboloids are further away from the cavity axis. At a distance of 4.5 m from the cavity axis the seismic signal is already absent.

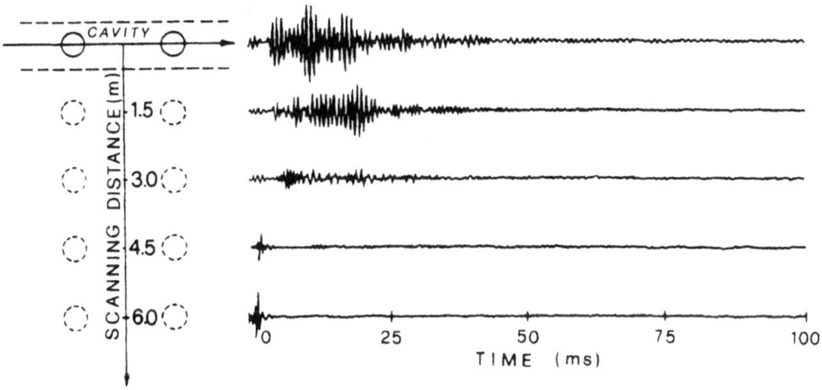

Fig.9. Amplitude versus time, for the acoustical signal reflected from the cavity as a function of the scanning distance.

Another test was made on a larger cavity about 3 m, deep in order to compare the receiving performance of the paraboloidal transducer with that of traditional geophones. First, the vertical axis of the cavity was prospected by insonifying the ground with the maximum energy (about 1000 joule) available on the transmitting source and by receiving with the paraboloidal transducer. Then, the latter paraboloid was replaced by 12 high-frequency geophones arranged along the circumference of the spot corresponding to the missing paraboloid and a successive acoustic pulse was sent into the ground in the same experimental conditions. Also in this case a 1000 - 2000 Hz band - pass filter was set in order to cut off surface waves.

The results are shown in Fig.10 which compares the responses of the echographic system and the geophones. While the paraboloidal transducer exhibits a seismogram (e) clearly attributable to P-waves reflected from cavity interface, the geophone system shows only a faint signal (g) due to electrical noise. The ringing evident in the seismogram (e) is very likely due to multiple reflections of the acoustic wave on the cavity interface which in this case is alongside the soil surface.

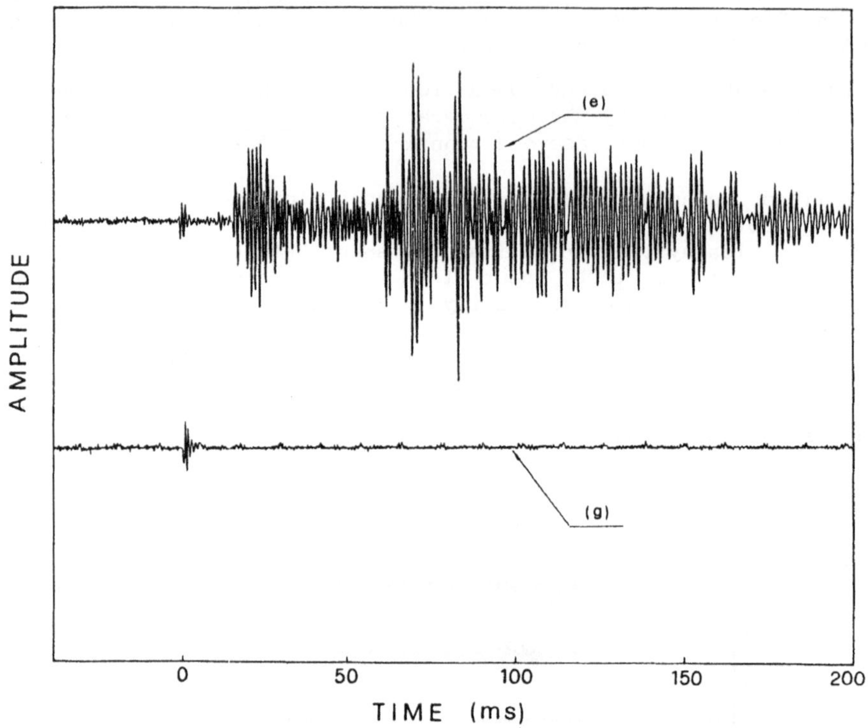

Fig.10. Comparison of the responses of the echographic apparatus (e) and geophone system (g) on a cavity, in the same experimental conditions.

CONCLUSIONS

Some important features characterize the present echographic system which can be considered a suitable means for high-resolution seismic prospecting. The system can generate wide frequency band pulses, with a variable frequency spectrum which can be adapted to different prospecting situations. This allows lower frequencies or higher frequencies to be used according to the need to explore anomalies to a great depth or near the surface.

Both the transmitting and the receiving paraboloidal transducers are equipped with suitable mechanisms to modify spark-gap and piezoelectric-ceramic positions around the respective foci, in order to optimize their performance. In particular, the spark electrode positioning in the transmitter makes it possible to change the acoustic beamwidth emitted, for suitable ratios of base diameter to wavelength.

In the receiving transducer, a mechanical device allows the piezoelectric - ceramic position to be changed near the focus, until optimal focusing of the seismic wave is achieved.

Optimal performance of the entire apparatus was only possible by a properly designed and assembled pair of paraboloidal transducers just like those described herein, since common reception devices, such as geophones, do not allow the P-wave reflected from underground anomalies to be distinguished from the prevailing surface waves.

The precise time control possible for this type of equipment allows a further interesting facility such as electronic beam forming and scanning when an array of these systems is employed in a seismic prospecting line.

The experimental results shown in the present paper, in conjunction with those reported in previous works by the same authors, prove that the novel seismic technique and the apparatus proposed can be used as well on land as in underwater prospecting, and with special interest can be applied in exploration of near surface zone where traditional technique fail.

ACKNOWLEDGMENTS

The authors wish to thank L.Pitolli for his technical assistance in measurements, P.Rossi, A.Lacchè and L.Birarelli for the collaboration in the field.

REFERENCES

Ari Ben-Menahem and Sarva Yit Singh,1981,Seismic Waves and Sources, Springer,New York.

Cannelli G.B.,D'Ottavi E. and Santoboni S.,1985, Shallow prospecting on land by means of a novel electroacoustical P-wave pulse generator, in: "Acoustical Imaging", Plenum,New York.

Cannelli G.B.,D'Ottavi E. and Santoboni S.,1987, Electroacoustic pulse source for high-resolution seismic explorations, Rev. Sci. Instrum. 58 (7).

Cannelli G.B. and D'Ottavi E., 1987, Italian National Patent n.48694-A/87 (11.12.87) and International Patent in progress.

Dobrin M.B.,1960, Introduction to Geophysical Prospecting, Mc Graw Hill, New York.

Knapp R.W., 1986, Geophone differencing to attenuate horizontal propagating noise, Geophysics, 51 (9).

Lombardi L.V.,1955, Notes on the use of multiple geophones, Geophysics, 20 (2).

Mac Quillin R., Bacon M. and Barclay W.,1980, An Introduction to seismic Interpretation, Graham and Trotman Ltd.,London.

GHOST AND SHADOW ZONE IN ULTRASONIC SECTOR SCAN B-MODE IMAGE

H. Yamada*, M. Nakamura*, Y. Tomikawa**, and M. Numata***

*Department of Electrical Engineering, Nihon University
**Department of Electrical Engineering, Yamagata University

1. Introduction

Recently, the B-mode image presentation based on the ultrasonic pulse echo method has been used widely in medical diagonosis.
This presentation does not necessarily show a true shape of a target, because of its assumption that the ultrasonic velocity and attenuation in the interior of media are constant. In practice, however the media are not uniform and cause various phenomena through reflection and refraction. This problem has not yet been fully analyzed.(1)(2)(3)
Consequently, we have studied various problems associated with ultrasonic fan beams striking on the surface of cylindrical double layers.
We successfully confirmed the occurrence of ghost and the existance of shadow zone, similarly to seismic one, by means of computer simulation and experiments.

2. Resolution and Distortion with respect to Images of Defects behind parallel Double layers in ultrasonic B-mode image using fan beam

2-1 Relationship between Ultrasonic Beam Ray and Image Response

The reflections and refractions at the boundory between two media with different acoustic impedances are well known, as analyzed by the ray theory.
We have simulated the distortion and resolution due to reflection, refraction and beam spread. A target reflector of square mesh was used. Square meshes were conceived to assess distortion in the center of the beam. In addition, resolution was assessed on the basis of whether the lines forming the meshes overlap one another.
The following conditions have been assumed for the simulation.
The parameters are velocity and thickness of parallel dual layer medium.
Fig.1 shows results of the above simulations. Fig.2 shows results of both the above simulations.

3. Singular Phenomena occurring when Ultrasonic Fan Beams strike concentric double layer

3-1 Analysis of Rays

We studied the propagation path of beams, whereby an ultrasonic fan beam strikes the medium with a sound velocity of C_1 (the first medium) from point P, in which a cylindrical object (the 2nd medium) with a radius of r

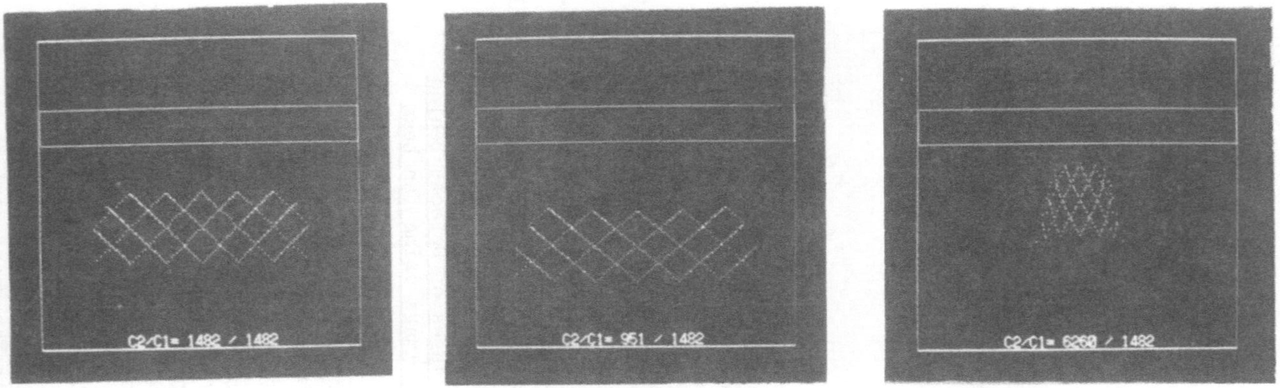

(a) Velocities change with respect to the liquid and solid layer and thickness of solid layers(t) is 30mm constants.

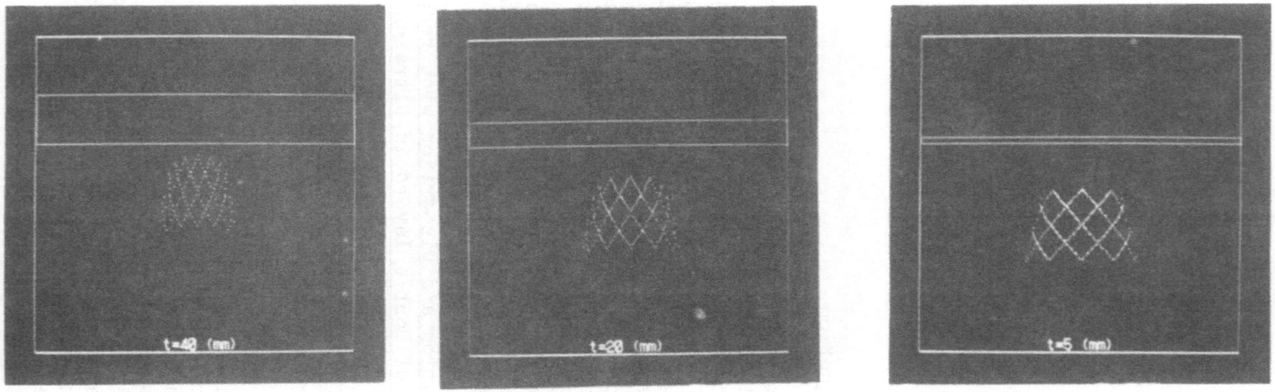

(b) Thickness changes, velocity ratio(n) is 3.97 constants, (5880/1482 (m/s) by applying the constants for water and steel.

Figure 1. Typical example of the distortion and resolution.

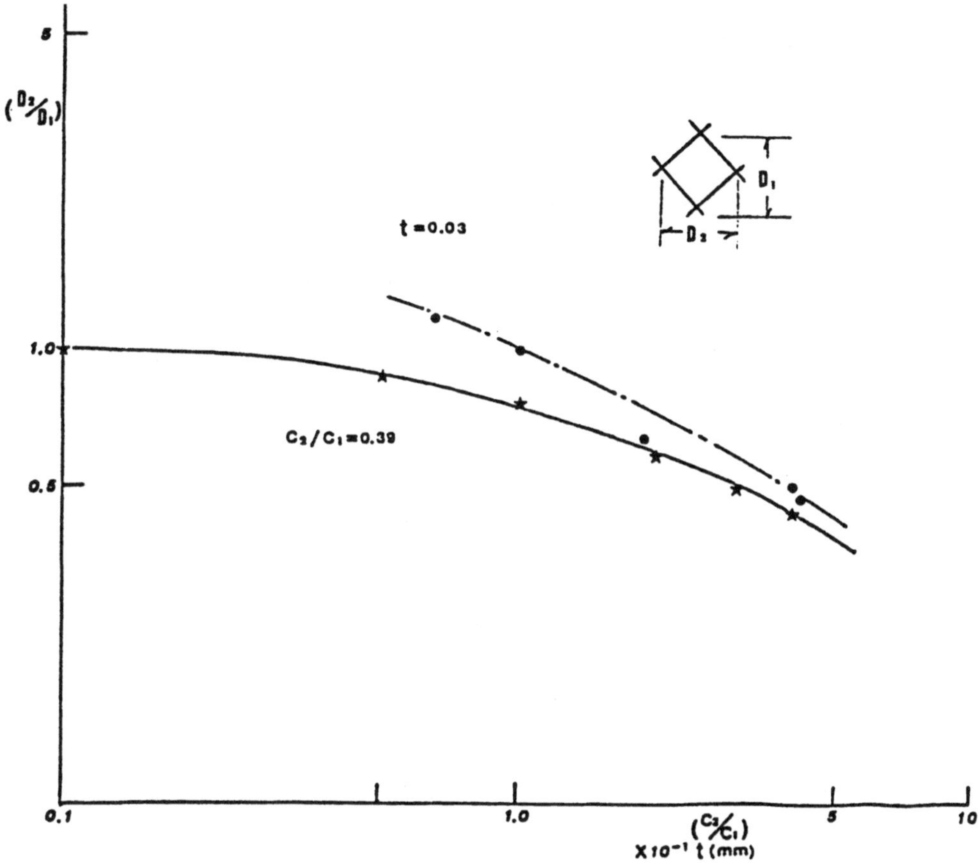

Figure 2. Results of the distortion obtained from the simulation with the velocity ratio(n) and solid layer thickness(t) as parameters.

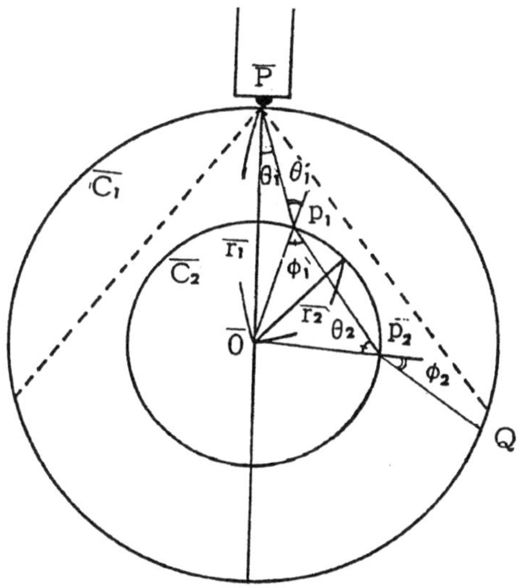

Figure 3. The schematic representation of the propagation path of beams and notation.

and a sound velocity of C_2 is present, as shown in Fig.4 .
The beams, having struck the surface of the 1st medium from point P at angle of θ_1' strike the surface of the 2nd medium at an angle of θ_1.
The beams refract at points p_1, and p_2, leading to point Q.
The refraction angles at points p_1, and p_2 and ϕ_1 and ϕ_2 respectively, are sought by using Snell's Law.
Therefore, we conducted simulation taking the different velocity ratio of each medium into consideration.

3-1-1 Where the sound velocity C_2 of the 2nd medium is lower than that of the 1st medium

Where the sound velocity C_2 of the 2nd medium is lower than that of the 1st medium, point Q (reached by the beam) comes to the right side of the circle in some cases and to its left side in others, as shown in Fig.4.(a),(b).
In Fig.4.(a), the ultrasonic beam striking the surface of the 1st medium from point P strikes the surface of the 2nd medium at an angle of θ_1' at point p_1 and is refracted at an angle of ϕ_1 .
The angle of ϕ_1 at which the ultrasonic beam is refracted is smaller than the incidence angle of θ_1', the inclinations of p_1 and p_2 are greater than that of P and p_1 and the ultrasonic beam further advances at an angle smaller than the incidence angle of the beam.
At p_2 the refraction angle of ϕ_2 is larger than θ_2 . Thus, the beam from point p_2 advances further to point Q (to the right side) at an angle smaller than that at which the beam passes through the 2nd medium.
Fig.4.(b) shows the case where point Q (reached by the beam) comes to the left side of the circle.

3-1-2 Where the sound velocity (C_2) of the 2nd medium is faster than that of the 1st medium

Where the sound velocity (C_2) of the 2nd medium is faster than that of the 1st medium, its conceivable propagation paths are as shown in Fig.5.(a) and (b).
In such a case, the refractions at point p_1 and p_2 are greater than the incidence angle and the beam spreads out as it advances, as shown in Fig.5.(a).
Where the incidence angle (θ_1') of the beam becomes somewhat larger, the beam is totally reflected and reaches point Q, as shown in Fig.5.(b) .

3-2 Simulation Method and Results

With the radii and sound velocities of the 1st and 2nd media determined, the propagation paths of the beam of light vertically striking the surface from point P were simulated.
The simulated propagation paths are shown in Figs.6 through 9.
Figs.6 and 7 show portions in which the directions of the beams of light striking the surfaces are reversed left to right and vice versa on arrival (a) and where the beams are focused at a single point, as shown in (b). The images resulting from these phenomena are such that the image of substances located in portion (a) are reversed left to right and vice versa. Namely, ghosts are formed. The images of substances in portion (b) are larger than actual size. The portion marked "Y", which is the center of the experiment, corresponds to portion (a). Fig.8 shows the case where, because of the presence of the 2nd medium, the portion behind it, shown as portion (c), is not struck by the beams.
For portion (c), therefore, no image is presented. Portion (c) is hence called a "Shadow Zone".
In Fig.9 C_2 is greater than C_1 and the beams spread out as they advance, and when their angles of incidence exceed a given value, they are totally reflected.

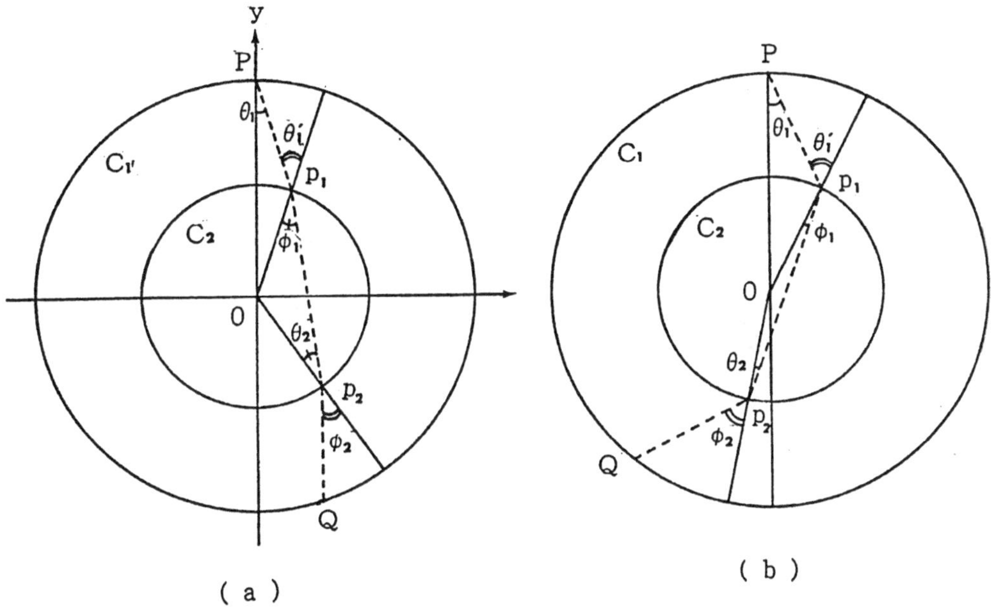

Figure 4. The sound velocity C_2 of the 2nd medium is lower than that of the 1st medium. (C_1), $(C_1 > C_2)$

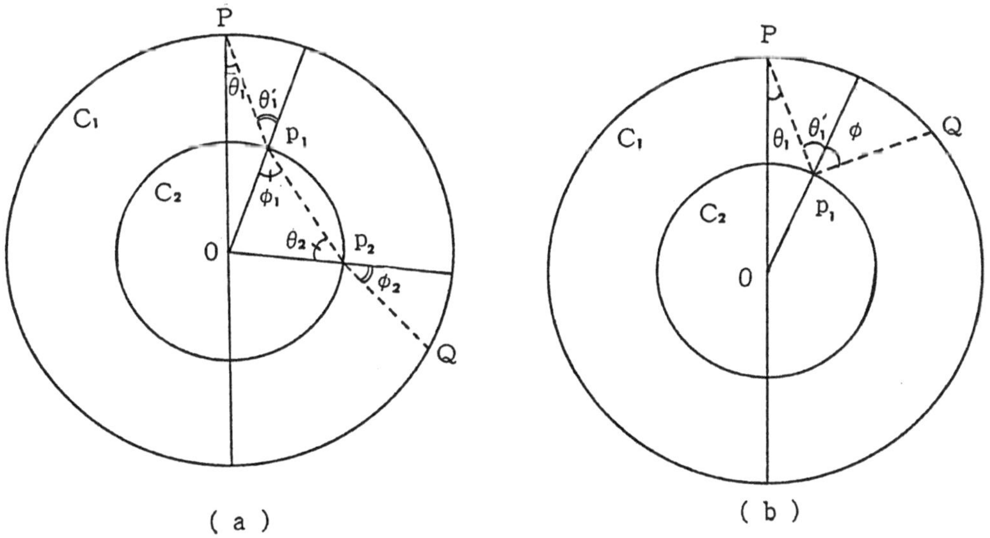

Figure 5. The sound velocity C_2 of the 2nd medium is faster than that of the 1st medium. (C_1), $(C_1 < C_2)$

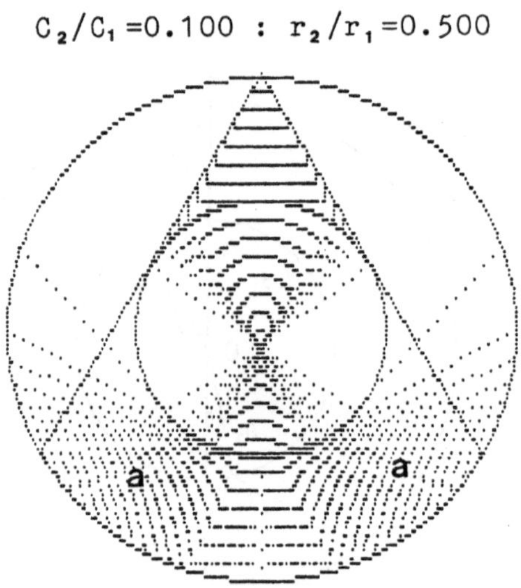

Figure 6.
The simulated propagation paths. Portion (a) is reversed left to rigth and vice versa.

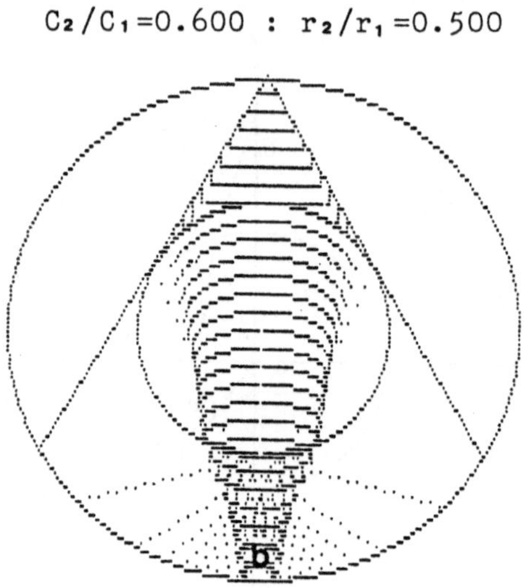

Figure 7.
Portion (b) is focused at single point.

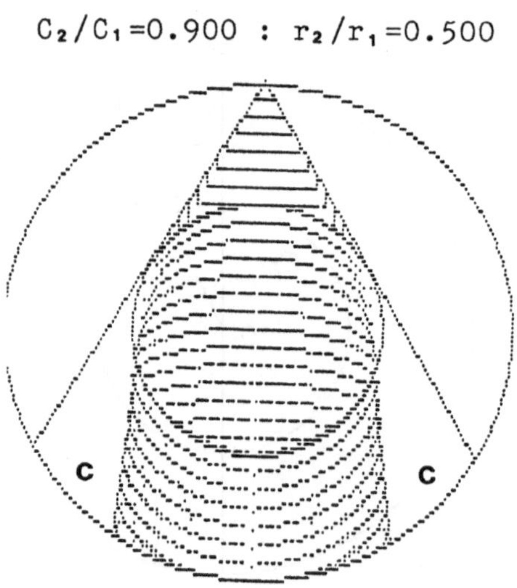

Figure 8.
Portion (c) is shadow zone.

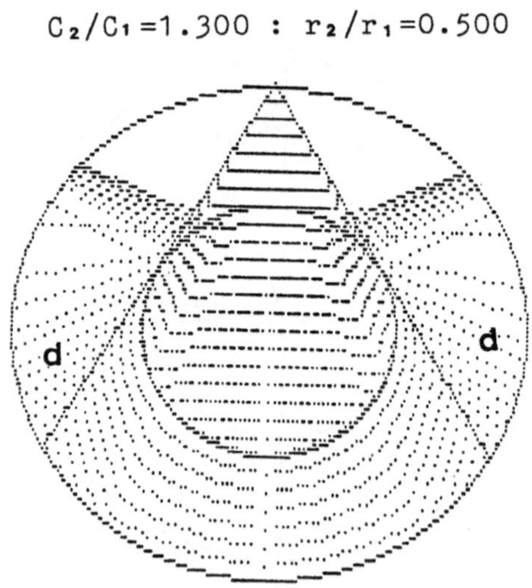

Figure 9.
Portion (d) total reflection and hence the object in this region is doubly displayed.

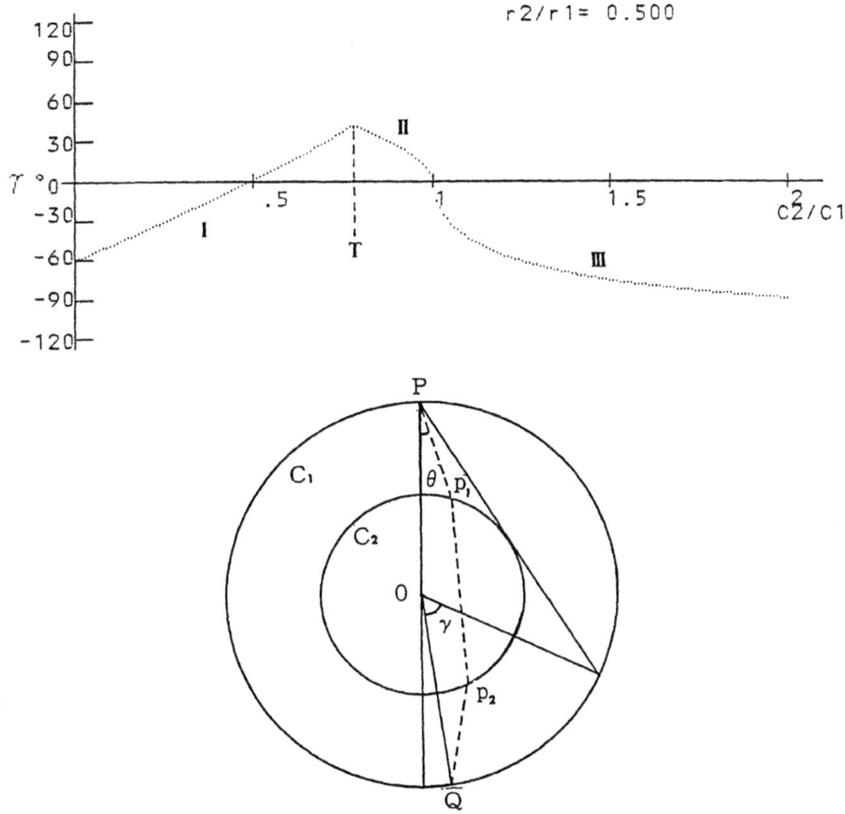

Figure 10. The angle of the shadow zone simulated changing the velocity ratio(n) from 0.01 to 2.00

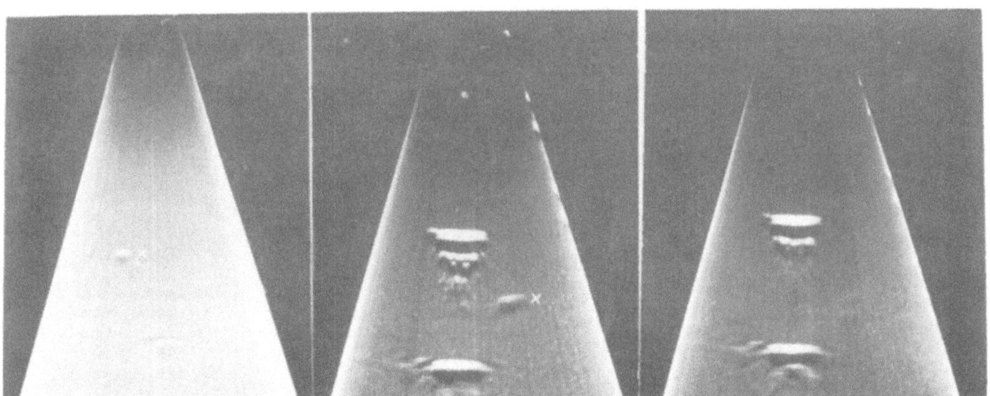

Figure 11. Results of these experiments.

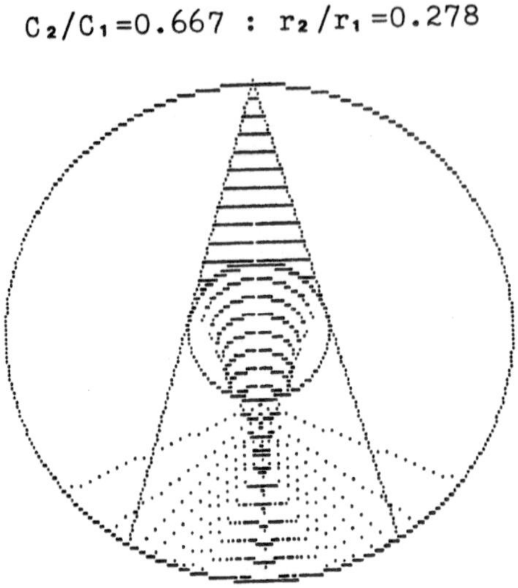

Figure 12. Simulation results under the same conditions as experiments.

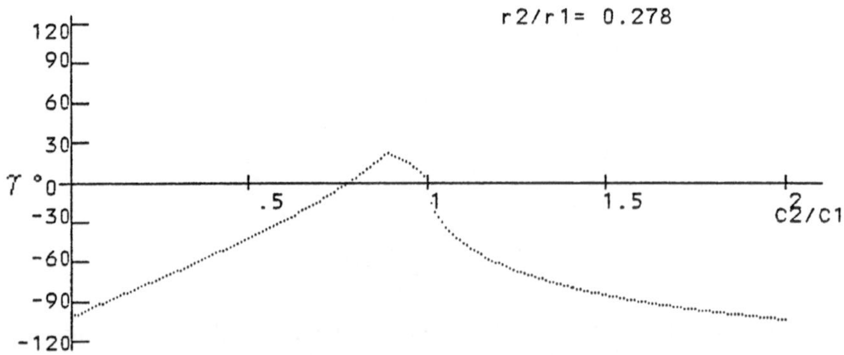

Figure 13. Results of the angle of the shadow zone, where the radial ratio of (r_2/r_1) is 0.278 and velocity ratio (C_2/C_1) is changed.

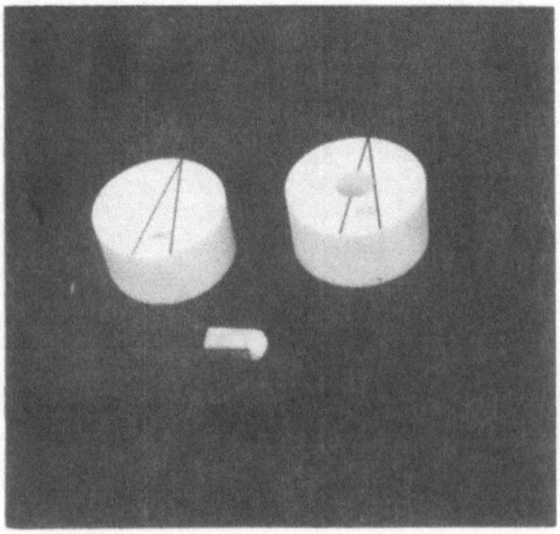

Photo 1.
Whole experimentation system.

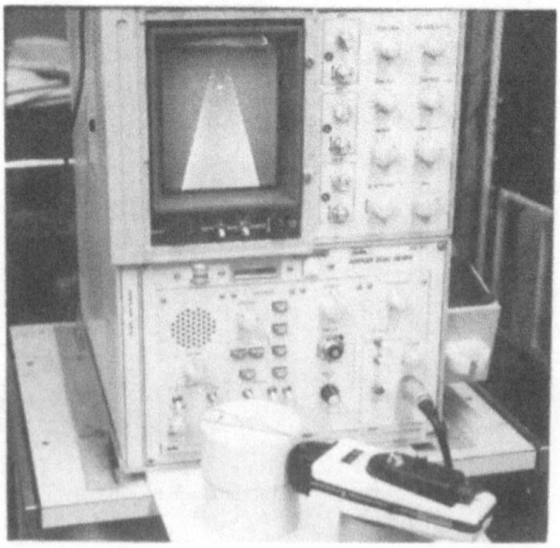

Photo 2.
Enlarged picture of the phantom

Overlapping parts of beams(d) appear and hence the object in this region (d) is doubly displayed.

4. Occurrence of Singular Phenomena and Confirming Ghosting in the B-mode Image

4-1 Development of Shadow Zone

Concerning the development of a Shadow Zone, using the radii given for the 1st and 2nd media, the angle of the Shadow Zone was sought while changing the velocity ratio (n) from 0.01 to 2.00. The results are shown in Fig.10 .
Where a Shadow Zone did not develop, the portion protruding from the tangent line from point P of the 2nd medium is assumed to be at a negative angle. To analytically seek the Shadow Zone, based on the radial ratio of (r_2/r_1) and with (C_2/C_1) as a variable, it can be divided into the following three portions and formulated as follows:

$$(\mathrm{I})\qquad \gamma = 4\sin^{-1}\left(\frac{r_2}{r_1}\right) - 2\cos^{-1}\left(\frac{C_2}{C_1}\right)$$

$$(\mathrm{II})\qquad \gamma = 2\cos^{-1}\left(\frac{C_2}{C_1}\right)$$

$$(\mathrm{III})\qquad \gamma = \sin^{-1}\left(\frac{C_1}{C_2}\right) - \sin^{-1}\left(\frac{r_2 C_1}{r_1 C_2}\right) - \cos^{-1}\left(\frac{r_2}{r_1}\right)$$

$$T = (c_2 / c_1) = \frac{\sqrt{r_1^2 - r_2^2}}{r_1}$$

In Fig.10, the scope (III) in which the velocity ratio is more than 1 corresponds to the case where C_2 the sound velocity of the 2nd medium, is faster than that of the 1st medium. Notably, also in the scopes (I and II) where the velocity is less than 1, the beams at the boundary on which a Shadow Zone is formed are not necessarily the same. Especially, in portion (I), there are cases where the angle becomes negative and also where it becomes positive.

4-2 Experiments to Confirm Ghosting in the B-mode Image

The image will be formed, as predicted by the results of the simulation, with the left and right sides reversed and the distance to the defect concerned being indicated as longer than it actually is.
To confirm the above, test samples of the same shape, containing foreign matter and those of different shapes were prepared and subjected to experiments using a Doppler Echo Graph (SSD 910 Type).
As a specimen, a cylindrical piece of synthetic resin (ABS resin) of 45 mm radius (r_1), with a water-filled hole having a radius (r_2) of 12.5 mm in its center, was used.
The sound velocity (C_1) for synthetic resin was taken as 2220 m/s and that for water (C_2), as 1480 m/s.
The experiments were conducted by using as reflector, i.e. a hole of 2.5mm radius drilled at a distance of 72 mm from a point struck by a beam coming in at an angle of 8° from point P.
The results of these experiments are shown in Fig.11. (a) through (c).
As may be seen from Fig.11.(a) through (c), the images of the defective portions in Fig.11.(a) and (b), marked X, show the left and right sides reversed. Fig.11.(c) is a photo of the defective portion shown in Fig.11.(b), into which water has been injected. Namely, the defective portion has become invisible.

Simulation result under the same conditions is shown in Figs.12 . Fig.13 shows the angle of the Shadow Zone, where the radial ratio of r/r is 0.278(=12.5/45) and the velocity ratio is changed. Photo 1 shows the whole experimentation system and Photo 2 shows an enlarged picture of the phantom to be checked.

5.Conclusion

We have attempted to substantiate the analysis concerning the propagation of ultrasonic beams through multilayer mediums, as shown in multilayer fan beam B-mode images by simulation based on the ray theory and phantom experimentations and have come to the following conclusion:
(1) A Shadow Zone can occur when fan beams strike the surface of a circular medium.
(2) A circular medium, on which the beams of light have the effect of a convex lens and converge or focus on a single spot, can be substantially deformed.
(3) Where the ultrasonic velocity ratio of the multilayer medium is small, the image may have its left and right sides reversed.
The occurrence of ghosting in the B-mode image has been confirmed by experiments using phantoms.
(4) The deformation of circular multilayer media due to fan beams striking their surfaces, associated ghosting and Shadow Zones can be presented by repetition using circular dual media and have thus been solved analytically.

References

(1) K.Soetanto,S.Ohtsuki,and M.Okujima;
"Shadow and distorted image behind a cylinder in the ultrasonic echogram",IEICE,US 83-27,pp.37-44,step.1983.(in japanese)
(2) M.Okujima,S.Ohtsuki and K.Soetanto;
"Echogram distortion behind cylinder by ultrasonic refraction", Proc. of JSUM,41-PE-27,pp.575-576,Dec.1982.(in japanese)
(3) K.Soetanto,K.Y.Zhou,S.Ohtsuki and M.Okujima;
"Ultrasonic tomograms of a liquid cylinder in scattering medium", Third International Congress on the Ultrasonic Examination of the Breast,No.61,pp.77,April 1980.

HOLOGRAPHIC SONAR USING ORTHOGONAL TRANSMITTING PULSES

Yasutaka Tamura and Takao Akatsuka

Department of Information Engineering
Yamagata University
Yonezawa 992, Japan

INTRODUCTION

The aim of this work is to develop an acoustical imaging system which permits high speed data acquisition with simple hardware. In the previous works [1,2], we have introduced the holographic sonar using transmitting pulses modulated by the system of orthogonal functions. The imaging system transmits uncorrelated pulses from multiple transmitters simultaneously, and detects the echo wave-front by the array of receivers. With the recorded waveforms of echo, beam-forming is accomplished indirectly.

Using the method, high speed data acquisition is possible since multiple sharp beams are synthesized with a single transmitting and receiving process. The complex beam-former using analog delay-lines is replaced with the digital signal processing system. The range resolving power is determined by the total band-width of the transmitting pulses. Hence, we can use the functions such as Walsh functions or frequency sifted sinusoidal waves for the modulation; the time-bandwidth product of received signal and the complexity of the system are reduced. We have demonstrated that the system has high spatial resolving power and wide view angle with a small number of elements.

In this paper, we report the 3-D holographic sonar which has circular array and the system of Walsh functions for the modulation. A high speed algorithm for the image reconstruction using the fast Walsh transform (FWT) is proposed.

Computer simulations and experimental results using airborne sensor array show the ability of the method.

PRINCIPLE OF BEAM-FORMING

Transmitting wavefront

The geometry for the beam-forming system is illustrated in Fig.1. The coaxial circular array which consists of N

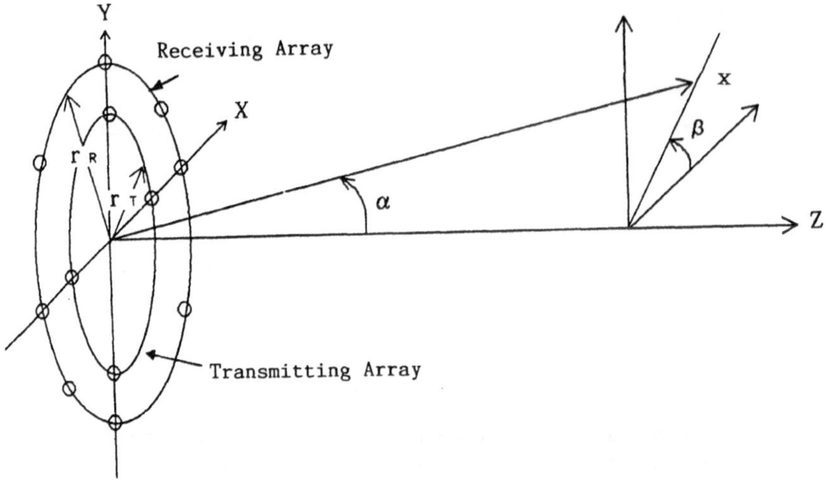

Fig.1 The geometry of the array.

transmitters and M receivers is on the x-y plane. The radii of the transmitting and receiving arrays are r_T and r_R respectively. The k-th transmitter is at the position ν_k and its coordinates of angle is ϑ_{Tk}, and i-th receiver μ_i and ϑ_{Ri}. Here, we assume that each element has omni-directional characteristics, the wave propagation speed C is constant and homogeneous in the medium, and the image plane is restricted in the paraxial region of the array.

The carrier signals of frequency f_o are multiplied by modulating sequences $m_k(t)$ of period T, (k=1, 2, ... ,N), and the modulated signals are fed into the transducers. The driving signals for the k-th transducer at the position ν_k is

$$u_k(t) = m_k(t)e^{j2\pi f_o t} \tag{1}$$

Here we assume that $u_k(t)$ k=1,2,...,N are orthogonal to each other, and the following equation holds:

$$\varphi_{ij}(\tau) = \int_{-\infty}^{\infty} u_i(t)u_j(t-\tau)^* dt \tag{2}$$

$$= \begin{cases} \varphi_{ii}(\tau) & \text{for } i=j \\ 0 & \text{for } i \neq j \end{cases} \tag{3}$$

where φ_{ij} and φ_{ii} are the cross-correlation functions and the auto-correlation functions of the transmitting pulses. The orthogonal transmitting pulses construct the wavefront of omni-directional energy propagation. Furthermore, short duration time of the pulses is preferable from the view-point of high speed data acquisition.

<u>Procedure of beam-forming</u>

The transmitting wavefront is reflected by objects, and the field is detected by the circular receiver array on the x-y plane and digitized. Suppose the receiving signal detected by the i-th receiver at the position μ_i is $r_i(t)$, then we can obtain the output $s(\mathbf{x})$ of the beam-former corresponding to the position $\mathbf{x}$ by the following operation:

$$s(\mathbf{x}) = \left| \sum_{i=1}^{M} \int_{-\infty}^{\infty} r_i(t) \sum_{k=1}^{N} u_k(t-\tau(\nu_k;\mathbf{x})-\tau(\mu_i;\mathbf{x}))^* dt \right|^2$$

$$= \left| \int_{-\infty}^{\infty} \sum_{i=1}^{M} r_i(t+\tau(\mu_i;\mathbf{x})) \sum_{k=1}^{N} u_k(t-\tau(\nu_k;\mathbf{x}))^* dt \right|^2 \tag{4}$$

where

$$\tau(\nu;\mathbf{x}) = \frac{|\mathbf{x}-\nu|}{C} \tag{5}$$

is the delay time due to the path of the element at ν to a position $\mathbf{x}$ in the wave field. For the position in the paraxial region, this equation is approximated as

$$\tau(\nu;\mathbf{x}) = \frac{(|\mathbf{x}|^2+|\nu|^2-2\mathbf{x}\cdot\nu)^{\frac{1}{2}}}{C} \tag{6}$$

$$\approx \frac{R+\dfrac{r^2}{2R} - \alpha\cdot r\cos(\beta-\vartheta)}{C} \tag{7}$$

where, (R,α,β) and (r,ϑ) are the polar coordinates of the point in image space and the element in the x-y plane as illustrated in Fig.1.

The operation (4) means that the image of the objects $s(\mathbf{x})$ is calculated by matching operation between the received echo field and the assumed one from a point target at the position concerned.

Spatial resolution

The point spread function (PSF) of the beam-former is the response of the imaging system to noise-free echo signals caused by a point target. Generaly, the PSF is given as follows:

$$h(\mathbf{x};\mathbf{x}_o) = \sum_{i=1}^{M} \int_{-\infty}^{\infty} \hat{r}_i(\mathbf{x}_o;t) \sum_{k=1}^{N} u_k(t-\tau(\mu_i;\mathbf{x})-\tau(\nu_k;\mathbf{x}))^* dt \tag{8}$$

where

$$\hat{r}_i(\mathbf{x}_o;t) = \sum_{k=1}^{N} u_k(t-\tau(\nu_k;\mathbf{x}_o)-\tau(\mu_i;\mathbf{x}_o)) \tag{9}$$

is the echo waveform due to the point target at $\mathbf{x}_o$. Since the Eq.(3) is still valid, Eq. (8) is rearranged as the next.

$$h(\mathbf{x};\mathbf{x}_o) = \sum_{i=1}^{M} \sum_{k=1}^{N} \varphi_{kk}(\tau(\nu_k;\mathbf{x})+\tau(\mu_i;\mathbf{x})-\tau(\nu_k;\mathbf{x}_o)-\tau(\mu_i;\mathbf{x}_o)) \tag{10}$$

When $|\mathbf{x}-\mathbf{x}_o|$ is small, PSF is represented as the function of differences of range and azimuth angles $(\Delta R,\Delta\alpha,\Delta\beta)$ as next.

$$h(\Delta R,\Delta\alpha,\Delta\beta) = \sum_{i=1}^{M} \sum_{k=1}^{N} \varphi_{kk}\left(\frac{2\Delta R-\Delta\alpha(r_T\cos(\Delta\beta-\vartheta_{Tk})+r_R\cos(\Delta\beta-\vartheta_{Ri})}{C}\right) \tag{11}$$

The resolution of the range δR is evaluated as the first zero point of

$$h_R(\Delta R) = h(\Delta R,0,0) = \sum_{i=1}^{M}\sum_{k=1}^{N} \varphi_{kk}\left(\frac{2\Delta R}{C}\right) \tag{12}$$

$h_R(\Delta R)$ is depend on the peak of the summation of auto-correlation functions. The sharpness of the peak is determined by the total frequency band-width B of all pulses. And the resolution limit for range is given by

$$\delta R = \frac{C}{2B} \tag{13}$$

We can evaluate the azimuthal resolution through the behavior of the PSF in the ΔR =0 plane. For small τ, correlation function φ_{kk} is approximated as:

$$\varphi_{kk}(\tau) = \cos 2\pi f_o \tau \tag{14}$$

when the band-widths of the pulses are not so large compared with their center frequency. Then the function $h(0,\Delta\alpha,\Delta\beta)$ is rearranged as

$$\begin{aligned} h_\alpha(\Delta\alpha) &= h(0,\Delta\alpha,\Delta\beta) \\ &\approx \sum_{i=1}^{M}\sum_{k=1}^{N} \cos\left[\frac{2\pi f_o \Delta\alpha (r_T \cos(\Delta\beta-\vartheta_{Tk}) + r_R \cos(\Delta\beta - \vartheta_{Ri}))}{C}\right] \\ &\approx J_o\left(\frac{2\pi f_o r_T \Delta\alpha}{C}\right)\cdot J_o\left(\frac{2\pi f_o r_R \Delta\alpha}{C}\right) \end{aligned} \tag{15}$$

where J_0 is the Bessel function of 0-th order. Here we assume that the numbers of the elements N and M are sufficiently large, and approximate the summation with the integration. Then the resolution of azimuth can be evaluated as first zero point of h_α, and it can be given by

$$\delta\alpha = \frac{1.2}{\pi r}\cdot\frac{C}{f_o} \tag{16}$$

where $r = \max(r_T, r_R)$ is the radius of the array.

Modulating function

We have proposed two modulating techniques for transmitting pulses, i.e. a set of Walsh functions and frequency shifted sinusoidal waves [1]. In this paper, we discuss the use of Walsh functions.

Walsh functions-Now we consider the Walsh functions which have binary value +1 and -1. The modulating functions are generated from the Hadamard matrix of N degree [W_{ij}], and are synchronized with a clock signal of period = Δt , like

$$\begin{aligned} m_k(t) &= W_{kj}\ ,\ (j-1)\Delta t \leqq t < j\Delta t \\ &= 0\ ,\ t<0\ ,\ t>T = N\Delta t\ . \end{aligned} \tag{17}$$

The modulating signals almost satisfy the orthogonal condition when the relative delay times are small enough. The summation of the autocorrelation functions becomes

$$\begin{aligned} \sum_{k=1}^{N} \varphi_{kk}(\tau) &= \int u_o(t) u_o(t-\tau)^* dt \\ u_o(t) &= \sum_{k=1}^{N} u_k(t)\ . \end{aligned} \tag{18}$$

The summation is approximately equal to the auto-correlation function of a rectangular pulse u_0 whose width is equal to the clock period of the Walsh-sequences (=T/N). The system has such a PSF as is similar to the PSF of the system using the rectangular pulse. Then, the resolution of range is determined by the clock period or the total frequency band width of the Walsh-sequences. The resolution of azimuth is determined by the ratio of array size to wave-length. But the orthogonal characteristics of the Walsh functions are violated when differential delay becomes large. Therefor the PSF must be examined by a numerical method, as we will present in later sections of this paper.

HIGH SPEED ALGORITHM FOR IMAGE FORMATION

The operation (4) is arranged as

$$s(\mathbf{x}) = \left| \sum_{i=1}^{M} \sum_{k=1}^{N} \Psi_{ik}(\tau(\nu_k;\mathbf{x})+\tau(\mu_i;\mathbf{x})) \right|^2 , \qquad (19)$$

where

$$\Psi_{ik}(\tau) = \int_{-\infty}^{\infty} r_i(t) u_k(t-\tau)^* dt \qquad (20)$$

is a correlation function between i-th received signal and k-th transmitting pulse. Then image can be calculated from the cross-correlation functions Ψ_{ik} .

The proposed high speed algorithm is based on the fact that the Walsh transform of the transmitting waveforms with respect to the index k yields the identical waveforms except for their delay times. With the Walsh transform of u_k with respect to k , we obtain

$$\sum_{k=1}^{N} W_{jk} u_k(t) = u_o(t-(j-1)\Delta t) , \qquad (21)$$

where u_o is the summed waveform of all transmitting pulses. Then, we obtain

$$\sum_{k=1}^{N} W_{jk} \Psi_{ik}(\tau) = \Psi_{io}(\tau-(j-1)\Delta t) \qquad (22)$$

$$\Psi_{io}(\tau) = \int_{-\infty}^{\infty} r_i(t) u_o(t-\tau)^* dt . \qquad (23)$$

From the eq.(22), The cross-correlation functions are calculated with inverse Walsh transformation:

$$\Psi_{ik}(\tau) = \sum_{j=1}^{N} W_{jk} \Psi_{io}(\tau-(j-1)\Delta t) \qquad (24)$$

Thus, we first compute the cross-correlation functions between received signals and the single reference waveform u_0 instead of u_k . Next, we obtain the cross-correlation functions with FWT. The procedure is illustrated in Fig.2.

SIMULATIONS AND EXPERIMENTAL SYSTEM

In this section, several results obtained by the computer simulation are presented to confirm the principle. And an experimental system using an airborne sensor array is proposed.

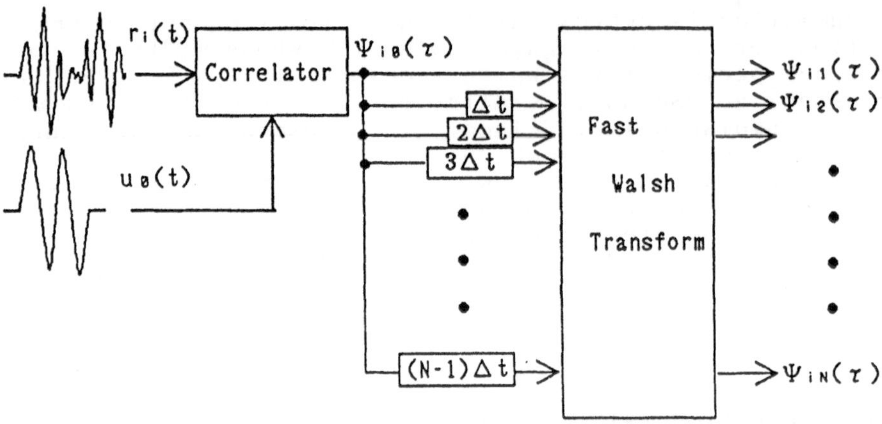

Fig.2. The procedure of high speed algorithm using FWT

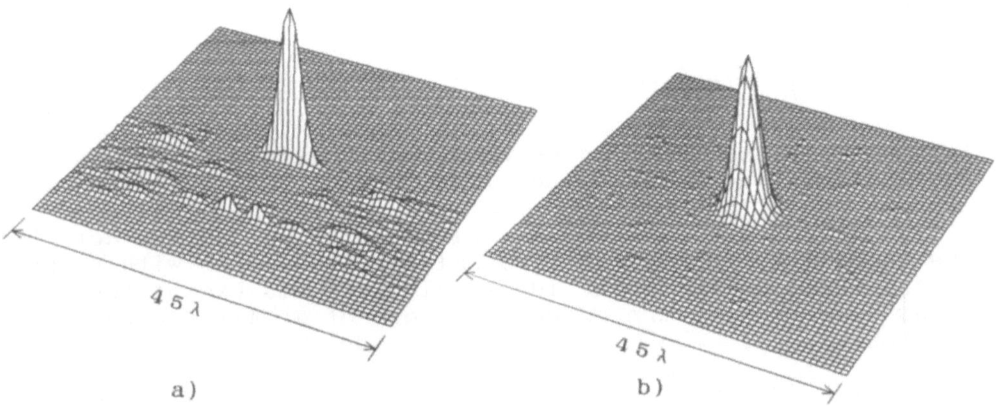

Fig.3. Results of computer simulation for beam-former using Walsh functions.
a) PSF in cross-sectional plane
b) PSF in focal plane

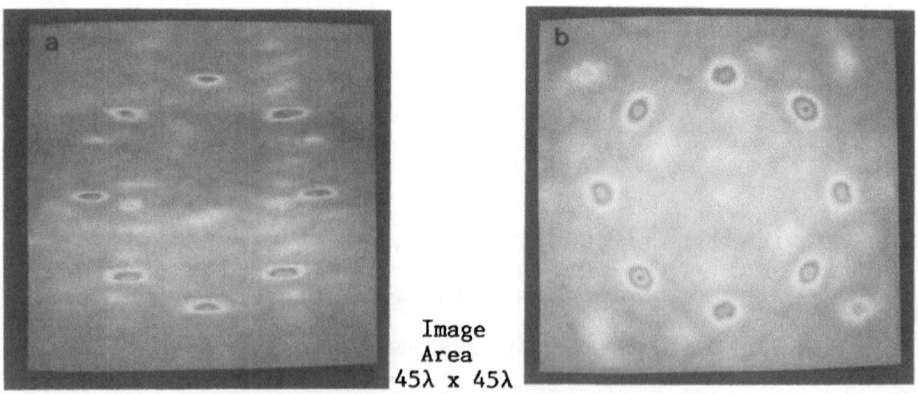

Fig.4. The simulated image of multiple point targets.
a) Cross sectional image
b) Focal plane image

Computer simulation

The computer simulation of the beam-forming system is performed to confirm the principle. In the computer simulation, a transmitter array of 16 elements and a receiver array of 8 elements is assumed. The radii of the transmitting and receiving arrays are 9 and 13 wavelengths of carrier wave respectively, and each transmitter is driven by the carrier signal modulated with the Walsh functions of binary order.

Calculated PSF at the point of 147 wavelengths apart from the array is illustrated in Fig.3. Clock width Δt is selected to be equal to two times as long as the carrier period. The calculation is performed according to the Eq.(19). The distorsion of the image and undesired peaks are considered to be caused by the violation of the orthogonal condition of the pulses.

The simulated image of multiple point targets is illustrated in Fig.4.

Experimental system

Fig. 5 shows the block diagram of experimental imaging system using the techniques described above. The array of the system consists of 16 ultrasound speakers and 8 microphones (Panasonic EAC-2M01A and EAC-1M01A). The frequency band of the elements is from 40kHz to 57kHz, and the frequency of the carrier wave is 50kHz (Wavelength λ =6.9mm). The arrangement of the array is identical to the simulation. Fig. 6 shows the view of the array. In the experimental system, digital circuits generate the carrier waves and modulating sequences and control the driving switch devices. Transmitting and receiving are repeated eight times because a single A/D converter is available. Digitized echo data are transferred to a mini-computer system. Image forming is performed by the computer.

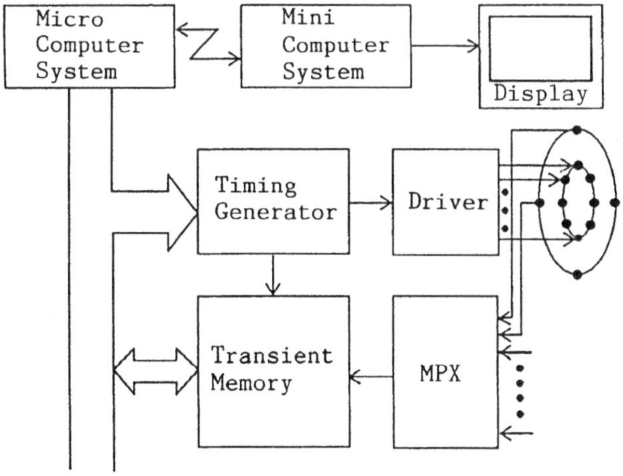

Fig.5. Block diagram of the experimental system

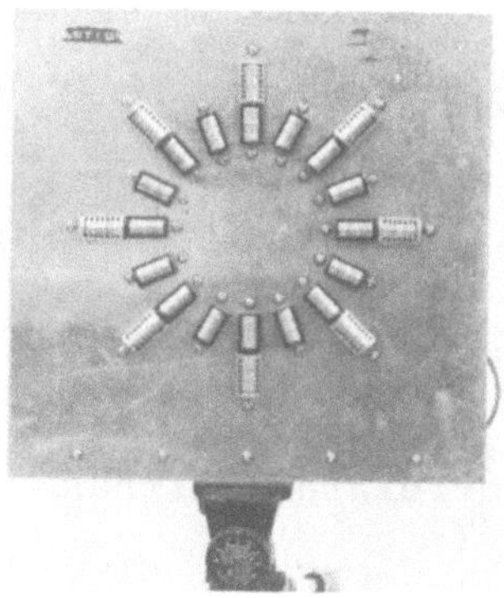

Fig.6. View of the array.

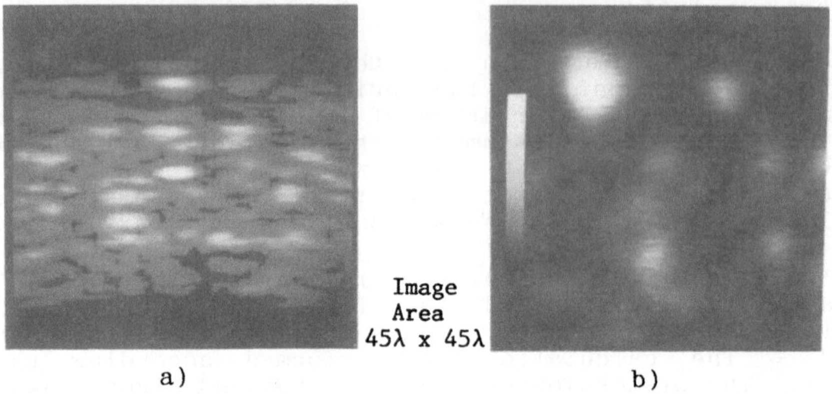

Fig.7. The images obtained by the experimental system
a) Cross sectional image of a steel rod
b) Focal plane image of a paper disk

Image obtained by the experimental system Fig.7 shows the images obtained by the experimental system. The clock period is selected to be two times as long as the carrier period, and the targets are a steel rod of 15mm D and a paper disk of 38mm D at the position of 1m apart from the array.

From the result of the experiment, it is expected that the system can construct the image with transmitting pulses of 0.64msec duration time when simultaneous digitization is available. And the lateral and range resolutions of the system are about 30mm and 10mm respectively at the distance of 1m apart from the array.

CONCLUSIONS

The data handling technique for ultrasound echo imaging system using sensor array is developed and discussed. The proposed method utilizes the indirect and numerical techniques. It is expected that the analog circuits will be replaced with simple switching hardware and data processing software using this method. The principle of the method is confirmed by the computer simulations and the experimental results.

Problems to be solved in future are mainly concerned with the improvement of beam-pattern and development of the real-time imaging system.

ACKNOWLEDGEMENT

We would like to thank Mr. Masanobu Takahashi and Mr. Teiichi Ohtaki for their considerable assistance with the experimental system and the computer programs.

REFERENCES

[1]Y.Tamura and T. Akatsuka, A digital beam-former using uncorrelated ultrasound pulses, Proc. of SPIE, 768, 85/92 (1987)
[2]Y. Tamura et al., Ultrasound holographic imaging system using high speed scanning with pulses modulated by Walsh functions, Trans. of SICE, 24-3, 213/219 (1988)

DIFFERENTIAL FREQUENCY RESPONSE AND ITS APPLICATION TO A STEP FREQUENCY SONAR

J. Nakayama, T. Goto and K. Iizuka(*)

Department of Electronics, Kyoto Institute of Technology
Matsugasaki, Kyoto 606, Japan
(*) Department of Electrical Engineering
University of Toronto, Toronto, Canada M5S-1A4

ABSTRACT

This paper introduces the concept of the differential frequency response as a new scheme of a short range CW sonar. It is shown that, by use of the differential frequency response, the divergence loss of sound wave can be compensated and a weak return signal from a target in the relatively far region can be enhanced. The concept is applied to a step frequency sonar locating targets in water, of which experimental results are described.

INTRODUCTION

A conventional sonar/radar transmits a narrow RF pulse and receives an echo backscattered from a target to estimate the time domain response in the baseband frequencies, from which the location and other properties of the target may be obtained. As is well known, however, the time domain response $R(\tau)$ may be obtained from the frequency response $H(f)$ by the Fourier transformation,

$$R(\tau) = \mathrm{Re}\{ \int_0^\infty H(f)\, e^{2\pi jf\tau}\, df\}, \tag{1}$$

where Re stands for the real part and τ is propagation delay time.

On the basis of this relation, Iizuka introduced the step frequency method[1] as a scheme of CW radar, which was successfully applied to a UHF imaging radar[2] and to an optical radar locating faults in optical fibers[3]. In the step frequency method, a continuous carrier with frequency f is transmitted and the phase and amplitude of the return signal from targets are measured to get the frequency response $H(f)$. Repeating this with different carrier frequencies, one gets a set of sampled data of $H(f)$, which is Fourier-transformed by a DFT algorithm to get the time domain response. Because the carrier frequencies can be made stable and accurate, the step frequency method gives accurate locations of targets. Another advantage is that the non-ideal performance of the

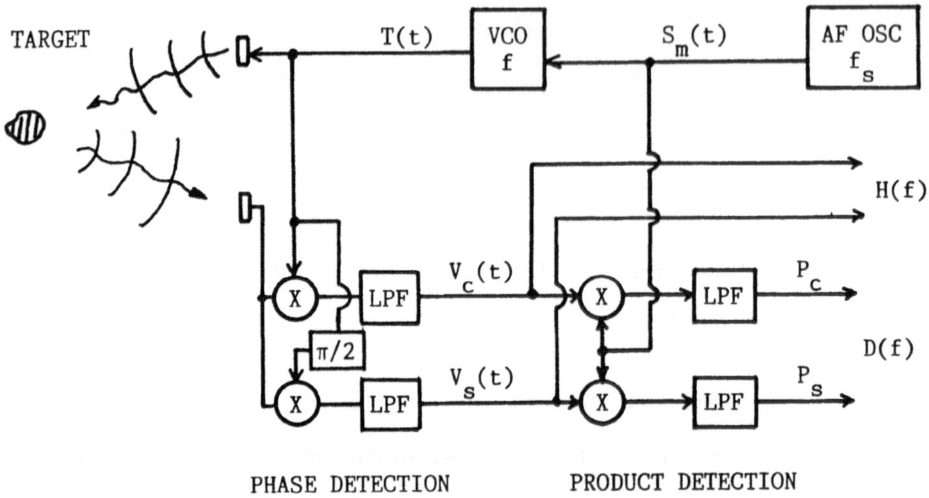

FIG. 1 Measurement of the differential frequency response D(f). the transmitter is made up of the AF (audio-frequency) oscillator and a VCO (voltage-controlled oscillator) to generate a frequency modulated carrier $T(t) = \sin[\,2\pi ft - (f_d/f_s)\cos(2\pi f_s t + \phi)\,]$ with the carrier frequency f, the modulation signal frequency f_s, and the frequency deviation f_d. the receiver consists of the quadrature phase detector followed by a pair of product detectors. when the frequency modulation is off, $V_c(t)$ and $V_s(t)$ become DC signals proportional to the real and imaginary parts of the frequency response H(f).

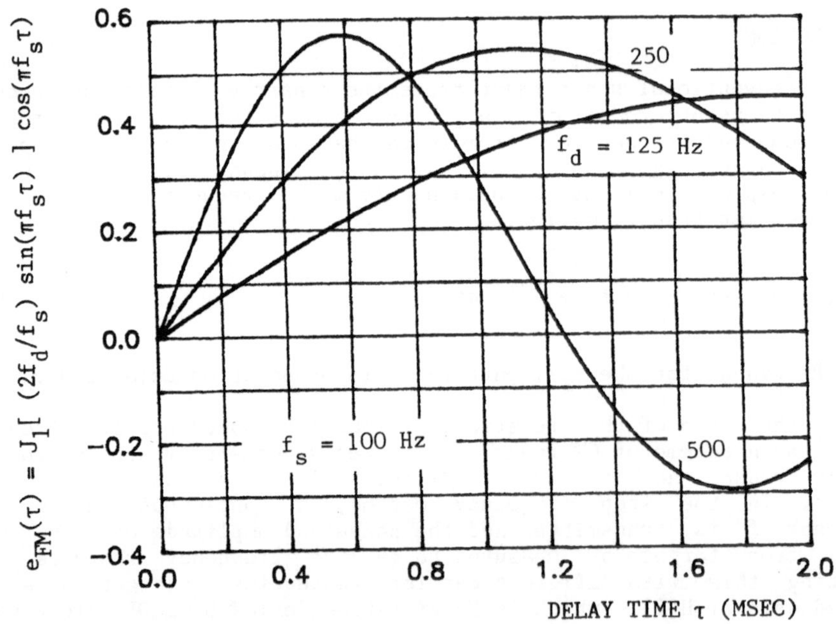

FIG. 2 The enhancement factor by the frequency modulation technique. f_s is the modulation signal frequency and f_d the frequency deviation.

transducers, electronic circuit can be easily corrected by multiplying the H(f) data by the appropriate correction factors before processing. It is difficult, however, to get a wide dynamic range, because the step frequency method is operated in CW mode, and because the unwanted signal due to the mutual coupling between the transmitter and receiver cannot be eliminated by a time-gating technique as in the impulse sonar/radar.

To overcome such a drawback in the step frequency method, we introduce the concept of the differential frequency response. Generally speaking, there is no need to obtain the time domain response for locating targets. In many practical cases, an enhanced time domain response $e(\tau)R(\tau)$ is more useful, where the enhancement factor $e(\tau)$ is an increasing function of τ. This is because the factor $e(\tau)$ compensates the divergence loss of sound wave, and enhances a weak return signal from a target in the far region. Furthermore, if $e(\tau) = 0$ for $\tau = 0$, the enhancement factor suppresses the unwanted leakage signal due to the mutual coupling between the transmitter and receiver. In the conventional pulse sonar/radar, such an enhanced time domain response is usually obtained by increasing the receiver gain against the delay time τ. But this also increases the receiver noise, so that the signal to noise ratio cannot be improved in principle. However, our differential frequency response may give an enhanced time domain response without increasing the system noise.

The idea of the differential frequency response is based on a simple mathematical relation,

$$\tau R(\tau) = -\frac{1}{\pi} \operatorname{Im}\{ \int_0^{\infty} \frac{dH(f)}{df} e^{2\pi j f \tau} df \}, \qquad (2)$$

which can be easily obtained by (1) and the integration by part. Here, Im stands for the imaginary part and $H(f) \to 0$ as $f \to \infty$ was implicitly assumed. Comparing (1) with (2), one finds that an enhanced time domain response $\tau R(\tau)$ appears in the left hand side in (2), instead of the time domain response $R(\tau)$. This relation suggests a new scheme of CW sonar/radar, that is, if one measures the derivative $dH(f)/df$, instead of H(f), and calculate the Fourier integral (2), one obtains an enhanced time domain response $\tau R(\tau)$.

DIFFERENTIAL FREQUENCY RESPONSE BY THE FREQUENCY MODULATION TECHNIQUE

The measurement of the derivative $dH(f)/df$ is, however, impractical in the exact sense. This is because $dH(f)/df = [H(f + \Delta f)-H(f)]/\Delta f$ as $\Delta f \to 0$ by the definition and because such a subtraction does not give any meaningful results for measured data $H(f + \Delta f)$ and H(f). Using a frequency modulation technique illustrated in figure 1, however, we can measure the differential frequency response D(f), which well approximates the derivative $dH(f)/df$ in a certain case.

In the frequency modulation technique, we transmit a sinusoidally frequency-modulated carrier T(t),

$$T(t) = \sin[\, 2\pi ft - (f_d/f_s) \cos(2\pi f_s t + \phi)], \qquad (3)$$

$$S_m(t) = \sin(2\pi f_s t + \phi), \qquad (4)$$

where f is the carrier frequency, f_d is the frequency deviation, $S_m(t)$ is the sinusoidal modulation signal, f_s is the modulation signal frequency, and ϕ is the phase constant.

The received signal $R_c(t)$ is first processed by means of quadrature phase detector in order to obtain the in-phase component $V_c(t)$ and quadrature phase component $V_s(t)$. If the modulation is off and $f_d = 0$, these components are DC signal proportional to the real and imaginary parts of the frequency response H(f), namely,

$$V_c(t) + j\,V_s(t) = H(f)\ ,\ (\ f_d = 0\ ). \tag{5}$$

When the modulation is on and $f_d \neq 0$, the instantaneous frequency in (3) becomes $f + f_d \sin(2\pi f_s t + \phi)$. Thus we could replace f in (5) by the instantaneous frequency to get,

$$V_c(t) + jV_s(t) = H[\ f + f_d \sin(2\pi f_s t + \phi\ )]$$

$$= H(f) + \frac{dH(f)}{df} f_d \sin(2\pi f_s t + \phi\ ) + \cdots\cdots\ , \tag{6}$$

where f_d and f_s are assumed to be much less than the carrier frequency f. Thus, to get the derivative dH(f)/df, we employ a product detection, where $V_c(t)$ and $V_s(t)$ are multiplied by the modulation signal $S_m(t)$ and then time-averaged by means of a pair of low-pass filters. In terms of the filtered signals, P_c and P_s, we put,

$$D(f) = P_c + j\,P_s, \tag{7}$$

which we call the differential frequency response. The differential frequency response D(f), however, is different from dH(f)/df because of what follows. As is easily seen in (6), P_c and P_s by the product detection may include higher order derivatives of H(f). Furthermore, in the total process of sound propagation both from the transmitter to a target and from the target to the receiver, the carrier and the modulation signal (actually, side bands of frequency modulated carrier) are both delayed. In (6), however, only the time delay of the carrier is taken into consideration, but the delay of the modulation signal is neglected.

A detailed analysis of the frequency modulation technique gives the enhancement factor $e_{FM}(\tau)$,

$$e_{FM}(\tau) = J_1[(2f_d/f_s) \sin(\pi f_s \tau)\ ] \cos(\pi f_s \tau), \tag{8}$$

where $\sin(\pi f_s \tau)$ and $\cos(\pi f_s \tau)$ describe the delay effect of the modulation signal, and J_1 is the Bessel function of the first order. As (2) shows, the enhancement factor form the derivative dH(f)/df is proportional to τ. As is illustrated in Fig. 2, however, the enhancement factor $e_{FM}(\tau)$ is proportional to τ only when τ is small enough. It becomes maximum about at a delay time τ_m,

$$\tau_m = \frac{1}{\pi f_s} \arcsin\left[\frac{1.841 \cdot f_s}{2\,f_d}\right], \tag{9}$$

since $J_1(x)$ takes its first maximum value at x = 1.841. For example, when f_s = 100 Hz and f_d = 250 Hz, we obtain τ_m = 1.2 msec, which corresponds to the one way distance L = 90 cm for underwater sound. Since $e_{FM}(\tau)$ decreases for $\tau > \tau_m$ and has zeros, the frequency modulation technique enhances return signals from short range targets whose delay time is less than τ_m.

We note that a complete linearity of the receiver has been assumed in the above analysis. Also, a wide receiver bandwidth was implicitly assumed so that no FM-AM conversion takes place in the whole system. In order to get the enhancement factor (8), we have taken the fundamental component with f_s in $V_s(t)$ and $V_c(t)$ in the product detection. If we take higher order harmonics in $V_s(t)$ and $V_c(t)$ instead of the fundamental component, however, we may get a higher order enhancement factor. This problem, however, will be omitted here.

APPLICATION TO A STEP FREQUENCY SONAR

We discuss an application of the differential frequency response to a short range sonar. However, we also describe the step frequency method using the frequency response H(f) for comparison. Let us consider a case where D(f) is sampled for equally spaced N frequencies f_n,

$$f_n = f_0 + n\,\Delta f, \quad (\,n = 0,\ 1,\ 2, \cdots, N-1), \tag{10}$$

where Δf is the frequency increment. Once the data $D(f_n)$ is obtained, it has to be Fourier-transformed according to (2). However, f_o is positive and the number of N of carrier frequencies is finite in the experiment below, so that the time domain response obtained from such data is regarded as a band limited signal without a DC component. Thus, we conveniently use the concept of the complex time domain response. We calculate the intensity $I(\tau)$ of the envelop of the complex time domain response by the DFT algorithm,

$$I(\tau_k) = \left| \exp(2\pi j f_o \tau_k) \sum_{n=0}^{N-1} D(f_n)\, W_n\, C_n \exp(2\pi jnk/N) \right|^2 , \tag{11}$$

where W_n is weight for pulse waveform synthesis and C_n is a factor which compensates unwanted frequency dependence of the transducers and the receiver. The τ_k is a quantized delay time relating to one-way distance L_k,

$$\tau_k = \frac{k}{N\,\Delta f}, \qquad L_k = \frac{1}{2}\, C\, \tau_k, \quad (\,k = 0,1,2,\cdots, N-1\,) \tag{12}$$

where C = 1500 m/sec is the sound velocity in water. The range resolution (quantized distance) becomes

$$\Delta L = \frac{C}{2N\Delta f}, \tag{13}$$

which is inversely proportional to the synthesized bandwidth $N\Delta f$. Because of the periodic nature of DFT, the maximum detectable range without ambiguity becomes

$$L_{max} = \frac{(N-1)\,C}{2N\Delta f}. \tag{14}$$

As Eq. (12) describes, the delay time τ_k is determined by Δf. Since the frequency synthesis technique always gives a highly accurate Δf, the delay time and the distance obtained by the differential frequency method become highly accurate. In an impulse sonar, the range resolution is

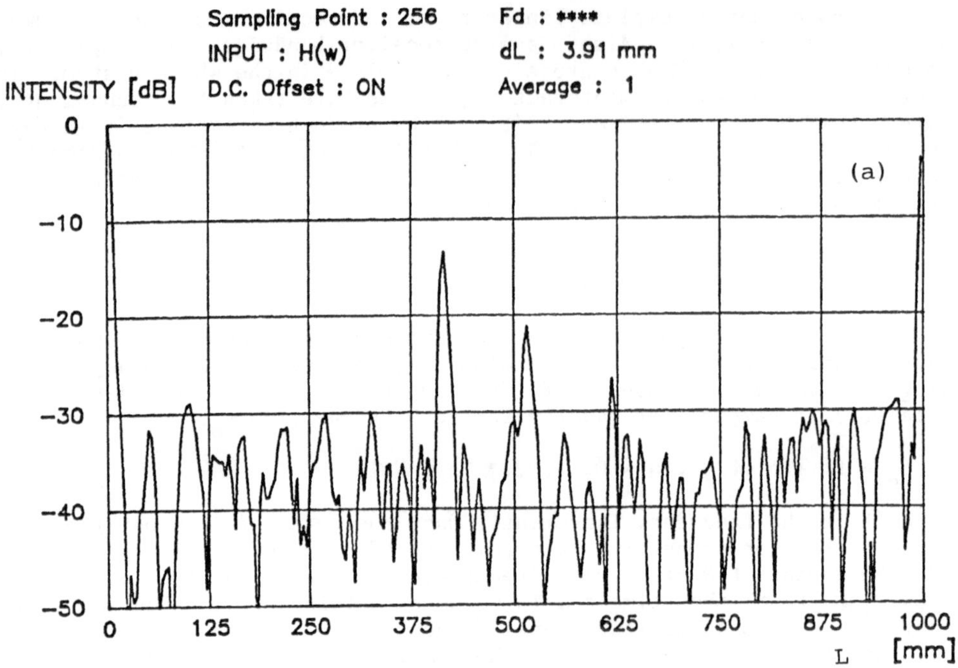

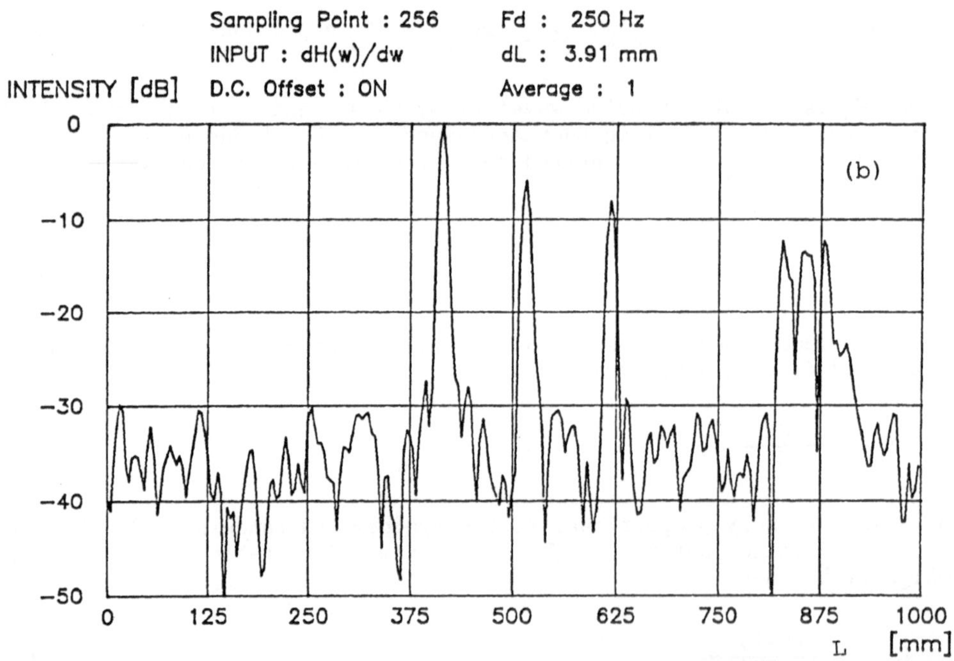

FIG. 3. Relative intensity $I(\tau)$ as a function of one-way distance L for a three-point target. (a) reconstruction by H(f) and (b) reconstruction by D(f). three targets are detected as three sharp peaks at L = 414.0, 515.6 and 617.2 mm. the largest peak at L = 0 in (a) is caused by the mutual coupling between transmitter and receiver. the sound absorber covering the water tank wall is detected as peaks at L = 828.1-878.9 mm in (b).

inversely proportional to the receiver bandwidth, so that the receiver bandwidth should be wide enough to get a high range resolution, But a wide frequency bandwidth means a high thermal noise level, and hence the range resolution and the system bandwidth are trade-off relation in an impulse sonar. On the other hand, the receiver measuring D(f) can have a very narrow bandwidth, which reduces much the thermal noise. Actually, the resolution is determined by the synthesized bandwidth $N\Delta f$ in the differential frequency response method.

ULTRASOUND EXPERIMENTS

In order to determine the feasibility of the differential frequency response D(f), we carried out an ultrasound experiment locating targets in a water tank with absorbing walls. For a one-dimensional imaging, the transmitting transducer as well as the receiving one is made long (10mm) and narrow (0.6mm) in the aperture so as to form a thin flat beam illuminating a uniformly long target. For comparison, however, we measured D(f) as well as H(f) at carrier frequencies with

$$f_o = 2.2 \text{ MHz}, \quad \Delta f = 750 \text{ Hz}, \quad N = 256, \tag{15}$$

so that the range resolution ΔL and the maximum range L_{max} become ΔL = 3.91 mm and L_{max} = 996 mm by (13) and (14). In the measurement of D(f), we set f_s the modulation signal frequency and f_d the frequency deviation as

$$f_s = 100\text{Hz}, \; f_d = 250\text{Hz}. \tag{16}$$

Using these parameters, we carried out experiments, results of which are described below.

For a three-point-target consisting of three brass rods with a diameter 3 mm, placed with equally spacing 100mm on a line, we measured H(f) first and then DFT processed to get intensity $I(\tau_k)$ against one-way distance L shown in Fig. 3(a), where C_n = 1 and the weight W_n in (11) was taken as 4-term Blackmann-Harris window[4]. The three rods are reconstructed as three peaks at L = 414.0, 515.6 and 617.2 mm. The distance between these peaks are 101.6 mm. Since the spacing of rods were 100 mm, so that the distance errors are 101.6-100 mm = 1.6 mm, which is less than the resolution 3.91 mm. Thus, the step frequency method using H(f) gives an accurate locations of targets. We may see, however, that the peak level for the far rod at L = 617.2 mm is only about 5dB larger than the noise level. Furthermore, there is the largest peak at L = 0, which is due to the unwanted mutual coupling between the transmitter and the receiver.

The same target was measured by the differential frequency method and DFT processed by (11) to get $I(\tau_k)$ shown in Fig. 3(b), which should be compared with Fig. (a). As Fig. (b) shows, there are three major peaks corresponding to the three rods. The locations of peaks are again 414.0, 515.6 and 617.2 mm, which are exactly same as the result obtained by H(f). However, the largest peak at L = 0 in Fig. (a) disappears in Fig. (b), in which the largest peak locates at L = 414.0 mm. This means that the differential frequency response suppresses the unwanted leakage signal due to the mutual coupling between transmitter and receiver. The peak at L = 617.2 mm becomes about 20 dB higher than the noise level. The peaks appeared about L = 828.1-878.9 mm are due to the reflection from the sound absorber covering the water tank wall. These peaks can not be obtained by the frequency response method, because they are smaller than the noise level. Seeing these facts, we may conclude that the differential frequency response actually enhances a weak return signal from a target in the relatively far region and suppresses the unwanted mutual coupling component appearing at L = 0.

CONCLUSIONS

Introducing the concept of differential frequency response, we have demonstrated, theoretically and experimentally, that the differential frequency response enhances a weak return signal from a target in the relatively far region and suppresses the unwanted leakage signal due to the mutual coupling between the transmitter and receiver.

The differential frequency response can be applied to a two/three dimensional imaging by means of multifrequency hologram matrix[5,6]. Such a two-dimensional imaging method will be studied near future.

Acknowledgment

This work was partially supported by the Grant-In-Aid for Scientific Research from the Ministry of Education of Japan.

References

1. K. Iizuka and A. P. Freundorfer, "Detection of Nonmetaric Object by a Step Frequency Radar", Proc. IEEE 71, 276 (1983)
2. K. Iizuka and A. P. Freundorfer, K. H. Wu, H. Mori, H. Ogura, and V. Nguyen, "Step-frequency radar", J. Appl. Phys. 56, 2573,(1984)
3. J. Nakayama, K. Iizuka and J. Nielsen, "Optical Fiber Fault Locator by the Step Frequency Method", Appl. Optics 26, 440, (1987)
4. F. J. Harris, " On the Use of Windows for Harmonic Analysis with the Discrete Fourier Transform", Proc. IEEE 66, 51 (1971).
5. J. Nakayama, H. Ogura, T. Miyashita and T. shibayama, "Imaging by means of Multifrequency Hologram Matrix", Trans. IECE, 62-B, 1081, (1979).
6. T. Miyashita, J. Nakayama, and H. Ogura, "Acoustical Imaging by means of Multifrequency Hologram Matrix", Acoustical Imaging, Vol. 9, 23, (1980).

A METHOD FOR IMPLEMENTING DOT REFLECTIVE ARRAYS

Bogdan Kamiński and Jerzy Kapelewski

Institute of Technical Physics
00-908 Warsaw 49, Kaliski St.
Poland

INTRODUCTION

In recent times there is a continuously growing interest in using the surface acoustic wave (SAW) devices technology for providing acoustical images of various objects located near the crystal surface as well as of some physical effects of a surface origination. This interest is due to peculiar possibilities provided by very character of this sort of acoustic waves which are distinctive for its relative easiness to be externally controlled with particular capability for detecting and examining shallow targets.

Up to now, a number of useful devices, basing on the use of surface waves, were elaborated. Basically, one can distinguish two groups of them. The first one, involves devices employing reflective arrays of peculiar geometry to focus the corresponding acoustic wave (either of the surface type or of a bulk type as converted from a SAW with focussing outside the substrate). In the second group, the emphasis is paid on signal processing function of SAW devices used for performing a sort of spectrum analysis of the acoustic waves as reflected from the objects in examination. This group can be exemplified by, e.g., the ultrasonic pulse Doppler flowmeter capable to provide images of the blood flow velocity accross vessel network [P.Tortoli et al, 1983]. The medical diagnostic instrumentation should be regarded as a major domain of practical applications for this sort of devices.

SAW REFLECTIVE ARRAY DEVICES

In the conventional interdigital version of SAW devices the transducers play the dual role of performing transduction and providing overall frequency response as required for a given use. In the reflective devices both of these functions are separated, the first one being performed by the system of transducers, and the second one by a reflection array, thus giving designer a large margin of freedom to optimize transducers for minimum insertion loss. The reflective array

This work was partially supported by the Problem CPBP.01.08

provides a way to fold the propagation path, giving rise to an effective reduction of the length of the substrate needed to achieve a given specification of parameters. In contrast with interdigital transducers the electrodes of which are connected to a common bus, the acoustic reflectors are electrically independent with respect to each other, rendering system to be essentially nonsensitive to local defects (e.g,broken or joined reflectors) arising in the fabrication process. This virtue is of particular importance for devices with long delay times. Additional advantage of using reflective array is its capability to virtually eliminate spurious bulk waves, which are known to be difficult to avoid in conventional SAW devices. These advantages are responsible for great utility of reflective-type devices, especially in constructing system with large value of delay time-bandwidth product, suitable for providing real time spectrum analysis (compressive receivers).

The reflective arrays currently used in SAW devices are composed either of grooves or of metallic stripes placed on the substrate surface together with a system of transducers. The version with grooves is more frequently used as it provides a good uniformity of the acoustic beam, for weighting can be realised here by varying the depth of grooves without any resorting to apodization. It has, however, serious disadvantages lying mainly in the complexity of the fabrication process (an additional step for providing the etching of the array is required, the array must be precisely aligned with respect to the positions of transducers, the weighting can not be incorporated into the mask). In turn, the version with strips has only a limited use, due to such drawbacks as beam nonuniformity introduced by an apodization, waveguiding effects and local currents which may arise along strips, piezoelectric shorting, etc.

All these disadvantages can be overcome, at the cost of some increasing in insertion loss, by using matallized dots as reflective elements.

THE TECHNIQUE OF MASK FABRICATION

Hitherto existing methods of implementation of SAW dot reflective arrays, although has led to very encouraging results, suffer, however, from a relatively high degree of technological complexity which renders it both time consuming and expensive. For instance, L.Solie (1977) has produced dot pattern on the mask by means of using succesive light flashes to get pictures of particular dots. In view of the large number of dots needed to achieve an acceptable level of loss (some 0,1-1,0 million for all the pattern) such a technique is not very useful, especially when a serial process is involved. The other approach, proposed by J.L.Thoss et al. (1981), consists in performing an interference pattern by overlaying two strip array with differing angles of inclination; the resulting arrays of dots is formed by the pattern of cross hatched regions where the strips do overlap. This technique requires, however, some additional proceeding to remove the pattern of residual segments of strips. The weighting being performed by incorporating changes into the strips width of the second grating pattern, imposes very demanding requirement on the accuracy level of the mask generation.

The method of mask fabrication proposed here does not suffer from this kind of shortcomings, being also distinctive from virtues of practical value such as technological simplicity, flexibility, and low cost. The reported proceeding consists in the following.

At first the corresponding mask of conventional stripe-type structure of 50 λ_o aperture and stripe-width suited to local value of $\lambda/4$ was produced. Next, according to the needed number of dots in each line (as given from a computer simulation), the additional mask was performed having opposite both the inclination angle and the sign of dispersion.

After exposing the latter as situated over the former and developing the mask, a "reticle" array was obtained. This array was subsequently used to form the picture of the dispersive delay line by incorporating the pattern of appropriate transducers. In the next stage, the mask thus obtained was given subject to a chemical processing, consisted in transferring the mask to a second photosensitive plate with the transducer being developed "negatively" and the reticle array "positively". A negative picture of the reticle array, introduced in such a way to the resulting mask, corresponds to the structure of dots required. The final stage of the procedure consists in implementing the corresponding metal structure on the crystalline substrate by means of the conventional one-stage photo-litography.

Randomization of dots distribution is effected here in a natural way as a result of the limited accuracy of the pattern generator used. It has proved to be quite sufficient for virtual elimination of detrimental coherence effects in the reflected acoustic beam.

THE DESCRIPTION OF THE DEVICE REPORTED

The reported version of the SAW dot reflective device involves two subsets realising the transmission and detection, respectively. The subset of transmission is composed of a pair of dot arrays of mutually opposite inclination angles. The transducers are made up of five-section capacitively weighted periodic systems of interdigital electrodes. Such a version has remarkable advantages over conventional ones in diminishing insertion losses, having a nearly rectangular frequency response and being able to generate a relatively flat SAW wavefront.

The dot reflective arrays (first proposed by L.Solie,1976) are composed of a number of parallel lines of dots, the weighting being effected by varying the dot density. As compared to more conventional versions of SAW reflective structures, that with dots is distinctive for its capability to eliminate many important sources of spurious effects, e.g., those resulted from the acoustic beam nonuniformity when the weighting is effected by means of apodization, the waveguide and local curents effects associated with strips, and in contrast to the version with grooves it has the advantage of one-stage photolitography.

The subset of detection differs from the proceeding one in

using a pair of three-section transducers to realize a coarse basic weighting for reducing sidelobe level.

With the foregoing method we have fabricated and studied a family of devices in consideration for several substrate materials and surface cuts (quartz YX and ST, lithium niobiate, bismuth germanium oxide - BGO). The configurations of the corresponding subset are illustrated in Fig.1.

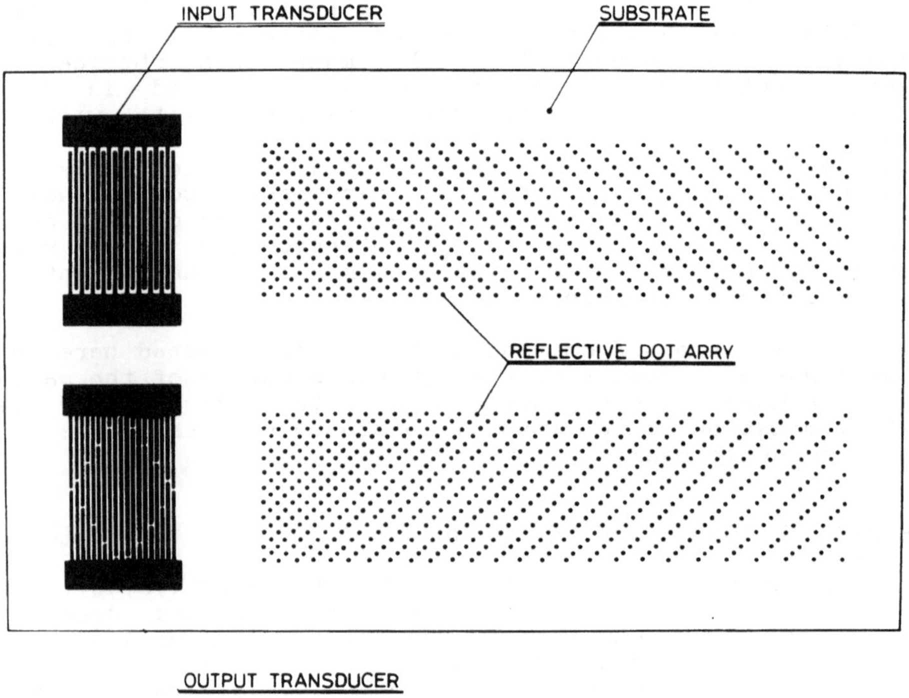

Fig. 1. Configuration of a SAW reflective dot array device.

PRELIMINARY RESULTS

The results presented in Fig.2-4 was obtained for the case of BGO (001) substrate and the (110) direction of the SAW as launched by the input transducer, with the central frequency (fo) of 30 MHz, the relative bandwidth (B) in the range of 4 MHz, and delay time (T) of 17 μs. The dots were composed of a copper layer (20 nm) overlaid by a layer of gold (300 nm).

It should be mentioned that, according to the knowledge of authors, this is a first attempt to implement a SAW dot reflective device on the BGO crystal.

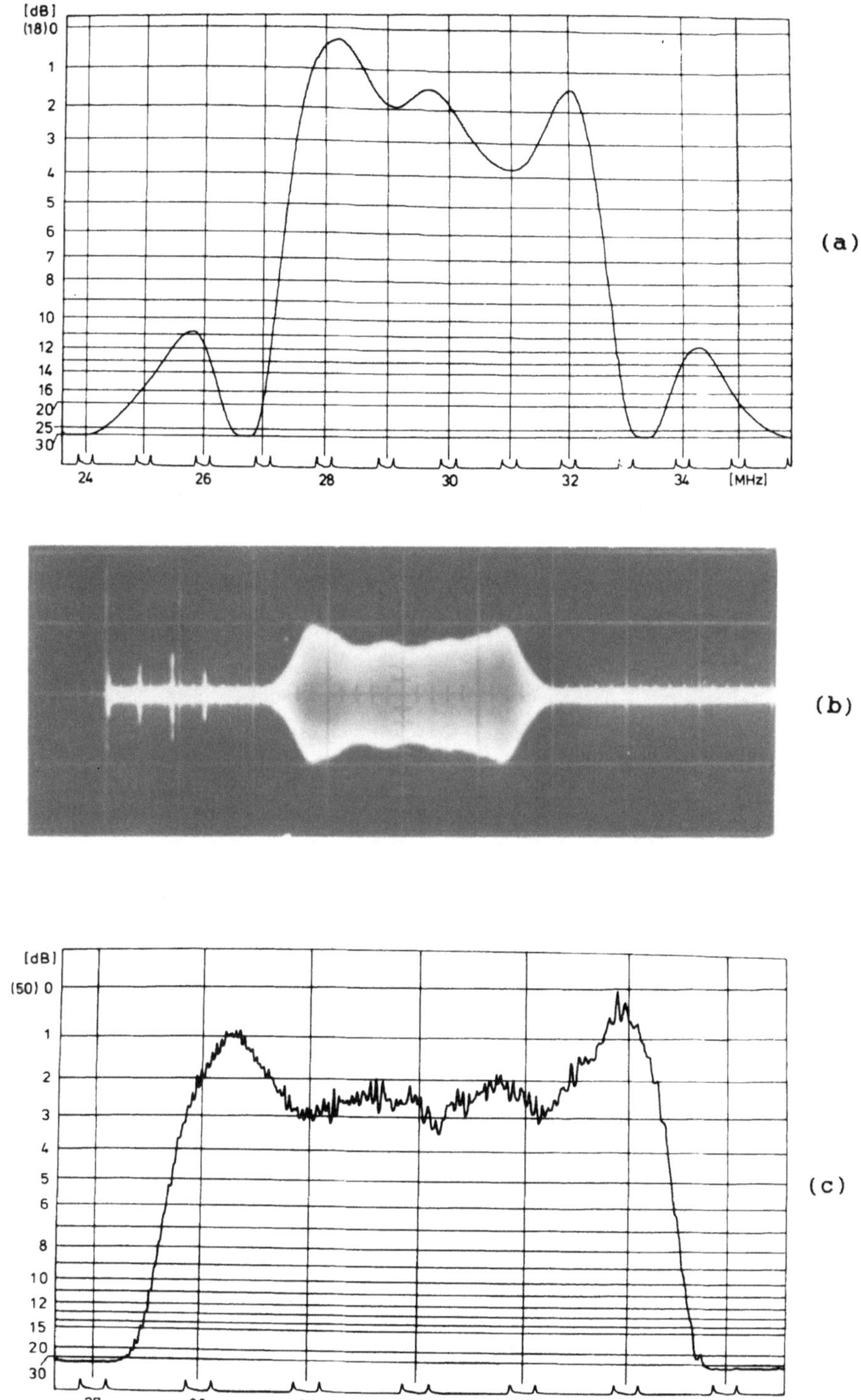

Fig. 2. Performance of the transmission subset.
(a) Frequency response of the line composed of five-section transducers (input and output ones); (b) impulse responseof the subset; (c) overall frequency response.

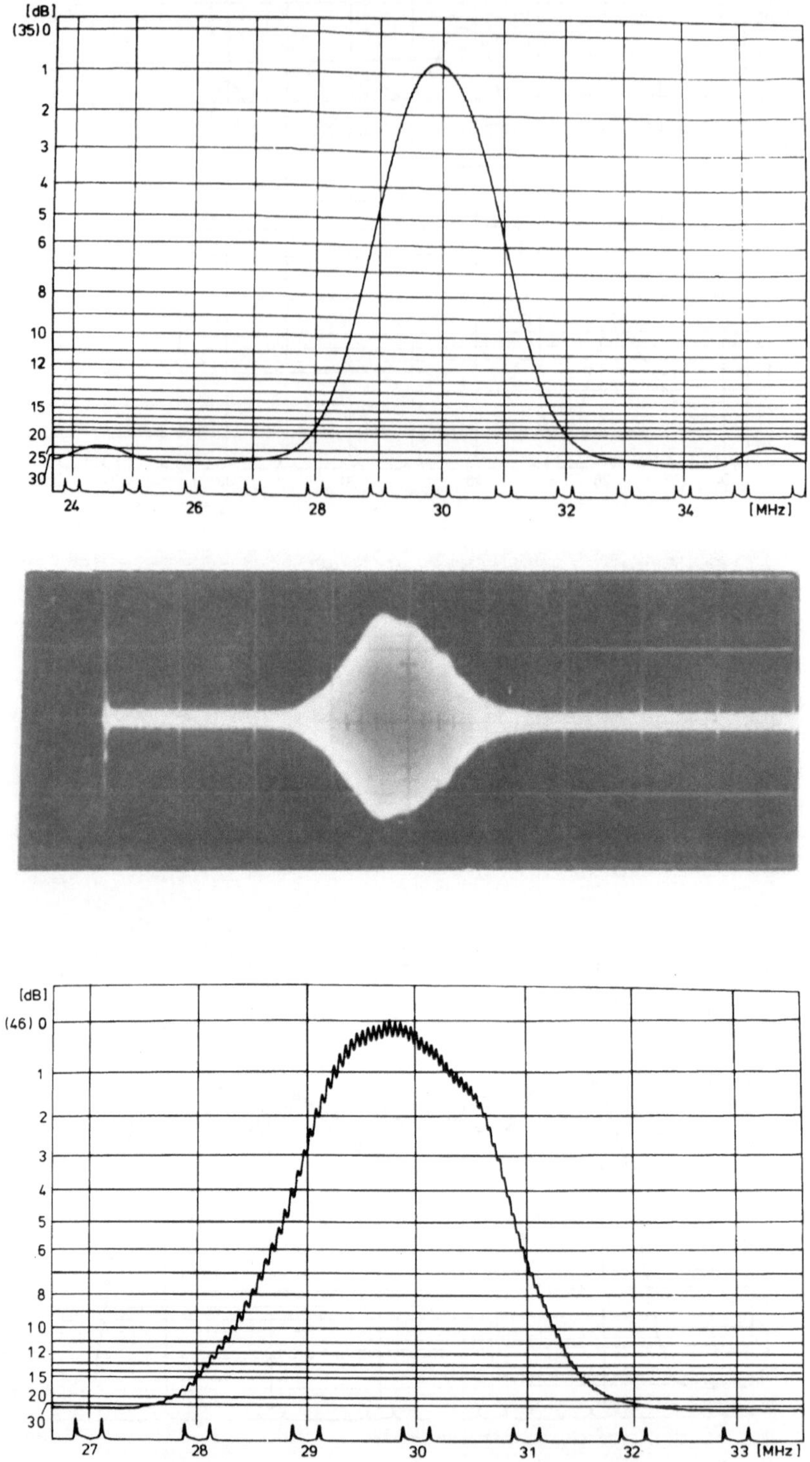

Fig. 3. Performance of the receiving subset.
(a) The weighting function as realized by a line three-sections transducers; (b) a impulse response of the subset; (c) the overall frequency response of the subset.

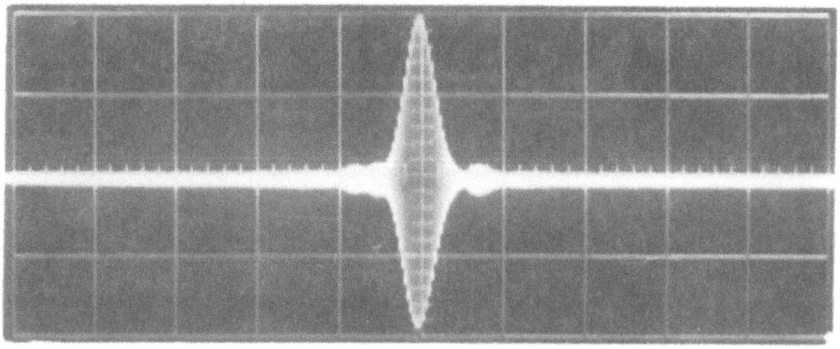

Fig. 4. The overall impulse response of the set exhibiting the effect of compression.

REFERENCES

Solie L.P.,1976, A SAW Filter Using a Reflective Dot Array, Proc. IEEE Ultrasonic Symposium, 309:312.

Solie L.P.,1976, Surface Acoustic Wave Reflective Dot Array, Appl. Phys. Lett., 28, 420:422.

Solie L.P.,1977, Reflective Dot Array Devices,Proc. IEEE Ultrasonic Symposium, 579:584.

Thoss J.L., Penumuri D., Thostenson M., 1981, Implementation of Reflective Array Matched Filters for Radar Applications, Proc. IEEE Ultrasonic Symposium, 63:68.

Tortoli P.,Manes G.F.,Atzeni C.,Boscaleri A., Pulse Doppler Flow Imaging Using a SAW CZT Processor, Proc. IEEE Ultra-Sonic Symposium,1059:1064.

PARTICIPANTS

Addison, Jr., R. C.
Rockwell International
PO Box 1085
1049 Camino Dos Rios
Thousand Oaks, CA 91360
USA

Akashi, N.
Dept. of Elec. Eng.
Tohoku Univ.
Sendai 980
Japan

Akiyama, I.
Dept. of Elec. Eng.
Sagami Institute of Tech.
Fujisawa 251
Japan

Aoki, Y.
Dept. of Information Eng.
Hokkaido Univ.
Kita-ku N13 W8 060
Japan

Attal, J
LAMM, USTL
Place Eugene Bataillon
34060 Montpellier, Cedex
France

Bereiter-Hahn, J
Cinematic Cell Research Group
Johann Wolfgang Goethe-University
D-6000 Frankfurt/M
FRG

Briggs, G. A. D.
Dept. of Metallurgy & Science
of Materials
Oxford Univ.
Parks Road, Oxford OX1 3PH
UK

Bussiere, J. F.
National Research Council Canada
Ind. Materials Research Inst.
75 De Mortagne Blvd.
Boucherville, Quebec J4B 6Y4
Canada

Cannelli, G. B.
Instituto di Acustica, CNR
00189 Rome
Italy

Cheeke, J. D. N.
Dept. of Physics
Univ. of Sherbrooke
Sherbrooke, Quebec
Canada J1K 2R1

Chen, Chin-Hsing
Dept. of Electrical Engineering
National Cheng Kung University
Tainan
Taiwan

Chen, Hou-Xia
Cardiovascular Institution
of Qingdao Medical College
10 Jiangsu Road, Qingdao
China

Cho, Y.
Research Institute of Elec. Com.
Tohoku Univ.
Sendai 980
Japan

Chubachi, N.
Dept. of Elec. Eng.
Tohoku Univ.
Sendai 980
Japan

Endoh, N.
Dept. of Med. Eng. & Cardiology
Research Institute
for Chest Diseases & Cancer
Tohoku Univ.
Sendai 980
Japan

Fukumoto, A.
Matsushita Research Institute
Tokyo, Inc.
Tama-ku 214
Japan

Gan, W. S.
Acoustical Services Pte Ltd
29 Telok Ayer St.
Singapore 0104
Singapore

Gilmore, R. S.
GE R & D Center
PO Box 43
Schenectady
NY 12301
USA

Gindre, M.
Laboratoire de Biophysique
(CNRS-UA 593)
24, rue du Faubourg
St. Jacques 75674, Paris Cedex 14
France

Gremaud, G.
EPFL
Institut de Genie Atomique
CH-1015-Lausanne
Switzerland

Ha, Kang-Lyeol
Dept. of Elec. Eng.
Tohoku Univ.
Sendai 980
Japan

Harumi, K.
Tokyo Univ. of Information Sci.
Chiba 280-01
Japan

Hashimoto, K.
Dept. of Elec. Eng.
Chiba Univ.
Chiba 260
Japan

Hatano, J.
Dept. of Materials
Science & Technology
Science Univ. of Tokyo
Shinjuku-ku 162
Japan

Hayakawa, Y.
Technical Research Lab.
Hitachi Kenki Co., Ltd.
Tsuchiura 300
Japan

Hill, C. R.
Physics Dept.
Royal Marsden Hospital
Sutton, Surrey SM2 5PT
UK

Hizama, M.
TOSHIBA Med. Eng. Lab.
Otawara 329-26
Japan

Hoppe, M.
Ernst Leitz Wetzlar GmbH
Postfach 2020
D-6330, Wetzlar 1
West Germany

Horiguchi, T.
C&C Systems Research Lab.
NEC Corp.
Miyamae-ku 213
Japan

Hoshimiya, T.
Tohoku Gakuin Univ.
Tagajo 985
Japan

Hwang, J. J.
ATL
PO Box 3003
22100 Bothell Hwy., S.E.
Bothell, WA 98041
USA

Iinuma, K.
TOSHIBA Med. Eng. Lab.
Otawara 329-26
Japan

Ikegami, M.
Government Ind. Development Lab.
Sapporo 004
Japan

Inoue, T.
Material Development Center
NEC Corp.
Miyamae-ku 213
Japan

Itoh, T.
Aloka Co., Ltd.
Mitaka 181
Japan

Iwashita, Y.
Dept. of Physics
Kagoshima Univ.
Kagoshima 890
Japan

Jhang, K. Y.
Graduate School at Nagatsuta
Tokyo Institute of Technology
Midori-ku 227
Japan

Justice, J. H.
Mobil R&D Corp.
Dallas Research Lab.
13777 Midway Road
Dallas, TX 75244
USA

Kadota, M.
K.K. Murata Ltd.
Nagaokakyo
Kyoto 617
Japan

Kanayama, K.
High Performance Materials &
Products Lab.
Mitsui Petrochemical Ind. Ltd.
Sodegaura-machi 299-02
Japan

Kapelewski, J.
Institute of Technical Physics
01-908 Warsaw 49, Kaliski St.
Poland

Kasahara, S.
School of Dentistry
Tohoku Univ.
Sendai 980
Japan

Kasai, C.
Aloka Co., Ltd.
Mitaka 181
Japan

Katakura, K.
Hitachi Ltd.
Meguro-ku 152
Japan

Kawai, T.
Yokogawa Elec. Corp.
Musashino 180
Japan

Kawano, K.
Dept. of Dental Radiology
Kagoshima Univ.
Higashi-gun 890
Japan

Khuri-Yakub, B. T.
Edward L. Ginzton Lab.
Stanford Univ.
Stanford, CA 94305
USA

Kojima, S.
Institute of Applied Physics
Univ. of Tsukuba
Tsukuba 305
Japan

Komatsuzaki, Y.
SIBER KIKAI K.K.
Siber Hegner Bldg.
Naka-ku 231
Japan

Kondo, M.
TOSHIBA R & D Center
Komukai-Toshibacho 1
Saiwai-ku 210
Japan

Kou, C.
Dept. of Elec. Com.
Shibaura Institute of Technology
Minato-ku 108
Japan

Koyama, K.
Dept. of Polymer Materials Eng.
Yamagata Univ.
Yonezawa 992
Japan

Kraegeloh, U. T.
SIBER KIKAI K.K.
Siber Hegner Bldg.
Naka-ku 231
Japan

Kubota, J.
Hitachi Research Lab.
Hitachi Ltd.
Hitachi 319-12
Japan

Kubota, M.
Dept. of Surgery
Tokai Univ. School of Medicine
Bohseidai
Isehara 259-11
Japan

Kulik, A.
EPFL
Institut de Genie Atomique
CH-1015 Lausanne
Switzerland

Kushibiki, J.
Dept. of Elec. Eng.
Tohoku Univ.
Sendai 980
Japan

Lee, H.
Dept. of Elec. & Computer Eng.
Univ. of Illinois
at Urbana-Champaign
Urbana, IL 61801
USA

Lees, S.
Dept. of Bioengineering
Forsyth Dental Center
140 Fenway
Boston, MA 02115
USA

Li, H. U.
Institute of Industrial Science
Univ. of Tokyo
Minato-ku 106
Japan

Li, L.
Research Lab. of Precision
Machinery & Electronics
Tokyo Institute of Technology
Midori-ku 227
Japan

Liu Ming-Yu
Dept. of Ultrasound
Heibei Tumor Hospital
Jiankang Road, Shijiazhuang,
Heibei
China

Mak, D. K.
Metals Technology Lab.
Canada Center
for Mineral & Energy Tech.
Ottawa, Ontario
Canada K1A 0G1

Maki, K.
Honda Electronics Co., Ltd.
Toyohashi 441-31
Japan

Mine, K.
SIBER KIKAI K.K.
Siber Hegner Bidg.
Naka-ku 231
Japan

Miyashita, T.
Dept. of Elec. Eng.
Kyoto Institute of Technology
Sakyo-ku 606
Japan

Miyazaki, M.
Honda Electronics Co., Ltd.
20 Oyamazuka, Oiwa-cho
Toyohashi 441-31
Japan

Mizuno, S.
Dept. of Elec. Com.
Shibaura Institute of Technology
Minato-ku 108
Japan

Mochizuki, T.
Research Institute
Aloka Co., Ltd.
Mitaka 181
Japan

Nagai, K.
Institute of Applied Physics
Univ. of Tsukuba
Tsukuba 305
Japan

Nakagawa, Y.
Ricoh R&D Center
Information Processing &
Electronics Research Center
Hino 191
Japan

Nakamoto, T.
Cooperation Center
for Science & Technology
Tokyo Institute of Technology
Meguro-ku 152
Japan

Nakamura, E.
SUMISHO MACHINERY TRADE CORP.
Kobun-Kosan Bldg.
Chiyoda-ku 101
Japan

Nakamura, K.
Dept. of Elec. Com.
Tohoku Univ.
Sendai 980
Japan

Nakamura, M.
Dept. of Elec. Eng.
Nihon Univ.
Narashino 275
Japan

Nakaso, N.
Fundamental Research Center
Toppan Printing Co., Ltd.
Kitakatsushika-gun 345
Japan

Nakayama, J.
Dept. of Electronics
Kyoto Institute of Technology
Matsugasaki
Kyoto 606
Japan

Nikoonahad, M.
Bio-Imaging Research Inc.
425 Barclay Blvd.
Lincolnshire, IL 60069
USA

Nomura, T.
Dept. of Elec. Com.
Shibaura Institute of Technology
Minato-ku 108
Japan

Nonaka, T.
Technical Research Lab.
Hitachi Construction Machinery
Tsuchiura 300
Japan

Nongaillard, B.
Lab. d'Opto-Acousto-Elec.
Univ. de Valenciennes
- Le Mont Houy
59326 Valenciennes, Cedex
France

Obata, M.
Dept. of Materials Processing
Tohoku Univ.
Sendai 980
Japan

Ohigashi, H.
Dept. of Polymer Materials Eng.
Yamagata Univ.
Yonezawa 992
Japan

Ohno, M.
Olympus Optical Co., Ltd.
Hachioji 192
Japan

Ohtsuki, S.
Research Lab. of Precision
Machinery & Electronics
Tokyo Institute of Technology
Midori-ku 227
Japan

Ohya, A.
Dept. of Elec. Eng.
Keio Univ.
Kohoku-ku 223
Japan

Okada, A.
SIBER KIKAI K.K.
Siber Hegner Bldg.
Naka-ku 231
Japan

Okada, H.
Dept. of Com. Eng.
Shibaura Institute of Technology
Minato-ku 108
Japan

Okamura, H.
Lightwave Com. Lab.
NTT Transmission Systems Lab.
Yokosuka 238-03
Japan

Okawai, H.
Research Institute
for Chest Diseases & Cancer
Tohoku Univ.
Sendai 980
Japan

Okuyama, D.
Dept. of Elec. Eng.
Akita Univ.
Akita 010
Japan

Oogaki, M.
Oki Elec. Industry Co., Ltd.
Minato-ku 108
Japan

Oohira, K.
Technical Research Institute
Toppan Printing Co., Ltd.
Kitakatsushika-gun 345
Japan

Park, S. B.
Dept. of Elec. Eng.
KAIST
PO Box 150
Cheongryang, Seoul
Korea

Pernod, P. J.
Institut Industriel du Nord
L.P.V.A., I.D.N.
B.P.48
59651 Villeneuve d'Ascq, Cedex
France

Saijo, Y.
Research Institute
for Chest Diseases & Cancer
Tohoku Univ.
Sendai 980
Japan

Sakurai, K.
Optoelectronics Research Lab.
NEC Corp.
Miyamae-ku 213
Japan

Sannomiya, T.
Dept. of Elec. Eng.
Tohoku Univ.
Sendai 980
Japan

Sato, T.
Graduate School at Nagatsuta
Tokyo Institute of Technology
Midori-ku 227
Japan

Sawaguchi, A.
Dept. of Elec. Eng.
The National Defense Academy
Yokosuka 239
Japan

Schwetlick, H.
Research Lab. of Precision
Machinery & Electronics
Tokyo Institute of Technology
Midori-ku 227
Japan

Shibayama, K.
Tamagawa Univ.
Machida 194
Japan

Shimizu, H.
Dept. of Elec. Com.
Tohoku Univ.
Sendai 980
Japan

Shimura, T.
Fujitsu Lab. Ltd.
Nakahara-ku
Kawasaki 211
Japan

Shingyouuchi, M.
Dept. of Elec. Eng.
Keio Univ.
Kohoku-ku
Yokohama 223
Japan

Sinton, A. M.
Dept. of Metallurgy & Science
of Materials
Oxford Univ.
Parks Road, Oxford OX1 3PH
UK

Sivers, E. A.
Bio-Imaging Research Inc.
425 Barclay Blvd.
Lincolnshire, IL 60069
USA

Somekh, M. G.
Dept. of Elec. & Electronic Eng.
Univ. College London
Torrington Place, London WC1E 7JE
UK

Spencer, S.
Dept. of Metallurgy & Science
of Materials
Oxford Univ.
Parks Road, Oxford OX1 3PH
UK

Sugawara, Y.
Dept. of Elec. Eng.
Tohoku Univ.
Sendai 980
Japan

Tabei, M.
Research Lab. of Precision
Machinery & Electronics
Tokyo Institute of Technology
Midori-ku 227
Japan

Takahashi, S.
NEC Corp., Ltd.
Takatsu-ku 213
Japan

Takamizawa, K.
TOSHIBA R & D Center
Komukai-Toshibacho 1
Saiwai-ku 210
Japan

Takeuchi, Y.
Yokogawa Medical Systems, Ltd.
Tachikawa 190
Japan

Tamura, Y.
Dept. of Information Eng.
Yamagata Univ.
Yonezawa 992
Japan

Tanaka, H.
Dept. of Elec. Com.
Tohoku Univ.
Sendai 980
Japan

Tanaka, H.
Matsushita Research Institute
Tokyo, Inc.
Tama-ku 214
Japan

Tanaka, M.
Dept. of Med. Eng. & Cardiology
Research Institute
for Chest Diseases & Cancer
Tohoku Univ.
Sendai 980
Japan

Tateoka, H.
Olympus Tech. Research Institute
Olympus Optical Co., Ltd.
Hachioji 192
Japan

Tokioka, M.
Central Research Lab.
Fuji Electric Co., Ltd.
Yokosuka 240-01
Japan

Tomikawa, Y.
Dept. of Elec. Eng.
Yamagata Univ.
Yonezawa 992
Japan

Tsukahara, Y.
Technical Research Institute
Toppan Printing Co., Ltd.
Kitakatsushika-gun 345
Japan

Uchida, M.
Dept. of Mathematical Eng.
Nihon Univ.
Narashino 275
Japan

Uchino, F.
Olympus Optical Co., Ltd.
Hachioji 192
Japan

Ueda, M.
Research Lab. of Precision
Machinery & Electronics
Tokyo Institute of Technology
Midori-ku 227
Japan

Umemura, S.
Central Research Lab.
Hitachi Ltd.
Kokubunji 185
Japan

Urbach, G.
Laboraroire de Biophysique
(CNRS-UA 593)
24, rue du Faubourg
St. Jacques 75674, Paris Cedex 14
France

Vetters, H. R.
Inst. for Materials Sci. & Eng.
Lesumer Heerstrasse 32
D-2820 Bremen 77
FRG

Wade, G.
Dept. of Elec. & Computer Eng.
Univ. of California
Santa Barbara, CA 93106
USA

Weglein, R. D.
6317 Drexel Avenue
Los Angeles, CA 90048
USA

Williams, P. G.
C-CORE
Bartlett Bldg.
Memorial Univ. of Newfoundland
St. Jone's
Canada AIB 3X5

Yamada, A.
Toyohashi Univ. of Technology
Toyohashi 440
Japan

Yamada, K.
Dept. of Elec. Com.
Tohoku Univ.
Sendai 980
Japan

Yamaguchi, M.
Dept. of Elec. Eng.
Chiba Univ.
Chiba 260
Japan

Yamakoshi, Y.
Graduate School at Nagatsuta
Tokyo Institute of Technology
Midori-ku 227
Japan

Yamamoto, A.
Dept. of Med. Eng. & Cardiology
Research Institute
for Chest Diseases & Cancer
Tohoku Univ.
Sendai 980
Japan

Yamanouchi, K.
Research Institute of Elec. Com.
Tohoku Univ.
Sendai 980
Japan

Yamashita, Y.
Dept. of Bio-Medical Eng.
Tokai Univ., School of Medicine
Isehara 259-11
Japan

Yan-Qing, Y.
Research Lab.
of Precision Machinery & Elec.
Tokyo Institute of Technology
Midori-ku 227
Japan

Yasuda, T.
Dept. of Elec. Com.
Shibaura Institute of Technology
Minato-ku 108
Japan

Yuan, Y.-Q.
Dept. of Radio Eng.
Nanjing Institute of Technology
China

Yuta, S.
Inst. of Information Sci. & Elec.
Univ. of Tsukuba
Tsukuba 305
Japan

Zhou, K. Y.
Research Lab.
of Precision Machinery & Elec.
Tokyo Institute of Technology
Midori-ku 227
Japan

Zieniuk, J. K.
IPPT-PAN
Institute of Fundamental Tech.
Research
Polish Academy of Sciences
00-049 Warsaw
Poland

AUTHOR INDEX

SUBJECT INDEX